Advances in Power Science and Technology

Advances in Power Science and Technology

Editors

Bo Yang
Zhijian Liu
Lin Jiang

MDPI

Basel • Beijing • Wuhan • Barcelona • Belgrade • Novi Sad • Cluj • Manchester

Editors
Bo Yang
Kunming University of Science and Technology
Kunming
China

Zhijian Liu
Kunming University of Science and Technology
Kunming
China

Lin Jiang
University of Liverpool
Liverpool
UK

Editorial Office
MDPI AG
Grosspeteranlage 5
4052 Basel, Switzerland

This is a reprint of articles from the Topic published online in the open access journals *Electricity* (ISSN 2673-4826), *Electronics* (ISSN 2079-9292), *Energies* (ISSN 1996-1073), *Processes* (ISSN 2227-9717), and *Sustainability* (ISSN 2071-1050) (available at: https://www.mdpi.com/topics/VT9C7KV1Q4).

For citation purposes, cite each article independently as indicated on the article page online and as indicated below:

Lastname, A.A.; Lastname, B.B. Article Title. *Journal Name* **Year**, *Volume Number*, Page Range.

ISBN 978-3-7258-1923-2 (Hbk)
ISBN 978-3-7258-1924-9 (PDF)
doi.org/10.3390/books978-3-7258-1924-9

Contents

About the Editors

Bo Yang

Prof. Bo Yang is a professor and doctoral supervisor at Kunming University of Technology. Served as a member of the Youth Working Committee of the 8th Council of the Chinese Electrotechnical Society. Executive Director of the Energy Information Society Technology Subcommittee and the Dynamic Power System Artificial Intelligence Application Technology Subcommittee of the IEEE PES Smart Grid and New Technology Committee (China). Senior Consulting Expert of the Yunnan Electric Power Industry Association. Co-Chairman of the EI Inclusion Conference ICPST 2023. Chairman of the ICPST 2024 Conference. Editor of three SCI journals, two EI journals, and more than ten SCI, EI, and Chinese core Special Issues. Received the title of Outstanding Young Talents in the first Yunnan Provincial Ten Thousand Talents Plan (at the provincial and ministerial level) and the Training and Visiting Program. Published two Chinese academic monographs, over 160 SCI journal papers (including 11 ESI papers and three hot topic papers), and 30 EI journal papers. Hosted 25 scientific research projects, including two National Natural Science Foundation projects. Selected as a highly cited scholar (Electrical Engineering) by Elsevier China from 2021 to 2023 and one of the top 2% globally influential scientists from 2022 to 2023. Received first prize in the technology category from the China Electric Power Enterprise Federation and second prize in the China Industry University Research Cooperation Innovation Achievement Award.

Zhijian Liu

Prof. Zhijian Liu is the Dean of the School of Electric Power Engineering at Kunming University of Technology and a Doctoral Supervisor. Mainly engaged in research on power system security and stability analysis, optimization, grid intelligence, intelligent detection of power equipment, and live working robots in power systems. Led projects with research funding of over CNY 2.3 million. Published over 50 papers in important academic journals and international conferences both domestically and internationally. Hosted or participated in 45 vertical and horizontal scientific research projects, including one National Natural Science Foundation, one Yunnan Provincial Natural Science Foundation, one Yunnan Provincial Department of Education Fund, and one School Youth Fund. Participated in one key project of the National Natural Science Foundation of China (ranked fourth) and one 863 Program project of the Ministry of Science and Technology (ranked fourth), and led research on one project of the Yunnan Provincial Development and Reform Commission (ranked second). Authorized 33 national invention and utility model patents and accepted over 10 patents. Received the first prize for technological invention in Yunnan Province in 2009 (ranked ninth), the first prize for scientific and technological progress in Yunnan Province in 2011 (ranked eighth), the first prize for teaching achievements in Yunnan Province in 2013 (ranked fourth), the special prize for teaching achievements in Kunming University of Technology in 2011 (ranked fourth), the first prize for school level teaching achievements in 2016 (ranked second), the 13th Wudaguan Excellent Teacher Award, and the provincial excellent master's thesis supervisor in 2017. The team leader of the 100 core courses and subject directions for graduate students.

Lin Jiang

Prof. Lin Jiang is a Reader at Liverpool University in the UK who has been researching the control and optimization of smart electricity and new energy generation for some time. Published over 250 mainstream papers, including over 180 SCI papers, of which 80 are from IEEE Transactions, and over

22 papers are selected as the top 1% of ESI papers (Google H-index 61, Selected Clarivate Web of Sciences, Cited by Scholars: 2019 (crossing filed), 2020/2021/2022 (Engineering)). Served as deputy editor for seven top journals, including *Control Engineering Practice*, *IEEE Trans Power Systems*, and *CSEE Journal of Energy and Power*.

energies

MDPI

Editorial

Exploring Sustainable Development of New Power Systems under Dual Carbon Goals: Control, Optimization, and Forecasting

Bo Yang [1,*], Jinhang Duan [1], Zhijian Liu [1] and Lin Jiang [2]

[1] Faculty of Electric Power Engineering, Kunming University of Science and Technology, Kunming 650500, China; 18787595370@163.com (J.D.); 12310010@kust.edu.cn (Z.L.)
[2] Department of Electrical Engineering and Electronics, University of Liverpool, Liverpool L69 3GJ, UK; l.jiang@liv.ac.uk
* Correspondence: yangbo_ac@outlook.com

In the context of achieving carbon neutrality, the substantial integration of high proportions of renewable energy sources has significantly impacted the dynamic characteristics of power systems, including frequency stability, voltage security, and synchronous stability, thereby posing formidable challenges to the secure and stable operation of power systems [1]. Various measures can be undertaken to enhance power system stability. For instance, enhancements to floating platforms can bolster the stability of offshore floating wind turbine units [2]. Furthermore, the introduction of novel devices can fulfill requirements for power density or improve power generation efficiency [3–5]. Nevertheless, the key to enhancing system stability lies in aspects such as grid control and optimization, planning and scheduling, and the accurate forecasting of renewable energy generation.

With the introduction of new energy sources, the accurate prediction of photovoltaic (PV) and wind power in power systems becomes increasingly critical. A deep learning-based comprehensive multi-site wind speed prediction model graph embedding-graph isomorphism-based gated recurrent unit demonstrates superior forecasting accuracy. Experimental results indicate that this model achieves a mean square error of 0.8457 m/s and a root mean square error (RMSE) of 0.9196 m/s in predicting wind speeds at a height of 10 m, thereby enhancing the reliability of wind speed forecasts and contributing to improved power generation efficiency in electric systems [6]. On the other hand, Fu et al. [7] proposed a novel hierarchical-improved variational mode decomposition–temporal convolutional network-gated recurrent unit multi-head attention PV power prediction mechanism, integrating improved variational mode decomposition, time convolutional network-gated recurrent unit architecture, and enriched multi-head attention mechanisms. Experimental results demonstrate a substantial decrease in RMSE and mean absolute error by 55.1% and 54.5%, respectively, compared to traditional methods, particularly evident during fluctuations in PV power under adverse weather conditions. The adoption of this approach effectively enhances the accuracy of prediction, which is crucial for ensuring the secure scheduling and stable operation of power systems. In renewable energy generation systems, the importance of energy storage systems is increasingly emphasized, and accurate battery modeling is essential for optimizing system control and enhancing power generation efficiency. However, due to insufficient data and noise interference, employing a generalized regression neural network for denoising and predicting real-time *V-I* data proves to be an effective strategy [8].

Control and optimization of the power grid are pivotal for enhancing the overall operational stability of the electricity system. Precise control of various grid components can typically be achieved through rational methodologies or the development of efficient controllers. For example, an optimized voltage feedforward control strategy is proposed to

Citation: Yang, B.; Duan, J.; Liu, Z.; Jiang, L. Exploring Sustainable Development of New Power Systems under Dual Carbon Goals: Control, Optimization, and Forecasting. *Energies* **2024**, *17*, 3909. https://doi.org/10.3390/en17163909

Received: 2 August 2024
Accepted: 7 August 2024
Published: 8 August 2024

reshape the phase characteristics of system output impedance, thereby significantly enhancing the system's adaptability to grid impedance. This approach addresses issues introduced by voltage feedforward and equivalent parallel virtual impedance correction, ultimately improving system stability [9]. Moreover, Zhao et al. [10] introduced a receding Galerkin optimal controller combined with a high-order sliding mode disturbance observer scheme. This enables precise tracking of large-scale load setpoints for boiler–turbine units under varying unknown disturbances, enhancing system robustness against interference and improving tracking capabilities within operational constraints. Additionally, Li et al. [11] presented adaptive controllers such as a dynamic model-based adaptive controller and an enhanced quasi-proportional-resonant controller, effectively mitigating harmonic and thrust pulsations while maintaining stability in linear induction motor systems. Furthermore, Zou et al. [12] introduced an intelligent, nonlinear robust controller using a chaos particle swarm gravity search optimization algorithm for diesel generator speed and excitation control. This approach significantly enhances dynamic accuracy and disturbance suppression capabilities compared to traditional proportional–integral–derivative methods. Given the multifaceted influences on grid components, establishing a rational simulation platform and devising effective control strategies remain critical despite the considerable challenges involved [13,14].

Effective control of the power grid not only enhances its stability but also closely correlates with economic benefits. A multi-stage segmented control model based on sensitivity analysis is proposed to address the stochastic nature of wind power generation output [15]. The results demonstrate that this control method achieves a balance between economic efficiency and safety by minimizing the risk of cascading failures in large-scale wind power systems. Additionally, Wang et al. [16] developed a controller for LED lights that utilizes simulation devices to achieve multi-channel intelligent sensing, dimming, and control, which is significant for alleviating global energy shortages. Furthermore, Akdeniz et al. [17] introduced a novel preventive control approach using non-operational vulnerability indices. The suggested decision support system is expected to assist power system operators in making critical decisions to adjust power system configurations in response to potential risks of cascading failures and emergencies. In practical emergency situations, this system is anticipated to reduce total operating costs by more than 20%.

The planning and scheduling of power systems significantly influence their reliability and stability. Effective grid planning and scheduling ensure stable operation under diverse load conditions, reduce operational costs, optimize power resource allocation, and enhance power supply efficiency [18]. Dong et al. [19] proposed a two-level optimization configuration method to mitigate challenges like voltage standard violations, excessive currents, and power imbalances resulting from integrating distributed PV systems into distribution grids. This method effectively balances grid capacity, minimizes active power losses, and reduces operating costs. In addition to the two-level optimization configuration, game theory finds wide application in economic dispatch within power systems. For example, the master–slave game model is employed to analyze the supply–demand interaction between 'grid-users' and retailers for implementing demand response strategies [20,21]. These interactive games allow participants to continually share their interest information to achieve a Nash equilibrium solution that aligns with their respective interests. He et al. [22] utilized cooperative game (CG) methods to dynamically coordinate wind, PV, and thermal power generation scheduling, optimizing energy complementary transmission systems' efficiency and facilitating the effective integration of new and traditional energy sources. Li et al. [23] utilized Stackelberg strategies to establish an equilibrium model for microgrid users and employed a multi-group genetic algorithm for iterative solutions. This model promotes user-centric energy utilization, enhances economic and system benefits, and effectively implements peak shaving and valley filling strategies. Current approaches to power system planning and scheduling are diverse, addressing uncertainties in renewable energy generation and decision-makers' varying risk attitudes. Wang et al. [24] introduced a novel adaptive evaluation model for power system grid planning based on GRA-TOPSIS integrated CG and improved cumulative prospect theory.

The proliferation of electrical systems has led to an increasing demand for effective system state management. An interval state estimation method has been proposed that considers measurement correlations within these systems [25]. This approach involves comparing the state estimation interval with the system's safe operating interval to ascertain whether the system's safe operational range adequately covers the estimated state interval. Such methodology enhances situational awareness among system administrators, facilitating precise adjustments and control of energy systems as needed. Moreover, further advancements in system management efficiency can be achieved through the development of an intelligent question–answering system tailored for electrical system regulations. For instance, an enhanced BERTserini algorithm based on the BERT model is introduced for intelligent interpretation of electrical regulations [26]. This approach obviates the manual organization of professional question–answer pairs, thereby significantly reducing labor costs compared to conventional question–answering systems. Furthermore, it demonstrates superior accuracy and faster response times in providing solutions.

All in all, the urgency to enhance the stability and efficiency of power systems is growing, particularly with the large-scale integration of renewable energy sources into the grid. Renewable energy sources, such as wind and solar power, exhibit intermittent and fluctuating characteristics that challenge grid stability. As the focus on energy efficiency intensifies, there is a pressing need to leverage artificial intelligence technology and dynamic pricing mechanisms to balance supply and demand, thereby improving efficiency while maintaining economic viability. Looking ahead, the enhancement of power system stability will increasingly depend on advancements in artificial intelligence technology and big data analytics. These developments will enable more accurate forecasting and responsive measures, as well as optimized scheduling and planning. Concurrently, the advancement of novel control technologies and systems will facilitate precise control, ensuring high efficiency, economic feasibility, stability, and environmental sustainability. Such measures are critical for advancing the sustainable development of power systems.

Author Contributions: B.Y.: conceptualization, methodology, and writing; J.D.: investigation and writing—review; Z.L.: investigation and writing—review; L.J.: supervision and writing—review and editing. All authors have read and agreed to the published version of the manuscript.

Funding: The authors received no specific funding for this study.

Conflicts of Interest: The authors declare no conflicts of interest.

References

1. Zhao, Y.; Chen, J. Collaborative optimization scheduling of multi-microgrids incorporating hydrogen-doped natural gas and P2G–CCS coupling under carbon trading and carbon emission constraints. *Energies* **2024**, *17*, 1954. [CrossRef]
2. Zhao, H.; Wu, X.; Zhou, Z. Exploring motion stability of a novel semi-submersible platform for offshore wind turbines. *Energies* **2024**, *17*, 2313. [CrossRef]
3. Yan, J.; Wu, H.; Fu, X.; Li, M.; Yu, Y. Modeling and suppression of conducted interference in flyback power supplies based on GaN devices. *Electronics* **2024**, *13*, 2360. [CrossRef]
4. Liu, H.; Wei, X.; Ai, J.; Yang, X. A layered parallel equaliser based on flyback transformer multiplexed for lithium-ion battery system. *Energies* **2024**, *17*, 754. [CrossRef]
5. Zidek, M.; Vanek, F.; Jezerska, L.; Prokes, R.; Gelnar, D. New equipment for determining friction parameters in external conditions: Measurements for the design. *Processes* **2023**, *11*, 3348. [CrossRef]
6. Wu, H.; Chen, H. Multi-site wind speed prediction based on graph embedding and cyclic graph isomorphism network (GIN-GRU). *Energies* **2024**, *17*, 3516. [CrossRef]
7. Fu, H.; Zhang, J.; Xie, S. A novel improved variational mode decomposition-temporal convolutional network-gated recurrent unit with multi-head attention mechanism for enhanced photovoltaic power forecasting. *Electronics* **2024**, *13*, 1837. [CrossRef]
8. He, P.; Zhou, X.; Liu, M.; Xu, K.; Meng, X.; Yang, B. Generalized regression neural network based meta-heuristic algorithms for parameter identification of proton exchange membrane fuel cell. *Energies* **2023**, *16*, 5290. [CrossRef]
9. Wang, S.; Cui, K.; Hao, P. Grid-connected inverter grid voltage feedforward control strategy based on multi-objective constraint in weak grid. *Energies* **2024**, *17*, 3288. [CrossRef]

10. Zhao, G.; Sun, Y.; Su, Z.-G.; Hao, Y. Receding Galerkin optimal control with high-order sliding mode disturbance observer for a boiler-turbine unit. *Sustainability* **2023**, *15*, 10129. [CrossRef]
11. Li, Y.; Dong, W.; Lv, C.; Wang, Z.; Zhang, Y. Control strategies of thrust ripple suppression for electromagnetic microgravity facility. *Electronics* **2024**, *13*, 1247. [CrossRef]
12. Zou, Y.; Xiao, B.; Qian, J.; Xiao, Z. Design of intelligent nonlinear H_2/H_∞ robust control strategy of diesel generator-based CPSOGSA optimization algorithm. *Processes* **2023**, *11*, 1867. [CrossRef]
13. Posada Contreras, J.; Rosas-Caro, J.C. Hardware-in-the-loop emulation of a SEPIC multiplier converter in a photovoltaic system. *Electricity* **2024**, *5*, 426–448. [CrossRef]
14. Wang, H.; Zhang, L.; Sun, Y.; Zou, L. Vibration scale model of a converter transformer based on the finite element and similarity principle and its preparation. *Processes* **2023**, *11*, 1969. [CrossRef]
15. Cheng, L.; Wang, T.; Wu, Y.; Gao, Z.; Ji, N. Hierarchical blocking control for mitigating cascading failures in power systems with wind power integration. *Energies* **2024**, *17*, 442. [CrossRef]
16. Wang, Z.; Zhang, D.; Li, J.; Zhang, W. An intelligent controller of LED street light based on discrete devices. *Energies* **2024**, *17*, 1838. [CrossRef]
17. Akdeniz, E.; Bagriyanik, M. A preventive control approach for power system vulnerability assessment and predictive stability evaluation. *Sustainability* **2023**, *15*, 6691. [CrossRef]
18. Ha, P.T.; Tran, D.T.; Phan, T.M.; Nguyen, T.T. Maximization of total profit for hybrid hydro-thermal-wind-solar power systems considering pumped storage, cascaded systems, and renewable energy uncertainty in a real zone, Vietnam. *Sustainability* **2024**, *16*, 6581. [CrossRef]
19. Dong, A.; Lee, S.-K. Research on a distributed photovoltaic two-level planning method based on the SCMPSO algorithm. *Energies* **2024**, *17*, 3251. [CrossRef]
20. Yang, Y.; Chen, T.; Yan, H.; Wang, J.; Yan, Z.; Liu, W. Optimization operation strategy for shared energy storage and regional integrated energy systems based on multi-level game. *Energies* **2024**, *17*, 1770. [CrossRef]
21. Hu, D.; Wang, T. Optimizing power demand side response strategy: A study based on double master-slave game model of multi-objective multi-universe optimization. *Energies* **2023**, *16*, 4009. [CrossRef]
22. He, W.; Liu, M.; Zuo, C.; Wang, K. Massive multi-source joint outbound and benefit distribution model based on cooperative game. *Energies* **2023**, *16*, 6590. [CrossRef]
23. Li, B.; Li, Y.; Li, M.-T.; Guo, D.; Zhang, X.; Zhu, B.; Zhang, P.-R.; Wang, L.-D. Research on optimal scheduling of multi-energy microgrid based on Stackelberg game. *Processes* **2023**, *11*, 2820. [CrossRef]
24. Wang, Y.; Yan, C.; Wang, Z.; Wang, J. Adaptability evaluation of power grid planning scheme for novel power system sonsidering multiple decision psychology. *Energies* **2024**, *17*, 3672. [CrossRef]
25. Huang, Y.; Feng, L. Interval state estimation of electricity-gas systems considering measurement correlations. *Energies* **2024**, *17*, 755. [CrossRef]
26. Gao, M.; Li, M.; Ji, T.; Wang, N.; Lin, G.; Wu, Q. Key technologies of intelligent question-answering system for power system rules and regulations based on improved BERTserini algorithm. *Processes* **2024**, *12*, 58. [CrossRef]

sustainability

MDPI

Article

A Preventive Control Approach for Power System Vulnerability Assessment and Predictive Stability Evaluation

Ersen Akdeniz [1,2,*] and Mustafa Bagriyanik [1]

1 Department of Electrical Engineering, Istanbul Technical University, Istanbul 34469, Turkey
2 Siemens Gamesa Renewable Enerji A.Ş, Izmir 35530, Turkey
* Correspondence: eakdeniz@itu.edu.tr

Abstract: Early detection of cascading failures phenomena is a vital process for the sustainable operation of power systems. Within the scope of this work, a preventive control approach implementing an algorithm for selecting critical contingencies by a dynamic vulnerability analysis and predictive stability evaluation is presented. The analysis was carried out using a decision tree with a multi-parameter knowledge base. After the occurrence of an initial contingency, probable future contingencies are foreseen according to several vulnerability perspectives created by an adaptive vulnerability search module. Then, for cases identified as critical, a secure operational system state is proposed through a vulnerability-based, security-constrained, optimal power flow algorithm. The modular structure of the proposed algorithm enables the evaluation of possible vulnerable scenarios and proposes a strategy to alleviate the technical and economic impacts due to prospective cascading failures. The presented optimization methodology was tested using the IEEE-39 bus test network and a benchmark was performed between the proposed approach and a time domain analysis software model (EMTP). The obtained results indicate the potential of analysis approach in evaluating low-risk but high-impact vulnerabilities in power systems.

Keywords: power system vulnerability assessment; preventive control; critical contingency selection; decision tree-based stability evaluation

Citation: Akdeniz, E.; Bagriyanik, M. A Preventive Control Approach for Power System Vulnerability Assessment and Predictive Stability Evaluation. *Sustainability* **2023**, *15*, 6691. https://doi.org/10.3390/su15086691

Academic Editors: Bo Yang, Zhijian Liu and Lin Jiang

Received: 23 March 2023
Revised: 11 April 2023
Accepted: 12 April 2023
Published: 15 April 2023

1. Introduction

Power system (PS) operation is a consumption–generation balancing act where operational costs aim to be minimized traditionally. In order to preserve the balance, system operators are obliged to take action in order to prevent cascading failures which might also lead to partial or total blackouts [1,2]. Thus, PS operators need to prepare emergency plans which requires a detailed analysis of their system including various aspects [3]. This phenomenon was also proven in recent blackout events, such as those in India and Turkey [4], where blackouts of whole electrical grids were caused due to operational failures, and the South Australian Transmission Grid failure which was due to insufficient analysis of vulnerabilities as a result of extreme weather conditions [5]. Any critical component failure may have negative impacts on system operational costs due to the ramp-up/down of generators or unserved energy penalties [6–9]. In practice, PS planners design the grid to be sufficient to cope with contingencies by allocating adequate reserves in generator production and transmission lines to provide a certain level of redundancy in case of pre-estimated critical contingencies [10,11]. However, as the system further expands from its original design, the implementation of additional reserves/capabilities is limited mostly by economic and environmental constraints, which weaken the hand of the PS operators (PSO) while keeping the system within the limits defined in power quality standards [12,13]. If a disturbance continues and required corrective action is not implemented, the system is expected to be drawn in an emergency state for which boundary limits are exceeded and as a result, the power system stability will be distorted [14–17]. Under these circumstances,

PSOs should take sufficient measures in order to maintain a secure operation, which might include load shedding and partial islanding as an alternative solution to prohibit a total blackout in the system [18–22].

In power system analysis, power flow optimization is known to be a critical topic attracting the attention of many researchers. Although several attempts have been made to model large-scale systems, none have proven to be completely successful [14,23]. The problem complexity arises not only from the problem size but also the complex non-linear dynamics of power systems. Most of the traditional OPF problems involve classical contingency constraints mainly relying on the electrical parameters of the system [17,18]. Steady-state security analysis is also a commonly used tool for determination of possible limit violations after credible outages [16,24]. However, as the grid complexity increases, the possible combinations of N-k contingencies drastically increase, which makes the analysis very time consuming, and makes real time analysis and online decision-making analysis almost unpractical [9,16]. Furthermore, as the cascading failure develops in a dynamic nature, steady state approximations become less reliable in predicting the behavior of the distorted system [25,26]. In such cases, transient stability analysis techniques and real-time transient models are implemented for obtaining a better system representation. However, as the dimension of the studied system enhances, implementation of time-domain analysis might become impractical due to grid complexity.

Most of the available work on contingency analysis studies uses exhaustive search and/or Monte Carlo simulation techniques to try and determine the worst contingencies [14,27,28]. For most of the N-2 contingency evaluation studies, electrically coupled pairs are generally used as a special case, such as after a line trip due to an initial fault, a neighboring or a parallel line dropping out of service due to overload, or a relay trip [17]. This type of approach provides a straightforward and reasonable way of determining the most probable contingencies in a classical manner. However, this approach might shadow contingencies due to new threats known as intentional attacks (cyber or physical) and adverse weather impacts.

Some recent studies [3,5,29] on power system vulnerability analysis are based mainly on topology and flow-based methods. Topology-based methods tend to be strongly dependent on grid topology, which sometimes does not discover the ongoing active phenomena inside. In flow-based approaches, energy balance equations and physical properties of the system are more important. However, as the system dimension expands, the solution complexity will also increase, which makes analysis inefficient. The techno-economic results of intentional attacks on transmission lines are demonstrated in [9], which indicates the necessity for PSOs to optimize their system to be resilient against such conditions. Critical equipment determination generally targets impacts due to single branch failure or randomly generated subsets. In [28], the proposed algorithm aims to select multiple contingency groups which can result in cascading failures. However, in these approaches there is a risk that the worst blackouts will not be detected and that high impact, small subsets will not be covered. In [7,8], the proposed methods are dependent on identifying over-limits and loss of loads but do not cover the complete process of disturbances. In [29], the impact of random line failures on grid vulnerability is studied, and it is concluded that small and large failures can induce similar performance loss in robustness. In [30], the overall vulnerability assessment of the power grid is made via structural and operative vulnerability indices defined for buses.

In our study, a different perspective of preventive control is presented by implementing a critical contingency selection algorithm through a dynamic vulnerability analysis module and a decision tree-based model stability evaluation with a multiple parameter knowledge base. In Section 2, the methodology of the solution approach is presented. In summary, for each critical scenario, an adaptive rescheduling and load shedding algorithm is used which considers operational and non-operational vulnerabilities for lines while bus loadings randomly changed within the $\pm 20\%$ range. The secure system state is derived from the decision tree (DT) evaluation module using key performance indices as prediction parameters. Then, the critical point for the system where stability deteriorates is detected.

At this point, a corrective security-constrained and genetic algorithm-based optimal power flow (CSC-OPF) is implemented to determine the secure system settings via optimizing the generator active power output and load shedding if necessary. In Section 3, the developed methodology is tested using the IEEE-39 bus network and a benchmark is made between the proposed approach and time domain analysis software (EMTP) model for showing the effectiveness of the method in the determination of the stability deviation point for the tested contingency sequence. The details of the CSC-OPF algorithm and a sample result of the secure system state transfer cost calculation is presented for a specific contingency case. In Section 4, a brief discussion on findings and future work is presented.

2. Methodology

Within the scope of this study, a vulnerability analysis tool is developed enabling the user to define various operational scenarios for the selected test cases via the developed graphical user interface. Initially, a user can choose the test case for analysis with the required operational constraints and later can calculate the performance indices defined in Table 1. Then, the program provides the user individual (operational (OPI), intentional attack (TAI), adverse weather (AWI)) and total vulnerability indices (TVI) and related line rankings corresponding to their relevant vulnerability type. As a result, the vulnerability analysis program (VAP) selects the most credible contingency cases, thus narrowing the possible contingency subsets.

Table 1. Performance indicators used as DT predictors.

Contingency-Based PIs	Severity-Based PIs
$PI_{P_c} = \sum_{l=1}^{n_{line}} \frac{w_P}{2} \left(\frac{P_l}{P_{l,max}} \right)^{2N}$	$PI_{P_s} = \sum_{i=1}^{n_{line}} w_p \frac{P_{loading} - P_{lim}}{1 - P_{cri}}$
$PI_{Vb_c} = \sum_{i=1}^{n_{bus}} \frac{w_b}{2} \left(\frac{\lvert V_i - V_{bc} \rvert}{\lvert V_{max} - V_{min} \rvert} \right)^{2N}$	$PI_{Vb_s} = \sum_{i=1}^{n_{bus}} w_p \frac{1 - V_{b,i}}{V_{cri}}$
$PI_{Qg_c} = \sum_{i=1}^{n_{gen}} \frac{w_g}{2} \left(\frac{Q_{g,i}}{Q_{g,max}} \right)^{2N}$	$PI_{Qg_s} = \sum_{i=1}^{n_{gen}} w_g \left(\frac{Q_g}{Q_{g,max}} - Q_{g,lim} \right)^{2N}$
$PI_{Pg_c} = \sum_{i=1}^{n_{gen}} \frac{w_g}{2} \left(\frac{P_g}{P_{g,max}} \right)^{2N}$	$PI_{Pg_s} = \sum_{i=1}^{n_{gen}} w_g \left(\frac{P_g}{P_{g,max}} - P_{g,lim} \right)^{2N}$
$PI_{Ang_c} = \sum_{l=1}^{n_{line}} w_{ang} \frac{\lvert \Delta\theta_i \rvert}{90}$	$PI_{Ang_s} = \sum_{l=1}^{n_{line}} w_{ang} \frac{\lvert \Delta\theta_i - \theta_{lim} \rvert}{45}$
$PI_{LoL} = \sum_{l=1}^{n_{bus}} w_{LoL} \frac{P_{LoL}}{P_{L\_tot}}$; $PI_{LoG} = \sum_{l=1}^{n_{gen}} w_{LoG} \frac{P_{LoG}}{P_{L\_totG}}$; $PI_{IB} = w_{LoL} \sum_{l=1}^{n_{bus}} isl_{bus}$	
$OPI_{ctg} = \sum \left(PI_{P_C} + PI_{Vb_C} + PI_{Qg_C} + PI_{Ang_C} + PI_{LoL} + PI_{LoG} + PI_{IB} \right)$	$OPI_{sev} = \sum \left(PI_{P_s} + PI_{Vb_s} + PI_{Qg_s} + PI_{Ang_s} + PI_{LoL} + PI_{LoG} + PI_{IB} \right)$

After this selection, with the help of a decision tree-based security analysis module, the impact of each selected contingency on system stability is evaluated. According to the results obtained from the CSC-OPF module, the program tries to optimize the generator's active power settings to obtain the best fitness value defined by overall cost function. If system limit constraints are still violated, load shedding is implemented for transferring the system to a more secure state which enables convergence in PF. The cost of the system for current operational status and the cost of transfer to a more secure state is calculated via the secure transfer cost (STC) calculation module, and if a feasible result is obtained, secure system transfer conditions are applied by the decision support module.

2.1. Contingency Selection

In contingency selection, direct and indirect methods are widely used [4]. The contingency impacts defined in terms of performance indices (such as active/reactive power and voltage level variations) are regarded as direct methods [5]. However, some critical contingencies were reported because of natural calamities, malicious attacks, or maloperations [6,18–20]. Thus, in order to represent real system conditions a broader approach is needed. Our approach also considers non-technical parameters, which have previously been defined as non-operational performance indices [31,32]. With the help of the proposed vulnerability evaluation module, critical contingencies were selected according to the total vulnerability evaluation (TVE) module where operational and non-operational constraints are used in the determination of the most vulnerable points in terms of operational, adverse weather, and intentional attack considerations. Thus, instead of investigating all mathematically possible combinations, such as those made in a brute-force contingency analysis, a narrowed contingency subset was obtained using a fuzzy interference system which reduced the analysis time drastically. Then, for each credible contingency a fitness value for an objective function was evaluated, which aims to keep the load shed at minimum level while maintaining low bus voltage deviations and transmission system losses. After determining the optimal system configurations for each contingency from the reduced subset, the optimum generator set points, and if necessary, other relevant control actions such as load shedding, were determined. By doing so, the system will be able to be survive after the occurrence of any critical pre-defined contingency, thus preventing cascading failures and partial blackouts in the system.

2.2. Decision Tree-Based Security Analysis

Decision trees for classification are an effective artificial intelligence tool for solving high-dimensional classification problems [20]. The principal motivation is to form a predictive model of the system that covers all possible operational scenarios [25]. The complicated classification problems are converted to a set of inequality equations composed of pre-defined predictor parameters or their linear combinations [17,23].

During the training process of decision trees, a minimum of 10 times more than the number of degrees of freedom model is required to cover all possible contingencies and operational scenarios [33]. If the training set is large enough, the quality of the obtained results will be good (the details of training and dataset creation are presented in Section 2.5.2). In this module, base scenario and contingency scenarios are studied using prediction results described as secure or insecure where scenarios are obtained via loading variations under line outage conditions considering 2 consecutive losses in the same time frame. The predictors are derived from contingency- and severity-based performance indices. Their combinations and parameter details are given in Table 1 and the methodology is described in [34].

In this work, the MATLAB R2021a statistical and machine learning toolbox [35] was used for the creation of classification trees in order to interpret the relationship between the prediction variable and target variable values, which was system stability in our case. Classification trees are the foundation for other, similar machine learning algorithms implementing different applications of decision trees. Classification trees, which were described first in [34], are used as a decision tree analysis method. In this approach, decision trees are used where each node becomes a split point for a predictor variable. The final convergence of the test network under various operational conditions is used as a stability indicator, and performance indicators (PI) described in Table 1 are used as predictor variables.

2.3. Main Algorithm for Decision Support

The proposed decision support tool given in Figure 1 performs as follows:

- Topological system information (i.e., switch status, line/bus outages) and electrical parameters are obtained to determine the initial operating point (OP) of the system.

- The environmental and weather forecast-related information is obtained and assumed to be changing in accordance with different zones defined for the test network.
- With this initial information, a total vulnerability evaluation of the system, including operational, environmental, and adverse weather-related indices, is made in order to select the most vulnerable parts of the system [34]. The probable and possible risks are combined to provide a more comprehensive contingency analysis of the system under study.
- From each module located in the basic vulnerability module/evaluation (BVE), a stack based on TVR is formed for which the size is determined according to the system operation requirements set by TSO.
- From each stack, total vulnerability ranking (TVR) modules provide separate subsets of vulnerable lines.
- Then, a pre-check of system security and stability is made via the critical contingency check module considering load variations at time step ($\Delta t = \tau$). If any critical contingency which requires immediate action is detected, the base critical contingency check (Base C3) module for rescheduling is applied in order to shift the system to a more secure operating point (OP_0′). Otherwise, the system is kept around the initial operating point while updating results obtained from BVE.
- After occurrence of the initial contingency, the test system is transferred to a new operating state (if the system satisfies basic N-1 requirements, if not it is expected that the results from the BVE are transferred to a secure operating point) which is described as disturbed operating point-1 (OP_1).

Then, the proposed standard vulnerability evaluation module (SVE) determines an updated ranking for the operational constraints to detect the most vulnerable parts of the system according to operational/electrical parameters, and the first subset of operational performance ranking (OPR) is provided.

- Finally, a comprehensive check of system security and stability is made via the critical contingency check (C3) module while updating the information received from SVE and considering the load variation uncertainty. Similarly, if any critical contingency is detected from N-1 contingency scenarios for which the candidates are prepared by SVE, the rescheduling and load shed (RLS) algorithm first tries rescheduling the available generators. If the reserve generation capability is not enough, then the proposed algorithm sheds the load according to the selection methodology proposed in the next section.

2.4. C3-RLS Algorithm

The proposed critical contingency check module is a DT-based system security evaluation method. The knowledge base of the module produced by the scenario generation module for which the operator can easily define various system operating conditions including load variations and pre-outaged components.

The flow process for the C3-RLS module is described in Figure 2.

- The standard vulnerability evaluation module online monitors the system status and other weather- and intrusion-related information. As the system conditions change above the predefined limits, the output stacks of the related vulnerability ranking modules are updated automatically. Each module presents a separate critical line list accordingly.
- Then, all candidates are stacked in the critical contingency pool to be analyzed via the proposed decision tree stability (DTS) evaluation module. In the DTS module, a stability analysis is carried out for which details are presented in part 2.5. Initially, generator re-dispatch values are determined for the foreseen instability. If rescheduling is not enough, loads are shed according to the indices computed via the power flow contribution matrix. The optimal values are calculated via the corrective security-constrained AC optimal power flow (CSC-OPF) algorithm.

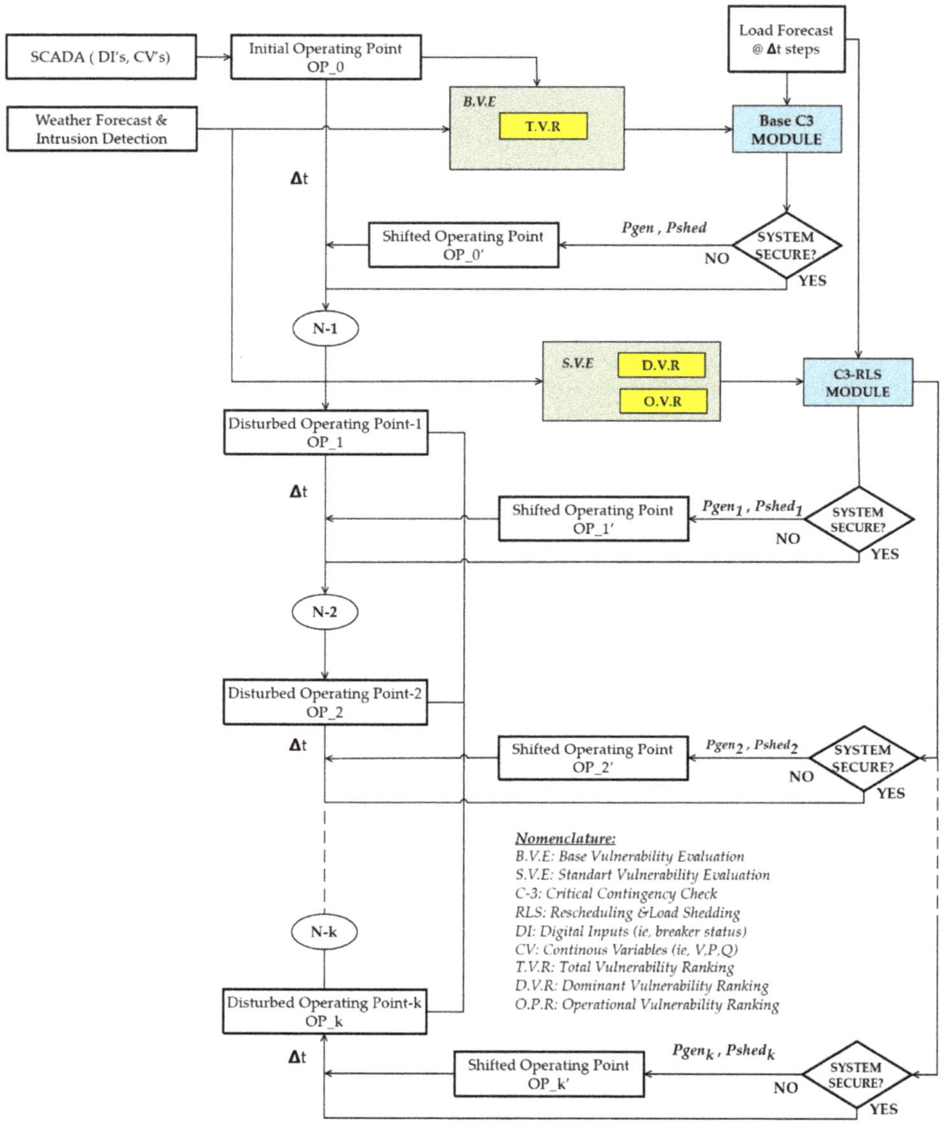

Figure 1. Proposed N-k contingency evaluation algorithm based on BVE, SVE, and C3-RLS modules.

In addition to operational costs based on generator active power, market prices related to the ramp up and down of generators, load shed, or unserved customer penalty costs are also foreseen in this module.

The knowledge-based decision support tool provides the best what-if operational scenarios and tries to optimize and transfer the electrical test system to a more secure state.

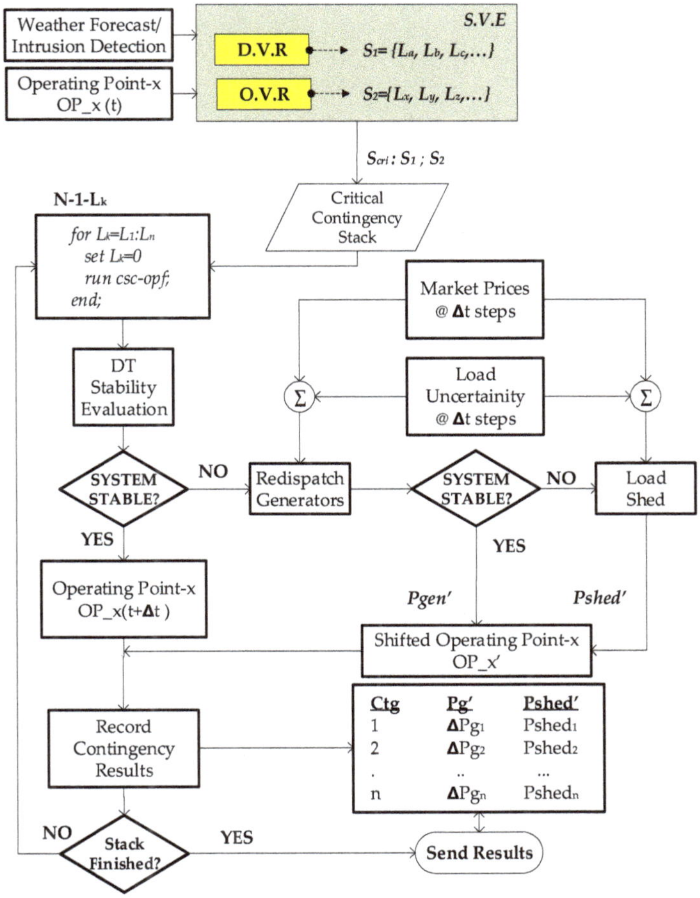

Figure 2. Proposed C3-RLS module flow chart.

2.4.1. Corrective Security-Constrained OPF Algorithm

The implemented corrective security-constrained optimal power flow (CSC-OPF) is based on an optimization approach using a genetic algorithm (GA) for finding the best combination of corrective actions and MATPOWER as power system equation solver. According to the objective function defined in (1), the GA implements an iterative search aiming minimize the costs for each system state change from u_0 to u_k due to the kth contingency:

$$\min_{u_0 \to u_k} \{f(u_k) + z(u_k)\} \tag{1}$$

$$z(u_k) = \left(w_{LoL} \times p_{LoL,k} + w_{Vb} \times \Delta V_{b,k} + w_{Vb_{ct},k} \times \Delta V_{b_{ct},k} + w_{P_L,k} \times \frac{P_{L,k}}{P_{L,0}} \right) \tag{2}$$

subject to:

$$g(x_0,\ u_0, y_0) = 0$$

$$h(x_0,\ u_0, y_0) \geq 0$$

$$g(x_k,\ u_k, y_k) = 0$$

$$h(x_k,\ u_k, y_k) \geq 0$$

$$|u_k - u_0| \leq \Delta u^{max}$$

where $f(u_k)$ is the system operating cost, $z(u_k)$, defined in (2), is the performance fitness function for the system state change, $g(x,u)$ represent equality constraints for power flow equations, and $h(x,u)$ includes system inequality constraints for which the details are described explicitly in [36].

2.4.2. Load Shed Selection

In case of generation inadequacy where the existing reserves cannot meet load demand requirements, a load shedding action is employed according to the protection strategy of the power system. The type of load (critical or uncritical) is one of the most common approaches in selection of the first group of loads to be separated from the system. If the power system operator has a bunch of loads to be selected, one of the proper shedding sequence methods is defined according to the load participation matrix [7]. For a pre-defined amount of load separation, those which alleviate branch flows the most are selected for decreasing the stress level of the transmission system. The quantity of the load shed is defined in (3) and (4):

$$\Delta P_{shed} = \alpha_{shed} \times \Delta V_{avg} \tag{3}$$

$$\Delta V_{avg} = \frac{1}{\tau} \int_{t_0}^{t_k} (V(t_k) - V(t_0))dt \tag{4}$$

where α_{shed} is the empirical factor relating the amount of load to be shed to the average voltage drop for the time frame of state change.

2.5. *Knowledge Base and DT Rules Generation*

2.5.1. Creation of Knowledge Base

With the help of the developed n-K contingency case generator tool given in Figure 3 and according to topological and pre-defined environmental configuration, datasets can be created by load scaling iterations for representing several variations in loadings for any IEEE test system whose data is available in MATPOWER. For each scenario the performance indices are calculated, and the convergence result of load-flow is defined as output flag (0/1) to be used in the DT analysis tool.

2.5.2. Decision Tree Formation

For the creation of the learning dataset to be used as a base input for DT, the performance indices defined in Table 1 were calculated for base (no contingency), N-1, and N-2 contingency cases under ±20% load changes for a 100-case per system configuration, which creates 108,200 cases for the IEEE-39 bus network available in MATPOWER 7.1 [37]. In order to create the optimal DT for security evaluation, the MATLAB statistical and machine learning toolbox was used. The obtained DT for each studied case is given in Figure 4 and the DT rules defining the security boundaries are presented in Table 2, respectively:

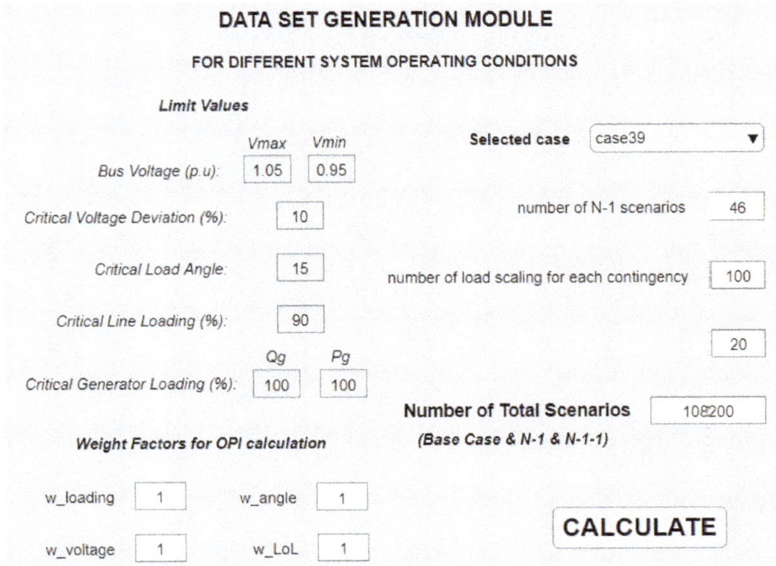

Figure 3. Dataset generation module developed in MATLAB-GUI.

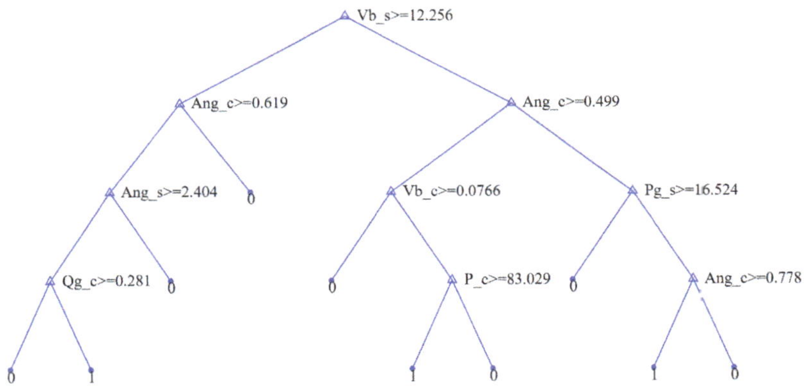

Figure 4. Classification tree for IEEE-39 security boundary definition (0: fail, 1: success).

2.6. Secure State Transfer Cost Calculation

For each selected contingency, the RLS module computes the proposed values for the generators, and if necessary, the loads to be shed. Defining the initial generator active power values as $P_{g,i}$, the proposed values are denoted as $P'_{g,i}$. The generic generation cost function given in (5) is as follows:

$$C_{gen} = a * P^2{}_{g,i} + b * P_{g,i} + c; \tag{5}$$

For which the relevant coefficients are defined in the MATPOWER case data file (i.e.; a = 0.01, b = 40, c = 0). The cost of the secure system transfer (C_{SST}) is defined as (6):

$$C_{SST} = C_{gen} * \sum_i^{n_g} Pg_i + C_{up} * \sum_i^{n_g} \Delta_{up} Pg_i + C_{down} * \sum_i^{n_g} \Delta_{down} Pg_i + C_{shed} * \sum P_{shed} \tag{6}$$

where C_{gen} is the base generation cost/MWh, C_{up} is the ramp-up cost/MWh, C_{down} is the ramp-down cost/MWh, and C_{shed} is the load shed penalty cost/MWh. The cost coefficients were determined heuristically according to the generic market costs available in [38], which can be modified via the developed GUI of the CSC-OPF module as given in Figure 5. Operational unit costs are defined as multiples of the standard production cost (SPC) which is atually the average hourly production cost defined in terms of $/MWh.

Table 2. DT rules defining security boundary for IEEE-39.

Order	Rules	Result
1	Vb_s < 12.256 and Ang_c ≥ 0.619	INSECURE
2	Vb_s < 12.256 and Ang_c < 0.619 and Ang_s ≥ 2.404	INSECURE
3	Vb_s < 12.256 and Ang_c < 0.619 and Ang_s < 2.4 and Qg_c < 0.281	INSECURE
4	Vb_s < 12.256 and Ang_c < 0.619 and Ang_s < 2.4 and Qg_c ≥ 0.281	SECURE
5	Vb_s ≥ 12.256 and Ang_c < 0.499 and Vb_c < 0.0766	INSECURE
6	Vb_s ≥ 12.256 and Ang_c < 0.499 and Vb_c > 0.076 and P_c < 83.029	SECURE
7	Vb_s ≥ 12.256 and Ang_c < 0.499 and Vb_c > 0.076 and P_c > 83.029	INSECURE
8	Vb_s ≥ 12.256 and Ang_c > 0.499 and Pg_s < 16.524	INSECURE
9	Vb_s ≥ 12.256 and Ang_c > 0.499 and Pg_s ≥ 16.524 and Ang_c < 0.778	SECURE
10	Vb_s ≥ 12.256 and Ang_c > 0.499 and Pg_s ≥ 16.524 and Ang_c ≥ 0.778	INSECURE

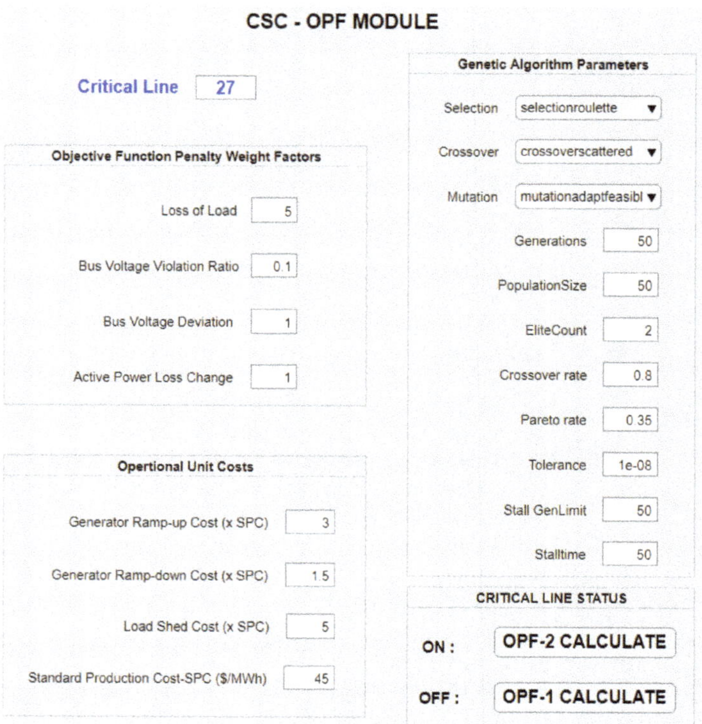

Figure 5. CSC-OPF module GUI.

3. Simulation and Results

The proposed analysis approach was tested on an IEEE-39 bus test network [11] for which implemented analysis tool developed in MATLAB. The system performance indices for IEEE-39 were calculated as given in Figure 6, in which OPI indicates operational performance impact indices, TAI indicates terrorist attack impact indices, AWI indicates the adverse weather impact indices, and TVI represents total vulnerability indices of the respective line outage obtained via fuzzy inference evaluation of three initially calculated indices for which the formulations are already defined in [34]. According to these rankings, the TSO analyst can choose the best-fitting vulnerability scenario that they would like to analyze depending on the specific vulnerability type. If they have no specific vulnerability interest they can choose the traditional analysis type, which is OPI-based analysis, or can choose all vulnerabilities included in a broader and possibilistic sense, which is TVI-based analysis.

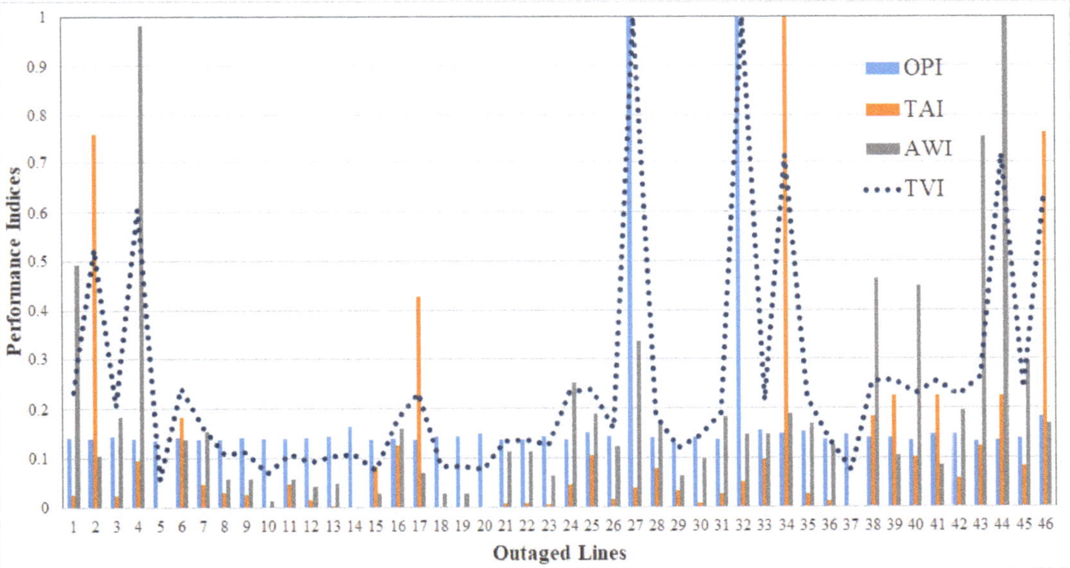

Figure 6. Performance indices distribution for IEEE-39 test network.

From the performance indices ranking, the total vulnerability ranking of 10 of the most critical lines is obtained. The contingency evaluation results for the IEEE-39 bus test system is obtained from the decision support module. The user interface is given in Figure 7 and the summary of the results is given in Table 3. This module enables the user to monitor two consecutive line outage contingencies at the same time.

3.1. Contingency Ranking and Stability Evaluation

In order to analyze higher N-k, where $k \geq 3$ the user can define these contingencies in the base case scenario definition, i.e., for analyzing $k = 5$, the base case conditions should be set to N-3 system conditions.

For the IEEE-39 test network, the selected operational performance indices-based contingencies, the obtained important PIs, and the loss of load (LoL) variations are given in Figure 8. It is observed that the IEEE-39 bus test system stability deteriorates after the N-4 contingency level. At this point, the specific contingency case must be benchmarked in a time domain analysis software to validate the critical contingency detection approach.

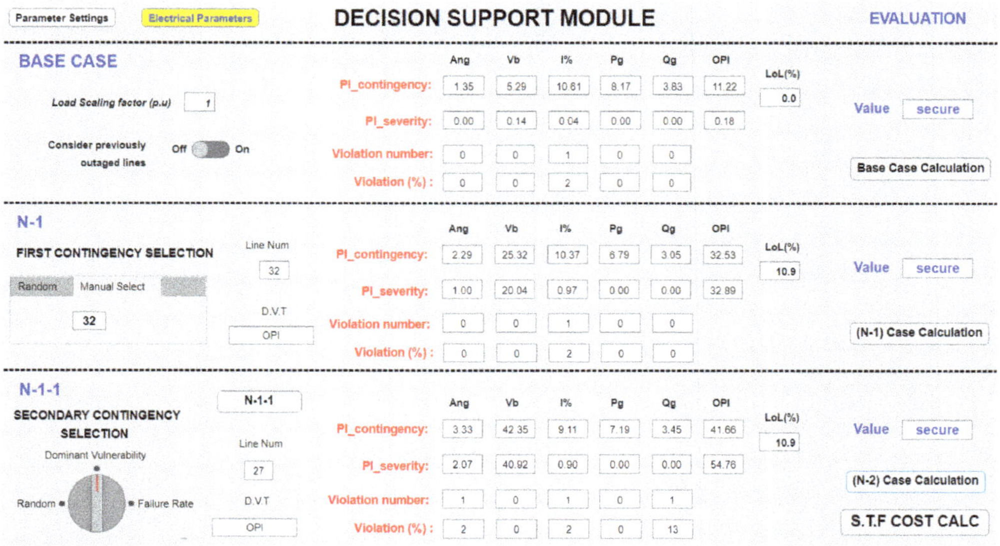

Figure 7. Decision support module GUI.

Table 3. Worst contingencies for IEEE-39 based on TVI ranking.

Order	Line ID	From–to Bus	DT Evaluation	Vulnerability Type	Isolated Bus Number
1	27	16–19	Insecure	OPI	4
2	32	19–20	Insecure	OPI	2
3	44	26–29	Secure	TAI	-
4	34	20–34	Insecure	TAI	1
5	46	28–29	Insecure	OPI	-
6	4	2–25	Insecure	AWI	-
7	2	1–39	Secure	TAI	-
8	43	26–28	Secure	AWI	-
9	39	23–36	Insecure	TAI	1
10	38	23–24	Secure	TAI	-

The foreseen cascading failure sequence was modeled in EMTP software [39] to obtain the dynamic behavior of the system based on generator active power outputs and rotational speeds. The connectivity diagram of the test model created within the developed interface with the most critical contingencies indicated in accordance with the dominant vulnerability types (OPI, AWI, TAI), as shown in Figure 9. For simulating the component outages, disconnectors are added to the relevant branches.

In order to simulate the consecutive cascading failure behavior, each component outage is assumed to occur at each 200 ms, respectively. The simulated behavior for the transient system model is modeled in an EMTP environment. Throughout the simulation, the active power output of the generators is given in Figure 10 and the rotational speed variation of the generators are presented in Figure 11.

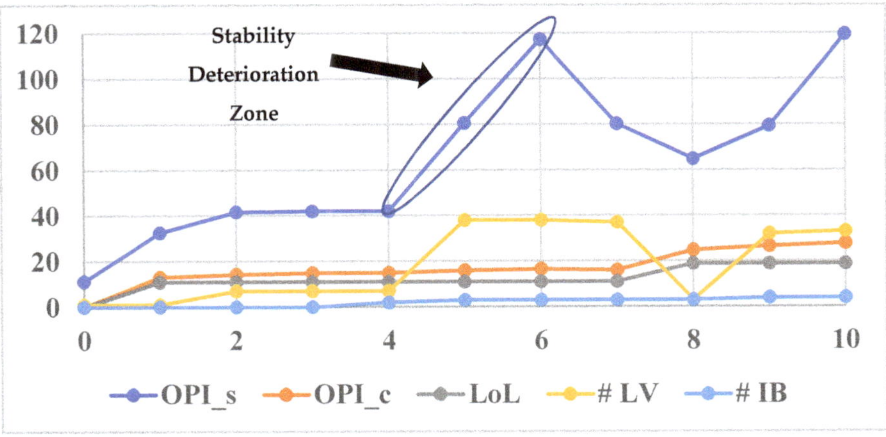

Figure 8. Performance indicators with respect to N-k contingencies (k = 1 to 10).

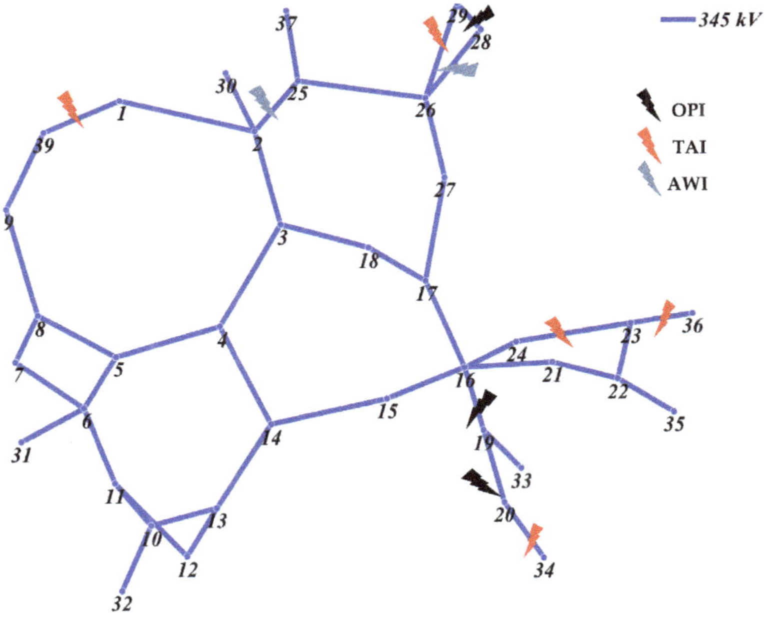

Figure 9. IEEE-39 bus system connectivity diagram indicating most vulnerable points according to OPI, AWI, and TAI indices.

As can be seen in Figure 9, the system stability starts to substantially deteriorate after the fourth (N-4) contingency, which corresponds to the 0.8 to 1.2 s interval zone shown in Figures 10 and 11. It is seen that after the sixth contingency, the severity related operational performance (OPI_s) and the number of limit violation indices (#LV) begin to change substantially. We can also observe slight changes in similar critical performance indicators, such as the related contingency operational performance (OPI_c), loss of load (LoL), and the number of isolated buses (#IB). It can be concluded that although the proposed algorithm is based on a steady state analysis tool where operational scenarios are defined in accordance with several N-k contingencies and load variations, by coupling it with decision tree evaluation using performance indices as predictive parameters, the proposed algorithm predicts good results for the detection of instability. It is also worth noting that the critical point where the generator angles start to deviate from original values (around t = 1.2 s), as seen in Figure 11, coincides with the N-6 contingency case shown in Figure 9 where deterioration of stability is clear.

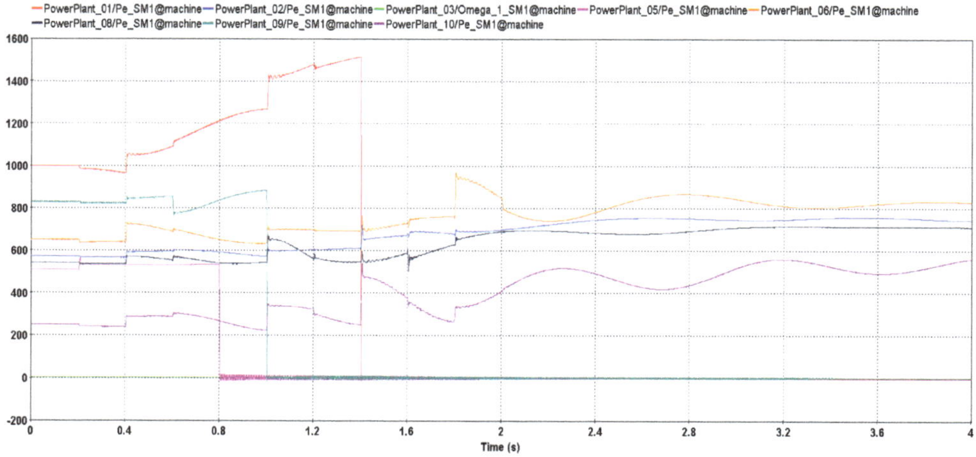

Figure 10. Active power variation of generators during foreseen cascaded failures.

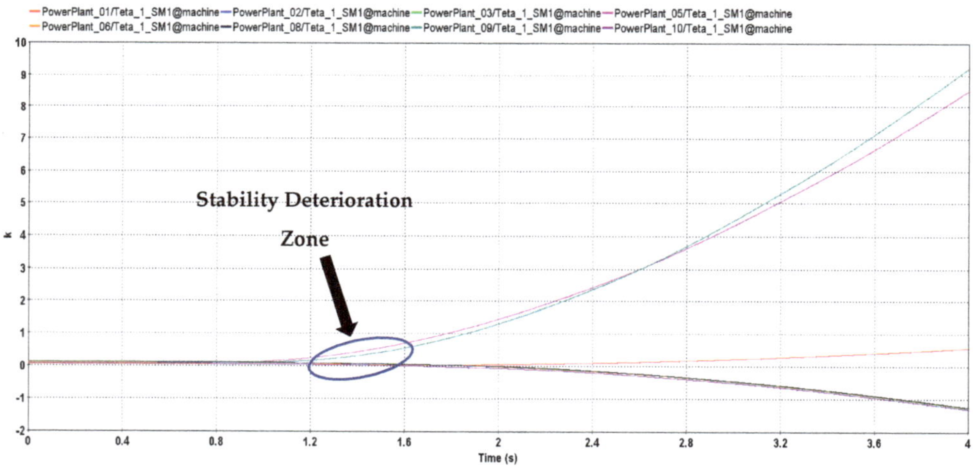

Figure 11. Generator rotational speed deviations during cascading failure.

3.2. *Secure System State Transfer*

For the N-k contingency analysis implemented according to the total vulnerability ranking, it was found that six contingencies resulted in an insecure state, and four of those also resulted in a bus isolated from the system. The line loadings and bus voltage deviations, being the most vulnerable cases, create a basis for improved constraints for the CSC-OPF problem. The genetic algorithm search aims to minimize the fitness objective applied to solve this corrective security-constrained OPF problem function, defined in Equation (1). Generator active power outputs are considered as main variables for optimization, which are represented as real numbers in the search space but are limited by the minimum and maximum limits of generators' active power. For candidate selection, the roulette selection type was used and scattered crossover and adaptive mutation were applied to the candidates that formed the search space. The GA performance is very dependent on the predefined crossover (P_c) and mutation (P_m) probabilities. The GA performance for various crossover and mutation probabilities was tested for P_c = 0.6 to 0.9 and P_m = 0.001 to 0.01 and the ideal results, which have a faster convergence behavior, were obtained for the following algorithm parameters; N_{gen}: 50, N_{size}: 50, P_c: 0.8, and P_m: 0.01. As the system size is relatively small, it is observed that the proposed GA fitness function (defined in Equation (1)) reaches an almost optimum solution after the 30th generation. The mean and best possible solution candidates reached the same fitness value, as it can be seen from Figure 12.

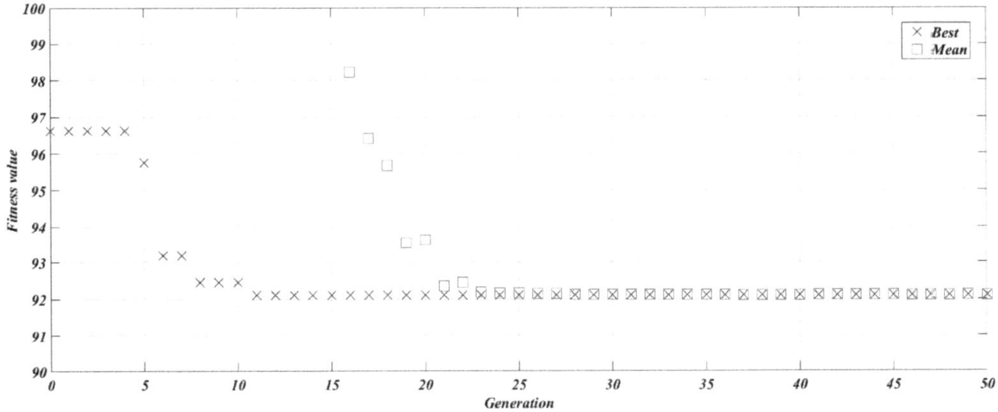

Figure 12. Convergence of VB-OPF algorithm for IEEE-39 bus test system.

As a numeric example, after the N-k contingency is selected, which includes outage of most vulnerable line (line-27/connecting buses 16 and 19), the CSC-OPF module calculates the new settings for generators as given in Table 4.

Table 4. New operational set points for generators after most credible contingency.

Gen-ID	P_{set}	P_{init}	P_{min}	P_{max}
1	667.48	250	0	1040
2	414.06	677.87	0	646
3	375.82	650	0	725
4	594.78	632	0	652
5	655.75	650	0	687
6	400.3	560	0	580
7	285.46	540	0	564
8	653.54	830	0	865
9	1040.9	1000	0	1100

Accordingly, the cost comparison of the systems when secure system transfer conditions are applied or not is shown in Figure 13. It is observed that by transferring the system to the proposed secure operating point using the CSC-OPF algorithm, although generation costs are slightly increased by avoiding the unserved energy (load shed) cost, the overall system operational cost is decreased by 20.4% for this specific contingency case.

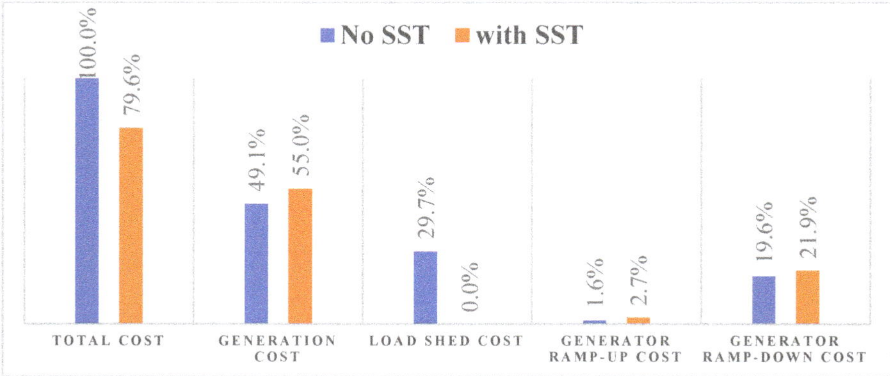

Figure 13. Cost comparison of SST operation with base (not applied) case in IEEE-39 network for the applied contingency scenario.

4. Results and Conclusions

In this study, a new preventive control approach is presented which implements critical contingency selection using a practical vulnerability search analysis and a decision tree-based stability evaluation module with a multi-variant parameter knowledge base. Using non-operational vulnerability indices, the proposed decision support system is expected to assist PSOs in making critical decisions for transferring the power system configuration to be resilient against the possible and probable contingencies which may lead to cascaded failures. By using the proposed methodology, the cascading failure withstand level of the system is expected to improve by transferring the system to a more secure operating point. In this study, it is also shown that although generator operational costs are slightly increased, the unserved energy costs due to load shedding can be alleviated, and as a result, more than 20% of the total system operation cost can be reduced in the specific contingency case. Furthermore, the validity of the stability deterioration detection approach was shown with the benchmark made using a commercial PSA software tool. Using this approach, we proposed a method that anticipates system stability during consecutive failures using the results obtained from steady state AC-OPF analysis, combined with decision trees, using presented performance indices iteratively. By implementing the proposed approach, users can understand several equipment outage scenarios based on vulnerability rankings, visualize the impacts of resulting outages, and estimate the operational cost of systems where a secure system transfer is considered or not.

The main contributions of this paper are as follows: (1) it introduces an approach for critical contingency selection considering operational and non-operational vulnerabilities jointly; (2) it studies a decision tree-based power system stability evaluation with a multi-parameter knowledge base where a corrective security-constrained and genetic algorithm-based OPF algorithm is used; and (3) it provides a contingency evaluation tool for system operators to quantify the impacts due to specific component outages arising from several vulnerabilities.

With the help of such knowledge-based decision support tools providing direct consolidated information from internal and external vulnerabilities, the load dispatch center's resilience capability is expected to be improved and PSOs will be capable of testing their disaster scenarios and understand the impacts of possible corrective maneuvers beforehand. In future work, the proposed algorithm will be tested on a more realistic network model with daily and seasonal load variations imposed.

Author Contributions: Methodology, E.A. and M.B.; Software, E.A.; Validation, E.A.; Investigation, E.A. and M.B.; Data curation, E.A.; Writing – original draft, E.A.; Writing – review & editing, M.B.; Supervision, M.B. All authors have read and agreed to the published version of the manuscript.

Funding: This research received no external funding.

Institutional Review Board Statement: Not applicable.

Informed Consent Statement: Not applicable.

Data Availability Statement: No new data were created.

Conflicts of Interest: The authors declare no conflict of interest.

Nomenclature

Indices	
l	line index
i	bus index
g	generator index
Constants	
n_{line}	total number of lines
n_{bus}	total number of buses
n_{gen}	total number of generators
N	formulation parameter (i.e., 2)
w_P	overloading weight factor
w_b	bus voltage weight factor
w_g	generator active/reactive power weight factor
w_{ang}	angle weight factor weight factor
w_{Vb}	bus voltage deviation weight factor
$\boldsymbol{w_{LoL}}$	loss of load weight factor
$\boldsymbol{w_{LoG}}$	loss of generation weight factor
$w_{Vb,ct}$	bus voltage violation weight factor
w_{PLoss}	active power loss change weight factor
$P_{l\text{-}max}$	max active power of lth line
$P_{g,max}$	max. active power of ith generator
$Q_{g,max}$	max. reactive power of ith generator
ΔP_{shed}	amount of load shed
Variables	
P_l	active power of lth line
V_i	voltage level of ith bus
V_{bc}	base case voltage of ith bus
V_{min}	minimum voltage of ith bus
V_{max}	maximum voltage of ith bus
Q_g	reactive power of ith generator
P_g	active power of ith generator
ΔV_b	total bus voltage deviation in p.u
$\Delta V_{b\text{-}ct}$	number of bus voltage limit violation
isl_{bus}	total islanded bus number
$P_{L,0}$	base case power loss
$P_{L;k}$	k^{th} contingency case power loss
PI_P	active power performance index
PI_V	voltage performance index
PI_{Pg}	generator active power performance index
PI_{Qg}	generator reactive power performance index
PI_{Ang}	load angle performance index
PI_{LoL}	percentage of load shed p. index
PI_{LoG}	percentage of lost generation p. index
PI_{IB}	isolated bus number performance index
OPI_{ctg}	contingency based operational p.index
OPI_{sev}	severity based operational p.index
ΔV_{avg}	average voltage drop

References

1. Andersson, G.; Donalek, P.; Farmer, R.; Hatziargyriou, N.; Kamwa, I.; Kundur, P.; Martins, N.; Paserba, J.; Pourbeik, P.; Sanchez-Gasca, J.; et al. Causes of the 2003 major grid blackouts in North America and Europe, and recommended means to improve system dynamic performance. *IEEE Trans. Power Syst.* **2005**, *20*, 1922–1928. [CrossRef]
2. Gao, B.; Shi, L. Modeling an Attack-Mitigation Dynamic Game-Theoretic Scheme for Security Vulnerability Analysis in a Cyber-Physical Power System. *IEEE Access* **2020**, *8*, 30322–30331. [CrossRef]
3. Mohammadi, F.; Rashidzadeh, R. Impact of stealthy false data injection attacks on power flow of power transmission lines-A mathematical verification. *Int. J. Electr. Power Energy Syst.* **2022**, *142*, 108293. [CrossRef]
4. Abedi, A.; Gaudard, L.; Romerio, F. Power flow-based approaches to assess vulnerability, reliability, and contingency of the power systems: The benefits and limitations. *Reliab. Eng. Syst. Saf.* **2020**, *201*, 106961. [CrossRef]
5. Sperstad, I.B.; Kjølle, G.H.; Gjerde, O. A comprehensive framework for vulnerability analysis of extraordinary events in power systems. *Reliab. Eng. Syst. Saf.* **2020**, *196*, 106788. [CrossRef]
6. Mohammadi, F. Emerging Challenges in Smart Grid Cybersecurity Enhancement: A Review. *Energies* **2021**, *14*, 1380. [CrossRef]
7. Donde, V.; López, V.; Lesieutre, B.; Pinar, A.; Yang, C.; Meza, J. Severe multiple contingency screening in electric power systems. *IEEE Trans. Power Syst.* **2008**, *23*, 406–417. [CrossRef]
8. Rocco, C.M.; Ramirez-Marquez, J.E.; Salazar, D.E.; Yajure, C. Assessing the vulnerability of a power system through a multiple objective contingency screening approach. *IEEE Trans. Reliab.* **2011**, *60*, 394–403. [CrossRef]
9. Sadeghian, O.; Mohammadi-Ivatloo, B.; Mohammadi, F.; Abdul-Malek, Z. Protecting Power Transmission Systems against Intelligent Physical Attacks: A Critical Systematic Review. *Sustainability* **2022**, *14*, 12345. [CrossRef]
10. Tan, S.; Guerrero, J.M.; Xie, P.; Han, R.; Vasquez, J.C. Brief Survey on Attack Detection Methods for Cyber-Physical Systems. *IEEE Syst. J.* **2020**, *14*, 5329–5339. [CrossRef]
11. Nezamoddini, N.; Mousavian, S.; Erol-Kantarci, M. A risk optimization model for enhanced power grid resilience against physical attacks. *Electr. Power Syst. Res.* **2017**, *143*, 329–338. [CrossRef]
12. Li, Z.; Shahidehpour, M.; Alabdulwahab, A.; Abusorrah, A. Analyzing Locally Coordinated Cyber-Physical Attacks for Undetectable Line Outages. *IEEE Trans. Smart Grid* **2018**, *9*, 35–47. [CrossRef]
13. Bi, W.; Zhang, K.; Li, Y.; Yuan, K.; Wang, Y. Detection Scheme Against Cyber-Physical Attacks on Load Frequency Control Based on Dynamic Characteristics Analysis. *IEEE Syst. J.* **2019**, *13*, 2859–2868. [CrossRef]
14. Abedi, A.; Gaudard, L.; Romerio, F. Review of major approaches to analyze vulnerability in power system. *Reliab. Eng. Syst. Saf.* **2019**, *183*, 153–172. [CrossRef]
15. Pandit, M.; Srivastava, L.; Sharma, J. Cascade fuzzy neural network based voltage contingency screening and ranking. *Electr. Power Syst. Res.* **2003**, *67*, 143–152. [CrossRef]
16. Fu, J.; Wang, L.; Hu, B.; Xie, K.; Chao, H.; Zhou, P. A Sequential Coordinated Attack Model for Cyber-Physical System Considering Cascading Failure and Load Redistribution. In Proceedings of the 2018 2nd IEEE Conference on Energy Internet and Energy System Integration (EI2), Beijing, China, 20–22 October 2018; pp. 1–6. [CrossRef]
17. Arroyo, J.M.; Fernández, F.J. Application of a genetic algorithm to n-K power system security assessment. *Int. J. Electr. Power Energy Syst.* **2013**, *49*, 114–121. [CrossRef]
18. Stott, B.; Jardim, J.; Alsac, O. DC power flow revisited. *IEEE Trans. Power Syst.* **2009**, *24*, 1290–1300. [CrossRef]
19. Liu, X.; Li, Z. Revealing the Impact of Multiple Solutions in DCOPF on the Risk Assessment of Line Cascading Failure in OPA Model. *IEEE Trans. Power Syst.* **2016**, *31*, 4159–4160. [CrossRef]
20. Yan, J.; Tang, Y.; He, H.; Sun, Y. Cascading Failure Analysis with DC Power Flow Model and Transient Stability Analysis. *IEEE Trans. Power Syst.* **2015**, *30*, 285–297. [CrossRef]
21. He, H.; Huang, S.; Liu, Y.; Zhang, T. A tri-level optimization model for power grid defense with the consideration of post-allocated DGs against coordinated cyber-physical attacks. *Int. J. Electr. Power Energy Syst.* **2021**, *130*, 106903. [CrossRef]
22. Rios, M.; Kirschen, D.; Jayaweera, D.; Nedic, D.; Allan, R. Value of security: Modeling time-dependent phenomena and weather conditions. *IEEE Trans. Power Syst.* **2002**, *17*, 543–548. [CrossRef]
23. Aliyana, E.; Aghamohammadia, M.; Kiab, M.; Heidaric, A.; Shafie-khahd, M.; Catalão, J.P.S. Decision tree analysis to identify harmful contingencies and estimate blackout indices for predicting system vulnerability. *Electr. Power Syst. Res.* **2020**, *178*, 106036. [CrossRef]
24. Liao, W.; Salinas, S.; Li, M.; Li, P.; Loparo, K.A. Cascading Failure Attacks in the Power System: A Stochastic Game Perspective. *IEEE Internet Things J.* **2017**, *4*, 2247–2259. [CrossRef]
25. Diao, R.; Vittal, V.; Sun, K.; Kolluri, S.; Mandal, S.; Galvan, F. Decision tree assisted controlled islanding for preventing cascading events. In Proceedings of the 2009 IEEE/PES Power Systems Conference and Exposition, Seattle, WA, USA, 15–18 March 2009; pp. 1–8. [CrossRef]
26. Zio, E.; Golea, L.R.; Rocco, S.C.M. Identifying groups of critical edges in a realistic electrical network by multi-objective genetic algorithms. *Reliab. Eng. Syst. Saf.* **2012**, *99*, 172–177. [CrossRef]
27. Cuadra, L.; Salcedo-Sanz, S.; Del Ser, J.; Jiménez-Fernández, S.; Geem, Z.W. A Critical Review of Robustness in Power Grids Using Complex Networks Concepts. *Energies* **2015**, *8*, 9211–9265. [CrossRef]

28. Henneaux, P.; Ciapessoni, E.; Cirio, D.; Cotilla-Sanchez, E.; Diao, R.; Dobson, I.; Gaikwad, A.; Miller, S.; Papic, M.; Pitto, A.; et al. Benchmarking quasi-steady state cascading outage analysis methodologies. In Proceedings of the 2018 IEEE International Conference on Probabilistic Methods Applied to Power Systems (PMAPS), Boise, ID, USA, 24–28 June 2018. [CrossRef]
29. Zeraati, M.; Aref, Z.; Latify, M.A. Vulnerability Analysis of Power Systems Under Physical Deliberate Attacks Considering Geographic-Cyber Interdependence of the Power System and Communication Network. *IEEE Syst. J.* **2018**, *12*, 3181–3190. [CrossRef]
30. Yan, J.; He, H.; Zhong, X.; Tang, Y. Q-Learning-Based Vulnerability Analysis of Smart Grid Against Sequential Topology Attacks. *IEEE Trans. Inf. Forensics Secur.* **2017**, *12*, 200–210. [CrossRef]
31. Rocchetta, R. Enhancing the resilience of critical infrastructures: Statistical analysis of power grid spectral clustering and post-contingency vulnerability metrics. *Renew. Sustain. Energy Rev.* **2022**, *159*, 112185. [CrossRef]
32. Du, H.; Lin, T.; Li, Q.; Fu, X.; Xu, X.; Cheng, J. Transmission expansion planning for power grids considering resilience enhancement. *Electr. Power Syst. Res.* **2022**, *211*, 108218. [CrossRef]
33. Breiman, L.; Friedman, J.; Olshen, R.A.; Stone, C.J. *Classification and Regression Trees*; Wadsworth: Belmont, CA, USA, 1984.
34. Akdeniz, E.; Bagriyanik, M. A knowledge based decision support algorithm for power transmission system vulnerability impact reduction. *Int. J. Electr. Power Energy Syst.* **2016**, *78*, 436–444. [CrossRef]
35. The MathWorks Inc. (2021a). Statistics and Machine Learning Toolbox Documentation, Natick, Massachusetts: The MathWorks Inc. Available online: https://www.mathworks.com/help/stats/index.html (accessed on 11 April 2023).
36. Zhu, J. *Optimization of Power System Operation*; IEEE Press Series on Power Engineering; John Wiley & Sons: Hoboken, NJ, USA, 2015.
37. Zimmerman, R.D.; Murillo-Sánchez, C.E.; Thomas, R.J. MATPOWER: Steady-State Operations, Planning, and Analysis Tools for Power Systems Research and Education. *IEEE Trans. Power Syst.* **2011**, *26*, 12–19. [CrossRef]
38. EPİAŞ Transparency Platform for Electricity Markets. Available online: https://seffaflik.epias.com.tr/transparency/piyasalar/gop/ptf.xhtml (accessed on 17 May 2021).
39. Electromagnetic Transients Program (EMTP®). Available online: https://www.emtp.com/ (accessed on 11 April 2023).

energies

MDPI

Article

Optimizing Power Demand Side Response Strategy: A Study Based on Double Master–Slave Game Model of Multi-Objective Multi-Universe Optimization

Diandian Hu [1] and Tao Wang [1,2,*]

[1] School of Mathmatics and Physics, North China Electric Power University, Baoding 071003, China
[2] Hebei Key Laboratory of Physics and Energy Technology, North China Electric Power University, Baoding 071003, China
* Correspondence: wangtao@ncepu.edu.cn

Abstract: In the pilot provinces of China's electricity spot market, power generation companies usually adopt the separate bidding mode, which leads to a low willingness of demand-side response and poor flexibility in the interaction mechanism between supply and demand. Based on the analysis of the demand response mechanism of the power day-ahead market with the participation of power sales companies, this paper abstracted the game process of the "power grid-sales company-users" tripartite competition in the electricity market environment into a two-layer (purchase layer/sales layer) game model and proposed a master–slave game equilibrium optimization strategy for the day-ahead power market under the two-layer game. The multi-objective multi-universe optimization algorithm was used to find the Pareto optimal solution of the game model, a comprehensive evaluation was constructed, and the optimal strategy of the demand response was determined considering the peak cutting and valley filling quantity of the power grid, the profit of the electricity retailers, the cost of the consumers, and the comfort degree. Examples are given to simulate the day-ahead electricity market participated in by the electricity retailers, analyze and compare the benefits of each market entity participating in the demand response, and verify the effectiveness of the proposed model.

Keywords: power demand response; master–slave game; demand response strategy; multi-objective multi-universe optimization

Citation: Hu, D.; Wang, T. Optimizing Power Demand Side Response Strategy: A Study Based on Double Master–Slave Game Model of Multi-Objective Multi-Universe Optimization. *Energies* **2023**, *16*, 4009. https://doi.org/10.3390/en16104009

Academic Editor: François Vallée

Received: 6 April 2023
Revised: 28 April 2023
Accepted: 4 May 2023
Published: 10 May 2023

1. Introduction

The new round of reform in the power sector has promoted the diversified development of market players, and the power demand-side response system has been gradually improved [1]. The electricity retailer has upgraded its revenue model from relying solely on price differentials to offering load integration and diversified comprehensive energy services. By balancing the varying demands of user groups and participating in demand-side response programs, the company not only actively cooperates with the peak-shaving and valley-filling measures of the power system but also increases its revenue. As the electricity market reform continues to deepen and the establishment of the electricity spot market gradually improves, the implementation model of demand response in the power sector is shifting from the "demand-side bidding + fixed compensation price" model to a market-oriented "demand-side bidding + maximum price limit" model [2,3]. The ability to optimize resource allocation by fully considering the demand side is crucial [4]. In this context, exploring the supply–demand interaction between the "grid–user–retailer" triad under market competition mechanisms and studying demand-side response strategies have become a pressing research topic.

Demand response subsidy pricing is an incentive measure established by grid companies to encourage consumers to reduce electricity consumption during peak periods, thereby reducing the burden on the power grid. Currently, there have been significant developments in the field of research related to demand-side resource prices [5] and the formulation of incentive mechanisms [6]. The traditional optimization theory system, which relies on single-agent decision-making, is unable to effectively address the challenges in actual power demand response management.Some artificial intelligence technology [7] and game theory methods [8] are applied to solve the power demand side response strategy [9]. Among them, there are two main types of applications for game theory: (1) The goal of studying demand-side electricity consumption behavior based on demand response optimized equipment is to discover the best strategy for arranging the power consumption of each device. Reference [10] presents a multi-layer game model that involves power companies, multiple home power management centers, and multiple devices within a household. This paper also proposed a response mechanism for managing household power load in a smart grid environment; (2) Investigate the game between the electricity demand side and the electricity supply company [11,12]. In reference [13], a game between electricity consumers and the power grid to determine the optimal price set by the grid, and users adjusting their optimal power consumption based on the price. Reference [14] thoroughly investigated decision-making behavior in power demand response management from the perspective of multi-population evolutionary game theory. At the same time, considering the master–slave game [15] very suitable for analyzing sequential decision-making problems in competitive environments [16], the basic theory of the modern engineering game has found extensive application in the domain of power system control and decision-making. Reference [17] tackles the issue of demand response subsidy pricing set by power grid companies. The research established a master–slave game model to illustrate the interaction between grid operators and multiple stakeholders. In reference [18], a master–slave game model was constructed with the aim of maximizing the interests of both users and load aggregator businesses. The optimal compensation pricing strategy for the load aggregator was obtained by solving the model, and the elasticity of user electricity consumption was analyzed to optimize user response. In the future, the demand response market will become more and more perfect, and the demand response model needs to consider the game of more market players.

In this paper, the response mechanism of the power market under the participation of retail electricity companies was analyzed, considering a more complete range of demand response participants in the market. With the goals of peak shaving and valley filling on the grid side, maximizing the profit of retail electricity companies and minimizing the electricity cost and ensuring the electricity comfort of users, the game process of "power grid-sales e-commerce-user" in the competitive environment of the power market was abstracted into a two-layer (power purchase layer/power sales layer) master–slave game model. After analyzing and solving the model, we proposed a new solution by introducing the multi-objective multiverse optimizer (MOMVO) algorithm [19]. This method converts the objective functions and constraints in the model into multiple fitness functions and seeks the Pareto front of the model. Finally, in the interactive game, by continuously interacting the respective benefit information between each subject, the Nash equilibrium solution that satisfies its own benefits is obtained, that is, the distribution of electricity and subsidy prices in which each market entity participates in demand response. By simulating the electricity market in which electricity sales companies participate, the benefits of demand response from various market entities were analyzed and compared, and the validity of the proposed model was confirmed.

2. Demand Response Revenue Model for Individual Market Players

In the event of power shortage (surplus) in the real-time balancing market, the power grid initiates a demand response mechanism for peak shaving (valley filling). Retail electricity companies participate in the demand response by purchasing electricity based on the electricity usage patterns of their customers, and then selling the electricity to them. This helps balance the demand and supply of electricity in the market. Upon receiving the incentive notification from the retail electricity company to reduce the load or the signal of rising electricity prices, the electricity users change their inherent electricity consumption patterns [20]. To guarantee the stability of the electricity network and regulate the increase of electricity costs, it is imperative to respond to power supply by reducing or shifting the electricity load during a specific time period. In response to the demand response programs issued by the grid, decisions are made by the demand side with the aim of maximizing the benefits of participation.

2.1. Demand Response Revenue Model for Grid

The benefits of the grid release of the peak-shaving demand response mainly include reduced power generation costs and avoidable transmission and distribution capacity costs; costs mainly include reduced electricity sales gains and compensatory costs for publishing demand responses.

(1) Reduced power generation costs:

$$C_f(Q) = aQ^2 + bQ + c, \tag{1}$$

$$\Delta C_f(Q) = C_f(Q) - C_f(Q'), \tag{2}$$

where a, b, c is the cost factor of power generation. Generally, $a > 0$ and $2a$ represent the slope of the marginal cost price curve, b is the starting marginal cost of the unit, c is the loss when the unit does not contribute, Q is the amount of electricity generated before demand response, Q' is the amount of power generated after demand response, and $\Delta C_f(Q)$ is the reduced power generation cost of the grid company.

(2) It can avoid transmission and distribution capacity costs.

The avoidable transmission and distribution capacity cost can be addressed through the avoidable transmission and distribution capacity per unit cost c_g and the actual ΔQ of avoided transmission and distribution capacity, as shown in Equation (3).

$$C_{con} = c_g \cdot \Delta Q. \tag{3}$$

Considering transmission and distribution losses, user participation in demand response does not directly correspond to load reduction on the grid side. The actual avoided transmission and distribution capacity ΔQ can be calculated as follows:

$$\Delta Q = \frac{1}{1-\alpha}. \tag{4}$$

Among them, Q_t stands for the total response of users during the time interval t, and α denotes the coefficient of transmission and distribution losses in the grid.

(3) Publish demand response subsidy fees.

To incentivize retail electricity companies to participate in demand response, the grid provides them with a compensation price. The compensation cost $C_{r,g}$ can be calculated as follows:

$$C_{r,g} = \sum_{t=1}^{T} \sum_{k \in N} r_k^t q_k^t, k = 1, 2, \ldots, N, t = 1, 2, \ldots, T. \tag{5}$$

Here, T represents the set of time periods for demand response, N denotes the set of sales companies, r_k^t is the subsidy unit price that the power grid provides to electricity retailer $k(k = 1, 2, \ldots, N)$ at period $t(t = 1, 2, \ldots T)$, and q_k^t refers to the response quantity of electricity retailer k during time period t.

(4) Reduced income from electricity sales.

$$B_{s,g} = \sum_{t=1}^{T} \sum_{k \in N} q_k^t \lambda_t, k = 1, 2, \ldots, N, t = 1, 2, \ldots, T. \tag{6}$$

Among them, B_s, g denotes the reduction in electricity sales revenue for the power grid, q_k^t is the response quantity of the electricity retailer k at time t, and λ_t stands for the spot electricity price in the market at time t.

Equation (7) depicts the revenue function of the power grid when it engages in demand response.

$$U_G = C_{con} + \Delta C_f - C_{r,g} - B_{s,g}. \tag{7}$$

2.2. *Demand Response Revenue Model for Electricity Retailers*

The cost of participating in pre-day peak shaving response of the grid is mainly comprised of reduced electricity sales revenue and allowance fees issued to users, while the income mainly consists of the response subsidies obtained.

(1) Demand response subsidies obtained.

The electricity retailer k participates in the day-ahead peak load reduction demand response program and receives subsidy payments, denoted by $\overline{B}_{r,k}$, as shown in Equation (8). Here, r_k^t is the subsidy unit price received by electricity retailer k during time period t, and q_k^t is the response amount of electricity sales company k during time period t.

$$\overline{B}_{r,k} = \sum_{t=1}^{T} r_k^t q_k^t, k \in N, t = 1,2,\ldots,T. \tag{8}$$

(2) Reduced power purchase costs.

The electricity retailer k participates in demand response and lowers the cost of procuring electricity from the grid, denoted by $\overline{C}_{g,k}$.

$$\overline{C}_{g,k} = \sum_{t=1}^{T} q_k^t \lambda_k^t, k \in N, t = 1,2,\ldots,T. \tag{9}$$

(3) Reduced electricity sales revenue.

This paper considered the signed time-of-use electricity pricing contract between the electricity sales company and users, where the electricity sales company determines the time-of-use periods and prices for the 24 h in a day in advance. The reduced electricity sales revenue $\overline{R}_{s,k}$ for the electricity sales company k is shown in Equation (10).

$$\overline{R}_{s,k} = \sum_{t=1}^{T} \sum_{i \in M_k} \widetilde{q}_i^t \widetilde{\lambda}_k^t, k \in N, i \in M_k, t = 1,2,\ldots,T. \tag{10}$$

Among them, $\widetilde{\lambda}_k^t$ is the live electricity rate of the electricity retailer k corresponding to the user in the t period, M_k represents the collection of users under the electricity sales company k, and $\widetilde{q}_i^t$ is the response volume of the user $i(i = 1,2,\ldots,M_k)$ during the t period.

(4) Published user response subsidy fees.

The subsidy price published by the electricity retailer to the user i during the t period is recorded as $\widetilde{r}_i^t$, then the response subsidy fee $\overline{D}_{r,k}$ issued by electricity retailer k to user is shown in Formula (11).

$$\overline{D}_{r,k} = \sum_{t=1}^{T} \sum_{i \in M_k} \widetilde{q}_i^t \widetilde{r}_i^t, k \in N, i \in M_k, t = 1,2,\ldots,T. \tag{11}$$

(5) Penalty fees.

When the actual response amount of the users under the electricity retailer k is lower than 95% of the declared response amount or higher than 105% of the declared response amount by the electricity retailer, the penalty fee $\overline{F}_{\omega,k}$ will be as shown in Formula (12), where ω_k is the penalty price of the electricity retailer k.

$$\overline{F}_{\omega,k} = \begin{cases} (q_k^t \cdot 95\% - \sum_{i \in M_k} \widetilde{q}_i^t) \cdot \omega_k, \sum_{i \in M_k} \widetilde{q}_i^t < q_k^t \cdot 95\%, \\ (\sum_{i \in M_k} \widetilde{q}_i^t - q_k^t \cdot 105\%) \cdot \omega_k, \sum_{i \in M_k} \widetilde{q}_i^t > q_k^t \cdot 105\%, \end{cases} k \in N, i \in M_k, t = 1,2,\ldots T. \quad (12)$$

The income of the electricity sales company k participating in the demand response $\overline{U}_{S,k}$ is as in Equation (13):

$$\overline{U}_{S,k} = \overline{B}_{r,k} + \overline{C}_{g,k} - \overline{R}_{s,k} - \overline{D}_{r,k} - \overline{F}_{\omega,k}, k \in N. \quad (13)$$

2.3. User Demand Response Revenue Model

(1) Response costs.

The cost of user response refers to the losses incurred by users reducing electricity consumption. In this paper, the user response cost $\widetilde{C}_{q,i}$ is expressed as a quadratic function, as shown in Equation (14).

$$\widetilde{C}_{q,i} = \beta_i (\overline{q}_i)^2 + \gamma \overline{q}_i, i \in M \quad (14)$$

$$\overline{q}_i = \sum_{t=1}^{T} \widetilde{q}_i^t, i \in M, t = 1,2,\ldots,T, \quad (15)$$

where β_i and γ_i are the user i response cost factors, both constants greater than 0, and the response amount of the user is recorded as $\overline{q}_i$.

(2) Response subsidy fees received by users.

The subsidy price that user i receives during the t session is $\widetilde{r}_i^t$, and the response subsidy fee that user i receives is $\widetilde{B}_{dr,i}$.

$$\widetilde{B}_{dr,i} = \sum_{t=1}^{T} \widetilde{r}_i^t \widetilde{q}_i^t, i \in M, t = 1,2,\ldots,T. \quad (16)$$

(3) Reduced power purchase costs.

In peak periods, users have the option to engage in demand response by either decreasing their consumption or shifting their usage, meaning that the demand response quantity of users includes both load reduction and load transfer.

Among them, the load transfer quantity for demand response scheduling by the user is mainly based on the demand elasticity theory. According to the demand principle of economics, the elasticity coefficient of electricity price may be characterized as the proportionate alteration in load demand resulting from the proportionate alteration in electricity price over a certain period, and the price elasticity coefficients are represented by Equation (17).

$$\begin{cases} e^{t_1,t_2} = \dfrac{\Delta L_{t_1} / L_{t_1}^0}{\Delta P_{t_2} / P_{t_2}^0} \\ e^{t_1,t_2} \le 0, t_1 = t_2; e^{t_1,t_2} \ge 0, t_1 \neq t_2. \end{cases} \quad (17)$$

Among them, $L_{t_1}^0$, ΔL_{t_1} are the initial electricity load at the user's t_1 moment and the load change before and after demand response, respectively. $P_{t_2}^0$, ΔP_{t_2} are the initial electricity price at t_2 and the price change before and after demand response, respectively. When $t_1 = t_2$, e^{t_1,t_2} is called the self-elastic response coefficient, the increase in electricity price will cause a decrease in the user's electricity demand, and its value is negative; when $t_1 \neq t_2$, then e^{t_1,t_2} is called the cross-elasticity coefficient, and the increase in the price at t_2 will cause users to transfer the load to the t_1 period, where the electricity price is lower, with a positive value.

According to the time-of-use electricity price contract signed between the selling company and the user, this paper considered that the user's transferable electricity occurs in the two periods of the highest and lowest real-time electricity prices. As a result of the varying response characteristics of different users towards electricity prices, the corresponding elasticity coefficients [21] are also different, so the corresponding elasticity coefficient is also different, and the elastic coefficient matrix of user i is:

$$E_i = \begin{pmatrix} e_i^{ff} & e_i^{fg} \\ e_i^{gf} & e_i^{gg} \end{pmatrix}. \tag{18}$$

Among them, the superscripts f and g, respectively, correspond to peak and valley periods, and the elements in the matrix correspond to the self-elasticity coefficient and the cross-elasticity coefficient of each time period. By deducing price-based demand response, the demand quantity of each user and the total load demand of each time period can be obtained.

$$L_i = L_i^0 + \begin{pmatrix} L_i^{f,0} & 0 \\ 0 & L_i^{g,0} \end{pmatrix} E_i \begin{pmatrix} \Delta P_i^{gf} / P_i^{f,0} \\ \Delta P_i^{fg} / P_i^{g,0} \end{pmatrix}. \tag{19}$$

Among them, L_i^0, L_i represent the power consumption before and after load transfer when user i participates in demand response, respectively. $L_i^{f,0}$, $L_i^{g,0}$ indicate the electricity consumption during the peak and trough periods before user i participates in demand response, respectively; $P_i^{f,0}$, $P_i^{g,0}$ indicate that user i participates in the power purchase price during the peak and trough periods before responding, respectively; ΔP_i^{fg} indicates the electricity purchase price differential during the user's peak-valley period; ΔP_i^{gf} represents the electricity purchase price differential between the user's trough and peak hours.

The reduced cost of power purchase for user i is denoted as $\widetilde{G}_{k,i}$ and is expressed as follows:

$$\widetilde{G}_{k,i} = \sum_{t=1}^{T} \widetilde{\lambda}_k^t \widetilde{q}_i^t + \Delta P_i^{fg} \Delta L_i, k \in N, i \in M, \tag{20}$$

where $\widetilde{\lambda}_k^t$ represents the electricity price of the electricity retailer k corresponding to the user during time period t, and ΔL_i represents the load transfer quantity of user i participating in the demand response.

In summary, the demand response benefit $\widetilde{U}_{k,i}$ of user i is:

$$\widetilde{U}_{k,i} = -\widetilde{C}_{k,i} + \widetilde{B}_{dr,i} + \widetilde{G}_{k,i}, k \in N, i \in M. \tag{21}$$

One of the constraints is that the response quantity $\widetilde{q}_i^t$ of user i during time period t cannot exceed the maximum response capacity h_i^t of any user during the same period, while it must also be higher than the minimum response quantity d_i^t. Therefore, the response quantity must satisfy the condition: $d_i^t \leq \widetilde{q}_i^t \leq h_i^t$.

3. Master–Slave Game Model and Solution of Demand-Side Response

3.1. Master–Slave Game Model

In 1952, Stackel-berg proposed the concept of the leader–follower game, where the leader has a strategic advantage and occupies a dominant or advantageous position, while the follower makes decisions following the leader. In real life, there are many specific examples of leader–follower games, such as the game between central and local governments, between a company and their subsidiaries, etc.

This paper considered the competition relationship between the three parties of "power grid–electricity retailer–user" and constructs a two-layer master–slave game model, as depicted in Figure 1. The higher-level demand response model involves the grid, which acts as a leader in the game. The grid publishes demand responses and takes into account constraints based on market electricity prices and electricity sales. The subsidy unit price of each electricity sales company is set with the aim of maximizing the grid's own income. Each electricity retailer in the lower-level model acts as a follower. After receiving the demand response information of the grid, they optimize their internal response volume and subsidy price to the user with the aim of maximizing their own revenue. The grid then adjusts the subsidy price according to the response strategy of the electricity retailer. This process creates a leader–follower sequential game and constitutes a Stackelberg game relationship [22], as well as a non-cooperative game relationship between each electricity seller.

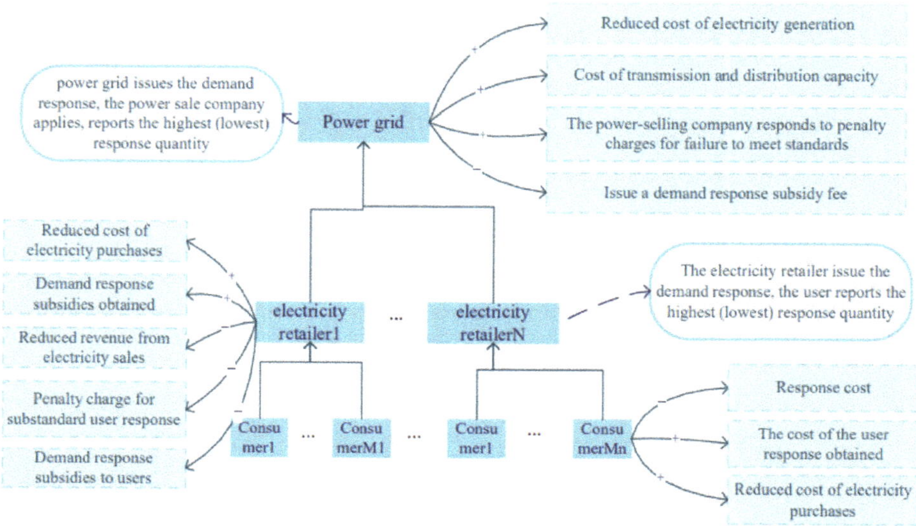

Figure 1. Tripartite demand response decision model of "power grid, electricity retailers and consumers".

In the lower-level model, the electricity sales company plays the role of a leader, publishes demand responses, combines its own power purchase constraints, and sets the subsidy unit price of each user with the goal of maximizing its personal income. Each user also acts as a follower. After receiving the demand response information from the electricity retailer, they optimize the response volume and power purchase period in order to maximize their own income. The electricity retailer then adjusts the subsidy price again based on the user's response strategy. This creates a leader–follower sequential game, constituting a Stackelberg game relationship. Each user also constitutes a non-cooperative game relationship in this model.

3.2. Game Model Solving

The game flow is as follows: The game model proposed earlier transforms the demand-side electricity purchasing decision problem into a multi-objective optimization problem, enabling more comprehensive and scientific purchasing decisions. The MOMVO algorithm [23] is a global search optimization algorithm with strong convergence speed, fast convergence rate, and good robustness, and it has been widely applied in many fields. In this paper, the game model was combined with the MOMVO algorithm by converting the objective functions and constraints in the master–slave game model into multiple fitness functions, and using the MOMVO algorithm to solve it. This method not only improves the efficiency of solving game models but also provides a new method for addressing optimization problems with multiple objectives.

The MOMVO algorithm utilizes the multiverse theory [24], which regards each optimal solution as a universe, to find all optimal solutions by simulating the interaction and variation of universes. The algorithm includes steps such as population initialization, individual evaluation, individual selection, individual evolution, and determining stopping conditions, and ultimately outputs all the found optimal solutions. This algorithm boasts several benefits, including potent global search capability, swift convergence speed and exceptional robustness, and has found extensive application across various domains.

The MVO algorithm is based on the multiverse theory's three main concepts: white hole, black hole, and wormhole. It establishes a mathematical model for optimization with candidate solutions defined as universes and their fitness measured by expansion rates. Each iteration uses black holes as candidate solutions, selecting better universes as white holes via roulette wheel selection. Black and white holes exchange their contents while some black holes can use wormholes to search for the best universes. The algorithm's internal loop structure's logical flow is illustrated in Figure 2.

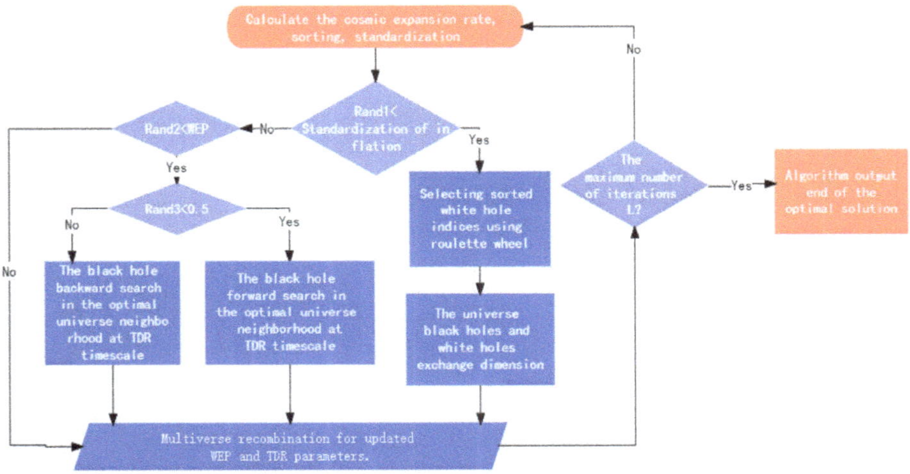

Figure 2. The algorithm's internal loop structure's logical flow.

In Figure 2, black holes update dimensions using two mechanisms.

(1) Based on sorted normalized expansion rates, white hole indices are selected using the roulette wheel selection principle and black holes exchange dimensional information with the selected white hole.

(2) When $Rand2 < WEP$, black holes travel through wormholes and update their dimensions using TDR parameters in the optimal universe neighborhood. The iteration Formula (22) is used, with j representing the specific dimension of the optimized problem.

$$X_j^i = \begin{cases} BestX_j - TDR(ub_j - lb_j) \cdot rand + lb_j, & Rand3 \geq 0.5 \\ BestX_j + TDR(ub_j - lb_j) \cdot rand + lb_j, & Rand3 < 0.5 \end{cases} \tag{22}$$

$$TDR = 1 - (\frac{1}{Len})^{\frac{1}{6}} \tag{23}$$

$$WEP = WEP_{min} + \frac{len}{Len} \cdot (WEP_{max} - WEP_{min}). \tag{24}$$

In Formulas (22)–(24), len and Len represent the current and maximum iterations, ub and lb represent the boundaries of the problem, $BestX$ represents the position of the optimal universe, and WEP and TDR are important parameters of the Multiverse Optimization Algorithm for the probability of wormhole existence and the travel distance rate. Formula (23) indicates that the TDR parameter of the Multiverse Optimization Algorithm is a concave decreasing function during the iteration, decreasing rapidly at first and gradually slowing down, while the WEP parameter increases linearly.

MOMVO is the multi-objective version of MVO, designed to store the best non-dominant solutions. To select the best solutions from the archive, a tunnel is established between solutions using the leader selection mechanism. In this approach, the crowding distance between each solution in the archive is initially selected, and the number of solutions in its neighborhood is used as a measure of coverage or diversity. MOMVO uses roulette wheel selection to improve the distribution of solutions across all objectives, favoring solutions with fewer individuals in the archive. The following equation is used to achieve this improvement.

$$P_r = h/N_m. \tag{25}$$

Using a constant parameter h greater than 1 and kept constant, the equation favors solutions near the i-th solution while reducing the fitness of hypercubes with more particles as a form of fitness sharing. This equation provides high probability solutions for regions with fewer solutions, improving other areas and attracting solutions to these regions with fewer individuals in the archive. It ultimately increases the coverage of the obtained Pareto optimal front.

Archives can only accommodate a limited number of non-dominated solutions and can become full during the optimization process. Therefore, a mechanism is needed to remove unnecessary solutions from the archive. An unnecessary solution is one that is surrounded by many solutions and thus requires cleanup to save space. The inverse Equation (26) is used to discard unwanted solutions from the archive and provide high probability for the MOMVO algorithm.

$$P_r' = N_m/h. \tag{26}$$

To quantify convergence, this paper selected the generational distance (GD) [25] and inverted generational distance (IGD) [26] proposed by Veldhuizen in 1998. These performance metrics serve to quantify the distribution of Pareto optimal solutions obtained. The corresponding mathematical equations for these performance measures are as follows:

$$GD = \frac{\sqrt{\sum_{i_1=1}^{no} s_{i1}^2}}{no} \tag{27}$$

$$IGD = \frac{\sqrt{\sum_{i_2=1}^{nt} (s_{i2}')^2}}{nt}, \tag{28}$$

where no, nt indicate the count of Pareto optimal solutions, and s_{i1}, s_{i2}' represent the Pareto optimal solution, and reference concentration is the most close to the true Pareto optimal solutions of m Euclidean distance. Please note that the Euclidean distances are calculated respectively in their target space.

In order to assist decision makers in selecting the optimal solution from the Pareto frontier, this paper modified the weights of individual indicators using a comprehensive evaluation index method and integrated them into a single objective function. The considered indicators included grid revenue, electricity retailer revenue, user revenue, peak load (valley filling) reduction, user electricity consumption comfort, and participation response satisfaction. The changes in indicator weights are related to the fairness of demand response and the optimal strategy chosen by the decision maker.

The electricity consumption comfort level u of users represents the sum of the standard deviations of the response volume in which all users are involved in the demand response, as in Equation (29). User satisfaction v represents the sum of the standard deviations of the unit price of subsidies obtained by all users participating in demand response, as shown in Equation (30), where M is a collection of all users, and $i = 1, 2, ..., M$.

$$u = \sum_{t=1}^{T} \sum_{i \in M} \sqrt{(\vec{q}_i^t) - \overline{\vec{q}}_i^t)^2}, t = 1, 2, ..., T, i \in M \tag{29}$$

$$v = \sum_{t=1}^{T} \sum_{i \in M} \sqrt{(\tilde{r}_i^t - \bar{\tilde{r}}_i^t)^2}, t = 1, 2, ..., T, i \in M. \quad (30)$$

In the model, the weight coefficient matrix of each index is formulated as $l = (l_{U_G}, l_{U_S}, l_{U_E}, l_Q, l_u, l_v)$, where $l_{U_G}, l_{U_S}, l_{U_E}, l_Q, l_u, l_v$ represent the weights of grid, electricity retailer, user revenue, peak load (valley filling) reduction, user electricity consumption comfort, and participation response satisfaction, respectively. The process of determining the indicator weights in this paper involved employing the CRITIC method, as expounded in [27], which is an objective weighting method that is superior to entropy weighting and standard deviation weighting methods. The CRITIC method takes into account the conflict and contrast intensity among indicators, and can simultaneously consider the size and correlation of indicator variability, using the inherent properties of the data for scientific evaluation.

4. Example Analysis

The reform of China's electricity spot market is in its early stages, with limited marketization and trading volume, making it difficult to obtain real-time electricity price data due to the need for mature market mechanisms and trading platforms for accurate price determination and real-time data acquisition. Therefore, this paper used real-time electricity price data from established electricity markets, such as PJM, which can be more practical and reliable. This paper selected data from one day in the PJM market to verify and analyze the model, which has a total of 24 time periods. In the power grid's demand response implementation phase, there could be multiple electricity retailers who are interested in taking part in the demand response program. This paper considered one power grid company, two electricity retailers, and three users participating in the demand response released by the electricity retailers. It was assumed that the total maximum response capacity of the users can meet the power grid's maximum response demand.

4.1. Parameter Settings

The user's time-of-use pricing schedule was set in intervals as shown in Table 1, while the utility company's time-of-use pricing is presented in Table 2. The differentiated elasticity coefficient for users was also set simultaneously, as shown in Table 3, while the response cost coefficient for different users is presented in Table 4.

Table 1. Time-of-use price contract time segment of users.

Type	Period of Time
Valley period	1, 2, 3, 4, 5, 6, 23, 24
Normal period	7, 8, 13, 14, 15, 16, 17, 18
Peak period	9, 10, 11, 12, 19, 20, 21, 22

Table 2. Time-of-use price of electricity retailers.

	Valley Period	Normal Period	Peak Period
electricity retailers A ($/MWh)	28.26	34.42	38.33
electricity retailers B ($/MWh)	26.21	33.65	39.21

The cost-per-unit of avoiding transmission and distribution capacity in the power grid was set at 2 $/MW, while the transmission and distribution loss coefficient was $\alpha = 0.04$, and the cost coefficient for power generation was $a = 0.06$, $b = 20$, and $c = 5$. The originally planned power output capacity for the day was $Q = 6000$ MW·h, and Figure 3 illustrates the live electricity rate prediction value. The grid plans to release a peak-shaving demand response project during the high-load peak period, which lasts for two hours (19:00–21:00).

Table 3. Customers differentiation elasticity.

Type of Load	Coefficient of Self-Elasticity	Cross Elastic Coefficient		
		Peak-Normal	Peak-Valley	Normal-Valley
user1	−0.05	0.02	0.04	0.02
user2	−0.12	0.02	0.03	0.11
user3	−0.38	0.03	0.23	0.03
user4	−0.56	0.02	0.04	0.02
user5	−0.14	0.15	0.03	0.18
user6	−0.23	0.03	0.08	0.06

Table 4. Response cost coefficient of different consumers (unit: $(MW·h)$^{-1}$).

User	Response Cost Factor β	Response Cost Factor γ
user1	0.2	60
user2	0.2	62
user3	0.2	58
user4	0.2	58
user5	0.2	60
user6	0.2	56

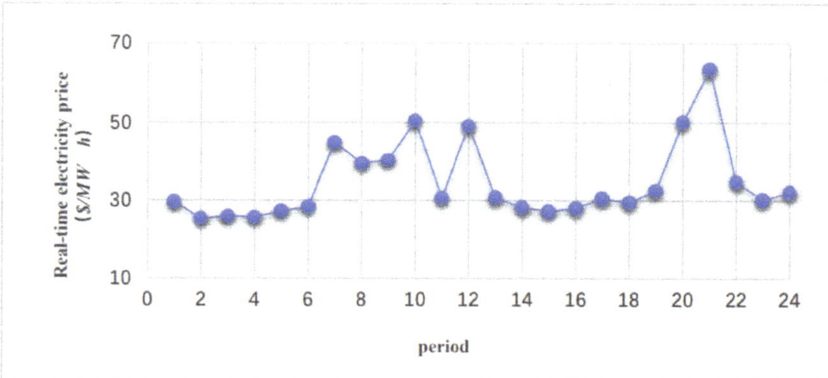

Figure 3. Real-time electricity price forecast value of power grid.

4.2. Scene Settings

This paper proposed a master–slave game model and used Python for simulation programming to obtain the equilibrium solution of the game between the power grid, retail electricity companies, and users. Four scenarios were set up for comparative analysis to verify the effectiveness of the optimization strategy, with different algorithms used to solve the game model's optimal solution while considering the impact of price elasticity on user demand response.

- Scenario 1 did not consider price elasticity on user demand response and used the MOMOV algorithm to solve the game model's optimal solution.
- Scenario 2 considered price elasticity and used the MOMOV algorithm.
- Scenario 3 considered price elasticity and used the MOPSO algorithm to solve the game model's optimal solution.
- Scenario 4 considered price elasticity and used the NSGA-II algorithm to solve the game model's optimal solution.

4.3. Comparative Analysis

When using the MOMOV algorithm for solving, Len was set to 100, while WEP linearly increased from 0.2 to 1 and TDR decreased concavely from 0.6 to 0. According to Scenario 1, the model converged in the 16th iteration. The optimal response of the retail electricity companies showed a convergence trend as the iteration increased, as shown in Figure 4. Table 5 shows the optimal response of the retail electricity companies and their corresponding subsidy price. Figure 5 shows the optimal response and subsidy price of users during the 20:00–21:00 period. It can be seen that user 4 has the highest response and subsidy price at 27.16 MW and 18.53 USD/MW, respectively, while user 2 has the lowest response and subsidy price at 18.85 MW and 13.37 USD/MW, respectively. When the user response is high, a corresponding high subsidy price can be obtained to stimulate users to increase their response.

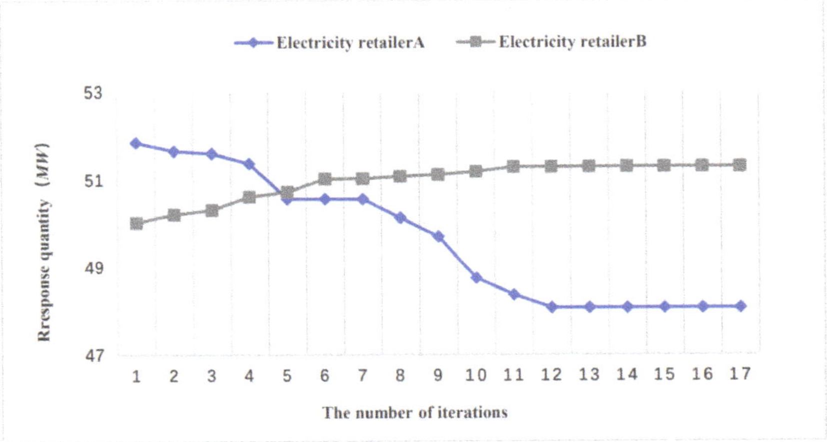

Figure 4. Convergence process of the optimal response quantity of the electricity retailers in Scenario 1.

Table 5. The optimal response quantity and subsidy unit price of electricity retailers.

Electricity Retailers	Period	Response Amount ($/MW)	Subsidy Unit Price ($/MW·h)
electricity retailers A	19:00–20:00	48.044	24.64
	20:00–21:00	49.068	25.07
electricity retailers B	19:00–20:00	51.417	27.09
	20:00–21:00	51.057	26.35

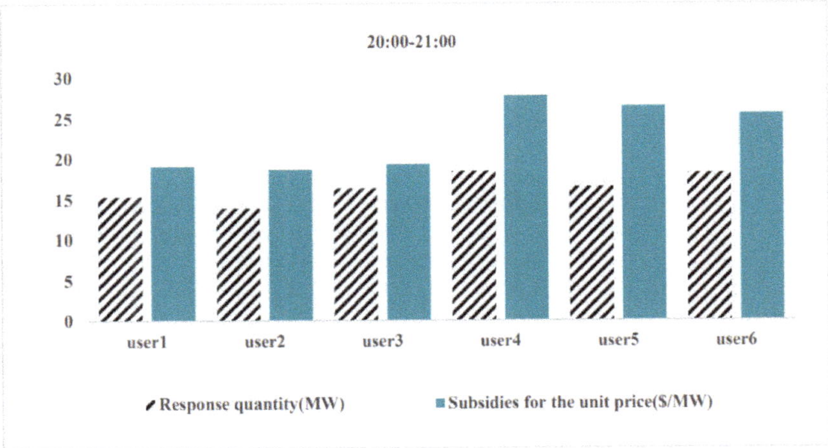

Figure 5. The optimal response quantity and subsidy unit price of users during 20:00–21:00.

In Scenario 2, considering the elasticity of users, the optimal response of the power company converged in the 18th iteration as shown in Figure 6. The optimal response of Retail Electricity Company A was 48.04 MW, and the optimal response of Retail Electricity Company B was 50.87 MW. At this time, the power grid's revenue was USD 3315.33 and the revenues of the two retail electricity companies were USD 2324.60 and USD 2420.37, respectively.

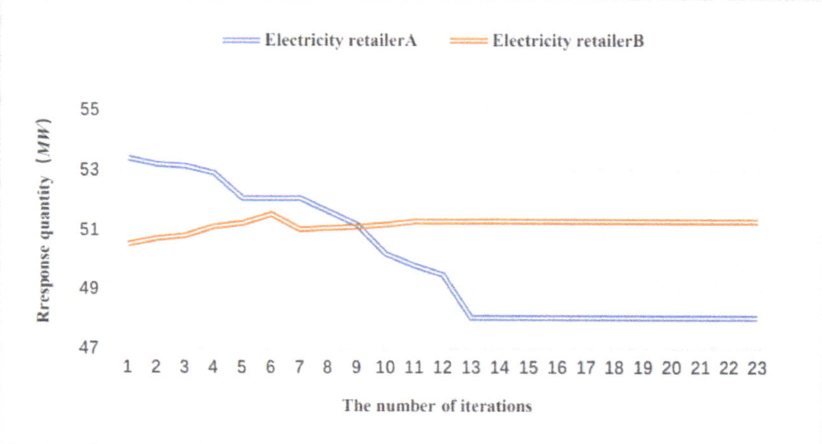

Figure 6. Convergence process of the optimal response quantity of the electricity retailers in Scenario 2.

Figure 7 shows the comparison of load before and after demand response by electricity users. It is evident from the figure that there is a significant decrease in the load during peak hours and an increase in electricity consumption during off-peak hours. Among all users, user 6 had the highest decrease in load during the 19:00–20:00 period, reaching 15.97%, while user 1 had the highest increase in electricity consumption, reaching 6.09%. By participating in demand response, users reduce their electricity costs and have a more stable load curve, reducing the losses caused by the extreme imbalance of power generation in the power grid while considering the price elasticity of demand response. User 4 had the highest price elasticity of demand response, resulting in the highest electricity consumption transfer, which was 1.13%. Through participation in demand response, users can reduce their power purchase costs while also making their load curve smoother, thereby reducing the losses caused by extreme power generation imbalance in the power grid.

In both Scenario 1 and Scenario 2, the MOMOV algorithm was used, but Scenario 2 took into account the elasticity coefficients of the users. The results of the two scenarios are shown in Table 6. The table shows that, in Scenario 2, the total user response increased by 2.89% and the power grid's benefits increased by 2.39%. It is evident that, in the process of maximizing its own benefits, the power grid company may sacrifice user benefits, which reduces the enthusiasm of users to participate in response. However, when considering the elasticity coefficients of users, the transfer of load reduces the user's response cost and increases their response enthusiasm, while optimizing the regulation of the power grid.

Table 6. Demand response in different scenarios.

Scenario	Response Cost Factor Grid Revenue ($)	Total Response (MW)	Load Transfer Amount (MW)
Scenario 1	3267.31	193.06	0
Scenario 2	3345.33	198.64	13.36

Figure 7. Load change before and after users participate in demand response.

In Scenario 2, when adjusting the profit weight l_{U_G} of the power grid while ensuring that other weight proportions remain unchanged, the changes in the profits of the power grid and electricity retailers are shown in Figure 8. When the profit weight l_{U_G} of the power grid increased from 0.1 to 0.4 under optimal decision-making, the profit of the power grid increased by 12.46%, while the profits of electricity retailer A and B decreased by 8.71% and 7.02%, respectively. In the game, as the profit weight of the power grid increased and the response subsidies received by electricity retailers decreased, their profits decreased.

With all other parameter values being equal, the results of solving Scenarios 2, 3, and 4 were compared. In other words, the effectiveness of the MOMOV, MOPSO, and NSGA-II algorithms for solving the game model was compared when taking into account the elasticity coefficients of users. The model was solved using each algorithm, independently run 20 times. The average and standard deviation of the IGD indicator were calculated, as shown in Table 7. It can be seen that the MOMOV algorithm performed the best in solving the model, with the smallest IGD indicator and standard deviation, indicating that this algorithm is more stable and effective than the other two algorithms. Therefore, the experimental findings serve as a testament to the efficacy of the MOMOV algorithm in addressing the decision variable contribution target analysis method for this particular model.

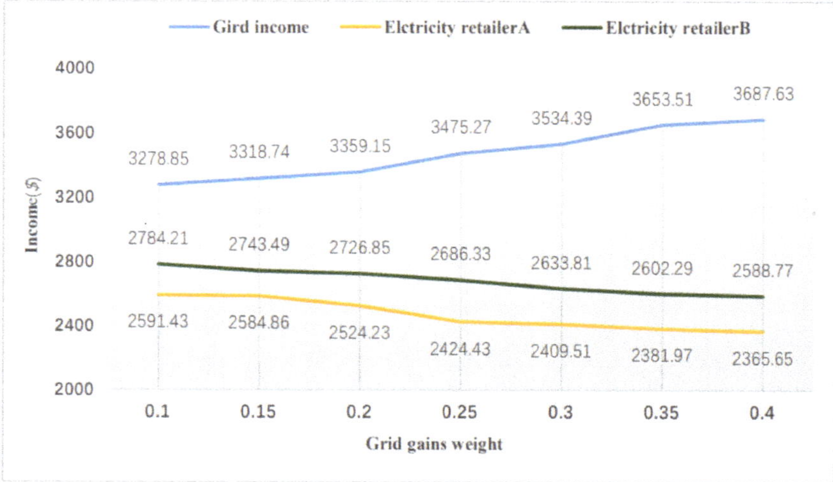

Figure 8. Response income with the participation of power grid and electricity retailers with the change of income weight of power grid.

Table 7. Results of the multi-objective algorithms (using GD, IGD) on the model.

MOPSO		NSGA-II		MOMOV	
GD	IGD	GD	IGD	GD	IGD
4.18×10^{-3}	1.32×10^{-2}	3.27×10^{-3}	2.12×10^{-2}	1.60×10^{-4}	4.10×10^{-3}

5. Conclusions

This paper addressed the demand response of the grid and proposed a game-theoretic model among grid, electricity retailers, and multiple users employing the master–slave game approach. Through the analysis and solution of the model, game equilibrium was achieved, and the corresponding strategy for power grid companies to set demand response subsidy prices was obtained. The simulation results of the example showed that the demand response game model constructed can achieve established goals, and the use of MOMOV for solving the model was superior to other optimization algorithms, which to some extent verified the rationality of the article. The subsidy price formulation strategy proposed in this paper not only considers user participation in demand response and reducing direct response subsidies, but also considers that users adjust their own electricity consumption under the influence of the price elasticity coefficient, and shift peak electricity consumption to valley hours. While users gain benefits, it has contributed positively to the process of reducing peak demand and filling low demand periods for the power system and effectively utilizing funds to improve the overall revenue of the grid in a competitive market environment. This has positive implications, while also providing a new train of thought for future demand response subsidized prices.

Author Contributions: Writing—original draft, D.H.; Writing—review & editing, T.W. All authors have read and agreed to the published version of the manuscript.

Funding: This research received no external funding.

Data Availability Statement: Publicly available datasets were analyzed in this study. This data can be found here: https://www.pjm.com (accessed on 5 April 2023).

Conflicts of Interest: The authors declare no conflict of interest.

References

1. Zhang, G.; Xue, S.; Fan, M.; Zhang, H. Design of market-oriented trading mechanism for demand response to China's electricity market. *Electr. Power Constr.* **2021**, *42*, 132–140.
2. Song, Y.; Bao, M.; Ding, Y.; Shao, C.; Shang, N. Review of Key Points and Related Suggestions of China's Electricity Spot Market Construction under New Power Reform. *Proc. CSEE* **2020**, *40*, 3172–3187.
3. He, Q.; Ai, Q. Bidding Strategy of Electricity Market with Demand Response virtual power plant in the environment of power sales side liberalization. *Electr. Power Constr.* **2019**, *3*, 2–6.
4. Wang, T.; Xu, K.; Zhu, Y. Source-grid-load multilayer game economic dispatching strategy of active distribution network. *Power Syst. Prot. Control* **2018**, *46*, 10–19.
5. Wang, H.; Xu, H.; Zhao, W. Optimization of multi-agent trading decisions in electricity market under uncertain demand and quota system. *Power Grid Clean Energy* **2021**, *37*, 34–44.
6. Sun, L.; Wang, Y.; Ding, J.; Ding, M.; Wang, P. Robust Planning Strategy for Generation and Transmission Considering Incentive Demand-Side Response. *Electr. Power Constr.* **2022**, *43*, 63–72.
7. Sun, Q.; Wang, X.; Wang, Y.; Zhang, Y.; Liu, Z.; He, J. Integrated energy market transaction optimization decision based on multi-agent Nash-Q reinforcement learning. *Power Syst. Autom.* **2021**, *45*, 10.
8. Cai, G.; Jing, Y.; Huang, N.; Yang, D.; Pan, X.; Shang, W. Optimal Scheduling of Charge and Discharge of Large-scale Electric Vehicles Based on Multi-agent Two-Layer Game under Power Demand Response Mechanism. *Proc. CSEE* **2023**, *43*, 14.
9. Lu, Q.; Zhang, Y. Demand response strategy of game between power supply and power consumption under multi-type user mode. *Int. J. Electr. Power Energy Syst.* **2022**, *134*, 107348. [CrossRef]
10. Dai, Y.; Sun, X.; Li, L.; Gao, H. Real-time Demand Response Mechanism of Smart Grid Residential Power Based on Multi-layer Game. *Oper. Res. Manag.* **2021**, *30*, 11–17.
11. Dai, Y.; Sun, X.; Qi, Y.; Leng, M. A real-time, personalized consumption-based pricing scheme for the consumptions of traditional and renewable energies. *Renew. Energy* **2021**, *180*, 452–466. [CrossRef]
12. Apostolopoulos, P.A.; Tsiropoulou, E.E.; Papavassiliou, S. Demand response manage-ment in smart grid networks: A two-stage game-theoretic learning-based approach. *Mob. Netw. Appl.* **2021**, *26*, 548–561. [CrossRef]
13. Wu, C.; Gao, B.; Tang, Y.; Wang, Q. Master–slave Game Based Bilateral Contract Transaction Model for Generation Companies and Large Consumers. *Autom. Electr. Power Syst.* **2016**, *40*, 56–62.
14. Cheng, L.; Yin, L. Behav-ioral decision-making in power demand-side response management: A multi-population evolutionary game dynamics perspective. *Int. J. Electr. Power Energy Syst.* **2021**, *129*, 106743. [CrossRef]
15. Lu, Q.; Lv, S.; Leng, Y. A Nash-Stackelberg game approach in regional energy market considering users' integrated demand response. *Energy* **2019**, *175*, 456–470. [CrossRef]
16. Motalleb, M.; Siano, P.; Ghorbani, R. Networked Stackelberg competition in a demand response market. *Appl. Energy* **2019**, *239*, 680–691. [CrossRef]
17. Dai, Y.; Gao, H.; Gao, Y.; Yuan, G. Real-time pricing mechanism in smart grid with forecasting update of power demand. *Autom. Electr. Power Syst.* **2018**, *42*, 58–63.
18. Lin, K.; Wu, J.; Liu, D.; Li, D. Energy management optimization of micro energy grid based on hierarchical Stackelberg game theory. *Syst. Technol.* **2019**, *43*, 973–983.
19. Mirjalili, S.; Jangir, P.; Mirjalili, S.Z.; Saremi, S.; Trivedi, I.N. Optimization of problems with multiple objectives using the multi-verse optimization algorithm. *Knowl.-Based Syst.* **2017**, *134*, 50–71. [CrossRef]
20. Lin, G.; Lu, S.; Guo, K.; Gao, Z.; Feng, X. Pricing Mech-anism of Demand Response Subsidy for Power Grid Companies Based on master–slave Game. *Power Syst. Autom.* **2020**, *44*, 59–67.
21. Xue, W.; Zhao, X.; Li, Y.; Mu, Y.; Tan, H.; Jia, Y.; Wang, X.; Zhao, H.; Zhao, Y. Research on the Optimal Design of Seasonal Time-of-Use Tariff Based on the Price Elasticity of Electricity Demand. *Energies* **2023**, *16*, 1625. [CrossRef]
22. Zhang, H.; Zhang, S.; Cheng, H.; Zhang, X.; Gu, Q. A Review of the Application of Stackelber Game in Electricity Market. *Trans. China Electrotech. Soc.* **2022**, *37*, 3250–3262.
23. Kumari, K.; Kumar, N.C.; Ranjan, N.M. Electric vehicles management enabling G2V and V2G in smart distribution system for maximizing profits using MOMVO. *Int. Trans. Electr. Energy Syst.* **2019**, *29*, e12013.
24. Haseeb, A.; Waleed, U.; Ashraf, M.M.; Siddiq, F.; Rafiq, M.; Shafique, M. Hybrid Weighted Least Square Multi-Verse Optimizer (WLS–MVO) Framework for Real-Time Estimation of Harmonics in Non-Linear Loads. *Energies* **2023**, *16*, 609. [CrossRef]
25. David, A.; VanVeldhuizen Gary, B. Lamont. Evolutionary Computation. In *Evolutionary Computation*; Massachusetts Institute of Technology Press: Cambridge, MA, USA, 1999.
26. Sierra, M.R.; Coello, C.A.C. Improving PSO-Based Multi-Objective Optimization Using Crowding, Mutation and—Dominance. In Proceedings of the Evolutionary Multi-Criterion Optimization, Guanajuato, Mexico, 9–11 March 2005; pp. 505–519.
27. Zhao, S.; Tang, S. Comprehe-nsive Evaluation of Transmission Network Planning Scheme Based on Improved Analytic Hierarchy Method, CRITIC Method and Approximation Ideal Solution Ranking Method. *Electr. Power Autom. Equip.* **2019**, *39*, 143–148+162.

processes MDPI

Article

Design of Intelligent Nonlinear H_2/H_∞ Robust Control Strategy of Diesel Generator-Based CPSOGSA Optimization Algorithm

Yidong Zou [1,†], Boyi Xiao [2,†], Jing Qian [2] and Zhihuai Xiao [1,*]

1 School of Power and Mechanical Engineering, Wuhan University, Wuhan 430072, China; yidongzou@whu.edu.cn
2 Faulty of Metallurgical and Energy Engineering, Kunming University of Science and Technology, Kunming 650000, China
* Correspondence: xiaozhihuai@126.com
† These authors contributed equally to this work.

Abstract: In today's human society, diesel generators (DGs) are widely applied in the human energy and electricity supply system due to its technical, operational, and economic advantages. This paper proposes an intelligent nonlinear H_2/H_∞ robust controller based on the chaos particle swarm gravity search optimization algorithm (CPSOGSA), which controls the speed and excitation of a DG. In this method, firstly, establish the nonlinear mathematical model of the DG, and then design the nonlinear H_2/H_∞ robust controller based on this. The direct feedback linearization and the H_2/H_∞ robust control theory are combined and applied. Based on the design of the integrated controller for DG speed and excitation, the system's performance requirements are transformed into a standard robust H_2/H_∞ control problem. The parameters of the proposed solution controller are optimized by using the proposed CPSOGSA. The introduction of CPSOGSA completes the design of an intelligent nonlinear H_2/H_∞ robust controller for DG. The simulation is implemented in MATLAB/Simulink, and the results are compared with the PID control method. The obtained results prove that the proposed method can effectively improve the dynamic accuracy of the system and the ability to suppress disturbances and improve the stability of the system.

Keywords: diesel generator; speed and excitation control; robust control; intelligent optimization

Citation: Zou, Y.; Xiao, B.; Qian, J.; Xiao, Z. Design of Intelligent Nonlinear H_2/H_∞ Robust Control Strategy of Diesel Generator-Based CPSOGSA Optimization Algorithm. *Processes* **2023**, *11*, 1867. https://doi.org/10.3390/pr11071867

Academic Editor: Hsin-Jang Shieh

Received: 25 May 2023
Revised: 12 June 2023
Accepted: 16 June 2023
Published: 21 June 2023

1. Introduction

The diesel generator (DG) is a prominent energy and power equipment extensively employed within the human industrial system [1]. Its versatility is especially crucial in regions characterized by weak grids and high costs associated with long-distance power transmission. Furthermore, in specific power supply scenarios such as military emergency power supply and disaster relief power supply, the reliance on traditional large power grids for providing adequate and timely power becomes impractical [2–4]. The wide range of applications of diesel generators is shown in Figure 1. In such contexts, DGs and their associated systems prove to be highly effective and reliable alternatives. However, to ensure the optimal performance and operational stability of DG systems, it is essential to develop advanced control strategies that can effectively regulate and optimize their operation.

The system consisting of DGs and the various types of loads connected can be considered a micro-grid (MG) [5]. The MG is an effective way to accelerate renewable energy development and solve the shortcomings of large grids [6]. The complementation of photovoltaic panels, wind turbines, and DGs in MG is one of the development trends of power generation and distribution in the future power system [7]. With the continuous improvement and perfection of DG technology, more and more people will also benefit from it, especially far away from the large grid. Therefore, the DG research work has progressive significance for the energy supply of human society.

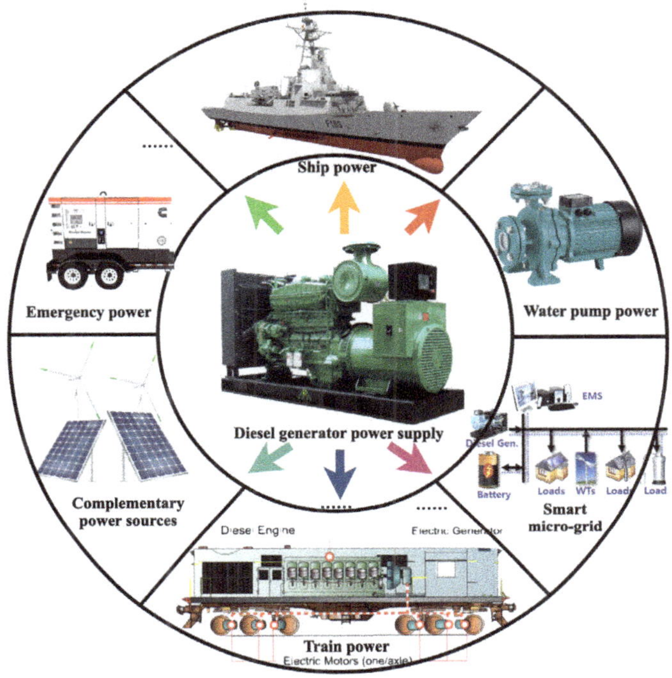

Figure 1. Widely used energy power device—diesel generator.

In many MGs, the DG is typically configured as the primary power source to ensure the safe and stable operation of the MG system [8,9]. Considering that DGs often contribute significantly to the overall capacity of most MGs, their control performance plays a critical role in determining the stability of the entire MG [10]. As a result, research focused on DG control is of fundamental importance. However, based on the current state of research, the comprehensive disturbance characteristics, such as voltage and frequency, specific to distributed MGs are rarely considered in the conventional DG control studies [11,12]. To address this gap, it is crucial to develop advanced control strategies that account for the intricate dynamics and disturbance characteristics associated with distributed MGs. By considering factors such as fluctuating loads, intermittent renewable energy sources, and varying grid conditions, the control of DGs can be significantly improved to ensure the optimal performance, stability, and resilience of the entire MG system.

The speed control system, comprising the diesel engine (DE) and the governor, plays a crucial role in determining the frequency of the synchronous generator. The frequency stability of the MG system relies heavily on the characteristics of the DE governor system [13,14]. However, when there are sudden changes in the load within the MG system, the power flow and output power of each generator also experience abrupt variations, disrupting the power balance between the generator and the prime mover. This leads to the occurrence of unbalanced torque and relative motion between the rotors of the generators [15]. The resulting frequency fluctuations not only impact the system's users but also detrimentally affect the overall operation of the MG system. Consequently, it becomes necessary to adjust the prime mover's frequency [16,17]. While much of the existing research has focused on optimizing the speed controller, there are limitations to the conventional approaches. For instance, Valenzuela et al. [18] proposed a conventional PI speed controller with reduced bandwidth to suppress excitation at the resonant frequency, but this approach leads to slow dynamic performance, which is undesirable. Mahmoud et al. [19] analyzed optimal tuning methods for the PI/PID controller to enhance system performance. However, this technique proves ineffective when the inertia of the driving machine is comparable to that of the load device [20]. In addition to these conventional approaches, advanced control strategies, such as sliding mode controller [21], model predictive control [22], fuzzy logic [23], neural networks [24,25], and optimization algorithms [26], have been proposed for speed controller optimization. These solutions exhibit improved dynamic properties and robustness against parameter changes. However, these works have not adequately considered the effects of the excitation regulation process. Moreover, in extreme cases, these methods may not be applicable to DG systems. In light of these limitations, it is essential to explore alternative control strategies that address the challenges associated with speed control in DG systems.

The balance between the mechanical power of the prime mover and the electromagnetic power of the generator is crucial to ensure stable operation. Failure to achieve this balance can result in rotor oscillations, electromagnetic transient processes, and dynamic excitation regulation caused by voltage and current fluctuations at the generator terminal [27,28]. Since the majority of electrical equipment in DG systems comprises inductive loads, the load current has a demagnetizing effect on the synchronous generator. Changes in current magnitude and power factor directly impact the terminal voltage, necessitating the use of an excitation control device for the synchronous generator [29]. Furthermore, to enhance power supply reliability and achieve cost-effective electricity generation, the parallel operation of multiple generator sets is often employed in certain systems under different working conditions. To maintain stability within the parallel system, the allocation of reactive power between the generators must be properly managed, a task regulated by excitation control mechanisms [30,31]. In the context of excitation control research, Patel et al. [32] proposed an alternative method based on command filtered backstepping principles to control the terminal voltage and stabilize the DG through the excitation system. In addition to the aforementioned approach, researchers have explored the integration of various advanced control techniques into excitation control. Examples include the decentralized improved cooperative excitation controller [33], the excitation controller based on linear matrix inequality (LMI) [34], robust control strategies [35,36], and model predictive control [37], all of which aim to improve system performance. However, it is worth noting that some controllers developed using these methods rely on linear approximations and bounded operating ranges. Consequently, their performance may degrade significantly in the presence of large disturbances [38].

The dynamic process of DGs in the system includes both the electromagnetic dynamic process and dynamic mechanical process, and the two are related to each other [39]. The comprehensive control of speed and excitation can be used as an effective measure to improve the transient stability of the MG system [40]. However, most of the existing DG strategies only consider one of these methods while ignoring the dual control functions of voltage regulation and prime mover speed regulation. In the above studies, stable islanded MGs based on nonlinear control techniques for speed regulation and excitation robust controllers for DGs have been studied [41]. However, the speed regulation and robust excitation controller they designed is not intelligent; that is, the user manually selects the parameters of the robust controller, and no optimization is used to optimize the adjustable parameters.

This paper proposes a robust control method for DG speed and excitation based on the intelligent CPSOGSA, which is used in island MG with multiple load types. The CPSOGSA optimizes the parameters of the robust controller, and the CPSOGSA is introduced for the robust control of DG speed and excitation. The simulation results show that the method effectively suppressed oscillations under severe disturbances and uncertain conditions. The main innovations of this paper can be divided into the following categories:

- Based on the theory of direct feedback linearization, a nonlinear speed and excitation robust controller of a DG is designed;

- The intelligent CPSOGSA is applied to optimize the dynamic output function parameters of the robust controller, thus introducing the CPSOGSA for this problem solving multi-objective mixed H_2/H_∞ robust controller;
- The proposed method effectively suppresses frequency and voltage oscillations under various load disturbances and uncertainties;
- Excellent damping efficiency, especially low overshoot, steady-state error, and settling time.

This paper is organized as follows: In Section 2, the isolated MG structure supplied by DG studied in this paper is introduced, and the dynamic mathematical model of the DG is established in detail, respectively. Then, in Section 3, a speed and excitation robust controller is designed for the MG system's primary power source-DG. Section 4 presents the CPSOGSA and its application to robust controller optimization problems. Simulation studies are conducted in Section 5 to demonstrate the advantages of the new method. Finally, Section 6 concludes this paper.

2. Modeling of Diesel Generator and Its Loads

This section briefly introduces the nonlinear mathematical model of DG, which mainly consists of two parts: speed regulation and the excitation system of DG.

2.1. Modeling of Speed Control Part

As illustrated in Figure 2, the diesel engine in the DG serves as a power machine that converts the heat energy from diesel fuel into the mechanical energy required by the generator [42]. However, it is important to note that the main driving output torque of the diesel engine, denoted as M_1, is a pulsating torque [43]. This pulsating torque can be mathematically decomposed into a Fourier series, represented by the following equation:

$$M_1 = M_p + \sum_{f=1}^{\infty} M_f^0 sin(f\omega_0 t + \varphi_f) \tag{1}$$

where M_1 is the DE pulsating torque; M_p is the average torque of one cycle; M_f^0 is the fth harmonic torque amplitude; ω_0 is the harmonic torque angular velocity; and φ_f is the fth harmonic torque initial phase angle.

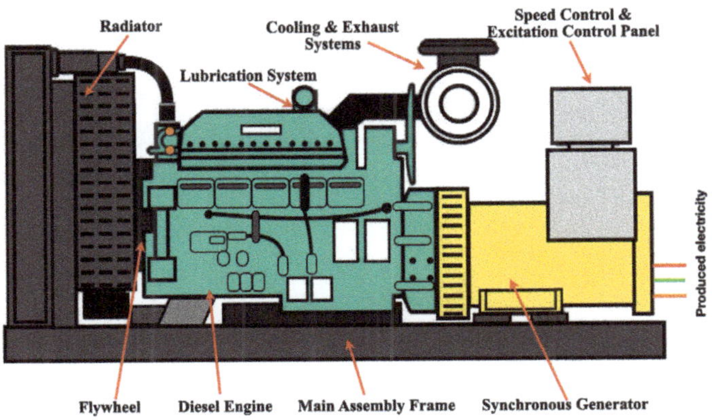

Figure 2. Principle diagram of DE speed regulation system.

As shown in Figure 3, a flywheel with a significant moment of inertia is installed on the main shaft of the diesel to keep the output torque of the diesel within the allowable range , which can make the flywheel rotation unevenness between (0.0033, 0.0067) [44]. During operation, the enormous flywheel torque ensures that the output torque of the diesel is uniform. Therefore, the output torque's unevenness can be ignored in the diesel's regular operation. In other words, the first constant in Equation (1) can be retained, and the rest of the harmonic torque composed of amplitude, phase, and frequency can be ignored, and Equation (1) is derived

$$M_1 = M_p \tag{2}$$

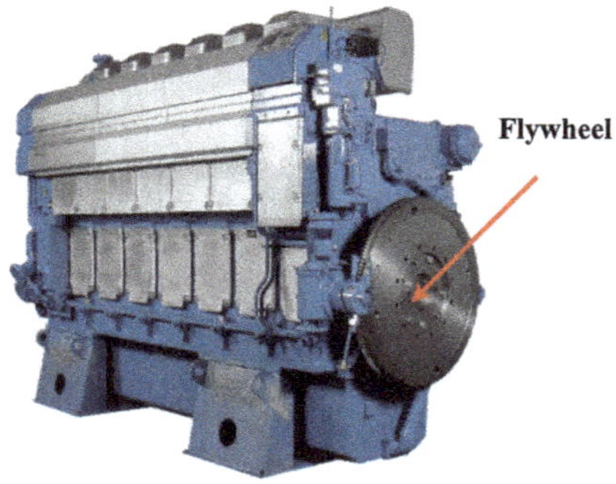

Figure 3. Flywheel of diesel.

The driving torque and rotating speed characteristics of DE are smooth curves, approximating several linear segments [45]. Divide the driving torque and rotating speed characteristics of the diesel into m sections; then, each section can be written as

$$M_{1j} = k_j n_d + b_j \tag{3}$$

where n_d is the DE rotational speed, and the constant value of k_j and b_j is determined by the positions of different segments ($j = 1, 2, \ldots, m$).

It can be seen from the adjustment characteristics of the diesel that the relationship between the main driving torque M_1 of the DE and the output throttle displacement L of the actuator presents a specific linear relationship [46]. At the same time, the driving torque of the diesel has a certain lag, so the adjustment characteristic of the diesel can be expressed as

$$M_1 = \frac{M_1^e}{L_e - L_0}(L(t - T_d) - L_0) \tag{4}$$

where M_1^e is the torque at rated oil supply; L_e is the rated stroke of the DE actuator; L_0 is the no-load stroke of the DE actuator; and T_d is the DE torque lag time.

Through the above analysis and derivation, the rotating speed characteristic and adjustment characteristic of the integrated the DE can be obtained as the expression of the driving torque M_1:

$$M_1 = k_j n_d + b_j + \frac{M_1^e}{L_e - L_0}(L(t - T_d) - L_0) \tag{5}$$

When the DE is running at the rated speed, the segment described in Equation (5) can be described as

$$M_1 = k_1 n_d + b_1 + \frac{M_1^e}{L_e - L_0}(L(t - T_d) - L_0) \tag{6}$$

Define $d_1 = b_1 - (M_1^e L_0)/(L_e - L_0)$; then, Equation (6) can be written as

$$M_1 = k_1 n_d + d_1 + \frac{M_1^e}{L_e - L_0} L(t - T_d) \tag{7}$$

The rotational speed, the number of cylinders, and the number of strokes of the diesel determine the torque lag time T_d of the diesel [47]. Generally, the following equation can be used to estimate the value range

$$\frac{15}{n_d} < T_d < \frac{15}{n_d} + \frac{60\tau}{n_d l_s} \quad (8)$$

where τ is the DE stroke coefficient and l_s is the number of DE cylinders.

The diesel parameters studied in this paper are the value of rated speed n_d = 1500 r/min, the number of cylinders is 16, and the stroke coefficient $\tau = 2$. According to Equation (8), the value range of T_d can be calculated as $0.01s < T_d < 0.015$. The speed transition time of the diesel is 2 s, and T_d is very small compared with it and can be ignored. Therefore, Equation (7) can become

$$M_1 = k_1 n_d + d_1 + \frac{M_1^e}{L_e - L_0} L \quad (9)$$

where L is the throttle actuator displacement.

The relationship between the rotational speed of the diesel n_d and the angular velocity of the crank shaft ω_g is

$$n_d = \frac{60\omega_g}{2\pi} \quad (10)$$

where ω_g is the angular velocity of the crankshaft shaft.

Equation (9) can be written as

$$M_1 = \frac{60k_1\omega_g}{2\pi} + d_1 + \frac{M_1^e}{L_e - L_0} L \quad (11)$$

The dynamic motion equation of the main shaft of the DG set can be expressed as

$$\dot{\omega}_g = \frac{1}{J}(M_1 - M_2 - M_d) \quad (12)$$

where J is the DE rotational moment of inertia; ω_g is the DE spindle angular velocity; M_1 is the DE output shaft torque; M_2 is the DE load torque; and M_d is the generator damping torque.

The damping torque M_d is generated by the damping winding of the synchronous generator, and its characteristic is that it is proportional to the electrical angular speed of the generator rotor, which can be calculated by Equation (13).

$$M_d = Dp\omega_g \quad (13)$$

where D is the damping factor and p is the number of synchronous generator pole pairs.

Substituting Equations (11) and (13) into Equation (12), we obtain

$$J\dot{\omega}_g + Dp\omega_g = \frac{60k_1\omega_g}{2\pi} + d_1 + \frac{M_1^e}{L_e - L_0} L - M_2 \quad (14)$$

General variables are marked per unit value in the research and analysis of electric power systems. In order to unify the form, Equation (14) can be standardized as per the unit value. This paper converts the mechanical angular speed into electrical angular speed and then standardizes the power, torque, and angular speed to study the speed response of the diesel. This paper defines the rated apparent power S_B of the DG as the reference value of power and defines $\omega_0 = 100\pi$ as the reference value of the angular speed. Therefore, in standardized form, Equation (14) can be written as

$$\frac{J\omega_{g0}^2}{S_B}\dot{\omega} = \frac{60k_1\omega_{g0}^2 - 2\pi Dp\omega_{g0}^2}{2\pi S_B}\omega + \frac{d_1\omega_{g0}}{S_B} + \frac{a\omega_{g0}}{S_B} L - M_2 \quad (15)$$

where ω_{g0} is $100\pi/p$; $a = \frac{M_1^e}{L_e - L_0}$.

Ignoring the no-load torque and the winding loss, M_2 is equal to the output torque of the synchronous generator. In the stability analysis or control system design, for the convenience of calculation, it is considered that the angular speed ω does not change much during the transient process and is approximately equal to the synchronous speed, that is, $\omega \approx 1$. Therefore, Equation (15) can be derived as

$$\frac{J\omega_{g0}^2}{S_B}\dot{\omega} = \frac{60k_1\omega_{g0}^2 - 2\pi Dp\omega_{g0}^2}{2\pi S_B}\omega + \frac{d_1\omega_{g0}}{S_B} + \frac{a\omega_{g0}}{S_B}L - P_e \tag{16}$$

The output power P_e of the salient-pole synchronous generator can be described as

$$P_e = \frac{E_q'U}{X_d'}\sin\delta + \frac{U^2}{2}\frac{X_d' - X_q}{X_d'X_q}sin(2\delta) \tag{17}$$

where E_q' is the q-axis transient potential; U is the DG terminal voltage; and X is the generator winding reactance.

The relationship between the generator rotor angle δ and the electrical angular velocity ω can be described as

$$\dot{\delta} = (\omega - 1)\omega_0 \tag{18}$$

where ω is the nominal value, δ and t are the actual values, and $\omega_0 = 100\pi$.

From Equations (16)–(18), the mathematical model of the electrical transient process of DGs is

$$\begin{cases} \dot{\delta} = (\omega - 1)\omega_0 \\ \dot{\omega} = \frac{T_b}{T_a}\omega + \frac{1}{T_a}c_1 + \frac{c_2}{T_a}L - \\ \quad \frac{1}{T_a}\frac{E_q'U}{X_d'}\sin\delta - \frac{1}{T_a}\frac{U^2}{2}\frac{X_d' - X_q}{X_d'X_q}\sin(2\delta) \end{cases} \tag{19}$$

in the above, the definitions of T_a, T_b, c_1, c_2, and a are as follows:

$$\begin{cases} T_a = \frac{J\omega_{g0}^2}{S_B} \\ T_b = \frac{60k_1\omega_{g0}^2 - 2\pi Dp\omega_{g0}^2}{2\pi S_B} \\ c_1 = \frac{d_1\omega_{g0}}{S_B} \\ c_2 = \frac{a\omega_{g0}}{S_B} \\ a = \frac{M_1^e}{L_e - L_0} \end{cases}$$

2.2. Modeling of Excitation Control Part

During the operation of DGs, excitation control is the most basic and indispensable. The synchronous generator is the control object of the excitation control system [46]. Studying the dynamic characteristics of the excitation system is inseparable from the analysis of the dynamic characteristics of the synchronous generator. The mathematical model of the electromagnetic transient process of the DG includes the stator voltage balance equation of the generator and the transient electromagnetic equation of each winding of the rotor [48].

The standard form of the transient electromagnetic equation of the synchronous generator can be described as

$$\begin{cases} \dot{E}'_q = \dfrac{1}{T_{d0}}E_{fd} - \dfrac{1}{T_{d0}}E'_q - \dfrac{X_d - X'_d}{T_{d0}} \\ \dot{E}''_q = \dfrac{1}{T_{d0}}\dfrac{X''_d - X_l}{X'_d - X_l}E_{fd} + \left(\dfrac{1}{T''_{d0}} - \dfrac{X''_d - X_l}{X'_d - X_1}\dfrac{1}{T_{d0}}\right)E'_q \\ -\dfrac{1}{T''_{d0}}E''_q - \left(\dfrac{X'_d - X''_d}{T''_{d0}} + \dfrac{X''_d - X_l}{X'_d - X_l}\dfrac{X_d - X'_d}{T_{d0}}I_d\right) \\ \dot{E}''_d = -\dfrac{1}{T''_{q0}}E''_d + \dfrac{X_q - X''_q}{T''_{q0}}I_q \\ U_d = -RI_d + \omega X''_q I_q + \omega E''_d \\ U_q = -RI_q - \omega X''_d I_d + \omega E''_q \\ U = \sqrt{U_d^2 + U_q^2} \end{cases} \tag{20}$$

where U is the stator winding terminal voltage; U_d and U_q are the d-axis and q-axis components of the stator winding terminal voltage; R is the stator winding resistance; X is the winding reactance; I is the winding current; T is the winding time constant; E''_d is the d-axis subtransitory potential; E'_q is the q-axis transient potential; E''_q is the q-axis subtransitory potential; and E_{df} is the excitation winding voltage.

By combining Equations (19) and (20), a nonlinear mathematical model of the DG is obtained. This model reflects the interaction and mutual influence of speed and voltage and more accurately describes the change law of speed and voltage.

3. Design of Nonliear H_2/H_∞ Synthetic Controller

3.1. Multi-Objective State-Feedback Theory

The control system shown in Figure 4 defines the closed-loop transfer function $T_\infty(s)$ from disturbance w to H_∞ control performance output vector z_∞, and the closed-loop transfer function $T_2(s)$ from disturbance w to H_2 control performance output vector z_2. The control goal is to design a state feedback control law $u = Kx$ to stabilize the closed-loop system progressively. In addition, it needs to:

- Keep the RMS gain (H_∞ norm) of $T_\infty(s)$ below a certain specified value $\gamma_0 > 0$;
- Maintain the H_2 norm (LQG cost) of $T_2(s)$ below a certain specified value $v_0 > 0$;
- Minimized the form of H_2/H_∞ trade-off standard $\alpha T_2(s) + \beta T_\infty(s)$;
- Place the closed-loop pole in the designated area of the open left half-plane.

The mixed H_2/H_∞ standard considers both the interference suppression aspect (RMS gain from w to z_∞) and the LQG aspect (H_2 norm from w to z_2). In addition, the closed-loop pole can be forced into a specific part of the stable half-plane to obtain a well-damped transient response. Such control problems are called H_2/H_∞ control design problems [48,49].

The linear system shown in Figure 4 can be expressed as:

$$\begin{cases} \dot{x} = \boldsymbol{A}x + \boldsymbol{B}_1 w + \boldsymbol{B}_2 u \\ z_\infty = \boldsymbol{C}_1 x + \boldsymbol{D}_{11} w + \boldsymbol{D}_{12} u \\ z_2 = \boldsymbol{C}_2 x + \boldsymbol{D}_{21} w + \boldsymbol{D}_{22} u \end{cases} \tag{21}$$

where the first term in Equation (21) is a dynamic linear state-space differential equation and the second and third terms are its dynamic performance evaluation signals.

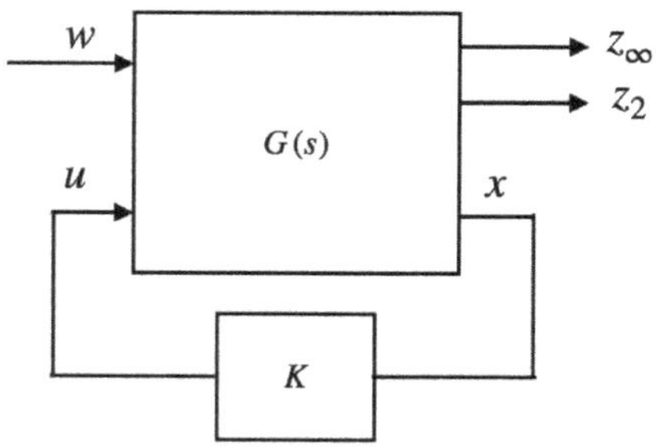

Figure 4. State-feedback control.

The closed-loop system formed after adding the controller Kx can be described in the form of state-space as

$$\begin{cases} \dot{x} = A_{cl}x + B_1 w \\ z_\infty = C_{cl1}x + D_{11}w \\ z_2 = C_{cl2}x + D_{21}w \end{cases} \tag{22}$$

where $A_{cl} = A + B_2K$; $C_{cl1} = C_1 + D_{12}K$; $C_{cl2} = C_2 + D_{22}K$.

Taken separately, the two design goals have the following LMI formulation:

The H_∞ performance: the closed-loop RMS gain from w to z_∞ does not exceed γ if and only if there exists a symmetric matrix X_∞ such that

$$\begin{cases} \begin{bmatrix} A_{cl}X_\infty + X_\infty A_{cl}^T & B_1 & X_\infty C_{cl1}^T \\ B_1^T & -\gamma I & D_{11}^T \\ C_{cl1}X_\infty & D_{11} & -\gamma I \end{bmatrix} < 0 \\ X_\infty > 0 \end{cases} \tag{23}$$

The H_2 performance: the closed-loop H_2 norm of $T_2(s)$ does not exceed v if there exist two symmetric matrices X_2 and Q such that

$$\begin{cases} \begin{bmatrix} A_{cl}X_2 + X_2A_{cl}^T & B_1 \\ B_1^T & -I \end{bmatrix} < 0 \\ \begin{bmatrix} Q & C_{cl2}X_2 \\ X_2C_{cl2}^T & X_2 \end{bmatrix} > 0 \\ Trace(Q) < v^2 \end{cases} \tag{24}$$

The mixed H_2/H_∞ performance: the pole of the closed-loop system is located in the left half-open complex plane, and the performance index $\alpha\|T(s)_\infty\|_\infty^2 + \beta\|T_2(s)\|_2^2$ is minimized.

To solve the mixed objective, the two sets of conditions add up to a non-convex optimization problem with variables Q, K, X_∞, and X_2. In order to facilitate the processing of this problem in the LMI framework, we seek a single Lyapunov matrix $X := X_\infty = X_2$ to force the solution of the two goals. With the change of variable $Y := KX$, this leads to the multi-objective state feedback synthesis problem, which the following sub-optimal LMI formula can express:

Minimize $\alpha\gamma^2 + \beta Trace(\boldsymbol{Q})$ over $\boldsymbol{Y}$, $\boldsymbol{X}$, $\boldsymbol{Q}$, and γ^2 satisfying:

$$\begin{cases} \begin{bmatrix} \boldsymbol{AX} + \boldsymbol{XA}^T + \boldsymbol{B}_2\boldsymbol{Y} + \boldsymbol{Y}^T\boldsymbol{B}_2^T & \boldsymbol{B}_1 & \boldsymbol{XC}_1^T + \boldsymbol{Y}^T\boldsymbol{D}_{12}^T \\ \boldsymbol{B}_1^T & -\gamma\boldsymbol{I} & \boldsymbol{D}_{11}^T \\ \boldsymbol{C}_1\boldsymbol{X} + \boldsymbol{D}_{12}\boldsymbol{Y} & \boldsymbol{D}_{11} & -\gamma\boldsymbol{I} \end{bmatrix} < 0 \\ \begin{bmatrix} \boldsymbol{Q} & \boldsymbol{C}_2\boldsymbol{X} + \boldsymbol{D}_{22}\boldsymbol{Y} \\ \boldsymbol{XC}_2^T + \boldsymbol{Y}^T\boldsymbol{D}_{22}^T & \boldsymbol{X} \end{bmatrix} < 0 \\ Trace(\boldsymbol{Q}) < v_0^2 \\ \gamma_0 < \gamma_0^2 \end{cases} \tag{25}$$

Denoting the optimal solution by $(\boldsymbol{X}^*, \boldsymbol{Y}^*, \boldsymbol{Q}^*, \gamma^*)$, the corresponding state-feedback gain is given by:

$$\boldsymbol{K}^* = \boldsymbol{Y}^*(\boldsymbol{X}^*)^{-1} \tag{26}$$

this gain guarantees the worst-case performances:

$$\begin{cases} \|T(s)_\infty\|_\infty \leq \gamma^* \\ \|T(s)_2\|_2 \leq \sqrt{Trace(\boldsymbol{Q}^*)} \end{cases} \tag{27}$$

The solution of the mixed H_2/H_∞ controller is usually configured with $\alpha = 0.5$ and $\beta = 0.5$ to obtain [46]. The robust controller obtained by this method can improve and optimize the control performance of the related system.

3.2. Design of Speed Controller

The schematic diagram of the DG integrated control system based on the nonlinear H_2/H_∞ integrated controller is shown in Figure 5. The DG integrated control system comprises a diesel, generator, nonlinear H_2/H_∞ integrated controller, actuator, fuel supply mechanism, and exciter. The nonlinear H_2/H_∞ integrated controller consists of two parts: one part is the nonlinear H_2/H_∞ speed controller, and the other is the nonlinear H_2/H_∞ voltage controller. The nonlinear H_2/H_∞ integrated controller should consider the coupling effect of speed and voltage, comprehensively control the speed and voltage, and control the mutual influence between the two to a minimum. Further, it should improve the stability of the frequency and voltage of the independent MG system.

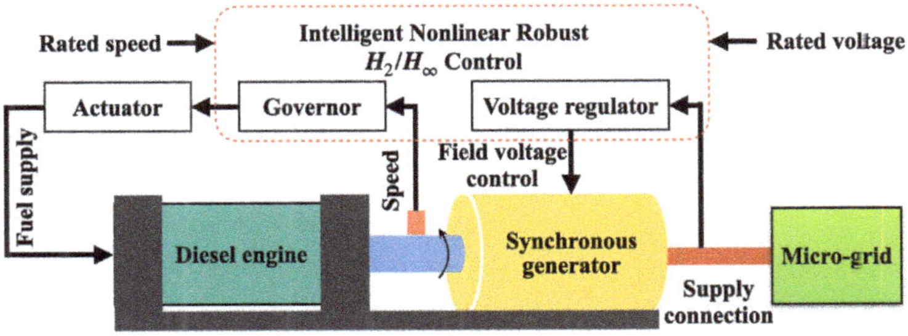

Figure 5. Principle diagram of diesel-generator set synthetic control system.

The differential equation expression of the actuator is:

$$\dot{L} = -\frac{1}{T_1}L + \frac{K_1}{T_1}u_1 \tag{28}$$

where T_1 is the actuator time constant and K_1 is the actuator gain.

The expression of the differential equation of the exciter is

$$\dot{E}_{fd} = -\frac{1}{T_2}E_{fd} + \frac{K_2}{T_2}u_2 \tag{29}$$

where T_2 is the exciter time constant and K_2 is the exciter gain.

By combining Equations (19) and (28), we can obtain the nonlinear mathematical model of the DG's speed control system:

$$\begin{cases} \dot{\delta} =(\omega - 1)\omega_0 \\ \dot{L} = -\frac{1}{T_1}L + \frac{K_1}{T_1}u_1 \\ \dot{\omega} =\frac{T_b}{T_a}\omega + \frac{1}{T_a}c_1 + \frac{c_2}{T_a}L - \frac{1}{T_a}\frac{E_q'U}{X_d'}\sin\delta - \\ \quad \frac{1}{T_a}\frac{U^2}{2}\frac{X_d' - X_q}{X_d'X_q}\sin(2\delta) \end{cases} \tag{30}$$

Equation (30) has apparent nonlinear terms. In order to be able to apply the above mixed H_2/H_∞ control theory, this paper uses the direct feedback linearization method to linearize the nonlinear mathematical model of the DG speed system described in Equation (30) to obtain its linear model and then obtain the corresponding nonlinear speed control law.

Three variables (X_1, X_2, and X_3) are defined in this paper, which can be expressed as:

$$\begin{cases} X_1 =\delta \\ X_2 =(\omega - 1)\omega_0 \\ X_3 =\frac{T_b}{T_a}\omega + \frac{1}{T_a}c_1 + \frac{c_2}{T_a}L - \frac{1}{T_a}\frac{E_q'U}{X_d'}\sin\delta \\ \quad \frac{1}{T_a}\frac{U^2}{2}\frac{X_d' - X_q}{X_d'X_q}\sin(2\delta) \end{cases} \tag{31}$$

According to the definitions of X_1, X_2 and X_3, the nonlinear differential equations described by Equation (30) can be rewritten as:

$$\begin{cases} \dot{X}_1 =X_2 \\ \dot{X}_2 =\omega_0 X_3 + d_1 W \\ \dot{X}_3 =\frac{T_b}{T_a}X_3 + \frac{c_2K_1}{T_aT_1}u_1 - \frac{c_2}{T_aT_1}L - \\ \quad \frac{E_q'U\omega_0}{T_aX_d'}\cos\delta(\omega - 1) - \\ \quad \frac{U^2\omega_0(X_d' - X_q)}{T_aX_d'X_d}\cos(2\delta)(\omega - 1) \end{cases} \tag{32}$$

where d_1W is the disturbance signal set when the mixed H_2/H_∞ control method is adopted, including equivalent disturbances caused by disturbance torque and modelling errors.

For speed control, this paper defines virtual control variables V as:

$$\begin{aligned} V =&\frac{c_2K_1}{T_aT_1}u_1 - \frac{c_2}{T_aT_1}L - \frac{E_q'U\omega_0}{T_aX_d'}\cos\delta(\omega - 1) - \\ &\frac{U^2\omega_0(X_d' - X_q)}{T_aX_d'X_q}\cos(2\delta)(\omega - 1) \end{aligned} \tag{33}$$

By introducing virtual control variables V, Equation (30) is described by linear state-space differential equations as:

$$\dot{X} = \boldsymbol{A}_s X + \boldsymbol{B}_{1s} W + \boldsymbol{B}_{2s} V \tag{34}$$

where X, $\boldsymbol{A}_s$, $\boldsymbol{B}_{1s}$ and $\boldsymbol{B}_{2s}$ are defined as follows:

$$X = \begin{bmatrix} X_1 \\ X_2 \\ X_3 \end{bmatrix}; \boldsymbol{A}_s = \begin{bmatrix} 0 & 1 & 0 \\ 0 & 0 & \omega_0 \\ 0 & 0 & \dfrac{T_b}{T_a} \end{bmatrix};$$

$$\boldsymbol{B}_{1s} = \begin{bmatrix} 0 \\ d_1 \\ 0 \end{bmatrix}; \boldsymbol{B}_{2s} = \begin{bmatrix} 0 \\ 0 \\ 1 \end{bmatrix}.$$

Add the dynamic performance evaluation signal described by Equation (35) to this system

$$\begin{cases} Z_{\infty s} = \boldsymbol{C}_{1s} X + \boldsymbol{D}_{11s} W + \boldsymbol{D}_{12s} V \\ Z_{2s} = \boldsymbol{C}_{2s} X + \boldsymbol{D}_{21s} W + \boldsymbol{D}_{22s} V \end{cases} \tag{35}$$

where

$$\boldsymbol{C}_{1s} = \begin{bmatrix} q_{11} & 0 & 0 \\ 0 & q_{12} & 0 \\ 0 & 0 & q_{13} \end{bmatrix}; \boldsymbol{D}_{12s} = \begin{bmatrix} 0 \\ 0 \\ r_1 \end{bmatrix};$$

$$\boldsymbol{C}_{2s} = \begin{bmatrix} q_{21} & 0 & 0 \\ 0 & q_{22} & 0 \\ 0 & 0 & q_{23} \end{bmatrix}; \boldsymbol{D}_{22s} = \begin{bmatrix} 0 \\ 0 \\ r_2 \end{bmatrix};$$

$$\boldsymbol{D}_{11s} = \boldsymbol{D}_{21s} = \begin{bmatrix} 0 \\ 0 \\ 0 \end{bmatrix}.$$

$\boldsymbol{C}_{1s}$, $\boldsymbol{C}_{2s}$, $\boldsymbol{D}_{11s}$, $\boldsymbol{D}_{12s}$, $\boldsymbol{D}_{21s}$, and $\boldsymbol{D}_{22s}$ are weighting matrices, $q_{ij} > 0$ $(i = 1,2; j = 1,2,3)$ and $r_i > 0$ $(i = 1,2)$ are weighting coefficients. A controller that meets the optimal performance requirements can be obtained by selecting different weighting coefficients. The performance requirements include the independent power system stability, frequency adjustment accuracy, and low energy consumption of the speed control system.

From Equations (34) and (35), the augmented controlled object based on the mixed H_2/H_∞ control theory can be obtained as

$$\boldsymbol{P}_s = \begin{bmatrix} \boldsymbol{A}_s & \boldsymbol{B}_{1s} & \boldsymbol{B}_{2s} \\ \boldsymbol{C}_{1s} & \boldsymbol{D}_{11s} & \boldsymbol{D}_{12s} \\ \boldsymbol{C}_{2s} & \boldsymbol{D}_{21s} & \boldsymbol{D}_{22s} \end{bmatrix} \tag{36}$$

Bring the augmented controlled object $\boldsymbol{P}_s$ into the framework shown in Figure 4 and meet the design requirements of the multi-objective H_2/H_∞ state-feedback controller at the same time, and then the H_2/H_∞ state feedback controller can be obtained as:

$$\begin{aligned} V &= \boldsymbol{F}_s X \\ &= \begin{bmatrix} f_{1s} & f_{2s} & f_{3s} \end{bmatrix} \begin{bmatrix} X_1 \\ X_2 \\ X_3 \end{bmatrix} \\ &= f_{1s} X_1 + f_{2s} X_2 + f_{3s} X_3 \end{aligned} \tag{37}$$

From Equations (33) and (37), the speed control law u_1 can be solved as:

$$\begin{aligned} u_1 =& \frac{1}{K_1}L + \frac{T_a T_1}{c_2 K_1}\boldsymbol{F}_s X + \frac{E_q' U \omega_0 T_1}{c_2 K_1 X_d'}\cos\delta(\omega-1) \\ &+ \frac{U^2 \omega_0 T_1 (X_d' - X_q)}{c_2 K_1 X_d' X_q}\cos(2\delta)(\omega-1) \end{aligned} \tag{38}$$

Substituting X_1, X_2, and X_3 into Equation (38), the DE nonlinear H_2/H_∞ speed control law in the usable form shown in Equation (39) can be obtained.

$$\begin{aligned} u_1 =& \frac{1}{K_1}L + \frac{T_a T_1}{c_2 K_1} f_{1s}\delta + \frac{T_a T_1}{c_2 K_1} f_{2s}(\omega-1)\omega_0 + \\ & \frac{T_b T_1}{c_2 K_1} f_{3s}\omega - \frac{T_1}{c_2 K_1}\frac{E_q' U}{X_d'} f_{3s}\sin\delta - \\ & \frac{T_1}{c_2 K_1}\frac{U^2}{2}\frac{X_d' - X_q}{X_d' X_q} f_{3s}\sin(2\delta) + \frac{T_1}{c_2 K_1} f_{3s} c_1 + \\ & \frac{T_1}{K_1} f_{3s} L + \frac{E_q' U \omega_0 T_1}{c_2 K_1 X_d'}\cos\delta(\omega-1) + \\ & \frac{U^2 \omega_0 T_1 (X_d' - X_q)}{c_2 K_1 X_d' X_q}\cos(2\delta)(\omega-1) \end{aligned} \tag{39}$$

3.3. Design of Excitation Control Controller

By combining Equations (20) and (29) of the first three terms, one can obtain the nonlinear mathematical model of the synchronous generator voltage regulation system, which is shown in Equation (40).

$$\begin{cases} \dot{E}_{fd} = -\dfrac{1}{T_2}E_{fd} + \dfrac{K_2}{T_2}u_2 \\ \dot{E}_q' = \dfrac{1}{T_{d0}}E_{fd} - \dfrac{1}{T_{d0}}E_q' - \dfrac{X_d - X_d'}{T_d}I_d \\ \dot{\delta} = (\omega - 1)\omega_0 \\ \dot{\omega} = \dfrac{T_b}{T_a\omega_0}\omega + \dfrac{1}{T_a\omega_0}c_1 + \dfrac{c_2}{T_a\omega_0}L - \\ \dfrac{1}{T_a\omega_0}\dfrac{E_q' U}{X_d'}\sin\delta - \dfrac{1}{T_a\omega_0}\dfrac{U^2}{2}\dfrac{X_d' - X_q}{X_d' X_q}\sin(2\delta) \end{cases} \tag{40}$$

Since the damper winding of the synchronous generator has a weak influence on the excitation control, the fourth and fifth terms of Equation (20) can be ignored when designing the voltage controller. Regarding the electromagnetic power P_e as an external disturbance, Equation (40) becomes

$$\begin{cases} \dot{E}_{fd} = -\dfrac{1}{T_2}E_{fd} + \dfrac{K_2}{T_2}u_2 \\ \dot{E}_q' = \dfrac{1}{T_{d0}}E_{fd} - \dfrac{1}{T_{d0}}E_q' - \dfrac{X_d - X_d'}{T_{d0}}I_d \\ \dot{\delta} = \omega - 1 \\ \dot{\omega} = \dfrac{T_b}{T_a\omega_0}\omega + \dfrac{1}{T_a\omega_0}c_1 + \dfrac{c_2}{T_a\omega_0}L - \dfrac{1}{T_a\omega_0}P_e \end{cases} \tag{41}$$

Since E_q' is not easy to measure, it is not conducive to designing a state feedback controller. Therefore, the voltage error ΔU is selected as the state variable, so this paper converts Eq' in Equation (41) into ΔU by a constant c_3. According to the relationship between variable data:

$$U = E_q' - c_3\delta \tag{42}$$

Through the relationship between U and the voltage U_0 at the stator winding end of the generator, it can be derived:

$$E_q' = U_0 + \Delta U + c_3\delta \tag{43}$$

According to the relationship described in Equation (43), the state-space differential equation expressed by Equation (41) can be rewritten as

$$\begin{cases} \dot{E}_{fd} = -\dfrac{1}{T_2}E_{fd} + \dfrac{K_2}{T_2}u_2 \\ \Delta\dot{U} = \dfrac{1}{T_{d0}}E_{fd} - \dfrac{1}{T_{d0}}\Delta U - \dfrac{1}{T_{d0}}c_3\delta - \\ \qquad c_3\omega + (c_3 - \dfrac{1}{T_{d0}}) - \dfrac{X_d - X_d'}{T_{d0}}I_d \\ \dot{\delta} = \omega - 1 \\ \dot{\omega} = \dfrac{T_b}{T_a\omega_0}\omega + \dfrac{1}{T_a\omega_0}c_1 + \dfrac{c_2}{T_a\omega_0}L - \dfrac{1}{T_a\omega_0}P_e \end{cases} \tag{44}$$

Equation (44) is written as a linear state-space differential equation in the form

$$\dot{X}' = \boldsymbol{A}_e X' + \boldsymbol{B}_{1e} W' + \boldsymbol{B}_{2e} u_2 \tag{45}$$

where

$$X'(t) = \begin{bmatrix} E_{fd} \\ \Delta U \\ \delta \\ \omega \end{bmatrix}; W' = \begin{bmatrix} I_d \\ L \\ P_e \\ 1 \end{bmatrix}; \boldsymbol{B}_{2e} = \begin{bmatrix} \dfrac{K_2}{T_2} \\ 0 \\ 0 \\ 0 \end{bmatrix};$$

$$\boldsymbol{A}_e = \begin{bmatrix} -\dfrac{1}{T_2} & 0 & 0 & 0 \\ \dfrac{1}{T_{d0}} & -\dfrac{1}{T_{d0}} & -\dfrac{c_3}{T_{d0}} & -c_3 \\ 0 & 0 & 0 & 1 \\ 0 & 0 & 0 & \dfrac{T_b}{T_a\omega_0} \end{bmatrix};$$

$$\boldsymbol{B}_{1e} = \begin{bmatrix} 0 & 0 & 0 & 0 \\ -\dfrac{X_d - X_d'}{T_{d0}} & 0 & 0 & c_3 - \dfrac{1}{T_{d0}} \\ 0 & 0 & 0 & -1 \\ 0 & \dfrac{c_2}{T_a\omega_0} & -\dfrac{1}{T_a\omega_0} & \dfrac{c_1}{T_a\omega_0} \end{bmatrix}.$$

Similarly, the dynamic performance evaluation signal of the system described in Equation (45) is defined as

$$\begin{cases} Z_{\infty e} = \boldsymbol{C}_{1e}X' + \boldsymbol{D}_{11e}W' + \boldsymbol{D}_{12e}u_2 \\ Z_{2e} = \boldsymbol{C}_{2e}X' + \boldsymbol{D}_{21e}W' + \boldsymbol{D}_{22e}u_2 \end{cases} \tag{46}$$

where

$$C_{1e} = \begin{bmatrix} q_{14} & 0 & 0 & 0 \\ 0 & q_{15} & 0 & 0 \\ 0 & 0 & q_{16} & 0 \\ 0 & 0 & 0 & q_{17} \end{bmatrix}; D_{12e} = \begin{bmatrix} r_3 \\ r_4 \\ r_5 \\ r_6 \end{bmatrix};$$

$$C_{2e} = \begin{bmatrix} q_{24} & 0 & 0 & 0 \\ 0 & q_{25} & 0 & 0 \\ 0 & 0 & q_{26} & 0 \\ 0 & 0 & 0 & q_{27} \end{bmatrix}; D_{22e} = \begin{bmatrix} r_7 \\ r_8 \\ r_9 \\ r_{10} \end{bmatrix};$$

$$D_{11e} = D_{21e} = \begin{bmatrix} 0 & 0 & 0 & 0 \\ 0 & 0 & 0 & 0 \\ 0 & 0 & 0 & 0 \\ 0 & 0 & 0 & 0 \end{bmatrix}.$$

C_{1e}, C_{2e}, D_{11e}, D_{12e}, D_{21e}, and D_{22e} are weighting matrices, $q_{ij} > 0\ (i = 1,2; j = 4,5,6,7)$, and $r_i > 0 (i = 3, 4, \ldots, 10)$ are weighting coefficients. The voltage controller that meets the optimal performance is solved by combining different weighting coefficients. The performance includes the system stability, voltage regulation accuracy, and low energy loss of the excitation system.

From Equations (45) and (46), the augmented controlled object P_e used to solve the mixed H_2/H_∞ robust controller can be obtained, which can be expressed as:

$$P_e = \begin{bmatrix} A_e & B_{1e} & B_{2e} \\ C_{1e} & D_{11e} & D_{12e} \\ C_{2e} & D_{21e} & D_{22e} \end{bmatrix} \tag{47}$$

Similarly, the augmented controlled object P_e is brought into the solution framework shown in Figure 4, and the voltage H_2/H_∞ state feedback controller of the DG is obtained to satisfy the LMI. The nonlinear H_2/H_∞ voltage control law of the DG is

$$\begin{aligned} u_2 &= F_e X' \\ &= \begin{bmatrix} f_{1e} & f_{2e} & f_{3e} & f_{4e} \end{bmatrix} \begin{bmatrix} E_{fd} \\ \Delta U \\ \delta \\ \omega \end{bmatrix} \\ &= f_{1e} E_{fd} + f_{2e} \Delta U + f_{3e} \delta + f_{4e} \omega \end{aligned} \tag{48}$$

Combining the Equations (39) and (48) derived above, the DG's nonlinear robust H_2/H_∞ integrated controller is obtained.

$$\begin{cases} u_1 = \dfrac{1}{K_1} L + \dfrac{T_a T_1}{c_2 K_1} f_{1s} \delta + \dfrac{T_a T_1}{c_2 K_1} f_{2s} (\omega - 1) \omega_0 + \\ \dfrac{T_b T_1}{c_2 K_1} f_{3s} \omega - \dfrac{T_1}{c_2 K_1} \dfrac{E'_q U}{X'_d} f_{3s} \sin \delta - \\ \dfrac{T_1}{c_2 K_1} \dfrac{U^2}{2} \dfrac{X'_d - X_q}{X'_d X_q} f_{3s} \sin(2\delta) + \dfrac{T_1}{c_2 K_1} f_{3s} c_1 + \\ \dfrac{T_1}{K_1} f_{3s} L + \dfrac{E'_q U \omega_0 T_1}{c_2 K_1 X'_d} \cos \delta (\omega - 1) + \\ \dfrac{U^2 \omega_0 T_1 (X'_d - X_q)}{c_2 K_1 X'_d X_q} \cos(2\delta)(\omega - 1) \\ u_2 = f_{1e} E_{fd} + f_{2e} \Delta U + f_{3e} \delta + f_{4e} \omega \end{cases} \tag{49}$$

It can be found from Equation (49) that the designed nonlinear control law considers the coupling effect of the DG's speed and voltage and comprehensively controls the two, which can improve the dynamic accuracy of the DG's speed and voltage. Further, it can improve the stability of the MG system connected to the DG.

4. Artificial Hybrid PSOGSA with Chaotic Maps Approach and Its Application to Multi-Objective H_2/H_∞ Robust Problem

4.1. The Proposed CPSOGSA Algorithm

The artificial intelligence CPSOGSA algorithm is a hybrid particle swarm optimization and gravitational search algorithm combined with chaotic mapping [50]. Moreover, the CPSOGSA is superior and more robust than other methods (such as MSA, GWO, WOA, and so on) [51]. Therefore, this paper also adopts CPSOGSA to optimize the robust controller. This subsection provides the necessary description of the algorithm.

4.1.1. Particle Swarm Optimization

The PSO consists of three crucial operators, which are the inertia weight vector w_p, the personal particle best (pbest), and the global particle best (gbest). The inertia weight gives diversification (exploration) capability, and <pbest, gbest> provides intensification (exploitation) power to PSO, respectively [52]. The velocity $v_i(k)$ and position $p_i(k)$ of each particle in the PSO algorithm are updated as follows:

$$\begin{cases} v_i(k+1) = w_p v_i(k) + c_1 \times \text{rand} \times (\text{pbest}_i - p_i(k)) \\ + c_2 \times \text{rand} \times (\text{gbest}_i - p_i(k)) \\ p_i(k+1) = p_i(k) + v_i(k+1) \end{cases} \tag{50}$$

where c_1 and c_2 are the acceleration coefficient and rand is a uniform random variable between O and 1.

4.1.2. Gravitational Search Algorithm

GSA is an optimization algorithm inspired by Newton's theories of gravity and motion. The operation steps of GSA are briefly summarized as follows [53].

Positions of N agents constituting the initialization population P_i in GSA are listed in Equation (51). The best and worst fitness values are computed for each iteration. The best and worst fitness values are defined as in Equation (52) for the problem.

$$P_i = \left(p_i^1, \ldots, p_i^d, \ldots, p_i^n\right), \quad i = 1, 2, \ldots, N \tag{51}$$

where p_i^d is the ith agent of the dth dimensions.

$$\begin{aligned} \text{best}(k) &= \min \text{fit}_i(k) \quad i \in \{1, 2, \ldots, N\} \\ \text{worst}(k) &= \max \text{fit}_i(k) \quad i \in \{1, 2, \ldots, N\} \end{aligned} \tag{52}$$

where $\text{fit}_i(k)$ denotes the fitness value of the ith individual.

The gravitational constant $G(k)$ for each iteration and inertial masses are calculated as follows:

$$\begin{cases} G(k) = G_0 e^{\left(-\alpha \frac{k}{K}\right)} \\ M_{ai} = M_{pi} = M_{ii} = M_i \quad i = 1, 2, 3, \ldots, N \\ m_i(k) = \dfrac{\text{fit}_i(k) - \text{worst}(k)}{\text{best}(k) - \text{worst}(k)} \\ M_i(k) = \dfrac{m_i(k)}{\sum_{j=1}^{N} m_j(k)} \end{cases} \tag{53}$$

where G_0 and α are the initial value of gravitational constant and the specified constant; k and K are the index and the total number of iterations; M_{ai} is the active mass at kth iteration for the ith agent; M_{pi} is the passive mass at kth iteration for the ith agent; M_{ii} is the the inertial mass at kth iteration for the ith agent; and M_i is the mass at kth iteration for the ith agent.

The force $F_{ij}^d(k)$ acting on the ith mass from the jth mass, the total force $F_i^d(k)$, and the acceleration value $\sigma_i^d(k)$ at iteration k are calculated as follows:

$$\begin{cases} F_{ij}^d(k) = G(k)\dfrac{M_{pi}(k) \times M_{aj}(k)}{R_{ij}(k)+\varepsilon}\left(p_j^d(k) - p_i^d(k)\right) \\ F_i^d(k) = \sum\limits_{j\in(k)\text{best},j\neq i} \text{rand}_j F_{ij}^d(k) \\ \sigma_i^d(k) = \dfrac{F_i^d(k)}{M_{ii}(k)} \end{cases} \tag{54}$$

where R_{ij} is the euclidean distance between the ith mass and the jth mass.

The velocity $V_i^d(k)$ and position $p_i^d(k)$ of the ith agent at the dth dimension of the next iteration are updated as follows:

$$\begin{cases} V_i^d(k+1) = \text{rand} \times V_i^d(k) + \sigma_i^d(k) \\ p_i^d(k+1) = p_i^d(k) + V_i^d(k+1) \end{cases} \tag{55}$$

4.1.3. Modified Hybrid Particle Swarm Optimization and Gravitational Search Algorithm with Chaotic Maps

The hybrid PSOGSA combines the advantages of PSO and GSA algorithms. The algorithm can be explained a follows:

$$\begin{cases} V_i(k+1) = w_p \times V_i(k) + c_1' \times \text{rand} \times \sigma_i(k) \\ \quad + c_2' \times \text{rand} \times (\text{gbest} - p_i(k)) \\ p_i(k+1) = p_i(k) + V_i(k+1) \end{cases} \tag{56}$$

When the chaotic map is used in the optimization algorithm, the optimization algorithm can quickly converge to the optimal solution and get rid of the optimal local solution [54]. Therefore, the one-dimensional and irreversible chaotic mapping method with this advantage is applied to the PSOGSA optimization algorithm so that the performance of PSOGSA can be further improved. The chaotic map equation gives these characteristics of the chaotic map. In [55], ten chaotic mapping methods are given to improve the performance of the hybrid PSOGSA algorithm, and the change curve of the chaotic map with the number of iterations is shown in Figure 6. From the figure, it can be found that there is no random value in any chaotic map.

In the proposed CPSOGSA algorithm, the chaotic map method is used to improve the local search ability of GSA. Equation (54) calculates the total force, while a random number between 0 and 1 is used as a weight. In other words, this parameter will impact the local search ability of the algorithm. If this situation is considered, using chaotic map methods instead of the random number could increase the convergence ability of the PSOGSA algorithm in an optimization process. Although each chaotic map follows the order in the Table, Equation (57) is applied instead of using a random parameter for calculating the total force in Equation (54). The application of chaotic map methods to the total force value is shown in Equation (57). After the total force value is calculated, the acceleration value is determined using Equation (53), while Equation (56) updates the velocities and positions of all agents in the population. In summary, the flowchart of the proposed CPSOGSA is shown in Figure 7.

$$F_i^d(k) = \sum_{j\in k\text{best},j\neq i} C(k)F_{ij}^d(k) \tag{57}$$

where $C(k)$ is the chaotic sequence constant for the kth iteration.

In this paper, the Tent chaos mapping is used to calculate the value of $C(k)$ in the algorithm, and its computational mathematical expression is shown as follows:

$$C(k+1) = \begin{cases} 2C(k) & C(k) < 0.5 \\ 2(1 - C(k)) & \text{else} \end{cases} \tag{58}$$

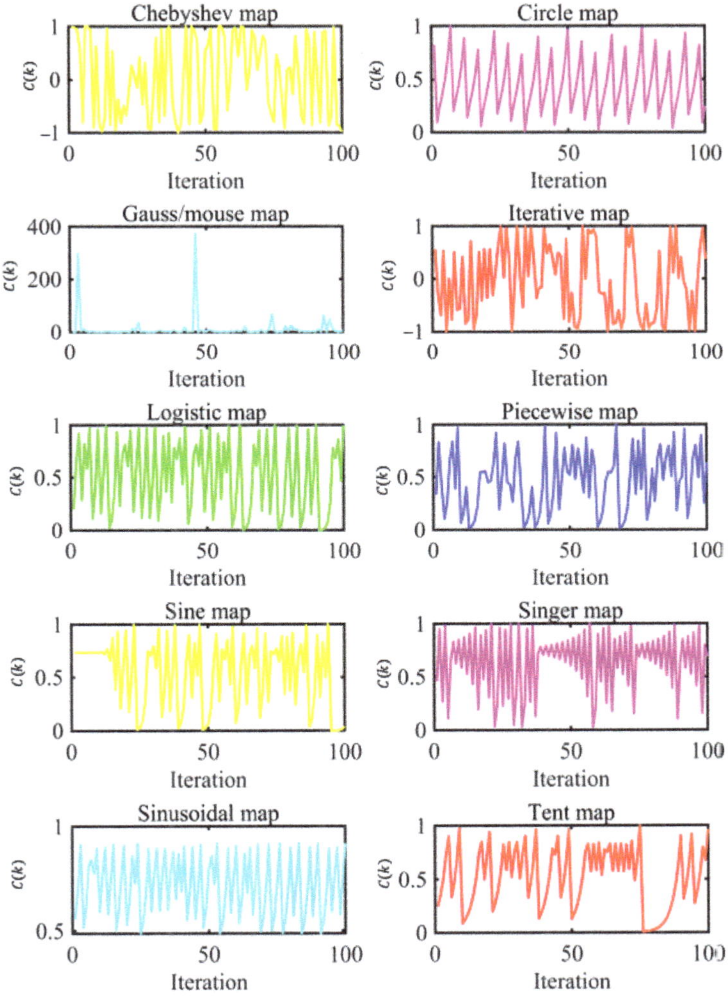

Figure 6. Curves of chaotic maps.

4.2. Applying CPSOGSA to Multi-Objective Robust H_2/H_∞ Problem

In subsection 3.2, the dynamic performance evaluation function and its weighting matrix for solving the robust controller of DG speed are proposed. The value of the weighting coefficient in these weighting matrices dramatically influences the performance of the speed controller. Therefore, these parameters need to be sufficient. and there are ten weighting coefficients for solving the speed robust controller, including the values of $q_{ij} > 0(i = 1,2; j = 1,2,3)$, $r_i(i = 1,2)$. This paper uses the CPSOGSA algorithm to optimize and adjust these parameters. In optimizing the speed controller, the objective optimization function of CPSOGSA is OF_1. The function described by Equation (59) is used to calculate the fitness value passed in the optimization process, aiming to minimize the absolute value of the change in the absolute value of the DG speed during this period.

$$OF_1 = \int |\Delta\omega|^2 dt \tag{59}$$

where $\Delta\omega$ is the speed deviation of DG.

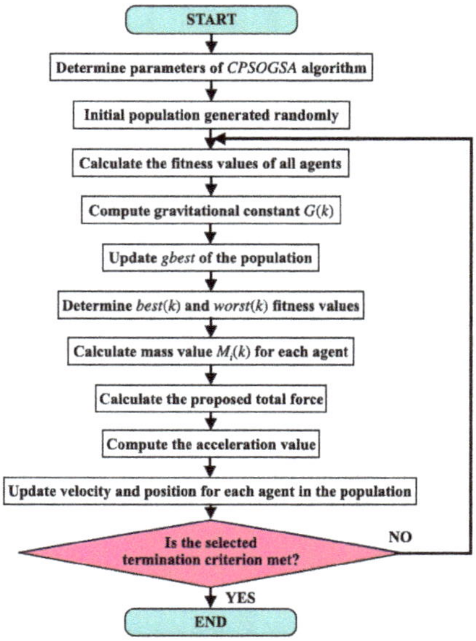

Figure 7. The flowchart of proposed CPSOGSA.

Similarly, CPSOGSA is used to adjust the dynamic performance evaluation function of the DG excitation controller proposed in subsection 3.3 and the weighting coefficients in the weighting matrix. For the solution of the excitation controller, the weighting coefficients include $q_{ij}(i = 1,2; j = 4,5,6,7), r_i(i = 3,4,\ldots,10)$. In the optimization process, the objective optimization function of CPSOGSA is OF_2. The function described by Equation (60) is used to calculate the fitness value of the excitation controller during the optimization process, aiming to minimize the absolute value of the DG voltage in time.

$$OF_2 = \int |\Delta U|^2 dt \tag{60}$$

where ΔU is the voltage deviation of DG.

In conclusion, these parameters need to be adequately tuned. The number of parameters to be adjusted is 16. These parameters are optimally adjusted using the CPSOGSA algorithm, whose overall optimal objective function is considered to be Equation (61) as the weighted sum of OF_1 and OF_2 given by Equations (59) and (59). Therefore, the optimization objective can be described as

$$OF = min\{OF_1 + OF_2\} \tag{61}$$

5. Numerical Study

The MG system composed of DG is shown in Figure 8. Among them, the rated power of the DE is 1250 kW, the rated speed is 1500 r/min, the rated torque is 11.9 kNm, and the moment of inertia is 71.82, while the rated voltage of the generator is 380 V, the rated current is 2310 A, the power factor is 0.8, and the rated frequency is 50 Hz. In addition, there are static and dynamic loads in this MG. The static load is mainly composed of electric heaters and lighting equipment, which accounts for a small proportion of the total loads; the dynamic loads refers to PMSM and driving devices, such as washing machines, water pumps, and fans, which account for a large proportion of the total load. Some parameters are also used for simulation, as shown in Table 1.

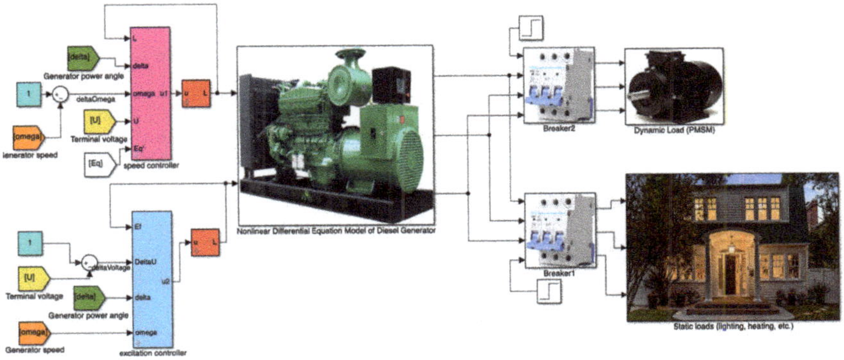

Figure 8. Diesel generator and its microgrid Simulink model.

Table 1. Relevant parameters for simulation.

T_a	T_b	c_1	c_2	X_d	X'_d
1.136	−0.492	1.515	−0.215	2.053	0.213
X_q	T_1	R	T_{d0}	K_1	d_1
1.003	0.050	0.011	2.200	0.200	0.1

5.1. Simulation Parameters

The selection of the relevant coefficients in the weighting function is a difficult point in the multi-objective H_2/H_∞ control and must be selected through repeated trials. After each trial selection, use the LMI toolbox to find the state feedback coefficient, substitute it into the simulation model, and then conduct the characteristic test to obtain the best comprehensive performance index. Simulations are carried out in MATLAB/Simulink and its programming environment interface. The parameters generated by the CPSOGSA algorithm are brought into the weighting function, and the state feedback control matrix $\boldsymbol{F}_s$ of the speed regulation part of the DG and the state feedback control matrix $\boldsymbol{F}_e$ of the excitation control part are calculated by the LMI and sent to Simulink. Simulink then iteratively evaluates the performance of the DG based on the adjusted parameters. In general, the flowchart of Figure 9 can be thought of as an optimization process. The number of iterative optimizations in this paper is 100.

As shown in Figure 10, under the optimization objective described in Equation (61), the fitness value (OF value) curve changes in different algorithm iterations 100 times . The optimization range of each parameter is $[1, 1e-3]$. It is not difficult to find from the figure that the CPSOGSA algorithm only needs about ten iterations to complete the optimization. On the contrary, the traditional algorithms are insufficient in terms of convergence and optimization efficiency. For example, the particle swarm algorithm PSO needs nearly 80 generations to converge, while the gravitational search algorithm falls into the local optimum at the beginning, and after introducing the chaotic sequence, GSA can avoid falling into the local optimum.

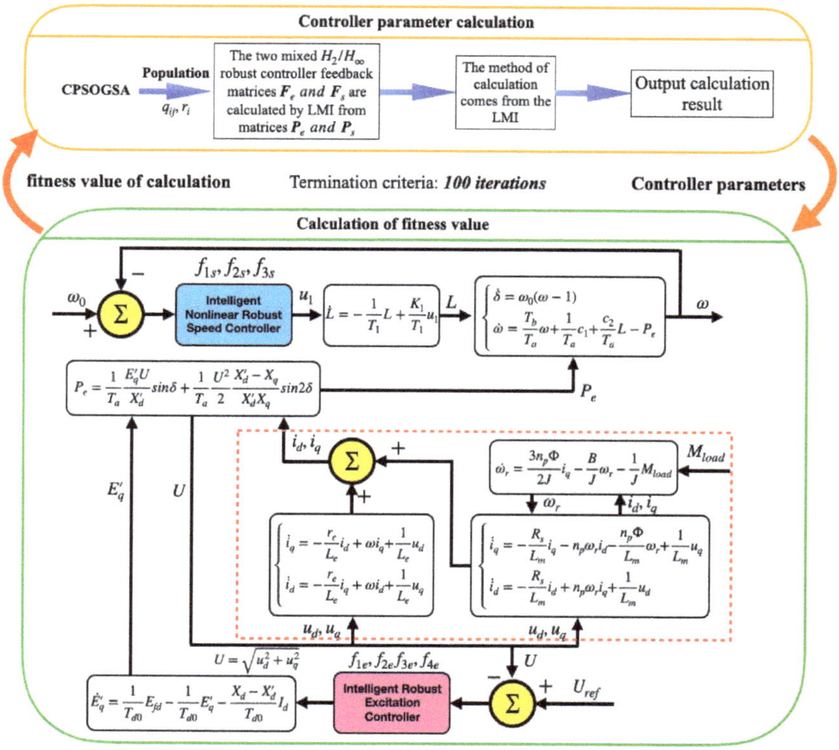

Figure 9. The flowchart of optimization process.

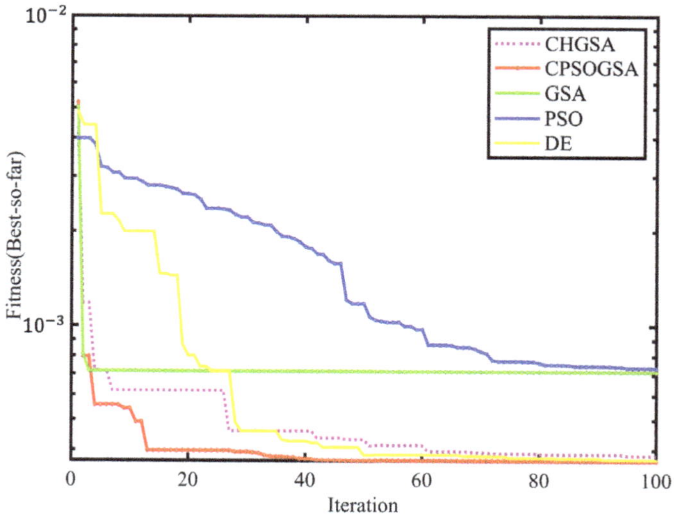

Figure 10. Iterative graph of the optimization process.

For the design of the speed control part of the nonlinear integrated controller based on the mixed H_2/H_∞ control theory, after the optimization of the CPSOGSA optimization algorithm, the weighting matrixs in Equation (35) are

$$C_{1s} = C_{2s}\begin{bmatrix} 0.4788 & 0 & 0 \\ 0 & 0.0015 & 0 \\ 0 & 0 & 0.1105 \end{bmatrix}$$

$$D_{12s} = D_{22s}\begin{bmatrix} 0 \\ 0 \\ 0.0011 \end{bmatrix}$$

Combining these matrices and using the linear matrix inequality toolbox in MATLAB, find the feedback coefficients in Equation (37) as $f_{1s} = -1.9188$, $f_{2s} = -1.8738$, and $f_{3s} = 35.604$.

Similarly, for the weighting matrix in Equation (46), these matrices are

$$C_{1e} = C_{2e} = \begin{bmatrix} 0.7817 & 0 & 0 & 0 \\ 0 & 3.824 \times 10^3 & 0 & 0 \\ 0 & 0 & 0.0256 & 0 \\ 0 & 0 & 0 & 0.1270 \end{bmatrix}$$

$$D_{12e} = D_{22e} = \begin{bmatrix} 0.1038 \\ 0.1143 \\ 0.1043 \\ 0.1417 \end{bmatrix}$$

Using the LMI toolbox in Matlab, obtain the coefficients of state-feedback controller in Equation (48): $f_{1e} = -2.227$, $f_{2e} = -7.516 \times 10^3$, $f_{3e} - 8.328 \times 10^{-4}$, and $f_{4e} = -9.495 \times 10^{-5}$.

5.2. Simulation Results

In the MATLAB/SIMULINK environment, we conduct simulation experiments on a DG system equipped with a mixed H_2/H_∞ robust integrated controller to verify the effectiveness of the controller described by Equation (49). The simulation parameters relevant to our study are presented in Table 1. During the comparative simulations, we employe specific parameter values for the PID speed controller: $K_p = 16.1313$, $K_i = 13.4692$, and $K_d = 0$. Similarly, the PID excitation controller is configured with the following parameters: $K_p = 50$, $K_i = 10$, and $K_d = 0$. The PID parameters for the speed control section are determined through optimization using the CPSOGSA algorithm, taking into account the desired system performance and stability. On the other hand, the parameters for the excitation control section are chosen based on prior research findings and practical expertise. By employing this combined approach, we aim to achieve an optimal balance between the speed control and excitation control of the DG system. The CPSOGSA optimization process effectively adjusts the PID parameters for the speed control section, ensuring an improved system response and improved system performance. Meanwhile, the predefined values for the excitation control parameters provide a solid foundation based on previous knowledge in the field.

Firstly, the comparison simulation experiment of DG no-load starting characteristics is carried out. When no load is connected, the DG start-up process is controlled by the controller and PID controller proposed in this paper, respectively. The simulation results are shown in Figures 11 and 12. It can be seen from the speed curve diagram in Figure 11 that from 0 to the establishment of the speed of 1 p.u, the proposed controller is about 3 s faster than the PID controller. From the voltage change curve in Figure 12, it can be seen that from 0 to 1 p.u of the machine terminal voltage, our controller is 10 s faster than the PID controller and has better start-up performance. It should be noted that the dead zone of the actuator is not limited in the simulation in this paper. Therefore, the speed of the

DG described in the simulation results reaches the rated value extremely fast under the proposed controller action.

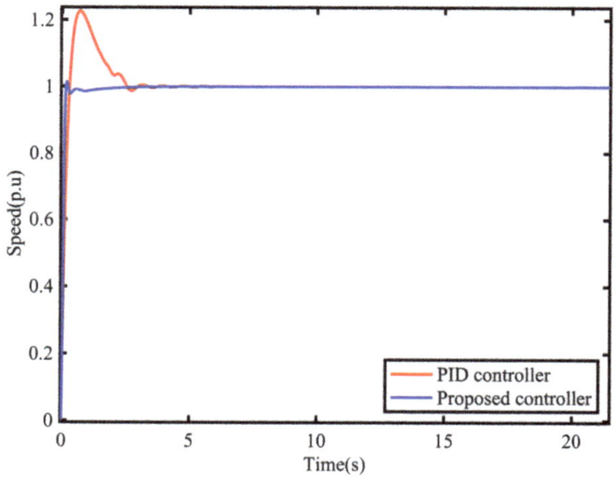

Figure 11. Variation curve of speed at no-load start-up.

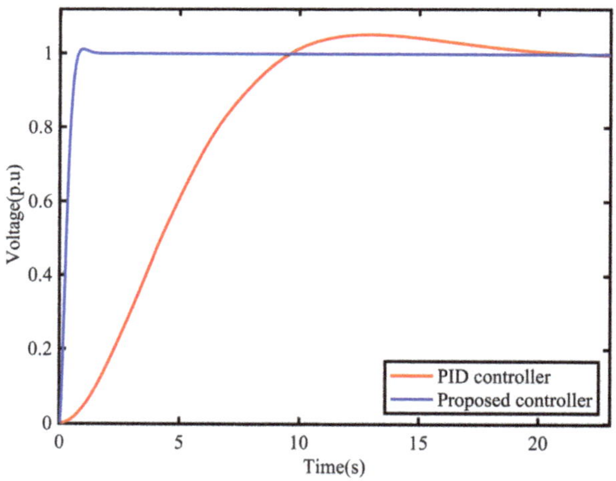

Figure 12. Variation curve of voltage at no-load start-up.

Then, the computer simulation of the designed system with sudden static loads is carried out. Figures 13 and 14 show the dynamic characteristic curves of the speed and voltage of the system after sudden static loads, and the two variables are per unit value. Among them, the blue curve represents the action result of the intelligent nonlinear H_2/H_∞ integrated controller, and the red curve represents the action result of the conventional PID controller.

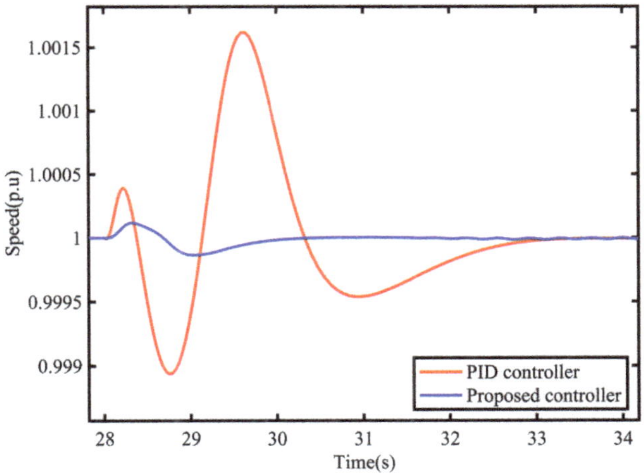

Figure 13. Speed response of system on suddenly increasing static load.

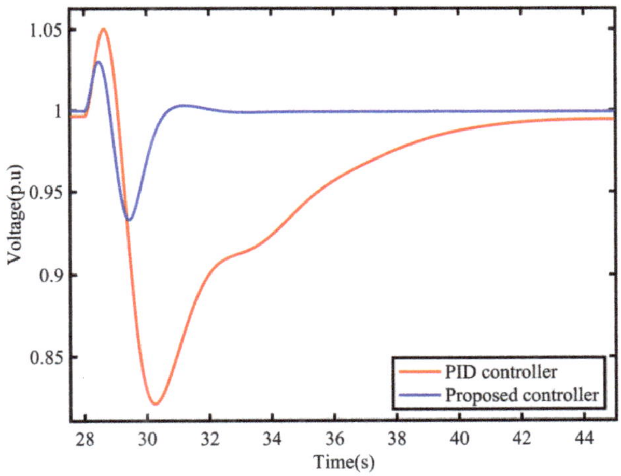

Figure 14. Voltage dynamic response of system on suddenly static load.

From the simulation results, the dynamic performance of the proposed controller is excellent. The dynamic regulation rate is minimal, the stabilization time is about 2 s, the dynamic voltage regulation rate is only about 0.05, and the stabilization time is 4 s. Then, when the conventional PID controller is applied, the dynamic regulation rate of the system is significant, and the stabilization time takes 5 s. The dynamic regulation rate of the system exceeds 0.15, and the stabilization takes a long time. It can be seen through computer simulation that after using the proposed controller, the rate and voltage rate are both decreased, and the stabilization time is shortened.

Finally, a simulation is performed that introduces a simulation of dynamic loads such as PMSM in the system. The dynamic loads greatly influences the grid voltage, especially when the grid is unloaded. The simulation runs for 55 s to connect to a PMSM with a size of 90 kW. The simulation results are shown in Figures 15 and 16. From the situation of the speed change in Figure 15, when the controller proposed in this paper is used, the system shows very little oscillation, while when the PID controller is used, the number of speed oscillations of the system exceeds five times. Figure 16 shows the response process of the terminal voltage under this working condition, and the controller proposed in this paper also shows better performance.

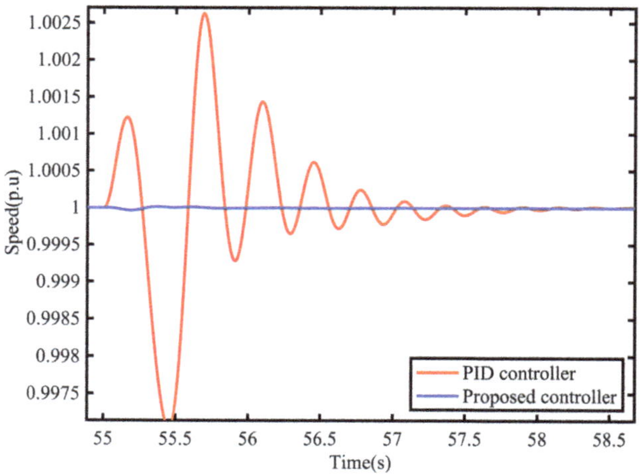

Figure 15. Speed response of system on suddenly increasing dynamic load.

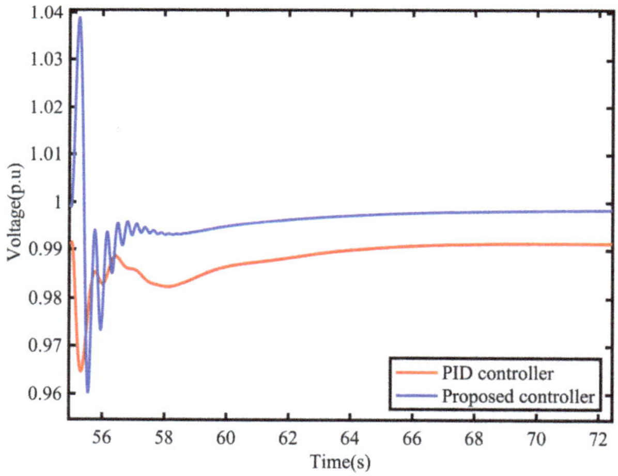

Figure 16. Voltage dynamic response of system on suddenly dynamic load.

In this paper, the intelligent nonlinear integrated controller effectively improves the dynamic accuracy of speed control and excitation control. The improvement of dynamic accuracy weakens the mutual influence and interaction between speed and voltage and solves the comprehensive control problem of speed and voltage. The method improves the stability of the MG system.

6. Conclusions

This paper firstly establishes a nonlinear dynamic model for a diesel generator (DG) and designs speed and excitation controllers for the DG. In the design process, we adopt the direct feedback linearization method and the multi-objective robust control theory. However, there are certain difficulties in determining the weighting parameters. To address this issue, we introduce the artificial intelligence CPSOGSA optimization algorithm to optimize the relevant weighting parameters that affect the controller performance. To validate the effectiveness of the proposed controller in improving a microgrid (MG) system powered by a DG, we create an MG model driven by a DG in Simulink and conduct corresponding simulation experiments. The simulation results demonstrate that compared to the traditional PID controller, the proposed controller effectively enhances the system's dynamic accuracy and disturbance suppression capability, thereby reducing fluctuations during load transients and operational condition changes. As a result, the stability of the MG system supplied by the DG is significantly improved. In summary, the nonlinear dynamic model and optimized controller proposed in this article exhibit excellent performance in an MG system driven by a DG, exerting positive effects on system stability and response characteristics. These research findings provide valuable references and guidance for the further optimization and enhancement of the application of DGs in MG systems.

Author Contributions: Conceptualization, Y.Z.; methodology, Y.Z. and B.X.; software, B.X.; validation, Y.Z. and B.X.; formal analysis, Y.Z.; investigation, Y.Z.; resources, Z.X.; data curation, Z.X. and J.Q.; writing—original draft preparation, Y.Z. and B.X.; writing—review and editing, Y.Z. and B.X.; visualization, J.Q.; supervision, Z.X.; project administration, Z.X. and J.Q.; and funding acquisition, Z.X. and J.Q. All authors have read and agreed to the published version of the manuscript.

Funding: This research was funded by the National Natural Science Foundation of China (No. 51869007 and 51979204).

Institutional Review Board Statement: Not applicable.

Informed Consent Statement: Not applicable.

Data Availability Statement: Data sharing is not applicable to this article as no datasets were generated or analysed during the current study.

Conflicts of Interest: The authors declare no conflict of interest.

References

1. Muresan, C.I.; Birs, I.; Ionescu, C.; Dulf, E.H.; De Keyser, R. A review of recent developments in autotuning methods for fractional-order controllers. *Fractal Fract.* **2022**, *6*, 37. [CrossRef]
2. Belboul, Z.; Toual, B.; Kouzou, A.; Mokrani, L.; Bensalem, A.; Kennel, R.; Abdelrahem, M. Multiobjective optimization of a hybrid PV/Wind/Battery/Diesel generator system integrated in microgrid: A case study in Djelfa, Algeria. *Energies* **2022**, *15*, 3579. [CrossRef]
3. Amiryar, M.E.; Pullen, K.R. Assessment of the carbon and cost savings of a combined diesel generator, solar photovoltaic, and flywheel energy storage islanded grid system. *Energies* **2019**, *12*, 3356. [CrossRef]
4. Hemeida, A.M.; Omer, A.S.; Bahaa-Eldin, A.M.; Alkhalaf, S.; Ahmed, M.; Senjyu, T.; El-Saady, G. Multi-objective multi-verse optimization of renewable energy sources-based micro-grid system: Real case. *Ain Shams Eng. J.* **2022**, *13*, 101543. [CrossRef]
5. Hasankhani, A.; Hakimi, S.M. Stochastic energy management of smart microgrid with intermittent renewable energy resources in electricity market. *Energy* **2021**, *219*, 119668. [CrossRef]
6. Hakimi, S.M.; Hasankhani, A.; Shafie-khah, M.; Catalão, J.P. Stochastic planning of a multi-microgrid considering integration of renewable energy resources and real-time electricity market. *Appl. Energy* **2021**, *298*, 117215. [CrossRef]
7. Hu, J.; Shan, Y.; Guerrero, J.M.; Ioinovici, A.; Chan, K.W.; Rodriguez, J. Model predictive control of microgrids–An overview. *Renew. Sustain. Energy Rev.* **2021**, *136*, 110422. [CrossRef]
8. Marqusee, J.; Becker, W.; Ericson, S. Resilience and economics of microgrids with PV, battery storage, and networked diesel generators. *Adv. Appl. Energy* **2021**, *3*, 100049. [CrossRef]
9. Bouchekara, H.R.; Javaid, M.S.; Shaaban, Y.A.; Shahriar, M.S.; Ramli, M.A.; Latreche, Y. Decomposition based multiobjective evolutionary algorithm for PV/Wind/Diesel Hybrid Microgrid System design considering load uncertainty. *Energy Rep.* **2021**, *7*, 52–69. [CrossRef]
10. Guo, Y.; Lei, X.; Wang, Q. Capacity coordination planning of isolated microgrid and battery swapping station based on the quantum behavior particle swarm optimization algorithm. *Int. Trans. Electr. Energy Syst.* **2021**, *31*, e12804. [CrossRef]
11. Rameshar, V.; Sharma, G.; Bokoro, P.N.; Çelik, E. Frequency Support Studies of a Diesel–Wind Generation System Using Snake Optimizer-Oriented PID with UC and RFB. *Energies* **2023**, *16*, 3417. [CrossRef]
12. Ahmed, M.; Meegahapola, L.; Vahidnia, A.; Datta, M. Stability and control aspects of microgrid architectures–A comprehensive review. *IEEE Access* **2020**, *8*, 144730–144766. [CrossRef]
13. McGowan, D.J.; Morrow, D.J.; Fox, B. Multiple input governor control for a diesel generating set. *IEEE Trans. Energy Convers.* **2008**, *23*, 851–859. [CrossRef]

14. Zou, Y.; Qian, J.; Zeng, Y.; Ismai, S.; Dao, F.; Feng, Z.; Nie, C.; Mei, H. Optimized Robust Controller Design Based on CPSOGSA Optimization Algorithm and H-two/H-infty Weights Distribution Method for Load Frequency Control of Micro-Grid. *IEEE Access* **2021**, *9*, 162093–162107. [CrossRef]
15. Qian, J.; Guo, Y.; Zou, Y.; Yu, S. Hamiltonian Modeling and Structure Modified Control of Diesel Engine. *Energies* **2021**, *14*, 2011. [CrossRef]
16. Latif, A.; Hussain, S.S.; Das, D.C.; Ustun, T.S. Double stage controller optimization for load frequency stabilization in hybrid wind-ocean wave energy based maritime microgrid system. *Appl. Energy* **2021**, *282*, 116171. [CrossRef]
17. Ali, H.; Magdy, G.; Xu, D. A new optimal robust controller for frequency stability of interconnected hybrid microgrids considering non-inertia sources and uncertainties. *Int. J. Electr. Power Energy Syst.* **2021**, *128*, 106651. [CrossRef]
18. Valenzuela, M.A.; Bentley, J.M.; Lorenz, R.D. Evaluation of torsional oscillations in paper machine sections. *IEEE Trans. Ind. Appl.* **2005**, *41*, 493–501. [CrossRef]
19. Al-Suod MM, S.; Oleksandr, U. Optimization of the Speed Controller in Gas Diesel Device Including in the Autonomous Electric Power System. *WSEAS Trans. Circuits Syst.* **2019**, *18*, 135–140.
20. Iwanski, G.; Bigorajski, Ł.; Koczara, W. Speed control with incremental algorithm of minimum fuel consumption tracking for variable speed diesel generator. *Energy Convers. Manag.* **2018**, *161*, 182–192. [CrossRef]
21. Afzal Thoker, Z.; Ahmad Lone, S. Dynamic performance improvement of wind-diesel power system through robust sliding mode control of hybrid energy storage system. *Wind. Eng.* **2022**, *46*, 1065–1079. [CrossRef]
22. Wang, R.; Li, X.; Ahmed, Q.; Liu, Y.; Ma, X. Speed control of a marine engine using predictive functional control based PID controller. In Proceedings of the 2018 Annual American Control Conference (ACC), Milwaukee, WI, USA, 27–29 June 2018; IEEE: Piscataway, NJ, USA, 2018; pp. 3908–3914.
23. Tran, T.A. The optimization of marine diesel engine rotational speed control process by fuzzy logic control based on particle swarm optimization algorithm. *Future Internet* **2018**, *10*, 99. [CrossRef]
24. Tran, T.A.; Haidara, G. A research on marine diesel engine speed controller by fuzzy logic control theory based on experimental investigation. *World* **2019**, *17*, 19. [CrossRef]
25. Ding, Y.; Shi, W.; Zhang, Y.; Guo, H. Design of Neural Network Speed Controller for Marine Diesel Generator Set. *Int. Core J. Eng.* **2021**, *7*, 586–592.
26. Asgari, S.; Suratgar, A.A.; Kazemi, M.G. Feedforward fractional order PID load frequency control of microgrid using harmony search algorithm. *Iran. J. Sci. Technol. Trans. Electr. Eng.* **2021**, *45*, 1369–1381. [CrossRef]
27. He, Y.; Fan, A.; Wang, Z.; Liu, Y.; Mao, W. Two-phase energy efficiency optimisation for ships using parallel hybrid electric propulsion system. *Ocean Eng.* **2021**, *238*, 109733. [CrossRef]
28. Li, J. *Design and Application of Modern Synchronous Generator Excitation Systems*; John Wiley & Sons: Hoboken, NJ, USA, 2019.
29. Chakraborty, C.; Basak, S.; Rao, Y.T. Synchronous generator with embedded brushless synchronous exciter. *IEEE Trans. Energy Convers.* **2019**, *34*, 1242–1254. [CrossRef]
30. Krishnamurthy, S.; Jahns, T.M.; Lasseter, R.H. The operation of diesel gensets in a CERTS microgrid. In Proceedings of the 2008 IEEE Power and Energy Society General Meeting-Conversion and Delivery of Electrical Energy in the 21st Century, Pittsburgh, PA, USA, 20–24 July 2008; IEEE: Piscataway, NJ, USA, 2008; pp. 1–8.
31. Gayatri, M.T.; Parimi, A.M.; Kumar, A.P. A review of reactive power compensation techniques in microgrids. *Renew. Sustain. Energy Rev.* **2018**, *81*, 1030–1036. [CrossRef]
32. Patel, R.; Hafiz, F.; Swain, A.; Ukil, A. Nonlinear excitation control of diesel generator: A command filter backstepping approach. *IEEE Trans. Ind. Inform.* **2020**, *17*, 4809–4817. [CrossRef]
33. Zhao, P.; Yao, W.; Wen, J.; Jiang, L.; Wang, S.; Cheng, S. Improved synergetic excitation control for transient stability enhancement and voltage regulation of power systems. *Int. J. Electr. Power Energy Syst.* **2015**, *68*, 44–51. [CrossRef]
34. Ramakrishnan, K. Delay-dependent stability of networked generator-excitation control systems: An LMI based approach. *IFAC-PapersOnLine* **2016**, *49*, 431–436. [CrossRef]
35. Berkoune, K.; Sedrine, E.B.; Vido, L.; Le Ballois, S. Robust control of hybrid excitation synchronous generator for wind applications. *Math. Comput. Simul.* **2017**, *131*, 55–75. [CrossRef]
36. Pramanik, M.A.; Roy, T.K.; Ghosh, S.K.; Anower, M.S.; Mahmud, M.A. Robust partial feedback linearizing excitation controller design for higher-order synchronous generator in smib systems to improve the transient stability. In Proceedings of the 2021 IEEE Texas Power and Energy Conference (TPEC), College Station, TX, USA, 2–5 February 2021; IEEE: Piscataway, NJ, USA, 2021; pp. 1–6.
37. Orchi, T.F.; Roy, T.K.; Mahmud, M.A.; Oo, A.M. Feedback linearizing model predictive excitation controller design for multimachine power systems. *IEEE Access* **2017**, *6*, 2310–2319. [CrossRef]
38. Roy, T.K.; Mahmud, M.A.; Shen, W.; Oo, A.M.; Haque, M.E. Robust nonlinear adaptive backstepping excitation controller design for rejecting external disturbances in multimachine power systems. *Int. J. Electr. Power Energy Syst.* **2017**, *84*, 76–86. [CrossRef]
39. Mobarra, M.; Rezkallah, M.; Ilinca, A. Variable speed diesel generators: Performance and characteristic comparison. *Energies* **2022**, *15*, 592. [CrossRef]
40. Mobarra, M.; Tremblay, B.; Rezkallah, M.; Ilinca, A. Advanced control of a compensator motor driving a variable speed diesel generator with rotating stator. *Energies* **2020**, *13*, 2224. [CrossRef]

41. Huang, M.L.; Song, K.M.; Wei, Z.D. Nonlinear H-two/H-infinity synthetic controller for diesel-generator set. *Control Theory Appl.* **2011**, *28*, 885–893.
42. Issa, M.; Ibrahim, H.; Lepage, R.; Ilinca, A. A review and comparison on recent optimization methodologies for diesel engines and diesel power generators. *J. Power Energy Eng.* **2019**, *7*, 31. [CrossRef]
43. Ramstedt, M. Cylinder-by-Cylinder Diesel Engine Modelling: A Torque-Based Approach. Master's Thesis, Linköping University, Department of Electrical Engineering, Linkoping, Sweden, 2004.
44. Zhang, Y.; Zhang, X.; Qian, T.; Hu, R. Modeling and simulation of a passive variable inertia flywheel for diesel generator. *Energy Rep.* **2020**, *6*, 58–68. [CrossRef]
45. Huang, M.; Wang, C. Nonlinear mathematical model of diesel generator sets in marine power stations. *J. Harbin Eng. Univ.* **2006**, *27*, 15–19.
46. Huang, M.L.; Wei, Z.D.; Song, K.M. Synchronous generator state feedback H-infty regulator for marine power plants. *J. Power Syst. Autom.* **2011**, *23*, 15–20.
47. Huang, M.; Wang, C. Simulation study of H-infty governor for diesel engines in marine power stations. *J. Electr. Eng. Technol.* **2006**, *10*, 125–129.
48. Zou, Y.; Hu, W.; Xiao, Z.; Wang, Y.; Chen, J.; Zheng, Y.; Qian, J.; Zeng, Y. Design of intelligent nonlinear robust controller for hydro-turbine governing system based on state-dynamic-measurement hybrid feedback linearization method. *Renew. Energy* 2023, *in press*.
49. Dao, F.; Zou, Y.; Zeng, Y.; Qian, J.; Li, X. An intelligent CPSOGSA-based mixed H-two/H-infty robust controller for the multi-hydro-turbine governing system with sharing common penstock. *Renew. Energy* 2023, *in press*.
50. Li, L.; Qian, J.; Zou, Y.; Tian, D.; Zeng, Y.; Cao, F.; Li, X. Optimized Takagi–Sugeno Fuzzy Mixed H-two/H-infty Robust Controller Design Based on CPSOGSA Optimization Algorithm for Hydraulic Turbine Governing System. *Energies* **2022**, *15*, 4771. [CrossRef]
51. Duman, S.; Li, J.; Wu, L.; Guvenc, U. Optimal power flow with stochastic wind power and FACTS devices: A modified hybrid PSOGSA with chaotic maps approach. *Neural Comput. Appl.* **2020**, *32*, 8463–8492. [CrossRef]
52. Rather, S.A.; Bala, P.S. Constriction coefficient based particle swarm optimization and gravitational search algorithm for multilevel image thresholding. *Expert Syst.* **2021**, *38*, e12717. [CrossRef]
53. Rather, S.A.; Bala, P.S. A hybrid constriction coefficient-based particle swarm optimization and gravitational search algorithm for training multi-layer perceptron. *Int. J. Intell. Comput. Cybern.* **2020**, *13*, 129–165. [CrossRef]
54. Xie, S.; Zeng, Y.; Qian, J.; Yang, F.; Li, Y. CPSOGSA Optimization Algorithm Driven Cascaded 3DOF-FOPID-FOPI Controller for Load Frequency Control of DFIG-Containing Interconnected Power System. *Energies* **2023**, *16*, 1364. [CrossRef]
55. Rather, S.A.; Bala, P.S. Hybridization of constriction coefficient-based particle swarm optimization and chaotic gravitational search algorithm for solving engineering design problems. In *Applied Soft Computing and Communication Networks: Proceedings of ACN 2019*; Springer: Singapore, 2020; pp. 95–115.

sustainability

MDPI

Article

Receding Galerkin Optimal Control with High-Order Sliding Mode Disturbance Observer for a Boiler-Turbine Unit

Gang Zhao, Yuge Sun, Zhi-Gang Su * and Yongsheng Hao

The National Engineering Research Center of Power Generation Control and Safety, School of Energy and Environment, Southeast University, Nanjing 210018, China; zhaogang@seu.edu.cn (G.Z.)
* Correspondence: zhigangsu@seu.edu.cn

Abstract: The control of the boiler-turbine unit is important for its sustainable and robust operation in power plants, which faces great challenges due to the control unit's serious nonlinearity, unmeasurable states, variable constraints, and unknown time-varying lumped disturbances. To address the above issues, this paper proposes a receding Galerkin optimal controller with a high-order sliding mode disturbance observer in a composite scheme, in which a high-order sliding mode disturbance observer is first employed to estimate the lumped disturbances based on a deviation form of the mathematical model of the boiler-turbine unit. Subsequently, under the hypothesis of state constraint, a receding Galerkin optimal controller is designed to compensate the lumped disturbances by embedding their estimates into the mathematically based predictive model at each sampling time instant. With the help of an interpolation polynomial, Gauss integration, and nonlinear solvers, an optimal control law is then obtained based on a Galerkin optimization algorithm. Consequently, disturbance rejection, target tracking, and constraint handling performance of a controlled closed-loop system are improved. Some simulation cases are conducted on a mathematical boiler-turbine unit model to demonstrate the effectiveness of the proposed method, which is supported by the quantitative result analysis, such as tracking and disturbance rejection performance indexes.

Keywords: Galerkin optimal control; boiler-turbine unit; high-order sliding mode disturbance observer

Citation: Zhao, G.; Sun, Y.; Su, Z.-G.; Hao, Y. Receding Galerkin Optimal Control with High-Order Sliding Mode Disturbance Observer for a Boiler-Turbine Unit. *Sustainability* **2023**, *15*, 10129. https://doi.org/10.3390/su151310129

Academic Editors: Bo Yang, Zhijian Liu and Lin Jiang

Received: 9 May 2023
Revised: 16 June 2023
Accepted: 22 June 2023
Published: 26 June 2023

1. Introduction

The boiler-turbine unit is the core of a thermal power plant as it produces the motive power to drive the generator. Thus, it is imperative to maintain safe and stable operation of the boiler-turbine unit. The research on boiler-turbine control methods is directly relevant to sustainability as it enables the optimization of power plant energy systems, leading to improved energy efficiency, sustainable operation, and reduced environmental impact. The control goal of the boiler-turbine unit is to meet the load demand of the electric power grid while also maintaining stable operation parameters [1]. Thereafter, control methods usually contribute to a more sustainable approach to power generation and daily automation operations. However, in practice, the boiler-turbine units are multi-input, multi-output (MIMO) nonlinear systems with multiple physical constraints and unmeasurable disturbances simultaneously. Moreover, the boiler-turbine units are now required to run over large-scale loads to adapt renewable energy power generation [2]. These all bring great challenges to the design of controllers for the boiler-turbine units with proper control performance.

Among the existing advanced controller methods applied for boiler-turbine units, the proportional-integral-differential (PID) [3] is extensively applied because of its robustness and ease of tuning. However, the control performance of PID is not satisfactory under a large-scale tracking scenario. To deal with the problem, some other advanced control approaches have been investigated in tracking conditions in recent years, including nonlinear control [4,5], gain scheduling control [6], adaptive control [7], robust control [8], and so on.

The emphasis on robust control shifts to enhancing the robustness index of the control system with plant uncertainties and external disturbances. In [9–11], fault-tolerant control methods can enhance the control system's tolerance by adaptively adjusting the control parameters through the information of fault diagnosis systems. Nevertheless, these control methods can't deal with the variable constraints directly. This obstacle leads to the investigation of several optimal controllers by minimizing a pre-defined performance index [12–17], in which model predictive control (MPC) is the prevailing one owing to its powerful abilities in addressing multiple variables, physical constraints, and tracking conditions.

Most of the aforementioned optimal controllers were designed based on either the black-box nonlinear model identified from the running data of the unit or linear models obtained by linearizing the unit's mathematical model. Thus, the nonlinearity contained in the information is discarded and not fully explored. Recently, more attention has been paid to designing optimal controllers based on the unit's mathematical model directly, and the pseudospectral (PS) method is an efficient one to solve the state- and control-constrained nonlinear optimal control problem (OCP) [18,19]. The main idea of the PS method is to transform the optimal control problem (OCP) into a nonlinear programming problem (NLP) [20], then use the NLP solver to solve out the optimal control sequence. In order to reduce the discretization error and obtain more accurate solutions, the Galerkin optimal control method [21,22] was proposed, in which the weak integral formulation is introduced to discretize the differential equations.

Considering the given merits, an adaptively receding Galerkin optimal controller (ARGOC) was proposed for a boiler-turbine unit with some unmeasured states and variable constraints in our previous work [23]. In the ARGOC, a state observer is first designed to estimate the key unmeasured state. Then, a receding Galerkin optimal controller is constructed by sufficiently taking information estimated and measured at each sampling time into account and borrowing the basic idea of receding optimization and feedback correction strategies from MPC. Thereafter, an independent model is embedded into the receding Galerkin controller structure to estimate and thus eliminate the constant disturbances in the output channels.

Nevertheless, the ARGOC fails when confronted with time-varying and state channel disturbances (i.e., lumped disturbances), which are more common in the boiler-turbine units than constant disturbances at the output channel. Under these circumstances, it is imperative to design a receding Galerkin optimal-based controller to deal with time-varying lumped disturbances. There are some difficulties in designing such a controller. The first is how to estimate the lumped disturbances in the presence of the unmeasurable state variable, fluid density, in the drum of the unit. In recent years, the disturbance observer (DO) [24–26] and the extended state observer (ESO) [27–29] have been proven to be efficient in estimating lumped disturbances in control systems. The main advantage of DO and ESO is that the disturbance observer can estimate the lumped disturbances without distinguishing whether they are external or internal. The time-varying disturbance reduction in the state channel is more focused on in this paper. Further, the disturbance observer convergence rate and accuracy, as well as the observer structure described above, still have improvement space.

The second difficulty lies in how to compensate the disturbance estimation into the receding Gakerkin optimal-based controller in an active manner. There are several disturbance compensation mechanisms in the optimal control field. In [30,31], a real-time updated equilibrium according to the disturbances and set-points is calculated first to transfer the aim of disturbance rejection and set-point tracking to the updated equilibrium. In [32], a disturbance model is added to the predictive model to compress the adverse effect of lumped disturbances. These disturbance rejection methods are not compensated by the disturbance observer in the manner of a model-predictive-based optimization process.

Motivated by the above statements, we aim to propose a receding Galerkin optimal control with a high-order sliding mode disturbance observer for a nonlinear boiler-turbine unit in a composite manner. The composite controller comprises a Galerkin optimal control-based feedback control part and a DO-based feedforward control part. The main contributions of the composite controller are summarized as follows:

(1) A high-order sliding mode disturbance observer is designed to estimate the lumped disturbances based on the derived deviation form of the mathematical model of the boiler-turbine unit, which aims at the MIMO system and additionally addresses the observer gain tuning of the unit.

(2) The estimates of the lumped disturbances are integrated into the predictive model and then feedforward compensated in the Galerkin optimal control-based feedback channel. As will be anticipated, the proposed composite controller exhibits superior time-varying lumped disturbance rejection and target tracking performance for a boiler-turbine unit.

The rest of the paper is organized as follows: The mathematical model of a disturbed boiler-turbine unit and problem statement are given in Section 2. Section 3 presents the methodology of the proposed composite controller. Simulation studies are carried out in Section 4 to verify the superiority of our proposed controller in time-varying lumped disturbance rejection and set-point tracking. The last section concludes this paper.

2. Disturbed Boiler-Turbine Unit and Problem Statement

In this paper, a 160 MW oil-fired drum-type boiler-turbine unit is considered 1, whose flow diagram is summarized in Figure 1. The mathematical model of this unit has been established in the form of

$$\begin{cases} \dot{x}_1 = f_1(x,u) := -a_1 u_2 x_1^{1.125} + a_2 u_1 - a_3 u_3 + \bar{d}_1(t), \\ \dot{x}_2 = f_2(x,u) := (b_1 u_2 - b_2) x_1^{1.125} - x_2 + \bar{d}_2(t), \\ \dot{x}_3 = f_3(x,u) := (c_1 u_3 - (c_2 u_2 - c_3) x_1)/c_4 + \bar{d}_3(t), \end{cases}$$
$$\begin{cases} y_1 = x_1, \\ y_2 = x_2, \\ y_3 = h(x,u) := 0.05(0.13073 x_3 + 100\alpha_{cs} + (q_e/9 - 67.975)), \end{cases} \tag{1}$$

where the state variables x_1, x_2 and x_3 signify drum pressure (kg/cm^2), electrical output (MW), and fluid density in the drum (kg/cm^3), respectively. The outputs are the drum steam pressure (y_1), electrical output (y_2), and drum water level (y_3). The controllable inputs u_1, u_2, and u_3 are the valve positions for fuel, steam, and feedwater flow, respectively. All the valve positions are normalized into [0, 1]. The coefficient α_{cs} and evaporation rate of steam q_e (kg/s) are defined as

$$\alpha_{cs} = \frac{(1 - 0.001538 x_3)(0.8 x_1 - 25.6)}{x_3(1.0394 - 0.0012304 x_1)} \tag{2}$$

$$q_e = (0.854 u_2 - 0.147) x_1 + 45.59 u_1 - 2.514 u_3 - 2.096 \tag{3}$$

$\bar{d}_i(t)$, $i = 1,2,3$ are unknown uncertainties or disturbances; referring to 33, parameters a_i, b_i, c_i do not have the physical meaning as the previous parameter, which is to be identified by the operating data, as shown in Table 1.

Table 1. Model Parameters of Boiler–Turbine Unit.

$a_1 = 0.0018$	$b_1 = 0.073$	$c_1 = 141$
$a_2 = 0.9$	$b_2 = 0.016$	$c_2 = 1.1$
$a_3 = 0.15$	$b_3 = 0.1$	$c_3 = 0.19$
		$c_4 = 85$

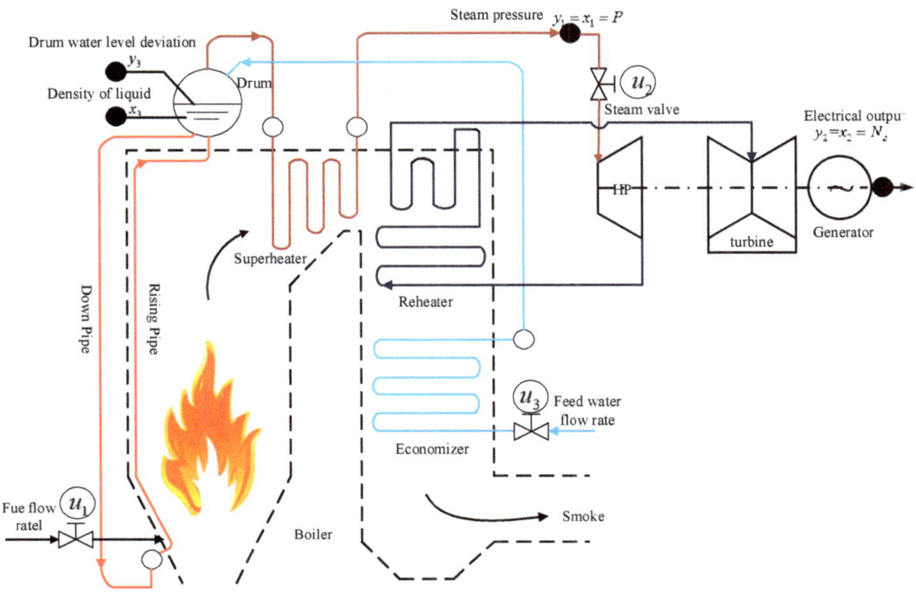

Figure 1. Structure of a 160 MW boiler-turbine unit in a thermal power plant.

Table 2 gives some typical operating points of the boiler-turbine unit in the absence of disturbances. The main control task is to regulate the outputs of the unit to track large-scale setpoints or references.

Table 2. Typical Operating Points of Boiler–Turbine Unit.

	#1	#2	#3	#4	#5	#6	#7
x_{1d}	75.6	86.4	97.2	108	118.8	129.6	135.4
x_{2d}	15.27	36.65	50.52	66.65	85.06	105.8	127
x_{3d}	299.6	324.4	385.2	428	470.8	513.6	556.4
u_{1d}	0.156	0.209	0.271	0.34	0.418	0.505	0.600
u_{2d}	0.483	0.552	0.621	0.69	0.759	0.828	0.897
u_{3d}	0.183	0.256	0.34	0.433	0.543	0.663	0.793
y_{3d}	−0.97	−0.65	−0.32	0	0.32	0.64	0.98

In order to implement the receding Galerkin optimal controller, all the states should be known in advance. However, the state variable x_3, that is the fluid density in the drum, cannot be measured directly. In this case, we adopt the available y_3 to design the controller instead. To simplify the controller design, the last three terms in y_3 can be viewed as a disturbance term, $w(t)$, that is

$$w(t) := 0.05(100\alpha_{cs} + q_e/9 - 67.975). \tag{4}$$

Therefore, the first derivative of y_3 is

$$\dot{y}_3 = c_0\dot{x}_3 + \dot{w}(t) = c_0[c_1 u_3 - (c_2 u_2 - c_3)x_1]/c_4 + c_0\bar{d}_3(t) + \dot{w}(t), \tag{5}$$

where constant $c_0 = 0.00654$.

It is observable from (1) that there is a rate of change constraint on the control inputs, which is expected to be introduced into the system dynamics for improving control performance. Define $v_1 = \dot{u}_1/c_1$, $v_2 = \dot{u}_2/c_2$, $v_3 = \dot{u}_3/c_3$ with expansion coefficients c_1, c_2,

c_3. Then, by given the setpoints/references y_{id} and their first derivatives $\dot{y}_{id}$ and defining outputs errors $z_i = y_i - y_{id}$, $i = 1, 2, 3$, dynamics are retrofitted as

$$\begin{aligned}
\dot{z}_1 &= -a_1 u_2 (z_1 + y_{1d})^{1.125} + a_2 u_1 - a_3 u_3 - \dot{y}_{1d} + d_1, \\
\dot{z}_2 &= (b_1 u_2 - b_2)(z_1 + y_{1d})^{1.125} - b_3 (z_2 + y_{2d}) - \dot{y}_{2d} + d_2, \\
\dot{z}_3 &= c_0 [c_1 u_3 - (c_2 u_2 - c_3)(z_1 + y_{1d})]/c_4 - \dot{y}_{3d} + d_3, \\
\dot{u}_1 &= c_1 v_1, \\
\dot{u}_2 &= c_2 v_2, \\
\dot{u}_3 &= c_3 v_3.
\end{aligned} \tag{6}$$

where the lumped disturbances are

$$d_i(t) := \begin{cases} \bar{d}_i(t) & i = 1, 2, \\ c_0 \bar{d}_3(t) + \dot{w}(t) & i = 3, \end{cases} \tag{7}$$

which satisfies the following assumption.

Assumption 1. *The lumped disturbances d_i satisfy $\left| d_i^{(j)} \right| \le \varsigma_i$, where $j = 1, 2, \ldots, L$ with a positive integer $L \le 3$ and positive constants ς_i, $i = 1, 2, 3$.*

Correspondingly, for safety consideration, model (1) should satisfy the following constraints [1, 5], which are restated as

$$\begin{aligned}
&70 \le x_1 \le 150,\ 10 \le x_2 \le 190,\ 0 \le u_1, u_2, u_3 \le 1, \\
&|v_1| \le 0.007/c_1, |v_2| \le 0.02/c_2, |v_3| \le 0.05/c_3, \\
&h(x, u) - 1 \le 0, h(x, u) + 1 \ge 0.
\end{aligned} \tag{8}$$

Dynamics (6)–(8) are utilized to design the composite controller, to which we turn next.

3. Main Results: Method

This section includes a concise and precise description of the proposed method design, their interpretation, as well as results that can be drawn.

3.1. High-Order Sliding Mode Disturbance Observer Design

A high-order sliding mode disturbance observer developed in [33–36] is normally used for the single-input-single-output (SISO) n-th order differential dynamics, while in this research it has tentatively been employed for the MIMO system (6) to estimate the unknown lumped disturbances as

$$\begin{aligned}
\dot{\xi}_0^i &= v_0^i + f_i(z, u, y_{id}, \dot{y}_{id}), \\
v_0^i &= -\lambda_0^i L_i^{\frac{1}{L+1}} |\xi_0^i - z_i|^{\frac{L}{L+1}} \mathrm{sign}(\xi_0^i - z_i) + \xi_1^i, \\
\dot{\xi}_1^i &= v_1^i, \\
v_1^i &= -\lambda_1^i L_i^{\frac{1}{L}} |\xi_1^i - v_0^i|^{\frac{L-1}{L}} \mathrm{sign}(\xi_1^i - v_0^i) + \xi_2^i, \\
&\vdots \\
\dot{\xi}_l^i &= v_l^i, \\
v_l^i &= -\lambda_l^i L_i^{\frac{1}{L+1-l}} \left| \xi_l^i - v_{l-1}^i \right|^{\frac{L-l}{L+1-l}} \mathrm{sign}\left(\xi_l^i - v_{l-1}^i \right) + \xi_l^i, \\
&\vdots \\
\dot{\xi}_L^i &= v_L^i, \\
v_L^i &= -\lambda_L^i L_i \mathrm{sign}(\xi_L^i - v_{L-1}^i),
\end{aligned} \tag{9}$$

where L_i, λ_l^i, $i = 1,2,3$, $l = 0,1,\ldots,L$ are observer coefficients and sign $(\cdot)$ is a sign function. Accordingly, $\varsigma_0^i, \varsigma_1^i, \varsigma_2^i, \ldots, \varsigma_L^i$ are, respectively, the estimates $\hat{z}_i, \hat{d}_i, \hat{\dot{d}}_i, \ldots \hat{d}_i^{(L-1)}$.

Define observer errors as $e_0^i = \varsigma_0^i - z_i, e_l^i = \varsigma_l^i - d_i^{(l-1)}$, l = 1, 2, 3, with $d_i^0 = d_i$ and $d_i^{(l-1)}$ the (l-1)th derivative of d_i. The observer error dynamics can be derived from (6) and (9) as

$$\begin{aligned}
\dot{e}_0^i &= -\lambda_0^i L_i^{1/4} \left|e_0^i\right|^{3/4} \mathrm{sign}\left(e_0^i\right) + e_1^i, \\
\dot{e}_1^i &= -\lambda_1^i L_i^{1/3} \left|e_1^i - \dot{e}_0^i\right|^{2/3} \mathrm{sign}\left(e_1^i - \dot{e}_0^i\right) + e_2^i, \\
\dot{e}_2^i &= -\lambda_2^i L_i^{1/2} \left|e_2^i - \dot{e}_1^i\right|^{1/2} \mathrm{sign}\left(e_2^i - \dot{e}_1^i\right) + e_3^i, \\
\dot{e}_3^i &\in -\lambda_3^i L_i \mathrm{sign}\left(e_3^i - \dot{e}_2^i\right) + [-\varsigma_i, \varsigma_i].
\end{aligned} \tag{10}$$

It can be concluded from 34–36 that the observer error is finite-time stable. That is, there exists a finite time t^*, such that $e_0^i \equiv 0, e_1^i \equiv 0, e_2^i \equiv 0, e_3^i \equiv 0$ when $t > t^*$.

Remark: The nature of HOSM is a differentiator that uses known information from system (6) to mathematically reconstruct an unknown disturbance. HOSM is the preliminary part for the whole composite controller design, which has significance for the reduction of disturbances caused by model uncertainties and coupling terms in (6). With this foreshadowing, the estimate information of all disturbances is embedded into the model (9) and restored as the nominal one for the rolling optimization process to calculate the control law.

3.2. Galerkin Optimal Control Design

The optimal control problem based on mathematical model (6) can be described as follows: determine the state-control function pair, $t \to (z,u) \in R^{N_z} \times R^{N_u}$ to minimize the following cost function

$$\begin{aligned}
&\min_{z,u,v} J_{BT} = \int_{t_0}^{t_f} \left[z^{\mathrm{T}} P z + v^{\mathrm{T}} Q v\right] dt \\
&s.t. \quad (9),\ (10),
\end{aligned} \tag{11}$$

where $z = [z_1,\ z_2,\ z_3]^{\mathrm{T}}, v = [v_1,\ v_2,\ v_3]^{\mathrm{T}}, P = diag\{p_1,\ p_2,\ p_3\}, Q = diag\{q_1,\ q_2,\ q_3\}$.

To solve the optimal control problem (11), a Galerkin optimal method is first designed to transform (11) into a nonlinear programming problem (NLP), which can then be computed by several sequential quadratic programming (SQP) software packages such as SNOPT [37] with high computational efficiency. To be specific, the Galerkin optimal method is performed in the following four steps: approximating state and control variables, discretizing the system dynamics and variable constraints, and integrating the cost function via interpolation polynomial [38,39] and Gauss integration [23] on a series of Legendre-Gauss-Lobatto (LGL) nodes. The LGL nodes are calculated as the roots of

$$\xi(\tau) = \left(1 - \tau^2\right) \dot{L}_N(\tau), \tag{12}$$

where $L_N(\tau)$ is the Nth order Legendre polynomial defined by

$$L_N(\tau) := \frac{(-1)^N}{2^N N!} \frac{d^N}{d\tau^N} \left(1 - \tau^2\right)^N. \tag{13}$$

Totally, there are $(N+1)$ LGL nodes in τ-space as $\{\tau_i\}_{i=0}^N$ $(\tau_0 = -1 < \tau_1 < \tau_2 < \cdots < \tau_N = 1)$. By converting the real-time domain $t \in [\, t_0, t_f \,]$ into a closed interval $\tau \in [-1,1]$ according to

$$\tau = \frac{2t - (t_f + t_0)}{t_f - t_0}, \tag{14}$$

one can get corresponding point $\{t_i\}_{i=0}^{N}$, $\left(t_0 = t^{N_0} < t^{N_1} < t^{N_2} < \cdots < t^{N_N} = t_f\right)$ in the real-time domain.

With the help of LGL nodes, the Galerkin method approximates the state and control variables by the Nth-order Lagrange interpolation polynomial defined on LGL nodes as follows:

$$z(\tau) \approx \sum_{j=0}^{N} \phi_j^N(\tau) \cdot z^{N_j}, \tag{15}$$

$$u(\tau) \approx \sum_{j=0}^{N} \phi_j^N(\tau) \cdot u^{N_j}, \tag{16}$$

where z^{N_j} and u^{N_j} are the state and control variables at the LGL nodes τ_j. $\phi_j^N(\tau)$ is the N-th order Lagrange interpolation basis function defined by $\phi_j^N(\tau) = \prod_{i=0.i\neq j}^{N} \frac{\tau-\tau_i}{\tau_j-\tau_i}$.

Using (15) and (16), the differential equation can be approximated using the following integral formulation

$$\int_{-1}^{1} \psi_i(\tau)\left(\dot{z}(\tau) - \frac{t_f - t_0}{2} f(z(\tau), u(\tau))\right) d\tau = 0 \tag{17}$$

with test functions $\psi_i(\tau)$. Equation (17) can be further rewritten as (18) when $\psi_i(\tau)$ is denoted as the basis function $\phi_j^N(\tau)$,

$$\sum_{j=0}^{N} \underbrace{\int_{-1}^{1} \phi_i^N(\tau)\dot{\phi}_j^N(\tau)d\tau}_{D_{ij}} \cdot z^{N_j} - \underbrace{\frac{t_f - t_0}{2}\int_{-1}^{1} \phi_i^N(\tau) f(z(\tau), u(\tau))d\tau}_{\Delta_i} = 0. \tag{18}$$

For simplicity, the D_{ij} and Δ_i can be approximated as

$$D_{ij} \approx \sum_{k=0}^{N} \phi_i^N(\tau_k)\dot{\phi}_j^N(\tau_k)\omega_k \approx \dot{\phi}_j^N(\tau_i)\omega_i = A_{ij}\omega_i \tag{19}$$

$$\Delta_i \approx \frac{t_f - t_0}{2} f(z(\tau_i), u(\tau_i))\omega_i \tag{20}$$

where A_{ij} is the Legendre differentiation matrix calculated by

$$A_{ij} = \begin{cases} \frac{L_N(\tau_i)}{L_N(\tau_j)} \frac{1}{\tau_i - \tau_j}, & i \neq j, \\ \frac{-N(N+1)}{4}, & i = j = 0, \\ \frac{N(N+1)}{4}, & i = j = N, \\ 0, & i = j \in [1, \cdots, N-1]. \end{cases} \tag{21}$$

ω_i, $i = 0, 1, \ldots, N$, are the quadrature weights, and the LGL version of quadrature weights is utilized in the present paper, which is calculated as

$$\omega_i = \frac{2}{N(N+1)[L_N(\tau_i)]}, i = 0, 1, \ldots, N. \tag{22}$$

With the D_{ij} in (19) and Δ_i in (20), the dynamics (18) can thus be finally simplified as

$$\begin{aligned} &\sum_{j=0}^{N} D_{ij} \cdot z_k^j - \Delta_{ki} = 0, \quad i = 0, 1, \cdots, N, \quad k = 1, \ 2, \ 3, \\ &\sum_{j=0}^{N} D_{ij} \cdot u_k^j - \Delta_{(k+3)i} = 0, \quad i = 0, 1, \cdots, N, \quad k = 1, \ 2, \ 3, \end{aligned} \tag{23}$$

where

$$\begin{aligned}
\Delta_{1i} &= \tfrac{t_f-t_0}{2}\left(-a_1u_2^i(z_1^i+y_{1d})^{1.125}+a_2u_1^i-a_3u_3^i-\dot{y}_{1d}+\xi_0^1\right)w_i,\\
\Delta_{2i} &= \tfrac{t_f-t_0}{2}\left((b_1u_2^i-b_2)(z_1^i+y_{1d})^{1.125}-b_3(z_2^i+y_{2d})-\dot{y}_{2d}+\xi_0^2\right)w_i,\\
\Delta_{3i} &= \tfrac{t_f-t_0}{2}\left(c_0\left[c_1u_3^i-(c_2u_2^i-c_3)(z_1^i+y_{1d})\right]/c_4-\dot{y}_{3d}+\xi_0^3\right)w_i,\\
\Delta_{4i} &= \tfrac{t_f-t_0}{2}(c_1v_1^i)w_i,\\
\Delta_{5i} &= \tfrac{t_f-t_0}{2}(c_2v_2^i)w_i,\\
\Delta_{6i} &= \tfrac{t_f-t_0}{2}(c_3v_3^i)w_i.
\end{aligned}$$

The variable constraints (10) can be discretized as

$$\begin{aligned}
&-10\le z_1^i\le 10,\quad -10\le z_2^i\le 10,\quad -1\le z_3^i\le 1,\quad i=0,1,\cdots,N\\
&0\le |u_1^i|\le 1,\quad 0\le |u_2^i|\le 1,\quad 0\le |u_3^i|\le 1,\quad i=0,1,\cdots,N\\
&|v_1^i|\le 0.007/c_1, |v_2^i|\le 0.02/c_2, |v_3^i|\le 0.05/c_3,\quad i=0,1,\cdots,N
\end{aligned} \tag{24}$$

In a relatively easy way, the cost function (11) is approximated according to the Gauss–Lobatto integration rule as follows:

$$\begin{aligned}
J &= \int_{t_0}^{t_f}\left[z^{\mathrm{T}}Pz+v^{\mathrm{T}}Qv\right]dt\\
&\approx \tfrac{t_f-t_0}{2}\sum_{j=0}^{N}\left[p_1\left(z_1^{N_j}\right)^2+p_2\left(z_2^{N_j}\right)^2+p_3\left(z_3^{N_j}\right)^2+q_1\left(v_1^{N_j}\right)^2+q_2\left(v_2^{N_j}\right)^2+q_3\left(v_3^{N_j}\right)^2\right]w_j.
\end{aligned} \tag{25}$$

Therefore, the continuous optimal control problem (13) can be constructed as

$$\begin{aligned}
J &= \tfrac{t_f-t_0}{2}\sum_{j=0}^{N}\left[p_1\left(z_1^{j}\right)^2+p_2\left(z_2^{j}\right)^2+p_3\left(z_3^{j}\right)^2+q_1\left(v_1^{j}\right)^2+q_2\left(v_2^{j}\right)^2+q_3\left(v_3^{j}\right)^2\right]w_j,\\
\text{s.t.}\ &\left\|\sum_{j=0}^{N}D_{ij}\cdot z_k^j-\Delta_{ki}\right\|_\infty\le\delta^N,\quad i=0,1,\cdots,N,\quad k=1,\ 2,\ 3,\\
&\left\|\sum_{j=0}^{N}D_{ij}\cdot u_k^j-\Delta_{(k+3)i}\right\|_\infty\le\delta^N,\quad i=0,1,\cdots,N,\quad k=1,\ 2,\ 3.\\
&-10\le z_1^i\le 10,\quad -10\le z_2^i\le 10,\quad -1\le z_3^i\le 1,\quad i=0,1,\cdots,N\\
&0\le |u_1^i|\le 1,\quad 0\le |u_2^i|\le 1,\quad 0\le |u_3^i|\le 1,\quad i=0,1,\cdots,N\\
&|v_1^i|\le 0.007/c_1, |v_2^i|\le 0.02/c_2, |v_3^i|\le 0.05/c_3,\quad i=0,1,\cdots,N,
\end{aligned} \tag{26}$$

where δ^N is a constant tolerance used to guarantee the feasibility of the NLP (26).

3.3. Receding Galerkin Optimal Control Design with High-Order Sliding Mode Disturbance Observer

It is evident that the Galerkin method interpreted in Section 3.2 is only feasible for the stabilization problem rather than the tracking problem. To deal with the tracking problem, the useful information at each sampling instant should be taken into account, including the information of states, outputs, and references. To address this problem, a receding version of Galerkin's optimal control strategy with a high-order sliding mode disturbance observer in a composite way is proposed by borrowing the basic idea from model predictive control (MPC) and is explained as follows:

(i). At current time instant t_k, the lumped disturbances of system (9) are estimated by the high-order sliding mode disturbance observer (11) as $\xi_0^1,\ \xi_0^2,\ \xi_0^3$.

(ii). Let the current state $z(t_k)$ and control $u(t_k)$ be the initial conditions, that is, $z_{0,k}=z(t_k)$, $u_{0,k}=u(t_k)$. Then embed the obtained $\xi_0^1,\ \xi_0^2,\ \xi_0^3$ into the nonlinear mathematical model based-predictive model, the optimal discrete state and control sequences $\left\{z^{j,k}\right\}_{j=0}^{N}$ and $\left\{u^{j,k}\right\}_{j=0}^{N}$ can be acquired by minimizing J_{BT} in (29) through the Galerkin optimal control method over the prediction horizon $[t_0, t_f]$: = $[t_k, t_k + \Delta T]$,

$$\begin{aligned} \min_{\{x^{j,k},u^{j,k},v^{j,k}\}} & J_{BT} = \tfrac{\Delta T}{2}\sum_{j=0}^{N}\left[(z^{j,k})^{\mathrm{T}}P(z^{j,k}) + (v^{j,k})^{\mathrm{T}}Qv^{j,k}\right]w_j, \\ s.t. & \left\|\sum_{j=0}^{N} D_{i,j}z^{j,k} - \Delta_i\right\|_{\infty} \le \delta^N, \\ & \|z^{0,k} - z(t_k)\|_{\infty} \le \delta^N, \\ & \|u^{0,k} - u(t_k)\|_{\infty} \le \delta^N, \end{aligned} \tag{27}$$

$$\begin{aligned} & -10 \le z_1^i \le 10, \quad -10 \le z_2^i \le 10, \quad -1 \le z_3^i \le 1, \quad i = 0,1,\cdots,N \\ & 0 \le |u_1^i| \le 1, \quad 0 \le |u_2^i| \le 1, \quad 0 \le |u_3^i| \le 1, \quad i = 0,1,\cdots,N \\ & |v_1^i| \le 0.007/c_1, |v_2^i| \le 0.02/c_2, |v_3^i| \le 0.05/c_3, \quad i = 0,1,\cdots,N, \end{aligned}$$

where ΔT is the length of horizon. D_{ij} and Δ_i are shown in (19) and in (20). To solve the discrete nonlinear programming problem (29), a nonlinear programming solver such as SNOPT is usually used 39.

(iii). Apply the optimal control law $u^{1,k}$ on the unit and repeat the above operations in steps (i) and (ii) at the coming time instant t_{k+1}. It is worth mentioning that the $u^{1,k}$ is a composite control law which contains the compensation of the lumped disturbances.

To sum up, the scheme of the proposed composite control method can be illustrated in Figure 2.

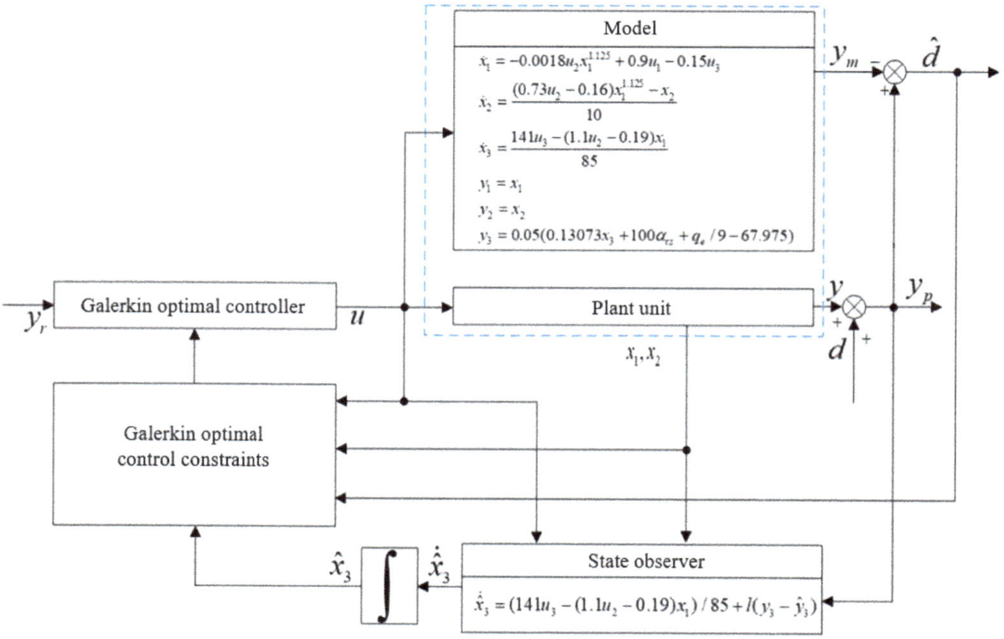

Figure 2. Block diagram of the receding Galerkin optimal controller with high-order sliding mode disturbance observer.

4. Simulations

In this section, some simulation cases are conducted to validate the performance of the proposed composite controller for the oil-fired drum-type boiler-turbine unit (1). For implementation purposes, a nonlinear programming solver SNOPT is adopted. Additionally, some controller parameters should be preset such as disturbance observer parameters (λ_l^i, L_i) and controller parameters (N, c, P, Q).

4.1. Parameters Assignment

As remarked in 1 and 35, the faster the convergence of disturbance observer should be, the larger λ_l^i and L_i are usually required. In fact, if the disturbance observer converges too fast, it will lead to the violation of the constraints imposed on control inputs. With these in mind, we select $\lambda_0^i = 1.5,\ \lambda_1^i = 0.1,\ \lambda_2^i = \lambda_3^i = 0.01,\ L_i = 2,\ i = 1,2,3$. Moreover, we choose $N = 20$, $P = diag\{1, 1, 2000\}$, $Q = diag\{2, 1, 2\}$, $c = [0.001, 0.001, 0.001]$ according to 23.

4.2. Case 1: Wide-Range Load Tracking without Lumped Disturbances

In this case, we intend to validate the tracking performance of the proposed composite controller without lumped disturbances. The change process of the working condition point is as in Table 3:

Table 3. The change process of the working condition point.

Time Period (s)	0–400	400–1000	1000–1400	1400–2400	2400–2700	2700–3200	3200–3500
Working condition	#2	#2 to #6	#6	#6 to #1	#1	#1 to #3	#3

From Table 3, the whole simulation time period is 3500 s, and the lumped disturbances are set to zero. With the initial conditions $z_1(0) = 5$, $z_2(0) = 10$, $z_3(0) = 0.1$, and the parameters assignment as Section 4.1. Implementing the composite controller on the unit (1) resulted in simulation results Figures 3 and 4, respectively.

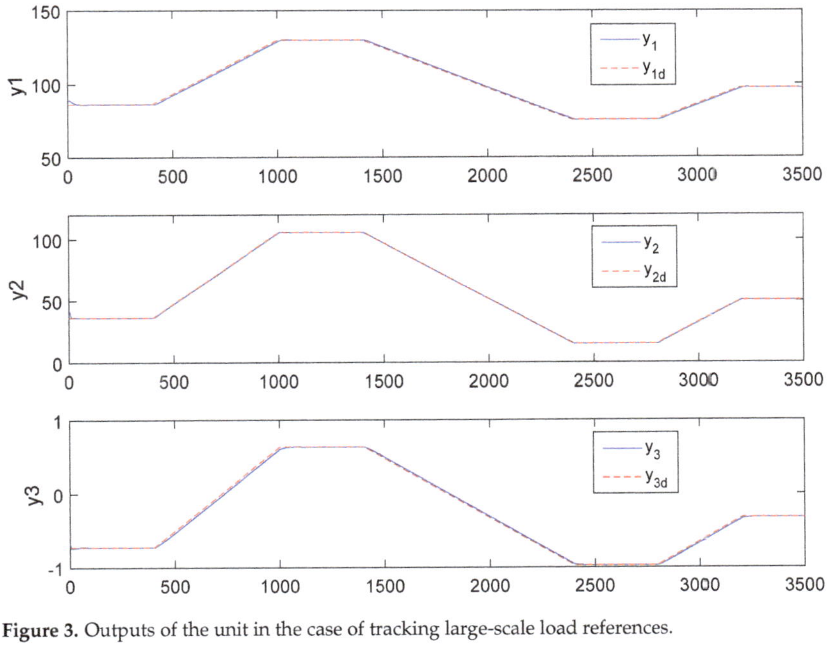

Figure 3. Outputs of the unit in the case of tracking large-scale load references.

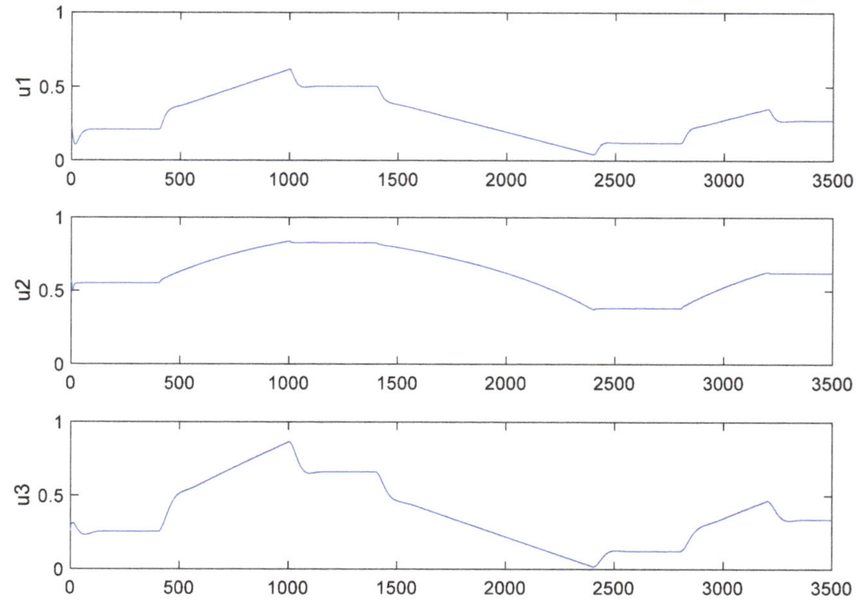

Figure 4. Control inputs of the unit in the case of tracking large-scale load references.

Figure 3 shows the output responses under control inputs in Figure 4. The x-axis is the time with its unit in seconds. It can be observed that the outputs y_i can track large-scale references y_{id} rapidly without deviation. Meanwhile, Figure 4 illustrates that all the control inputs u_i strictly meet the constraints.

4.3. *Case 2: Wide-Range Load Tracking with Lumped Disturbances*

In Case 1, the boiler-turbine unit is just regulated to track large-scale references without lumped disturbances. In this case, we aim to run the unit to track large-scale references in the presence of lumped disturbances $d_i(t)$ defined by

$$\begin{aligned} d_1 &:= -0.0002(z_1 + y_{1d})^{9/8}u_2 - 0.1u_1 + 0.02u_3, \\ d_2 &:= (0.01u_2 + 0.004)(z_1 + y_{1d})^{9/8} - 0.01(z_2 + y_{2d}) + 0.2\sin(0.01\pi t), \\ d_3 &:= 0.001u_3 - \left(2 \times 10^{-5}u_2 + 5 \times 10^{-6}\right)(z_1 + y_{1d}). \end{aligned} \tag{28}$$

The whole simulation time period is 3500 s, and the lumped disturbances (28) occur at the 400th second and vanish at the 3200th second. The initial conditions and parameter assignments are the same as those in Case 1. Figures 5–7 illustrate the estimates of lumped disturbances defined in (28), the response curves of the outputs y_i, and the control inputs u_i.

It can be seen in Figure 5 that the lumped disturbances can be estimated by the high-order sliding mode disturbance observer (11) with high accuracy. Based on this, Figure 6 exhibits that the outputs y_i can track large-scale references y_{id} rapidly in most cases, except for the period when drastic disturbances begin to happen or vanish. Meanwhile, it can be seen that the nominal performance can be reserved when the lumped disturbances vanish at the 2300th second. Furthermore, Figure 7 shows that the control inputs u_i change drastically due to the large-scale changes of references and the drastic lumped disturbances during time period (400, 3200), but the control inputs can be guaranteed inside the boundary [0, 1]. These all illustrate the excellent disturbance rejection and set-point tracking performance of the proposed composite control method.

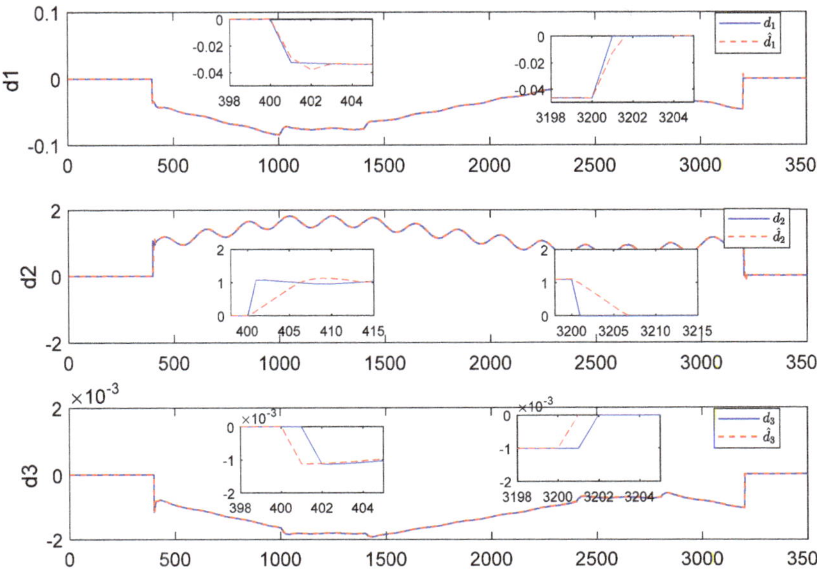

Figure 5. Estimates of lumped disturbances.

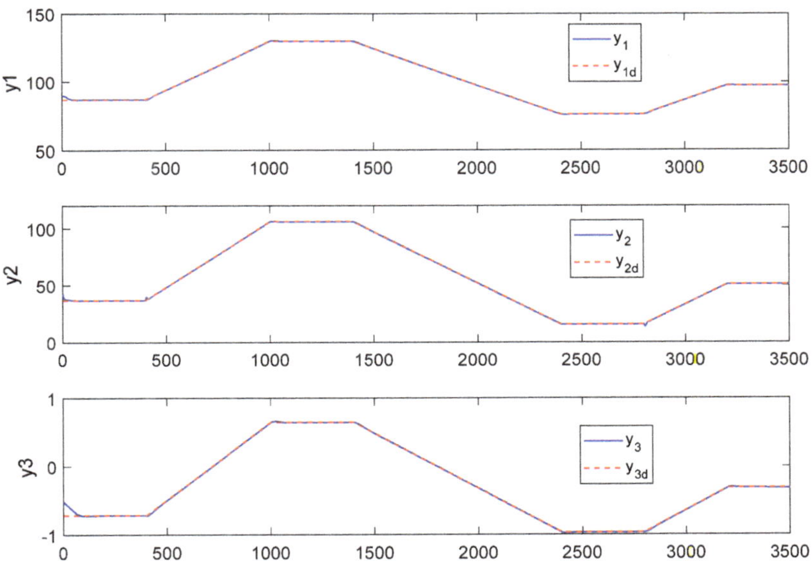

Figure 6. Output trajectories in presence of lumped disturbances.

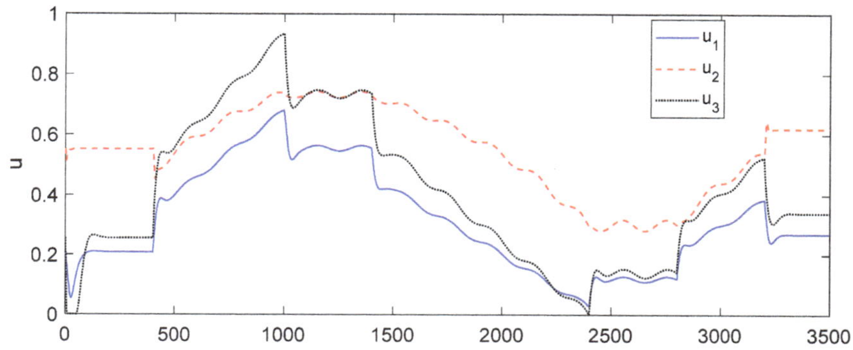

Figure 7. Control inputs curves.

4.4. Case 3: Control Performance Comparison under Wide-Range Load Tracking Conditions

To further verify the control performance of our proposed composite controller, a state feedback controller developed in 1 is carried out as the comparative method by considering its anti-disturbance ability.

The whole simulation time period is 3500 s. The lumped disturbances, initial conditions, and parameters assignment are set the same as those of Case 2. Corresponding simulation results are drawn in Figures 8 and 9.

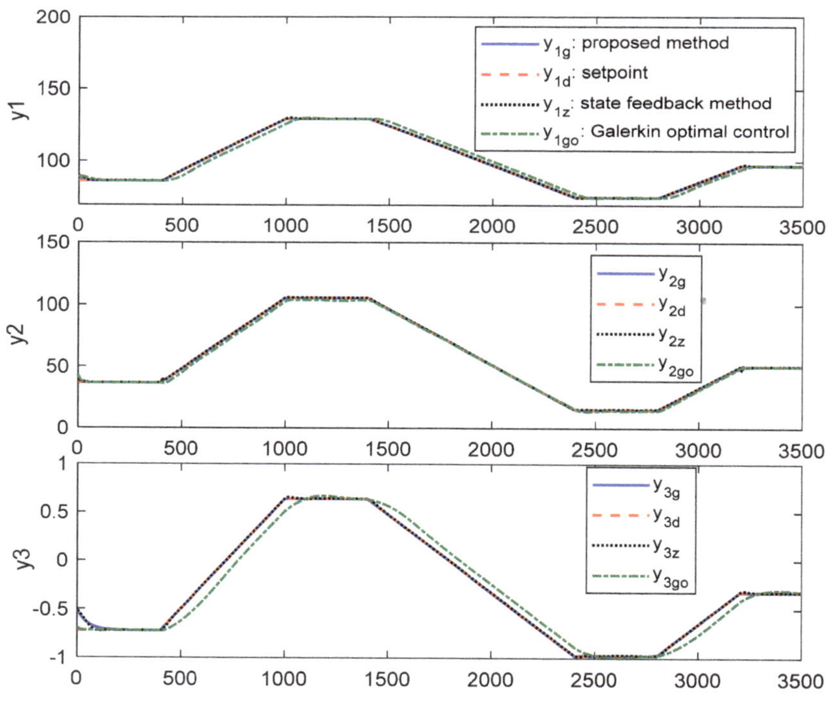

Figure 8. Control outputs in presence of lumped disturbances.

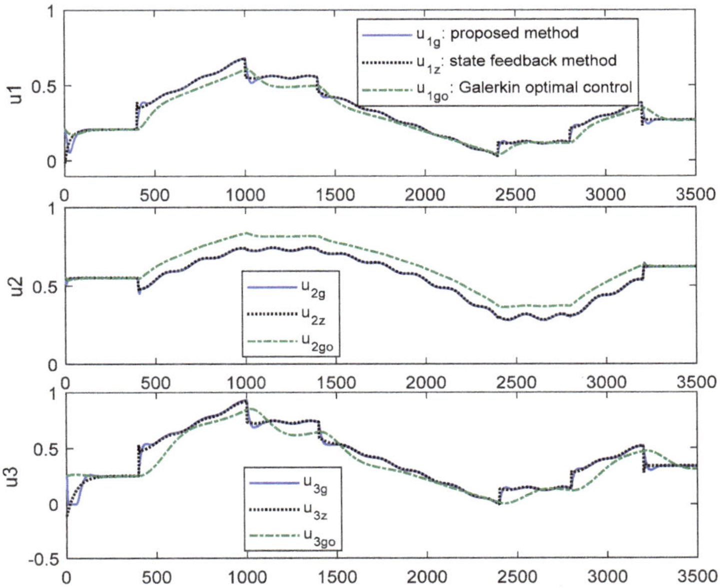

Figure 9. Control inputs in presence of lumped disturbances.

Figure 8 shows the output responses and Figure 9 gives the corresponding control inputs. Among them, y_{id}, $y_{ig}(u_{ig})$ and $y_{iz}(u_{iz})$ are setpoints, the outputs (inputs) produced by the proposed composite controller, and the outputs (inputs) produced by the comparative controller.

It can be noted from Figures 8 and 9 that the outputs of the two controllers can track the setpoints well with similar control inputs curves when the load changes over a wide range, indicating that the two control methods have similar disturbance rejection capability and can be adapted to large range load tracking scenarios. Moreover, for more clear comparisons, the following four-time frame are selected: (1) 1~100 s: the tracking process is of the set value under the initial deviation condition. (2) 390 s-480 s: the lumped disturbances occur and load starts to rise. (3) 990~1080 s: the lumped disturbances and the load begins to decline. (4) 3190~3280 s: the lumped disturbances disappear and the load begins to stabilize. The corresponding outputs and control inputs are illustrated in Figures 10–15.

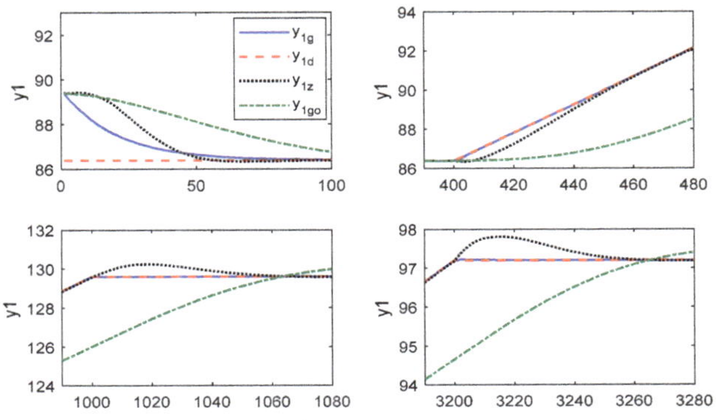

Figure 10. The comparison results of y_1 between the two controllers.

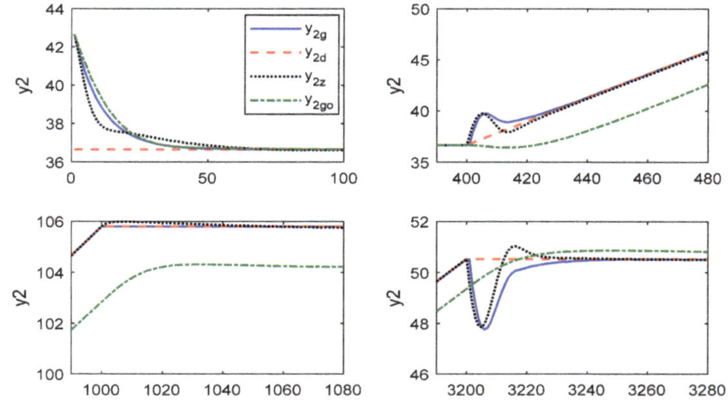

Figure 11. The comparison results of y_2 between the two controllers.

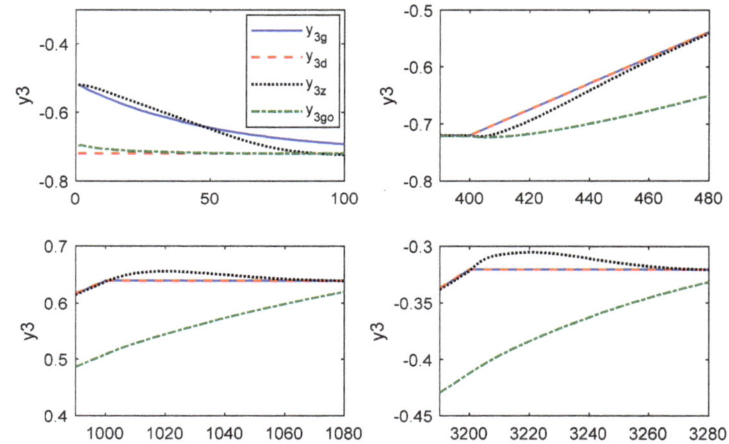

Figure 12. The comparison results of y_3 between the two controllers.

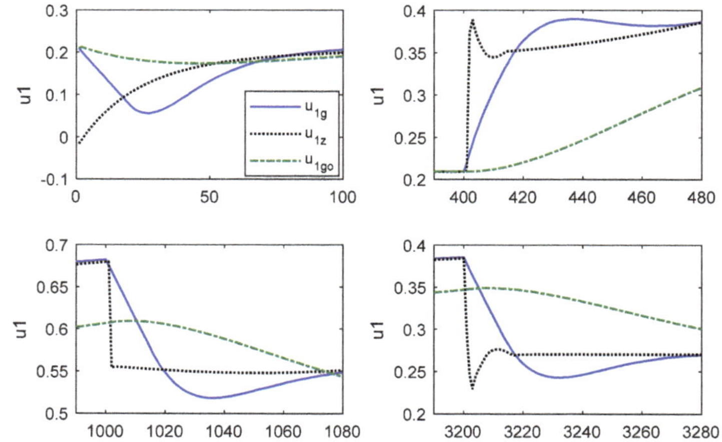

Figure 13. The comparison results of u_1 between the two controllers.

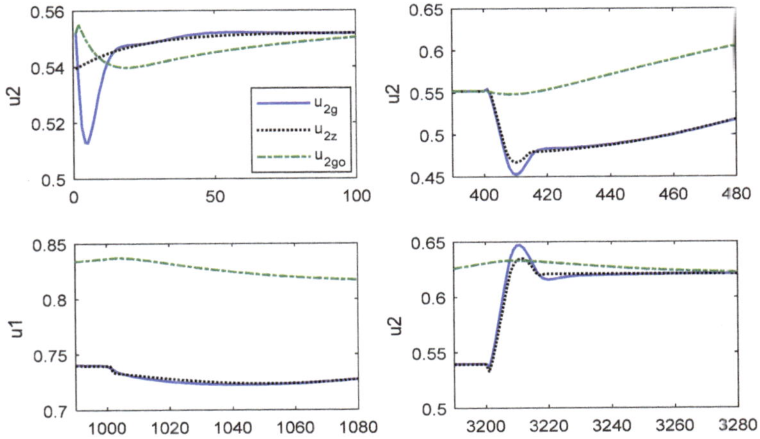

Figure 14. The comparison results of u_2 between the two controllers.

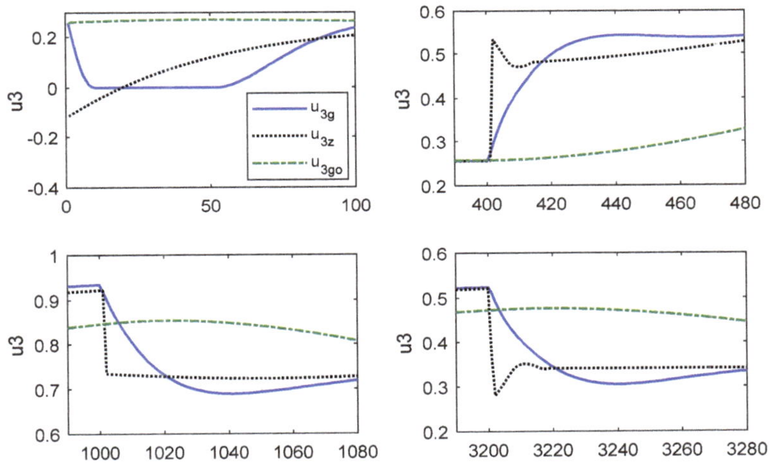

Figure 15. The comparison results of u_3 between the two controllers.

It can be concluded from Table 4 and Figures 10–15 that (i) the tracking rate of the comparative controller (y_{iz}) is faster than that of the proposed composite controller (y_{ig}), which means the rate of change of u_{iz} is higher than that of u_{ig}; (ii) When the deviation of initial condition setting is large, the initial stage of u_{3z} will be less than zero, which does not meet the constraint. However, the situation will not occur on the u_{3g} produced by the proposed composite controller due to the consideration of constraint during optimization; and (iii) the rate of change of the control input obtained by the comparative method is significantly higher than that of the proposed composite controller when the lumped disturbances occur, and the load starts to rise. This is due to the fact that the constraints of the input and output rates are taken into account in the controller design of the proposed control method while the constraints on control inputs are guaranteed by selecting reasonable state feedback controller's parameters of the comparative method. As a result, it is difficult to meet all constraints under the condition of severe disturbances of the comparative method. The IAE value of y_{3go} at all time frames are larger than the proposed method, because the disturbance rejection ability is weaker than the proposed one. Overall, the proposed method has a lower IAE index in each time frame compared with the other two methods.

Table 4. Quantitative integral absolute error (IAE) indexes of the simulation outputs.

$(\mathbf{IAE}_{y_1},\mathbf{IAE}_{y_2},\mathbf{IAE}_{y_3})$	Time Frame (1)	Time Frame (2)	Time Frame (3)	Time Frame (4)
Proposed method	(61.098, 63.04, 8.73)	(0.376, 26.6, 0.00105)	(0.0346,0.0472,0.000258)	(0.466, 29.64, 0.0027)
Galerkin control	(171.77, 72.39, 0.472)	(181.15, 219.023, 5.38)	(138.84, 165.61, 6.97)	(97.867, 39.399, 4.63)
State feedback control	(90.676, 60.343, 8.053)	(22.1, 25.217, 0.866)	(21.73, 6.45, 0.689)	(19.738, 25.095, 0.631)

According to the above comparison results, both controllers can fast-track the large-scale load. But the method proposed in [1] will result in control inputs that do not meet the constraints if the parameters are not selected properly or the disturbances are too severe. On the contrary, the presented receding Galerkin optimal controller with high-order sliding mode disturbance observer integrates the advantages of the high-order sliding mode disturbance observer and the receding Galerkin optimal control method, making it superior in constraint satisfaction, disturbance rejection, and large-scale load tracking.

5. Conclusions

In this paper, a receding Galerkin optimal controller with a high-order sliding mode disturbance observer is proposed to improve the control performance of a boiler-turbine unit. The presented controller can be directly designed based on the nonlinear mathematical unit model. More precisely, a high-order sliding mode disturbance observer is first employed to estimate the lumped disturbances and unmeasurable deviation states. Next, the estimate of the lumped disturbances is feedforward compensated in the receding optimization process. Then, based on the traditional Galerkin optimal control method, the idea of receding optimization is proposed to deal with lumped disturbances, variable constraints, and large-scale load tracking at the same time. Simulation results have shown that the proposed controller can regulate the boiler-turbine unit to track large-scale load set-points and meet the variable constraints in the presence of various types of unknown disturbances.

6. Annexe

All the necessary variables of controlled boiler–turbine unit by proposed method are listed in Table 5 for the easy of query.

Table 5. Variable of Controlled Boiler–Turbine Unit.

Variable	$a_1 \sim a_3$	$b_1 \sim b_3$	$c_1 \sim c_4$	α_{cs}
Name	identified parameter	identified parameter	identified parameter	coefficient
	x_1	x_2	x_3	y_1
	drum pressure	electrical output	fluid density in the drum	drum steam pressure
	y_2	y_3	u_1	u_2
	electrical output	drum water level	valve positions for fuel	valve positions for steam
	u_3	q_e	$\overline{d}_i(t),\ i = 1,2,3$	$x_{1d} \sim x_{3d}$
	valve positions for feedwater flow	evaporation rate of steam	unknown disturbances	setpoint signal for state
	$u_{1d} \sim u_{3d}$	y_{3d}	$w(t)$	$d_1(t) \sim d_3(t)$
	setpoint signal for input	setpoint signal for output	disturbance term	lumped disturbance
	$v_1 \sim v_3$	$z_1 \sim z_3$	$\zeta_0^i \sim \zeta_L^i$	J_{BT}
	virtual control law	states/outputs error	HOSMO estimate	cost function
	$L_N(\tau)$	z^{N_j}	u^{N_j}	
	Nth order Legendre polynomial	state variables at the LGL nodes	control variables at the LGL nodes	

Author Contributions: Conceptualization, Z.-G.S. and G.Z.; methodology, Z.-G.S.; software, G.Z.; validation, G.Z. and Y.H.; formal analysis, Y.S.; investigation, G.Z.; resources, Z.-G.S. and G.Z.; data curation, Y.H.; writing—original draft preparation, G.Z.; writing—review and editing, Y.S.; visualization, G.Z. and Y.S.; supervision, Z.-G.S.; project administration, Z.-G.S.; funding acquisition, Z.-G.S. All authors have read and agreed to the published version of the manuscript.

Funding: This research was funded by the National Natural Science Foundation of China, grant number 52076037.

Institutional Review Board Statement: Not applicable.

Informed Consent Statement: Not applicable.

Data Availability Statement: The data that support the findings of this study are available on request from the corresponding author, upon reasonable request.

Conflicts of Interest: The authors declare no conflict of interest.

References

1. Su, Z.G.; Zhao, G.; Yang, J.; Li, Y.G. Rejection of Nonlinear Boiler-Turbine Unit Using High-Order Sliding Mode Observer. *IEEE Trans. Syst. Man Cybern. -Syst.* **2018**, *50*, 5432–5443. [CrossRef]
2. Xu, Z.Y.; Qu, H.N.; Shao, W.H.; Xu, W.S. Virtual power plant based pricing control for wind/thermal cooperated generation in China. *IEEE Trans. Syst. Man Cybern. -Syst.* **2016**, *46*, 706–712. [CrossRef]
3. Zhang, S.; Taft, C.W.; Bentsman, J.; Hussey, A.; Petrus, B. Simultaneous gains tuning in boiler/turbine PID-based controller clusters using iterative feedback tuning methodology. *ISA Trans.* **2012**, *51*, 609–621. [CrossRef]
4. Yang, S.; Qian, C.; Du, H. A genuine nonlinear approach for controller design of a boiler–turbine system. *ISA Trans.* **2012**, *51*, 446–453. [CrossRef]
5. Fang, F.; Wei, L. Backstepping-based nonlinear adaptive control for coal-fired utility boiler–turbine units. *Appl. Energy* **2011**, *88*, 814–824. [CrossRef]
6. Chen, P.C.; Shamma, J.S. Gain-scheduled ℓ1-optimal control for boiler-turbine dynamics with actuator saturation. *J. Process Control.* **2004**, *14*, 263–277. [CrossRef]
7. Ghabraei, S.; Moradi, H.; Vossoughi, G. Multivariable robust adaptive sliding mode control of an industrial boiler–turbine in the presence of modeling imprecisions and external disturbances: A comparison with type-I servo controller. *ISA Trans.* **2015**, *58*, 398–408. [CrossRef]
8. Sariyildiz, E.; Oboe, R.; Ohnishi, K. Disturbance observer-based robust control and its applications: 35th anniversary overview. *IEEE Trans. Ind. Electron.* **2020**, *67*, 2042–2053. [CrossRef]
9. Morales, J.Y.R.; Mendoza, J.A.B.; Torres, G.O.; Vázquez, F.D.J.S.; Rojas, A.C.; Vidal, A.F.P. Fault-Tolerant Control implemented to Hammerstein–Wiener model: Application to Bio-ethanol dehydration. *Fuel* **2022**, *308*, 121836. [CrossRef]
10. Ortiz Torres, G.; Rumbo Morales, J.Y.; Ramos Martinez, M.; Valdez-Martínez, J.S.; Calixto-Rodriguez, M.; Sarmiento-Bustos, E.; Torres Cantero, C.A.; Buenabad-Arias, H.M. Active Fault-Tolerant Control Applied to a Pressure Swing Adsorption Process for the Production of Bio-Hydrogen. *Mathematics* **2023**, *11*, 1129. [CrossRef]
11. Morales, J.Y.R.; López, G.L.; Martínez, V.M.A.; Vázquez, F.D.J.S.; Mendoza, J.A.B.; García, M.M. Parametric study and control of a pressure swing adsorption process to separate the water-ethanol mixture under disturbances. *Sep. Purif. Technol.* **2020**, *236*, 116214. [CrossRef]
12. Chen, P.C. Multi-objective control of nonlinear boiler-turbine dynamics with actuator magnitude and rate constraints. *ISA Trans.* **2013**, *52*, 115–128. [CrossRef]
13. Li, Y.; Shen, J.; Lee, K.Y.; Liu, X. Offset-free fuzzy model predictive control of a boiler–turbine system based on genetic algorithm. *Simul. Model. Pract. Theory* **2012**, *26*, 77–95. [CrossRef]
14. Kong, X.; Liu, X.; Lee, K.Y. Nonlinear multivariable hierarchical model predictive control for boiler-turbine system. *Energy* **2015**, *93*, 309–322. [CrossRef]
15. Ławryńczuk, M. Nonlinear predictive control of a boiler-turbine unit: A state-space approach with successive on-line model linearisation and quadratic optimisation. *ISA Trans.* **2017**, *67*, 476–495. [CrossRef]
16. Wu, X.; Shen, J.; Li, Y.; Lee, K.Y. Data-driven modeling and predictive control for boiler–turbine unit using fuzzy clustering and subspace methods. *ISA Trans.* **2014**, *53*, 699–708. [CrossRef]
17. Liu, X.; Kong, X. Nonlinear fuzzy model predictive iterative learning control for drum-type boiler–turbine system. *J. Process Control* **2013**, *23*, 1023–1040. [CrossRef]
18. Yang, S.; Qian, C. Real-time optimal control of a boiler-turbine system using pseudospectral methods. In Proceedings of the 19th Annual Joint ISA POWID/EPRI Controls and Instrumentation Conference and 52nd ISA POWID Symposium, Rosemont, IL, USA, 12–14 May 2009; Volume 477, pp. 166–177.
19. Elnagar, G.; Kazemi, M.A.; Razzaghi, M. The pseudospectral Legendre method for discretizing optimal control problems. *IEEE Trans. Autom. Control* **1995**, *40*, 1793–1796. [CrossRef]

20. Biegler, L.T. Nonlinear programming: Concepts, algorithms, and applications to chemical processes. In *Society for Industrial and Applied Mathematics*; SIAM: Philadelphia, PA, USA, 2010. [CrossRef]
21. Boucher, R.; Kang, W.; Gong, Q. Galerkin optimal control for constrained nonlinear problems. In Proceedings of the 2014 American Control Conference, Portland, OR, USA, 4–6 June 2014; pp. 2432–2437.
22. Boucher, R.; Kang, W.; Gong, Q. Galerkin optimal control. *J. Optim. Theory Appl.* **2016**, *169*, 825–847. [CrossRef]
23. Zhao, G.; Su, Z.G.; Zhan, J.; Zhu, H.; Zhao, M. Adaptively receding Galerkin optimal control for a nonlinear boiler-turbine unit. *Complexity* **2018**, *2018*, 8643623. [CrossRef]
24. Xu, B. Disturbance observer-based dynamic surface control of transport aircraft with continuous heavy cargo airdrop. *IEEE Trans. Syst. Man Cybern. Syst.* **2016**, *47*, 161–170. [CrossRef]
25. Yang, J.; Liu, X.; Sun, J.; Li, S. Sampled-data robust visual servoing control for moving target tracking of an inertially stabilized platform with a measurement delay. *Automatica* **2022**, *137*, 110105. [CrossRef]
26. Li, S.; Yang, J.; Chen, W.H.; Chen, X. *Disturbance Observer-Based Control: Methods and Applications*; CRC Press: Boca Raton, FL, USA, 2014.
27. Han, J. From PID to active disturbance rejection control. *IEEE Trans. Ind. Electron.* **2009**, *56*, 900–906. [CrossRef]
28. Madonski, R.; Łakomy, K.; Stankovic, M.; Shao, S.; Yang, J.; Li, S. Robust converter-fed motor control based on active rejection of multiple disturbances. *Control Eng. Pract.* **2021**, *107*, 104696. [CrossRef]
29. Gao, Z. Scaling and bandwidth-parameterization based controller tuning. *ACC* **2003**, 4989–4996.
30. Gao, X.H.; Wei, S.; Wang, M.; Su, Z.G. Optimal disturbance predictive and rejection control of a parabolic trough solar field. *Int. J. Robust Nonlinear Control*, 2023; *in press*. [CrossRef]
31. Zhang, F.; Wu, X.; Shen, J. Extended state observer based fuzzy model predictive control for ultra-supercritical boiler-turbine unit. *Appl. Therm. Eng.* **2017**, *118*, 90–100. [CrossRef]
32. Zeng, L.; Li, Y.; Liao, P.; Wei, S. Adaptive disturbance rejection model predictive control and its application in a selective catalytic reduction denitrification system. *Comput. Chem. Eng.* **2020**, *140*, 106963. [CrossRef]
33. Bell, R.D.; Åström, K.J. *Dynamic models for Boiler-Turbine-Alternator Units: Data Logs and Parameter Estimation for a 160 MW Unit*; Research Reports TFRT-3192; Department of Automatic Control, Lund Institute of Technology (LTH): Lund, Sweden, 1987.
34. Levant, A. Higher-order sliding modes, differentiation and output-feedback control. *Int. J. Control* **2003**, *76*, 924–941. [CrossRef]
35. Shtessel, Y.B.; Shkolnikov, I.A.; Levant, A. Smooth second-order sliding modes: Missile guidance application. *Automatica* **2007**, *43*, 1470–1476. [CrossRef]
36. Li, S.; Sun, H.; Yang, J.; Yu, X. Continuous finite-time output regulation for disturbed systems under mismatching condition. *IEEE Trans. Autom. Control* **2014**, *60*, 277–282. [CrossRef]
37. Gill, P.E.; Murray, W.; Saunders, M.A. SNOPT: An SQP algorithm for large-scale constrained optimization. *SIAM Rev.* **2005**, *47*, 99–131. [CrossRef]
38. Zhan, J.; Su, Z.; Hao, Y. Optimal control for a boiler-turbine system via Galerkin method. In Proceedings of the 2017 2nd International Conference on Power and Renewable Energy (ICPRE), Chengdu, China, 20–23 September 2017; pp. 776–780.
39. Gong, Q.; Kang, W.; Ross, I.M. A pseudospectral method for the optimal control of constrained feedback linearizable systems. *IEEE Trans. Autom. Control* **2006**, *51*, 1115–1129. [CrossRef]

processes

Article

Vibration Scale Model of a Converter Transformer Based on the Finite Element and Similarity Principle and Its Preparation

Hao Wang [1], Li Zhang [1,*], Youliang Sun [1,2] and Liang Zou [1]

[1] School of Electrical Engineering, Shandong University, Jinan 250061, China; 202120603@mail.sdu.edu.cn (H.W.); youliang.s@163.com (Y.S.); zouliang@sdu.edu.cn (L.Z.)
[2] Shandong Electric Power Equipment Co., Ltd., Jinan 250061, China
* Correspondence: zhleee@sdu.edu.cn

Abstract: A similarity criterion suitable for studying the vibration characteristics of converter transformers is proposed based on comprehensive consideration of geometric dimensions, electric field, magnetic field, force field, sound field, and coupling field interface interactions. By comparing the magnetic field, stress, displacement, sound field distribution, and vibration characteristics of the scale model of the converter transformer with the initial model, the reliability of the similarity criterion was determined. Based on the vibration similarity criterion of the converter transformer, a prototype of the proportional model was designed and manufactured, and vibration signals under no-load and load conditions were tested. These signals correspond to the vibration signals of the iron core and winding in the finite element model, respectively. Through comparative analysis, the reliability of the prototype and the vibration similarity model of the converter transformer has been proven, which can provide an accurate and effective laboratory research platform for in-depth research on the vibration and noise of the converter transformer and equipment protection.

Keywords: vibration; similarity model; converter transformer; finite element; prototype preparation

Citation: Wang, H.; Zhang, L.; Sun, Y.; Zou, L. Vibration Scale Model of a Converter Transformer Based on the Finite Element and Similarity Principle and Its Preparation. *Processes* **2023**, *11*, 1969. https://doi.org/10.3390/pr11071969

Academic Editors: Bo Yang, Zhijian Liu and Lin Jiang

Received: 12 June 2023
Revised: 26 June 2023
Accepted: 27 June 2023
Published: 29 June 2023

1. Introduction

In recent years, with the acceleration of human economic and social development, energy and environmental issues have become more severe, and the imbalance between electricity load and energy base has become increasingly serious. The rapid development of high-voltage direct current transmission technology has made great contributions to solving this problem [1–7]. Converter transformers, which withstand mixed AC and DC voltages during operation, are key equipment in DC transmission systems. The harmonic and DC components they bear during operation are also higher than ordinary power transformers, and the risks of core saturation and resonance effects are higher, resulting in more serious vibration and noise problems [8–12]. Researchers have been committed to studying the generation mechanism and vibration suppression of converter transformer vibration, but they are still troubled by its variable testing environment, numerous electric/magnetic field analyses, and complex vibration propagation processes. The mechanism of internal vibration generation and propagation in converter transformers still needs further research, and the effectiveness of vibration suppression is limited. The safety and reliability of converter stations need to be further improved [13–16].

The large volume and expensive price of converter transformers make it impractical to conduct in-depth and effective vibration research on them in laboratory environments. Establishing a large-scale experimental vibration platform for power equipment is an effective method to solve these problems. It can not only increase the feasibility of experimental mechanism research, but also reduce the computational time and burden of the experiment. However, the consistency between the operation and vibration characteristics of the equipment after proportional deceleration and the actual operating equipment is currently a challenging issue [17,18]. The similarity model method based on similarity theory can effectively solve the problems of economy and timeliness of the test, and can be used to verify new technologies, design, and manufacturing difficulties that have not been considered in the process of testing, and predict the performance indicators of the entire system [19,20]. From the above, it can be seen that, based on the principle of similarity, reducing the equipment volume proportionally is a feasible solution to solve the problem of difficulty in conducting internal vibration characteristics research on large-capacity converter transformers. However, currently, most proportional model designs for power devices consider only geometric dimensions and a small number of electrical parameters with similar derivation, and lack consideration of electromagnetic and mechanical parameters of vibration [21]. For the moment, there is no similarity theory proposed to study the vibration characteristics of converter transformers, nor any related research on designing and compiling scale models of converter transformers based on similarity methods. How to derive the similarity criteria for converter transformers comprehensively and reasonably and how to modify them in practical applications need to be discussed in detail.

The deep application of finite element technology provides a convenient and effective research platform for the analysis and research of large-scale power equipment involving complex internal structures and multi-physical field interface effects. Reference [22] analyzed the monopole ionization field in DC wires through full multi-grid measurements. Reference [23] constructed a generator power prediction model based on the finite element method and compared the results with neural network algorithms, proving the effectiveness of the model. Reference [24] established a two-dimensional physical model of the transformer and quantitatively analyzed the direction and magnitude of the force on the transformer winding under low load and overload. Reference [25] established a finite element model of a three-phase five-column converter transformer, applied different DC components to the model, studied the changes in magnetic flux and force of the converter transformer under different DC components, and quantified its loss characteristics.

From the above research, it can be seen that the application of finite element technology in the coupling field effect of transformers has become very mature. However, the finite element model of the converter transformer used in current research is too idealized in the modeling process, without refining and distinguishing between the core lamination structure and the winding form. This integrated core and winding structure can quickly obtain uniformly distributed magnetic field morphology and calculation results for vibration characteristics, meeting the operational requirements of composite ideal transformers. However, electromagnetic and vibration calculation results that require details and consider coupling boundary effects cannot be convincing [17,26,27]. Considering the significant differences in structure and nonlinear characteristics between the actual operation of the converter transformer and the ideal transformer, especially when calculating the internal structure coupling field interface interaction and nonlinear force loss under different external excitation conditions, using the ideal converter transformer model can cause unacceptable calculation errors [28,29]. Therefore, when using the finite element method for coupling calculation of the field interface of the converter transformer, the design cannot be blindly simplified [30].

Firstly, based on the similarity principle, the similarity criteria of the electromagnetic field and structural mechanics were derived. Secondly, based on the structure and operating parameters of a single-phase dual-winding current transformer in operation, the contribution of the oil tank structure, the lamination form of the iron core, the entanglement and winding form of the winding, and the force characteristics of the pad between the winding cakes were fully considered. Furthermore, in order to verify the reliability of the similarity criterion, a similarity model of the converter transformer was established based on the physical model of the converter transformer mentioned above. A comprehensive comparison was made between the electromagnetic field, mode of vibration, force, and displacement distribution characteristics of the scale model and the initial model of the converter transformer. Finally, a proportional prototype of a converter transformer with a similarity coefficient of 1:5 was designed and prepared based on the similarity criterion. The basic experimental items and vibration characteristics were analyzed, further verifying the reliability and engineering practicality of the similarity criterion.

2. Similarity Principle

2.1. Geometric Similarity

Based on the principle of similarity, the full size of the converter transformer should be proportionally reduced, that is, the overall geometric length l of length, width, height, and radius should be changed to $l' = kl$. So, the corresponding overall geometric area S becomes $S' = k^2S$, and the overall volume V becomes $V' = k^3V$.

2.2. Similarity of Structural Mechanics

The quality and volume of the model have a significant impact on the overall modal distribution of the model, which can easily alter the original resonance effect of the model, thereby affecting the final vibration distribution characteristics of the model. Therefore, it is necessary to discuss the structural parameters such as the natural frequency of the model. For mass M, due to the material density of scale models ρ, the model volume remains unchanged and decreases to three times the original k^3, so M' becomes:

$$M' = k^3 M \tag{1}$$

For the stiffness coefficient K, it shall meet the following requirements before and after similarity:

$$K' = kK \tag{2}$$

So. the natural frequency $\omega_0{}'$ of scale models becomes:

$$\omega_0{}' = \sqrt{\frac{kK}{k^3 M}} = \frac{1}{k}\sqrt{\frac{K}{M}} \tag{3}$$

2.3. Similarity of Electromagnetic Field

Supposing that the current density J' of the scale model of the converter transformer is equal to the current density J of the initial model, to ensure the controllability of excitation and the controllability of reducing winding losses:

$$J' = \frac{I'}{S'} = \frac{I}{S} = J \tag{4}$$

The sectional area of similar winding $S' = k^2S$, then the current strain of similar winding is:

$$I' = k^2 I \tag{5}$$

The integral form of Ampere's law is:

$$\oint Bdl = \mu_0 NI \tag{6}$$

where B is the magnetic flux density, N is the number of turns, and μ_0 is the vacuum permeability.

The magnetic flux density is one of the main indexes considered in the design of the converter transformer core, which should be selected in the linear area of the B-H curve, and the magnitude of the magnetic flux density affects the working characteristics and vibration characteristics of the core. Based on the operation characteristics of the converter transformer, the core material generally works between 1.5–1.8 T. As the amplitude of I decreases after the model is reduced, in order to reach the linear working area of the hysteresis loop more easily, the number of coil turns is increased from N to $k^{-1}N$, and the magnetic flux density obtained after similarity should be:

$$B' = B \tag{7}$$

It is worth emphasizing that through the electromagnetic field similarity criterion, the working current density and magnetic field intensity of the scale model are consistent with the prototype, which can effectively ensure the similarity of the working characteristics of the converter transformer before and after scaling.

The same winding material is used before and after the similarity, so the winding conductivity remains unchanged, but at the same time, the length of the winding decreases to k times, and the cross-sectional area of the winding decreases to k^2 times. Therefore:

$$R' = \frac{1}{\sigma} \cdot \frac{l'}{S'} = \frac{1}{\sigma} \cdot \frac{kl}{k^2 S} = \frac{R}{k} \tag{8}$$

Based on Ohm's circuit law, the proportional model has the following voltage:

$$U' = I'R' = kU \tag{9}$$

Assuming that the operating frequency of the power supply is unchanged before and after the similarity, we can judge whether the similar process between the above parameters is balanced by the relationship between the main magnetic flux and the operating voltage when the converter transformer works normally:

$$U = 4.44fNBS \tag{10}$$

According to Formulas (4)–(9), $U' = kU$, $N' = k^{-1}N$, $B' = B$, $S' = k^2S$, and $f' = f$ are brought into Formula (10). The equation is balanced on both sides of the equals sign, and the similarity criterion is correct.

2.4. Similarity of Solid Mechanics

According to the principle of the vibration mechanics model, the winding is regarded as an elastic mass. The gasket and insulation structure together are equivalent to the spring damping element, that is, the winding is equivalent to the rigid body structure with mass m, the gasket between windings is equivalent to the spring element K, and the insulation block and compression structure are equivalent to the spring K_B and K_H respectively. Assuming that the damping coefficient between windings is C, the governing equation of winding vibration can be obtained as follows [31]:

$$M\frac{d^2z}{dt^2} + C'\frac{dz}{dt} + K'z = F + Mg \tag{11}$$

where M is the mass matrix, C' is the damping matrix, K' is the elastic matrix, z is the displacement, g is the acceleration of gravity.

The force acting on a moving charge in a magnetic field is called the Lorentz force, which is the force exerted by the magnetic field on the moving charge in the wire. The Ampere force is the macroscopic manifestation of the Lorentz force, which is the force acting on the charged wire in the magnetic field. When alternating current is applied, a leakage of magnetic field that varies with the power frequency is generated around the winding. The moving charge in the winding is subjected to the action of the alternating magnetic field. During the positive and negative half cycles of the power supply, they are subjected to opposing forces, resulting in two vibrations occurring within one power frequency. The leakage flux through the winding is divided into axial and radial components. The axial leakage flux interacts with the winding current to generate radial electromagnetic force F_x, and the radial leakage flux interacts with the winding current to generate axial electromagnetic force F_z, which can be expressed as:

$$F_x = iB_x \cdot 2\pi r \tag{12}$$

$$F_z = iB_z \cdot 2\pi r \tag{13}$$

After similarity, the winding radius $r = kr$, magnetic flux density $B' = B$, and winding current $i' = k^2i$. Therefore, we obtain the electromagnetic force $F_l' = k^4F_l$ on the winding of the converter transformer after similarity.

The vibration model of silicon steel sheet under alternating magnetic field is as follows:

$$D_0\left(\frac{\partial^4 v}{\partial x^4} + 2\frac{\partial^4 v}{\partial x^2 \partial y^2} + \frac{\partial^4 v}{\partial y^4}\right) = \frac{\omega \sin 2\omega t}{\Delta l_x}\int_v \frac{1}{2}E\lambda_Z^3 dV \tag{14}$$

where D_0 is the bending stiffness of silicon steel sheet, v is the Poisson's ratio of the material, x and y are the transverse and longitudinal dimensions of two-dimensional silicon steel sheet, respectively, ω is angular frequency, t is the time, l_x is the elongation, E is the elastic modulus of the material, λ_Z is the axial magnetostriction coefficient, V is the volume element.

Nowadays, the gaps between silicon steel sheets in the iron cores used in converter transformers are becoming smaller and smaller, and the stacking method is more advanced. The Maxwell stress between silicon steel sheets can be ignored, so the main contribution of iron core vibration is the magnetostrictive effect of ferromagnetic materials under alternating electric fields. The magnetostrictive force caused by the magnetostrictive effect in both parallel and vertical directions can be expressed as:

$$F_c = F_{cmax}sin2\omega t = 0.5\nabla(H^2\tau\partial\frac{\partial\mu}{\partial\tau}) \tag{15}$$

where F_c is the magnetostrictive force and F_{cmax} is the amplitude of the electromagnetic force caused by magnetostriction. Therefore, the force magnitude F_c' on the iron core changes to $F_c' = kF_c$.

3. Finite Element Physical Model and Its Similarity Process

3.1. Finite Element Model of Electric-Magnetic-Force Multi-Field Coupling for Converter Transformer

According to a 500 kV converter transformer in operation (ZZDFPZ-415000/500-800), a three-dimensional physical model of the converter transformer with multi-field coupling was established using the finite element simulation platform, as shown in Figure 1. The model was simplified according to the design drawings and was designed as a single-phase double winding structure with two main columns and two side columns. Fixture replacement can be neglected with fixed constraints, but the winding entanglement structure, distribution form of winding pads, and iron core lamination form that have a significant impact on the vibration characteristics of the model cannot be omitted. The material property parameters of the model are listed in Table 1.

Table 1. The material property parameters of the model.

Name	Iron Core	Winding	Spacer	Insulating Oil
Density (kg/m^3)	7870	8940	1117	895
Young's modulus (GPa)	6	12.6	1	-
Poisson's ratio	0.45	0.34	0.38	-
Conductivity (S/m)	1.12×10^7	5.99×10^7	1	1
Relative permittivity	1	1	1	-

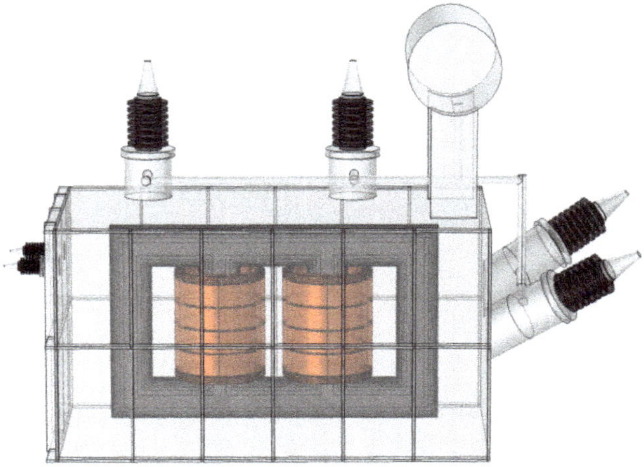

Figure 1. Finite element model of multi-field coupling for converter transformer.

Considering the time cost, the computational complexity and convergence speed of the model are optimized, and some details are simplified without affecting the vibration characteristics. The core material is the main contribution to the nonlinearity of the converter transformer. It is necessary to define the values of *B* and *H* at each location based on the *B*-*H* curve of the silicon steel sheet to ensure the reliability of the model. The *B*-*H* curve used for the definition is shown in Figure 2.

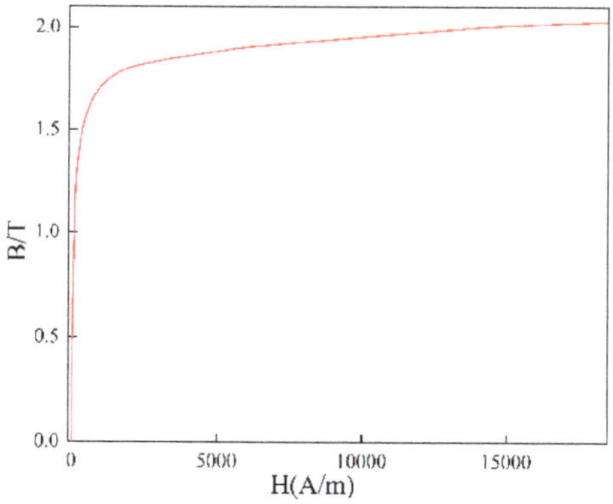

Figure 2. Determination of B-H curve of silicon steel sheet corresponding to reference converter transformer.

The design of the winding structure should consider not only the arrangement between each turn of the winding and the position of the cushion blocks, but also the trajectory of current flow in the winding cake, which is one of the main difficulties in finite element modeling. The windings of the high-voltage transformers currently in operation are mostly wound continuously, as shown in Figure 3. By using the field–circuit coupling method, as shown in Figure 4, the specified conduction path of excitation is set according to the node order to achieve a winding mode of entangled–continuous–entangled.

Figure 3. Entangled winding-wire turn arrangement.

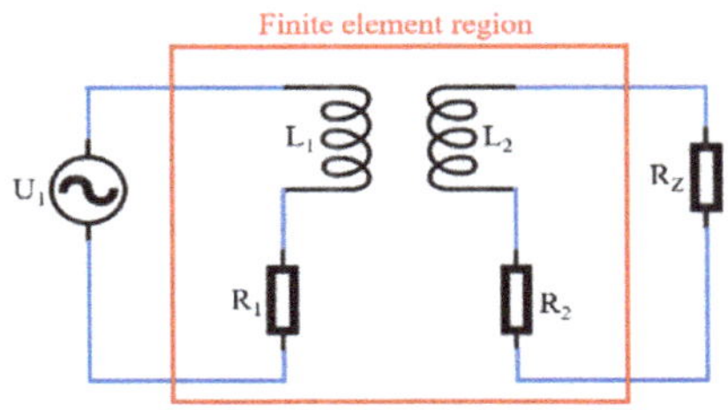

Figure 4. Field–circuit coupling method.

The current flow sequence of the winding is shown in Figure 5. The overall appearance structure of the winding is shown in Figure 6, and the geometric parameters of each part of the winding are listed in Table 2.

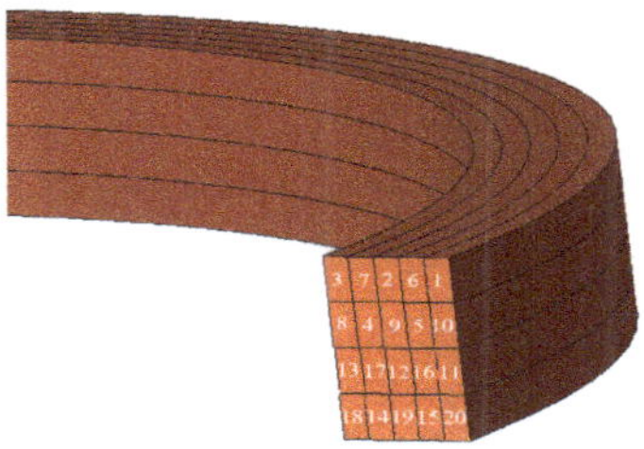

Figure 5. Current sequence of entangled winding.

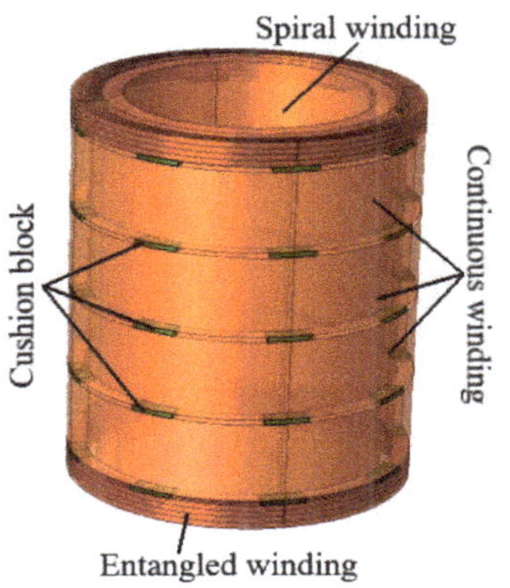

Figure 6. Single-phase winding model.

Table 2. Geometry parameters of initial model and scale model.

Parameters	Initial Model	Similitude Model
Yoke length (m)	6.72	1.344
Core height (m)	3.6	0.72
Winding wire cake height (m)	1.764	0.353
Winding radius at grid side (m)	0.664	0.133
Radius of valve side winding (m)	0.468	0.094

In order to save time and computational resources, the physical model of the converter transformer was divided into 1/2 models along the longitudinal section. The 1/2 model was meshed, and the winding, iron core, and pad were subjected to ultra-fine free tetrahedral structural dissection. Other areas were subjected to plane triangle and sweep operations.

3.2. Vibration Scale Model of Converter Transformer

Based on the similarity criterion, a 1/5 scale model of the converter transformer was established, and the geometric parameters are shown in Table 2. The magnetic flux density distribution, stress distribution and vibration characteristics of the converter transformer model and its scale model were calculated, and the calculation results are compared and analyzed.

4. Analysis and Discussion on Simulation Results of Finite Element Scale Model of Converter Transformer

4.1. Similarity of Modal Shapes

Mode of vibration was the first structural characteristic to be verified after the converter transformer geometry transformation. In order to improve the universality of modal analysis, in addition to the initial model and 1/5 similarity model, a 1/2 similarity model was also established and analyzed, and the first six natural frequency distributions of the model core and winding under three geometric sizes were obtained, which are listed in Tables 3 and 4, respectively. By comparing the modal shapes of the models with three geometric sizes, it was found that the modal distributions of the windings and iron cores under each size were roughly the same, and the natural frequency showed an inverse trend to the similarity coefficient. The law follows the derivation of Formula (3). In order to reduce the generation of resonance effects during operation, mode has become one of the main parameters considered in the transformer design process. The natural frequencies of the core and winding are both staggered from the power frequency, avoiding resonance, indicating that the structural design of the model is relatively reasonable.

Table 3. Natural frequency distribution of the core.

Natural Frequency	Initial Model (Hz)	1/2 Model (Hz)	1/5 Model (Hz)
First order	77.46	155.06	373.64
Second order	114.55	229.28	563.93
Third order	138.76	277.58	673.93
Fourth order	240.83	481.99	1203.96
Fifth order	271.62	548.29	1358.53
Sixth order	316.01	632.02	1593.26

Table 4. Natural frequency distribution of the winding.

Natural Frequency	Initial Model (Hz)	1/2 Model (Hz)	1/5 Model (Hz)
First order	77.53	155.22	387.62
Second order	88.78	196.67	428.94
Third order	115.64	229.58	573.14
Fourth order	138.79	277.58	693.93
Fifth order	168.62	337.24	842.95
Sixth order	240.78	481.57	1203.57

The winding and core modes of the initial model, the 1/5 model, and the 1/2 model are shown in Figure 7a–c and Figure 7d–f, respectively. As the model scale decreases proportionally, the modal distribution pattern of the model does not change much, and the natural frequency changes of each size of model follow the load similarity criterion.

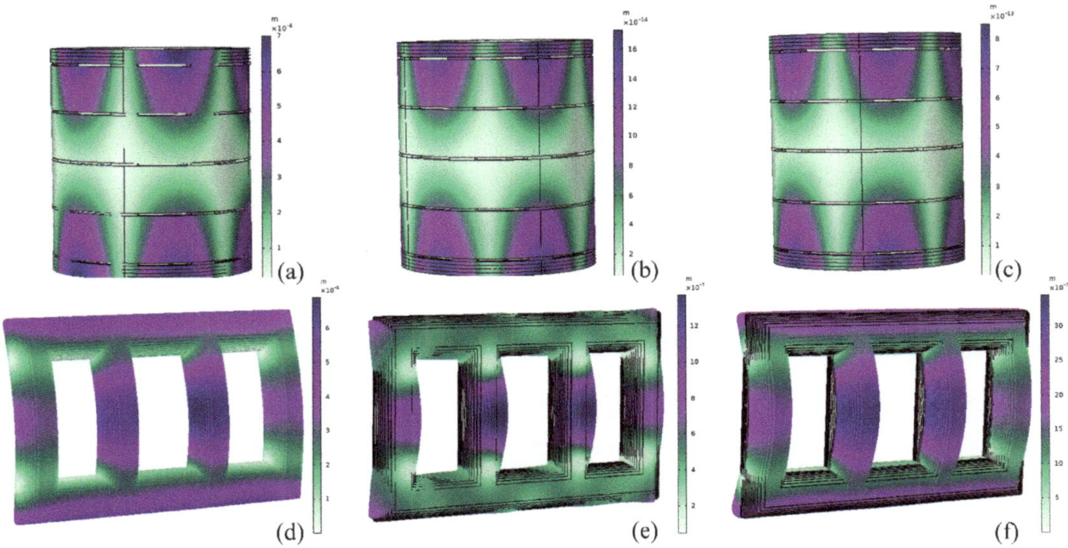

Figure 7. Modal distribution of iron core and winding: (**a**) original model winding; (**b**) 1/2 model winding; (**c**) 1/5 model winding; (**d**) Original model iron core; (**e**) 1/2 model iron core; (**f**) 1/5 model iron core.

4.2. Distribution Characteristics of Electromagnetic Fields in Similar Processes

The changes of the magnetic field intensity distribution of the core and winding in the process of model similarity were calculated and compared, and are depicted in Figures 8 and 9. As depicted in Figure 8, the change of core magnetic flux density distribution in the process of similarity can be ignored, and the amplitude of magnetic flux density basically remained unchanged, which is consistent with the conclusion of the similarity criteria. The leakage flux density distributions of the initial model and 1/5 scale model are shown in Figure 9. Through comparison, it was found that the model did not change the spatial distribution trend of the leakage flux in the similar process, and the amplitude variation law of the leakage flux density of both models meets the proposed electromagnetic field similarity criterion.

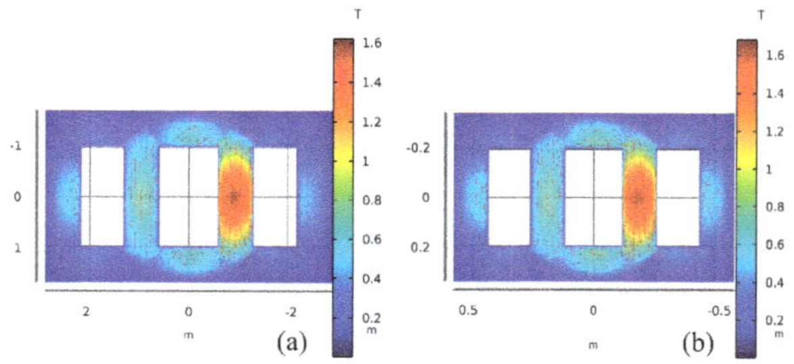

Figure 8. Magnetic flux density distribution of (**a**) initial model and (**b**) scale model.

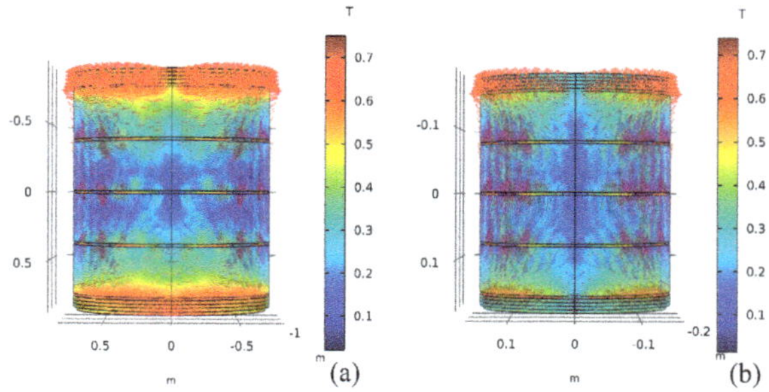

Figure 9. Leakage flux density of (**a**) initial model and (**b**) scale model.

4.3. Stress and Displacement Distribution Characteristics in Similar Processes

Figure 10 compares the stress distribution characteristics of the iron core during similar processes. Comparing the two figures in Figure 10, it can be found that there is a difference in stress amplitude between the initial model and the scale model, but the stress distribution characteristics of the two are basically the same. Before and after similarity, the areas with greater stress on the iron core are distributed in the middle of the main column. The four points with the maximum force amplitude are distributed in the corner of the core window near the winding, similar to the analysis of magnetic field distribution. The stress threshold of the iron core in the scale model is roughly equal to the initial model, which is mainly affected by the magnetostrictive force of the magnetic domain under alternating magnetic fields. According to Formula (11), it is proportional to the square of the voltage. Since the voltage remains constant during similar processes, the force on the solid iron core remains unchanged. The distribution and amplitude change of the force on the core meet the similarity criterion [32].

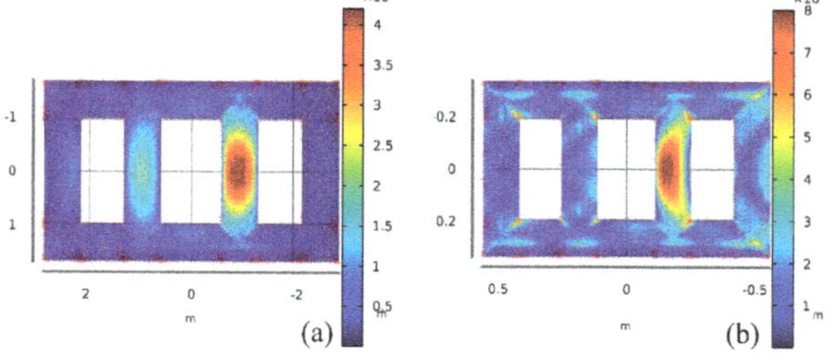

Figure 10. Stress distribution of (**a**) iron core of initial model and (**b**) scale model.

The converter transformer winding stress distribution is depicted in Figure 11. The stress distribution of the winding in the similar model is basically the same as the initial model, and the direction of action of the stress points is also the same. The stress amplitude of the winding in the scale model is about k^3 times that of the initial model. The amplitude of the numerical winding before and after the similarity keeps an appropriate corresponding relationship in the order of magnitude.

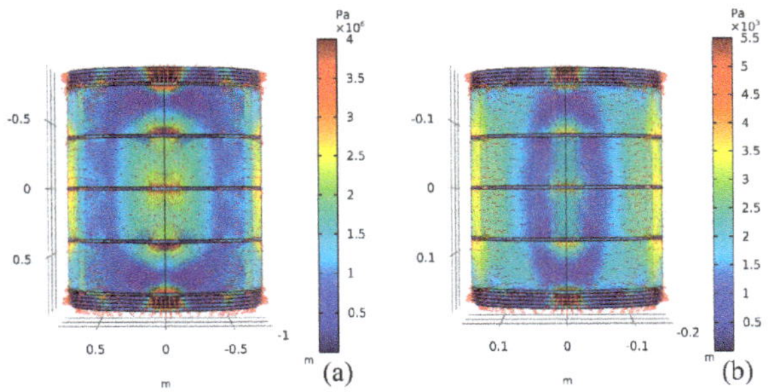

Figure 11. Force distribution of winding of (**a**) initial model and (**b**) scale model.

Figures 12 and 13 depict the displacement distribution characteristics of the initial model and the proportional model iron core and winding, respectively. Comparing the two figures in Figure 12, it can be found that the displacement distribution characteristics and deformation trend of the iron core do not change much during similar processes, and the change trend of displacement amplitude is the same as that of stress. The main contribution of core displacement comes from the magnetostrictive effect of the magnetic domains inside the core under alternating magnetic fields, which is proportional to the square of the voltage. During similar processes, this law remains unchanged.

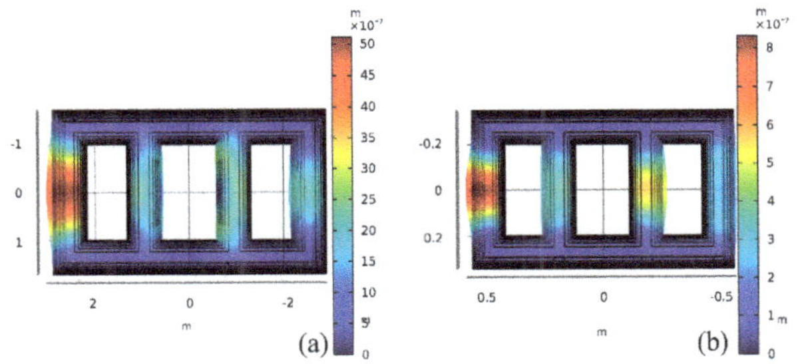

Figure 12. Core displacement distribution of (**a**) initial model and (**b**) scale model.

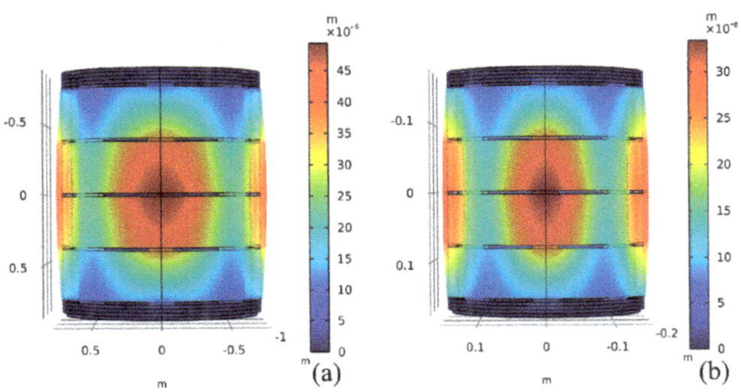

Figure 13. Winding displacement distribution of (**a**) initial model and (**b**) scale model.

From Figure 13, it can be seen that the deformation trend of the windings in the initial model and scale models is the same. The main contribution of winding displacement comes from the combined Lorentz force on electrons in an alternating magnetic field, and the leakage flux density depends on the current, so its variation pattern is the same as that of current in similar processes. Figure 13 shows a trend of outward stretching in both the original and scale models, showing maximum values of winding shape variables in the middle. The radial electromagnetic force of the winding exerts a stretching effect on it, causing its deformation direction to expand outward.

A hemispherical air domain with a radius of 10 m was added to the converter transformer model before and after scaling, and a perfect matching layer with a thickness of 1 m was set up to conduct frequency domain analysis of the sound field. As shown in Figure 14, before scaling, the sound pressure distribution of the converter transformer was between ±64 Pa, and after scaling, the sound pressure distribution was between ±33 Pa. The sound pressure frequencies were mainly 100, 200, and 300 Hz, and the 100 Hz component was the largest. The sound pressure distribution shows that the lateral sound pressure value is greater than the front and greater than the top. The maximum sound pressure levels before and after scaling appear at 68 dB and 23.2 dB, respectively. The sound pressure distribution shows attenuation outward along the center of the hemisphere, with the fastest attenuation above, indicating that the sound pressure distribution characteristics before and after scaling were normal.

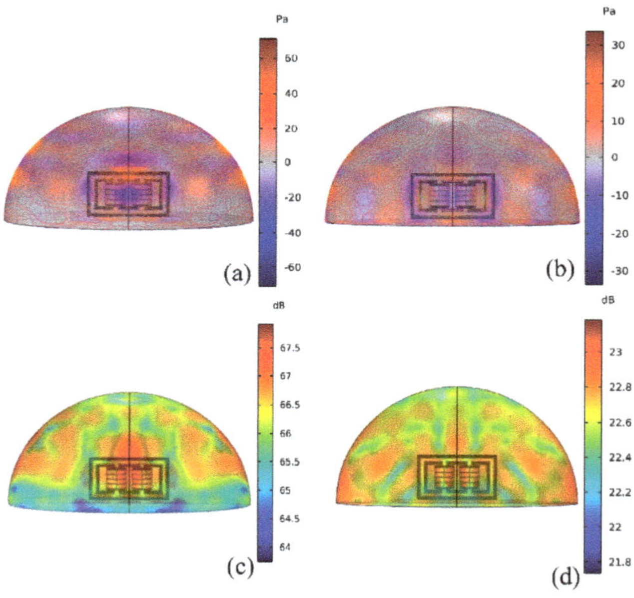

Figure 14. (**a**) Initial model sound pressure distribution; (**b**) Sound pressure distribution of Scale model; (**c**) Initial model sound pressure level distribution; (**d**) Scale model sound pressure level distribution.

5. Preparation of Scale Prototype

According to the similarity criterion, a scale prototype of converter transformer with a scale of 1:5 was designed and manufactured. The prototype was designed as single-phase double winding, on load voltage regulation, with a rated voltage ratio of 100/30 kV. The relevant design parameters are shown in Table 5.

Figure 15 shows the design drawing of the scale prototype converter transformer. The iron core diameter is 204/208 mm, and the design type is two main columns and two side columns. The six-step stacking method is adopted. The step amount is 9 mm, and the lamination form is 1, 3, 5, 2, 4, and 6. The casting structure is selected as the positioning method, and the casting ring/locating nail is directly connected with the clamp.

Table 5. Design parameters of scale prototype.

Parameter	Name/Value	Parameter	Name/Value
Model	ZZD-800/100-30	Rated voltage ratio	100/30 kV
Impedance	>2%	Impedance deviation	+2.5%
Temperature rise	Winding hot spot 78 K	Temperature rise	Hot spot of iron core and metal parts 80 K
Insulation level	Valve side neutral point LI60AC35 DC40	Insulation level	Meshwork side LI60AC35
Load loss	Pk ≤ 15 kW	No load loss	Po ≤ 2 kW Io ≤ 2%
Connection group	Ii0 (YNY0)	Noise level	55 dB (A)

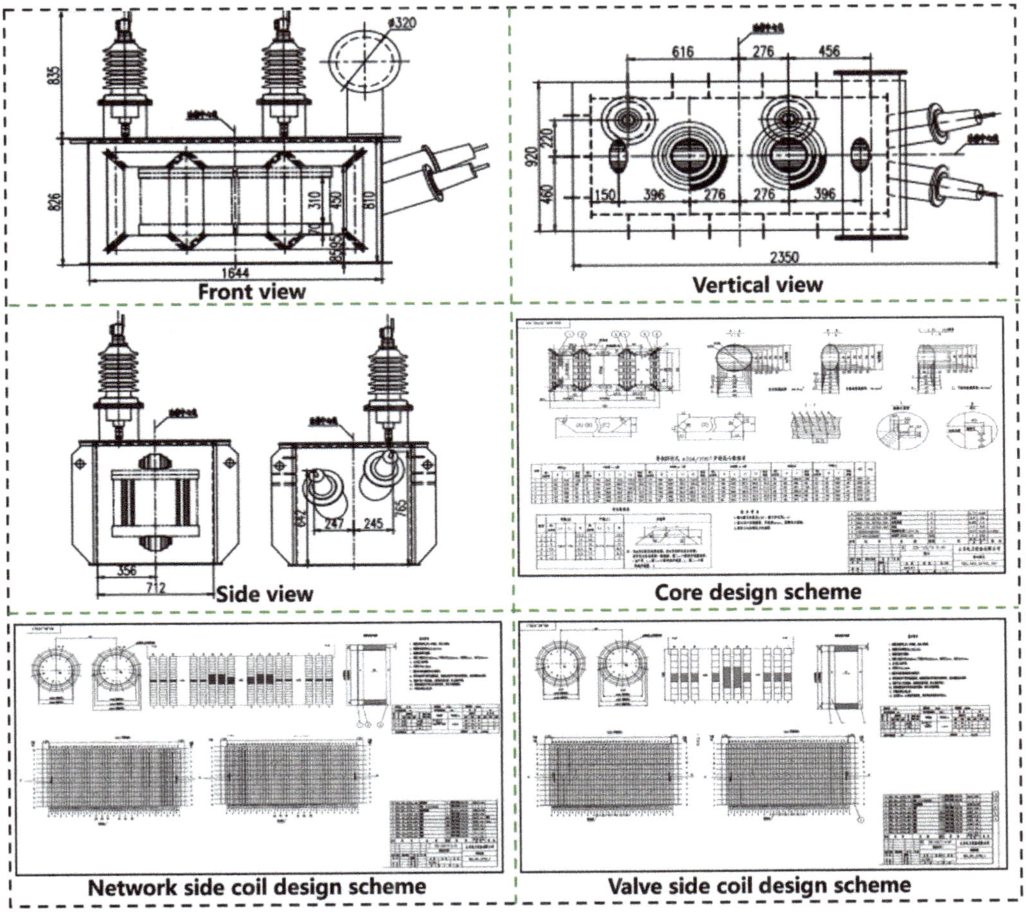

Figure 15. Design drawing of scale prototype.

After the reduction of the converter transformer, considering the spatial distribution, it is impractical to raise the number of winding turns to several. When designing the converter transformer scale prototype, we tried to use another way to remove the part of the equivalent winding that had not increased. Because the magnetic field distribution is mainly affected by the number of turns, namely Formulas (6) and (10), when other parameters were fixed, we replaced the silicon steel sheet material (30QG120) to approximately replace the effect of winding turns on *B*. The coil structure has been designed as a valve net, two-column type, with column 1 winding to the right and column 2 winding to the left. The inner diameter of the valve side coil is 260 mm, the outer diameter is 302 mm, the radial dimension is 21 mm, and the reactance height is 320 mm. The inner diameter of the coil at the grid side is 398 mm, the outer diameter is 454 mm, the radial size is 28 mm, and the reactance height is 320 mm, all of which are divided into 12 grades. The preparation process and physical images of the proportional prototype are shown in Figure 16a–c, and the corresponding converter transformer of the prototype is shown in operation in Figure 16d.

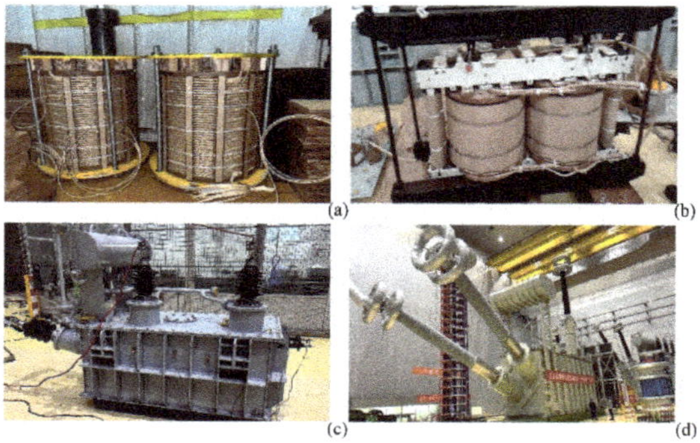

Figure 16. Prototype preparation and vibration characteristics testing: (**a**) Winding preparation; (**b**) Assembly of iron core winding; (**c**) Complete prototype appearance; (**d**) Refer to converter transformer.

Since the angle correlation between the vibration signal loss and its transmission time is stronger than that of the sensor, the measuring point was set as far as possible perpendicular to the sound source (winding and iron core). In view of the position of the low-voltage bushing above the oil tank and the distribution of iron core fasteners above the oil tank, only two measuring points above the oil tank, namely point 10 and point 11, were selected. Figure 17 depicts the distribution of measuring points of the converter transformer scale prototype.

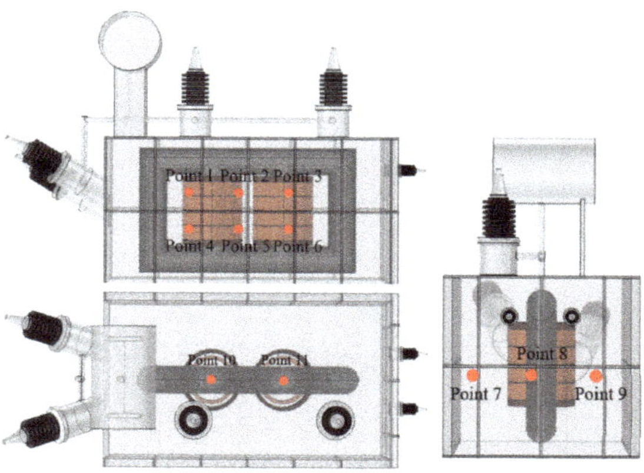

Figure 17. Distribution of measuring points.

The vibration signals of each measuring point of the prototype were processed by FFT, and the frequency domain diagrams of no-load and load vibration signals of each measuring point were obtained, as shown in Figure 18a,b. The frequency domain signals of the vibration acceleration at the corresponding points of the finite element proportional model under no-load and load conditions were extracted and are shown in Figure 18c,d, in order to compare and analyze the simulated and measured signals more intuitively.

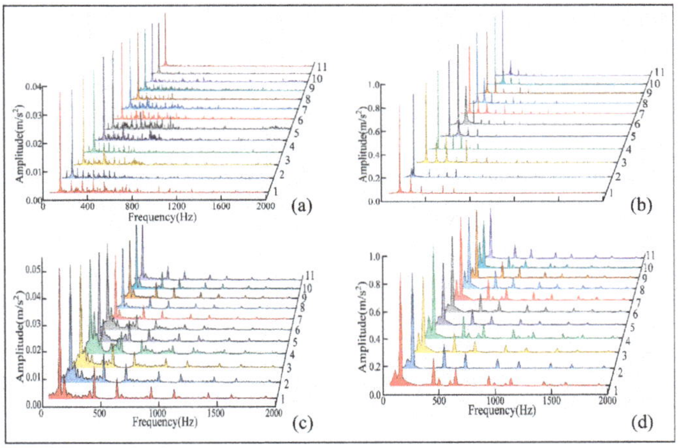

Figure 18. Frequency domain diagram of vibration signal of prototype and simulation model: (**a**) prototype no-load; (**b**) prototype load; (**c**) simulated no-load; (**d**) simulated load.

As shown in Figure 18, the vibration acceleration signal distribution of the finite element scale model based on the similarity principle was basically the same as the actual prepared scale prototype. Both no-load and load conditions, regardless of the main frequency or amplitude range, exhibited a certain degree of similarity. Among them, the similarity between the average acceleration amplitude of each measuring point under load and the simulation results exceeded 80%, and the similarity between the average acceleration amplitude of each measuring point under no load and the simulation results was close to 90%, proving that the standards proposed in this article are reliable for guiding product design and preparation. At the same time, sound pressure testing was conducted on the prototype of the scale prototype of the converter transformer, and the average sound pressure level was 24 dB, which is basically consistent with the scale simulation model, and proves that the criteria proposed in this paper are reliable for guiding the design and preparation of products.

6. Conclusions

1. Different from the traditional geometric similarity criteria in previous studies, the similarity criteria suitable for studying the vibration mechanism of converter transformers were derived by comprehensively considering the electromagnetic field and structural mechanics.

2. Considering the coupling effect of electric field, magnetic field, and solid mechanics, and the contribution of the core lamination mode, entangled–continuous–entangled winding structure, and gasket on vibration characteristics, a finite element model of a converter transformer was established.

3. The modal transformation laws of similar pre- and post-converter transformers were derived and verified, providing theoretical support for the study of resonance avoidance mechanisms based on scale models.

4. Based on the similarity criterion, a finite element similarity model of the converter transformer was established, and the electromagnetic and vibration characteristics of the

model during the similarity process were compared, verifying the reliability of the similarity criterion.

5. The proportional model prototype of the converter transformer was designed and prepared, and basic experimental projects and vibration characteristics analyses were conducted on the prototype. The results were compared with simulation research to further verify the reliability of the similarity criterion and provide a reference for the design of the proportional model prototype of high-voltage power equipment.

6. This scale model can serve as an experimental platform for studying the vibration mechanism of converter transformers, making it possible to analyze the vibration characteristics of internal components of high-power equipment in a laboratory environment. At the same time, more convenient testing and experimental scheme layout greatly saves time and computational costs, and has great engineering application value.

The vibration similarity model for converter transformers proposed in this paper can provide a reliable laboratory research platform for studying the mechanism and suppression of vibration and noise in converter transformers, it can save a lot of time and computational costs for conducting relevant vibration mechanism experiments, and has certain reference value for experimental research on proportional models of other power equipment.

Author Contributions: Conceptualization, L.Z. (Li Zhang); methodology, L.Z. (Liang Zou); software, L.Z. (Liang Zou); validation, Y.S.; formal analysis, H.W.; investigation, Y.S.; resources, L.Z. (Li Zhang); data curation, L.Z. (Li Zhang); writing—original draft preparation, H.W.; writing—review and editing, H.W.; visualization, Y.S.; supervision, L.Z. (Li Zhang); project administration, L.Z. (Li Zhang); funding acquisition, Y.S. All authors have read and agreed to the published version of the manuscript.

Funding: This work was supported by the Key R&D Program of Shandong Province (2021CXGC010210).

Institutional Review Board Statement: Not applicable.

Informed Consent Statement: Not applicable.

Data Availability Statement: Not applicable.

Conflicts of Interest: The authors declare that they have no known competing financial interests or personal relationships that could have appeared to influence the work reported in this paper.

References

1. Zhang, P.; Li, L.; Cheng, Z.; Tian, C.; Liu, Y. Simulation and experimental comparison of core vibration of shunt reactor and transformer model. *J. Electrotech.* **2018**, *33*, 9.
2. Makmud, M.Z.H. Optimal Design of Corona Ring for 132 kV Insulator at High Voltage Transmission Lines Based on Optimisation Techniques. *Energies* **2023**, *16*, 778.
3. Kell, D. HVDC to Grow Rapidly. *Nat. Gas Electr.* **2015**, *31*, 11–18. [CrossRef]
4. Ghilman, R.; Al-Majali, H.D. Performance analysis of multi-level high voltage direct current converter. *Int. J. Electr. Comput. Eng.* **2022**, *12*, 1368–1376.
5. Mohan, M. A comprehensive review of DC fault protection methods in HVDC transmission systems. *Prot. Control Mod. Power Syst.* **2021**, *6*, 1–20.
6. Yang, B.; Liu, B.; Zhou, H.; Wang, J.; Yao, W.; Wu, S.; Shu, H.; Ren, Y. A critical survey of technologies of large offshore wind farm integration: Summary, advances, and perspectives. *Prot. Control Mod. Power Syst.* **2022**, *7*, 233–264. [CrossRef]
7. Husin, H.; Zaki, M. A critical review of the integration of renewable energy sources with various technologies. *Prot. Control Mod. Power Syst.* **2021**, *6*, 37–54.
8. Acarolu, H.; Márquez, G.; Pedro, F. High voltage direct current systems through submarine cables for offshore wind farms: A life-cycle cost analysis with voltage source converters for bulk power transmission. *Energy* **2022**, *249*, 123713. [CrossRef]
9. Zhao, Y.; Crossley, P. Impact of dc bias on differential protection of converter transformers. *Int. J. Electr. Power Energy Syst.* **2020**, *115*, 105426.1–105426.10. [CrossRef]
10. Liu, X.; Wu, J.; Jiang, F.; Wang, Y.; Zhang, C.; Hui, Y. Electromagneto-mechanical numerical analysis and experiment of transformer influenced by DC bias considering core magnetostriction. *J. Mater. Sci. Mater. Electron.* **2020**, *31*, 16420–16428. [CrossRef]
11. Weiser, B.; Pfützner, H.; Anger, J. Relevance of magnetostriction and forces for the generation of audible noise of transformer cores. *IEEE Trans. Magn.* **2000**, *36*, 3759–3777. [CrossRef]

12. Jiang, P.; Zhang, Z.; Dong, Z.; Wu, Y.; Xiao, R.; Deng, J.; Pan, Z. Research on distribution characteristics of vibration signals of ±500 kV HVDC converter transformer winding based on load test. *Int. J. Electr. Power Energy Syst.* **2021**, *132*, 107200. [CrossRef]
13. Weiser, B.; Hasenzagl, A.; Booth, T.; Pfützner, H. Mechanisms of noise generation of model transformer cores. *J. Magn. Magn. Mater.* **1996**, *160*, 207–209. [CrossRef]
14. Bilgundi, S.K.; Sachin, R.; Pradeepa, H.; Nagesh, H.B.; Kumar, M.V.L. Grid power quality enhancement using an ANFIS optimized PI controller for DG. *Prot. Control Mod. Power Syst.* **2022**, *7*, 26–39. [CrossRef]
15. Kitagawa, W.; Ishihara, Y.; Todaka, T.; Nakasaka, A. Analysis of structural deformation and vibration of a transformer core by using magnetic property of magnetostriction. *Electr. Eng. Jpn.* **2010**, *172*, 19–26. [CrossRef]
16. He, J.; Chen, K.; Li, M.; Luo, Y.; Liang, C.; Xu, Y. Review of protection and fault handling for a flexible DC grid. *Prot. Control Mod. Power Syst.* **2020**, *5*, 151–165. [CrossRef]
17. Guillod, T.; Kolar, J.W. Medium-frequency transformer scaling laws: Derivation, verification, and critical analysis. *CPSS Trans. Power Electron. Appl.* **2020**, *5*, 18–33. [CrossRef]
18. Zhang, P.; Li, L.; Cheng, Z.; Tian, C.; Han, Y. Study on vibration of iron core of transformer and reactor based on maxwell stress and anisotropic magnetostriction. *IEEE Trans. Magn.* **2019**, *55*, 9400205. [CrossRef]
19. Stephen, A.S.; Ross, C. Scale model studies of AC substation electric fields. *IEEE Trans. Power Appar. Syst.* **1979**, *98*, 926–939.
20. Brubaker, M.A.; Lindgren, S.R.; Frimpong, G.; Walden, J.M. Streaming electrification measurements in a 1/4-scale transformer model. *IEEE Trans. Power Deliv.* **1999**, *14*, 978–985. [CrossRef]
21. Piantini, A.; Janiszewski, J.M.; Borghetti, A.; Nucci, C.A.; Paolone, M. A Scale Model for the Study of the LEMP Response of Complex Power Distribution Networks. *IEEE Trans. Power Deliv.* **2007**, *22*, 710–720. [CrossRef]
22. Li, Q.; Wang, X.; Zhang, L.; Lou, J.; Zou, J. Modelling methodology for transformer core vibrations based on the magnetostrictive proper-ties. *IET Electr. Power Appl.* **2012**, *6*, 604–610. [CrossRef]
23. Zou, L.; Gong, P.; Zhang, L.; Zhao, T.; Li, Q. Scale reduction test and model simplification of space magnetic field of dry-type air core reactor. *High Volt. Technol.* **2014**, *40*, 8.
24. Pu, Z.H.; Ruan, J.J.; Du, Z.Y.; Zhang, Y.D. Analysis of voltage distribution characteristics in hvdc converter transformer winding based on the reduced-scale model. *IEEE Trans. Magn.* **2014**, *50*, 1–5. [CrossRef]
25. Zhang, X.; Liu, X.; Guo, F.; Xiao, G.; Wang, P. Calculation of DC Bias Reactive Power Loss of Converter Transformer via Finite Element Analysis. *IEEE Trans. Power Deliv.* **2020**, *36*, 751–759. [CrossRef]
26. Yao, Y.; Koh, C.S.; Ni, G.; Xie, D. 3-D nonlinear transient eddy current calculation of online power transformer under DC bias. *IEEE Trans. Magn.* **2005**, *41*, 1840–1843.
27. Wang, T.; Xie, Q.; Zhang, Y.; Du, Z.; Ruan, J.; Tan, D.; Zhu, L. Research on scaling relationship of power transformer based on similarity theory. *New Technol. Electr. Energy* **2015**, *34*, 5.
28. Moses, A.J.; Anderson, P.I.; Phophongviwat, T. Localized Surface Vibration and Acoustic Noise Emitted from Laboratory-Scale Transformer Cores Assembled from Grain-Oriented Electrical Steel. *IEEE Trans. Magn.* **2016**, *52*, 1–15. [CrossRef]
29. Shilyashki, G.; Pfützner, H.; Hamberger, P.; Aigner, M.; Kenov, A.; Matkovic, I. Spatial distributions of magnetostriction, displacements and noise generation of model transformer cores. *Int. J. Mech. Sci.* **2016**, *118*, 188–194. [CrossRef]
30. Garcia, B.; Burgos, J.C.; Alonso, A. Transformer Tank Vibration Modeling as a Method of Detecting Winding Deformations-Part I: Theoretical Foundation. *IEEE Trans. Power Deliv.* **2006**, *21*, 157–163. [CrossRef]
31. Baravati, P.R.; Moazzami, M.; Hosseini, S.M.H.; Mirzaei, H.R.; Fani, B. Achieving the exact equivalent circuit of a large-scale transformer winding using an improved detailed model for partial discharge study. *Int. J. Electr. Power Energy Syst.* **2022**, *134*, 107451. [CrossRef]
32. Gao, Y.; Muramatsu, K.; Hatim, M.J.; Fujiwara, K.; Ishihara, Y.; Fukuchi, S.; Takahata, T. Design of a Reactor Driven by Inverter Power Supply to Reduce the Noise Considering Electromagnetism and Magnetostriction. *IEEE Trans. Magn.* **2010**, *46*, 2179–2182. [CrossRef]

energies

MDPI

Article

Generalized Regression Neural Network Based Meta-Heuristic Algorithms for Parameter Identification of Proton Exchange Membrane Fuel Cell

Peng He [1], Xin Zhou [1], Mingqun Liu [1], Kewei Xu [1], Xian Meng [1] and Bo Yang [2,*]

[1] Electric Power Science Institute, Yunnan Power Grid Co., Ltd., Kunming 650000, China; hepeng08@yn.csg.cn (P.H.); zhouxin@dlyjy.yn.csg.cn (X.Z.); lmq1997@tom.com (M.L.); kewei_xu@outlook.com (K.X.); fxz4939@163.com (X.M.)

[2] Faculty of Electric Power Engineering, Kunming University of Science and Technology, Kunming 650500, China

* Correspondence: yangbo_ac@outlook.com

Abstract: An accurate parameter extraction of the proton exchange membrane fuel cell (PEMFC) is crucial for establishing a reliable cell model, which is also of great significance for subsequent research on the PEMFC. However, because the parameter identification of the PEMFC is a nonlinear optimization problem with multiple variables, peaks, and a strong coupling, it is difficult to solve this problem using traditional numerical methods. Furthermore, because of insufficient current and voltage data measured by the PEMFC, the precision rate of cell parameter extraction is also very low. The study proposes a parameter extraction method using a generalized regression neural network (GRNN) and meta-heuristic algorithms (MhAs). First of all, a GRNN is used to de-noise and predict the data to solve the problems in the field of PEMFC, which include insufficient data and excessive noise data of the measured data. After that, six typical algorithms are used to extract the parameters of the PEMFC under three operating conditions, namely high temperature and low pressure (HTLP), medium temperature and medium pressure (MTMP), and low temperature and high pressure (LTHP). The last results demonstrate that the application of GRNN can prominently decrease the influence of data noise on parameter identification, and after data prediction, it can greatly enhance the precision rate and reliability of MhAs parameter identification, specifically, under HTLP conditions, the *V-I* fitting accuracy achieved 99.39%, the fitting accuracy was 99.07% on MTMP, and the fitting accuracy was 98.70%.

Keywords: PEMFC; GRNN; MhAs; parameter identification; data processing; HTLP; MTMP; LTHP

Citation: He, P.; Zhou, X.; Liu, M.; Xu, K.; Meng, X.; Yang, B. Generalized Regression Neural Network Based Meta-Heuristic Algorithms for Parameter Identification of Proton Exchange Membrane Fuel Cell. *Energies* **2023**, *16*, 5290. https://doi.org/10.3390/en16145290

Academic Editor: Djaffar Ould-Abdeslam

Received: 5 June 2023
Revised: 27 June 2023
Accepted: 6 July 2023
Published: 10 July 2023

1. Introduction

With the rapid development of technology and continuous economic growth, the demand for various fossil fuels and electricity is increasing day by day. The existing problem is that the energy conversion efficiency of traditional fossil energy is relatively low, and it causes huge environmental pollution, bringing the greenhouse effect, rising sea levels, acid rain, and other thorny environmental problems [1,2]. In addition, the massive development and utilization of traditional non-renewable energy will also cause the global energy crisis [3]. In this context, countries around the world have begun to vigorously develop clean energy and renewable energy. The proton exchange membrane fuel cell (PEMFC) is widely used because of the advantages of high energy density, high power generation efficiency, starting at a low temperature and a long working life [4,5].

With the widespread application of the PEMFC, precise modeling of batteries is crucial for optimizing the control of cell systems and improving cell power generation efficiency. Currently, there are many models for the PEMFC, including three-dimensional steady-state models [6] and electrochemical steady-state models [7]. Among them, electrochemical stability models can predict the state of batteries very well, it is beneficial for the safe and stable operation of the PEMFC and cell management. Accurate cell models rely on accurate internal model parameters, so accurate parameter identification of PEMFC batteries is a prerequisite for establishing accurate and reliable cell models. However, because the PEMFC parameter identification is a nonlinear problem with multiple variables, multiple peaks, a strong coupling, and limited *V-I* data measured by the cell, it is arduous to use traditional numerical analysis methods for parameter identification, for example, the least squares method, gradient descent method, and the identification results are not ideal [8]. However, meta-heuristic algorithms (MhAs) are widely used in the field of PEMFC parameter extraction due to their low initial value requirements and global search ability, which can avoid falling into local optimum [9].

A study [10] proposed an identification method based on adaptive focusing particle swarm optimization (AFPSO). Compared with particle swarm optimization (PSO), AFPSO has a stronger global search capability and faster optimization speed. Final experimental results also demonstrate that the obtained results have high fitting accuracy with the data obtained from experimental testing, and can effectively identify the parameters of the cell. In work [11], a PEMFC parameter estimation study based on an extended Kalman filter (EKF) was proposed. By constructing a semi-mechanistic and semi-empirical PEMFC model, based on the characteristics of the existing sensor signals in the model system, the EKF was used to extract parameters of the cell model. Specifically, this method estimates the parameters during PEMFC off-design operation, and it is more in line with the actual application of batteries. Work [12] uses an extreme learning machine (ELM) to identify cell parameters, where under the actual operating conditions, the measured current and voltage data will inevitably have exception values, that is, noise data, which will affect the identification accuracy of parameters in the model. Therefore, in this study, it is proposed to use ELM to train the data, then perform noise reduction processing, then use algorithms for parameter identification. The results obtained also prove that the data after noise reduction processing is used for parameter extraction, and the identification accuracy is significantly improved. In the literature [13], a PEMFC parameter identification study of improved chicken swarm optimization (ICSO) was proposed. In this paper, the author introduced a Tent mapping strategy to initialize the population, which can improve the uniformity and ergodicity of the population. Secondly, set adaptive inertia weights on the feeding speed of individual chickens, which can improve the optimization efficiency of individual hens, and the Levy flight strategy were introduced to randomly update the chicken position, greatly improving the algorithm's global search ability. Finally, by comparing the parameter identification results obtained by ICSO with those obtained by other heuristic algorithms, it was proven that the ICSO algorithm has better parameter identification accuracy and a stronger model generalization ability. Reference [14] proposed an improved method based on a differential evolution algorithm, which is unique in that it references a probability selection model, which assigns a selection probability for every individual in the evolutionary population regarding their performance. In the work, to verify the effectiveness of algorithm, standard test functions were also used for testing. The experimental results showed that after the algorithm improvement, high data fitting accuracy can be achieved, and the parameters of the cell can be identified very accurately. In literature [15], a novel method based on the Levenberg Marquardt backpropagation (LMBP) algorithm was proposed. The neural network was designed based on the PEMFC model, and the LMBP algorithm was used for parameter identification. The LMBP is a variant of the Newton method, which combines the steepest descent method with the Gaussian Newton method and iteratively calculates using the Jacobian matrix, greatly improving computational efficiency. The final experimental results in the study also indicate that neural networks have higher fitting accuracy compared to heuristic algorithms, and the speed of parameter identification research through LMBP is much faster than that of heuristic algorithms. SOA is a swarm intelligence optimization algorithm that simulates the random search behavior of human beings. The SOA algorithm optimizes the parameters of the PEMFC model, and then compares the results with those of other algorithms, proving that the algorithm has good fitting accuracy, and it can significantly improve and enhance the accuracy of the PEMFC model parameters [16]. In reference [17], a PEMFC parameter identification method based on Bayesian regularization neural network (BRNN) was proposed. BRNN is used to de-noise data and MhAs are used to identify parameters, and the results are compared with other heuristic algorithms. The extraction results of BRNN data de-noising are more accurate than the original data results, and the results obtained are more stable with fewer outliers.

Overall, current research on PEMFC parameter identification mainly utilizes the MhAs method [18–20], and most of the research focuses on algorithm improvement to improve the accuracy and speed of parameter extraction. Only a few studies consider the impact of the data itself on the identification results. However, the study proposes MhAs based on a generalized regression neural network (GRNN) for PEMFC parameter extraction, which trains the GRNN, predicting and de-noising the data, fully considering the insufficient measured data and the impact of noise data on the final identification results, and conducting parameter identification research on the PEMFC under three operating conditions, namely high temperature and low pressure (HTLP), medium temperature and medium pressure (MTMP), and low temperature and high pressure (LTHP) [17]. The last results demonstrate that after data processing, its identification accuracy is higher and its performance is better. This study provides a new approach to the identification of PEMFC parameters, and its contributions and innovations can be summarized as follows:

1. Established the PEMFC model and conducted parameter identification research on the model under three operating conditions;
2. Considering the influence of insufficient data volume and noise data, a GRNN was used to de-noise and predict the measured *V-I* data, and the final results fully demonstrate its excellent robustness when applied to PEMFC parameter extraction under various operation conditions;
3. Based on the data processed by a GRNN, six typical heuristic algorithms were compared for their effectiveness in PEMFC parameter identification. The results demonstrate that after data processing, accuracy can be greatly improved.

The structure of the remaining part is as follows: Section 2 is the modeling of the PEMFC, mainly introducing the internal chemical mechanism of PEMFC power generation and its cell model, and then establishing an objective function for the model. Section 3 mainly displays the application of GRNN-MhAs in PEMFC parameter identification research, which involves using a GRNN for data de-noising and prediction processing, and then using MhAs for parameter identification. Section 4 mainly displays the parameter identification results obtained by six algorithms under three working conditions. Section 5 is the discussion section. Section 6 provides some important conclusions obtained from this research, as well as some prospects for future PEMFC parameter identification research.

2. PEMFC Modeling

Establishing the PEMFC model is beneficial to conduct in-depth research on the parameter identification of a cell. This section mainly introduces the basic principles and mathematical models of the mechanism of the PEMFC.

2.1. The Mechanism of the PEMFC

In principle, the PEMFC is equivalent to a reverse device for water electrolysis. A typical PEMFC is composed of an anode, a cathode, and a proton exchange membrane. The anode is the site of hydrogen fuel oxidation, the cathode is the site of oxidant reduction, and both poles contain catalysts to accelerate electrode electrochemical reactions [21–23].

In addition, the electrochemical reaction mechanism of the PEMFC is shown in Equations (1)–(3) [24].

Anode side:

$$H_2 \rightarrow 2H^+ + 2e^- \tag{1}$$

Cathode side:

$$\frac{1}{2}O_2 + 2H^+ + 2e^- \rightarrow H_2O \tag{2}$$

Overall chemical reaction,

$$H_2 + \frac{1}{2}O_2 \rightarrow H_2O \tag{3}$$

2.2. Mathematical Model of the PEMFC

The model introduced in this section is only one kind of cell model, namely the 0-D model. Note that many other multi-dimensional models exist. Considering the impact of some losses in electrochemical reactions on the output characteristics of the PEMFC, the output voltage is as follows [25]:

$$V_{est} = E_{nernst} - V_{act} - V_{ohm} - V_{con} \tag{4}$$

where V_{act}, V_{ohm} and V_{con}, respectively, represent activation voltage loss (V), ohmic voltage loss (V), and concentration voltage loss (V); E_{nernst} is the thermodynamic electromotive force (V); E_{nernst} can be expressed as [26]:

$$E_{nernst} = \frac{\Delta G}{2F} + \frac{\Delta S}{2F}(T_k - T_{ref}) + \frac{RT}{2F}\left[\ln(P_{H_2}) + \frac{1}{2}(P_{O_2})\right] \tag{5}$$

where ΔG and ΔS represent changes in free Gibbs energy and entropy, respectively, the value of ΔG is 228,170 J/mol; F represents a constant (96,485.3383 C/mol); R is the universal gas constant (8.314 J/(K·mol)); T_k and T_{ref}, respectively, represent the actual temperature and reference temperature; T_k has a value of 353.15 K under HTLP operating conditions, 333.15 K under MTMP operating conditions, and 313.15 K under operating conditions; P_{H_2} and P_{O_2} denote the partial pressure of hydrogen (atm) and oxygen (atm), which can be expressed as [27]:

$$P_{H_2} = 0.5 \times RH_a \times P_{H_2O}^{sat} \times \left[\left(\frac{RH_a \times P_{H_2O}^{sat}}{P_a} \times \exp\left(\frac{1.635\left(\frac{i_{cell}}{A}\right)}{T_k^{1.334}}\right)\right)^{-1} - 1\right] \tag{6}$$

$$P_{O_2} = RH_c \times P_{H_2O}^{sat} \times \left[\left(\frac{RH_a \times P_{H_2O}^{sat}}{P_c} \times \exp\left(\frac{4.192\left(\frac{i_{cell}}{A}\right)}{T_k^{1.334}}\right)\right)^{-1} - 1\right] \tag{7}$$

where RH_a and RH_c are the relative humidity of the vapor, the values of RH_a and RH_c are both 1 under HTLP operating conditions, 2 under MTMP operating conditions, and 3 under operating conditions; P_a and P_c the inlet pressure of the anode and cathode (atm), respectively; i_{cell} is the output current (A); A is the effective activation area, the value of ΔG is 50.6 cm^2; $P_{H_2O}^{sat}$ is the saturation pressure (atm), which is as follows:

$$\log_{10}\left(P_{H_2O}^{sat}\right) = 2.95 \times 10^{-2} \times T_c - 9.19 \times 10^{-5} \times T_c^2 + 1.44 \times 10^{-7} \times T_c^3 - 2.18 \tag{8}$$

$$T_c = T_k - 273.15 \tag{9}$$

In addition, the activation voltage loss V_{act} can be expressed as:

$$V_{act} = \varepsilon_1 + \varepsilon_2 T_k + \varepsilon_3 T_k \ln(C_{O_2}) + \varepsilon_4 T_k \ln(i_{cell}) \tag{10}$$

where ε_1, ε_2, ε_3, and ε_4 are semi-empirical coefficients; C_{O_2} denote the concentration of oxygen catalyzed by the anode catalyst surface (mol/cm^3), which is shown below:

$$C_{O_2} = \frac{P_{O_2}}{5.08 \times 10^6 \times e^{\left(\frac{-498}{T_k}\right)}} \tag{11}$$

In addition, the ohmic voltage V_{ohm} loss is as follows [28]:

$$V_{ohm} = i_{cell}(R_m + R_c) \tag{12}$$

where R_m and R_c are the electron transfer resistance and proton exchange membrane equivalent resistance (Ω), R_m can be expressed as:

$$R_m = \rho_m \left(\frac{l}{A}\right) \tag{13}$$

where l is the thickness of the proton exchange membrane, the value of l is 178 μm; ρ_m represents the resistivity (Ω·cm), which can be expressed as:

$$\rho_m = \frac{181.6 \times \left[1 + 0.03 \times \left(\frac{i_{cell}}{A}\right) + 0.062 \times \left(\frac{T_k}{303}\right)^2 \left(\frac{i_{cell}}{A}\right)^{2.5}\right]}{\left[\lambda - 0.643 - 3 \times \left(\frac{i_{cell}}{A}\right)\right] \exp\left[4.18 \times \left(\frac{T_k - 303}{T_k}\right)\right]} \tag{14}$$

where λ is the water content.

In addition, the concentration voltage V_{con} loss can be expressed as:

$$V_{con} = -b \ln\left(\ln \frac{J}{A \times J_{max}}\right) \tag{15}$$

where b is the parameter coefficient (V); J is the current density (A/cm^2); J_{max} is the maximum current density, the value of J_{max} is 1.5 A/cm^2.

Finally, it is clear from Equations (4)–(15) that the PEMFC needs to identify seven unknown parameters, namely ε_1, ε_2, ε_3, ε_4, λ, R_c, b.

2.3. Objective Function

This study utilizes RMSE to measure the accuracy of extraction results. It can effectively reflect the accuracy of the calculated value, that is, the degree of deviation between the calculated value and the actual value. Therefore, RMSE is defined as the objective function, as follows:

$$\text{RMSE}(x) = \sqrt{\frac{1}{N}\sum_{i=1}^{N}[V_{act}(i) - V_{est}(i)]^2}, x = [\varepsilon_1, \varepsilon_2, \varepsilon_3, \varepsilon_4, \lambda, R_c, b] \tag{16}$$

where N is the quantity of data; V_{act} and V_{est} represent the measured voltage and calculated voltage.

Furthermore, the constraints of key parameters are as follows:

$$\text{s.t.}\begin{cases} \varepsilon_{i,min} \le \varepsilon_i \le \varepsilon_{i,max} \\ \lambda_{min} \le \lambda \le \lambda_{max} \\ R_{c,min} \le R_c \le R_{c,max} \\ b_{min} \le b \le b_{max} \end{cases}, \forall i \in \{1,2,3,4\} \tag{17}$$

3. GRNN-MhAs for PEMFC Parameter Extraction

3.1. Principle of GRNN

A GRNN is a special form of nonlinear regression feedforward neural network, belonging to the branch of radial basis function (RBF). GRNN is based on non-parametric regression and follows the principle of maximum probability to obtain the network output [29]. The GRNN model inherits the good nonlinear approximation function of RBF neural network. The algorithm of the GRNN model has a fast convergence speed, a small amount of calculation, and can be well handled in the face of fewer training samples. It has been widely applied in structural analysis, control decision-making, system identification, and other aspects, especially in dealing with a curve fitting.

As shown in Figure 1, the GRNN model consists of four function layers, namely the input layer, pattern layer, summation layer, and output layer [29]. The network input $X = [x_1, x_2, \ldots, x_n]^T$, and its output is $Y = [y_1, y_2, \ldots, y_k]^T$.

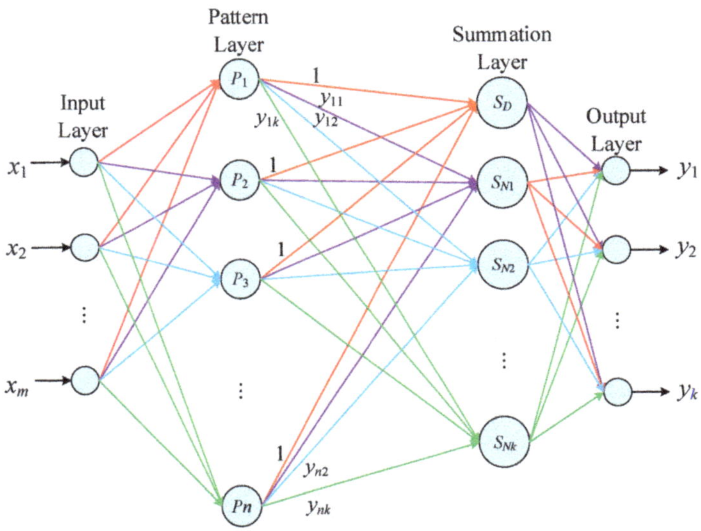

Figure 1. Structure diagram of the GRNN.

The GRNN adopts the idea of nonlinear regression analysis. Let x, y be random variables, let X be the real observation value, $g(x, y)$ be the joint probability density function, and the regression of y for x is determined by the following Equation (18):

$$E(y|X) = y(X) = \frac{\int_{-\infty}^{+\infty} y g(X, y) \mathrm{d}y}{\int_{-\infty}^{+\infty} g(X, y) \mathrm{d}y} \quad (18)$$

The function $g(x, y)$ can be obtained by nonparametric estimation of the observation samples of x and y, as the Equations (19) and (20) show:

$$g(X, y) = \frac{1}{n(2\pi)^{\frac{p+1}{2}} \sigma^{p+1}} \sum_{i=1}^{n} \exp[-d(X, x_i)] * \exp\left[-(y - y_i)^2\right] \quad (19)$$

$$d(X, x_i) = \sum_{j=1}^{p} \left[\frac{x_{0j} - x_{ij}}{\sigma}\right]^2 \quad (20)$$

where σ is called the smoothing factor. Bring Equations (19) and (20) into Equation (18), and because $\int_{-\infty}^{+\infty} x e^{-x^2} \mathrm{d}x = 0$, simplifying Equation (18) can be shown as follows:

$$y(X) = \frac{\sum_{i=1}^{n} y \exp\left[-(y - y_i)^2\right]}{\sum_{i=1}^{n} \exp[-d(X, x_i)]} \quad (21)$$

Obviously, in Equation (21) above, when the input training samples are determined, the training of neural networks is essential to determine the smoothness factor σ. The process only requires adjusting the smoothness factor σ to change the transfer function.

Based on the above principles, the basic operation process of a GRNN is as follows [30]:

Step 1 Input Layer: The number of neurons is equal to the dimension m of the input vector $X = [x_1, x_2, \ldots, x_n]^{\mathrm{T}}$ in the learning sample, and directly transfers the input variables to the pattern layer.

Step 2 Model Layer: The number of neurons in the model layer is equal to the number of learning samples n, and the neurons correspond to the learning samples one by one. Assuming a function which is shown in the following formula:

$$D_i^2 = (X - X_i)^{\mathrm{T}} * (X - X_i) \tag{22}$$

where D_i^2 represents the square of the Euclid distance between the input variable of the i-th neuron and its learning sample X. In mode layer, Gaussian function is chosen as the activation kernel function, and the transfer function can be expressed as:

$$p_i = \exp\left[-\frac{D_i^2}{2\sigma^2}\right],\ i = 1, 2, \ldots, n \tag{23}$$

where σ is a smoothing parameter.

Step 3 Summation Layer: about the GRNN, two types of neurons are used for summation in the summation layer. Among them, the first type corresponds to the dimension k of the output vector, with a total of k nodes. The connection weight between the i-th neuron in the pattern layer and the j-th molecular summation neuron in the summation layer is the j-th element of the output sample Y_i, there is a transfer function as follows:

$$S_{Nj} = \sum_{i=1}^{n} y_{ij} p_i,\ j = 1, 2, \ldots, k \tag{24}$$

The second type only has one neuron S_D. Perform arithmetic summation on all neurons in the pattern layer, and another transfer function is as follows:

$$S_D = \sum_{i=1}^{n} p_i \tag{25}$$

Step 4 Output Layer: Each neuron divides the output of the summation layer, and the output of the j-th neuron corresponds to the estimation result $\hat{Y}(X)$ is as follows:

$$y_j = \frac{S_{Nj}}{S_D},\ j = 1, 2, \ldots, k \tag{26}$$

In summary, in the training process of the GRNN, only the smoothing parameters need to be adjusted σ to change the transfer function to obtain regression estimates.

3.2. Parameter Extraction Process

The conventional process for parameter identification of PEMFC based on a GRNN and MhAs is mainly divided into three sectors: data collection, data preprocessing, and optimization parameter extraction, and the specific process is shown in Figure 2.

The concrete process can be expressed as follows: collect actual cell voltage and current data, and then the GRNN model is trained for data prediction and data noise reduction to obtain the predicted data and de-noising data. Finally, six heuristic algorithms were used to optimize and iterate the PEMFC data, and the final parameter identification results were obtained. Note that this work uses RMSE to measure the size of error, and the steps are shown in Figure 3.

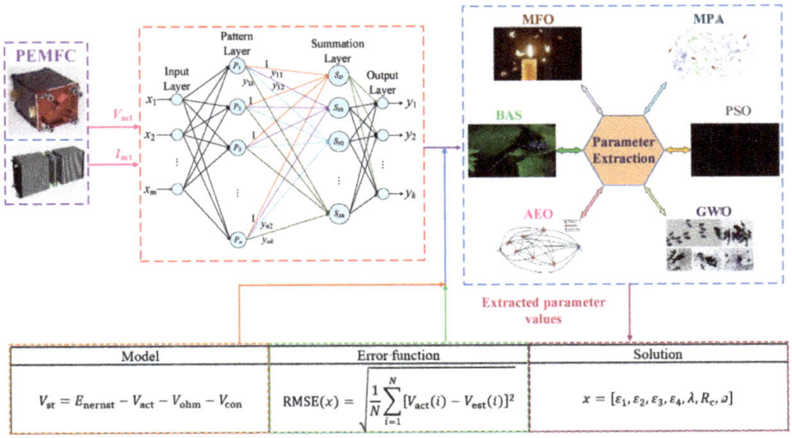

Figure 2. Schematic diagram of PEMFC parameter identification structure.

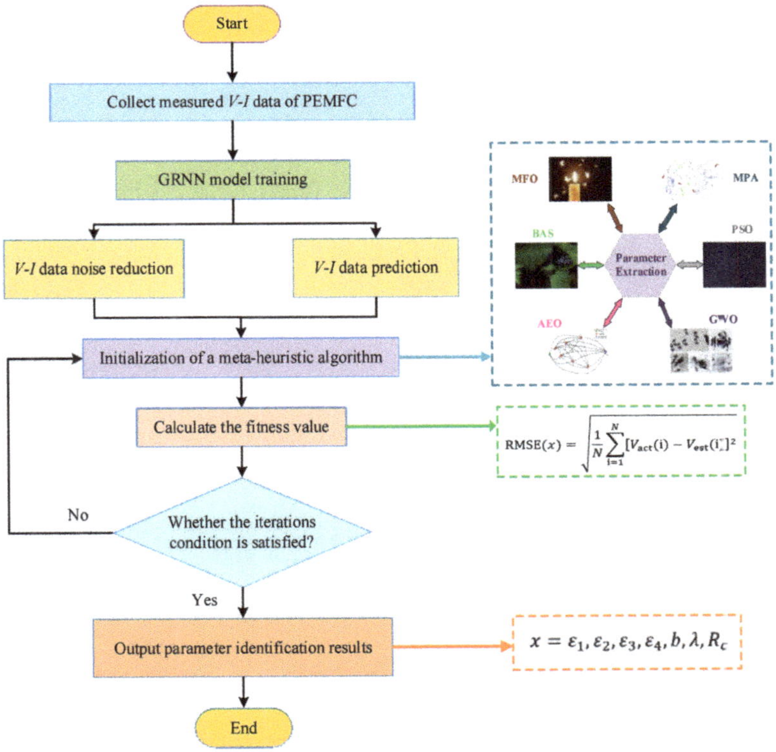

Figure 3. The extraction steps of GRNN-MhAs for the PEMFC.

4. Case Studies

In this part, a GRNN and six typical MhAs were used to extract the parameters of the PEMFC model, respectively, moth fire optimization (MFO) [31], PSO [32], beetle antennae search (BAS) [33], grey wolf optimization (GWO) [34], marine predator algorithm (MPA) [35], and artificial ecosystem-based optimization (AEO). Then, the operating conditions were set up according to the actual working conditions of the cell, namely HTLP,

MTMP, and LTHP. Due to the phenomenon of noise and insufficient available data, a GRNN was used to preprocess, de-noise and predict the 25 pairs of current and voltage data extracted from the cell. Finally, 145 sets of data were predicted and used for parameter identification research under multi-data. In this study, the parameters of PEMFC are shown in Table 1.

Table 1. Model and algorithms parameter settings.

Types	Parameters	Value
PEMFC	PEM effective area	50.6 cm^2
	PEM thickness	178 μm^2
Algorithms	Maximum iterations	500
	Run times	10

Remark 1. *The cell data in this study comes from experimental data provided by the cell manufacturer. The reason why a GRNN is used to process the V-I data of the PEMFC in this research is due to the inevitable impact of noise data in the measurement data. In addition, due to the loss of measured data, to verify the robustness of the GRNN applied to PEMFC parameter recognition, as well as the difficulty in measuring the V-I data of the PEMFC during actual operation, and due to battery aging and other phenomena, the difference between the measured data and the data from the battery factory is significant, which has a significant impact on the final parameter identification results.*

4.1. GRNN for V-I Data Preprocessing

4.1.1. GRNN for V-I Data De-noising

Small fluctuations may affect experimental data, similarly, the PEMFC is inevitably affected by noise when used in different environments. Undoubtedly, irregular changes in multiple variables can affect the inaccurate parameter identification of the PEMFC.

Therefore, to minimize the effect of the noise condition on the accuracy of the calculation results as much as possible, this paper adopts a GRNN [31]. The results obtained by de-noising the original data obtained under three operating conditions using the GRNN are shown in Figures 4–6.

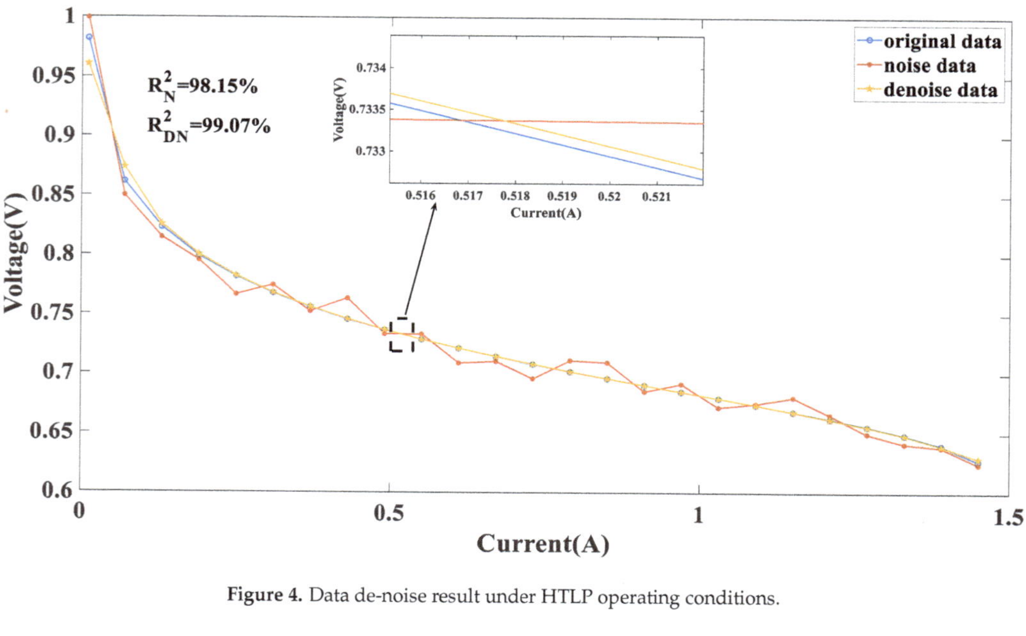

Figure 4. Data de-noise result under HTLP operating conditions.

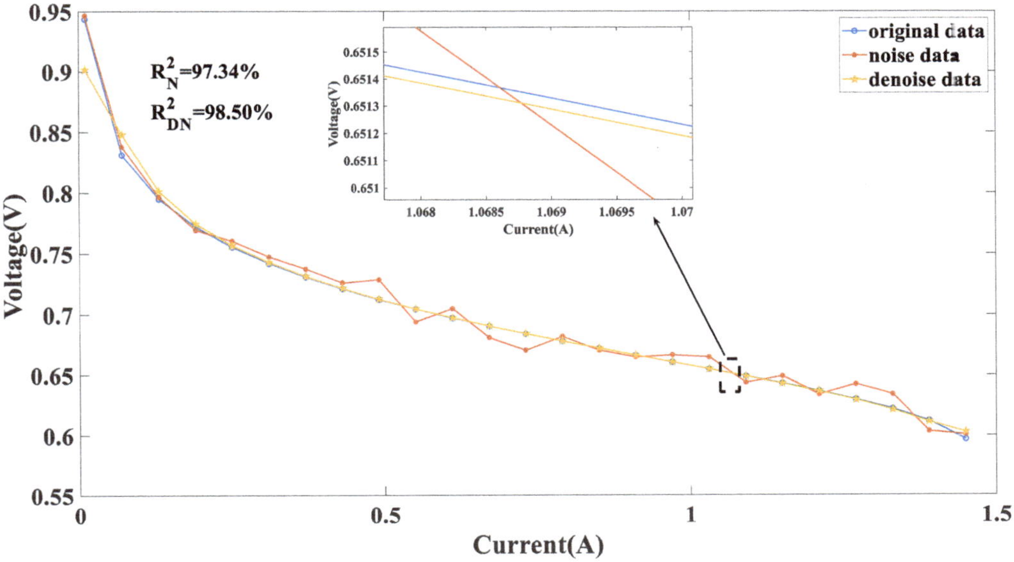

Figure 5. Data de-noise result under MTMP operating conditions.

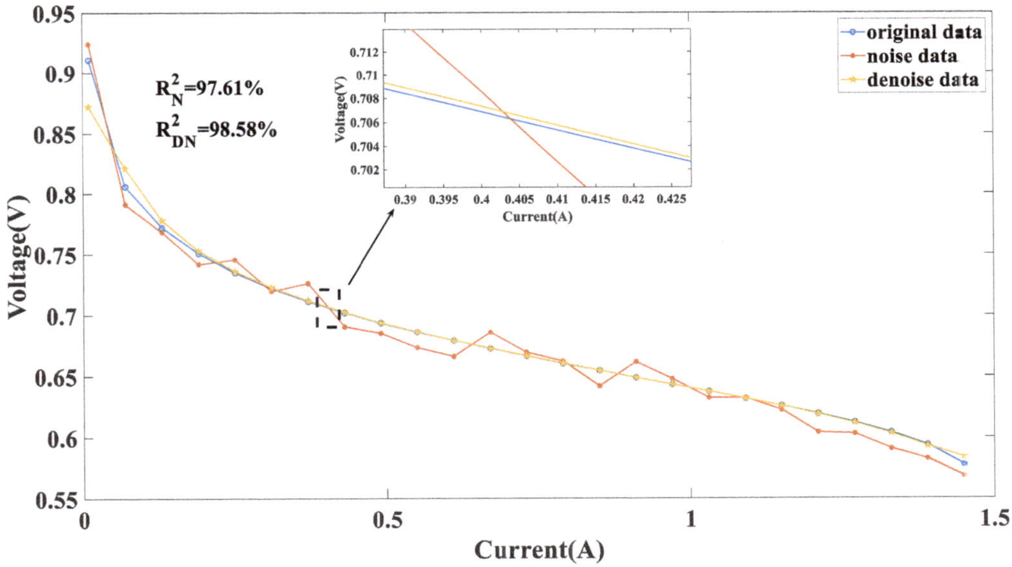

Figure 6. Data de-noise results under LTHP operating conditions.

4.1.2. GRNN for V-I Data Prediction

The parameter identification of PEMFC essentially relies on the most primitive current and voltage data, and the accuracy of the final identified parameters largely depends on the original data. However, actual data are difficult to obtain.

Therefore, this study uses existing data to train the GRNN model, then performs data prediction, expands the data volume, and improves the accuracy of identification parameters. The results obtained by data prediction of the original data obtained under three operating conditions using the GRNN are shown in Figures 7–9.

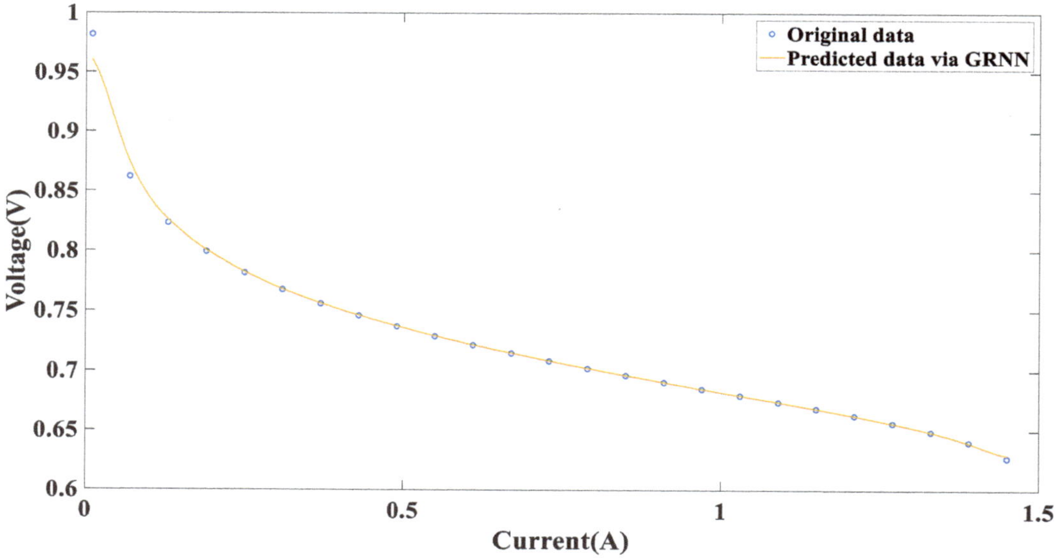

Figure 7. Data prediction result under HTLP operating conditions.

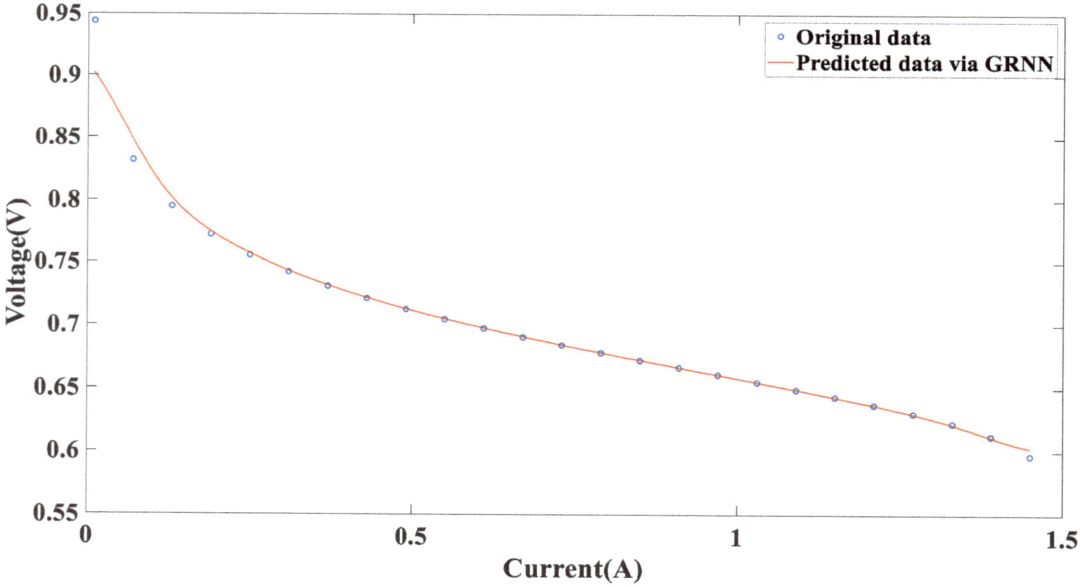

Figure 8. Data prediction result under MTMP operating conditions.

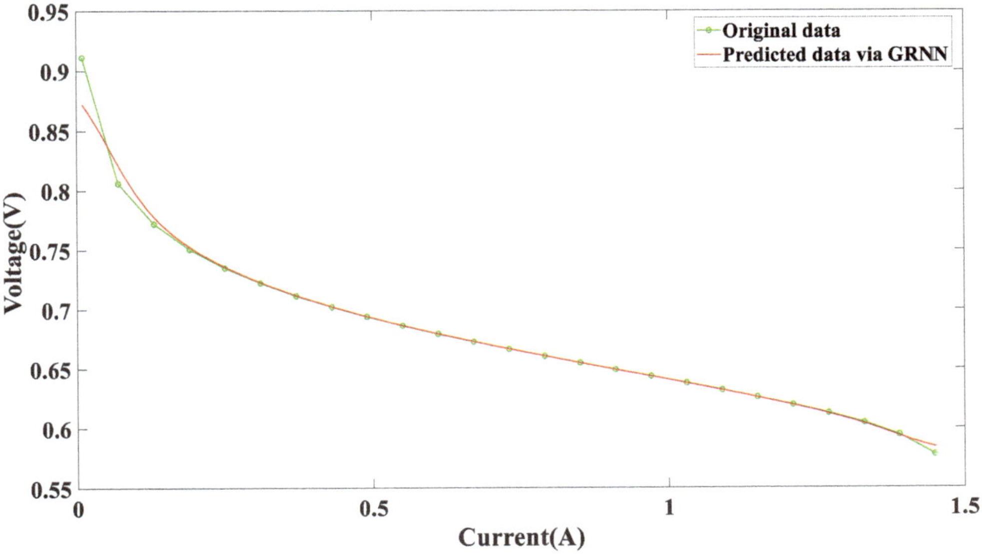

Figure 9. Data prediction result under LTHP operating conditions.

4.2. PEMFC Parameter Extraction of HTLP

4.2.1. Noised Data

Table A1 of Appendix A shows the statistics of the results of parameter extraction of noise and noise reduction data, respectively, by six algorithms under HTLP, where the symbol 'N' denotes the results obtained from noised data and 'DN' denotes the results obtained from de-noised data. From Table A1 of Appendix A, it is obvious that after data noise reduction, the RMSE is lower than that obtained from noised data. After data noise reduction, the RMSE of the PSO algorithm and the BAS algorithm has a magnitude of the minus second power of ten, while the RMSE of the other four algorithms has a magnitude of the minus third power of ten. The MPA algorithm exhibits the most significant decrease of 82.10%, whereas the BAS algorithm demonstrates a comparatively smaller reduction of 42.62%.

In addition, Figure 10 shows the RMSE convergence curves obtained by six algorithms trained on two datasets. The results obtained based on data de-noising have smaller errors than those obtained from noised data. The special process is that the RMSE obtained by six algorithms on de-noised data is lower than that obtained from noised data.

In order to acquire the visual impact of the two different training data, the boxplot illustrates the distribution of RMSE obtained by MhAs which is presented in Figure 11. It can be seen from the figure that after data de-noising, the RMSE corresponding to each algorithm in the boxplot decreased to a certain extent. However, after data de-noising, the upper and low bounds of the boxplot of PSO and BAS changed significantly, shrinking toward the RMSE median. In addition, MPA, AEO, GWO, and MFO have superior performance compared with other algorithms. This fully shows that GRNN data noise reduction can improve the stability of MhAs in parameter identification.

Figure 12 presents the *V-I* characteristic curves based on high-temperature and low-pressure obtained by the GRNN fitting the MPA algorithm under noise reduction data conditions. It can be seen that the curve of fitting data almost coincides with the curve of actual data and the error measured by RMSE is equal to 99.39%, which demonstrates the parameter identification effect is in line with expectations.

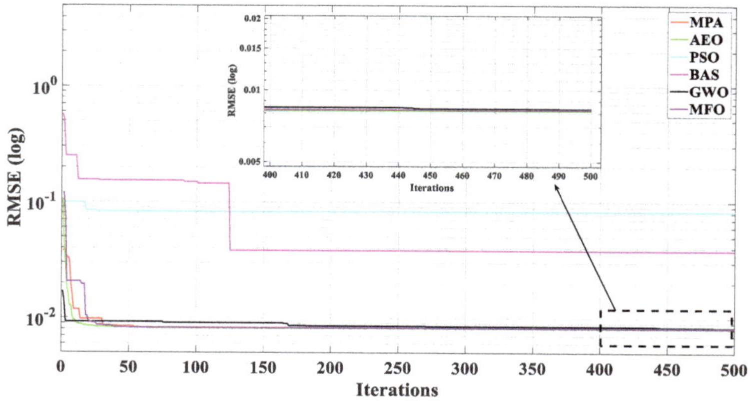

(a)

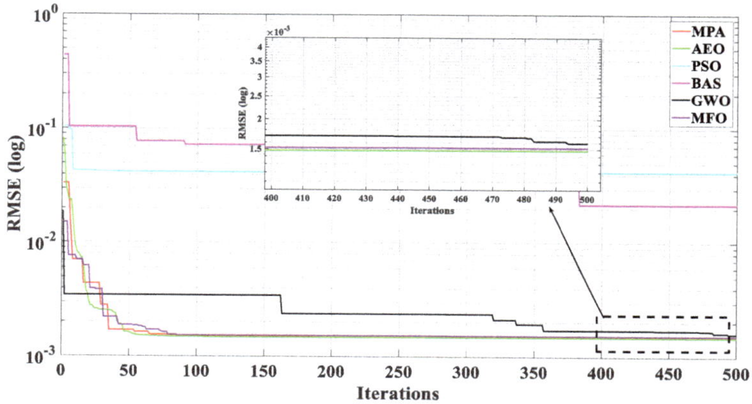

(b)

Figure 10. Convergence curves of RMSEs obtained by MhAs on noise data and de-noised data under HTLP. (**a**) noise data and (**b**) de-noised data.

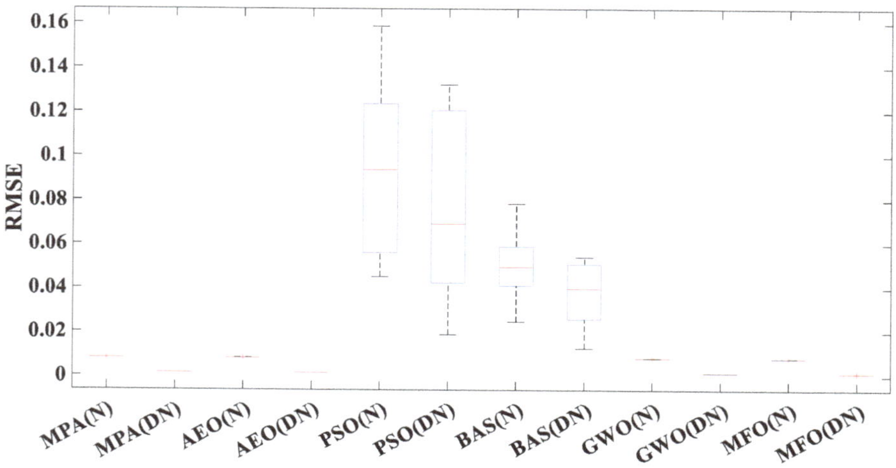

Figure 11. Boxplot of RMSEs obtained by MhAs on noise data and de-noised data under HTLP.

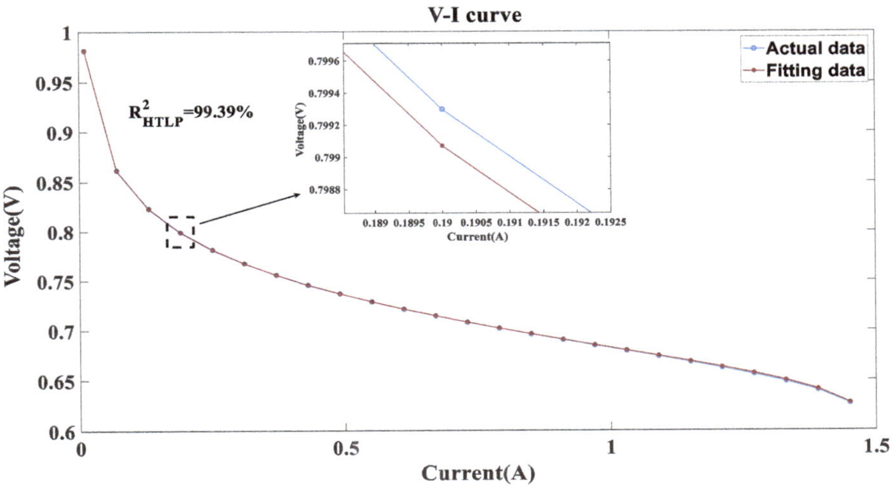

Figure 12. GRNN for *V-I* curve fitting based on de-noised data under HTLP of MPA.

4.2.2. Insufficient Data

Table A2 of Appendix A shows the statistics of the results of parameter extraction of insufficient and predicted data, respectively, by six algorithms under HTLP, where the symbol 'O' denotes the source data and 'P' denotes the predicted data. From Table A2 of Appendix A, it can be obtained by observation that after data prediction, the RMSE achieved by the five algorithms is lower than that obtained from predicted data, except for PSO algorithm. After data prediction, the RMSE of the PSO algorithm and the BAS algorithm has a magnitude of the minus second power of ten, while the RMSE of the other four algorithms has a magnitude of the minus fourth power of ten. The GWO algorithm exhibits the most significant decrease of 66.66%, whereas the MPA algorithm demonstrates a comparatively smaller reduction of 28.58%.

Figure 13 describes the RMSE convergence curves obtained by six algorithms on two datasets, with most algorithms having lower RMSE obtained from predicted data, and only RMSE based on multi-data of PSO being larger than RMSE based on low data. In addition, compared with other algorithms, MPA, AEO, and MFO can quickly acquire a smaller RMSE and have great stability.

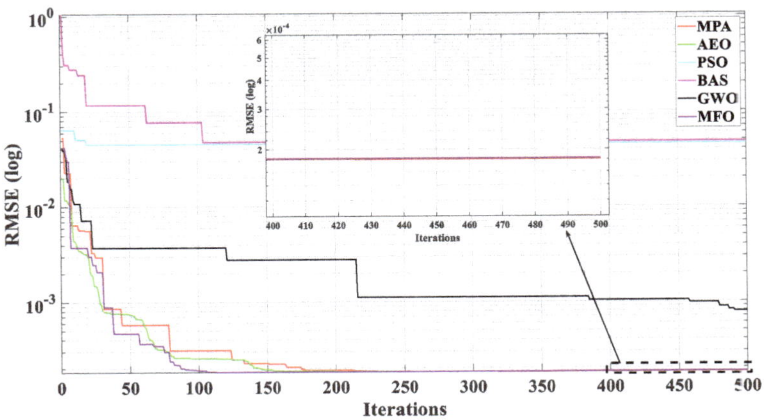

(a)

Figure 13. *Cont.*

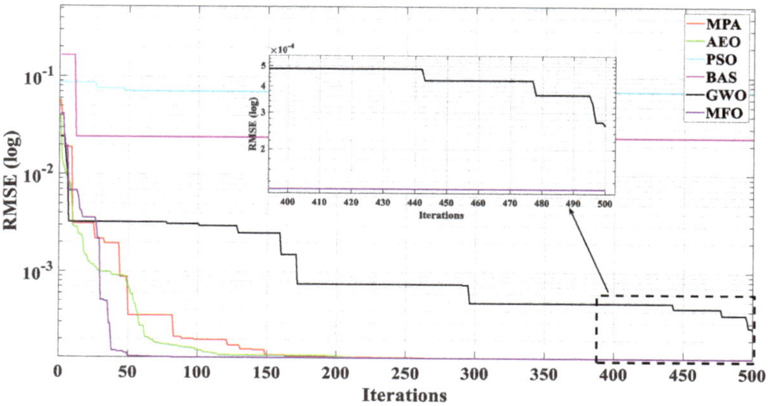

(b)

Figure 13. Convergence curves of RMSEs obtained by MhAs on original data and predicted data under HTLP. (**a**) original data and (**b**) predicted data.

The boxplot illustrates the distribution of RMSE obtained by MhAs which is presented in Figure 14. It can be obtained by observation that except for the PSO, the RMSE obtained based on predicted data are lower than the RMSE obtained from original data. On the contrary, the RMSE of PSO has increased. In addition, MPA, AEO, GWO, and MFO have superior performance compared with other algorithms.

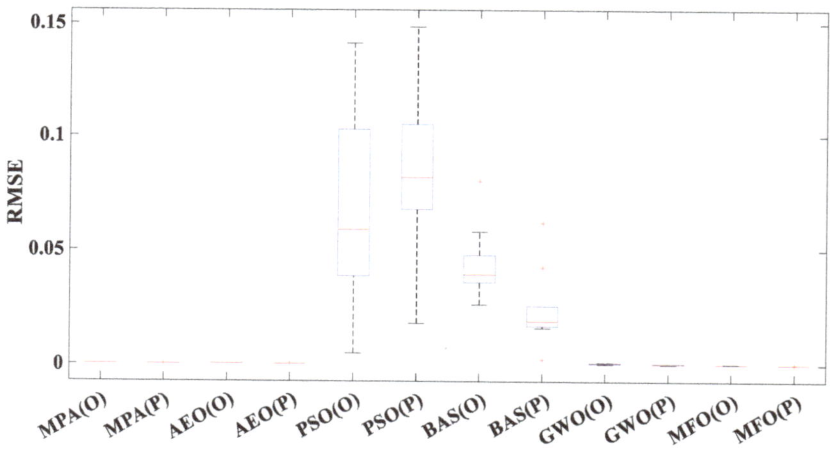

Figure 14. Boxplot of RMSEs obtained by MhAs on original data and predicted data under HTLP.

4.3. *PEMFC Parameter Extraction of MTMP*

4.3.1. Noised Data

Table A3 of Appendix A shows the statistics of the results of parameter extraction of noise and noise reduction data, respectively, by six algorithms under MTMP. From Table A3 of Appendix A, it can be seen that after data noise reduction, the RMSE obtained by the six algorithms is lower than that obtained from noised data. After data noise reduction, the RMSE of the PSO algorithm and the BAS algorithm has a magnitude of the minus second power of ten, while the RMSEs of the other third algorithms have a magnitude of the minus third power of ten. The GWO algorithm exhibits the most significant decrease of 66.53%, whereas the BAS algorithm demonstrates a comparatively smaller reduction of 25.09%.

Figure 15 describes the RMSE convergence curves obtained by six algorithms under noise and noise reduction data conditions. It can be obtained by observation that the RMSE based on de-noised data of parameter identification results has decreased. The special process is that the RMSE obtained by six algorithms on de-noised data is lower than that obtained from noised data.

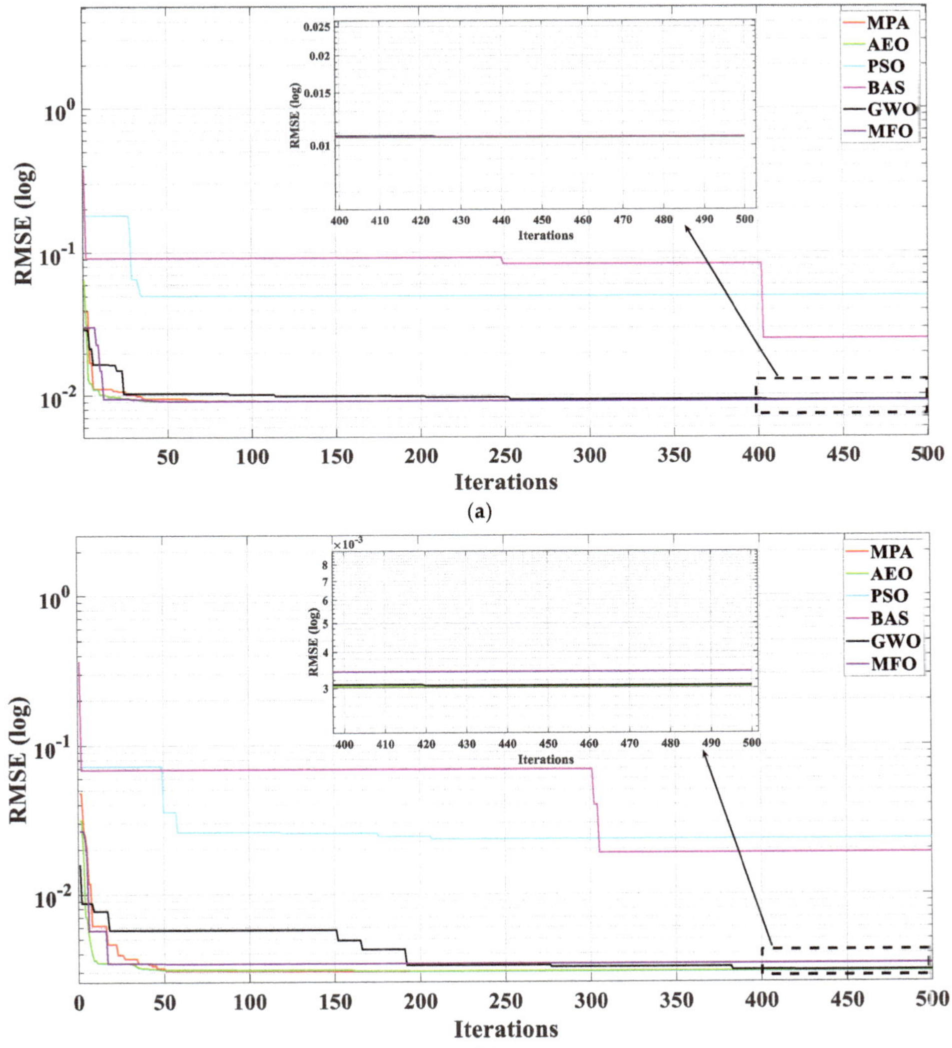

Figure 15. Convergence curves of RMSEs obtained by MhAs on noise data and de-noised data under MTMP. (**a**) noise data and (**b**) de-noised data.

Figure 16 describes the RMSE distribution boxplot obtained by six algorithms. It can be obtained by observation that except for the BAS, the RMSE obtained from predicted data has decreased. On the contrary, the upper and low bounds of BAS have increased. Also, there are a few outliers in the boxplot of MFO and PSO. In addition, MPA, AEO, and GWO have superior performance compared with other algorithms. This fully shows that GRNN data noise reduction can improve the stability of MhAs in parameter identification.

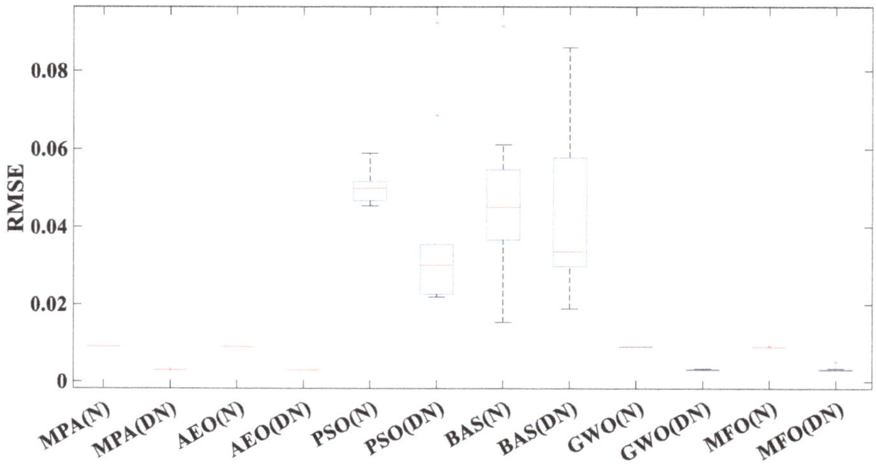

Figure 16. Boxplot of RMSEs obtained by MhAs on noise data and de-noised data under MTMP.

Figure 17 presents the *V-I* characteristic curves based on medium-temperature and medium-pressure obtained by the GRNN fitting the GWO algorithm under noise reduction data conditions. It can be obtained by observation that the curve of fitting data almost coincides with the curve of actual data and the error measured by RMSE is equal to 99.07%, which demonstrates the parameter identification effect is in line with expectations.

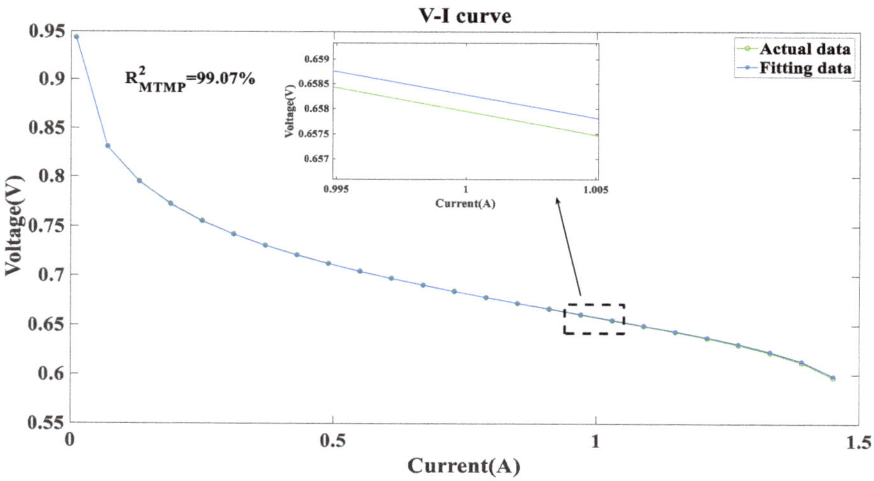

Figure 17. GRNN for *V-I* curve fitting based on de-noised data under MTMP of GWO.

4.3.2. Insufficient Data

Table A4 of Appendix A shows the statistics of the results of parameter extraction of insufficient and predicted data, respectively, by six algorithms under MTMP. From Table A4 of Appendix A, it is obvious that by data prediction, the RMSE obtained by the four algorithms is lower than that obtained from predicted data, except for the MFO and PSO algorithms. After data prediction, the RMSE of the PSO algorithm and the BAS algorithm has a magnitude of the minus second power of ten, while the RMSE of other algorithms has a magnitude exceeding the minus fourth power of ten. The MPA algorithm exhibits the most significant decrease of 63.40%, whereas the BAS algorithm demonstrates a comparatively smaller reduction of 13.26%.

Figure 18 describes the RMSE convergence curves obtained by six algorithms on two datasets, with most algorithms having lower RMSE based on predicted data, and only RMSE based on prediction data of BAS and PSO being larger than RMSE based on low data. In addition, compared with other algorithms, MPA and AEO can quickly acquire a smaller RMSE and have great stability.

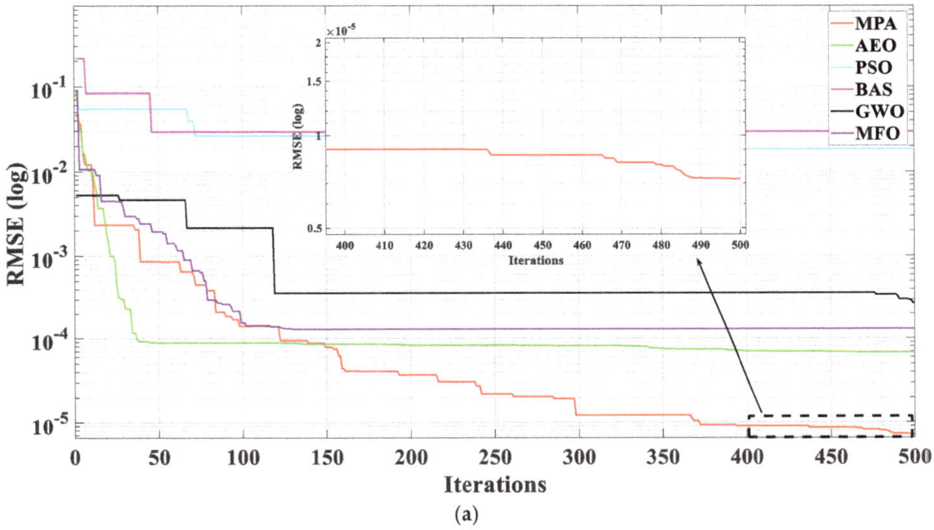

Figure 18. Convergence curves of RMSEs obtained by MhAs on original data and predicted data under MTMP. (**a**) original data and (**b**) predicted data.

Figure 19 describes the RMSE distribution boxplot obtained by six algorithms. It can be obtained by observation that the RMSE obtained based on predicted data has decreased. In addition, MPA has superior performance compared with other algorithms.

This fully shows that GRNN data noise reduction can improve the stability of MhAs in parameter identification.

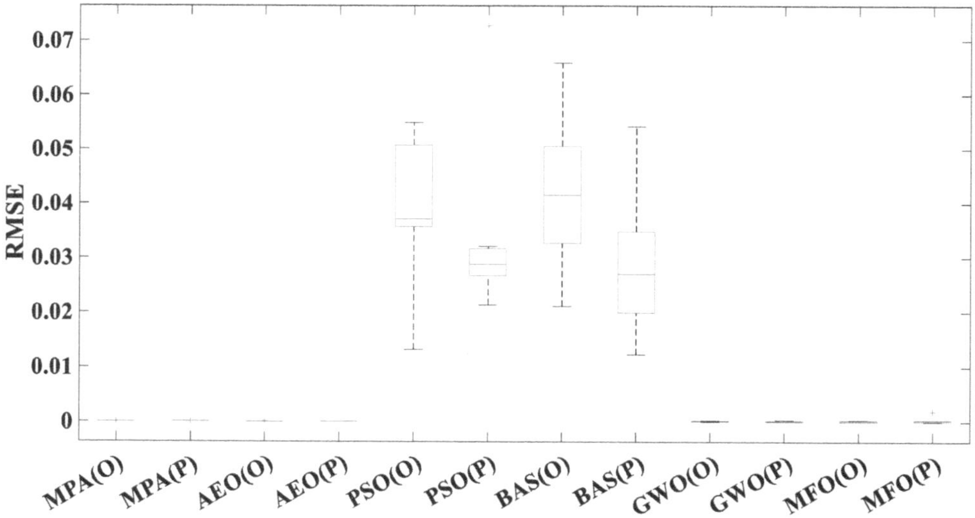

Figure 19. Boxplot of RMSEs obtained by MhAs on original data and predicted data under MTMP.

4.4. *PEMFC Parameter Extraction of LTHP*

4.4.1. Noised Data

Table A5 of Appendix A shows the statistics of the results of parameter extraction of noise and noise reduction data, respectively, by six algorithms under LTHP. From Table A5 of Appendix A, it can be obtained by observation that after data de-noising, the RMSE obtained by the five algorithms is lower than that obtained from noised data, except for the BAS algorithm. In particular, the MFO algorithm exhibits the most significant decrease of 73.85%, whereas the PSO algorithm demonstrates a comparatively smaller reduction of 26.21%. After data noise reduction, the RMSE of the PSO algorithm and the BAS algorithm has a magnitude of the minus second power of ten, while the RMSE of the other four algorithms has a magnitude of the minus third power of ten.

Figure 20 describes the RMSE convergence curves obtained by six algorithms under noise and de-noised data conditions. It can be obtained by observation that most of the RMSE based on de-noised data of identification results have decreased, while the RMSE of the BAS has increased after data noise reduction.

The boxplot illustrates the distribution of RMSE obtained by MhAs which is presented in Figure 21. It can be obtained by observation that except for the BAS, the RMSE obtained from predicted data has decreased. On the contrary, the upper and low bounds of PSO and the upper bound of BAS have increased. In addition, MPA, AEO, and GWO have superior performance compared with other algorithms. This fully shows that GRNN data noise reduction can improve the stability of MhAs in parameter identification.

Figure 22 shows the *V-I* characteristic curves based on low-temperature and high-pressure obtained by GRNN fitting the MFO algorithm under noise reduction data cases. It can be obtained by observation that the curve of fitting data almost coincides with the curve of actual data and the error measured by RMSE is equal to 98.70%, which demonstrates the parameter identification effect is in line with expectations.

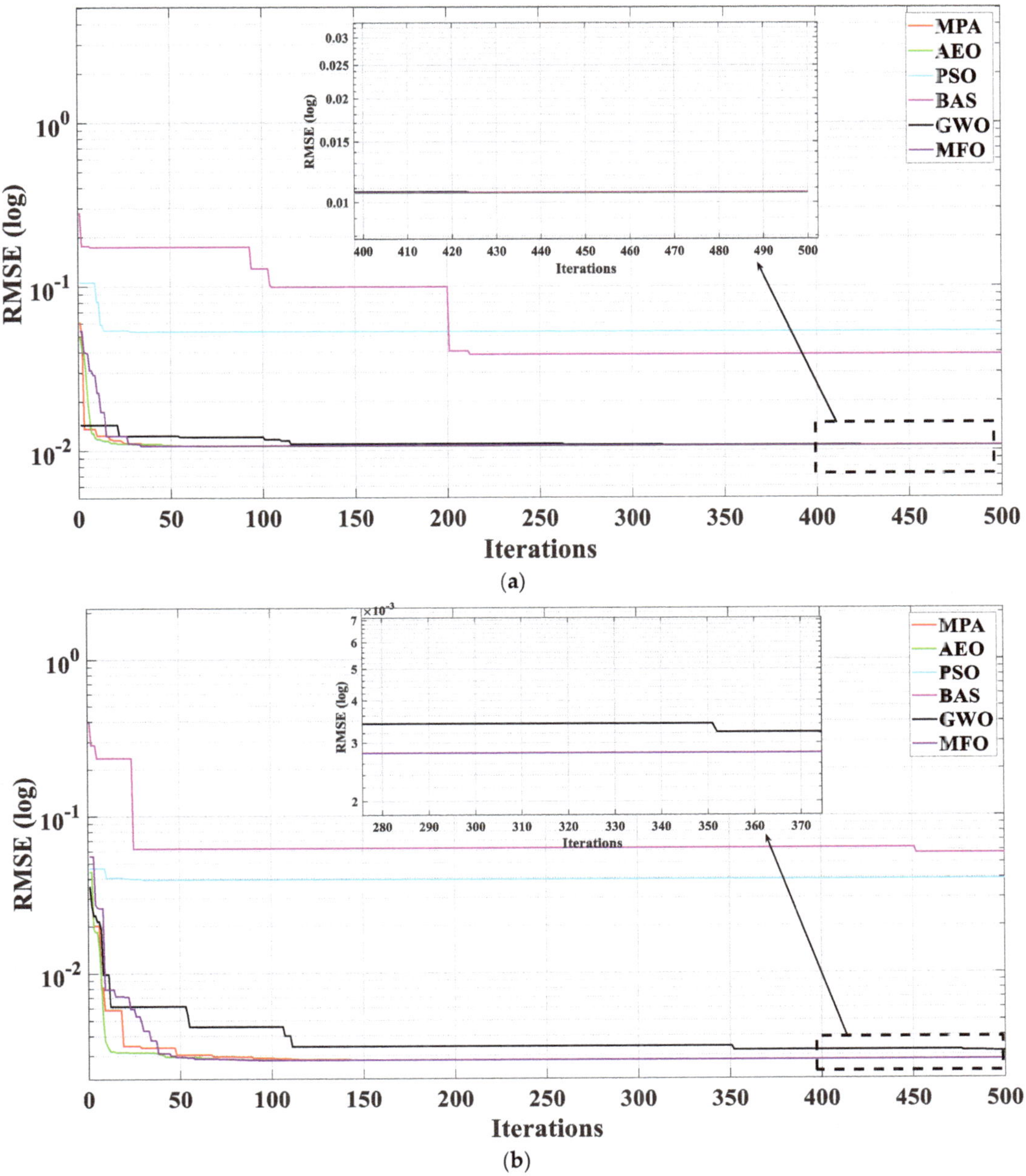

Figure 20. Convergence curves of RMSEs obtained by MhAs on noise data and de-noised data under LTHP. (**a**) noise data and (**b**) de-noised data.

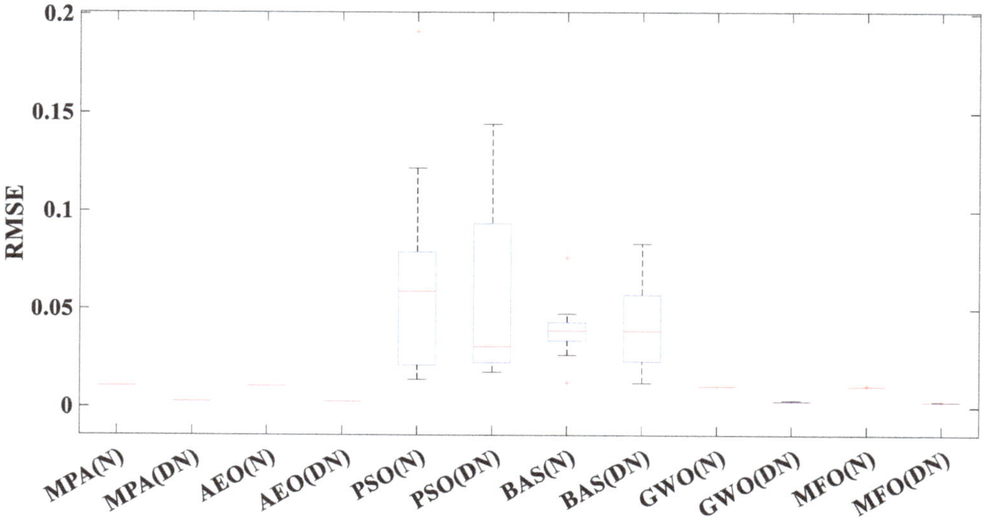

Figure 21. Boxplot of RMSEs obtained by MhAs on noise data and de-noised data under LTHP.

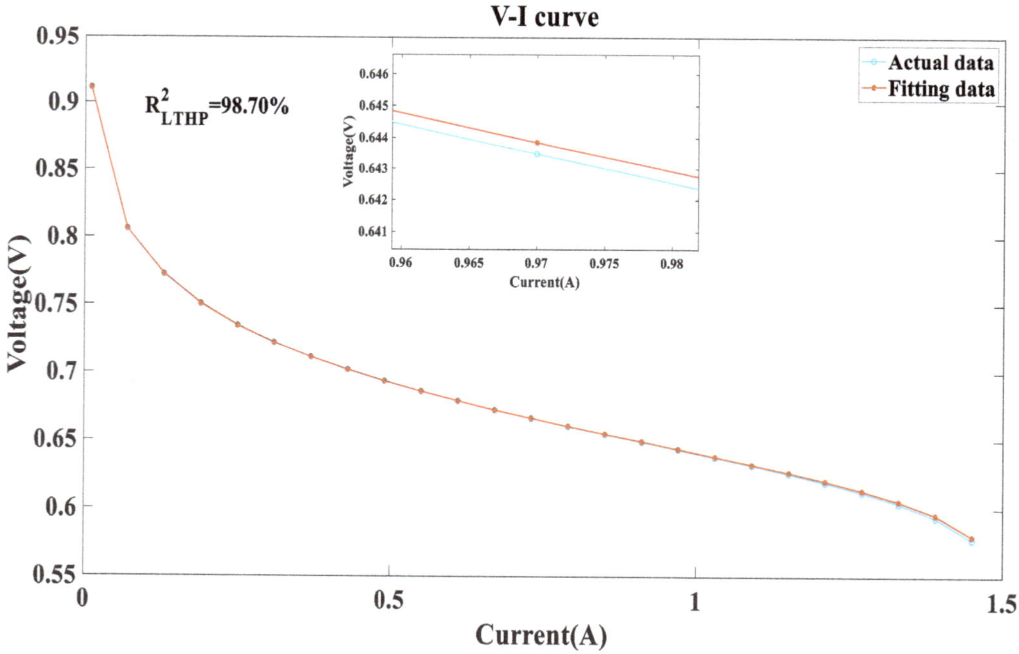

Figure 22. GRNN for *V-I* curve fitting based on de-noised data under LTHP of MFO.

4.4.2. Insufficient Data

Table A6 of Appendix A shows the statistics of the results of parameter extraction of insufficient and predicted data, respectively, by six algorithms under LTHP. From Table A6 of Appendix A, it is obvious that after data prediction, the RMSE obtained by the six algorithms is lower than that obtained from predicted data. After data prediction, the RMSE of the PSO algorithm and the BAS algorithm has a magnitude not exceeding the minus third power of ten, while the RMSE of the other algorithm has a magnitude exceeding the minus fourth power of ten. The MFO algorithm exhibits the most significant decrease of 92.69%, whereas the PSO algorithm demonstrates a comparatively smaller reduction of 43.01%.

Figure 23 describes the RMSE convergence curves obtained by six algorithms on two datasets, with all algorithms having lower RMSE based on predicted data.

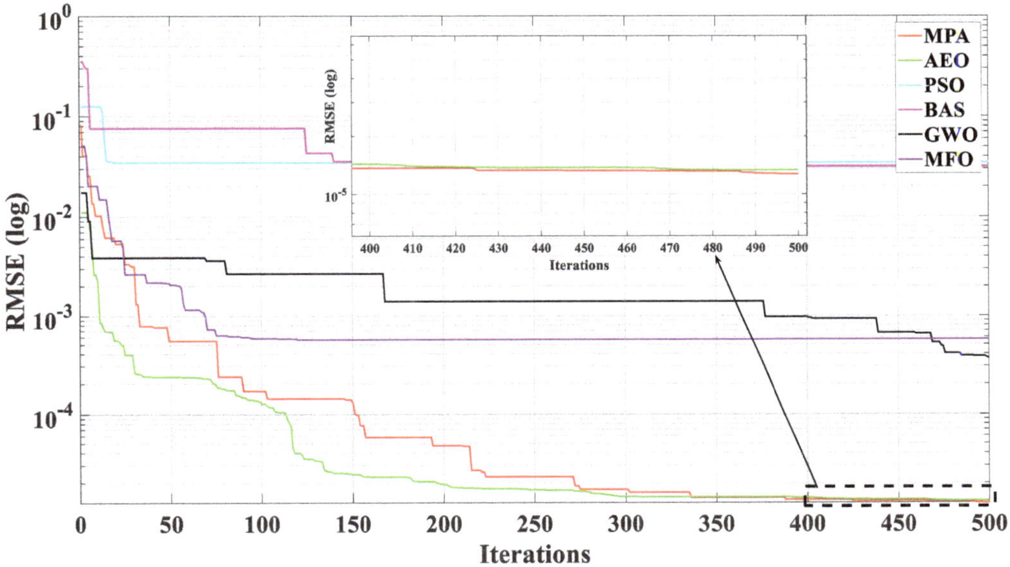

(**a**)

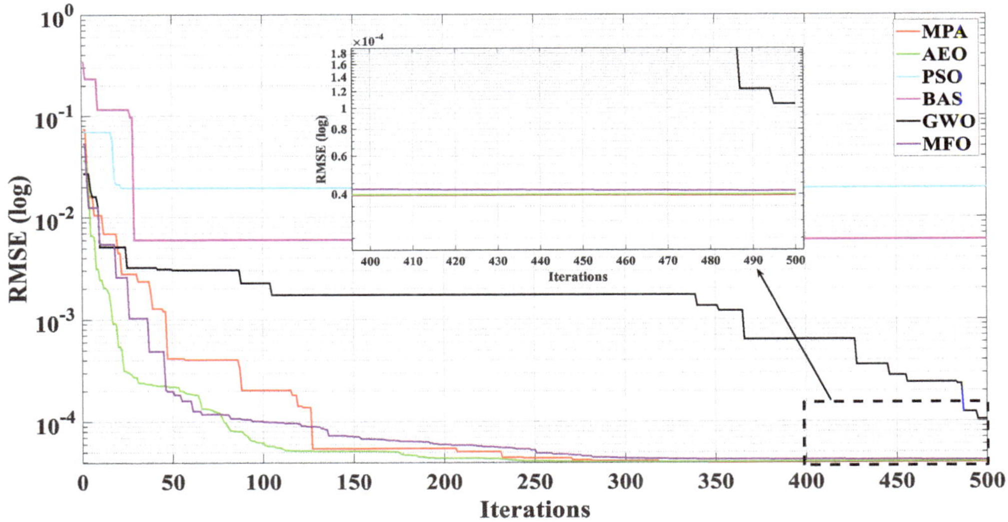

(**b**)

Figure 23. Convergence curves of RMSEs obtained by MhAs on original data and predicted data under LTHP. (**a**) original data and (**b**) predicted data.

The boxplot illustrates the distribution of RMSE obtained by MhAs which is presented in Figure 24. It can be obtained by observation that except for the BAS and PSO, the RMSE of other algorithms obtained from predicted data has decreased. However, the lower bound RMSE of PSO and the upper bound RMSE of BAS have increased. In addition, MPA, AEO, and MFO have superior performance compared with other algorithms. This fully shows that GRNN data noise reduction can improve the stability of MhAs in parameter identification.

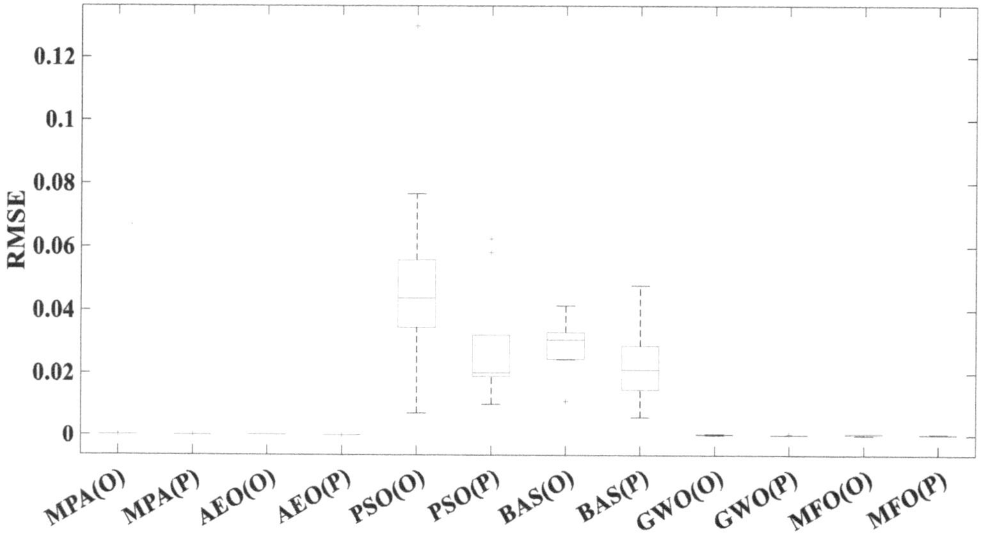

Figure 24. Boxplot of RMSEs obtained by MhAs on original data and predicted data under LTHP.

5. Discussions

Table 2 summarizes the research related to PEMFC parameter identification in recent years. It can be seen from the statistical comparison results that most studies have not simultaneously considered the impact of noise data and insufficient data volume on the final parameter extraction accuracy. The research conducted in this study precisely compensates for the shortcomings in this area and provides excellent guidance for the research on PEMFC parameter extraction direction. However, through the research in this study, it can be found that in using heuristic algorithms, due to their unique parameter random search ability, some algorithms have abnormal numerical accuracy in the results when extracting parameters. For example, under MTMP conditions, after data prediction and parameter extraction using the BAS algorithm, the identification accuracy showed abnormalities, after data de-noising, and the accuracy decreased by 13.26%. Overall, the method proposed in this research is not only suitable for PEMFC parameter identification but also for photovoltaic (PV) and solid oxide fuel cell (SOFC) parameter identification. Through this study, it has been fully demonstrated that it has extremely good performance in the field of parameter identification. All experimental results in this article are based on the data in Tables A7–A9, where Table A7 represents the *V-I* data of PEMFC under HTLP, Table A8 represents the *V-I* data under MTMP, and Table A9 represents the *V-I* data under LTHP.

Table 2. Summary of research on parameter identification of some PEMFCs in recent years.

Approach	Year	Cell Type	Data Process			Operating Condition	
			Data De-Noise	Data Prediction	No	Consider	No Consider
BRNN-MhAs [17]	2021	Ballard-Mark-V PEMFC	✓			✓	
IDE [14]	2021	N.P.	✓				✓
LMBP [15]	2021	Ballard-Mark-V PEMFC			✓	✓	
ICSO [13]	2023	N.P.			✓	✓	
ELM-MhAs [12]	2023	Ballard-Mark-V PEMFC	✓			✓	

Note. N.P.: Not provided.

Additionally, the study did not take into account the impact of changes in temperature and other factors on the identification results during the actual operation of the PEMFC, and the shortcomings of this study are that although the overall accuracy can be improved after data processing, it has not improved much. Further research is needed in this direction in the future. Additionally, the research did not consider the specific impact and role of the identified parameters on the cell itself [36–38].

6. Conclusions and Prospect

This study proposes a parameter identification method for the PEMFC using GRNN and MhAs. The original cell *V-I* data are processed using GRNN, which includes data de-noising and data prediction. In addition, six typical heuristic algorithms were used to extract parameters of the PEMFC under three operating conditions: HTLP, MTMP, and LTHP. Then, the obtained results were compared with the results extracted from the original data, and the results show that using GRNN to process the data can markedly enhance the precision rate of final identification, specifically, after data prediction, the accuracy of the MFO algorithm has been improved by 92.69% under LTHP conditions. And after data de-noising processing, it is obvious that it can improve the stability of parameter identification results. Finally, by substituting the identified parameters into the model, the fitting accuracy of *V-I* data obtained under all three operating conditions was very high. Specifically, under HTLP conditions, the *V-I* fitting accuracy achieved 99.39%, the fitting accuracy was 99.07% on MTMP, and the fitting accuracy was 98.70%. All in all, after processing the PEMFC data using GRNN and using MhAs for cell parameter extraction, the efficiency, accuracy, and stability of the final identification results of PEMFC parameter identification can be greatly improved. This study provides a novel approach to the field of PEMFC parameter identification.

In the end, this study provides significant guidance for future research on PEMFC parameter extraction. However, future research on this aspect should pay more attention to the impact of data analysis on the final identification results. In addition, consideration should also be given to the impact of the identified parameters on the internal mechanism of the cell itself. In general, further research should be conducted on the internal characteristics of the cell, such as its state of charge and health, through the identified parameters.

Author Contributions: P.H.: writing the original draft and editing. X.Z., M.L., K.X.: conceptualization. X.M., B.Y.: visualization and contributed to the discussion of the topic. All authors have read and agreed to the published version of the manuscript.

Funding: This research received no external funding.

Data Availability Statement: Not applicable.

Conflicts of Interest: The authors declare that they have no known competing financial interest or personal relationships that could have appeared to influence the work reported in this paper.

Appendix A

Table A1. Parameters identification results of noise data and de-noised data based under HTLP and MhAs.

State	Algorithms	Data	Identified Parameters							RMSE
			ε_1	ε_2	ε_3	ε_4	λ	$R_c(\Omega)$	b(V)	
HTLP	MPA	N	−0.9071	2.5000×10^{-3}	3.6000×10^{-5}	-1.0000×10^{-4}	23.0000	1.0000×10^{-4}	0.0136	8.2100×10^{-3}
		DN	−0.9437	3.4000×10^{-3}	9.8000×10^{-5}	-1.0000×10^{-4}	13.5231	8.0000×10^{-4}	0.0136	$\mathbf{1.4792 \times 10^{-3}}$
	AEO	N	−0.9384	2.5000×10^{-3}	3.6000×10^{-5}	-1.0000×10^{-4}	18.1835	1.0000×10^{-4}	0.0136	8.2182×10^{-3}
		DN	−0.9142	2.8000×10^{-3}	6.1300×10^{-5}	-1.5836×10^{-4}	13.4758	8.0000×10^{-4}	0.0136	$\mathbf{1.4798 \times 10^{-3}}$
	PSO	N	−0.9610	2.9000×10^{-3}	3.6000×10^{-5}	-2.6000×10^{-4}	17.8012	1.0000×10^{-4}	0.0220	8.3356×10^{-2}
		DN	−0.9882	3.0000×10^{-3}	3.6000×10^{-5}	-2.6000×10^{-4}	14.1474	1.0000×10^{-4}	0.0693	$\mathbf{4.2543 \times 10^{-2}}$
	BAS	N	−1.0078	3.2000×10^{-3}	5.5900×10^{-5}	-1.6467×10^{-4}	17.2423	1.5174×10^{-4}	0.0217	3.8371×10^{-2}
		DN	−0.9756	3.0000×10^{-3}	7.1700×10^{-5}	-1.1658×10^{-4}	13.9875	2.4255×10^{-4}	0.0156	$\mathbf{2.2016 \times 10^{-2}}$
	GWO	N	−0.8561	2.7000×10^{-3}	6.1100×10^{-5}	-1.7406×10^{-4}	19.4860	6.5726×10^{-4}	0.0136	8.3379×10^{-3}
		DN	−0.8714	2.8000×10^{-3}	7.0143×10^{-5}	-1.5764×10^{-4}	11.3971	4.8605×10^{-4}	0.0136	$\mathbf{1.5914 \times 10^{-3}}$
	MFO	N	−0.9520	2.7000×10^{-3}	4.3300×10^{-5}	-1.7414×10^{-4}	23.0000	8.0000×10^{-4}	0.0136	8.2942×10^{-3}
		DN	−0.8531	3.1000×10^{-3}	9.3100×10^{-5}	-1.5858×10^{-4}	12.7877	1.0000×10^{-4}	0.0136	$\mathbf{1.5251 \times 10^{-3}}$

Table A2. Parameters identification results of original data and predicted data based under HTLP and MhAs.

State	Algorithms	Data	Identified Parameters							RMSE
			ε_1	ε_2	ε_3	ε_4	λ	$R_c(\Omega)$	b(V)	
HTLP	MPA	O	−0.9788	2.6836×10^{-3}	3.6000×10^{-5}	-1.7361×10^{-4}	23.0000	1.0000×10^{-4}	0.0136	1.8172×10^{-4}
		P	−0.9040	2.4849×10^{-3}	3.6873×10^{-5}	-1.7361×10^{-4}	23.0000	3.2975×10^{-4}	0.0136	$\mathbf{1.2978 \times 10^{-4}}$
	AEO	O	−1.0234	3.3728×10^{-3}	7.4400×10^{-5}	-1.7358×10^{-4}	22.9704	1.0000×10^{-4}	0.0136	1.8329×10^{-4}
		P	−0.9936	3.0606×10^{-3}	5.8834×10^{-5}	-1.7361×10^{-4}	22.9999	3.0556×10^{-4}	0.0136	$\mathbf{1.2979 \times 10^{-4}}$
	PSO	O	−0.9918	2.9998×10^{-3}	5.3800×10^{-5}	-1.8100×10^{-4}	20.4025	2.9781×10^{-4}	0.0136	$\mathbf{4.5757 \times 10^{-2}}$
		P	−0.9975	3.1380×10^{-3}	6.0205×10^{-5}	-1.0567×10^{-4}	21.0376	7.1037×10^{-4}	0.1105	7.1117×10^{-2}
	BAS	O	−0.9646	2.7871×10^{-3}	4.7900×10^{-5}	-1.2795×10^{-4}	11.8778	7.2134×10^{-4}	0.0315	4.7789×10^{-2}
		P	−1.0520	3.2208×10^{-3}	6.9550×10^{-5}	-1.1996×10^{-4}	16.9202	1.0746×10^{-4}	0.0233	$\mathbf{2.4037 \times 10^{-2}}$
	GWO	O	−1.0150	3.3641×10^{-3}	7.5400×10^{-5}	-1.7248×10^{-4}	14.8613	1.8872×10^{-4}	0.0136	7.8535×10^{-4}
		P	−1.1592	3.9697×10^{-3}	8.8869×10^{-5}	-1.7324×10^{-4}	21.1633	2.9474×10^{-4}	0.0137	$\mathbf{2.6180 \times 10^{-4}}$
	MFO	O	−0.8531	2.5174×10^{-3}	4.9000×10^{-5}	-1.7360×10^{-4}	23.0000	1.0000×10^{-4}	0.0136	1.8301×10^{-4}
		P	−0.8539	3.2393×10^{-3}	9.8000×10^{-5}	-1.7361×10^{-4}	23.0000	2.6234×10^{-4}	0.0136	$\mathbf{1.2979 \times 10^{-4}}$

Table A3. Parameters identification results of noise data and de-noised data based under MTMP and MhAs.

State	Algorithms	Data	Identified Parameters							RMSE
			ε_1	ε_2	ε_3	ε_4	λ	$R_c(\Omega)$	b(V)	
MTMP	MPA	N	−0.9356	2.5684×10^{-3}	3.6000×10^{-5}	-1.7905×10^{-4}	23.0000	1.0000×10^{-4}	0.0136	9.0391×10^{-3}
		DN	−0.9416	3.3689×10^{-3}	9.8000×10^{-5}	-1.4121×10^{-4}	10.0000	8.0000×10^{-4}	0.0149	$\mathbf{3.0236 \times 10^{-3}}$
	AEO	N	−1.0197	3.0243×10^{-3}	5.0100×10^{-5}	-1.7902×10^{-4}	22.9999	1.1388×10^{-4}	0.0136	9.0398×10^{-3}
		DN	−0.9375	2.5467×10^{-3}	4.1900×10^{-5}	-1.4125×10^{-4}	10.0000	8.0000×10^{-4}	0.0149	$\mathbf{3.0294 \times 10^{-3}}$
	PSO	N	−0.9882	3.0804×10^{-3}	3.6000×10^{-5}	-2.6000×10^{-4}	14.1474	1.0000×10^{-4}	0.0693	5.0083×10^{-2}
		DN	−0.9547	3.3380×10^{-3}	9.8000×10^{-5}	-9.5400×10^{-5}	13.7242	8.0000×10^{-4}	0.0292	$\mathbf{2.3422 \times 10^{-2}}$
	BAS	N	−0.9699	3.0822×10^{-3}	8.2000×10^{-5}	-9.9500×10^{-5}	19.7473	4.3042×10^{-4}	0.0369	2.5185×10^{-2}
		DN	−0.9584	2.7603×10^{-3}	4.5600×10^{-5}	-1.7596×10^{-4}	11.8635	7.0335×10^{-4}	0.0272	$\mathbf{1.8866 \times 10^{-2}}$
	GWO	N	−1.1160	3.7093×10^{-3}	7.7473×10^{-5}	-1.7869×10^{-4}	22.9456	6.0307×10^{-4}	0.0136	9.1580×10^{-3}
		DN	−1.0319	2.8337×10^{-3}	4.2200×10^{-5}	-1.4112×10^{-4}	10.0000	4.7206×10^{-4}	0.0151	$\mathbf{3.0655 \times 10^{-3}}$
	MFO	N	−0.9990	3.1232×10^{-3}	6.1200×10^{-5}	-1.7904×10^{-4}	23.0000	1.0000×10^{-4}	0.0136	9.0445×10^{-3}
		DN	−1.1997	3.5568×10^{-3}	5.7400×10^{-5}	-1.4348×10^{-4}	10.0000	1.0000×10^{-4}	0.0136	$\mathbf{3.4203 \times 10^{-3}}$

Table A4. Parameters identification results of original data and predicted data based under MTMP and MhAs.

State	Algorithms	Data	Identified Parameters							RMSE
			ε_1	ε_2	ε_3	ε_4	λ	$R_c(\Omega)$	b(V)	
MTMP	MPA	O	−0.9872	3.6025×10^{-3}	9.7600×10^{-5}	-1.7220×10^{-4}	16.5370	7.9936×10^{-4}	0.0158	7.2400×10^{-6}
		P	−0.8886	3.3117×10^{-3}	9.8000×10^{-5}	-1.7224×10^{-4}	17.0972	7.9999×10^{-4}	0.0158	$\mathbf{2.6500 \times 10^{-6}}$
	AEO	O	−0.8531	2.9435×10^{-3}	7.9900×10^{-5}	-1.7234×10^{-4}	19.3780	8.0000×10^{-4}	0.0161	6.8300×10^{-5}
		P	−1.0941	3.8739×10^{-3}	9.4200×10^{-5}	-1.7224×10^{-4}	17.0795	8.0000×10^{-4}	0.0158	$\mathbf{3.1200 \times 10^{-5}}$

Table A4. *Cont.*

State	Algorithms	Data	Identified Parameters							RMSE
			ε_1	ε_2	ε_3	ε_4	λ	$R_c(\Omega)$	b(V)	
MTMP	PSO	O	−0.9523	3.3678×10^{-3}	9.8000×10^{-5}	-9.5400×10^{-5}	18.0146	8.0000×10^{-4}	0.0373	$\mathbf{1.8500 \times 10^{-2}}$
		P	−1.0564	3.5663×10^{-3}	9.8000×10^{-5}	-9.5400×10^{-5}	21.0946	8.0000×10^{-4}	0.0174	2.1842×10^{-2}
	BAS	O	−0.9972	3.5814×10^{-3}	9.2100×10^{-5}	-1.2458×10^{-4}	19.1842	1.6857×10^{-4}	0.0408	2.9545×10^{-2}
		P	−0.9651	3.3046×10^{-3}	9.2400×10^{-5}	-1.2678×10^{-4}	18.4227	7.8429×10^{-4}	0.0436	$\mathbf{2.5628 \times 10^{-2}}$
	GWO	O	−0.9400	2.5785×10^{-3}	3.6500×10^{-5}	-1.7195×10^{-4}	10.1534	1.0110×10^{-4}	0.0137	2.6819×10^{-4}
		P	−0.9093	2.8226×10^{-3}	5.9800×10^{-5}	-1.7183×10^{-4}	11.3025	3.9408×10^{-4}	0.0142	$\mathbf{2.2363 \times 10^{-4}}$
	MFO	O	−0.9032	2.4639×10^{-3}	3.6300×10^{-5}	-1.7252×10^{-4}	23.0000	8.0000×10^{-4}	0.0163	$\mathbf{1.2940 \times 10^{-4}}$
		P	−0.8551	3.0497×10^{-3}	8.6800×10^{-5}	-1.7204×10^{-4}	10.3446	1.0000×10^{-4}	0.0136	2.8621×10^{-4}

Table A5. Parameters identification results of noise data and de-noised data based under LTHP and MhAs.

State	Algorithms	Data	Identified Parameters							RMSE
			ε_1	ε_2	ε_3	ε_4	λ	$R_c(\Omega)$	b(V)	
LTHP	MPA	N	−0.9243	3.3842×10^{-3}	9.8000×10^{-5}	-1.6452×10^{-4}	10.9407	8.0000×10^{-4}	0.0136	1.0637×10^{-2}
		DN	−0.8969	3.2076×10^{-3}	9.8000×10^{-5}	-1.4003×10^{-4}	10.0000	8.0000×10^{-4}	0.0150	$\mathbf{2.7961 \times 10^{-3}}$
	AEO	N	−1.0965	3.3347×10^{-3}	5.5484×10^{-5}	-1.6454×10^{-4}	10.9277	8.0000×10^{-4}	0.0136	1.0638×10^{-2}
		DN	−0.9823	3.3547×10^{-3}	8.9066×10^{-5}	-1.4016×10^{-4}	10.0000	7.9998×10^{-4}	0.0150	$\mathbf{2.7965 \times 10^{-3}}$
	PSO	N	−0.9883	3.0805×10^{-3}	3.6000×10^{-5}	-2.6000×10^{-4}	14.1474	1.0000×10^{-4}	0.0693	5.3206×10^{-2}
		DN	−0.9449	2.9247×10^{-3}	3.6000×10^{-5}	-2.6000×10^{-4}	18.8108	1.0000×10^{-4}	0.0474	$\mathbf{3.9261 \times 10^{-2}}$
	BAS	N	−0.8917	2.6943×10^{-3}	5.5447×10^{-5}	-1.6776×10^{-4}	13.2096	4.8206×10^{-4}	0.0181	$\mathbf{3.8512 \times 10^{-2}}$
		DN	−0.9934	3.2872×10^{-3}	6.4747×10^{-5}	-1.3804×10^{-4}	14.4718	3.5248×10^{-4}	0.0240	5.7263×10^{-2}
	GWO	N	−1.0674	3.0840×10^{-3}	4.4299×10^{-5}	-1.6480×10^{-4}	10.7232	2.5220×10^{-4}	0.0136	1.0681×10^{-2}
		DN	−1.0596	3.5278×10^{-3}	8.3867×10^{-5}	-1.4073×10^{-4}	14.3708	5.6604×10^{-4}	0.0176	$\mathbf{3.1399 \times 10^{-3}}$
	MFO	N	−1.1679	3.2876×10^{-3}	3.6000×10^{-5}	-1.6487×10^{-4}	10.7555	1.0000×10^{-4}	0.0136	1.0692×10^{-2}
		DN	−1.0772	3.7835×10^{-3}	9.8000×10^{-5}	-1.4003×10^{-4}	10.0000	8.0000×10^{-4}	0.0150	$\mathbf{2.7961 \times 10^{-3}}$

Table A6. Parameters identification results of original data and predicted data based under LTHP and MhAs.

State	Algorithms	Data	Identified Parameters							RMSE
			ε_1	ε_2	ε_3	ε_4	λ	$R_c(\Omega)$	b(V)	
LTHP	MPA	O	−0.9778	3.5659×10^{-3}	9.7853×10^{-5}	-1.7098×10^{-4}	16.4825	8.0000×10^{-4}	0.0169	3.9481×10^{-5}
		P	−0.9494	3.4339×10^{-3}	9.4921×10^{-5}	-1.7101×10^{-4}	17.0471	7.9376×10^{-4}	0.0170	$\mathbf{1.2754 \times 10^{-5}}$
	AEO	O	−0.9935	3.1003×10^{-3}	6.1264×10^{-5}	-1.7103×10^{-4}	15.4458	5.4652×10^{-4}	0.0167	3.9974×10^{-5}
		P	−0.9575	3.0169×10^{-3}	6.3508×10^{-5}	-1.7106×10^{-4}	16.0734	4.7360×10^{-4}	0.0169	$\mathbf{1.3427 \times 10^{-5}}$
	PSO	O	−0.9703	2.9028×10^{-3}	5.0162×10^{-5}	-1.6158×10^{-4}	15.0542	2.0792×10^{-4}	0.0328	3.4588×10^{-2}
		P	−0.9990	2.7318×10^{-3}	3.6000×10^{-5}	-1.4170×10^{-4}	10.1816	5.8666×10^{-4}	0.0420	$\mathbf{1.9711 \times 10^{-2}}$
	BAS	O	−0.9182	3.0406×10^{-3}	7.5044×10^{-5}	-1.7750×10^{-4}	13.8505	4.2091×10^{-4}	0.0301	3.1607×10^{-2}
		P	−0.9931	3.1320×10^{-3}	5.9052×10^{-5}	-1.7616×10^{-4}	15.8719	3.5783×10^{-4}	0.0192	$\mathbf{6.0350 \times 10^{-3}}$
	GWO	O	−1.1501	3.6883×10^{-3}	6.7517×10^{-5}	-1.7042×10^{-4}	10.4514	2.1721×10^{-4}	0.0146	3.6019×10^{-4}
		P	−1.0010	3.0434×10^{-3}	5.5508×10^{-5}	-1.7142×10^{-4}	17.3989	4.1713×10^{-4}	0.0172	$\mathbf{1.0451 \times 10^{-4}}$
	MFO	O	−0.9458	3.0558×10^{-3}	6.8896×10^{-5}	-1.7060×10^{-4}	10.0000	8.0000×10^{-4}	0.0136	5.6641×10^{-4}
		P	−0.9375	2.5788×10^{-3}	3.6958×10^{-5}	-1.7101×10^{-4}	16.5422	8.0000×10^{-4}	0.0169	$\mathbf{4.1393 \times 10^{-5}}$

Table A7. Original data under HTLP condition [39].

V	I
0.0100	0.9817
0.0700	0.8619
0.1300	0.8233
0.1900	0.7993
0.2500	0.7817
0.3100	0.7677
0.3700	0.7560

Table A7. *Cont.*

V	I
0.4300	0.7459
0.4900	0.7370
0.5500	0.7289
0.6100	0.7215
0.6700	0.7147
0.7300	0.7082
0.7900	0.7021
0.8500	0.6963
0.9100	0.6906
0.9700	0.6850
1.0300	0.6795
1.0900	0.6740
1.1500	0.6684
1.2100	0.6625
1.2700	0.6562
1.3300	0.6491
1.3900	0.6404
1.4500	0.6272

Table A8. Original data under MTMP condition [39].

V	I
0.0100	0.9439
0.0700	0.8314
0.1300	0.7950
0.1900	0.7723
0.2500	0.7556
0.3100	0.7422
0.3700	0.7310
0.4300	0.7213
0.4900	0.7126
0.5500	0.7047
0.6100	0.6975
0.6700	0.6907
0.7300	0.6843
0.7900	0.6782
0.8500	0.6723
0.9100	0.6665
0.9700	0.6608
1.0300	0.6551

Table A8. *Cont.*

V	*I*
1.0900	0.6493
1.1500	0.6434
1.2100	0.6371
1.2700	0.6302
1.3300	0.6223
1.3900	0.6124
1.4500	0.5969

Table A9. Original data under LTHP condition [39].

V	*I*
0.0100	0.9112
0.0700	0.8061
0.1300	0.7720
0.1900	0.7506
0.2500	0.7349
0.3100	0.7222
0.3700	0.7115
0.4300	0.7023
0.4900	0.6940
0.5500	0.6864
0.6100	0.6794
0.6700	0.6728
0.7300	0.6666
0.7900	0.6606
0.8500	0.6548
0.9100	0.6491
0.9700	0.6435
1.0300	0.6378
1.0900	0.6320
1.1500	0.6259
1.2100	0.6195
1.2700	0.6125
1.3300	0.6043
1.3900	0.5940
1.4500	0.5777

References

1. Yang, B.; Liu, B.Q.; Zhou, H.Y.; Wang, J.B.; Yao, W.; Wu, S.C.; Shu, H.C.; Ren, Y.X. A Critical Survey of Technologies of Large Offshore Wind Farm Integration: Summarization, Advances, and Perspectives. *Prot. Control Mod. Power Syst.* **2022**, *7*, 17. [CrossRef]
2. Yang, B.; Li, Y.L.; Li, J.L.; Shu, H.C.; Zhao, X.Y.; Ren, Y.X.; Li, Q. Comprehensive Summarization of Solid Oxide Fuel Cell: Control: A State-of-the-art Review. *Prot. Control Mod. Power Syst.* **2022**, *7*, 36. [CrossRef]
3. Joaquin, E.C.; Daniel, A.M.; Edwin, R.T.; Jan, M. A systematic review of real-time detection and classification of power quality disturbances. *Prot. Control. Mod. Power Syst.* **2023**, *8*, 3.
4. Guo, Y.Y.; Dai, X.D.; Jermsittiparsert, K.; Razmjooy, N. An Optimal Configuration for A Battery and PEM Fuel Cell-based Hybrid Energy System Using Developed Krill Herd Optimization Algorithm for Locomotive Application. *Energy Rep.* **2020**, *6*, 885–894. [CrossRef]
5. Ahmed, M.A.; Attia, A.E.; Gamal, M.S. Steady-State Modeling of Fuel Cells Based on Atom Search Optimizer. *Energies* **2019**, *12*, 1884.
6. Valencia, J.M.; Sierra, J.M.; Figueroa, S.J.; Diaz, S.E.; Meza, M. 3D CFD Modeling of A PEM Fuel Cell Stack. *Int. J. Hydrogen Energy* **2016**, *41*, 23425–23433. [CrossRef]
7. Pan, W.X.; Chen, J.Q.; Zhang, Y.; Gong, Y. Modeling of PEMFC and Application in Hybrid Power System. *Power Syst. Prot. Control* **2012**, *40*, 13–18.
8. Hu, J.; Su, J.H.; Du, Y.; Wan, H.N.; Shi, Y. Parameter Identification of PEMFC Stack Model based on Least Squares Method. *Acta Energiae Solaris Sin.* **2021**, *42*, 1–4.
9. Jones, D.D.; Mirrazavi, S.K.; Tamiz, M. Multi-objective Meta-heuristics: An Overview of the Current State-of-art. *Eur. J. Oper. Res.* **2002**, *137*, 1–9. [CrossRef]
10. Li, Q.; Chen, W.R.; Liu, S.K.; Lin, C.; Jia, J.B. Mechanism Modeling of Proton Exchange Membrane Fuel Cell Based on Adaptive Focusing Particle Swarm Optimization. *Proc. CSEE* **2009**, *29*, 119–124.
11. Zhou, S.; Wang, Y.J.; Jiang, L.W.; Jin, W. PEMFC State and Parameter Estimation Based on EKF with Tracking Factor. *J. Proj. Rocket. Missiles Guid.* **2019**, *39*, 45–50.
12. Rahul, K.; Ravita, L.; Rajesh, K. Parameters Extraction of PEMFC Model Using Evolutionary Based Optimization Algorithms. *Adv. Clean Energy Sustain.* **2022**, 443–451. [CrossRef]
13. Yang, Y.L.; Ling, M. Parameter Identification of Proton Exchange Membrane Fuel Cell Model based on Improved Chicken Swarm Optimization Algorithm. *Acta Energiae Solaris Sin.* **2023**, *44*, 269–278.
14. Xu, B. Parameter Optimal Identification of Proton Exchange Membrane Fuel Cell Model based on An Improved Differential Evolution Algorithm. *CIESC J.* **2021**, *72*, 1512–1520.
15. Yang, B.; Zeng, C.Y.; Wang, L.; Guo, Y.Y.; Chen, G.H.; Guo, Z.X.; Chen, Y.J.; Li, D.Y.; Cao, P.L.; Shu, H.C.; et al. Parameter Identification of Proton Exchange Membrane Fuel Cell via Levenberg-Marquardt Backpropagation Algorithm. *Int. J. Hydrogen Energy* **2021**, *46*, 22998–23012. [CrossRef]
16. Li, Q.; Chen, W.R.; Dai, C.H.; Jia, J.B.; Han, M. Proton Exchange Membrane Fuel Cell Model Optimization Based on Seeker Optimization Algorithm. *Proc. CSEE* **2008**, *28*, 119–124.
17. Yang, B.; Li, D.Y.; Zeng, C.Y.; Chen, Y.J.; Guo, Z.X.; Wang, J.B.; Shu, H.C.; Yu, T.; Zhu, J.W. Parameter Extraction of PEMFC via Bayesian Regularization Neural Network based Meta-Heuristic Algorithms. *Energy* **2021**, *228*, 120592. [CrossRef]
18. Liu, L.N.; Fathi, G. A Metaheuristic-based Methodology for Efficient System Identification of the Proton Exchange Membrane Fuel Cell Stacks. *Int. J. Hydrogen Energy* **2022**, *47*, 39626–39638. [CrossRef]
19. Abhishek, S.; Rizwan, A.K.; Abhinav, S.; Diwakar, K.; Shailendra, R. A Novel Opposition-Based Arithmetic Optimization Algorithm for Parameter Extraction of PEM Fuel Cell. *Electronics* **2021**, *10*, 2834.
20. Mohammed, Y.S.; Oscar, B.; Aissa, B. A Novel Adaptive PID Controller Design for a PEM Fuel Cell Using Stochastic Gradient Descent with Momentum Enhanced by Whale Optimizer. *Electronics* **2022**, *11*, 2610.
21. Beicha, A. Modeling and Simulation of Proton Exchange Membrane Fuel Cell Systems. *J. Power Sources* **2012**, *205*, 335–339. [CrossRef]
22. Huangpu, Y.G.; Shi, L.; Li, Y.R. Modeling, Simulation and Control of Proton Exchange Membrane Fuel Cell Systems. *J. Northwest. Polytech. Univ.* **2015**, *33*, 682–687.
23. Song, P.; Liu, C.H.; Sun, Y.H. Modeling and Simulation of 5kW Proton Exchange Membrane Fuel Cell. *Chin. Battery Ind.* **2009**, *14*, 382–385.
24. Zhong, Y.; Liu, L.; Hou, J. Simulation of Proton Exchange Membrane Fuel Cell Output Characteristics. *J. Heilongjiang Univ. Sci. Technol.* **2021**, *31*, 802–806.
25. Zhang, G.; Liu, G.J. Output Voltage Stability Control Technology for Proton Exchange Membrane Fuel Cells. *J. Power Supply* **2022**, *20*, 134–140.
26. Sun, Z.; Wang, N.; Bi, Y.R.; Srinivasan, D. Parameter Identification of PEMFC Model based on Hybrid Adaptive Differential Evolution Algorithm. *Energy* **2015**, *90*, 1334–1341. [CrossRef]
27. Sun, Z.; Cao, D.; Ling, Y.W.; Xiang, F.; Sun, Z.X.; Wu, F. Proton Exchange Membrane Fuel Cell Model Parameter Identification based on Dynamic Differential Evolution with Collective Guidance Factor Algorithm. *Energy* **2021**, *216*, 119056. [CrossRef]

28. Huang, Z.W.; Zhou, J.B.; Cui, Y.S. Performance Simulation of Proton Exchange Membrane Fuel Cell System Based on Fuzzy Logic. In Proceedings of the 2020 IEEE Conference on Telecommunications, Optics and Computer Science, Shenyang, China, 11–13 December 2020; pp. 382–384.
29. Xin, Z.L.; Jiang, Y.Q.; Zhong, W.; Zhang, H.Y.; Hou, J.W. The Prediction Model of Irregular Cavity Volume based on CNN-GRNN Neural Network. In Proceedings of the 2022 2nd International Conference on Algorithms, High Performance Computing and Artificial Intelligence, Guangzhou, China, 21–23 October 2022; pp. 758–761.
30. Wang, C.H.; Yuan, Y.; Liu, L.S.; Chen, K.N.; Wu, H.S. Earthquake Magnitude Prediction Based on Generalized Regression Neural Network Optimized by Principal Component Analysis. *Sci. Technol. Eng.* **2022**, *22*, 12733–12738.
31. Fan, L.P.; Xu, Y.W. Constant Pressure Control of Proton Exchange Membrane Fuel Cells Based on Neural Networks. *Comput. Appl. Chem.* **2018**, *35*, 896–902.
32. Jacob, S.L.K.; Madhu, R. An Efficient Approach for Anti-Jamming in IRNSS Receivers Using Improved PSO based Parametric Wavelet Packet Thresholding. *Prot. Control Mod. Power Syst.* **2022**, *3*, 21.
33. Yang, B.; Wang, J.B.; Yu, L.; Shu, H.C.; Yu, T.; Zhang, X.S.; Yao, W.; Sun, L.M. A Critical Survey on Proton Exchange Membrane Fuel Cell Parameter Estimation Using Meta-Heuristic Algorithms. *J. Clean. Prod.* **2020**, *265*, 121660. [CrossRef]
34. Hossam, S.S.; Gaber, M.; Abualkasim, B.; Lstvan, V. Adaptive Coordination Control Strategy of Renewable Energy Sources, Hydrogen Production Unit, and Fuel Cell for Frequency Regulation of a Hybrid Distributed Power System. *Prot. Control Mod. Power Syst.* **2022**, *7*, 34.
35. Majumder, P.; Lu, C.H.; Eldho, T.I. Two-Step Approach based Multi-Objective Groundwater Remediation in Highly Heterogeneous Media Using Enhanced Random Vector Functional Link Integrated with Evolutionary Marine Predator Algorithm. *J. Contam. Hydrol.* **2023**, *256*, 104201. [CrossRef] [PubMed]
36. Amina, Y.; Abdul, R.Y.; Muhammad, B.S.; Saba, Z.; Mudassar, R.; Robina, N.; Ridab, A.E.A.; Shaherbano, B. Fuel Cell Voltage Regulation Using Dynamic Integral Sliding Mode Control. *Electronics* **2022**, *11*, 2922.
37. Guo, Y.; Yang, D.F.; Zhang, Y.; Wang, L.C.; Wang, K. Online Estimation of SOH for Lithium-ion Battery based on SSA-Elman Neural Network. *Prot. Control Mod. Power Syst.* **2022**, *7*, 40. [CrossRef]
38. Xu, B.; Zhang, G.Y.; Li, K.; Li, B.; Chi, H.Y.; Yao, Y.; Fan, Z. Publisher Correction: Reactive Power Optimization of a Distribution Network with High-Penetration of Wind and Solar Renewable Energy and Electric Vehicles. *Prot. Control Mod. Power Syst.* **2023**, *8*, 1. [CrossRef]
39. Attia, A.E. Extracting Optimal Parameters of PEM Fuel Cells Using Salp Swarm Optimizer. *Renew. Energy* **2018**, *119*, 641–648.

energies

MDPI

Article

Massive Multi-Source Joint Outbound and Benefit Distribution Model Based on Cooperative Game

Wang He, Min Liu *, Chaowen Zuo and Kai Wang

School of Electrical Engineering, Guizhou University, Guiyang 550025, China; 13985899533@163.com (W.H.); 18748969164@163.com (C.Z.); wk791234348@163.com (K.W.)
* Correspondence: lium@gzu.edu.cn

Abstract: In light of the challenges posed by the widespread distribution of new energy sources in China and their distance from load centers, the power system must effectively integrate both new energy and thermal power transmission. To address this issue, we propose a dynamic coordinated scheduling model that combines wind, photovoltaic, and thermal power to optimize the profit of the energy complementary delivery system. Additionally, we present an improved ant lion optimization algorithm to investigate the coordinated scheduling and benefit distribution of these three power sources. This paper introduces a cooperative mode for benefit distribution and utilizes an enhanced Shapley value method to allocate the benefits of joint operation among the three parties. The distribution of benefits is based on the contribution of each party to the joint proceeds, considering the profit levels of joint outbound and independent outbound modes. Through our analysis, we demonstrate that the upgraded ant lion optimization algorithm facilitates finding the global optimal solution more effectively within the feasible zone. Furthermore, our suggested three-party combined scheduling model and profit-sharing approach are shown to be superior and feasible.

Keywords: cooperative game; enhanced Shapley value method; enhanced ant lion algorithm; profit distribution

Citation: He, W.; Liu, M.; Zuo, C.; Wang, K. Massive Multi-Source Joint Outbound and Benefit Distribution Model Based on Cooperative Game. *Energies* **2023**, *16*, 6590. https://doi.org/10.3390/en16186590

Academic Editors: Ferdinanda Ponci, Lin Jiang, Bo Yang and Zhijian Liu

Received: 5 July 2023
Revised: 10 August 2023
Accepted: 18 August 2023
Published: 13 September 2023

1. Introduction

Currently, there are evident challenges in the power system, including uneven load distribution, intricate path selection, and implementation complexities [1]. Consequently, the investigation of outbound dispatching operations necessitates the integration of various types of power supplies [2].

In Ref. [3], a static optimization method is proposed for configuring wind power transmission capacity, considering the convergence characteristics of wind farms of varying scales. Ref. [4] analyzes the fluctuation characteristics of a wind power transmission project in Jiu Quan and provides technical insights for planning and researching wind power base transmission projects. In Ref. [5], an equivalent power model for a photovoltaic power station is constructed, and a planning method for optimizing the configuration of its external tie line is proposed, taking into account the output convergence characteristics. To implement this novel approach, Ref. [6] efficiently and reliably transmits locally centralized grid-connected wind and solar power generation to the load center. The authors develop a wind–thermal capacity ratio calculation model, providing valuable insights for wind power transmission research. It is worth noting that most of the preceding literature focuses on capacity allocation for individual energy transmission, neglecting the combined integration of renewable energy sources [7–9].

Numerous scholars have dedicated their research efforts to the optimization of joint delivery scheduling involving wind power, thermal power, and other types of power supply, instead of considering wind power and photovoltaic power separately. Ref. [10] thoroughly examines the scheduling mechanism for the combined operation of wind power and hydropower, shedding light on the interdependent relationship between hydropower capacity and new energy sources. Furthermore, in the context of joint operation involving wind power, photovoltaic power, and hydropower, References [11,12] propose a comprehensive method for optimizing transmission capacity, taking into account factors such as power loss due to transmission restrictions, the potential adverse effects of bundled transmission on thermal power, and the construction costs of transmission and transformation projects. The above-mentioned literature primarily focuses on the integration of wind power and thermal power, as well as the technical and economic analyses of wind–thermal complementarity. Their collective aim is to address the challenges posed by multi-type energy joint large-scale delivery.

Ref. [13] introduces a control strategy designed to ensure system stability for wind–thermal bundled delivery coordination. In Ref. [14], a novel solar–wind–natural gas hybrid system is proposed, emphasizing synergistic complementarity to enhance the economics of new energy sources. Furthermore, Ref. [3] examines the present state of hydropower, wind power, and solar combined operation and identifies critical issues. To accommodate wind generation and PV grid hookup, thermal power units are required for frequent deep peaking and start–stop peaking [15], thereby freeing up grid space. Additionally, under the carbon emission trading mechanism, Ref. [16] develops a wind–solar–heat joint delivery scheduling optimization model. While the preceding literature thoroughly explores and proposes solutions for various aspects of multi-source systems, it nearly overlooks the willingness of thermal power units to participate as auxiliary peak shaving units in large-scale multi-source power joint delivery systems. The critical issue of profit distribution within these systems is scarcely addressed. As cooperative entities, thermal power units must adjust their start–stop status and output in real time to ensure the stable delivery of wind and photovoltaic units, inevitably affecting their own interests to some extent.

In cooperative games, players have the opportunity to form alliances, share earnings, and cooperate through side payments and payoffs [17]. For instance, the cooperative income points of power generation alliances involving wind–thermal systems [18], and the cooperative gain power allocation of wind–solar–hydro multi-agent energy systems [19] have been studied. In the realm of cooperation mode, References [20,21] explored the Owen value [22] and Shapley value [23] techniques to allocate profits based on cooperation, and they concluded that the replacement of power generation rights could serve as a viable cooperation mode for alliances comprising different scale generating units. Furthermore, Ref. [24] utilized the Shapley value allocation technique within a cooperative mode for comparison. Employing the Nash bargaining theory, [25] established a cooperative scheduling model for a wind–solar–hydrogen multi-energy system, aiming to achieve both individual benefits for members and overall benefits for the alliance. Additionally, Ref. [26] introduced regulations for on-grid electricity by bundling hydrogen energy storage on the wind power side, proposing a two-way electricity price compensation mechanism to encourage hydropower's participation in wind power consumption, thereby fostering positive growth within the cooperative alliance. In the context of cost allocation, Ref. [27] meticulously studied, analyzed, and compared various cooperative game theory approaches, ultimately adopting the Shapley method for an effective cost distribution to achieve the best allocation. Furthermore, Ref. [28] employed traditional cooperative game theory to allocate cascade hydropower compensation benefits, leading to fair and equitable outcomes.

The Shapley value allocation technique, as a classic method in cooperative game theory, is renowned for its fairness and reasonability. It serves as a powerful motivator for alliance members to engage in cooperation, fostering positive growth in their individual operational benefits. However, the symmetry inherent in the Shapley value assumes that all alliance members are of the same nature and possess consistent standings within the alliance, disregarding the influence of preferences, discursive power, and market policies among the diverse members. Consequently, the original Shapley value necessitates improvements to address this limitation, wherein the alliance members' weights are assigned differently.

This leads to the development of an enhanced Shapley value allocation strategy known as the M-Shapley value method [29].

The Ant Lion Optimizer (ALO) is a novel meta-heuristic algorithm, conceptualized by Australian scholar Seyedali Mirjalili, drawing inspiration from the natural hunting behavior of ant lions [30]. Nevertheless, within the ALO algorithm's iterative optimization process, the presence of ant lions with relatively low fitness levels poses a challenge. When ants traverse alongside ant lions, it could potentially diminish the search efficiency. Additionally, the ant's random walk process, though facilitating exploration ability, may also hinder the full utilization of discovered optimal solutions, consequently impacting the optimization performance and convergence efficiency of the ALO algorithm to a certain extent. To address these issues, this paper proposes an enhanced ant lion optimization algorithm (MALO) to address the mathematical model at hand. By integrating the vortex convergence method into the modeling, the improved algorithm not only enhances the utilization capability compared to the original algorithm, but also adheres to the biological mechanism of an ant lion preying on ants. This synergistic improvement not only enables better utilization of resources, but also aligns more closely with the natural predator–prey relationship observed in ant lions.

The main contributions of this paper are outlined below:

(1) An advanced three-party joint delivery operation model, integrating wind, photovoltaic, and thermal power, is established utilizing mixed-integer linear programming (MILP), and a novel improved ant lion algorithm is introduced to effectively solve this model;
(2) To address the mathematical model presented in this paper, an enhanced ant lion optimization algorithm (MALO) is proposed. By incorporating the vortex convergence method into the modeling, the improved algorithm not only enhances the utilization ability compared to the original algorithm, but also aligns seamlessly with the biological mechanism of an ant lion preying on ants;
(3) The incremental benefits derived from the joint operation of wind, photovoltaic, and thermal power are allocated using the M-Shapley value method, overcoming the limitations observed in the conventional Shapley value method. This leads to a more equitable and efficient benefit distribution strategy for the three-party joint operation.

The subsequent sections of this paper are structured as follows:

In Section 2, a comprehensive three-party joint delivery optimization operation model is established, employing the principles of MILP. Section 3 puts forward an enhanced ant lion optimization method dedicated to resolving the model developed in Section 2. To address the issue of incremental benefit distribution, Section 4 delves into the cooperative game theory and the M-Shapley value method. The numerical simulation results are presented and analyzed in Section 5, shedding light on their implications. Finally, Section 6 provides a concise summary encapsulating the key findings and conclusions of the entire discourse.

2. Joint Wind, Photovoltaic, and Thermal Power Delivery Model

2.1. Optimal Target Function

The rationale behind the alliance of wind, photovoltaic, and thermal power lies in the geographical distribution of wind energy resources in China, which are located far from the load centers. Additionally, the intermittent and volatile nature of renewable energy sources, such as wind and solar, poses potential risks to the stability of the vast power system. To address this challenge, the output of thermal power plants needs to be finely adjusted to complement and coordinate with the delivery of wind and photovoltaic energy, ensuring the overall power output remains relatively stable.

The joint delivery system's optimization goal can be achieved through the optimization of three key factors. Firstly, taking advantage of the favorable on-grid electricity prices for new energy sources, wind power and solar firms can maximize the utilization of wind energy. Secondly, by fully utilizing the peak shaving benefits of thermal power in the partnership, the system's overall efficiency can be enhanced. Thirdly, optimizing the scheduling combination of the alliance's units can lead to a reduction in power generation costs. It is important to note that wind power incurs no variable cost as it does not rely on fuel consumption. However, optimizing the three aforementioned factors can present conflicting objectives. If the primary aim is to minimize wind and photovoltaic curtailment, new energy resources would be utilized more efficiently. However, the unpredictable and intermittent nature of wind power may result in larger power fluctuations on the transmission line. On the other hand, if the optimization goal is to minimize power fluctuation on the transmission line or minimize coal consumption in thermal power units, it may contradict the original intention of establishing a clean energy joint delivery system.

To ensure the long-term and coordinated development of the alliance, a comprehensive approach incorporating the three criteria is adopted. The optimization model for the wind–thermal joint energy delivery system is built on the premise of satisfying the power fluctuation requirements of the transmission line while ultimately maximizing the benefits for the alliance.

The objective function for optimization is as follows:

$$\max F \tag{1}$$

$$F = \pi^c p^c + \pi^w p^w + \pi^s p^s - c^c - c^w - c^s \tag{2}$$

where F is the total profit of the combined system; π^c is the on-grid price of thermal power in the sending end area; π^w and π^s are the benchmark price of wind power and solar power in the sending end area, respectively; p^c, p^w, and p^s are the on-grid electricity of thermal power, wind power, and photovoltaic power generation, respectively; c^c, c^w, and c^s are the total power generation costs of thermal power units, wind farms, and solar electric fields, respectively.

The power generation cost of each power plant consists of two components: the real-time power generation cost and the fixed cost. The real-time cost is determined by the fuel consumption necessary for power generation, while the fixed cost encompasses operation and maintenance expenses as well as depreciation costs. Hence, the following components are involved:

$$c^c = c^{cv} + c^{cf} \tag{3}$$

$$c^w = c^{wv} + c^{wf} \tag{4}$$

$$c^s = c^{sv} + c^{sf} \tag{5}$$

where c^{cv} and c^{cf} are the real-time output cost and fixed cost of thermal power units, respectively; c^{wv} and c^{wf} are the real-time output costs of wind farms and solar electric fields, respectively, which can be set to zero because they do not consume traditional energy; c^{sv} and c^{sf} are the fixed costs of wind farm and solar electric field, respectively. Due to their high construction costs, they can be converted into constants according to the design operating years.

c^{cv} is expressed as:

$$c^{cv} = \lambda^c \sum_{t=1}^{T} \sum_{i=1}^{N} [u_{it} f_i(P_{it}^c) + u_{it}(1 - u_{i(t-1)}) SU_i] \tag{6}$$

$$f_i(P_{it}^c) = a_i + b_i P_{it}^c + c_i (P_{it}^c)^2 \tag{7}$$

where λ^c is the price of coal; u_{it} is a 0–1 variable, indicating the start–stop state of the unit—1 is the operating state and 0 is the shutdown state; SU_i is the starting consumption of the unit; a_i, b_i, and c_i are the fuel cost coefficients of unit output of thermal power units; P_{it}^c is the real-time output of the i-th unit in t period; $f_i(P_{it}^c)$ is the total coal consumption of unit i in t period.

2.2. Constraint Condition

To guarantee the secure and stable electricity supply, it is imperative to impose constraints on the technical parameters of the energy complementary system.

2.2.1. Maximum Transmission Power Constraint

$$P_t^{tr} \leq P_{\max}^{tr} \tag{8}$$

$$\sum_i^N P_{it}^c(1-\mu_i^c) + P_t^{re-w}(1-\mu^w) + P_t^{re-s}(1-\mu^s) = P_t^{tr} \tag{9}$$

where P_t^{tr} is the amount of electricity transmitted from the transmission line during the t period; P_t^w and P_t^s are the actual output of wind farm and photovoltaic electric field int t period, respectively; $P_{\max}^{tr}$ is the maximum transmission capacity of the transmission line; μ_i^c, μ^w, μ^s, and μ^s are the auxiliary power consumption rates of thermal power plants, wind farms, and solar electric fields, respectively.

2.2.2. Transmission Power Stability Constraint

To avert significant fluctuations in the transmission power of the energy complementary system and mitigate potential risks to the receiving end power system's stability, strict constraints are imposed on the variation range of the transmission power.

$$\Delta P^- \leqslant P_t^{tr} - P_{t-1}^{tr} \leqslant \Delta P^+ \tag{10}$$

where ΔP^- and ΔP^+ are the upper and lower limits of the fluctuation amplitude of transmission power, respectively, which are determined by the reserve capacity of the power system in the receiving area. The constraint requirement on transmission power fluctuations is inversely proportional to the magnitude of the peak regulation margin within the receiving area power grid. A larger peak regulation margin results in a lower constraint requirement for transmission power fluctuations, while a smaller peak regulation margin necessitates a higher constraint on transmission power fluctuations.

2.2.3. Thermal Power Unit Output Constraint

Equation (11) is the upper and lower limits of real-time output power of thermal power units. Equation (12) is the climbing constraint of the unit.

$$u_{it}P_i^{c-\min} \leqslant P_{it}^c \leqslant u_{it}P_i^{c-\max} \tag{11}$$

$$\eta_i^{c-} \leqslant P_{it}^c - P_{it-1}^c \leqslant P_i^{c-\min} \tag{12}$$

where $P_i^{c-\max}$ and $P_i^{c-\min}$ are the maximum and minimum output power of thermal power unit i, respectively. $P_i^{c-\min}$ and η_i^{c-} are the response speed limits of power increase and decrease in coal–thermal units, respectively.

2.2.4. Start–Stop Time Constraint of Thermal Power Unit

The time constraints for start-stop operations of thermal power units are mathematically represented by Equations (17) and (18):

$$\left(T_{i,t-1}^{\text{on}} - T_{\text{M},i}^{\text{on}}\right)(u_{i,t-1} - u_{i,t}) \geqslant 0 \tag{13}$$

$$\left(T_{i,t-1}^{\text{off}} - T_{\text{M},i}^{\text{off}}\right)(u_{i,t} - u_{i,t-1}) \geqslant 0 \tag{14}$$

where Formula (13) is the minimum start-up time constraint of unit i, $T_{i,t-1}^{\text{on}}$ is the running time of unit i at time $t-1$. $T_{\text{M},i}^{\text{on}}$ A is the shortest running time of unit i. Equation (14) is the minimum downtime constraint of unit i, $T_{i,t-1}^{\text{off}}$ is the downtime of unit i at time $i-1$. $T_{\text{M},i}^{\text{off}}$ is the minimum downtime of unit i.

2.2.5. Wind Farm and Solar Electric Field Output Constraints

Equations (15) and (16) are utilized to delineate the association betwixt the real potency of renewable energy and the forsaken wind and solar energy, as well as the projected potency.

$$P_t^{re-w} + P_t^{ab-w} = P_t^{pr-w} \quad (15)$$

$$P_t^{re-s} + P_t^{ab-s} = P_t^{pr-s} \quad (16)$$

where P_t^{re-w} and P_t^{ab-w} are the actual output and wind curtailment power of the wind farm in t period, respectively, P_t^{re-s} and P_t^{ab-s} are the actual output and abandoned photovoltaic power of the solar electric field in period t, respectively. P_t^{pr-w} and P_t^{pr-s} are the predicted output of wind farm and solar electric field, respectively.

2.2.6. Generation Reserve Constraint

By employing Equation (17), one can obtain the constraint conditions for power generation reserve.

$$\sum_{i=1}^{I} P_i^{c-max} \geqslant \gamma^c \sum_{i}^{I} P_{i,t}^{c}(1-\mu_i^c) + \gamma^{W} P_t^{re-W}(1-\mu^{W}) + \gamma^s P_t^{re-s}(1-\mu^s) \quad (17)$$

where γ^c, γ^W, and γ^s are the reserve coefficients of thermal power and wind power, respectively.

The dynamic optimization scheduling model for a large-scale multi-source joint distribution system has been presented in this study. Building upon this foundation, it is possible to develop a profit maximization scheduling model for independent delivery of wind power, as well as a coordinated scheduling model for wind power, thermal power, and photovoltaic power, through appropriate adjustments to the objective function and constraint conditions.

3. Improved ALO Algorithms

3.1. The Original Ant Lion Optimization Algorithm

The ant lion belongs to the ant lion family and undergoes two distinct stages in its life cycle: the larvae and the adult stages.

In the context of optimization algorithms, the ant lion algorithm mimics the hunting behavior of ant lion larvae in nature. As depicted in Figure 1, the ant lion creates a cone-shaped trap using its massive jaws before initiating the predation process. It positions itself at the bottom of the trap, patiently awaiting the arrival of ants and other small insects [31]. Once the prey falls into the trap, the ant lion skillfully throws sand at the edge of the pit, burying the trapped ants and subsequently preying on them. After consuming the prey, the ant lion disposes of the remaining remnants of the ants outside the pit, and then proceeds to modify the pit for future hunting endeavors.

Figure 1. Foraging behavior of ant lions.

There are six steps to use ALO to solve the optimization problem:

① The ant population moves in a random manner following the guidelines specified by Equation (18). However, to guarantee that the ants traverse the feasible region in a randomized fashion, certain adjustments are applied based on Equation (20):

$$X(t) = [0, cumsum(2r(t_1) - 1, \ldots, cumsum(2r(t_k) - 1)] \quad (18)$$

where $X(t)$ is the step set of ant random movement. k is the number of steps that move randomly. The mathematical expression of r is:

$$r = \begin{cases} 1, rand > 0.5 \\ 0, rand \leq 0.5 \end{cases} \quad (19)$$

where $rand$ is a random number between [0, 1].

$$X_i^t = \frac{(X_i^t - a_i) \times (d_i^t - c_i^t)}{(b_i - a_i)} + c_i \quad (20)$$

where a_i and b_i represent the maximum and maximum values of the random movement of the i-dimensional variable, respectively; c_i and d_i represent the minimum and maximum values of the i-dimensional variable in the t generation, respectively.

② As the ant moves in a random manner, it becomes subject to the influence of the trap set by the ant lion. This interaction is mathematically expressed as follows:

$$c^t = Al_j^t + c^t \quad (21)$$

$$d^t = Al_j^t + d^t \quad (22)$$

where c^t and d^t represent the minimum and maximum values of all variables at the t-th iteration, respectively. c_j^t and d_j^t represent the minimum and maximum values of the j-th ant in the t-th iteration, respectively. Al_j^t is the position of the j-th ant lion in the t-th iteration.

③ If an ant falls into the trap, the ant lion will promptly respond by hurling sand towards the trap's edge, effectively hindering the ant from escaping. As a consequence, the ant's activity range progressively diminishes over time. This phenomenon can be mathematically represented as follows:

$$c^t = \frac{c^t}{I} \quad (23)$$

$$d^t = \frac{d^t}{I} \quad (24)$$

$$I = \begin{cases} 1, t \leq 0.1T \\ 10^{\omega} \cdot \frac{t}{T}, t > 0.1T \end{cases} \quad (25)$$

where I is the proportional coefficient, T is the maximum number of iterations, ω is a constant representing the number of iterations that increase with the number of iterations.

④ Each ant can only be captured by a single ant lion, and the specific ant lion to capture an ant is determined using the Roulette Wheel approach. Ant lions that are more fit have a higher likelihood of capturing ants. To make this determination, the individual fitness of each ant lion is compared to the individual fitness of the ant. If the individual fitness of the ant surpasses that of the ant lion, the ant lion preys on the ant, and its position is then replaced with the position of the ant. This process concludes with the update of the ant lion's position. The mathematical formula representing this process is as follows:

$$Al_j^t = Ant_i^t, if : f(Ant_i^t) > f\left(Al_j^t\right) \quad (26)$$

where Ant_i^t is the position of the i-th ant at the t-th iteration. $f(Ant_i^t)$ is the fitness of the i-th ant, $f\left(Al_j^t\right)$ is the fitness of the j-th ant lion. In the single objective optimization problem, only the individual with the best fitness function is selected.

⑤ The elite strategy of the ant lion involves the calculation of the fitness for each individual ant lion. The ant lion that exhibits the highest fitness value emerges as the elite ant lion. This winning ant lion is designated as the elite member of the population. The mathematical expression representing this process is as follows:

$$f\left(Al^t_{elite}\right) = \min\left(f\left(Al^t_1\right), f\left(Al^t_2\right) \ldots f\left(Al^t_m\right)\right) \tag{27}$$

where $f\left(Al^t_{elite}\right)$ is the fitness of the elite ant lion.

⑥ The update of the ant population is performed to preserve the global diversity of ants after each iteration. This update is carried out based on the mathematical expression (28).

$$Ant^t_i = \frac{Al^t_{\text{elite}} + Al^t_{\text{select}}}{2} \tag{28}$$

where Al^t_{elite} is the elite ant lion position at the t-th iteration. Al^t_{select} is the ant lion position selected at the t-th iteration. The average value of the two is the ant individual in the new generation of ant population.

3.2. Ant Lion Algorithm Enhancement

3.2.1. Improve the Roulette Selection Values

The ALO method utilizes roulette selection to randomly choose ant lions for random wandering, aiming to enhance the diversity of the ant population. However, this approach may inadvertently select ant lions with low fitness, leading to a potential reduction in the algorithm's overall optimization efficiency. To address this issue, a constraint mechanism is imposed on the roulette fitness selection value to ensure that the selection process adheres to specific limitations. The constraint mechanism is as follows:

$$f\left(Al^t_j\right) \leqslant \frac{1}{N}\sum\nolimits_{i=1}^{N} f\left(Al^t_j\right) \tag{29}$$

If Al^t_j is less than the ant lion's average fitness value, it participates in roulette selection; otherwise, it does not.

3.2.2. Vortex Convergence

The random walk exhibits significant divergence, ensuring the algorithm's search capability, yet it fails to fully exploit the optimal solution. Moreover, the original program inadequately simulates the process of an ant lion collecting ants, merely updating the ant lion's position based on Formula (26). To enhance the algorithm's utilization capacity and more accurately emulate the scenario when ants are trapped in the ant lion's trap, the improved algorithm induces the ants to spiral towards the ant lion positioned at the trap's center. This behavior is simulated using the logarithmic spiral [32] (Figure 2), and the ants' positions are updated through the following formula:

$$Ant^{t+1}_{ij} = EAl^t_j + D^t_{ij} \cdot Ae^t \cdot \cos(2\pi t) \tag{30}$$

$$t = \left(-2 - \frac{t}{T}\right) \cdot rand + 1 \tag{31}$$

where EAl^t_j is the j-dimensional position of the ant lion capturing ants of the t-th generation. $D^t_{ij} = \left|Ant^t_{ij} - EAl^t_j\right|$ is the distance on dimension j between the ant i and the ant lion that captured it. A is a constant that determines the spiral's shape, it is 0.5 in this paper.

This enhancement offers two significant advantages. Firstly, it more accurately replicates the biological behavior of an ant lion preying on ants, thus enhancing the algorithm's biological plausibility. Secondly, the process of ants spiraling into the ant lion trap enhances the algorithm's utilization capacity, achieving a balance between search exploration and efficient utilization in the ant lion algorithm.

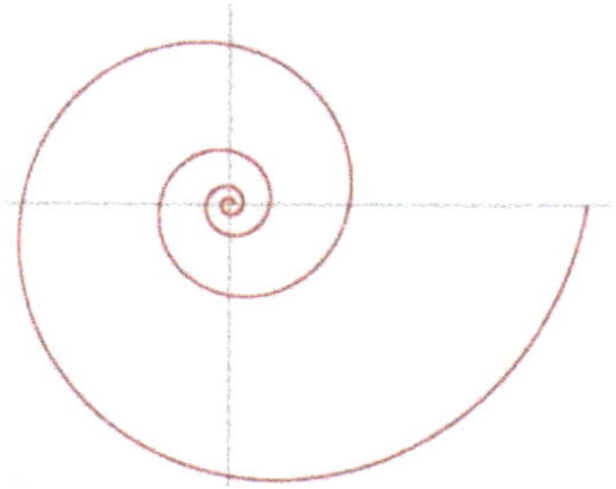

Figure 2. Logarithmic spiral.

3.3. Improved Antlion Algorithm Steps

In summary, the steps for the improved Ant Lion algorithm are outlined as follows:

(1) Input the original data and various constraints, determine the size and dimension of ants and ant lions, randomly initialize the positions of ants and ant lions within the feasible region, and calculate their respective fitness values;
(2) Select the ant lion with the highest fitness level in the ant lion population as the elite ant lion;
(3) Conduct random walks for the chosen ant lion, optimizing the Roulette Wheel approach and the current optimal ant lion, and then update the position of the ant using Formula (30);
(4) Calculate and compare the fitness of the ants after the position update with the fitness of the ant lion. If it is lower than the ant lion's fitness, the ant lion is replaced, and the finest fitness ant lion within the ant lion population becomes the new elite ant lion. The new step update formula generates a new population, calculates the fitness values, directly compares them to the fitness of the elite ant lion, and selects the value with the best fitness as the elite ant lion;
(5) Determine if the maximum number of iterations has been reached; if so, conclude the iteration; otherwise, continue with Step (3).

These steps have been enhanced to better imitate the ant lion's hunting behavior and strike a balance between exploration and utilization, resulting in a more efficient and effective Ant Lion algorithm.

4. Cooperative Game and Profit Distribution Model

4.1. Cooperation Game Model

Cooperative game theory is an effective mathematical modeling technique used to address the problem of benefit distribution arising from the collaborative efforts of multiple stakeholders [33,34]. When multiple stakeholders' actions collectively determine the outcome of a situation and their coordinated behavior leads to greater advantages, it is considered cooperation. The fundamental challenge in cooperative game research is to establish a method of allocating benefits that promotes cooperation among all stakeholders. Notably, fairly distributing the benefits of the alliance to each member is a crucial issue. Dissatisfaction with the distribution can hinder alliance formation and potentially lead to the failure of the alliance, even if it is initially established. Therefore, a key assumption in N-person cooperation is to engage in advance discussions to determine the structure of the alliance and the mechanism for distributing benefits resulting from cooperation [35]. $N = \{1, \dots n\}$ is the player set and 2^N is the power set of N, i.e., each and every subset S of N is regarded as a coalition and assigned a real value via the characteristic function v. Note that normally, there holds $v(\phi) = 0$ which is referred to as normalization in cooperative game theory. Obviously, any payoff allocation $x = (x_1, \dots, x_n)$ should normally meet two requirements: [36]

$$x_i \geq v(\{i\}) \quad (i = 1, 2, \cdots, \mathrm{n}) \tag{32}$$

$$\sum_{i=1}^{n} x_i = v(N) \tag{33}$$

where x_i denotes the share of the i-th member of the alliance. The Formula (32) is known as the individual rationality condition, stating that the income of each member must be greater than that of a single member. Similarly, Formula (33) is called efficiency.

4.2. Shapley Profit Distribution Model

The Shapley Value [23] is the most widely used point-valued solution concept for cooperative game problems. One key benefit of utilizing the Shapley value is its ability to allocate benefits based on the marginal contribution rate of alliance members. In other words, member i's share of the benefits is equal to the average of the marginal benefits they bring through their participation in the alliance. Suppose the cooperative game system is comprised of n members, with the set N= {1, 2, 3, ... n} serving as an indicator. The Shapley value $x(v)$ of a game $v \in G^N$ is the average of the marginal vectors of the game. The specific expression can be formulated as follows [23]:

$$x_i(v) = \sum_{S:i \notin S} \frac{|S|!(n-1-|S|)!}{n!}(v(S \cup \{i\}) - v(S)) \tag{34}$$

$$\frac{|S|!(n-1-|S|)!}{n!} = \frac{1}{n}\binom{n-1}{|S|}^{-1} \tag{35}$$

where for each $i \in N$, the i-th coordinate $x_i(v)$ of $x(v)$ is the expected payoff of player i according to this random procedure. S is a subset of N not containing i. Create a subset S with $i \notin S$ in the following way. To begin, randomly select a number from the urn containing potential sizes ranging from 0 to ($n-1$), where each number (i.e., size) has probability $\frac{1}{n}$ to be drawn. If size s is chosen, draw a set out of the urn consisting of subsets of N$\backslash\{i\}$ of size s, where each set has the same probability $\binom{n-1}{s}^{-1}$ to be drawn. If S is drawn, then one gives player i the amount $v(S \cup \{i\}) - v(S)$. Subsequently, it becomes evident that considering Equation (35), the expected payoff for player i within this stochastic procedure corresponds to the Shapley value assigned to player i in the game $v \in G^N$.

4.3. M-Shapley Profit Distribution Model

According to the Shapley value, all alliance members are considered to have the same nature, and their status within the alliance remains constant, without considering the influence of individual preferences, willingness to participate, and market policies among different members. However, to account for such factors and enhance the original Shapley value, a modified Shapley value allocation approach has been proposed [29]. In this modified approach, various weights are assigned to alliance members. Let us assume that the weight assigned by the alliance members is denoted as follows:

$$M_i = (M_1, M_2, \cdots, M_n) \tag{36}$$

Furthermore:

$$\sum_{i=1}^{n} M_i = 1 \tag{37}$$

Since the default members of the Shapley value strategy have the same affecting factors, that is, $\overline{M} = 1/n$, the distinction is:

$$\Delta M_i = M_i - \frac{1}{n} \tag{38}$$

Furthermore:

$$\sum_{i=1}^{n} \Delta M_i = 0 \tag{39}$$

When the difference value indicates the correction factor while taking into account the real influencing factors, the correction amount of member benefit distribution is:

$$\Delta x_i(v) = v(N) \times \Delta M_i \tag{40}$$

The updated M-Shapley value is as follows:

$$x_i'(v) = x_i(v) + \Delta x_i(v) \tag{41}$$

5. Example Analysis

5.1. Simulation Setup

The total installed capacity of the wind farm participating in energy delivery at the sending end area is considered to be 3300 MW. Table 1 presents the comparable usage efficiency of the wind farm at each phase of a typical day. The photovoltaic electric field is equipped with an equivalent usage efficiency of [19], with an auxiliary power consumption rate of 2%, and incurs a fixed cost of 8 million yuan. Additionally, there are six thermal power units serving as supplementary power sources, and their parameters can be found in Ref. [18]. The power receiving area is expected to provide a specific amount of peak regulation reserve for the transmission system, while the technical constraints of the transmission line are as follows: $P_{\max}^{tr} = 200$ MW,$\Delta P^{+} = 100$ WM, $\Delta P^{-} = -100$ WM. In the sending end area, the on-grid price of wind power is 570 yuan/(MWh), the on-grid price of photovoltaic electricity is 950 yuan/(MW h), the on-grid price of thermal power is 320 yuan/(MWh), and the coal price is 600/t. The modified ant lion algorithm includes 100 ant populations and 100 ant lion populations, with the maximum number of iterations set to 200.

Table 1. Equivalent utilization rate of wind power plant.

Time Period	Utilization Ratio	Time Period	Utilization Ratio	Time Period	Utilization Ratio
1	0.41	9	0.22	17	029
2	0.59	10	0.15	18	0.25
3	0.71	11	0.09	19	0.18
4	084	12	o.22	20	0.16
5	0.69	13	0.23	21	0.18
6	0.57	14	0.16	22	0.29
7	0.49	15	0.28	23	0.36
8	0.41	16	0.35	24	0.46

5.2. Power Transmission Simulation Results

Figure 3 illustrates the real-time output scheduling results of each thermal power unit under the three-party joint delivery mode. It is evident that, to maintain the economic efficiency of the joint delivery system, the No.5 and No.6 units, characterized by lower capacity and higher coal consumption levels, play a more significant role in providing the overall system's peak shaving reserve.

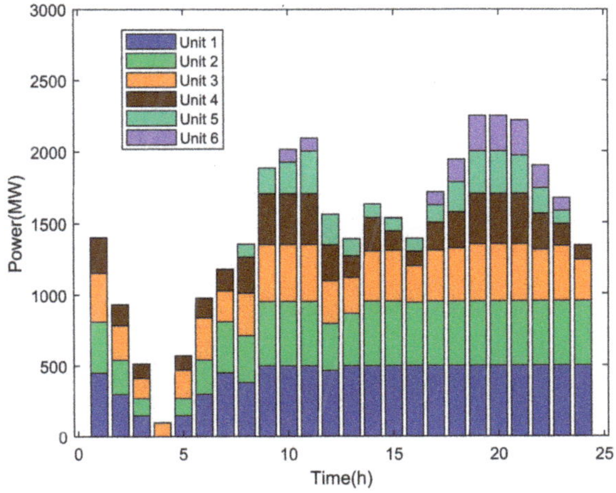

Figure 3. Real-time output of three-party combined thermal power unit.

Figure 4 depicts the power fluctuation of transmission lines under various transmission modes.

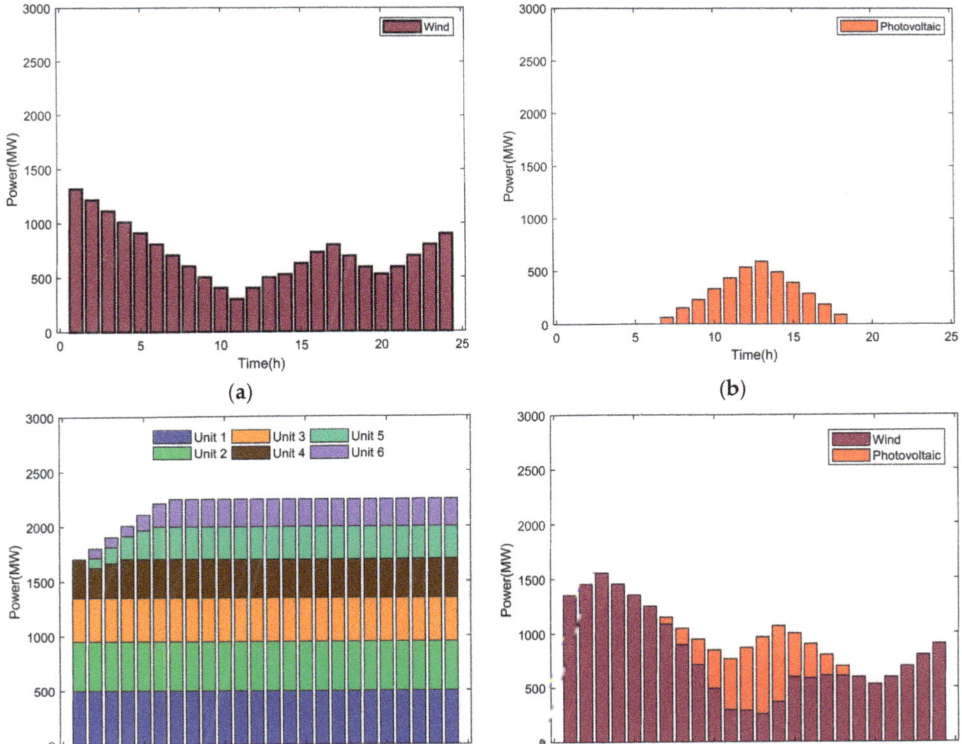

Figure 4. *Cont.*

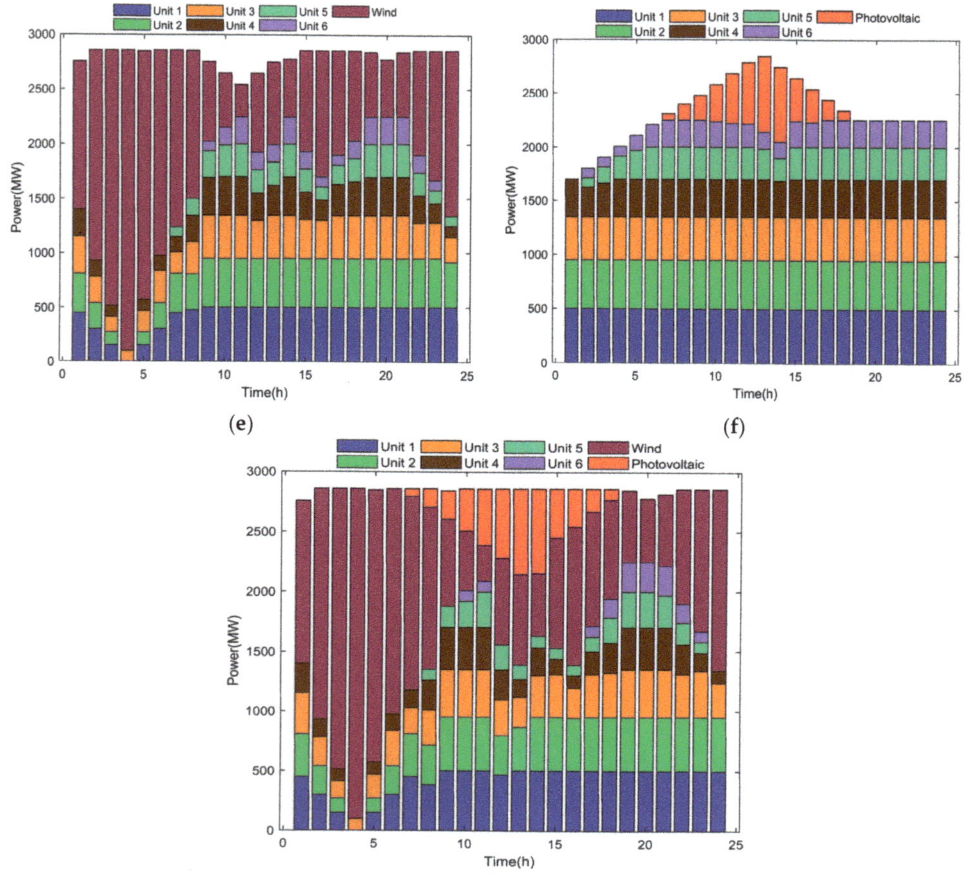

Figure 4. The system transmits power in several modes. (**a**) Independent wind power transmission. (**b**) Photovoltaic independent delivery. (**c**) Thermal power independent delivery. (**d**) Joint delivery of wind and photovoltaic. (**e**) Combined delivery of thermal and wind. (**f**) Photovoltaic and thermal power joint delivery. (**g**) Tripartite joint delivery.

The aforementioned operational results reveal significant variations in the total transmission power of the system between the independent transmission mode of wind power and photovoltaic power and the combined transmission mode of wind power and photovoltaic power, especially when the thermal power unit is not included in the combined transmission system. While independent delivery of thermal power units ensures stable delivered power, it contradicts the environmental trend of energy conservation and emission reduction, and the higher coal consumption cost diminishes the system's economic profit. Conversely, joint delivery operations involving the thermal power unit, such as the three-party joint delivery mode, the thermal–thermal joint mode, and the thermal–photovoltaic joint mode, result in relatively low power fluctuation and ensure the stability of the delivery system.

In Table 2, the wind and photovoltaic abandonment rates are presented for different operation modes. Notably, in the combined operation mode of wind, solar, and thermal power, wind power abandonment is only 1.6%, while solar electric field abandonment is 0%. This three-party cooperative delivery system, including thermal power units, significantly enhances the utilization rate of wind and solar energy, as evidenced by the lower rates of abandoned wind and photovoltaic energy compared to their respective rates in the independent operation mode of wind farms and solar electric fields.

Table 2. Comparison of wind and photovoltaic abandonment under three-party joint and independent modes.

Mode of Operation	Abandoned Wind Rate/%	Photovoltaic Curtailment Ratio/%
Tripartite Union	1.60	0
Wind power independence	37.82	-
Photoelectric independence	-	9.82

In Figure 5, the optimization iteration diagram showcases the process of solving the objective function using both the ant lion algorithm and the modified ant lion algorithm. From the graph, it is evident that the modified ant lion method has the potential to yield a superior solution, while also maintaining the option for further optimization, thanks to its improved balance between search and utilization.

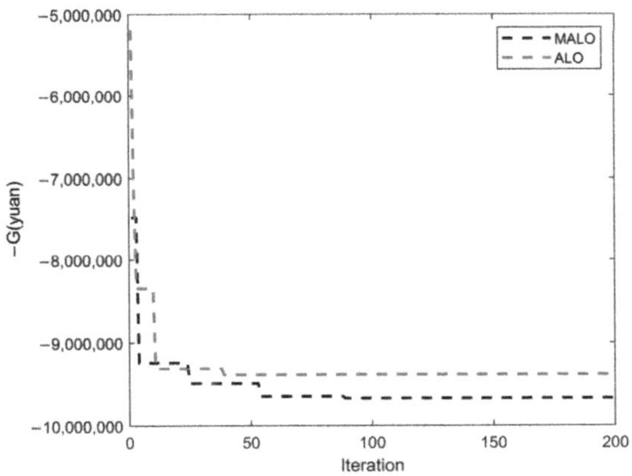

Figure 5. Comparison of ant lion algorithm and improved ant lion iteration diagram.

Table 3 provides a comprehensive comparison between ALO, MALO, and other literature in terms of profitability. Remarkably, the energy complementary system based on MALO achieves a profit of 9,669,273 yuan, surpassing the literature's profit by 154,621 yuan [19]. This result demonstrates the superior performance and effectiveness of the MALO method in optimizing the profitability of the system.

Table 3. Comparison of joint total profit of three parties.

Solving Algorithm	Profit	Literature
ALO	9,484,652/Yuan	--
LINGO optimizer	9,514,652/Yuan	Literature [19]
MALO	9,669,273/Yuan	--

In this research, we employ the enhanced ant lion optimization algorithm to determine the maximum profits under seven distinct operating modes, utilizing the proposed three-party joint delivery model. The specific profits for each mode are presented in Table 4. The results provide valuable insights into the profitability of the system under various scenarios, highlighting the effectiveness of the enhanced algorithm in optimizing the energy delivery process.

Table 4. Profit comparison under different modes.

Operating Mode	Tripartite Union	Combined Thermal and Wind Power	Combined Thermal Power and Photovoltaic	Wind Power, Photovoltaic Joint
Profit/Yuan	9,669,273	8,856,441	3,276,310	4,165,790
Operating mode	Thermal power independence	Wind power independence	Photoelectric independence	
Profit/Yuan	2,521,129	1,936,098	629,670	

The data presented in the analysis table clearly illustrate that the joint operation of the participating electric fields in the three-party system yields additional profits due to mutual cooperation. Notably, the income from joint operation significantly surpasses the sum of the incomes obtained from individual operations, making the profit generated by the three-party joint delivery the highest among all scenarios. This outcome underscores the substantial benefits of collaborative efforts and joint optimization in enhancing overall profitability.

The combined operation of thermal power units alongside wind power and solar energy in the energy complementary system plays a crucial role in significantly increasing the overall output power. When the income of wind farms and solar power plants is determined by the on-grid pricing of wind power and photovoltaic electricity, their revenues rise in tandem. However, as complementary power sources, thermal power units must adjust their output to match the levels of wind turbines to prioritize the delivery of wind and solar power. Consequently, the output power of thermal power units in the energy complementary system is limited, and the economic benefits of each unit become uncertain due to frequent power adjustments.

If the revenue of thermal power units is based on the benchmark price of thermal power, it becomes challenging for these units to achieve optimal returns, with some units even failing to break even. To properly reflect the synergistic value of thermal power units as complementary power sources, a fair and rational profit distribution mechanism is essential. This mechanism should distribute profits according to the contribution rate of wind farms, photovoltaic power plants, and thermal power units to the overall benefits of the energy complementary system.

To ensure cooperation among individuals participating in the energy complementary system, it is crucial to establish a profit distribution approach that encourages alliance cooperation by promising higher profit income compared to non-cooperation scenarios. In this context, the M-Shapley value distribution technique can effectively address the profit distribution challenge of an energy complementary system. By adopting the M-Shapley value, the system can achieve an equitable distribution of profits, considering the contributions of each participant to the overall benefits of the cooperative alliance. This approach encourages mutual cooperation and maximizes the profitability of the energy complementary system.

5.3. M-Shapley Value Allocation Strategy

The contribution of each participant to the alliance and the income of various alliance formation procedures must be considered when calculating the Shapley value. To simplify the concept, suppose that {1,2,3} represent a vast alliance of wind power, photovoltaic power, and thermal power. {1,2}, {1,3}, and {2,3} represent an alliance of three parties. {1}, {2}, and {3} denotes a coalition of one party. The profit distribution of the three power generation enterprises can be calculated using Formula (32), as follows:

$$x_1(v) = \frac{2!}{3!} \times (v\{1\}) + \frac{1!}{3!} \times (v\{1,2\} - v\{2\}) + \frac{1!}{3!} \times (v\{1,3\} - v\{3\}) + \frac{2!}{3!} \times (v\{1,2,3\} - v\{2,3\}) = 4,421,592.33$$

$$x_2(v) = \frac{2!}{3!} \times (v\{2\}) + \frac{1!}{3!} \times (v\{1,2\} - v\{1\}) + \frac{1!}{3!} \times (v\{2,3\} - v\{3\}) + \frac{2!}{3!} \times (v\{1,2,3\} - v\{1,3\}) = 1,356,481.17$$

$$x_3(v) = \frac{2!}{3!} \times (v\{3\}) + \frac{1!}{3!} \times (v\{1,3\} - v\{1\}) + \frac{1!}{3!} \times (v\{2,3\} - v\{2\}) + \frac{2!}{3!} \times (v\{1,2,3\} - v\{1,2\}) = 3,891,199.5$$

Profits can be obtained from three power generation firms, but the Shapley value is increased by taking into account the contribution of thermal power units to the system to ensure their willingness to participate in joint operation. Set M= (M1, M2, M3) = (0.33,0.333,0.34) for each of the three parties. And, according to Equations (38) and (40), the profit distribution correction for each member is as follows:

$$\Delta x_1(v) = v(1,2,3) \times \Delta M_1 = -32,230.91$$

$$\Delta x_2(v) = v(1,2,3) \times \Delta M_2 = -32,230.91$$

$$\Delta x_3(v) = v(1,2,3) \times \Delta M_3 = 64,461.82$$

The modified improved Shapley value is obtained by Formula (41):

$$x_1'(v) = x_1(v) \times \Delta x_1(v) = 4,389,361.42$$

$$x_2'(2) = x_2(v) \times \Delta x_2(v) = 1,324,250.26$$

$$x_3'(v) = x_3(v) \times \Delta x_3(v) = 3,955,661.32$$

Conclusively, the wind farm has yielded a substantial profit of 4,389,361.42 million yuan, while the photovoltaic power plant has recorded a commendable profit of 1,324,250.26 million yuan. Additionally, the thermal power plant has demonstrated its profitability with an impressive earning of 3,955,661.32 million yuan.

Following the implementation of the M-Shapley strategy for profit distribution, a noteworthy outcome emerges. Wind farm's share of the alliance's profit stands at 45.39%, a considerable decrease from the pre-M-Shapley value distribution level of 77.92%. Similarly, the photovoltaic power plant's share accounts for 13.70% of the alliance's profits, reflecting a decrease compared to the previous level of 15.87%. This adjustment aims to effectively capture the cooperative value of thermal power units within the energy complementary export system. A portion of the economic income generated by wind power units and photovoltaic power plants is allocated as a subsidy to offset the opportunity cost associated with thermal power. Nevertheless, despite the reduction in profit share, both wind farms and photovoltaic power plants experience significant enhancements when compared to their independent operation. Notably, the profit obtained by wind farms after M-Shapley value distribution amounts to a substantial 438,936,142,000 yuan, surpassing the independent operation profit of 1,936,098 yuan. Similarly, the profit attained by photovoltaic power plants after M-Shapley value distribution reaches an impressive 132,425,026,000 yuan, significantly exceeding the independent operation profit of 629,670 million yuan. Thus, there exist compelling justifications for wind farms and photovoltaic power plants to actively engage in the joint delivery system. Additionally, by amalgamating the distribution outcomes of the M-Shapley value method with the installed capacity and actual output of each electricity field, along with the profit levels associated with different combinations outlined in Table 5, it becomes evident that participants who contribute more to the alliance's interests secure greater profits through the M-Shapley value distribution. This serves as a means to attract their continued involvement in the alliance, thereby fostering the creation of even more substantial profits.

Table 5. Comparison of direct allocation and M-Shapley value allocation results.

Distribution Mode	Wind Power		Photovoltaic		Thermal Power	
	Profit/Ten Thousand Yuan	Profit Ratio/Ten Thousand Yuan	Profit/Ten Thousand Yuan	Profit Ratio/Ten Thousand Yuan	Profit/Ten Thousand Yuan	Profit Ratio/Ten Thousand Yuan
Direct distribution	7,534,297.52	77.92%	1,534,513.63	15.87%	600,461.85	6.21%
M-Shapley strategy	4,389,361.42	45.39%	1,324,250.26	13.70%	3,955,661.32	40.91%

6. Conclusions

To address the paradox of China's new energy distribution, which is concentrated in certain areas but far from the power load centers, this study employs cooperative game theory to explore the coordinated scheduling and benefit distribution of wind power, photovoltaic, and thermal power during delivery. The cooperative model can be divided into two sub-problems: cooperation optimization and benefit distribution. Based on the analysis, the following conclusions can be drawn.

This paper introduces a novel large-scale multi-source joint export model based on cooperative game theory, integrating new energy and thermal power generation concepts. The model is effectively solved using the enhanced ant lion optimization technique proposed in this study. The improved ant lion algorithm enhances the feasibility of obtaining the global optimal solution, making the formation of a grand alliance among wind power plants, solar power plants, and thermal power plants conceivable. Such an alliance promises tremendous economic and social benefits.

The enhanced ant lion optimization algorithm greatly facilitates the attainment of the global optimal solution. It is now possible to form a large alliance comprising a wind farm, a photovoltaic electric field, and a thermal power plant, leading to tremendous economic and social benefits.

A well-structured benefit distribution mechanism plays a pivotal role in the sustainable development of the wind, photovoltaic, and thermal energy delivery alliance. In this study, the M-Shapley value method is employed to allocate the benefits of the joint operation among the three parties. The key finding is that participants who contribute more to the alliance's benefits are rewarded with higher profits in the M-Shapley value approach, making alliance participation more attractive and enabling the creation of greater profits.

In the investigated multi-source joint delivery model, wind power, solar power, and thermal power all contribute to the power supply side. Moreover, the inclusion of additional elements, such as photothermal power plants and energy storage, can be explored further in this context. While addressing the benefit distribution among alliance members in this study, we assume all entities to be completely rational. However, in real-world scenarios, firms often exhibit "incomplete rationality" as they seek to maximize their own interests. Therefore, it becomes imperative to research and modify the preconditions for the distribution of alliance benefits, ensuring greater applicability of the benefit distribution scheme in real market settings.

Author Contributions: Conceptualization, methodology, software, writing—original draft preparation, writing—review and editing, visualization, W.H.; formal analysis, investigation, supervision, M.L.; supervision, validation, C.Z.; supervision, validation, K.W. All authors have read and agreed to the published version of the manuscript.

Funding: This research received no external funding.

Data Availability Statement: The research data has been described in Section 5, along with the corresponding references. The data is derived from reference [19].

Conflicts of Interest: The authors declare no conflict of interest.

References

1. Cui, Y.; Li, C.; Zhao, Y. Source-grid-load multi-period optimal scheduling method considering wind-photovoltaic-photothermal combined DC transmission. *Proc. CSEE* **2022**, *42*, 559–573.
2. He, W.; Liu, M.; Wang, K.; Pan, B. Massive Multi-source Joint Outbound and Benefit Distribution Model Based on Cooperative Game. In Proceedings of the 2023 IEEE International Conference on Power Science and Technology (ICPST), Kunming, China, 5–7 May 2023; pp. 519–527.
3. Shen, J.; Wang, Y.; Cheng, C. Research status and prospect of water-wind-solar multi-energy complementary power generation scheduling problem. *Proc. CSEE* **2022**, *42*, 3871–3885.
4. Zhang, B.; Yu, J.; Huang, W. Challenges and thinking of large-scale new energy transmission power grid in Gansu. *Grid. Clean. Energy* **2020**, *36*, 81–89, 96.
5. Mei, H.; Gao, B.; Cao, Z. Optimal configuration of wind-solar-thermal-storage combined delivery system with CSP power station. *J. Solar. Energy* **2022**, *43*, 124–133.
6. Alassi, A.; Bañales, S.; Ellabban, O.; Adam, G.; MacIver, C. HVDC Transmission: Technology Review, Market Trends and Future Outlook. *Renew. Sustain. Energy Rev.* **2019**, *112*, 530–554. [CrossRef]
7. Li, Q.; Zhao, D.; Yin, J. Sediment Instability Caused by Gas Production from Hydrate-bearing Sediment in Northern South China Sea by Horizontal Wellbore: Evolution and Mechanism. *Nat. Resour. Res.* **2023**, *32*, 1595–1620. [CrossRef]
8. Wang, F.; Liu, X.; Jiang, B.; Zhuo, H.; Chen, W.; Chen, Y.; Li, X. Low-loading Pt nanoparticles combined with the atomically dispersed FeN4 sites supported by FeSA-N-C for improved activity and stability towards oxygen reduction reaction/hydrogen evolution reaction in acid and alkaline media. *J. Colloid Interface Sci.* **2023**, *635*, 514–523. [CrossRef]
9. Li, Q.; Zhang, C.; Yang, Y.; Ansari, U.; Han, Y.; Li, X.; Cheng, Y. Preliminary experimental investigation on long-term fracture conductivity for evaluating the feasibility and efficiency of fracturing operation in offshore hydrate-bearing sediments. *Ocean Eng.* **2023**, *281*, 114949. [CrossRef]
10. Xie, J.; Zhang, Y.; Pan, X. A short-term optimal scheduling model for wind- solar- hydro hybrid generation system with cascade hydropower considering regulation reserve and spinning reserve requirements. *IEEE Access* **2021**, *9*, 10765–10777. [CrossRef]
11. Zhang, G.; Zhu, Y.; Xie, T.; Zhang, K.; He, X. Wind Power Consumption Model Based on the Connection between Mid- and Long-Term Monthly Bidding Power Decomposition and Short-Term Wind-Thermal Power Joint Dispatch. *Energies* **2022**, *15*, 7201. [CrossRef]
12. Xu, D.; Guo, J.; Ding, X. Co-evolutionary genetic algorithm based joint economic dispatch of wind-solar-storage in distribution network. *New Technol. Electr. Energy* **2020**, *39*, 51–57.
13. Li, F. *Study on Improving Transient Stability of Wind-Thermal-Bundled System by Current-Limiting SSSC*; Northeast Electric Power University: Jilin, China, 2021.
14. Xu, J.; Liu, L.; Wang, F. Equilibrium strategybased economic- reliable approach for day ahead scheduling towards solar wind gas hybrid power generation system: A case study from China. *Energy* **2022**, *240*, 122728. [CrossRef]
15. Rana, M.M.; Atef, M.; Sarkar, M.R.; Uddin, M.; Shafiullah, G. A Review on Peak Load Shaving in Microgrid—Potential Benefits, Challenges, and Future Trend. *Energies* **2022**, *15*, 2278. [CrossRef]
16. Tan, Q.; Ding, Y.; Li, Y. Multi-objective optimization of wind-solar-thermal joint dispatch strategy considering economic-environmental balance. *Power Const.* **2020**, *41*, 129–136.
17. Contreras, J. A Cooperative Game Theory Approach to Transmission Planning in Power Systems. Ph.D. Thesis, University of California, Berkeley, CA, USA, 1997.
18. Tan, Z.; Song, Y.; Zhang, H.; Shang, J. Large-scale wind power and thermal power combined delivery system and its profit distribution model. *Power Syst. Auto* **2013**, *37*, 63–70.
19. Wu, D.; Yu, A. Profit allocation based on nucleolus theory in large-scale multi-source joint delivery coordination scheduling. *Power System Technol.* **2016**, *40*, 2975–2981.
20. Chen, C.; Wu, C.; Kang, K.; Lin, X.; Ma, Y.; Sui, Q.; Xu, H. Optimal Strategy of Distributed Energy Storage Two-layer Cooperative Game Based on Improved Owen-Value Method. *Proc. CSEE* **2022**, *42*, 3924–3936.
21. Ye, G.; Gao, F. Coordinated Optimization Scheduling of Data Center and Electricity Retailer Based on Cooperative Game Theory. *CPSS Trans. Power Electr. App.* **2022**, *9*, 273–282. [CrossRef]
22. Owen, G. Values of Games with a Priori Unions. In *Mathematical Economics and Game Theory*; Springer: Berlin/Heidelberg, Germany, 1977; Volume 141, pp. 76–88.
23. Shapley, L.S. *A Value for n-Person Games. Contributions to the Theory of Games II*; Princeton University Press: Princeton, NJ, USA, 1953; pp. 307–317.
24. Xie, J.; Zhang, L.; Chen, X. Incremental benefit allocation for joint operation of multi-stakeholder wind-PV-hydro complementary generation system with cascade hydro-power: An Aumann-Shapley value method. *IEEE Access* **2020**, *8*, 68668–68681. [CrossRef]
25. Ma, T.; Pei, W.; Xiao, H. Cooperative operation method for wind-solar-hydrogen multi-agent energy system based on Nash bargaining theory. *Proc. CSEE* **2021**, *41*, 25–39.
26. Li, X.; Yang, Y. Joint optimal dispatch of wind power and cascade hydropower with hydrogen storage based on bi-directional electricity price compensation. *Power Syst. Technol.* **2020**, *44*, 3297–3305.
27. Shandilya, S.K.; Izonin, I.; Singh, K.K. Modeling and Comparative Analysis of Multi-Agent Cost Allocation Strategies Using Cooperative Game Theory for the Modern Electricity Market. *Energies* **2022**, *15*, 2352. [CrossRef]

28. Zhang, L.; Xie, J.; Chen, X. Cooperative game-based synergistic gains allocation methods for wind-solar-hydro hybrid generation system with cascade hydropower. *Energies* **2020**, *13*, 3890. [CrossRef]
29. Shi, H. *Research on the Mechanism of Wind-Thermal Energy Delivery and Benefit Distribution Based on Cooperative Game*; Xian University of Technology: Xi'an, China, 2019.
30. Mirjalili, R. The ant lion optimizer. *Adv. Eng. Soft.* **2015**, *83*, 80–98. [CrossRef]
31. Wang, Y.; Ni, Y.; Zheng, Y. Remaining useful life prediction of lithium-ion battery based on ALO-SVR. *China J. Electr. Eng.* **2021**, *41*, 1445–1457.
32. Liu, S.; Gao, Z.; Li, M. A closed-form logarithmic spiral method for seismic passive earth pressure in anisotropic sand. *Comput. Geotech.* **2022**, *152*, 105052. [CrossRef]
33. Xie, Y.; Yao, Y.; Wang, Y.; Cha, W.; Zhou, S.; Wu, Y.; Huang, C. A Cooperative Game-Based Sizing and Configuration of Community-Shared Energy Storage. *Energies* **2022**, *15*, 8626. [CrossRef]
34. Wu, W.; Zhu, J.; Chen, Y. Modified Shapley Value-Based Profit Allocation Method for Wind Power Accommodation and Deep Peak Regulation of Thermal Power. *IEEE Trans. Ind. App.* **2023**, *59*, 276–288. [CrossRef]
35. Devi, N.N.; Thokchom, S.; Singh, T.D.; Panda, G.; Naayagi, R.T. Multi-Stage Bargaining of Smart Grid Energy Trading Based on Cooperative Game Theory. *Energies* **2023**, *16*, 4278. [CrossRef]
36. Branzei, R.; Dimitrov, D.; Tijs, S. *Models in Cooperative Game Theory*; Springer Science & Business Media: Berlin/Heidelberg, Germany, 2008.

processes

Article

Research on Optimal Scheduling of Multi-Energy Microgrid Based on Stackelberg Game

Bo Li [1], Yang Li [2], Ming-Tong Li [1], Dan Guo [1], Xin Zhang [1], Bo Zhu [1], Pei-Ru Zhang [1] and Li-Di Wang [1,*]

[1] College of Information and Electrical Engineering, Shenyang Agricultural University, Shenyang 110065, China; 18842586001@163.com (B.L.); goldenflying123@syau.edu.cn (D.G.); mlzx662022@163.com (X.Z.); m15898315745@163.com (B.Z.); 17640499389@163.com (P.-R.Z.)
[2] State Grid Liaoning Electric Power Co., Ltd. Shenyang Power Supply Company, Shenyang 110065, China
* Correspondence: wanglidi@syau.edu.cn

Abstract: In recent years, rapid industrialization has driven higher energy demand, depleting fossil-fuel reserves and causing excessive emissions. China's "dual carbon" strategy aims to balance development and sustainability. This study optimizes microgrid efficiency with a tiered carbon-priced economy. A Stackelberg game establishes microgrid-user equilibrium, solved iteratively with a multi-population algorithm (MPGA). Comparative analysis can be obtained without considering demand response scenarios, and the optimization cost of microgrid operation considering price-based demand response scenarios was reduced by 5%; that is 668.95 yuan. In addition, the cost of electricity purchase was decreased by 23.8%, or 778.6 yuan. The model promotes user-driven energy use, elevating economic and system benefits, and therefore, the scheduling expectation of "peak shaving and valley filling" is effectively realized.

Keywords: energy; MPGA; stackelberg game; tiered carbon-priced economy

Citation: Li, B.; Li, Y.; Li, M.-T.; Guo, D.; Zhang, X.; Zhu, B.; Zhang, P.-R.; Wang, L.-D. Research on Optimal Scheduling of Multi-Energy Microgrid Based on Stackelberg Game. *Processes* **2023**, *11*, 2820. https://doi.org/10.3390/pr11102820

Academic Editors: Lin Jiang, Bo Yang and Zhijian Liu

Received: 22 August 2023
Revised: 7 September 2023
Accepted: 21 September 2023
Published: 24 September 2023

1. Introduction

With the rapid development of modern industry, the contradiction between energy consumption and the environment has become increasingly prominent, and the large emission of greenhouse gases has led to a more severe ecological environment and more serious global warming: according to the BP World Energy Statistical Yearbook 2022 [1], global primary energy consumption in 2021 showed a rebound trend, an increase of nearly 6%. This significantly reversed the sharp decline in energy consumption due to lockdowns in many countries around the world in 2020. Therefore, all countries are seeking energy transformation, and any successful and long-lasting energy transformation cannot be separated from the three elements: safety, burden, and low-carbon. Therefore, countries have formulated corresponding emission reduction targets, focusing on policy incentives and government economic investment as the main body to achieve excessive excess of traditional energy to low-carbon energy. In the daily production and consumption process led by electricity as a high-quality energy source, according to the theoretical guidance of "high energy, high energy, low energy and low-use, temperature-oriented, step use", the three-energy coupling of electricity–heat–gas are explored and improved energy conversion. The new way of rate, especially the utilization of thermal energy, has achieved results in the energy terminal. As a large country of energy use, China strives to achieve carbon peaks by 2030 and achieve carbon neutralization by 2060 [2]. In order to achieve the ambitions and commitments of decarburization and achieve promise, China must establish a relatively sound energy system, with the development of clean energy as the guide, and take the first step in the transformation of China's energy industry structure [3].

In recent years, many scholars have conducted a lot of research on the optimization of the comprehensive energy system. Zhou Nan and others introduced the time-sharing pricing strategy into the optimization scheduling of energy storage to maximize the annual profit and utilization rate of photovoltaic systems [4]. Li Xiulei et al. established a general model for energy storage and demand response optimization planning, and analyzed the impact of energy storage and demand response on goals in operational strategies. The end of the energy cycle in demand response is usually thermal energy, so the main coupling method is the electrical thermal component, and the load type mainly exists in the form of temperature-controlled loads. However, due to its inherent limitations, thermal energy is often strongly influenced by external weather factors and has a delay effect. Therefore, through research on modeling, optimization, and control of electrical thermal coupling components, its impact on power system stability can be reduced [5]. Wang Dan et al. provided an energy-saving design for temperature-controlled loads in buildings, while considering the limitations of user comfort to determine the operating method of the power plant, and established the optimal allocation model for energy-saving power plants [6]. Wang Chengshan et al. used a simplified equivalent thermodynamic model with first-order parameters and a state control model to smooth out renewable energy in microgrids by replacing energy storage technology with demand side and load response technology [7]. Tang Xiaoting et al. constructed a mathematical model that includes input and output energy balance constraints of energy hubs, system capacity constraints, and energy storage efficiency constraints of renewable energy generation technology [8]. Heiskanin et al. Proposed the Energy System Analysis Environmental Assessment Framework (EAFESA) [9]. This framework can minimize the shortcomings of the two models and maximize the combination of the two models to analyze the no climatic environmental impacts of energy scenarios based on the life cycle. Wu Yong et al. established a multi-objective optimization model for comprehensive planning of various energy storage capacities with the goal of minimizing economic costs and network carbon emissions, but did not consider the impact of different energy storage components on the system [10]. Zhang Xizheng searched algorithm (GSA) and particle swarm optimization (PSO) algorithm by combining gravitational, a hybrid modified GSA-PSO (MGSA-PSO) scheme is proposed to optimize the load dispatch of the microgrid containing electric vehicles. The load dispatch optimization are implemented and analyzed, including the unordered charging strategy, the ordered charging–discharging strategy, and the ordered charging–discharging strategy with distributed generations [11]. Zhang Feng et al. applied a robust ALO optimizer (ALO) algorithm for MPP tracking of solar photovoltaic system, designed the charge controller of the energy storage system, and designed the DC–AC converter to match the frequency of RES with the frequency of DG [12].

In summary, due to the continuous improvement of energy coupling in microgrids and the continuous reform of the electricity market, the mathematical models of equipment in microgrids have increased, and operation scheduling strategies have become more complex. Traditional control methods for loads need to be reformed. In addition, in the context of huge carbon emissions, carbon reduction and environmental benefits cannot be achieved in parallel. Therefore, this article conducts reasonable scheduling of devices in microgrids, seeks more reasonable and economical device matching methods, introduces a demand response mechanism, constructs a low-carbon economic optimization model for microgrids with tiered carbon prices, and establishes a Stackelberg game model to control user load, achieving the goal of "peak shaving and valley filling", which is of great significance for the overall energy utilization of microgrids.

Considering the multi-energy micro-network planning process under the carbon trading mechanism, there are many stakeholders during energy transactions: the distribution network–microcyllar network and user side. In order to meet the demands of various interests at the same time, the main game framework was introduced. At the same time, because there is an energy interaction between the distribution network and the microcontrollers, the relationship between the energy demand response and the user's side has the relationship between the micro grid and the user. That is to say, on the upper layer of the main game, the microblogs are maximized to the system's efforts to maximize the effort of the system according to the price signal given by the distribution network. To participate in the scheduling for the goal, and to obtain a balanced solution, micro-network response strategies through price demand can guide users to participate in micro-network scheduling can achieve energy transition. As the main body of energy use, the spontaneous use of the electricity price incentive signal transfer period or cut load based on the spontaneous operator's electricity price incentive signal can be used to achieve the goal of peak-cutting the valley of the load of the microblog.

Therefore, this article proposes a multi-energy microfinance low-carbon tone considering the main game, and the establishment of the overall micro-power grid.

This article does the following:

1. A low-carbon economic optimization model of microgrid with tiered carbon price was constructed. The carbon emissions involved in the operation of each equipment are finally traded through the carbon trading market, and the sensitivity analysis of the system is carried out by adjusting the ladder carbon trading parameters.
2. A Stackelberg game model with microgrid as the main body of the game and user response as the follower of the game was established, which further improved the level of load participation in the energy system and proved the existence of equilibrium solutions in the game. The model is iteratively solved by a variety of group algorithms, and immigration operators and artificial selection operators are added to the traditional genetic algorithm to prevent all individuals in the population from tending to the same state and stop evolution, and at the same time increase the memory population and improve the model calculation efficiency.

2. Optimization Model of Micro-Grid Low-Carbon Economy with Ladder Carbon Price

2.1. Stepped Carbon Trading Model

Carbon emission quota trading can limit the amount of carbon emissions according to specific emission industries. At the same time, the real-time price of the carbon market is determined by the carbon trading market, and the carbon emission quota is managed in a reasonable and efficient way [13].

2.1.1. Calculation Model of Carbon Emissions

The main carbon sources in microgrid are power consumption process, heat generation and power generation process of cogeneration unit, gas-fired boiler, and coal-fired unit. Because the coal-fired units are equipped with carbon capture power plants, it can help the microgrid consume part of CO_2. Therefore, when calculating the actual carbon emissions, the CO_2 absorbed by the carbon capture equipment needs to be removed. The total carbon emission model of microgrid is shown in Formula (1), and the total amount of gas purchased is shown in Formula (2).

$$E_{all} = \lambda_e \sum_{t=1}^{T} P_{e,buy} + \lambda_g \sum_{t=1}^{T} P_{g,buy} + Q_{CO_2}^{PGU} - \sum_{t=1}^{T} Q_{CO_2}^{CCS} \tag{1}$$

$$P_{g,buy} = P_{g,CHP} + P_{g,GB} \tag{2}$$

$P_{e,buy}$, $P_{g,buy}$ are purchase electricity and gas, λ_e, λ_g are carbon emission coefficients of electricity-consuming equipment and gas-consuming equipment, $Q_{CO_2}^{PGU}$ is emissions for coal-fired power plants CO_2 quantity, and $Q_{CO_2}^{CCS}$ is the amount of carbon capture equipment CO_2.

2.1.2. Carbon Decentralization Quota Model

The initial allocation of carbon emission rights in this paper mainly includes electricity purchase quota, cogeneration unit quota, gas boiler quota, and coal-fired power plant quota. The carbon emission quota model for power purchase in microgrid is shown in

Formula (3), and the model synthesis of cogeneration unit, gas boiler, and coal-fired power plant is shown in Formula (4).

$$E_e = \varepsilon_e \sum_{t=1}^{T} P_{e,buy} \tag{3}$$

$$E_x = \varepsilon_x \cdot P_x \tag{4}$$

ε_e, ε_x are carbon emission quotas for unit electric power and unit power consumed by different equipment.

2.1.3. Stepped Carbon Trading Model

Based on the above carbon emission model and carbon emission rights allocation model, a ladder carbon trading model is established, and a plurality of carbon emission rights purchase intervals are set in the ladder carbon trading mechanism. The initial carbon emission quota allocated by the system is removed from the total carbon emission, and the mathematical Formula is shown in Formula (5). At the same time, pricing with different gradients is carried out according to the net carbon emission of the system in different charging intervals, when the net carbon emission is less than a given interval length. In the internal time, only the transaction amount at the base price of carbon trading is paid. When the price is higher than a given interval length, the price in each step interval is fixed. Every step increase, the carbon trading price increases exponentially, the carbon emission right purchase ratio increases, and the corresponding price will also increase. According to this trading model, the mathematical model Formula of ladder carbon trading as shown in Formula (6) is obtained.

$$E = E_{all} - E_e - \sum_{i}^{N} E_x \tag{5}$$

$$C_{CO_2} = \begin{cases} c \cdot E, E \leq l \\ c \cdot (1+\lambda)(E-l) + cl, l \leq E \leq 2l \\ c \cdot (1+2\lambda)(E-2l) + c(2+\lambda)l, 2l \leq E \leq 3l \\ c \cdot (1+3\lambda)(E-3l) + c(3+3\lambda)l, 3l \leq E \leq 4l \\ c \cdot (1+4\lambda)(E-4l) + c(4+6\lambda)l, E \geq 4l \end{cases} \tag{6}$$

C_{CO_2} is carbon transaction costs, E is carbon emission of the system, i is equipment selection, including cogeneration units, gas-fired boilers and coal-fired power plants, N is the total number of device, λ is the price growth rate, l is the interval length, c is the base price of carbon trading.

2.2. *Microgrid Low-Carbon Economic Optimization Model with Step Carbon Price*

Based on the established carbon trading mechanism, a low-carbon optimization model of microgrid with ladder carbon trading mechanism is proposed. The model considers the lowest comprehensive operating cost of the whole microgrid from the economic level, implements the initial carbon emission quota form from the environmental level, and comprehensively considers the operating cost, system power consumption cost, gas consumption cost, and carbon trading cost of the whole system exceeding the carbon emission quota. The electric balance, thermal balance, gas balance and hydrogen balance in the overall operation of the system are taken as equality constraints, and the climbing constraints and output upper limit constraints of each device are taken as inequality constraints, so as to improve the economy and low carbon of the microgrid [14].

2.2.1. Objective Function

The overall objective function of microgrid not only minimizes the fuel cost and the cost of purchasing electricity and gas, but also supplements the environmental problems according to the carbon trading mechanism. Among them, the carbon transaction cost is the ladder carbon price cost considering exceeding the carbon emission quota. On the premise of ensuring the safety and reliability of the whole microgrid system, the environmental factors are quantified, and the output degree of each micro-source is optimized according to the load change, so as to obtain the minimum comprehensive cost of the system.

$$minF = min\sum_{t=1}^{T}\left(C_{PGU} + C_{e,\ buy} + C_{g,\ buy} + C_{CO_2}\right) \tag{7}$$

F is the minimum operate cost of the system, C_{PGU} is fuel costs, $C_{e,\ buy}$ is the cost of electricity purchase, $C_{g,\ buy}$ is gas purchase cost, and C_{CO_2} is carbon transaction costs.

a. Fuel cost

$$C_{pgu} = ap_{PGU}^2 + bp_{PGU} + c \tag{8}$$

b. Energy purchase cost

$$C_{e,\ buy} + C_{g,\ buy} = c_e\sum_{t=1}^{T} P_{e,\ buy} + c_g\sum_{t=1}^{T} P_{g,\ buy} \tag{9}$$

$P_{e,\ buy}$ is the total power consumption of the system, $P_{g,\ buy}$ is the total gas consumption of the system, c_e is real-time electricity prices, c_g is real-time gas price, and T takes 24 h.

2.2.2. System Operation Constraints

According to the law of conservation of energy, when the microgrid is running as a whole, the output energy of each form should always be equal to the input energy of each form to maintain the system operation. At the same time, the output energy of each equipment is kept within the rated power and cannot exceed the specified maximum output, so the related mathematical Formulas of balance constraint and imbalance constraint in the system are integrated, respectively.

The equality constraints are as follows:

a. Electric power balance

On the premise of ignoring the network loss, the discharge of wind power, cogeneration units, and power storage equipment, the purchase of electricity from the superior power grid and the overall output of hydrogen fuel cells are equal to the sum of the electric power consumed by the electric load and the charging of power storage equipment, as shown in Formula (10).

$$P_{WT}(t) + P_{CHP}(t) + P_{SB,\ dis}(t) + P_{e,\ buy}(t) + P_{HFC,\ e}(t) = P_{load}(t) + P_{SB,\ chr}(t) \tag{10}$$

$P_{WT}(t)$ is output power for wind energy, $P_{CHP}(t)$ is output power for electric energy of cogeneration unit, $P_{SB,\ dis}(t)$ is the discharge power for electric storage device, $P_{HFC,\ e}(t)$ is electricity production of hydrogen fuel cell, $P_{load}(t)$ is the electrical load, and $P_{SB,\ chr}(t)$ is charging the power storage equipment.

b. Thermal power balance

The input heat of the system is equal to the output heat; that is, the sum of the exothermic power of cogeneration unit, gas boiler, and heat storage equipment is equal to the thermal power consumed by heat storage equipment and heat load.

$$P_{HFC,\ h}(t) + H_{CHP}(t) + H_{GB}(t) + H_{TS,\ dis}(t) = H_{TS,\ chr}(t) + H_{Load}(t) \tag{11}$$

$H_{CHP}(t)$ the thermal energy output power of the cogeneration unit, $H_{GB}(t)$ is the heat generation of gas boilers, $H_{TS,\ dis}(t)$ is released for the heat storage device, $H_{TS,\ chr}(t)$ is the heat release of heat storage equipment, and $H_{Load}(t)$ is the heat load.

c. Natural gas power balance

Similarly, the output of natural gas is equal to the input, that is, the sum of gas purchase, gas storage tank, gas release, and gas conversion technology output is equal to the gas load, gas storage tank, gas storage, cogeneration unit, and gas boiler gas consumption.

$$G_{buy}(t) + G_{ES,\ dis}(t) + G_{MR}(t) = G_{load} + G_{ES,\ chr}(t) + G_{CHP}(t) + G_{GB}(t) \tag{12}$$

$G_{buy}(t)$ is the amount of gas purchased, $G_{ES,\ dis}(t)$ is the amount of gas vented by the gas receiver, $G_{MR}(t)$ is the amount of CH_4 dioxide produced by methanation of electricity-to-gas technology, G_{load} is gas load, $G_{ES,\ chr}(t)$ is the gas storage capacity of the gas storage equipment, $G_{CHP}(t)$ is the air consumption of the cogeneration unit, and $G_{GB}(t)$ is the amount of natural gas consumed by the gas boiler unit.

d. Hydrogen power balance

Because the two steps of electro-gas conversion technology in microgrid are modeled separately and equipped with hydrogen storage tank and hydrogen fuel cell, hydrogen energy balance is added.

$$P_{EL,\ H_2}(t) + P_{HS,\ dis}(t) = P_{H_2,\ MR}(t) + P_{H_2,\ HFC}(t) + P_{HS,\ chr}(t) \tag{13}$$

$P_{EL,\ H_2}(t)$ is the hydrogen production capacity of the electrolytic cell in the first step of converting electricity to gas, $P_{HS,\ dis}(t)$ is the hydrogen storage capacity of the hydrogen storage device, $P_{H_2,\ MR}$ is the hydrogen consumption of methanation reaction, $P_{H_2,\ HFC}$ is the hydrogen consumption of the hydrogen fuel cell, $P_{HS,\ chr}(t)$ is amount of hydrogen released for hydrogen storage equipment.

Inequality constraints are as follows:

a. Coal-fired units

During the operation of coal-fired units, it is necessary to ensure that the output is within the allowable range, that is, to ensure the output constraint.

$$P_{PGU}^{min} \leq P_{PGU} \leq P_{PGU}^{max} \tag{14}$$

P_{PGU}^{max} and P_{PGU}^{min} are the upper and lower limits of the output of coal-fired units.

The output adjustment of coal-fired units should be within the allowable range, that is, climbing constraint.

$$\Delta P_{PGU}^{min} \leq \Delta P_{PGU} \leq \Delta P_{PGU}^{max} \tag{15}$$

$$\Delta P_{PGU}^{min} \leq \Delta P_{PGU} \leq \Delta P_{PGU}^{max} \tag{16}$$

ΔP_{PGU}^{max} is the maximum upward climb power of the coal-fired unit, ΔP_{PGU}^{min} is the maximum downward climb power of the coal-fired unit, ΔP_{PGU} is the amount of power change, $P_{PGU}(t)$ is the power of the coal-fired unit at the time of t, $P_{PGU}(t-1)$ is the power of the coal-fired unit at the time of $t-1$.

b. Cogeneration unit

Because the cogeneration unit meets both the electric load and the heat load, it needs to meet both the electric output constraint and the heat output constraint.

$$P_{CHP,\ e}^{min} \leq P_{CHP,\ e} \leq P_{CHP,\ e}^{max} \tag{17}$$

$$H_{CHP,\ h}^{min} \leq H_{CHP,\ h} \leq H_{CHP,\ h}^{max} \tag{18}$$

$P_{CHP,\ e}^{max}$ and $P_{CHP,\ e}^{min}$ are the upper and low limits of the electric output of the cogeneration unit, $H_{CHP,\ h}^{min}$ and $H_{CHP,\ h}^{max}$ are the upper and lower limits of thermal output of cogeneration unit.

Climbing constraints are shown in Formulas (19) and (20).

$$\Delta P^{min}_{convert,\ e} \leq \Delta P_{convert,\ e} \leq \Delta P^{max}_{convert,\ e} \tag{19}$$

$$\Delta P_{convert,\ e} = P_{convert,\ e}(t) - P_{convert,\ e}(t-1) \tag{20}$$

$\Delta P^{max}_{convert,\ e}$ is the maximum upward climb power of that cogeneration unit, $\Delta P^{min}_{convert,\ e}$ is the maximum downward climb power of the cogeneration unit, $\Delta P_{convert,\ e}$ is the amount of power change under the pure condensation condition, $P_{convert,\ e}(t)$ is the electrical power of the cogeneration unit after the t time conversion, $P_{convert,\ e}(t-1)$ is the electrical power of the cogeneration unit after the $t-1$ time conversion.

c. Electro-gas conversion technology

The electrolyzer and methanation reaction should meet the hydrogen production output constraint and the CH_4 output constraint, respectively.

$$P^{min}_{EL,\ H_2} \leq P_{EL,\ H_2} \leq P^{max}_{EL,\ H_2} \tag{21}$$

$$G^{min}_{MR} \leq G_{MR} \leq G^{max}_{MR} \tag{22}$$

$P^{max}_{EL,\ H_2}$ and $P^{min}_{EL,\ H_2}$ are the upper and lower limits of hydrogen production by the electrolyzer, G^{max}_{MR} and G^{min}_{MR} are the upper and lower limits of the CH_4 quantity obtained by the reaction.

Climbing constraints are shown in Formulas (23)–(26).

$$\Delta P^{min}_{EL,\ H_2} \leq \Delta P_{EL,\ H_2} \leq \Delta P^{max}_{EL,\ H_2} \tag{23}$$

$$\Delta P_{EL,\ H_2} = P_{EL,\ H_2}(t) - P_{EL,\ H_2}(t-1) \tag{24}$$

$$\Delta G^{min}_{MR} \leq \Delta G_{MR} \leq \Delta G^{max}_{MR} \tag{25}$$

$$\Delta G_{MR} = G_{MR}(t) - G_{MR}(t-1) \tag{26}$$

$\Delta P^{max}_{EL,\ H_2}$ and ΔG^{max}_{MR} is the maximum upward climb power of that electrolytic cell and methanation reaction, $\Delta P^{min}_{EL,\ H_2}$ and ΔG^{min}_{MR} are the electrolyzer and methanation reaction large downward climb power, $\Delta P_{EL,\ H_2}$ and ΔG_{MR} are the amount of power change, $P_{EL,\ H_2}(t)$ and $G_{MR}(t)$ are the power of the electrolyzer and methanation reaction at the time of t, $P_{EL,\ H_2}(t-1)$ and $G_{MR}(t-1)$ are the power of the electrolyzer and methanation reaction at the time of $t-1$.

d. Gas-fired boiler unit

The output constraint and climbing constraint of gas-fired boilers are similar to those of coal-fired units, as shown in Formulas (27)–(29).

$$H^{min}_{GB} \leq H_{GB} \leq H^{max}_{GB} \tag{27}$$

$$\Delta H^{min}_{GB} \leq \Delta H_{GB} \leq \Delta H^{max}_{GB} \tag{28}$$

$$\Delta H_{GB} = H_{GB}(t) - H_{GB}(t-1) \tag{29}$$

H^{max}_{GB} and H^{min}_{GB} are the upper and lower limits of gas boiler output, ΔH^{min}_{GB} is the maximum upward climbing power of the gas boiler, ΔH^{min}_{GB} is the maximum downward climbing power of the gas boiler, ΔH_{GB} is the amount of power change, $H_{GB}(t)$ is the gas boiler power at the time of t, $H_{GB}(t-1)$ is the gas boiler power at the time of t.

e. Hydrogen fuel cell

Similarly, the output constraints and climbing constraints of hydrogen fuel cells are shown in Formulas (30)–(32).

$$P_{HFC,\,e}^{min} \le P_{HFC,\,e} \le P_{HFC,\,e}^{max} \quad (30)$$

$$\Delta P_{HFC,\,e}^{min} \le \Delta P_{HFC,\,e} \le \Delta P_{HFC,\,e}^{max} \quad (31)$$

$$\Delta P_{HFC,\,e} = P_{HFC,\,e}(t) - P_{HFC,\,e}(t-1) \quad (32)$$

$P_{HFC,\,e}^{max}$ and $P_{HFC,\,e}^{min}$ are the upper and lower limits of hydrogen fuel cell output, $\Delta P_{HFC,\,e}^{max}$ is the maximum upward climb power of the hydrogen fuel cell, $\Delta P_{HFC,\,e}^{min}$ is the minimum upward climb power of the hydrogen fuel cell, $\Delta P_{HFC,\,e}$ is the amount of power change, $P_{HFC,\,e}(t)$ is the hydrogen fuel cell power at the time of t, $P_{HFC,\,e}(t-1)$ is the hydrogen fuel cell power at the time of $t-1$.

3. Game Theory Basis

Gaming theory refers to the influence of the individual income of each participant in the case of the interdependence and interdependence of the participants. Therefore, because the income of all parties in the game receives the influence of multiple parties, each participant is considered to make rational judgments. Through the information obtained by themselves, they will give their strategies in real time to the overall feedback. The complete game consists of three basic elements:

3.1. *Participants*

Participants refer to the subjects that can Formulate strategies and make rational judgments in the overall game. Among them, the collection mathematics represents the participants (33).

$$N = \{1,\ 2,\ 3,\ \ldots,\ n\} \quad (33)$$

3.2. *Strategy Set*

Strategy is an important factor in the overall game. Participants have changed different strategies by collecting information, that is, the methods and means of maximizing their own interests to achieve their own interests. The number of strategies can be selected by themselves. Because the game is the mutual impact of the main body Formulation strategy, the sequence of the strategy Formulation has a huge impact on the results of the game. Among them, all participants Formulated the strategy collection mathematics indication (34).

$$S = \{S_1,\ S_2,\ S_3,\ \ldots S_n\} \quad (34)$$

3.3. *Effectiveness*

The effect refers to the benefits that the participating entities of each game are after the game. Among them, benefits can be positive income or negative benefits, that is, the maximum benefits to obtain. The goal of the participants is to maximize the benefit by adjusting the strategy, and the benefit collection mathematical representation of the participants is shown in the math indication (35).

$$u = \{u_1,\ u_2,\ u_3,\ \ldots u_n\} \quad (35)$$

After determining the basic three elements of the game, the establishment of a complete game is completed.

3.4. *Stackelberg Game*

Because the microgrid and the user side consider the price-based demand response, that is, there is no relevant agreement between the participants, users spontaneously change their energy consumption habits. Participants constantly adjust their own strategies to maximize their own interests during the scheduling process. There is still an obvious decision-making sequence between the microgrid and the client, with the microgrid acting first as the leader and the client following up as the follower. Among them, the leader first makes the appropriate decision, occupying the position of priority decision, and the follower makes the decision after receiving the signal from the leader. That is, microgrid is the leader to adjust the real-time price of distribution network, and the client is the follower to respond to the real-time price proposed by microgrid. This game type is called Stackelberg game. Stackelberg game is divided into dynamic non-cooperative game, also known as Stackelberg game. The model problems solved by the Stackelberg game mainly have the following characteristics [15]:

(1) There is no agreement among the participants, and they make their own decisions.
(2) The decision of each participant has an impact on the benefits of other participants.
(3) Because of the different market positions, there is a decision-making order among the participants. The leader first makes appropriate decisions according to the target benefit, and the followers make decisions on their own goals after receiving the leader's decision-making signal. There is a restrictive relationship between them.
(4) The final decision-making scheme of each participant needs the unanimous consent of all participants.

The mathematical definition model of Stackelberg game is as follows:

Let the leader's strategy set is X, the follower's strategy set is Y, the leader's utility function is $f : X \times Y \to R$, the utility function of the follower is $g \cdot X \times Y \to R$.

As a dynamic game, leaders send out decision signals $x_n \in X$, followers make decisions according to the leader's strategy, and the set of followers' balance points is $A(x)$. Therefore, the mechanism mapping is generated: $X \to A(Y)$. When the next leader makes a decision, in order to maximize their own interests at all times, they will consider the follower scheme, and the overall game optimization goal is $max f(x, y)$. Let (x^*, y^*) be the equilibrium point of the game, then the equilibrium condition is shown in Formula (4) [16].

$$\begin{cases} \max\limits_{x \in X} f(x^*, y^*(x^*)) \\ g(x^*, y^*) \geq g(x^*, y) \end{cases} \tag{36}$$

4. Multi-Energy Microgrid Model Based on Stackelberg Game

4.1. The Demand Response Type

For the classification of electric energy demand response, the response method is mainly used as a differentiation method, both of which require a contract with the energy supply company, one is the price-based demand response, including time-of-use electricity price, real-time electricity price and peak electricity price. Compared with the incentive type, the price type has a lower degree of change in the user's energy habits, and only needs to make voluntary adjustments to the price signal. The incentive-type response is to adjust the energy-using behavior strictly according to the load reduction calculation method and response time signed in the contract, and carry out the corresponding compensation mechanism for the adjusted energy-using behavior or impose corresponding fines for the part that does not meet the response requirements. Compared with the two corresponding methods, the current user acceptance of price-based demand response is higher, and the project implementation scope is wider, and this paper mainly studies price-based electricity price in the subsequent microgrid-user game stage.

The relationship model between electricity and electricity price is mainly divided into: electricity price elasticity matrix, user psychology model, exponential function fitting model, and statistical principle model [17]. Among them, the elasticity matrix of electricity and electricity prices is described as the change in electricity demand caused by the change in electricity price, and the elastic coefficient of electricity price is defined, and the calculation Formula is shown in Formula (37).

$$\varepsilon = \frac{\Delta P_e}{P_e}\frac{C_e}{\Delta C_e} \tag{37}$$

In the Formula, ΔP_e is the amount of electricity increase, and ΔC_e is the increase in electricity price.

The electricity price response mechanism is divided into single period and multi-period, that is, fixed ladder electricity price and real-time electricity price, due to the peak and valley hours and price of the ladder electricity price is fixed, users will only change energy demand in a certain period, and real-time electricity price will prompt users to change energy consumption behavior according to different time periods. Therefore, a multi-period response mechanism is introduced, in which the multi-period elasticity coefficient is divided into the self-elasticity coefficient of the user response in the current period and the cross-elasticity coefficient of the user response behavior in other periods, and the mathematical model of the self-elasticity coefficient and cross-elasticity coefficient obtained according to the above definition is shown in Formulas (38) and (39).

$$\varepsilon_{ii} = \frac{\Delta P_i}{P_i}\frac{C_i}{\Delta C_i} \tag{38}$$

$$\varepsilon_{ij} = \frac{\Delta P_i}{P_i}\frac{C_j}{\Delta C_j} \tag{39}$$

In the Formula, ε_{ii} is the self-elastic coefficient and ε_{ij} is the cross-elastic coefficient. i is the ith dispatch period and j is the jth dispatch period.

According to the definition of the above elasticity coefficient and the mathematical model, the user price response model of the overall period is obtained as shown in Formula (40) and the electricity price elasticity matrix as shown in Formula (41)

$$\begin{bmatrix} \frac{\Delta C_1}{C_1} \\ \frac{\Delta C_2}{C_2} \\ \vdots \\ \frac{\Delta C_n}{C_n} \end{bmatrix} = E \begin{bmatrix} \frac{\Delta P_1}{P_1} \\ \frac{\Delta P_2}{P_2} \\ \vdots \\ \frac{\Delta P_n}{P_n} \end{bmatrix} \tag{40}$$

$$E = \begin{bmatrix} \varepsilon_{11} & \varepsilon_{12} & \cdots & \varepsilon_{1n} \\ \varepsilon_{21} & \varepsilon_{22} & \cdots & \varepsilon_{2n} \\ \vdots & \vdots & \ddots & \vdots \\ \varepsilon_{n1} & \varepsilon_{n2} & \cdots & \varepsilon_{nn} \end{bmatrix} \tag{41}$$

ΔC_n is the change in electricity price at n moments after the demand response, ΔP_n is the load change at n times after the demand response, P_n is the load before the n time response.

4.2. Microgrid-User Stackelberg Game Structure

As the leader in the Stackelberg game, microgrids generate revenue by developing pricing strategies different from those of distribution networks, selling electricity to users at different energy purchase prices [18,19]. Users, as followers in the Stackelberg game relationship, formulate a decision plan to reduce the cost of electricity purchase, that is, increase the amount of load transfer to reduce the cost of electricity purchase, and achieve the maximization of the follower's benefits, that is, the maximum value of the follower's objective function, as shown in Formula (51). Then, the scheme obtained by changing the load transfer amount is transmitted to the microgrid, which makes the next optimal decision based on the user's response results. At this point, the output plan of the microgrid may not match the user's demand, thereby reducing the maximum profit of the microgrid. Therefore, at this point, the microgrid will redesign its output plan and energy prices to achieve the highest return on the microgrid, which is the maximum value of the leader's objective function, as shown in Formula (42). The Stackelberg game structure of microgrids is shown in Figure 1.

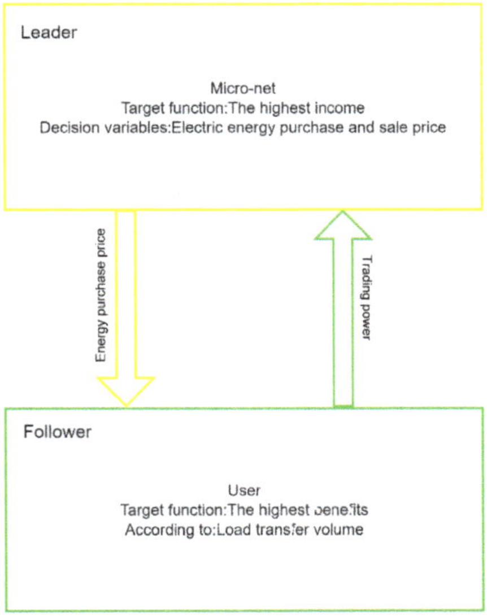

Figure 1. Microgrid Stackelberg game structure.

4.3. *Microgrid Revenue Model*

As the leader in the game, microgrid should achieve the optimal scheduling strategy of microgrid operation as a whole [20,21], set the price of electricity sales within 24 h in combination with the load change given by the user, realize the highest income, and construct the optimal operation income scheduling model of components of microgrid as a whole.

4.3.1. Objective Function

S is the minimum operating cost of the system in the game model, $C_{es,\ buy}$ is the cost of purchasing electricity from the superior power grid in the game model, $C_{eu,\ sell}$ is the electricity price income sold by microgrid to users, $C_{gs,\ buy}$ is the cost of buying gas, $C_{CO_2,\ s}$ is that transaction cost of carbon in the game model, $e_{es,\ buy}$ is the unit price of selling electricity for the distribution network, $e_{eu,\ sell}$ is the amount of electricity purchased for the microgrid, $P_{es,\ buy}$ is the electricity purchased by the user sold by the microgrid, P_{user} is the amount of power after the user responds.

$$minS = min\sum_{t=1}^{T}\left(C_{es,\ buy} - C_{eu,\ sell} + C_{gs,\ buy} + C_{CO_2,\ s}\right) \tag{42}$$

$$C_{es,\ buy} = e_{es,\ buy}P_{es,\ buy} \tag{43}$$

$$C_{eu,\ sell} = e_{eu,\ sell}P_{user} \tag{44}$$

4.3.2. Constraints

Due to the introduction of the demand response mechanism, the upper and lower limits of the microgrid's power purchase from the distribution network and the upper and lower limits of the microgrid's selling price to users are increased, in which the power balance is shown in Formula (47). Thermal power balance, natural gas power balance, and hydrogen power balance are shown in the Formulas (12)–(14). The operating models, contribution models, and climbing models of thermoelectric units are as follows (17)–(20). The two-stage operation model, the power model, and the climbing model of the electrical rotation technology are as follows (21)–(26). The power boiler crew and the climbing models are shown in the Formulas (27)–(29), respectively. Hydrogen fuel cell operation models, contribution models, and climbing models are shown in the Formulas (30)–(32).

$$P_{es,\ buy}^{min} \le P_{es,\ buy} \le P_{es,\ buy}^{max} \tag{45}$$

$$e_{eu,\ sell}^{min} \le e_{eu,\ sell} \le e_{eu,\ sell}^{max} \tag{46}$$

$P_{es,\ buy}^{min}$ and $P_{es,\ buy}^{max}$ are the upper and lower limits of the amount of electricity that the microgrid purchases from the distribution network, $e_{eu,\ sell}^{max}$ and $e_{eu,\ sell}^{min}$ are the upper and lower limits of electricity price sold by microgrid to users.

$$P_{WT}(t) + P_{CHP}(t) + P_{SB,\ dis}(t) + P_{e,\ buy}(t) + P_{HFC,\ e} = P_{user}(t) + P_{SB,\ chr}(t) \tag{47}$$

$P_{WT}(t)$ is output power for wind energy, $P_{CHP}(t)$ is output power for electric energy of cogeneration unit, $P_{SB,\ dis}(t)$ is discharge power for that electric storage device, $P_{HFC,\ e}(t)$ is electricity production of hydrogen fuel cell, $P_{SB,\ chr}(t)$ is charging the power storage equipment.

Due to the introduction of price-based demand response, the relevant constraints of demand response should be increased. First, it is required that the total amount of load in the whole dispatching period remains unchanged, as shown in Formulas (48) and (49). At the same time, in order to achieve the ultimate goal of demand response, it is necessary to control the load of each step in the scheduling between the upper and lower limits of the load value before response, as shown in Formula (50).

$$\textstyle\sum_{t=1}^{T} \Delta P_{user,\ t} = 0 \tag{48}$$

$$\Delta P_{user,\ t} = P_{user,\ t}^{0} - P_{user,\ t} \tag{49}$$

$$\begin{cases} P_{user,t}^{max} \le P_{user,t}^{0,max} \\ P_{user,t}^{min} \ge P_{user,t}^{0,min} \end{cases} \tag{50}$$

$\Delta P_{user,\ t}$ is the change in user electricity consumption before and after demand response, $P_{user,\ t}^{0}$ is the electricity consumption of users before demand response, $P_{user,\ t}^{0,\ max}$ and $P_{user,\ t}^{0,\ min}$ are the upper and lower limits of the demand response preload value, $P_{user,\ t}^{max}$ and $P_{user,\ t}^{min}$ are the upper and lower limits of the demand response afterload value.

4.4. User Benefit Model

User benefit refers to the reasonable adjustment of required energy according to its own demand for electricity and energy price in the process of demand response [22]. In this paper, the user satisfaction model is introduced to constrain the load variation. While considering the microgrid to guide users to adjust their own energy consumption period, the satisfaction of power consumption mode and expenditure satisfaction are used as constraints to participate in dispatching.

4.4.1. Objective Function

As a follower of the game, the user responds after the price signal is given by the microgrid. Considering the user's power consumption income and cost, the goal is to maximize the user's benefit, and the objective function is shown in Formula (51).

$$maxU = C_{UE} - C_{eu,\ sell} \tag{51}$$

$$C_{UE} = \alpha \log(1 + P_{user,\, t}) \tag{52}$$

U is that maximum benefit of user in the game model, C_{UE} is the benefit of users, α is energy preference coefficient for users.

4.4.2. Constraints

a. Satisfaction with electricity consumption mode

Before the introduction of electricity price demand response, users' electricity consumption periods were mainly selected according to their own preferences, and at this time, users' satisfaction with electricity consumption methods was the highest [23]. However, after the introduction of electricity price demand response, the user changes his own electricity preference to reduce electricity expenditure by responding to the price signal given by the microgrid. The mathematical model is shown in Formula (53) and the satisfaction constraint is shown in Formula (54).

$$M = 1 - \frac{\sum_{t=1}^{T} |\Delta P_{user,t}|}{\sum_{t=1}^{T} P_{user,t}^{0}} \tag{53}$$

$$M \geq M_{min} \tag{54}$$

M_{min} is the lower limit of user's satisfaction with electricity consumption.

b. Expenditure satisfaction

After the price-based demand response is implemented in the microgrid, users will adjust according to the real-time electricity price to ensure that the electricity expenditure will not have a great impact [24]. Therefore, expenditure satisfaction is usually used to measure the change in user expenditure. The mathematical model is shown in Formula (55), and the satisfaction constraint is shown in Formula (56).

$$N = 1 + \frac{\sum_{t=1}^{T} \left(e_{eu,sell}^{0} P_{user,t}^{0} - e_{eu,sell} P_{user,t} \right)}{e_{eu,sell} P_{user,t}} \tag{55}$$

$$N \geq N_{min} \tag{56}$$

In the Formula, $e_{eu,\, sell}^{0}$ is the unit price of electricity sales for microgrid before demand response, P_{user}^{0} is power consumption of users before demand response. N_{min} is the lower limit of user spend satisfaction.

4.5. Establishment and Proof of Microgrid-Client Stackelberg Game

As shown in Figure 2, the microgrid and the user constitute a dynamic non-cooperative game, and the game relationship constitutes a Stackelberg game Formula as shown in Formula (57).

$$G = \{(MGO \cup USER);\ \Phi_{MGO};\ \Psi_{USER};\ S_{MGO};\ U_{USER}\} \tag{57}$$

Formula (52) contains three elements of the game: participants, strategy sets, and utility.

1. Game participants: participants in the game of microgrid and users as the main slave, expressed in the form of set as follows $(MGO \cup USER)$.
2. Strategy set: The microgrid is the leader in the Stackelberg game and the optimization strategy is Formulated first, and the electricity price strategy proposed by the microgrid to the user is represented by the set Φ_{MGO}. The set of load adjustment strategies made by the user is represented by set Ψ_{USER}.
3. Utility: The cost set of the microgrid is represented by set S_{MGO}, and the benefit set of users is represented by U_{USER}. the cost collection of microgrid.

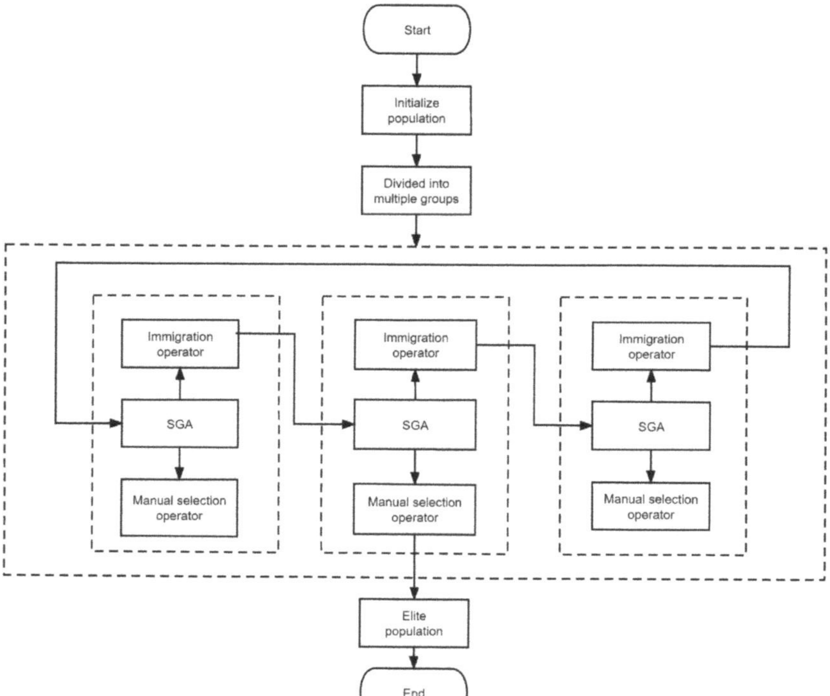

Figure 2. Simplified flowchart of multi-population genetic algorithm.

When the follower in the game responds to the leader's strategy, and the leader accepts the response, it shows that the upper and lower game as a whole has reached the equilibrium condition [24]. That is, when the user responds to the electricity price strategy proposed by the microgrid according to the optimized operation scheme, and the microgrid accepts the response strategy, the microgrid-user reaches the equilibrium condition. Make Φ^*_{MGO} a vector set representing all optimal strategies of a microgrid, and Ψ^*_{USER} a vector set representing all the response strategies of the client. To reach the Stackelberg equilibrium condition, Formula (58) must be satisfied.

$$\begin{cases} S_{MGO}(\Phi^*_{MGO}, \Psi^*_{USER}) \geq S_{MGS}(\Phi_x, \Phi^*_{n-x}, \Psi^*_{USER}) \\ U_{USER}(\Phi^*_{MGO}, \Psi^*_{USER}) \geq U_{USER}\left(\Phi^*_{MGO}, \Psi_y, \Psi^*_{n-y}\right) \\ \forall \Phi_x \in \Phi_{MGO} \\ \forall \Psi_y \in \Psi_{USER} \end{cases} \quad (58)$$

Φ_x is the optimal operation scheme of the microgrid, Ψ_y is the optimal response scheme of the user terminal, Φ^*_{n-x} is yeah, except Φ_x other strategies than, Ψ^*_{n-y} yeah, and except Ψ_y other strategies outside.

In the equilibrium state of Stackelberg game, neither party can obtain greater benefits by unilaterally proposing new strategies, and it is necessary to verify the existence and uniqueness of the equilibrium solution before solving it. The theorems for verifying the existence of the equilibrium solution are as follows:

1. The decision schemes of leaders and followers are all non-empty bounded convex sets;
2. After the top leaders make decisions, the followers have corresponding unique solutions;
3. After the lower followers respond, the leader has a unique solution.

Compared with the above definitions, the existence and uniqueness of Stackelberg equilibrium solution of microgrid-user Stackelberg game model established in this chapter are proved:

1. As shown in Formulas (46) and (50), the policy set Φ_{MGO} and Ψ_{USER} are non-null bounded convex set;
2. As shown in Formulas (52)–(54), each term in S_{MGO} is a linear or constant function with respect to $P_{es,\ buy}$ or P_{user}, then S_{MGO} is a concave function with respect to $P_{es,\ buy}$ and P_{user}.
3. As shown in Formulas (51) and (52), U_{USER} is a continuous function with respect to $P_{es,\ buy}$ and P_{user}.

5. The Solution Method of Game Model Based on Multi-Population Genetic Algorithm

The previous chapter analyzed the energy flow relationship between devices in the microgrid and the optimization functions of various game entities, and established a Stackelberg game model for the microgrid user. As the function to be solved is a large-scale linear programming problem, compared to other algorithms, using multiple population genetic algorithms can effectively reduce the complexity of the solution and improve the efficiency of the solution.

Genetic Algorithm (GA) is an adaptive global optimization probability search algorithm proposed by Professor Holland in the United States in 1975 that simulates the genetic evolution of biological organisms in nature. Organisms evolve through heredity, variation, and natural selection, and genetic algorithms are inspired by Darwin's theory of natural selection. The solution of the example corresponds to the chromosomes in the genetic process, and the set of all chromosomes is a population, and the individuals are eliminated between the populations according to the principle of "natural selection, survival of the fittest", and the selection between individuals in the corresponding population is in programming. Starting from the initialization of the population, the interval judgment of the fitness function is carried out for each generation of the population, and according to the designed fitness ratio, the appropriate strategy is selected to select the excellent individuals of the current population, and the selected excellent individuals are crossed and mutated to form a new population. Analogous to the evolution process of species, generation-by-generation, continuously enhancing the fitness of the population until the optimal solution is output after the desired conditions are completed. Since the genetic algorithm does not rely on gradient calculation, it has strong robustness and global optimization ability [24,25].

Multi-population genetic algorithm divides a single population into multi-threaded populations and adds immigration operators. In the evolution process of different populations, the migration operator introduces the optimal individuals to other populations every certain number of iterations, which realizes the information exchange between different populations and the balance of global and local search performance. Secondly, the elite population is established, and each generation of evolution selects the best individuals of other populations to join the elite population and save them through artificial selection operators, and no genetic operation is carried out to ensure that the best individuals are not destroyed, so all the optimal solutions produced by each evolution can be completely preserved [26–29]. The simplified process of improvement is shown in Figure 3.

The solution process for the whole game system is as follows:

(1) Initializing the operation parameters of the microgrid and the load data of the user terminal, and sending the electricity price strategy drawn up by the microgrid to the lower layer;
(2) Converting the maximum energy consumption benefit of the user terminal into a negative cost, feeding back according to the pricing signal of the microgrid, and feeding back the load signal to the upper-level dispatching;
(3) The microgrid solves the objective function through the feedback signal;

(4) Judging whether the game equilibrium solution is reached, and if so, outputting the result; otherwise, return to (2) to continue scheduling.

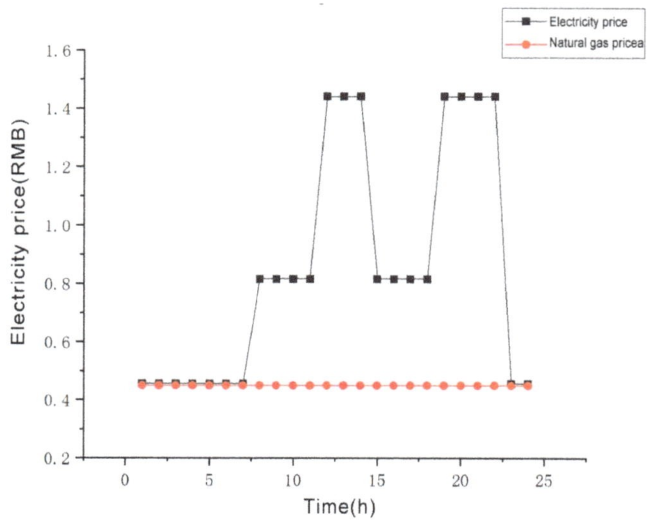

Figure 3. Electricity and gas prices.

6. Example Analysis

6.1. Basic Data

The reaction parameters of each equipment are shown in Table 1. The electricity price and natural gas price are shown in Figure 3, but considering the price-based user demand response, the initial tiered price is adopted as shown in Figure 4, and the new demand response parameter data are shown in Table 2 [30]. The typical electricity load, heat load, and gas load curves of the multi energy microgrid, as well as the predicted output power of wind power generation, are shown in Figure 5. Peak and valley time of use electricity prices are used for billing, with low peak periods ranging from 23.00 to 7.00, flat peak periods ranging from 8.00 to 11.00 and 15.00 to 18.00. Daytime peak periods are reached from 12.00 to 14.00, and nighttime peak periods are reached from 19.00 to 22.00.

Table 1. Operating parameters of each device.

Equipment	Value	Efficiency/Carbon Emissions Quota	Value
Carbon capture power plant contribution range/kw·h	[0, 200]	η_{CHP}	0.92
The range of thermal power union crew/kw·h	[0, 300]	η_{EL}	0.88
Thermoelectrician Unit is a thermal power ratio	1.8	η_{MR}	0.6
The range of electrolytic tank equipment/kw·h	[0, 500]	η_{HFC}	0.85
Methane reaction force range/kw·h	[0, 250]	ε_e	0.798
Hydrogen fuel cell contribution range/kw·h	[0, 250]	ε_h	0.985

Table 2. Demand Response Related Parameters.

Parameter Name	Value	Parameter Name	Value
M_{min}	0.9	ε_{ii}	−0.2
N_{min}	0.9	ε_{ij}	0.033

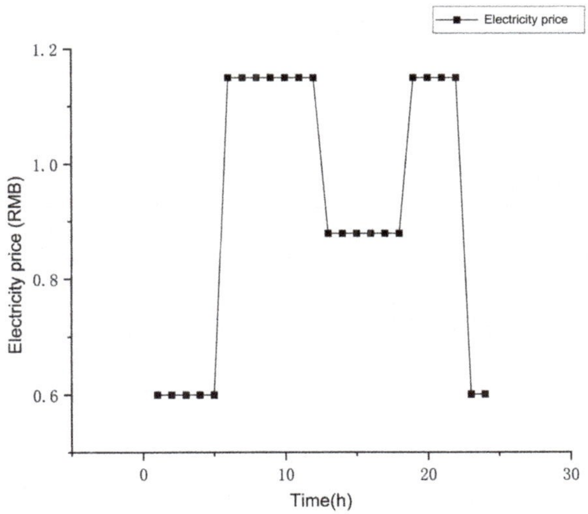

Figure 4. Initial electricity price.

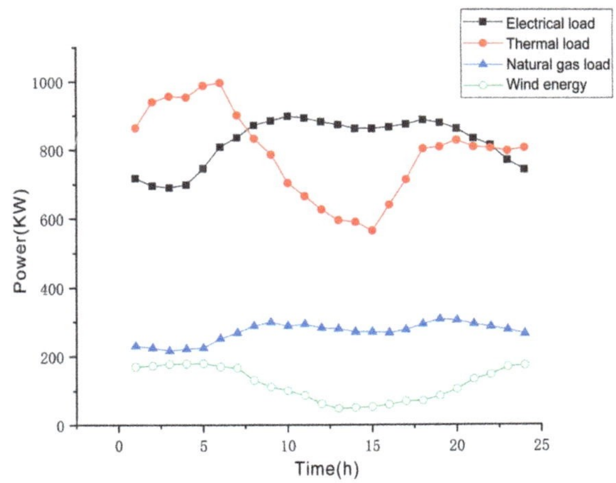

Figure 5. Typical daily electricity, heat, gas load forecasting, and wind power output forecasting.

6.2. Game Equilibrium Results

The lowest comprehensive cost of the microgrid is 11,667.044709 yuan, which takes 692.513986 s. Comparing Figure 6a with Figure 6b, the game process between the microgrid and the user terminal can be analyzed. When the number of iterations is 1–8 times, the results of the game between microgrid and user do not change, and the microgrid has the lowest return and the highest user benefit. When the number of iterations is between 8 and 14, the revenue of microgrid gradually increases, and the corresponding user income gradually decreases. At a time of 15–20 iterations, both microgrid and user benefits remain the same. When the number of iterations is between 21 and 24, the revenue of microgrid increases significantly, and the corresponding user revenue decreases sharply. When the number of iterations is between 25 and 30, the microgrid revenue and user revenue once again enter a stable state.

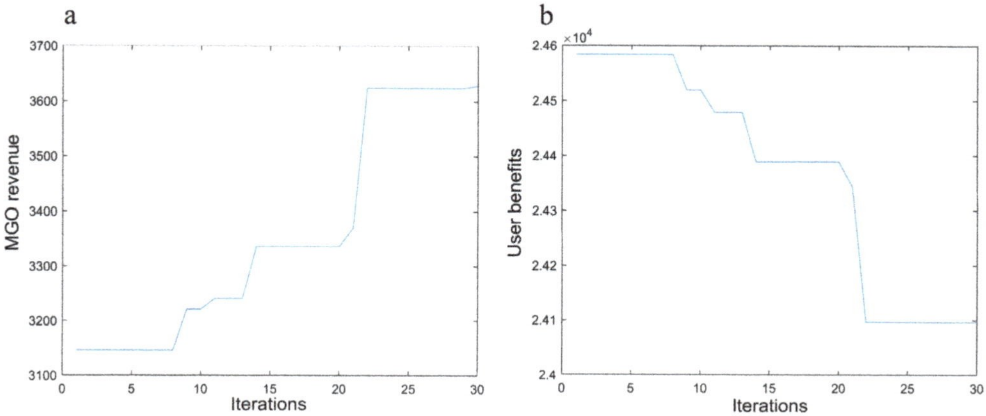

Figure 6. Game equilibrium comparison. (**a**) Microgrid revenue iteration. (**b**) User revenue iteration.

The electricity sales plan of microgrid to users obtained by game solution is shown in Figure 7, and the load after the price-based demand response of the user is shown in Figure 8. According to Figure 7, it can be seen that the purchase price of microgrid is lower than the transaction price between microgrid and users at any time, which is the result of microgrid's game in order to encourage users to respond and achieve the optimal overall economic dispatching level. The price signal sent by the microgrid between 12:30–18:00 and 0:00–1:00 and 2:00–5:00 at the peak time of electricity prices continues to be low, encouraging users to adjust the load during this period. According to Figure 8, it can be seen that after the introduction of price-based demand response, the user's load curve has changed greatly, and the load has been transferred according to the real-time price adjustment given by the microgrid.

In order to verify the optimization effect of the proposed multi-group optimization method on the model, the multi-group optimization algorithm (MPGA) is compared with particle swarm optimization (PSO) and standard genetic algorithm (GA), and the results are shown in Figure 9.

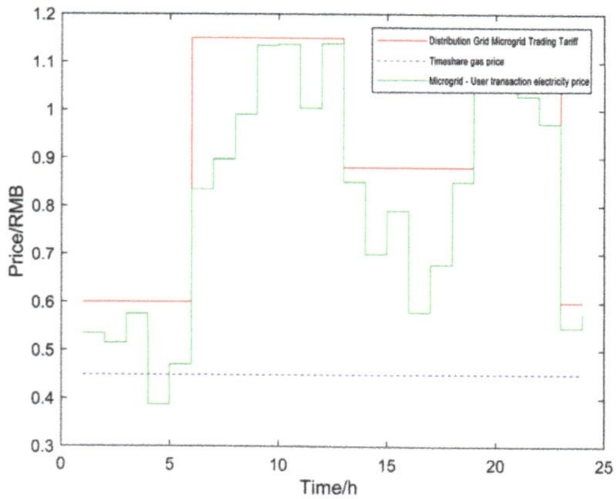

Figure 7. Electricity sales plan for microgrid users.

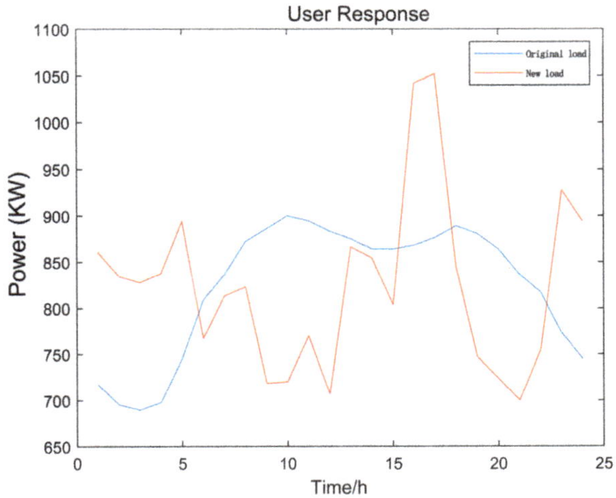

Figure 8. Electricity load of users after electricity price demand response.

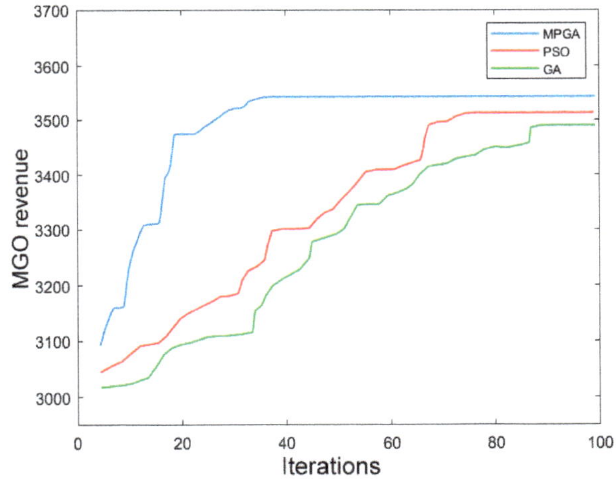

Figure 9. Algorithm convergence curve comparison chart.

As shown in Table 3, the total cost of microgrid obtained by MPGA algorithm is 11, 667.04 yuan, which reaches convergence in 31 times. The total cost of PSO algorithm is 12, 306.7 yuan, which reaches convergence in 74 times. The total cost of GA algorithm is 13, 368.8 yuan, which reaches convergence in 87 times. It is proved that the convergence speed and one-day expenditure cost of MPGA algorithm are the best, and the optimal solution is due to other algorithms.

Table 3. Comparison of optimization results of different algorithms.

Algorithm	Micro-Net Total Cost/RMB	Number of Iterations
MPGA	11,667.04	31
PSO	12,306.73	74
GA	13,368.85	87

6.3. Optimization of Operation Results

In order to further analyze the impact of price-based demand response on the system, two comparison scenarios are set up. The carbon transaction cost, carbon emission, electricity purchase cost, gas purchase cost, and total cost of microgrid in different scenarios are shown in Table 4.

Table 4. Optimization results for different scenarios.

Parameter	Consider Demand Response	Ignore Demand Response
Carbon emission/kg	4857.90	4733.93
Carbon trading cost/RMB	2303.95	2261.96
Power purchase cost/RMB	3268.39	4046.99
Gas purchase cost/RMB	6094.70	6027.04
Total cost/RMB	11,667.04	12,335.99

As shown in Table 3, compared with not considering demand response, in the scenario of considering demand response, the gas purchase cost has slightly increased, while the initial carbon emissions of natural gas related equipment are low, so the carbon emissions and carbon transaction costs have increased. However, based on a low-carbon model with a tiered carbon trading mechanism, the cost of electricity and total costs significantly decrease, but the increase in environmental costs is relatively small and within a reasonable range. Moreover, considering demand response can enable users to participate in the response, which is conducive to the consumption of renewable energy. The price is based on demand response and leads the user to transfer the load, so the power purchase cost is reduced. Through comprehensive calculation, the total cost of the scenario considering demand response is reduced, and the goal of economic optimal scheduling of the microgrid system is achieved.

The equipment scheduling level of the microgrid after considering the demand response is shown in Figures 10 and 11, and the equipment scheduling level of the microgrid without considering the demand response is shown in Figures 12 and 13.

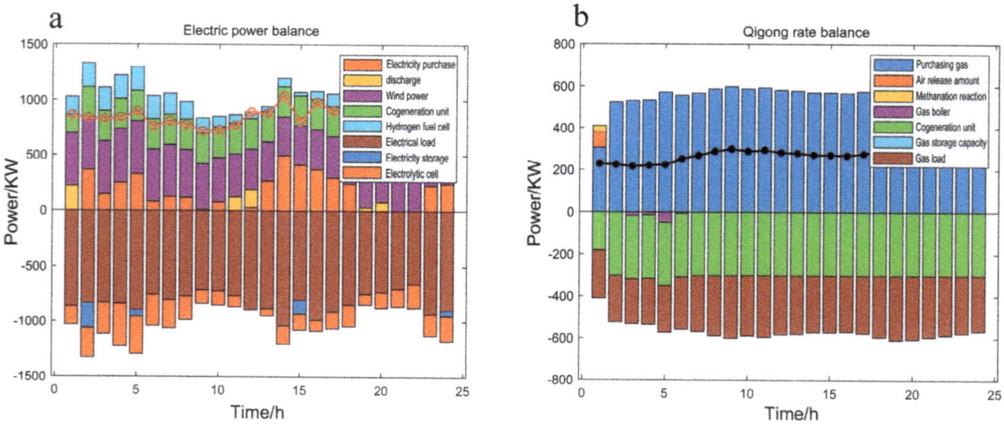

Figure 10. Device operation after demand response. (**a**) Power balance. (**b**) Natural gas energy balance.

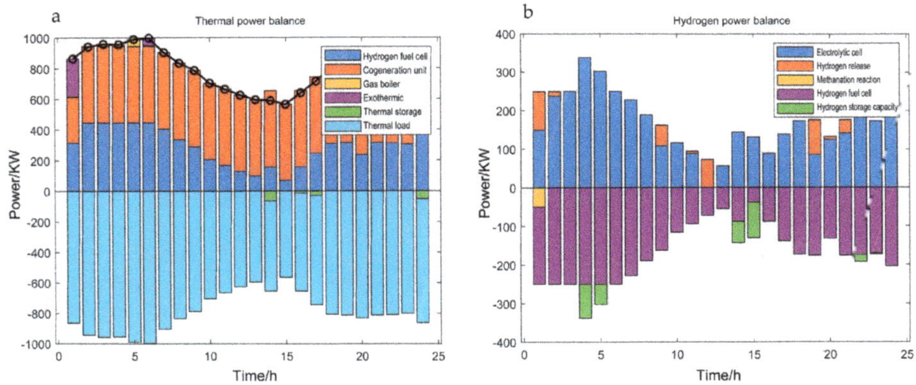

Figure 11. Device operation after demand response. (**a**) Thermal balance. (**b**) Hydrogen balance.

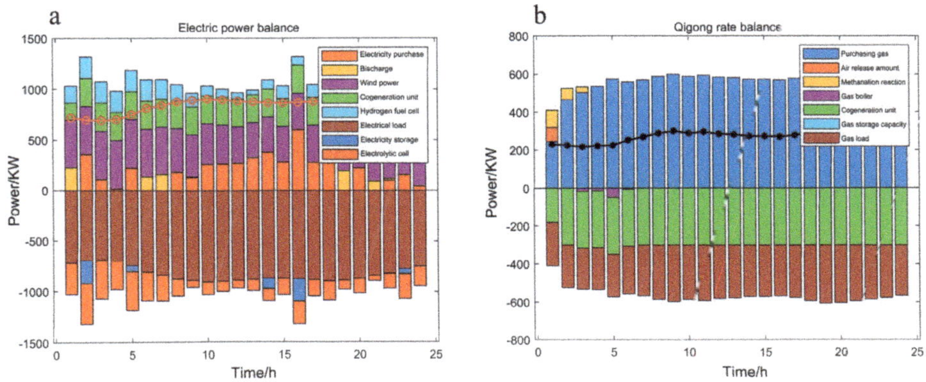

Figure 12. Device operation without demand response. (**a**) Power balance. (**b**) Natural gas energy balance.

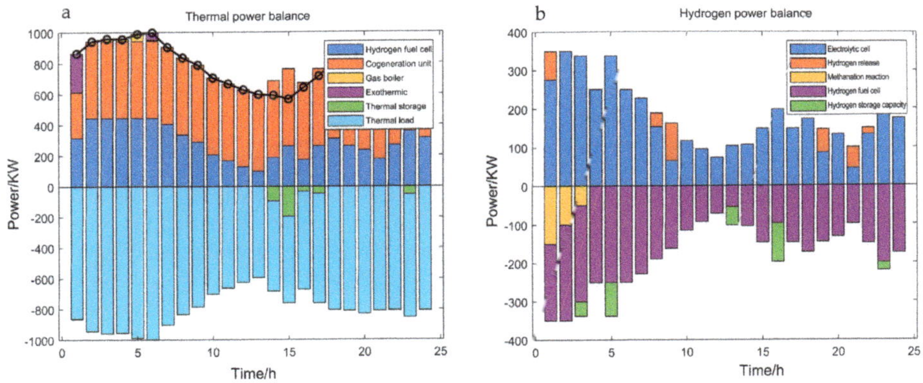

Figure 13. Device operation without demand response. (**a**) Thermal balance. (**b**) Hydrogen balance.

By comparing Figures 10a and 11a, it can be seen that the energy purchase period, without considering demand response, is mainly in the peak time of energy consumption, and the power purchase period in the micro grid scenario after considering demand response is mainly distributed in the peak and valley time of electricity price. By comparing the gas consumption of methanation reactions in Figures 10b, 11, 12b, and 13, it can be seen that, considering demand response scenarios, it is 26.4% and 16.7% of the gas consumption without considering demand response scenarios, respectively. Without considering demand response, methanation reactions account for a larger proportion of the gas consumption. This is because in this scenario, the microgrid does not consider user load transfer, and purchases during the low electricity price period for the first step of hydrogen production from electricity to gas. Therefore, the hydrogen production is more than considering demand response scenarios, and as a raw material for methanation reactions and hydrogen fuel cells, it relatively reduces the purchase of natural gas during this scheduling period.

7. Conclusions

This article further rationalizes the configuration of microgrids and analyzes the respective needs of the two stakeholders, microgrids, and users. Therefore, a low-carbon economic optimization model for microgrids with a tiered carbon trading mechanism is constructed to ensure system economy while reducing carbon emissions. In addition, on this basis, a Stackelberg game model with price-based demand response, led by microgrids and followed by users, was introduced. The microgrid benefits are determined as the optimal scheduling of device operation, while the user side benefits are determined as having the highest energy efficiency, with the lowest cost. Firstly, the three elements and types of game are proposed, and the applicable game type in this chapter is Stackelberg game. Secondly, a microgrid user Stackelberg game structure was established, and the goal of the micro grid game was to achieve optimal operational economy, as well as constraints due to the consideration of increased demand response. The game goal of the user end was proposed to maximize revenue, and a satisfaction model including user electricity consumption and electricity expenditure was analyzed as a constraint to participate in the game. Once again, a Stackelberg game model was established between the microgrid and users, and it was proven that there exists an equilibrium solution to the game. Finally, an example analysis was conducted to verify the feasibility of the model. The specific conclusions are as follows:

(1) Comparative analysis without considering demand response scenarios shows that the optimization cost of microgrid operation considering price-based demand response scenarios has decreased by 5%, which is 668.95 yuan. Among them, the power purchase cost has decreased by 23.8%, which is 778.6 yuan, the carbon emissions have increased by 17%, which is 83.96 kg, and the carbon trading cost has increased by 1.8%, which is 41.98 yuan. This proves that the introduction of demand response can improve the overall economic benefits of microgrids while slightly increasing environmental costs.

(2) After considering demand response, the selling price of microgrids is always lower than the purchase price from the distribution network, and the price reduction rate is relatively high during the initial load valley, encouraging users to adjust their load during the time period. After the demand response, the user's load curve underwent significant adjustments and transformations, and the corresponding load transfer was carried out according to the price signal of the microgrid, achieving the expected "peak shaving and valley filling" effect of microgrid scheduling.

(3) In the microgrid scheduling scenario considering price-based demand response, the electricity purchase period is mainly distributed during the low and flat peak periods of tiered electricity prices. Due to the transfer of user load, the energy pressure caused by user load is reduced.

Author Contributions: Conceptualization, B.L. and L.-D.W.; methodology, B.L, Y.L. and D.G.; software, B.L., M.-T.L. and Y.L.; validation, B.L., D.G. and X.Z.; formal analysis, M.-T.L., X.Z. and P.-R.Z.; investigation, Y.L., P.-R.Z. and X.Z.; resources, L.-D.W.; data curation, B.L., M.-T.L. and B.Z.; writing—original draft preparation, B.L.; writing—review and editing, B.Z., D.G. and L.-D.W.; supervision, L.-D.W.; project administration, L.-D.W.; funding acquisition, L.-D.W. All authors have read and agreed to the published version of the manuscript.

Funding: This article is supported by the Liaoning Provincial Scientific Research Funding Project, the comprehensive load modeling and electric heating control mechanism of the village-level traffic DC micro-network comprehensive load modeling (LJKMZ20221032).

Data Availability Statement: Data are availability on request from the authors.

Conflicts of Interest: The authors declare no conflict of interest. The funders had no role in the design of the study; in the collection, analyses, or interpretation of data; in the writing of the manuscript; or in the decision to publish the results.

Glossary

Symbol	Meaning
$P_{g,buy}$	Gas purchases
$P_{e,.buy}$	Purchase electricity
λ_e	Power-hungry devices
λ_g	Air consumption equipment
$Q_{CO_2}^{PGU}$	The amount of CO_2 emitted by coal-fired power plants
$Q_{CO_2}^{CCS}$	The amount of CO_2 captured by the carbon capture equipment
ε_e	Unit electrical power
ε_x	Carbon allowances per unit of power consumed by different devices
C_{CO_2}	Carbon trading costs
E	Net carbon emissions of the system
i	Equipment options, including cogeneration units, gas-fired boilers, and coal-fired power plants
N	Total number of devices
λ	Price growth rate
l	The length of the interval
c	Carbon trading base price
F	Minimum operating costs of the system
C_{PGU}	Fuel costs
$C_{e,\ buy}$	Electricity purchase costs
$C_{g,\ buy}$	The cost of purchasing gas
$P_{g,\ buy}$	Overall system air consumption
c_e	Real-time electricity prices
c_g	Real-time gas prices
T	Take 24 h
$P_{WT}(t)$	Wind energy output power
$P_{CHP}(t)$	The power output of the cogeneration unit
$P_{SB,\ dis}(t)$	Discharge power of power storage equipment
$P_{HFC,\ e}(t)$	Hydrogen fuel cells produce electricity
$P_{load}(t)$	Electrical load
$P_{SB,\ chr}(t)$	The amount of charge of the storage device
$H_{CHP}(t)$	Combined heat and power unit thermal energy output power
$H_{GB}(t)$	Gas boiler heat generation
$H_{TS,\ dis}(t)$	Heat release from heat storage equipment
$H_{TS,\ chr}(t)$	Heat storage equipment stores heat
$H_{Load}(t)$	Heat load
$G_{buy}(t)$	Gas purchases
$G_{ES,\ dis}(t)$	Air receiver outgassing
$G_{MR}(t)$	The amount of CH_4 produced by methanation in power-to-gas technology
G_{load}	Gas load
$G_{ES,\ chr}(t)$	Gas storage capacity for gas storage equipment
$G_{CHP}(t)$	Air consumption of a cogeneration unit
$G_{GB}(t)$	Gas boiler units consume natural gas
$P_{EL,\ H_2}(t)$	The amount of hydrogen produced by the electrolyzer in the first step of electro-to-gas

$P_{HS,\,dis}(t)$	Hydrogen storage equipment storage capacity
$P_{H_2,\,MR}$	Hydrogen consumption in methanation reactions
$P_{H_2,\,HFC}$	Hydrogen fuel cell hydrogen consumption
$P_{HS,\,chr}(t)$	The amount of hydrogen released by hydrogen storage equipment
P_{PGU}^{max}	Upper limit of coal-fired unit output
P_{PGU}^{min}	Lower limit of coal-fired unit output
ΔP_{PGU}^{max}	The maximum uphill climbing power of coal-fired units
ΔP_{PGU}^{min}	Maximum downhill climb power of coal-fired units
ΔP_{PGU}	The amount of power change
$P_{PGU}(t)$	t time coal-fired unit power
$P_{PGU}(t-1)$	$t-1$ time coal-fired unit power
$P_{CHP,\,e}^{max}$	Upper limit of electrical output of cogeneration units
$P_{CHP,\,e}^{min}$	Lower limit of electrical output of cogeneration units
$H_{CHP,\,h}^{min}$	Lower limit of thermal output of cogeneration units
$H_{CHP,\,h}^{max}$	Upper limit of thermal output of cogeneration units
$\Delta P_{convert,\,e}^{max}$	The maximum upward climb power of the cogeneration unit
$\Delta P_{convert,\,e}^{min}$	Maximum downhill climb power of cogeneration units
$\Delta P_{convert,\,e}$	Maximum downhill climb power of cogeneration units
$P_{convert,\,e}(t)$	The electrical power of the cogeneration unit after conversion at time t
$P_{convert,\,e}(t-1)$	The electrical power of the cogeneration unit after conversion at time $t-1$
$P_{EL,\,H_2}^{max}$	The upper limit of hydrogen production capacity of the electrolyzer
$P_{EL,\,H_2}^{min}$	The lower limit of hydrogen production by the electrolyzer
G_{MR}^{max}	The upper limit of the amount of CH_4 resulting from the reaction
G_{MR}^{min}	Lower limit of the amount of CH_4 resulting from the reaction
$\Delta P_{EL,\,H_2}^{max}$	The maximum uphill climb power of the electrolyzer
ΔG_{MR}^{max}	The maximum uphill climb power of the methanation reaction
$\Delta P_{EL,\,H_2}^{min}$	Maximum downhill climb power of the electrolyzer
ΔG_{MR}^{min}	Methanation reaction maximum downhill climb power
$P_{EL,\,H_2}(t)$	The amount of power change in the electrolyzer
$G_{MR}(t)$	The amount of change in methanation reaction power
$P_{EL,\,H_2}(t-1)$	Electrolyzer power at $t-1$ time
$G_{MR}(t-1)$	Methanation reaction power at time $t-1$
H_{GB}^{max}	Upper limit of gas boiler output
H_{GB}^{min}	Lower limit of gas boiler output
ΔH_{GB}^{max}	The maximum upward climbing power of the gas boiler
ΔH_{GB}^{min}	Maximum downhill climb power of gas boilers
ΔH_{GB}	The amount of change in power
$H_{GB}(t)$	Gas boiler power at time t
$H_{GB}(t-1)$	$t-1$ time gas boiler power
$P_{HFC,\,e}^{max}$	Upper limit of hydrogen fuel cell output
$\Delta P_{HFC,\,e}^{min}$	Hydrogen fuel cell minimum uphill climb power
$\Delta P_{HFC,\,e}$	The amount of change in power
$P_{HFC,\,e}(t)$	The power of the gas boiler at the time t
$P_{HFC,\,e}(t-1)$	The power of the gas boiler at the time $t-1$
$C_{es,\,buy}$	The cost of purchasing electricity to the upper grid in the game model
$C_{eu,\,sell}$	The electricity price revenue sold by the microgrid to the user
$C_{gs,\,buy}$	The cost of purchasing gas
$C_{CO_2,\,s}$	Carbon trading costs in game models
$e_{es,\,buy}$	The unit price of electricity sold in the distribution network
$e_{eu,\,sell}$	Electricity purchased by the microgrid
$P_{es,\,buy}$	The electricity purchased by the user is sold by the microgrid
P_{user}	The amount of power after the user responds
$P_{es,\,buy}^{min}$	The lower limit of the amount of electricity that the microgrid purchases from the distribution grid
$P_{es,\,buy}^{max}$	The upper limit of the amount of electricity that microgrids can purchase from the distribution grid

$e^{max}_{eu,\ sell}$	The upper limit of the electricity price that microgrid can sell to users
$e^{min}_{eu,\ sell}$	The lower limit of the electricity price sold by the microgrid to the user
$P_{CHP}(t)$	The output power of the electrical energy of the cogeneration unit
$P_{SB,\ dis}(t)$	The discharge power of the power storage device
$P_{HFC,\ e}(t)$	The power generation of hydrogen fuel cells
$P_{SB,\ chr}(t)$	The power to charge the storage device
$\Delta P_{user,\ t}$	Changes in user electricity consumption before and after demand response
$P^{0}_{user,\ t}$	The user's electricity consumption before the demand responds
$P^{0,\ max}_{user,\ t}$	The upper limit of the demand response preload value
$P^{0,\ min}_{user,\ t}$	The lower bound of the demand response preload value
$P^{max}_{user,\ t}$	The upper limit of the demand response afterload value
$P^{min}_{user,\ t}$	The lower bound of the demand response afterload value
C_{UE}	User's interests
α	User energy preference coefficient
M_{min}	The lower limit of user satisfaction with electricity consumption
P^{0}_{user}	Electricity consumption by users before demand response
N_{min}	Minimum consumer spend satisfaction
Φ_x	Optimal operation scheme of microgrid
Ψ_y	Optimal response on the user side
Φ^{*}_{n-x}	Other strategies except Φ_x
Ψ^{*}_{n-y}	Other strategies except Ψ_y

References

1. BP World Energy Statistical Yearbook (2022 Edition). [EB/OL]. Available online: https://www.bp.com/en/global/corporate/energy-economics/statistical-review-of-world-energy/downloads.html (accessed on 17 June 2022).
2. *Speech at the General Debate of the 75th United Nations General Assembly*; State Council of the People's Republic of China: Beijing, China, 2020.
3. Zhang, Y.; Zhang, N.; Dai, H. Analysis model construction and transformation path comparison of low-carbon development of China's power system. *China Power* **2021**, *54*, 1–11.
4. Zhou, N.; Fan, W.; Liu, N. Multi objective capacity optimization configuration of photovoltaic micro grid energy storage system based on demand response. *Power Grid Technol.* **2016**, *40*, 1709–1716.
5. Li, X.; Geng, G.; Ji, Y. Joint optimization planning of energy storage and demand side response in active distribution network. *Power Grid Technol.* **2016**, *40*, 199–206.
6. Wang, D.; Fan, M.; Jia, H. Demand Response of Home Temperature Control Load and Modeling of Energy Efficiency Power Plant Considering User Comfort Constraints. *Proc. CSEE* **2014**, *34*, 2071–2077.
7. Wang, C.; Liu, M.; Lu, N. Power fluctuation smoothing method of microgrid tie line using residential temperature control load control. *Proc. CSEE* **2012**, *32*, 36–43.
8. Tang, X.; Su, J.; Li, L. Optimization of economic dispatch of energy hub based on multi-energy storage microgrid. *Power Capacit. React. Power Compens.* **2021**, *42*, 182–187.
9. Heiskanen, E. The Institutional Logic of Life Cycle Thinking. *J. Clean. Prod.* **2022**, *10*, 427–437. [CrossRef]
10. Wu, Y.; Lv, L.; Xu, L. Comprehensive optimal configuration of multiple energy storage capacities of multi-energy microgrid considering electric/heat/gas coupling demand response. *Power Syst. Prot. Control* **2020**, *48*, 1–10.
11. Zhang, X.; Wang, Z.; Lu, Z. Multi-objective load dispatch for microgrid with electric vehicles using modified gravitational search and particle swarm optimization algorithm. *Appl. Energy* **2022**, *306*, 118018. [CrossRef]
12. Zheng, S.; Shahzad, M.; Asif, H.M.; Gao, J.; Muqeet, H.A. Advanced optimizer for maximum power point tracking of photovoltaic systems in smart grid: A roadmap towards clean energy technologies. *Renew. Energy* **2023**, *206*, 1326–1335. [CrossRef]
13. Victoria, M.; Zhu, K.; Brown, T.; Andresen, G.B.; Greiner, M. Early decarbonisation of the European Energy System pays off. *Nat. Commun.* **2020**, *11*, 6223. [CrossRef] [PubMed]
14. Chen, J.; Hu, Z.; Chen, Y. Thermoelectric optimization of integrated energy system considering stepped carbon trading mechanism and electricity to hydrogen. *Electr. Power Autom. Equip.* **2021**, *41*, 48–55.
15. Saboori, H.; Hemmati, R. Considering carbon capture and storage in electricity generation expansion planning. *IEEE Trans. Sustain. Energy* **2016**, *7*, 1371–1378. [CrossRef]
16. Chen, C.; Duan, S.; Cai, T.; Liu, B.; Hu, G. Optimal allocation and economic analysis of energy storage system in microgrids. *IEEE Trans. Power Electron.* **2011**, *26*, 2762–2773. [CrossRef]
17. Regufe, M.J. Current Developments of Carbon Capture Storage and/or Utilization–Looking for Net-Zero Emissions Defined in the Paris Agreement. *Energies* **2021**, *14*, 2406. [CrossRef]
18. Liu, Y. *Research on Collaborative Optimal Scheduling of Carbon Capture and Waste Incineration Virtual Power Plants*; East China Jiaotong University: Nanchang, China, 2021.

19. Kirschen, D.S.; Strbac, G.; Cumperayot, P.; de Paiva Mendes, D. Factoring the Elasticity of Demand in Electricity Prices. *IEEE Trans Power Syst.* **2000**, *15*, 612–617. [CrossRef]
20. Li, L.; Xue, Y.; Tian, L.; Yuan, X. Research on optimal configuration strategy of energy storage capacity in grid-connected microgrid. *Prot. Control Mod. Power Syst.* **2017**, *2*, 35. [CrossRef]
21. Ma, J.; Li, L.; Wang, H.; Du, Y.; Ma, J.; Zhang, X.; Wang, Z. Carbon Capture and Storage: History and the Road Ahead. *Engineering* **2022**, *14*, 33–43. [CrossRef]
22. Wei, F. Research on Integrated Energy System Planning and Operation Optimization Based on Multi-Objective Optimization and Dynamic Game Method. Bachelor's Thesis, South China University of Technology, Guangzhou, China, 2017.
23. Cheng, S.; Chen, Z.; Wang, R. Multi microgrid two-level coordinated optimal scheduling based on hybrid game. *Electr. Power Autom. Equip.* **2021**, *41*, 41–46.
24. Chalkiadakis, G. Cooperative game theory: Basic concepts and computational challenges. *IEEE Intell. Syst.* **2012**, *27*, 86–90. [CrossRef]
25. Makarov, Y.V.; Du, P.; Kintner-Meyer, M.; Jin, C.; Illian, H. Sizing energy storage to accommodate high penetration of variable energy resources. *IEEE Trans. Sustain.* **2012**, *3*, 34–40. [CrossRef]
26. Guan, X.; Xie, S.; Chen, G. Parameter optimization method for dynamic adjustment interval of multipopulation genetic algorithm. *Comput. Appl. Softw.* **2022**, *39*, 273–282+312.
27. Guan, X.; Xie, S.; Chen, G.; Qu, M. Modal parameter identification by adaptive parameter domain with multiple genetic algorithms. *J. Mech. Sci. Technol.* **2020**, *34*, 4965–4980.
28. Song, W.; Dong, W.; Kang, L. Group anomaly detection based on Bayesian framework with genetic algorithm. *Inf. Sci.* **2020**, *533*, 138–149. [CrossRef]
29. Sabyasachi, M.; Antonios, T. Autonomous Addition of Agents to an Existing Group Using Genetic Algorithm. *Sensors* **2020**, *20*, 6953.
30. Ma, Y.; Wang, H.; Hong, F.; Yang, J.; Chen, Z.; Cui, H.; Feng, J. Modeling and optimization of combined heat and power with power-to-gas and carbon capture system in integrated energy system. *Energy* **2021**, *236*, 121392. [CrossRef]

processes

MDPI

Article

New Equipment for Determining Friction Parameters in External Conditions: Measurements for the Design

Martin Zidek [1,2], Filip Vanek [2], Lucie Jezerska [1,*], Rostislav Prokes [1,2] and Daniel Gelnar [1]

1 ENET Centre, CEET, VSB—Technical University of Ostrava, 17. listopadu 15/2172, 708 33 Ostrava, Czech Republic; martin.zidek@vsb.cz (M.Z.); rostislav.prokes@vsb.cz (R.P.); daniel.gelnar@vsb.cz (D.G.)

2 Department of Mining Engineering and Safety, Faculty of Mining and Geology, VSB—Technical University of Ostrava, 17. listopadu 15/2172, 708 33 Ostrava, Czech Republic; filip.vanek.st@vsb.cz

* Correspondence: lucie.jezerska@vsb.cz

Abstract: Friction parameters such as the angle of internal friction and the external friction of soils (bulk materials) show the possibilities of further material use. These are, for example, possibilities for soil processing, handling, and storage. The determination of friction parameters is usually carried out under laboratory conditions. For the possibility of determining the properties of soils outside the laboratory in terms of immediate material response, a laboratory prototype was developed. The main objective for its development was to determine the effect of the shape of the friction surface when "sliding" on the soil. This was achieved with the help of validation equipment designed to measure, test, and validate the processes of raking, material piling, material transfer and removal, and tool movement or sliding on or in a material. It was found that by using an appropriate speed and normal load, the Jenike method can be applied to determine the angle of external friction over a shorter distance with an error of about 6–7.5% from the values measured on a calibrated shear machine. The results also showed that the method can be applied to detect the shear stresses that arise when a tool is plunged into a material, and thus predict the possible increase in energy loss during the process.

Keywords: innovative device; development; external friction; bulk material; soil

Citation: Zidek, M.; Vanek, F.; Jezerska, L.; Prokes, R.; Gelnar, D. New Equipment for Determining Friction Parameters in External Conditions: Measurements for the Design. *Processes* **2023**, *11*, 3348. https://doi.org/10.3390/pr11123348

Academic Editor: Farhad Ein-Mozaffari

Received: 31 October 2023
Revised: 27 November 2023
Accepted: 29 November 2023
Published: 1 December 2023

1. Introduction

For the transport, handling, processing, storage, and other processes with soils (bulk materials), it is important to determine the mechanical and physical properties. These properties are determined primarily to predict the process behavior of bulk/particle materials and to possibly find the critical points, of both a given equipment and the critical conditions of the bulk material [1]. In some cases, the mechanical properties, transport, and storage of the material can be improved by the slight addition of a suitable additive [2–4]. In others, pretreatment of the material, such as milling, to change the behavior of the process [5] or pre-homogenization of alloys to improve their mechanical properties [6] are sufficient. The selected properties are measured mainly in accredited laboratories on standardized equipment under laboratory conditions. However, many times these properties do not correlate with the properties of the material in its raw state, which is also closely related to the place of collection. It is important to observe the sampling method of the different samples and also to determine the possible sampling error rate [7]. Measuring the mechanical and physical properties of soils in real conditions is difficult due to environmental influences such as rain, sun, wind, and temperature. Sample deterioration occurs during collection and transport. The sample should have the same characteristics as when it was collected. Loose material can degrade, change temperature, absorb moisture, or become soft or compacted if improperly transported and stored [8]. A sample so affected may not have the same properties as when it was collected. It would be appropriate to measure the parameters at a given location and under given conditions (climatic, stress, and dynamic effects), which is only possible when portable equipment is available [9–11]. For soil shear tests, a simple portable device has been developed to determine friction parameters in clay soils under different climatic conditions [12]. The device works on the principle of pushing the measuring head into the soil, and when the desired value of the normal load is reached, the shear head rotates at a defined depth. For accurate determination of the properties of soils, bulk materials, or other powders under realistic conditions, this approach is the right choice.

The friction parameters (angle of internal friction and external friction) of the soils affect all processes where the soil is further manipulated. They are important indicators of possibilities for material use in a multitude of applications [13–16]. Friction parameters affect both the life of the tools in contact with the soil and the soil itself, in the form of possible intoxication by residues resulting from wear or degradative reactions in the soil. Experience shows where it is essential to determine the necessary parameters on site [17,18]. These include field experiments in snow conditions, where portable instruments can determine the instantaneous cohesion of the snow. This can be used to analyze how cohesion and friction affect the increase in avalanche pressure. Therefore, portable devices that determine the angle of external friction are very important with respect to the instantaneous response of the material. Therefore, this study focuses on the development of such a device that determines the external friction angle based on the dynamic motion and response of the tool. Similar issues are also being addressed by ski manufacturers. It is necessary to know the reactions and interactions that occur between the ski motion and the snow [19,20].

This study consists of several parts. First, the validation system (device) on which the measurements were performed and its calibration and methodologies are described. Of course, a characterization of the bulk material used is presented. In order to ensure accurate measurements, the calibration of the process under static as well as dynamic loading was performed and is described below. The experiment itself, where a set of determinations was carried out to analyze the possibility of determining the friction parameters of the soils, was carried out and is described in detail on two different soils and at three working speeds. The initial shear stress and the partial reaction of the material as the tool (friction surface) moved within it were monitored in the measurement line. The final part of the paper is devoted to a discussion of the above-mentioned issues.

2. Materials and Methods

This study describes the design of a device for determining the dynamic frictional properties of a soil. It highlights the problem of the sliding of a solid body over the soil and the effect of increased wall friction when the tool penetrates the soil under the application of a specified normal load.

2.1. Materials

The soil type was determined using the composition and particle size of the coarse sandy soil (>0.6 mm to 2 mm) according to the norm CSN EN ISO 14688-1 [19].

Two samples with different fractions were used in the study. Silica sand with a grain size of 0.3–1 mm (sample 1) and silica sand with a grain size of 1.4–2 mm (sample 2) were measured. PLA filament was used as the contact material, which was used in the 3D printing of the tool. The active parts of the tools were smoothed/sanded after printing (with 120 grit sandpaper). Other materials used were aluminum (strain gauge, validation device frame), clear polycarbonate (plastic plates), steel (rails, guides, fasteners), and chipboard (dividing plate, bottom). The following parameters were measured: angle of external friction, bulk density, volume density, and particle size distribution.

2.2. Devices

In this study, five devices and measurement methods were used to measure the mechanical and physical properties of the bulk material. A Jenike shear tester was used to measure the angle of external friction of the bulk material (Figure 1). The volume density of the bulk material was measured using an Anton Paar gas pycnometer, type Ultrapyc 5000 [21,22]. The bulk density was measured via an experiment of pouring bulk material into a measuring cylinder [23] and the particle size distribution was measured on a Camsizer particle size analyzer [24].

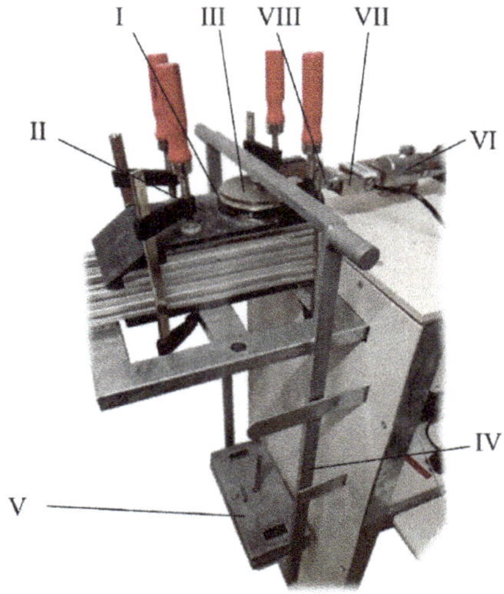

Figure 1. Jenike shear tester.

An alternative way of measuring the friction parameters and the penetration of the tool into the bulk material was measured on a validation system for pulling and pushing tools (hereafter validation system, Figure 2). Also, the apparatus was used to dynamically run the system. The validation system was patented by the VSB—Technical University of Ostrava [25].

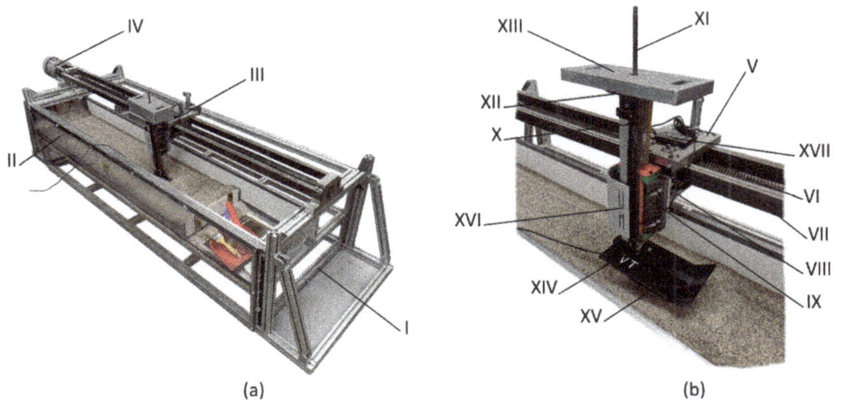

Figure 2. Validation system, (**a**) axonometric view, and (**b**) details of the device for alternative measurement of friction parameters.

The Jenike shear tester [26,27] uses the rectilinear movement of the measuring cell with the measured bulk material to determine the shear stress between the bulk material and the friction surface (angle of external friction). During the measurement, the bulk material is loaded into the shear cell/ring (I). The cell/ring (I) is then placed on the contact material (II) (shear surface of the tool). A lid (III) is placed on the bulk material, which is placed in the cell/ring (I), and contains a spike on the top. This spike is used to place a holder (IV) on which a weight (V) is placed to apply a normal load. The piston (VI) produces a straight-line movement, which is transferred to the lid (III) and cell/ring (I) via the tensiometer (VII) and thorn (VIII). The shear stress is measured using a tensiometer (VII). The normal and friction parameters are then entered into tables and evaluated by linearizing the set of measurements and reading the angle of the linearized curve and the horizontal axis of the graph. The working speed of the Jenike shear machine used in our laboratory is v = 25 mm min^{-1} and the volume of the shear cell/ring is S = 6.78 × 10^{-3} m^2. Normal tension was induced using 1–4 weights with a weight of 4.4 kg per weight.

The Anton Paar gas pycnometer Ultrapyc 5000 (Anton Paar GmbH, Graz, Austria) is a device for measuring the real/volume density of solid porous and bulk materials. It works on the principle of measuring the volume of a gas (most commonly helium) [28].

The validation system is primarily designed to provide the necessary data in the areas of material raking, material piling, material transfer and removal, and tool movement or sliding on or in a material. Figure 2 describes the validation system, which consists of a support structure (I) composed of aluminum profiles and clear polycarbonate plates, an adjustable bottom (II), a linear guide (III), and a drive (IV). The linear guide (III) consists of the travel (V), the threaded rod TR40x7 (Libor Muron PERFEKT, Ostrava, Czech Republic) (VI), and the guide rods (VII). A weldment (VIII) is attached to the travel (V) using fasteners, to which the flange carriage with ball guide HGW 45HC Z0H (HIWIN s.r.o., Brno, Czech Republic) (IX) is attached in the same way. The T-rail HIWIN HGR45T-0364 (HIWIN s.r.o., Brno, Czech Republic) (X) is inserted through the groove of the ball guide. A holder (XII) is attached to the head of the rail (X) by means of a rod (XI), on which a weight (XIII) is placed. A diaphragm force transducer (MEG30, 500 N, MEGATRON, s.r.o., Prague, Czech Republic) (XIV) is placed at the foot of the rail (X), on which is the skid/tool (XV). The normal load on the bulk material was set at 1–4 weights (XIII) (where one weight has a mass of 4.4 kg). In addition, a digital caliper (HMB Machines, Moordrecht, The Netherlands) (XVI) is attached to the device, where the values can be read from the panel (XVII), which is located on the travel (V). The drive (IV) is started using the Sinamics G110 AIN frequency converter.

After the drive (IV) is put into operation, the threaded rod TR40x7 (VI) starts to rotate. Using the RT40x7 nut, which is located in the linear guide (III), the rotary motion is converted into a straight-line motion. The linear guide (III) then moves back and forth in the axis of the threaded rod TR40x7 (VI). This motion is transferred from the weldment (VIII), the ball guide HGW 45HC Z0H (IX), the T-rail HIWIN HGR45T-0364 (X), and the diaphragm force transducer (MEG30, 500 N) (XIV) to the tool (XV), which moves parallel to the axis of the threaded rod TR40x7 (VI).

Figure 3 shows the basic geometrical parameters of the experiment, the length of the tool movement, its direction of movement, and the height of the material in the validation device. Three working speeds were used for the experiment. The tool start-up length was then run between points A and B and the measurement itself was run between points B and C.

The active element of the experiment that is in contact with the soil is the tool shown in Figure 4a. To maintain the same height and level of the soil, a raking plate was created (Figure 4b).

The steady-state shear friction values were then evaluated and read using a strain gauge attached to the tool.

For accurate measurement values, it was important to determine the individual weights of the components in the unfolded state (Mettler Toledo PG5002-SDR scale, Greifensee, Switzerland), as well as the weight of the overall assembly as a dynamic system. The actual weight of the dynamic motion was determined using a gauge created for this dynamic motion (Figure 5). On the ball guide rail (IX), the gauge structure (I) (made of aluminum profiles) was placed for dynamic operation, to which a diaphragm force transducer (MEG30, 500 N) (II) was attached in the lower part by means of a screw

(III). The dynamic effects of motion and load acting on the rail (V) are transmitted to the diaphragm transducer (II) by means of a pin (IV).

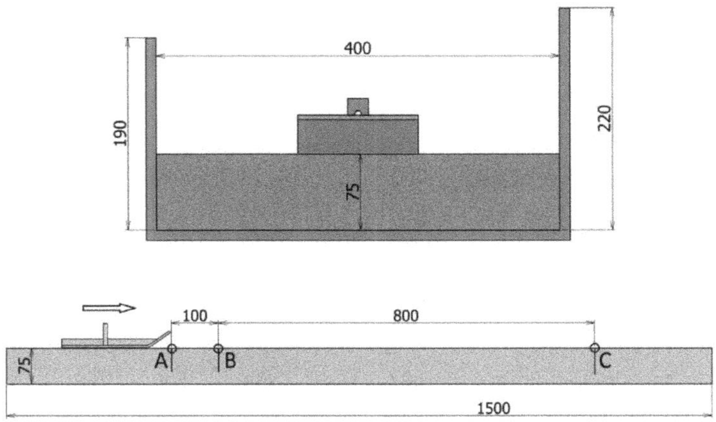

Figure 3. Basic geometric parameters for experimental measurements, trough width, wall height, height of material in the trough, direction, and length of movement (mm).

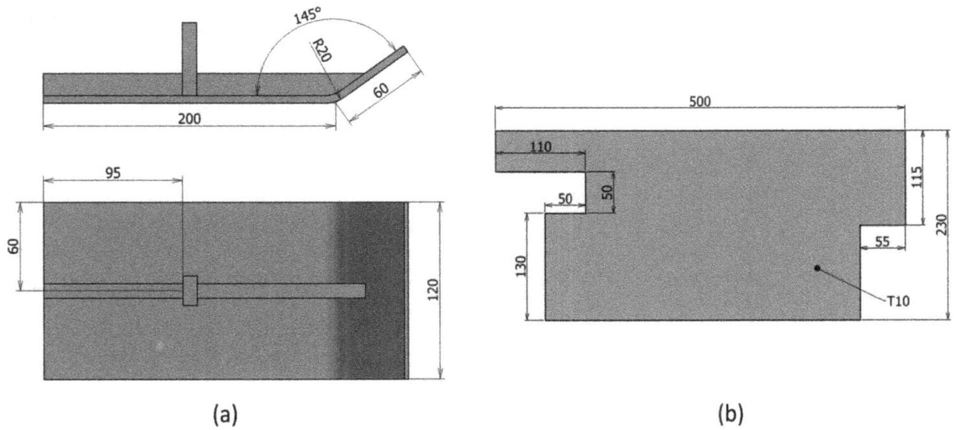

Figure 4. (**a**) Tool geometry, (**b**) rake plate geometry (mm).

During motion, the diaphragm transducer (III) senses the dynamic effects and normal forces that were inferred from the experiment from the movement over the soil. Measurements were made for the unloaded system and with weights 1–4.

The theoretical speed of the tool movement was set using the current frequency f [Hz] on the frequency converter (Siemens SINAMICS G110, SIEMENS, Munich, Germany). This value was calculated using the measured revolutions n_m (rpm) read by a revolution sensor (Voltcraft, DT-30LK, Conrad Electronic SE, Hirschau, Germany) on the shaft with thread pitch p (m) of the TR40x7 lead screw. The theoretical velocity v_t was then determined using Equation (1).

$$v_t = n_m \cdot p \left[\text{m}\cdot\text{s}^{-1}\right] \quad (1)$$

where p is the pitch of the lead screw and n_m is the measured revolutions of the lead screw.

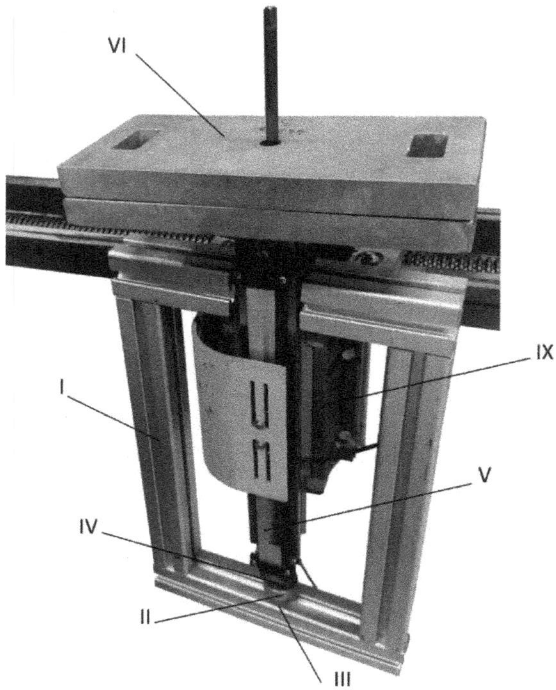

Figure 5. Equipment for measuring dynamic effects.

The measuring tools were 3D-printed with fused filament fabrication (FFF) on a Prusa I MKS3 printer (Prague, Czech Republic), and the filament used was PLA produced by Prusament. The printing method used was FFF, which uses a hot–cold–hot environment. The 3D printer head scoops and melts the PLA wire and deposits individual layers of material onto a pre-heated bed, where they are cooled to form the desired tool shape. This leads to better mechanical properties, where the adhesion between layers is strengthened, and prevents warping and curling of the printed parts. The G-codes for the prints were sliced in PrusaSlicer version 2.3.0 with a layer height of 0.20 (mm). The fillers were set at 20%. The filling pattern was chosen to be gyroid except for 100% of the filling, which is forced rectilinear. The PLA tools were printed at an extrusion temperature of 210 °C and a bed temperature of 60 °C.

2.3. Methods

This study is based on Jenike's method for determining wall friction. Iqbal T. [29], Chen P. [30], and resp. Teunou E. [31] describe the standard procedure, cell dimensions, loads, and conditions for measurement and evaluation using the Jenike method. On this basis, an innovative alternative method has been developed where, unlike the existing method, the friction surface (tool surface) moves over the bulk material, at a higher speed and on a longer trajectory.

The required soil (approximately 65–70 kg) was first loaded into the validation device. The surface of the soil was then leveled to the desired height using a rake plate (used before each measurement to compare the surface of the soil and to keep the material at the same height). Next, a diaphragm force transducer (MEG30, 500 N) and a skid/tool were attached to the rail. Before the actual slipping/sliding process of the skid/tool on the soil, the skid was carefully and slowly placed on the soil. Then, the diaphragm force transducer (MEG30, 500 N) was reset, and the measurement started.

During calibration, the diaphragm force transducer (MEG30, 500 N) was placed firmly on a stable nonmoving frame and successively loaded with 1 to 4 calibrated weights. It was then calibrated to these values using DEWESoft 7.1 software.

The normal forces were measured using a diaphragm force transducer (MEG30, 500 N), which was attached to the dynamic effects measuring device. These forces were induced by a weight placed on the bracket while the device was moving (Figure 5).

To determine the initial shear stress, the tool was in position A (Figure 3). Then, a normal load was applied to the tool using weights. After placing the weights on the holder, the diaphragm force transducer (MEG30, 500 N) was reset to zero. The device was then put into operation. The force against the sliding motion of the tool on the soil was measured using the diaphragm force transducer at the distance between points A and B (Figure 3). The average steady-state values were then used in the evaluation. The data were entered into tables and graphs. Measurements were made for 1–4 weights. These values were compared with the angle of external friction measurements using Jenike's straight-line method.

To ensure the same conditions for the next experiment, we proceeded as follows: The tool was placed on the soil at point A (Figure 3), the diaphragm force transducer was then reset to zero. The device was put into operation and set into motion. Without weights, the slide moved to point B. Once the tool passed point B (Figure 3), 1–4 weights were added to the bracket. The measurement section then ran between points B and C in Figure 3. At point C, the measurement was stopped.

The forces resisting the movement of the skid over the soil were sensed using a membrane force transducer. The DEWESoft measuring system was used for the evaluation.

The result was data that determined the magnitude of the resistance of the skid against movement on the soil at the chosen normal load. All experiments were performed 10 times due to possible errors and dynamic shocks. Inaccurate values were then remeasured, and the resulting data were statistically evaluated.

When comparing the data from the experiment and the data measured on the calibrated device, statistical evaluation method x_s [-] Equation (2) was used to determine the percentage deviation of the experiment from the measurements on the calibrated device.

$$x_s = ABS\left(\left(\frac{X_{exp}}{X_{kal}} \cdot 100\right) - 100\right) \tag{2}$$

where X_{exp} [-] is the parameters measured in the experiment on the validation device and X_{kal} [-] is the parameters measured on the calibrated devices.

3. Discussion and Results

In this section of the article, we will discuss the measured parameters of the soil and the results of the process of the tool sliding on the soil, the overall evaluation of the experiments, and the determination of the deviations and measurement errors. From the measured data, the deviations in the theoretically calculated and measured basic parameters were compared and determined. All parameters were statistically evaluated and are presented in the table and graphs below.

3.1. Characterization of Soil Using Standardized Methods

The following parameters were measured to compare the results and individual dependencies: particle size distribution, volume weight, bulk density, angle of external friction, and angle of repose (Table 1). These parameters were measured at BSC (Bulk Solid Center), Czech Republic, VŠB TU-Ostrava, CENET.

Table 1. Average values of the determined basic parameters.

Parameter	Value					
	Sample 1			Sample 2		
Particle size distribution x (mm)	Q3_10 (%)	Q3_50 (%)	Q3_90 (%)	Q3_10 (%)	Q3_50 (%)	Q3_90 (%)
	0.45	0.66	1.13	1.47	2.01	2.45
Volume weight ρ_v (kg·m^{-3})		2.64			2.64	
Bulk density ρ_s (kg·m^{-3})		1.49			1.51	
Angle of external friction φe (°)		26.74			25.4	
Angle of repose Ψ_s (°)		36.91			38.8	

3.2. Basic Parameters of Experimental Measurements of Dynamic Friction Parameters

First, it is important to point out the difference between the theoretical calculation and the experimental measurement. It is important to identify the difference and thus the possible bias for the theoretical calculations. The two values will always be different; this can be caused by many parameters, such as an inappropriate choice of drive, resistance to movement in the movement screw, a not-strong-enough device for the rail attachment, or dynamic shocks in the system.

This section contains all the necessary theoretical, calculated, and experimental basic parameters. These are mainly the parameters of the frequency converter, drive, screw line, rotational speed, and feed rate (Table 2), and also the normal loads, component weights, and frequency sensing parameters. The parameters were experimentally determined to be smeared for the smooth running and calibration of the measuring experimental device of tool sliding on the soil.

Table 2. Average measured line speed v_V.

Parametr	Frequency on the Inverter		
	5 Hz	10 Hz	15 Hz
Average angular speed on threaded rod (rpm)	141.49	288.62	436.64
Average linear line speed (m·s^{-1})	0.017	0.034	0.051

The frequency converter was used to set three transport speeds for frequencies (5 Hz, 10 Hz, 15 Hz). The speed of the screw line was read using a Voltcraft DT-30LK laser tachometer, directly on the threaded rod. The measured speeds for each frequency are shown in Table 2. Equation (1) was used to calculate the transport speed.

The individual components of the dynamic system were considered both individually and as a whole. They were also weighed both statically and dynamically in motion. The weights of the individual components of the linear-line travel (Figure 2b) are as follows: The rail weighs 3.81 kg, the diaphragm force transducer XIV 0.04 kg, the weight holder and axle XI and XII 0.42 kg, the drop/height device 0.14 kg, the slide/tool XVI 0.12 kg, and the weight XIII 4.43 kg. Subsequently, the total weight of the moving unit was determined for weights 1–4. As the rail moves in the line, there is friction and consequently a loss of normal load. The theoretical static loads and a comparison of the weights of the components individually weighted and weighted in the guideway are shown in Table 3. The in-line parameter is the weight of the moving components in the normal direction stored in the line on the validation device.

Table 3. Comparison of weights: individually weighted and weight in the guideway.

Parameter	Number of Weights			
	1	2	3	4
Individually weighted (kg)	8.95	13.37	17.8	22.22
Weight in the guideway (kg)	7.58	11.85	16.19	20.47
Decrease S (%)	15.32	11.38	9.07	7.88

Therefore, it can be predicted that the resistance of the ball guideway at rest absorbs approximately 8 to 15% of the applied normal load. The weight in the guideway's mean absolute deviation (MAD) ranged from 0.06 kg to 0.12 kg, the root mean square error (RMSE) ranged from 0.45 kg to 0.98 kg, and the coefficient of determination (R-squared) was 1.

3.3. Calibration of a Strain Gauge for Measuring Shear Parameters

The calibration of the diaphragm force transducer (MEG30, 500 N, Figure 2 (XIV) and 5 (II)) was carried out by successively loading the strain gauge with four weights from 4.43 kg to 17.6 kg. A holder was attached to the strain gauge for firm fixing. The measuring system and DEWESoft software were used for the evaluation. The values were read after the load curve had stabilized. This calibration was performed for both the strain gauge used for the shear readings and the strain gauge used on the device to read the dynamic effects during motion. The strain gauge was reset before each measurement (Figure 6).

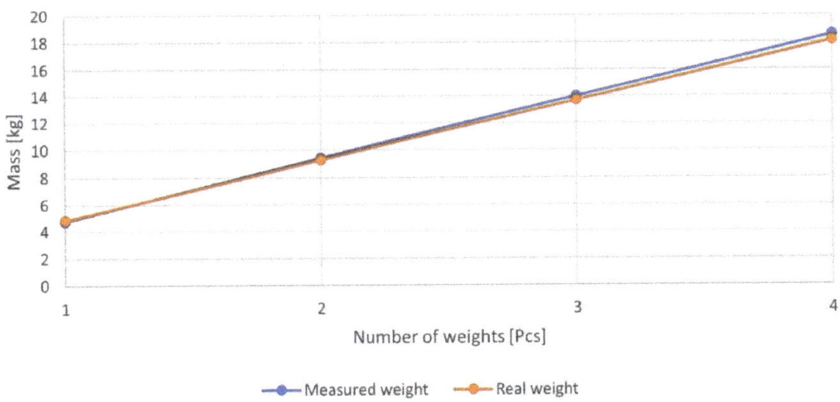

Figure 6. Calibration graph of diaphragm force transducer (MEG30, 500 N).

It was found that the deviation of the strain gauge from the real weight increases with increasing load, up to a maximum of 2.35%. MAD ranged from 0.06 N to 0.12 N, RMSE ranged from 0.09 N to 0.43 N, and R-squared was 1. The calibration of the strain gauge was successful.

3.4. Measurement of Weighted, Static, and Dynamic Normal Loads

The average shift times were 5 Hz/48 s, 10 Hz/23.80 s, and 15 Hz/15.53 s. For accurate measurements at such a dynamic motion, it was important to set and evaluate the normal load for our chosen parameters of tool movement over the soil. Therefore, four normal loads were compared, and different dynamic effects were applied in each.

Normal loads were compared, with loads applied to the soil from the rail (X), the diaphragm force transducer (XIV), the weight holder and axis (XI, XII), and the equipment for determining draft/height and weight; see Figure 5.

The normal load was measured in three ways: (I) The individual components were measured separately. The end weight was obtained by summing the partial values (weighted). (II) The second method was static weighting of the normal load, where the rail was placed in the line and loaded. The values were read using a calibrated balance directly on the equipment (static, see Section 3.2). (III) The third method was the reading of the normal load using a dynamic effect measuring device (Figure 5). The normal load values were read in straight line motion for 3 speeds of the screw linear guideway (Dyn. 5 Hz, Dyn. 10 Hz, Dyn. 15 Hz). For all three methods of reading the normal load, 1 to 4 weights were used. The resulting data are shown in Figure 7.

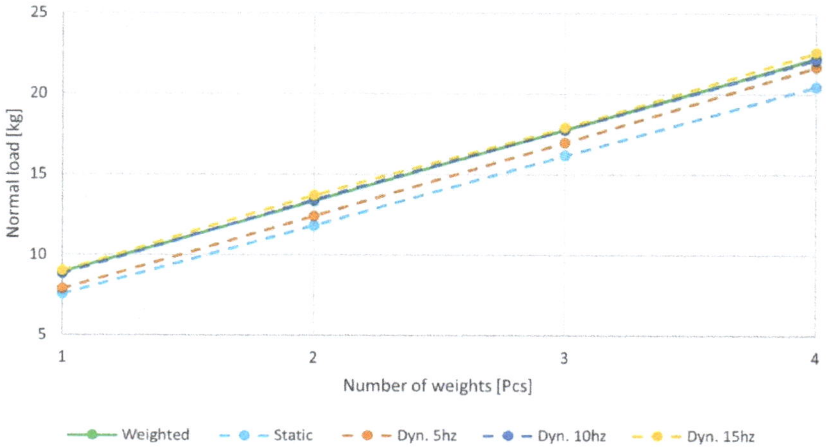

Figure 7. Comparison of weighted (I), static (II), and dynamic (III) normal loads.

For the values read statically (II), it is clear that the rail guide absorbs part of the normal load. For the data readings of the linear guideway motion (III), the normal load gradually increases compared to the static load to values equal to the weighted component data (I). For this experiment, this is the ideal condition where the losses that occurred in the static method (I) disappear. MAD ranged from 0.07 kg to 0.13 kg, RMSE ranged from 0.11 kg to 0.45 kg, and R-squared was 1.

3.5. *Determination of Shear Parameters of Soils on a Validation Stand*

When measuring the shear parameters, emphasis was placed on three important values. The first was the initial shear stress at start-up, the second was the shear stress generated between the soil and the tool after start-up, and the third was the steady-state shear stress after the measurement distance had been completed. As the tool sank/was immersed into the soil, it was important to account for the increasing cross-sectional area through which the tool was sliding on the soil.

When sensing the movement of the tool on the loose material, it was found that the tool came into contact the soil with an area of $S_1 = 0.024$ m^2 during the start-up, and after the start-up distance it sank by the measured value h_x, and therefore exerted an average shear area $S_x = 0.0265$ m^2. After traveling the measured runway, it sank to such an extent that it exerted a full shear area, and therefore a maximum shear area of $S_2 = 0.03$ m^2 was calculated at the end of the measurement.

Several sets of measurements were made. The first set (a) was for initial shear friction and angle of external friction, and the second set (b) was for steady-state shear forces during motion.

A total of 15 measurements were evaluated for each speed (5 Hz, 10 Hz, 15 Hz) and for individual weights (1–4 weights). Subsequently, erroneous measurements with interfering elements or fluctuations were removed. For illustration, Figure 8 shows the evaluation of the shear force data, induced by 1 weight at 10 Hz speed. In the results of this measurement, measurement numbers 6, 10, 11, 12, and 15 were removed due to measurement deviations.

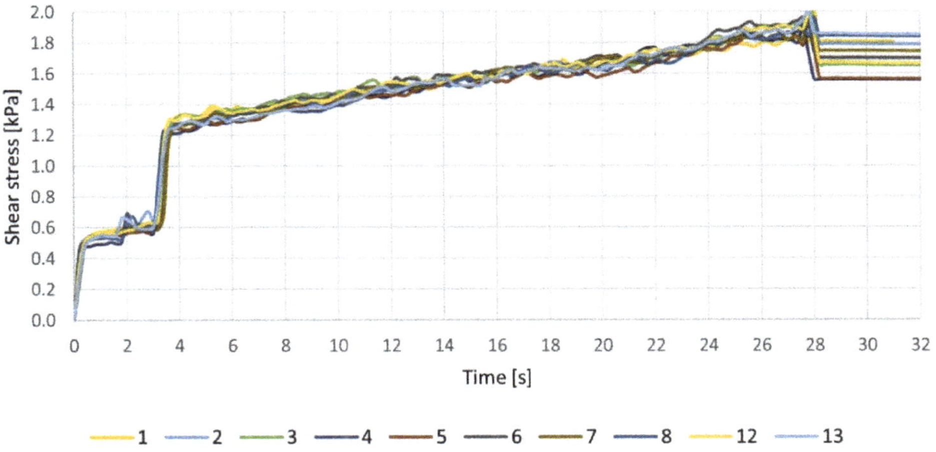

Figure 8. Shear stress versus time for 1 weight at 10 Hz.

Subsequently, the values of the 10 measurements were evaluated and entered into curves. For the given normal stresses, the given shear stresses were determined. For this example, the average value of the steady-state shear stress is 1.86 kPa.

For the first set of measurements (a), the initial shear stress, the tool was placed on the soil. It was then loaded with the required number of 1–4 weights and the strain gauge was set to zero. After starting, the shear force of the tool was then read. This was carried out for three transport speeds (5 Hz, 10 Hz, 15 Hz). The parameters were evaluated, and the initial shear stress and the angle of external friction when the tool was run over the soil were determined (Figure 9). This angle was then compared with the angle of external friction, which was determined via Jenike's method.

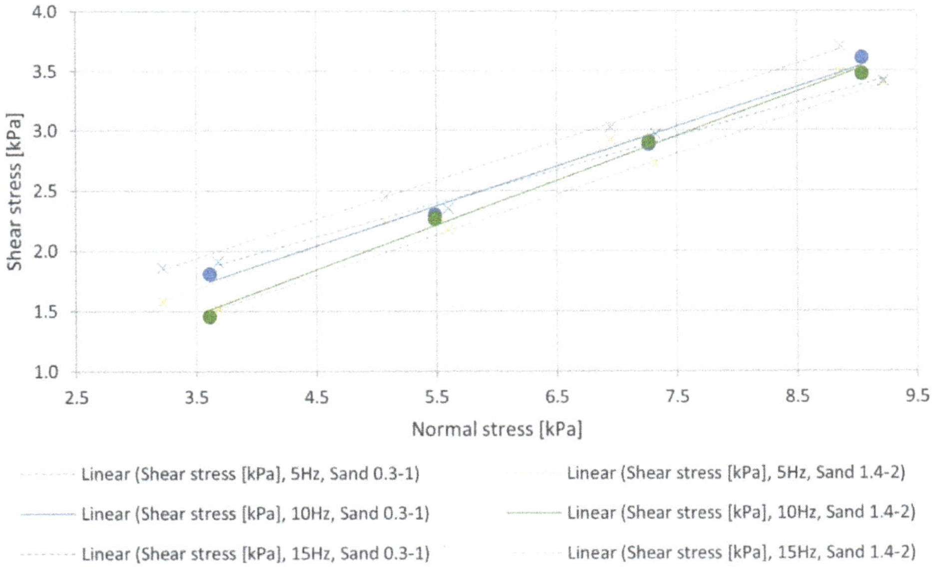

Figure 9. Evaluation of the initial shear stress for 2 tested materials and 3 transport speeds.

The second experiment (b) was performed as follows: The tool was placed on the soil, and the strain gauge was set to zero. The run-up of the device was carried out without weights, and these were added after traveling the run-up distance, which is 0.1 m (AB, Figure 3). Subsequently, the required number of weights was added, and the tool moved along the entire measuring path, which has a length of 0.8 m (BC, Figure 3). Measurements were taken for each speed (5 Hz, 10 Hz, 15 Hz) and the required number of weights 1–4 (Figure 10).

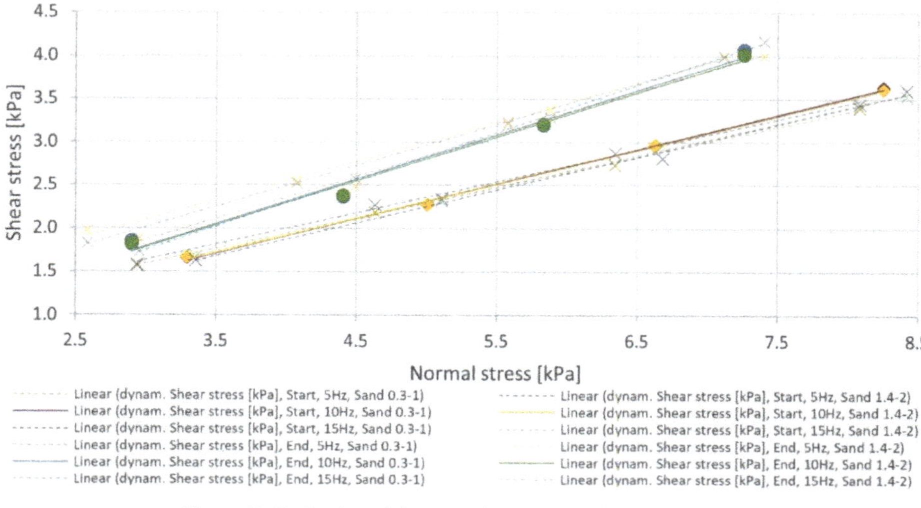

Figure 10. Evaluation of dynamic shear stresses for 2 tested materials and 3 transport speeds (start and end).

The points were interspersed with linear regression. The angle of external friction is defined as the angle given by a linear line with the *x*-axis Table 4.

Table 4. Evaluation of the angle of external friction (°) for 2 tested materials and 3 transport speeds.

Transport Speed	Beginning		Start		End	
	Sample 1	Sample 2	Sample 1	Sample 2	Sample 1	Sample 2
5 Hz	23.89	22.53	23.68	24.26	30.29	30.45
10 Hz	22.28	21.57	24.18	24.11	29.22	29.05
15 Hz	23.67	22.41	23.50	23.65	29.46	29.18

The measurement error variances were as follows: MAD was in the range of 0.67 N to 1.28 N; RMSE increased with the length of the offending path, at the beginning 0.59° to 1.01°, at the start 2.46° to 2.75°, and at the end 6.5 to 6.68°; and R-squared was in the range of 0.990 to 0.999.

3.6. Comparison and Evaluation of Measured Data and Experiments

After determining these parameters, it was predicted that the dynamic motion would affect the normal load. For this reason, a device was constructed to measure the dynamic effects (Figure 5). Both the dynamic effects measurement device and the validation system used (Figure 2) contain a diaphragm force transducer. Its accurate calibration was therefore crucial to the study. The calibration was carried out through successive loading of 1–4 weights. Then, these measured parameters were compared with those weighed on a calibrated Mettler Toledo PG5002-SDR balance. A comparison of the individual values can then be seen in Figure 6, where it can be seen that the measurement inaccuracy increases as weights are added gradually, up to a maximum of 2.35% of the calibrated measurement.

After calibration of the membrane force transducer on the validation and dynamic effects measuring device, it was possible to determine the dynamic effects of moving materials in motion. The speed of the moving linear line was determined according to the frequency applied in the frequency converter (5 Hz/0.017 $m \cdot s^{-1}$, 10 Hz/0.034 $m \cdot s^{-1}$, 15 Hz/0.051 $m \cdot s^{-1}$). The values of different measured normal loads are shown in Figure 7 for comparison. Here, it can be seen that the dynamic motion for the frequency of 10 Hz is closest to the weighted values, with a maximum deviation of 1.31% from the values measured on the calibrated balance.

To determine the initial shear stress, the first seconds of tool movement over the soil were evaluated at a given normal load (1–4 weights) and at three transport speeds. Figure 10 shows the linear interleaving of the points and a comparison of the different sets of measurements for 1–4 weights, the three transport speeds, and the two materials tested. The angle formed by the trend line and the x-axis then gives the given friction angle. The friction angle read at speed 2 with a frequency of 10 Hz agrees with the calibration measurement using the Jenike shear machine with the smallest deviation for both samples tested, with differences of 7.5% for sample 1 and 5.9% for sample 2.

The subsequent set was based on the addition of weights during the measurement and the shear stress response as the tool moved over the soil from location A to location B (Figure 3). To ensure repeatability and the same operating conditions in the experiment, the unloaded tool with the rail was placed on the soil. After resetting the strain gauge, the rail was put into a straight-line motion where it moved from point A to point B (Figure 3). In this way, identical initial measurement conditions were achieved. After passing through the AB path, the required number of weights (1–4 weights) was added. The tool traveled along the path BC (Figure 3), where the response of the tool movement over the soil was measured at a given normal load.

Table 4 evaluates the angles of external friction for the beginning shear stress, and the start and end of the measurement under dynamic normal loading. If we consider past measurements, the result was that the most effective tool movement speed was at 10 Hz. Therefore, the percentage difference from the initial stress at the beginning of the experiment was higher by 8.5% and 11.8% for sample 1 and sample 2, respectively. For the final phase of the experiment, the measured percentage deviation from the initial tension was 31.1% higher for sample 1 and 34.7% higher for sample 2.

Such a rapid increase in shear friction was caused by a negative phenomenon that occurred when the tool was sheared over the soil. As the tool moved, it gradually sank/was immersed into the soil. The tool pulled the soil in front of it, and thus more forces were exerted on the tool against the movement. This phenomenon resulted in larger shear cross-sections of the tool.

4. Discussion

This research has shown that by using appropriate input parameters, the initial shear stress/external friction can be determined and then this theory can be applied and used to determine the possible increase in shear stress when the tool is immersed in the soil. Based on this finding, further experiments will be carried out to allow a wider application of this measurement method and the subsequent development of a measurement system to determine the mechanical and physical properties of soils as a function of external conditions. This can make the transportation, treatment, and processing of soils more efficient in many industries. By measuring these parameters on site, sampling is avoided and degradation during transport to the laboratory is prevented.

Shear stress measurements have been found to be extremely dependent on the shear area. Therefore, a follow-up study is proposed to evaluate the relationship between frictional surface area and particle grain size and the effect on frictional parameters. In this way, we will be able to better understand the frictional properties of the frictional surface on the soil and derive frictional parameters more accurately. Evaluation of tool penetration into the bulk material or prediction of tool resistance/wear during movement in soil will also be part of future research.

5. Conclusions

This study was designed to validate the theory for the design of a device measuring basic friction parameters for subsequent possible alternative uses and applications. The theory is based on the Jenike shear test, which is most commonly performed under laboratory conditions. However, sometimes it is preferable/desirable to measure the material directly at the place of occurrence/sampling. For this reason, the dynamic motion of the tool on the soil (coarse sand soil) was tested and then compared with the measurement on the calibrated device.

The basic parameters such as the basic geometric dimensions of the trough (Figure 3), the tool, or rake plate (Figure 4) help to prepare the measurement, as well as the average angular velocity and the average linear conduction velocity at the inverter frequencies of 5 Hz, 10 Hz, and 15 Hz (Table 2). The required weights of the individual moving parts were determined. These parts were weighed both individually and as a whole on a calibrated balance. It was important to weigh these parts once already placed in the line. By comparing individual values for different loads, it was found that the lines in which the moving parts are placed absorb approximately 8% to 15% of the applied normal load (Table 3).

The motion of the device was dynamic. Therefore, it was checked whether dynamic shocks would not be generated when the skid moved. For this experiment, a device was created to determine the dynamic effects (Figure 5). The device sensed dynamic shocks for a defined normal load (1–4 weights) and for a defined transport speed (at frequencies of 5 Hz, 10 Hz, and 15 Hz). It was found that as the velocity of motion increased, the normal load stabilized and the absorption of the normal force exerted by the line was canceled out. After analysis, it was found that the smallest deviation in the normal load during dynamic motion (max. 1.3%) and the load weighted on the calibrated balance is at a frequency of 10 Hz. After calibration of the dynamic normal load, the device was ready.

After experiments where the tool was sliding over the soil, it was found that when comparing the initial shear stress of the validation device, the smallest percentage deviation of the measurement from the calibrated Jenike method at a frequency of 10 Hz was found to be 7.5% for sample 1 and 5.9% for sample 2. This was the second confirmation that the smallest deviations from calibrated measurements occur only at the frequency of 10 Hz.

Furthermore, the shear stresses generated when the tool was immersed in the soil were observed to determine possible negative effects. It was found that after traveling the BC distance where the measurements were made, the shear stress increased by up to 1/3 when the tool was immersed.

Author Contributions: L.J. and D.G. discovered the mechanisms. M.Z. wrote the main manuscript text. L.J. and M.Z. corrected the text. M.Z. and D.G. designed the experiments. M.Z. and F.V. performed the experiments. M.Z. and R.P. analyzed the data. All authors have read and agreed to the published version of the manuscript.

Funding: This research was funded by the project SP2023/094 Specific Research in Selected Areas of Energy Processes and REFRESH—Research Excellence For Region Sustainability and High-tech Industries, CZ.10.03.01/00/22_003/0000048.

Data Availability Statement: Data are contained within the article.

Conflicts of Interest: The authors declare no conflict of interest.

Nomenclature

Symbols

v_t	theoretical velocity	$(m \cdot s^{-1})$
n_m	measured revolutions of the lead screw	(s^{-1})
p	pitch of the lead screw	(m)
X_{exp}	parameters measured in the experiment	(-)
X_{kal}	parameters measured on the calibrated devices	(-)
x_s	statistical evaluation method	(-)
x	particle size distribution	(mm)
Q3_10	particle size in 10% of the cumulative curve	(%)
Q3_50	particle size in 50% of the cumulative curve	(%)
Q3_90	particle size in 90% of the cumulative curve	(%)
ρ_v	volume weight	$(kg \cdot m^{-3})$
ρ_s	bulk density	$(kg \cdot m^{-3})$
φ_e	angle of external friction	(°)
Ψ_s	angle of repose	(°)
S	decrease	(%)
S_1	area of contact between the tool and the bulk material when starting the tool	(m^2)
S_2	area of the tool after the run-up distance	(m^2)
S_x	area of the tool at the end of the path	(m^2)
hx	drop of the tool after the run-up distance	(m)

Abbreviations

PLA	polylactide fibres
FFF	fused filament fabrication
BSC	Bulk Solid Center
VŠB TU-Ostrava	VSB—Technical University of Ostrava

References

1. Deng, T.; Garg, V.; Pereira Diaz, L.; Markl, D.; Brown, C.; Florence, A.; Bradley, M.S.A. Comparative studies of powder flow predictions using milligrams of powder for identifying powder flow issues. *Int. J. Pharm.* **2022**, *628*, 122309. [CrossRef]
2. Kumar, P.; Subbarao, P.M.V.; Kala, L.; Vijay, V.K. Influence of physical. mechanical. and thermal properties of biomass pellets from agriculture residue: Pearl millet cob and mix. *Bioresour. Technol. Rep.* **2022**, *20*, 101278. [CrossRef]
3. Anda, R.; Chai, J.; Negami, T. Effect of chemical additives on the consolidation behaviours of mini PVD unit cells from macro to micro. *Geotext Geomembr.* **2023**, *51*, 199–208. [CrossRef]
4. Yu, T.; Zhao, Z.; Li, J. Effect of sintering temperature and sintering additives on the properties of alumina ceramics fabricated by binder jetting. *Ceram. Int.* **2023**, *49*, 9948–9955. [CrossRef]
5. Castellano, J.M.; Gómez, M.; Fernández, M.; Esteban, L.S.; Carrasco, J.E. Study on the effects of raw materials composition and pelletization conditions on the quality and properties of pellets obtained from different woody and non woody biomasses. *Fuel* **2015**, *139*, 629–636. [CrossRef]
6. Yu, H.; Dong, X.R.; Kang, S.M.; Yu, W.; Wang, Z.F.; Mu, J.Y.; Cui, X.W.; Li, J.H.; Yin, F.X.; Shin, K.S. Effect of the pre-homogenization on the precipitation behaviors. mechanical and corrosion properties of as-extruded MgY binary alloys. *Mater. Charact.* **2021**, *178*, 111307. [CrossRef]
7. Merkus, H.G. *Particle Size Measurements: Fundamentals, Practice, Quality*; Springer: New York, NY, USA, 2009; pp. 73–117, ISBN 978-1-4020-9015-8.
8. Sahoo, R. Degradation characteristics of steel making materials during handling. *Powder Technol.* **2007**, *176*, 77–87. [CrossRef]
9. Kang, X.; Sun, H.M.; Luo, H.; Dai, T.; Chen, R.P. A Portable Bender Element-Double Cone Penetration Testing Equipment for Measuring Stiffness and Shear Strength of In-Situ Soft Soil Deposits. *KSCE J. Civ. Eng.* **2020**, *24*, 3546–3560. [CrossRef]
10. Roy, A.K.; Camping, J.D. Development of a portable shear test fixture for low modulus porous (foam) materials. *Exp. Mech.* **2003**, *43*, 39–44. [CrossRef]
11. Aikins, K.A.; Ucgul, M.; Barr, J.B.; Jensen, T.A.; Antille, D.L.; Desbiolles, J.M.A. Determination of discrete element model parameters for a cohesive soil and validation through narrow point opener performance analysis. *Soil Tillage Res.* **2021**, *213*, 105123. [CrossRef]
12. Garciano, L.O.; Upadhyaya, S.K.; Jones, R.A. Measurement of soil parameters useful in predicting tractive ability of off-road vehicles using an instrumented portable device. *J. Terramechanics* **2010**, *47*, 295–305. [CrossRef]
13. Wang, Y.; Fan, Z.; Zhou, X.; Zeng, C.; Xu, P.; Xie, X.; Wang, X.; Zhang, M.; Su, Z.; Huang, Q. Friction properties of bulk isotropic pyrocarbon materials based on different composite microstructures. *J. Mater. Res. Technol.* **2022**, *21*, 4079–4092 [CrossRef]

14. Li, R.; Chen, O.; Ji, L.; Peng, X.; Huang, G. Based on internal friction theory: Investigating of Fe-based bulk amorphous alloys on mechanical properties with different Si content. *J. Non-Cryst. Solids* **2021**, *563*, 120813. [CrossRef]
15. Zegzulka, J. The angle of internal friction as a measure of work loss in granular material flow. *Powder Technol.* **2013**, *233*, 347–353. [CrossRef]
16. Zegzulka, J.; Gelnar, D.; Jezerska, L.; Ramirez-Gomez, A.; Necas, J.; Rozbroj, J. Internal Friction Angle of Metal Powders. *Metals* **2018**, *8*, 255. [CrossRef]
17. Kyburz, M.L.; Sovilla, B.; Gaume, J.; Ancey, C. Decoupling the Role of Inertia. Friction. and Cohesion in Dense Granular Avalanche Pressure Build-up on Obstacles. *J. Geophys. Res.-Earth* **2020**, *125*, e2019JF005192. [CrossRef]
18. Yoneyama, T.; Kitade, M.; Osada, K. Investigation on the ski-snow interaction in a carved turn based on the actual measurement. *Procedia Eng.* **2010**, *2*, 2901–2906. [CrossRef]
19. Federolf, P.; Roos, M.; Lüthi, A.; Dual, J. Finite element simulation of the ski–snow interaction of an alpine ski in a carved turn. *Sport Eng.* **2010**, *12*, 123–133. [CrossRef]
20. *CSN EN ISO 14688-1*; Geotechnical investigation and Testing-Identification and Classification of Soil-Part 1: Identification and Description. The Czech Office for Standards, Metrology and Testing: Prague, Czech Republic, 2018.
21. Torresani, E.; Carrillo, M.; Haines, C.; Martin, D.; Olevsky, E. Fabrication of powder components with internal channels by spark plasma sintering and additive manufacturing. *J. Eur. Ceram. Soc.* **2023**, *43*, 1117–1126. [CrossRef]
22. Lifton, J.J.; Tan, Z.J.; Goh, G.A.; Mutiargo, M. On the uncertainty of porosity measurements of additively manufactured metal parts. *Measurement* **2022**, *188*, 110616. [CrossRef]
23. Kalman, H.A.; Portnikov, D. Analyzing bulk density and void fraction: A. the effect of archimedes number. *Powder Technol.* **2021**, *381*, 477–487. [CrossRef]
24. Zegzulka, J.; Gelnar, D.; Jezerska, L.; Prokes, R.; Rozbroj, J. Characterization and flowability methods for metal powders. *Sci. Rep.* **2020**, *10*, 21004. [CrossRef] [PubMed]
25. Zidek, M.; Rozbroj, J.; Zegzulka, J.; Necas, J.; Marschalko, M.A. Validation System of Traction and Pressing Tools. Czech Republic. Patent 306578. Granted 01.02.2017. Registered 29.09.2016. 2017. Available online: https://isdv.upv.gov.cz/doc/FullFiles/Patents/FullDocuments/306/306578.pdf (accessed on 21 April 2023).
26. Yang, J.; Buettner, K.E.; Dinenna, V.L.; Curtis, J.S. Computational and experimental study of the combined effects of particle aspect ratio and effective diameter on flow behavior. *Chem. Eng. Sci.* **2022**, *255*, 117621. [CrossRef]
27. Salehi, G.; Barletta, D.; Poletto, M.A. Comparison between powder flow property testers. *Particuology* **2017**, *32*, 10–20. [CrossRef]
28. Anton Paar GmbH. Brochure Ultrapyc Series, I15IP006EN-B. 2020. Available online: https://www.labo.de/upload_weka/nwo/003/623/I15IP006EN_B_Brochure_UltrapycSeries__2__3623086.pdf (accessed on 18 July 2003).
29. Iqbal, T.; Fitzpatrick, J.J. Effect of storage conditions on the wall friction characteristics of three food powders. *J. Food Eng.* **2006**, *72*, 273–280. [CrossRef]
30. Chen, P.; Yuan, Z.; Shen, X.A.; Zhang, Y. Flow properties of three fuel powders. *Particuology* **2012**, *10*, 438–443. [CrossRef]
31. Teunou, E.; Fitzpatrick, J.; Synnott, E. Characterisation of food powder flowability. *J. Food Eng.* **1999**, *39*, 31–37. [CrossRef]

processes

MDPI

Article

Key Technologies of Intelligent Question-Answering System for Power System Rules and Regulations Based on Improved BERTserini Algorithm

Ming Gao [1,2], Mengshi Li [1], Tianyao Ji [1,*], Nanfang Wang [2], Guowu Lin [2] and Qinghua Wu [1]

1 College of Electric Power Engineering, South China University of Technology, Guangzhou 510640, China; 202011002583@mail.scut.edu.cn (M.G.); mengshili@scut.edu.cn (M.L.); wuqh@scut.edu.cn (Q.W.)
2 School of Electrical Engineering, Guangzhou City University of Technology, Guangzhou 510800, China; wnf2184649296@163.com (N.W.); 16676676716@163.com (G.L.)
* Correspondence: tyji@scut.edu.cn

Abstract: With the continuous breakthrough of natural language processing, the application of intelligent question-answering technology in electric power systems has attracted wide attention. However, at present, the traditional question-answering system has poor performance and is difficult to apply in engineering practice. This paper proposes an improved BERTserini algorithm for the intelligent answering of electric power regulations based on a BERT model. The proposed algorithm is implemented in two stages. The first stage is the text-segmentation stage, where a multi-document long text preprocessing technique is utilized that accommodates the rules and regulations text, and then Anserini is used to extract paragraphs with high relevance to the given question. The second stage is the answer-generation and source-retrieval stage, where a two-step fine-tuning based on the Chinese BERT model is applied to generate precise answers based on given questions, while the information regarding documents, chapters, and page numbers of these answers are also output simultaneously. The algorithm proposed in this paper eliminates the necessity for the manual organization of professional question–answer pairs, thereby effectively reducing the manual labor cost compared to traditional question-answering systems. Additionally, this algorithm exhibits a higher degree of exact match rate and a faster response time for providing answers.

Keywords: intelligent question-answering system; improved BERTserini algorithm; rules and regulations; information retrieval

Citation: Gao, M.; Li, M.; Ji, T.; Wang, N.; Lin, G.; Wu, Q. Key Technologies of Intelligent Question-Answering System for Power System Rules and Regulations Based on Improved BERTserini Algorithm. *Processes* **2024**, *12*, 58. https://doi.org/10.3390/pr12010058

Academic Editor: Peng Li

Received: 22 November 2023
Revised: 20 December 2023
Accepted: 22 December 2023
Published: 26 December 2023

1. Introduction

The intelligent question-answering system is an innovative information service system that integrates natural language processing, information retrieval, semantic analysis and artificial intelligence. The system mainly consists of three core parts, which are question analysis, information retrieval and answer extraction. Through these three parts, the system can provide users with accurate, fast and convenient answering services.

The representative systems of the intelligent question-answering system include:

(1) Rule-based algorithms (1960s–1980s). The question-answering system based on this pattern mainly relies on writing a lot of rules and logic to implement the dialogue. ELIZA [1], developed by Joseph Weizenbaum in the 1960s, was the first chatbot designed to simulate a conversation between a psychotherapist and a patient. PARRY [2] is a question-and-answer system developed in the 1970s that simulates psychopaths. The emergence of ELIZA and PARRY provided diverse design ideas and application scenarios for subsequent intelligent question-answering systems, thereby promoting the diversification and complexity of dialogue systems. However, the main problem of this model is its lack of flexibility and extensibility. It relies too much on rules or templates set by humans, and

consumes a lot of time and manpower. When the questions become complicated, it is difficult to obtain satisfactory answers through simple rules set by the model.

(2) Statistics-based algorithms (1990s–2000s). The question-answering system based on this model adopts the method of statistical learning to learn patterns and rules from a large number of dialogue data. Common algorithms include Vector Space Model [3] and Conditional Random Fields [4]. ALICE (Artificial Linguistic Internet Computer Entity) [5] is an open-source natural language processing project. The system in question is an open-domain question-answering platform capable of addressing queries across a multitude of subjects and domains. Jabberwacky [6] is an early intelligent chatbot employing machine learning and conversational models to enhance its responses continually. These systems are designed to train models that can learn the relationships between questions and answers present in the corpus. Therefore, these models can carry out more natural and smooth dialogue. However, the ability of context understanding and generalization ability is weak, so it is difficult to adapt to model sharing and transfer learning in various professional fields. Moreover, considering statistical models are trained on a large corpus, this kind of model may suffer from data bias when dealing with domain-specific problems and fail to provide accurate answers.

(3) Algorithms based on hybrid technology (2010s–early 2020s). The question-answering system, grounded on this model, can amalgamate diverse techniques encompassing rules, statistics, and machine learning. It leverages multiple input modalities, including speech, image, and text, to interoperate seamlessly. The overarching objective is to facilitate users in accomplishing specific tasks or goals within designated domains, such as booking, traveling, shopping, or ordering food. This synergistic integration of multifarious technologies and input modes fosters a more sophisticated and intelligent dialogue system. Typical question-answering systems based on the hybrid technology model include Apple's Siri [7], Microsoft's Cortana [8], Amazon's Alexa [9], Facebook's M [10], and Google's Google Assistant [11]. These systems are centered around artificial intelligence and natural language processing technology, aiming to furnish users with personalized and convenient information and services to cater to diverse needs.

The system built based on this pattern has stronger context understanding and personalized customization, but there are two shortcomings: first, the quality of dialogue in such a system is not stable; second, the generalization ability of the model is limited. It is difficult to realize model sharing, transfer learning and answer generation in professional fields. The training of this model requires excessive investment in computing and data resources, and its training and deployment speed is slow.

(4) Algorithms based on pre-trained language (2020s). The model is based on pre-trained language models such as BERT [12], GPT (Generative Pre-trained Transformer) [13], etc. These models are pre-trained on large-scale data and they learn rich language representation and context understanding skills to generate more natural, fluid, and accurate responses. In addition, through the supervised training on domain-specific question-answering datasets, the question-answering system can answer questions in specialized professional fields. Ref. [14] primarily investigates a question-answering system for agriculture. This system utilizes artificial intelligence technology and relevant datasets to provide farmers with information on topics such as weather, market prices, plant protection, and government plans. Ref. [15] proposed a TD-BERT model based on BERT. This model leverages the powerful semantic representation capabilities of BERT and integrates target information to enhance the accuracy of sentiment classification. Ref. [16] proposed a BERTserini algorithm which improves the exact match rate of the question-answering system. In comparison to the original BERT algorithm, the proposed method surpasses its processing byte limit and can provide accurate answers for multi-document long texts.

Although systems built on the BERTserini algorithm perform well on public datasets, there are some problems in the application in professional fields such as electrical power engineering. Considering the low exact match rate and poor answer quality, engineering applications of these models are challenging. The problems are mainly caused by the following aspects.

(1) Lack of model expertise: Language models such as BERT or GPT are usually pre-trained from large amounts of generic corpus collected on the Internet. However, the digital realm offers limited professional resources pertaining to industries like electrical power engineering. As a result, the model has insufficient knowledge reserve when dealing with professional question, which affects the quality of the answers; (2) Differences in document format: There are significant differences between the format of documentation in the electrical power engineering field and that of public datasets. The documents in the electrical power engineering field often exhibit unique formatting, characterized by an abundance of hierarchical headings. It is easy to misinterpret the title as the main content and mistakenly use it as the answer to the question, leading to inaccurate results; (3) Different scenario requirements: Traditional answering systems do not need to pay attention to the source of answers in the original document. However, a system designed for professional use must provide specific source information for its answers. If such information is not provided, there may arise doubts regarding the accuracy of the response. This further diminishes the utility of the application in particular domains.

This paper proposes an improved BERTserini algorithm to construct an intelligent question-answering system in the field of electrical power engineering. The proposed algorithm is divided into two stages:

The first stage is text segmentation. During this phase, the text is segmented and preprocessed. Firstly, a multi-document long text preprocessing method that supports rules and regulations text is proposed. This approach can accurately segment rules and regulations text and generate an index file of answer location information. By doing so, the system can better comprehend the structure of the regulation text, enabling it to locate the answer to the user's question more accurately. Secondly, through the FAQ [17] pre-module, high-frequency questions are intercepted for question pre-processing. This module matches and classifies user-raised questions based on a pre-defined list of common questions, intercepting and addressing high-frequency issues. This reduces the repetition of processing the same or similar problems and enhances the system's response efficiency. Finally, Anserini [18] is employed to extract several paragraphs highly relevant to user problems from multi-document long text. Anserini is an information-retrieval tool based on a vector space model that represents a user question as a vector and each paragraph in a multi-document long text as a vector. By calculating the similarity between the user problem vector and each paragraph vector, several paragraphs with high relevance to the user problem can be selected. These paragraphs serve as candidate answers for the system to further analyze and generate the final answer.

The second stage is the answer-generation and source retrieval stage. During this phase, the Chinese Bert model undergoes fine-tuning [19], which comprises two steps involving key parameter adjustments. This process enhances the model's comprehension of the relationship between the question and the answer, thereby improving the accuracy and reliability of the generated response. Subsequently, based on the input question, the Bert model extracts several candidate answers from the N paragraphs with the highest similarity to the question, as determined by Anserini. The user can then filter through these multiple relevant paragraphs to identify the answer that best aligns with their query. Finally, the candidate answers are weighted, and the highest-rated answer is outputted along with the chapter and position information of the answer in the original document. This approach facilitates users in quickly locating the most accurate answer while providing pertinent contextual information.

The improved BERTserini algorithm proposed in this paper has three main contributions.

(1) The proposed algorithm implements multi-document long text preprocessing technology tailored for rules and regulations text. Through optimization, the algorithm segments rules and regulations into distinct paragraphs based on its inherent structure and supports answer output with reference to chapters and locations within the document.

The effectiveness of this pretreatment technology is reflected in the following three aspects: first, through accurate segmentation, paragraphs that may include questions can be extracted more accurately, thus improving the accuracy of answer generation. Secondly, the original Bert model exhibits a limitation that it outputs the heading of rules and regulations text as the answer frequently. To address this issue, an improved BERTserini algorithm has been proposed. Finally, the algorithm is able to accurately give the location information of answers in the original document chapter. The algorithm enhances the comprehensiveness and accuracy of reading comprehension, generating answers to questions about knowledge and information contained in professional documents related to the field of electric power. Consequently, this leads to a marked improvement in answer quality and user experience for the question-answering system.

(2) The proposed algorithm optimizes the training of the corpus in the field of electrical power engineering and fine-tunes the parameters of the large language model. This method eliminates the necessity for the manual organization of professional question–answer pairs, knowledge base engineering, and manual template establishment in BERT reading comprehension, thereby effectively reducing labor costs. This enhancement significantly enhances the accuracy and efficiency of the question-answering system.

(3) The proposed algorithm has been developed for the purpose of enhancing question-answering systems in engineering applications. This algorithm exhibits a higher degree of exact match rate of questions and a faster response for providing answers.

The remaining sections of this article are organized as follows. Section 2 provides an introduction to the background technology of intelligent question-answering systems. Section 3 describes the procedural steps of an improved BERTserini algorithm. Section 4 presents the experimental results of the proposed algorithm and its implementation in engineering applications. Finally, Section 5 draws conclusions.

2. Background of the Technology

2.1. FAQ

Frequently Asked Questions (FAQs) are a collection of frequently asked questions and answers designed to help users quickly find answers to their questions [17]. The key is to build a rich and accurate database of preset questions, which consists of questions and the corresponding answers. They are manually collated from the target documents. The FAQ provides an answer that corresponds to the user's question by matching it with the most similar question.

2.2. BM25 Algorithm

The Best Match 25 (BM25) algorithm [18,19] was initially proposed by Stephen Robertson and his team in 1994 and applied to the field of information retrieval. It is commonly used to calculate the relevance score between documents and queries. The main logic of BM25 is as follows: Firstly, the query statement involves word segmentation to generate morphemes. Then, the relevance score between each morpheme and the search result is calculated.

Finally, by weighting summing the relevance scores of the morpheme with the search results, the relevance score between the retrieval query and the search result documents is obtained. The formula for calculating BM25 algorithm is as follows:

$$Score(D, Q) = \sum_{i}^{n} W_i \cdot R(q_i, D) \tag{1}$$

In this context, Q represents a query statement, q_i represents a morpheme obtained from Q. For Chinese, the segmented results obtained from tokenizing query Q can be considered as morpheme q_i. D represents a search result document. W_i represents the weight of morpheme q_i, and $R(q_i, D)$ represents the relevance score between morpheme q_i and document D. There are multiple calculation methods for weight parameter W_i, with

Inverse Document Frequency (IDF) being one of the commonly used approaches. The calculation process for IDF is as follows:

$$IDF(q_i) = \log(\frac{N - n(q_i) + 0.5}{n(q_i) + 0.5}) \quad (2)$$

In the equation, N represents the total number of documents in the index, and $n(q_i)$ represents the number of documents that contain q_i.

Finally, the relevance scoring formula for the BM25 algorithm can be summarized as follows:

$$Score(D, Q) = \sum_{i=1}^{n} IDF(q_i) \cdot \frac{f(q_i, D) \cdot (k_1 + 1)}{f(q_i, D) + k_1 \cdot \left(1 - b + b \cdot \frac{|D|}{avgdl}\right)} \quad (3)$$

where k_1 and b are adjustment factors, $f(q_i, D)$ represents the frequency of morpheme q_i appearing in document D, $|D|$ denotes the length of document D, and $avgdl$ represents the average length of all documents.

2.3. Anserini

Anserini [20] is an open-source information retrieval toolkit that supports various text-based information retrieval research and applications. The goal of Anserini is to provide an easy-to-use and high-performance toolkit that supports tasks such as full-text search, approximate search, ranking, and evaluation on large-scale text datasets. It enables the conversion of text datasets into searchable index files for efficient retrieval and querying. Anserini incorporates a variety of commonly used text retrieval algorithms, including the BM25 algorithm. With Anserini, it becomes effortless to construct a BM25-based text retrieval system and perform efficient search and ranking on large-scale text collections. The flowchart of the algorithm is illustrated in Figure 1.

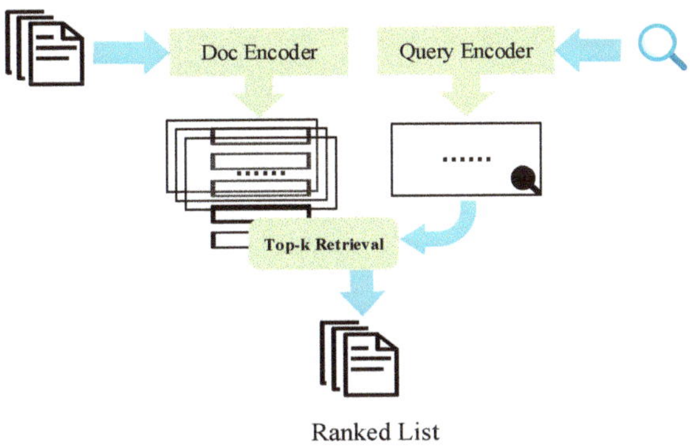

Figure 1. The flowchart of the Anserini algorithm.

2.4. BERT Model

Bidirectional Encoder Representations from Transformers (BERT) [12] is a pre-trained language model proposed by Google in 2018. The model structure is shown in Figure 2. In the model, E_i represents the encoding of words in the input sentence, which is composed of the sum of three word embedding features. The three word embedding features are Token Embedding, Position Embedding, and Segment Embedding. The integration of these three words embedding features allows the model to have a more comprehensive understanding of the text's semantics, contextual relationships, and sequence information, thus enhancing the BERT model's representational power. The transformer structure in the figure is represented as Trm. The T_i represents the word vector that corresponds to the trained word E_i.

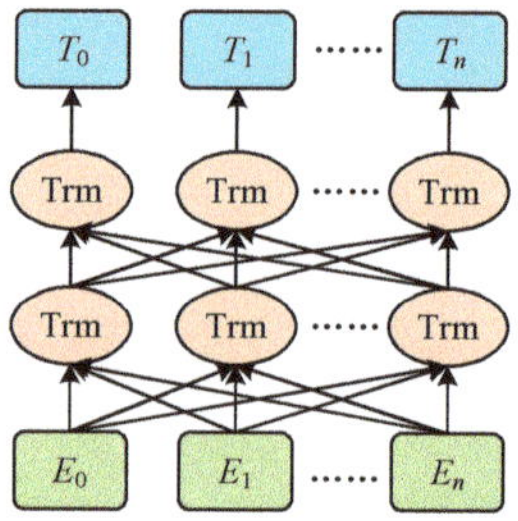

Figure 2. Architecture of BERT.

BERT exclusively employs the encoder component of the Transformer architecture. The encoder is primarily comprised of three key modules: Positional Encoding, Multi-Head Attention, and Feed-Forward Network. Input embeddings are utilized to represent the input data. Addition and normalization operations are denoted by "Add&norm". The fundamental principle of the encoder is illustrated in Figure 3.

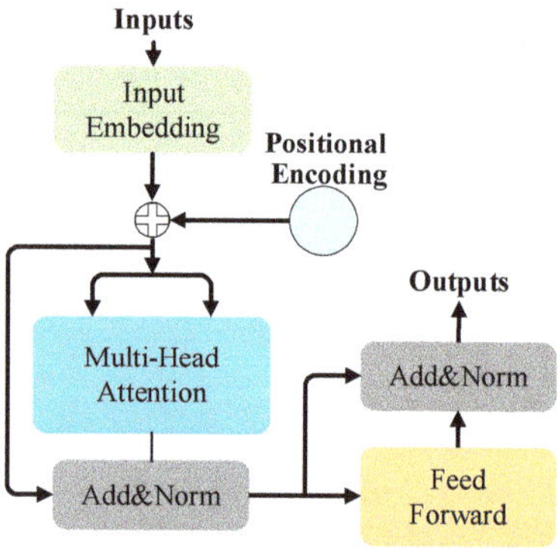

Figure 3. Transformer encoder principle.

In recent years, several Chinese BERT models have been proposed in the Chinese language domain. Among these, the chinese-BERT-wwm-ext model [21] released by the HIT·iFLYTEK Language Cognitive Computing Lab (HFL) has gained significant attention and serves as a representative example. This model, based on the original Google BERT model, underwent further pretraining using a total vocabulary of 5.4 billion words, including Chinese encyclopedia, news, and question-answer datasets. The model adopts the Whole Word Masking (wwm) strategy, which is an improvement tailored to Chinese language characteristics. In Chinese processing, as words are composed of characters, and a word may consist of one or more characters, it becomes necessary to mask the entire word rather than just a single character. The wwm strategy is designed to better understand and capture the semantics of Chinese vocabulary. In summary, this model is an improved Chinese version of BERT that, through whole-word masking, exhibits enhanced performance in Chinese language understanding.

2.5. BERTserini Algorithm

The architecture of BERTserini algorithm [16] is depicted in Figure 4. The algorithm employs the Anserini information extraction algorithm in conjunction with a pretraining BERT model. In this algorithm, the Anserini retriever is responsible for selecting text paragraphs containing the answer, which are then passed to the BERT reader to determine the answer scope. From Figure 4, it can be observed that BERTserini is an intelligent question-answering system that combines the BERT language model with the Anserini information retrieval system. It synergistically harnesses the powerful language understanding capabilities of BERT and the efficient retrieval functionalities of Anserini. This algorithm exhibits significant advantages over traditional algorithms. It demonstrates fast execution speed similar to traditional algorithms while also possessing the characteristics of end-to-end matching, resulting in more precise answer results. Furthermore, it supports extracting answers to questions from multiple documents. This algorithm is primarily applied to open-domain question-answering tasks, where the system needs to find answers to questions from a large amount of unstructured text.

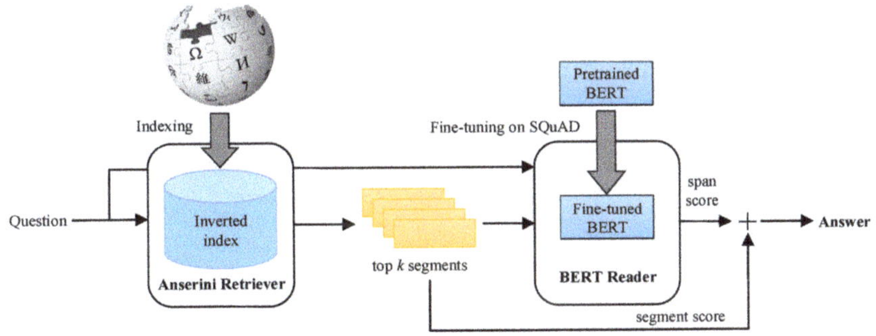

Figure 4. Architecture of BERTserini.

3. Improved BERTserini Algorithm

3.1. Algorithm Description

The improved BERTserini algorithm presented in this paper can be divided into two stages, and the flowchart is illustrated in Figure 5.

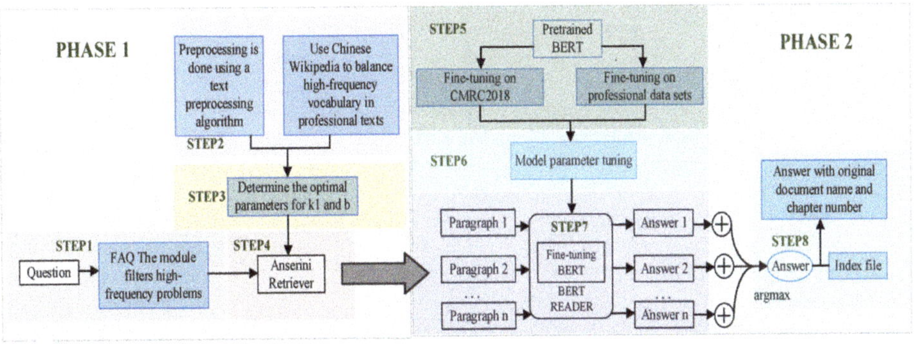

Figure 5. Flowchart of the proposed algorithm.

(1) Phase 1: Text Segmentation Stage

The first stage is text segmentation stage, which comprises two key components: (1) Question preprocessing: The FAQ module is utilized to intercept high-frequency questions in advance, thereby achieving question preprocessing. If the FAQ module cannot provide an answer that corresponds to the user's query, then the query is transferred to the subsequent stage of paragraph extraction. Anserini retrieval technology is utilized for paragraph extraction, enabling the rapid extraction of highly relevant paragraphs which are pertinent to user queries within multi-document long text. (2) Document preprocessing: Due to the high degree of keyword overlap in power regulation documents. The paper proposes a multi-document long text preprocessing method supporting regulation texts, which can accurately segment the regulation texts and support the retrieval and tracing of the answer chapters' sources.

STEP 1: The FAQ module filters out high-frequency problems.

The FAQ module is designed to pre-process questions by intercepting and filtering out high-frequency problems. To achieve this, the module requires a default question library that contains a comprehensive collection of manually curated questions and their corresponding answer pairs from the target document. By matching the most similar question to the user's inquiry, the FAQ module can efficiently provide an accurate answer based on the corresponding answer to the question.

The FAQ module employs ElasticSearch, an open-source distributed search and analysis engine, to match user queries in a predefined question library. ElasticSearch is built upon the implementation of Lucene, an open-source full-text search engine library released by the Apache Foundation, and incorporates Lucene's BM25 text similarity algorithm. This algorithm calculates similarity by evaluating word overlap between the user query's text and the default question library, as shown in (3).

The FAQ module will directly return the preset answer to the matched question if the BM25 score returned by ElasticSearch exceeds the predetermined threshold. In cases where the return score falls below this threshold, instead of returning an answer, the question is referred to subsequent steps.

STEP 2: Text preprocessing and document index generation.

This step involves two tasks.

The first task is due to the high overlap of professional terminology in similar regulatory texts. If Anserini is directly used to retrieve and calculate the paragraphs in professional documents, it may result in an issue where certain professional terms have lower weights W_i in Equation (2). The main reason for this is that if we assume q_i is a power industry term initially used as a retrieval keyword, its occurrence in multiple professional documents results in a larger value for $n(q_i)$ in the calculation of Equation (2). This value becomes essentially close to the total document count N, leading to a decrease in the calculated result of $IDF(q_i)$. The issue arising from this is that when retrieving the professional term q_i, the original expectation was to find paragraphs or documents strongly related to it. However, due to the decrease in $IDF(q_i)$, the probability of finding paragraphs or documents strongly associated with this professional term is actually reduced. Conversely, in this situation, some non-specialized terms may have relatively larger IDF values. This situation is exactly opposite to the intended calculation goal of the IDF algorithm. For keywords that possess strong discriminative power for document categories, the expectation is that documents containing such keywords should be relatively scarce in the corpus. Consequently, the IDF value for these keywords should be larger.

For example, "generator" is a professional term and keyword in power regulatory texts. However, due to its high frequency across multiple professional documents, the IDF value calculated according to Equation (2) may not be high. On the other hand, non-specialized terms like "tool" may have a higher IDF value because of their infrequent occurrence in professional documents. As a result, after inputting the retrieval query, Anserini calculates and retrieves documents that are not strongly related to the professional term, contrary to the intended outcome. Therefore, in the process of constructing the index file, besides incorporating regulatory texts, Chinese Wikipedia textual data has been included. This action increases the value of N, consequently enlarging the gap between and $n(q_i)$. This adjustment elevates the calculated IDF value for professional terms according to Equation (2), thereby mitigating the adverse effects caused by the high frequency of certain professional terms.

The second task Involves proposing a multi-document lengthy text preprocessing algorithm that supports regulatory texts. This algorithm accurately segments regulatory texts, retains information about the sections to which paragraphs belong, and generates an index file. The specific method is as follows:

Convert documents in .pdf or .docx format to plain text in .txt format.

Remove irrelevant information such as header/footer and page number.

Use regular expressions to extract the title number from the text (for example: Section 3.3.1), and match the title number to the text.

Use rules to filter out paragraphs in the text such as tables and pictures that are not suitable for machine reading comprehension.

Use Anserini to divide the text title number into words and index the corresponding text.

STEP 3: Determine the two parameters k1 and b.

The k1 and b parameters utilized in the Anserini module are empirically selected to determine the optimal parameters for this study. A specific methodology is employed, starting from 0.1 within their respective value ranges and incrementing by 0.05 to systematically explore all possible combinations of k1 and b values. The selection of the best k1 and b values is based on the accuracy assessment of the second stage Bert reading comprehension module.

STEP 4: Extract paragraphs and generate paragraph scores.

Based on the user's question, Anserini extracts relevant paragraphs from the preprocessed document by filtering out those that are not related to the query. It then matches the question with the paragraphs in the index and selects the top N paragraphs with the highest relevance to the question. This paragraph is evaluated using the BM25 algorithm, as specified in Equations (1)–(3), and is denoted by $S_{anserini}$.

(2) Phase 2: answer generation and source retrieval stage

The second stage is the answer-generation and source-retrieval stage. After undergoing two steps of fine-tuning and key parameter tuning, the model is capable of extracting accurate answers from N paragraphs based on the given question. Additionally, the model can output the chapter information of the answer in the original document according to the index file.

STEP 5: Select the appropriate Chinese Bert model and fine-tune it.

In this research, the Chinese-Bert-WWM-EXT Base model is chosen as the foundational framework. The initial step involves fine-tuning the model using the Chinese Open domain Question answering dataset (CMRC2018). Subsequently, a second round of fine-tuning is conducted by employing the training exam questions related to rules and regulations as specialized datasets.

STEP 6: Algorithm parameter tuning.

Based on the structural and characteristic features of regulatory documents, the following five crucial parameters of the improved BERTserini algorithm have been optimized:

paragraph_threshold. The paragraph threshold is employed to exclude paragraphs with Anserini scores below this specified limit, thereby conserving computational resources.

phrase_threshold. The answer threshold serves as a filter, excluding responses with a Bert reader score below the specified limit.

remove_title. Removes the paragraph title. If this item is True (y = True, n = False), paragraph headings are not taken into account when the Bert reader performs reading comprehension.

max_answer_length. The maximum answer length. The maximum length of an answer is allowed to be extracted when the Bert reader performs a reading comprehension task.

mu. Score weight is implemented to evaluate both the answer and paragraph using the Bert reader and Anserini extractor, subsequently calculating the final score value of the answer.

STEP 7: Extract the answers and give a reading comprehension score.

Bert is used to extract the exact answers to the question from the N paragraphs extracted by Anserini. The sum of the probability of starting and ending positions (logits) for each answer predicted by the model is used as the score of the answer generated by the Bert reading comprehension module. It can be expressed by the following equation:

$$S_{bert} = max(start\ logit) + max(end\ logit) \tag{4}$$

STEP 8: The candidate answers are scored by a comprehensive weighted score, rank the answers by score, output the answer with the highest score, and give the original document name and specific chapter information for the answer.

Use the following equation to calculate the overall weighted score of the answer:

$$S = (1 - \mu) \times S_{anserini} + \mu \times S_{bert} \tag{5}$$

The final score of the answer is calculated by the above formula. $S_{anserini}$ represents the BM25 score returned by the Anserini extracter, and S_{bert} represents the answer score returned by Bert. The answers are sorted by the calculated answer score, and the final output is the answer with the highest score. According to the index file, the original document name and chapter information are output together.

3.2. Main Innovations

(1) Multi-document long text preprocessing method which can process rules and regulations text and support answer provenance retrieval.

In this paper, a multi-document long text preprocessing method is proposed that facilitates answer provenance retrieval and can effectively process the rules and regulations text, which provides a technical path for the construction of intelligent question-answering system in specific professional fields. The innovation point of this method is reflected in STEP 2. This method divides the rules and regulations into chapters. The original document name of each paragraph and its chapter number information can be preserved. To address the issue of excessive frequency of certain proper nouns, the method incorporates text data from Chinese Wikipedia and performs balance processing. By incorporating a larger corpus, the frequency of a specific proper noun in the text can be effectively diminished, thereby mitigating its influence on the model. This innovative preprocessing method can improve the calculation effect of the subsequent reading comprehension module. The answer can be provided in the original document, including chapter and location information.

(2) Determination of optimal parameters of Anserini and improved BERTserini algorithm.

① Determination of the optimal parameters of Anserini. In STEP 3, the optimal parameters of Anserini are determined. All possible combinations of k1 and b are experimentally tried one by one. And the best value is selected according to the answer performance of the subsequent reading comprehension module questions. The determination of the optimal parameters of Anserini improves the performance of the intelligent question-answering system and the exact match of answers (EM).

② Determination of the optimal parameters of the improved BERTserini algorithm: In STEP 6, the optimal parameters of the improved BERTserini algorithm are determined. According to the structure and characteristics of regulation documents, five important parameters are optimized. Thus, the algorithm can determine the reasonable threshold of generating candidate answers when the Bert reading comprehension module performs the reading comprehension task. And the answer generation does not take into account the paragraph title and the optimal overall rating weight and other details that constitute high-quality questions and answers.

(3) Fine-tuning of multi-data sets for Bert reading comprehension model.

This step is illustrated in STEP 5. The Bert model is pre-trained using the CMRC2018 data, and a two-step fine-tuning was carried out using the existing rules and regulations exam questions. By making full use of data sets in different fields, the accuracy and generalization ability of the model are improved. This method achieves better results in question-answering system. At the same time, this method also reduces the time and labor cost required for the manual editing of question–answer pairs in traditional model training. It also significantly improves Bert's reading comprehension of rules and regulations.

(4) Clever use of FAQ.

The clever use of the FAQ is reflected in STEP1. In this paper, the existing rules and regulations are used to train and test questions, which constitutes the questions and answers pairs required by the pre-FAQ module to intercept some high-frequency questions. In this way, a low-cost FAQ module is constructed, which improves the answering efficiency of high-frequency questions, and also improves the exact match rate (EM) of the intelligent question-answering system.

4. Results Analysis of the Experiment

4.1. Data Description

4.1.1. Document Description

For the present study, a total of 30 documents including regulations, provisions, and operation manuals related to the theme of power safety are selected, such as a company power grid work regulations. The total size of the documents is 30.1 MB, and the intelligent system is required to preprocess all the content within the documents, perform machine reading comprehension, and efficiently answer questions.

4.1.2. Fine-Tuning Dataset Description

In this study, four datasets are experimented for fine-tuning the Bert model, which include Chinese Machine Reading Comprehension 2018 (CMRC2018) [22], Delta Reading Comprehension Dataset (DRCD) [23], Safety Procedure Test Item data set (SPTI), and a dataset generated through data augmentation based on documentations of a power grid company. The first two datasets are open-source. Among them, the CMRC2018 dataset contains a large amount of Chinese text. After fine-tuning, it can be adapted to specific domains or application scenarios, thereby improving performance. DRCD is also a Chinese machine reading comprehension dataset, primarily used to train and evaluate models in understanding Chinese texts and answering related questions. The text in the DRCD dataset is sourced from various authentic corpora, including Chinese Wikipedia, to ensure a simulation of real-world scenarios. Based on end-to-end manual evaluation, the results indicate that the model trained using CMRC2018 data performs the best in this study. Therefore, it has been selected as the fine-tuning training dataset. The dataset follows the format of the SQuAD dataset [24]. It consists of a total of 10,142 training samples, 3219 validation samples, and 1002 testing samples. The overall size of the dataset is 32.26 MB. The SPTI consists of 1020 training and examination questions related to electrical safety regulations.

4.1.3. BERT Model Description

In this study, the Chinese-BERT-wwm-ext model [25] released by the HFL is used for training.

4.1.4. Parameter Tuning Explanation for Improved BERTserini Algorithm

The parameter settings in this study are as follows. paragraph_threshold = 10, phrase_threshold = 0, remove_title = n (n = False, y = True), if remove_title = y, the paragraph titles will not be considered by the BERT reader algorithm during reading comprehension. max_answer_length = 50, mu = 0.6.

The parameter in the BM25 algorithm used in the Anserini module has a value range of (0–1), and the parameter has a value range of (0–3).

4.2. Document Preprocessing Performance

In accordance with the document pre-processing algorithm proposed, the document format output by Anserini is illustrated in Figure 6. Within this context, "text" denotes the output paragraphs obtained from Anserini, "paragraph_score" represents the specific score assigned to each paragraph, this is the $S_{anserini}$ mentioned in STEP 4 in Section 3.

Finally, "docid" indicates the name of the document along with the corresponding section information where the paragraph is situated.

[
{
"text": "Terms and definitions of electric power facilities. General term for the equipment related to power generation, transformation, transmission, distribution, and supply applied in the power system.",
"paragraph_score": 18.094900131225586,
"docid": "TAG%%××Electric Grid Limited Liability Company Power Safety Regulations. pdf%%3.3_1",
},
{
"text": "Management content and methods. The classification criteria for accident events. Accident classification. Accidents are classified into three categories: electrical personnel accidents, electrical equipment accidents, and electrical safety accidents. They are ranked in descending order of severity of consequences. The accident levels are classified as extremely serious, serious, significant, and general, with a total of four levels. For detailed criteria for accident classification, please refer to Appendix A.1.",
"paragraph_score": 13.659899711608887,
"docid": "TAG%%××Limited Liability Company Accident and Incident Management Measures.pdf%%5.1.1_1",
},
]

Figure 6. Document preprocessing of the Anserini module.

4.3. Question-Answering Performance

The comparison of the question-answering performance before and after the improvement of the BERTserini algorithm is presented in Table 1. It can be observed that the original BERTserini algorithm exhibits inaccuracies in extracting the start and end positions of answers when addressing power regulations and standards questions, and even results in incomplete sentences. Compared to the original BERTserini algorithm, the improved BERTserini algorithm proposed in this paper can accurately locate the paragraph containing the correct answer and perform precise answer extraction. Additionally, it removes specific details like paragraph headings during the answering process, adapting to the structural characteristics of professional domain regulatory texts. The answers to certain questions are more accurate and concise than manually generated standard answers.

Table 1. Comparison of question-answering performance before and after the improvement of the BERTserini algorithm.

Question	Standard Answer	Original BERTserini Algorithm			Improved BERTserini Algorithm		
		Answer	Whether Exact Match	Trace the Source and Results of the Answers	Answer	Whether Exact Match	Trace the Source and Results of the Answers
When should the audited units send the relevant documents and information to the professional audit teams?	The audited unit should send the relevant documents and information 5 working days before the audit.	The audited unit should send the relevant documents and information 5 working days before the audit.	Yes	No	5 working days before the audit.	Yes	Yes (× × Power Grid Co., LTD. Safety Production Risk Management System Audit Business Guide 5.1.5)
What is the key inspection content of quarterly safety production?	At the end of each quarter, the safety supervision department at all levels shall determine the key contents of supervision in the next quarter according to the annual work plan, seasonal characteristics, and key work arrangements.	Quarterly safety production focus supervision within.	No	No	At the end of each quarter, the safety supervision department at all levels shall determine the key contents of supervision in the next quarter according to the annual work plan, seasonal characteristics, and key work arrangements.	Yes	Yes (× × Power Grid Co., LTD. Safety Production Risk Management System Audit Business Guide 5.2.2)
What does correction mean?	Measures taken to eliminate nonconformities that have been identified.	Correction means the elimination of discrepancies that have been found.	No	No	Measures taken to eliminate nonconformities that have been identified.	Yes	Yes (× × Power Grid Co., LTD. Safety Zone Representative Management Service Guide 4.5)
What is the safety supervision department directly under the company responsible for?	To organize the preparation and issuance of the annual safety measure plan of the unit, supervise and evaluate the implementation of the safety measure plan of the unit and the power supply units at the county.	The safety supervision department of each unit directly under the company is responsible for organizing and compiling.	No	No	To organize the preparation and issuance of the annual safety measure plan of the unit, supervise and evaluate the implementation of the safety measure plan of the unit and the power supply units at the county.	Yes	Yes (× × Power Grid Co., LTD. Safety zone Representative Management Service Guide 5.1.2)
What is an electric utility?	A general term for equipment related to generation, transformation, transmission, distribution and supply used in power systems.	Utility applied to electricity.	No	No	A general term for equipment related to generation, transformation, transmission, distribution and supply used in power systems.	Yes	Yes (× × Power Grid Co., LTD. Power Safety Working Regulations 3.3)

4.4. Comparison of Different Algorithms

This paper uses the Exact Match rate (EM), Recall rate (R), and $F1$ score to measure the question-answering performances of different algorithms. Among them, EM represents the percentage of questions in the question-answering system where the answers provided are an exact match with the standard answers.

The specific calculation formula is as follows:

$$EM = \frac{1}{n_{samples}} \sum_{i=0}^{n_{samples}-1} I(\hat{y}_i = y_i) \tag{6}$$

where $n_{samples}$ represents the total number of samples. $I(x)$ is an indicator function that takes the value of 1 when $\hat{y}_i$ is identical to y_i, and 0 otherwise. As can be seen from the formula, a higher EM value indicates a higher exact match.

Recall rate is to determine the proportion between the number of questions accurately answered by the question-answering system and the total number of questions.

The calculation formula is as follows:

$$R = \frac{TP}{TP + FN} \tag{7}$$

where TP signifies the accurate count of questions answered correctly by the question-answering system. Conversely, FN denotes the incorrect count of questions that were responded to inaccurately by the system.

The calculation formula for the F1 score is as follows:

$$F1 = \frac{2(EM \times R)}{EM + R} \tag{8}$$

In this study, Algorithm 1, as shown in Table 2, used the original BERTserini algorithm, adopting the algorithmic steps from reference [17] to construct the intelligent question-answering system. In Algorithm 2, an additional pre-processing algorithm is incorporated based on Algorithm 1. Fine-tuning is conducted using the CMRC2018 dataset, and parameter optimization for the BERTserini algorithm is performed. Algorithm 3 is an extension of Algorithm 2, incorporating the SPTI dataset for fine-tuning. Algorithm 4 is an improved version of the Bertserini algorithm, which is based on Algorithm 3. In addition to incorporating the SPTI dataset for fine-tuning, a pre-processing FAQ module based on short-text similarity calculation is added to filter out frequently asked questions. This module enables more efficient and effective preprocessing of the questions.

Table 2. Comparison of different algorithms.

Algorithm	Content	EM		R		F1	
		Value	Percentage of Improvement	Value	Percentage of Improvement	Value	Percentage of Improvement
Algorithm 1	Original BERTserini	0.261	——	0.453	——	0.331	——
Algorithm 2	Document preprocessing + Original BERTserini +Fine-Tuning (CMRC2018) + Parameter tuning	0.502	48%	0.783	42.1%	0.615	46.1%
Algorithm 3	Document preprocessing + Original BERTserini + Fine-Tuning (CMRC2018) + Parameter tuning + Fine-Tuning (SPTI)	0.702	62.8%	0.919	50.7%	0.796	58.4%
Algorithm 4	Document preprocessing + Original BERTserini + Fine-Tuning (CMRC2018) + Parameter tuning + Fine-Tuning (SPTI) + FAQ	0.856	69.5%	0.976	53.6%	0.912	63.7%

Note: In the table "Percentage of improvement" is calculated based on the values of Algorithm 1 as a reference point. It represents the relative increase in evaluation value (EM, R, and F1) achieved by Algorithm 2, Algorithm 3, and Algorithm 4 compared to Algorithm 1.

As shown in Table 2, the EM value for Algorithm 1 is only 0.261, indicating poor performance. After adopting Algorithm 3, the EM value reaches 0.702, representing an improvement of 62.8%. After adopting the proposed Algorithm 4, the EM value reaches 0.856, demonstrating the best performance. In comparison to Algorithm 1, the proposed algorithm achieved an improvement of 69.5% in terms of EM value, 53.6% in terms of R value, and 63.7% in terms of F1 value. These results demonstrate a practical level of engineering advancement.

4.5. Engineering Application

An intelligent question-answering system for power regulations and standards is constructed based on the proposed improved BERTserini algorithm and experimental data presented in this paper, the UI interface of intelligent question-answering system based on improved BERTserini algorithm, as shown in Figure 7. The English explanation of UI interface in intelligent question-answering system based on improved BERTserini algorithm is shown in Figure 8.

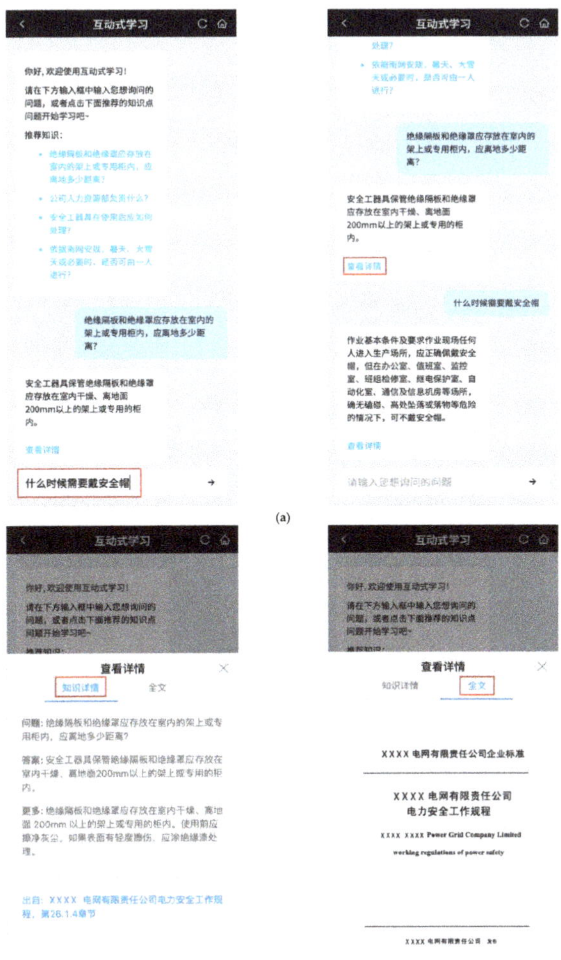

Figure 7. UI interface of intelligent question-answering system based on improved BERTserini algorithm. (**a**) Multi-turn interactive question-answering interface. (**b**) Knowledge details page. (**c**) Full-text source page.

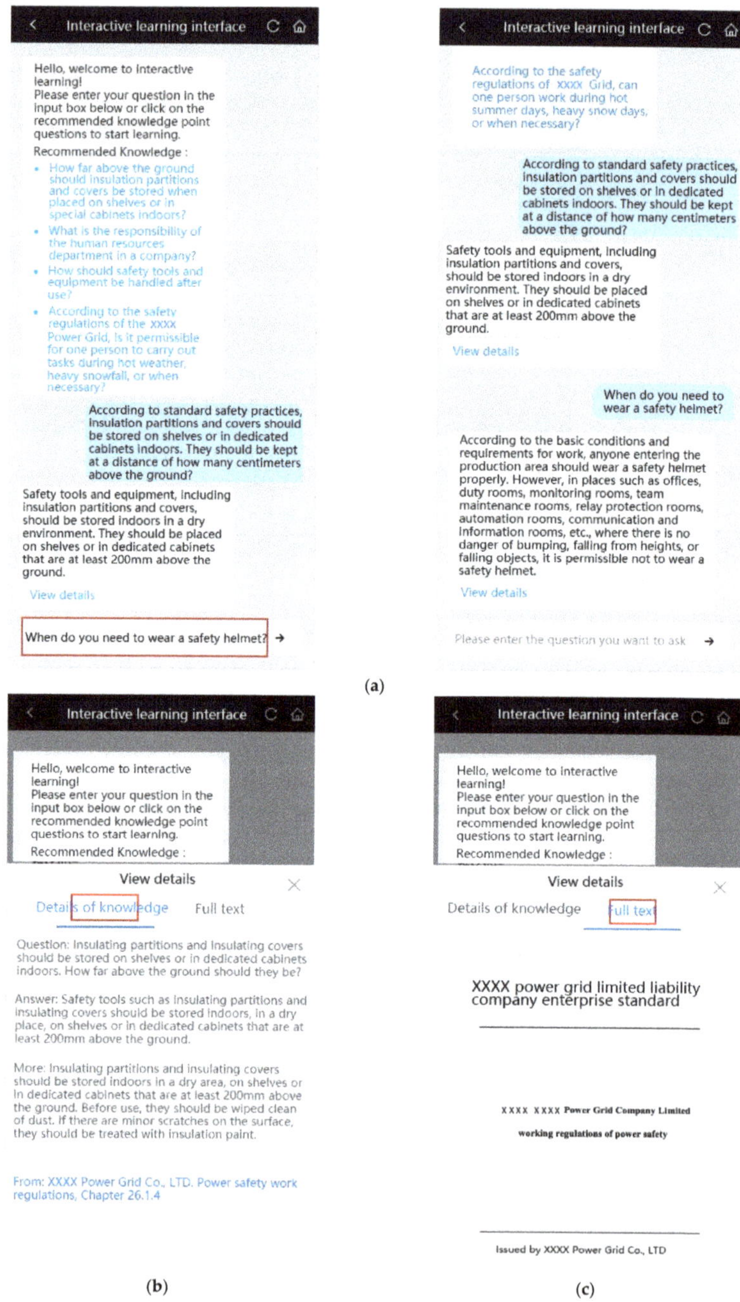

Figure 8. English explanation of UI interface in intelligent question-answering system based on improved BERTserini algorithm. (**a**) Multi-turn interactive question-answering interface. (**b**) Knowledge details page. (**c**) Full-text source page.

The system provides users with a multi-turn interactive question-answering interface on the topic of power safety, as illustrated in Figures 7a and 8a. Users can ask questions by either voice input or manual input. After sending the question, they will receive the system's response within 400 ms. Clicking on the "view details" link below the answer will cause the system to pop up a window displaying the source of the answer, including the name of the original document and the chapter number, as shown in Figures 7b and 8b. Clicking on the "full text" link allows users to view the content of the original document where the answer is located, as shown in Figures 7c and 8c.

5. Conclusions

The improved BERTserini algorithm proposed in this paper is designed for intelligent question-and-answer processing of power regulation documents. In comparison to the original BERTserini algorithm, this approach offers the following advantages:

(1) The improved BERTserini algorithm supports multi-document long text preprocessing for rules and regulations. This algorithm is capable of answering documents containing 30+ rules and regulations with a length of 30M+ bytes. This addresses the issue in the original BERTserini algorithm where document titles of regulatory documents were erroneously output as answers. Furthermore, it accurately provides the document name and chapter/page number information for answers that the original BERTserini algorithm could not identify. These enhancements significantly enhance the quality of answers and user experience in the question-answering system.

(2) The improved BERTserini algorithm proposed in this paper underwent two rounds of fine-tuning using the CMRC2018 and the specialized dataset SPTI. Algorithm parameters were also optimized. The intelligent question-answering system built upon it demonstrates a more precise answer generation capability compared to the original BERTserini algorithm when addressing domain-specific questions.

(3) The improved BERTserini algorithm proposed in this paper significantly enhances the exact match rate for intelligent question-answering in the domain of regulatory texts. Experimental data indicate that, compared to the original BERTserini algorithm, the exact match rate has increased by 69.5%, the R-value has improved by 53.6%, and the F1-value has risen by 63.7%. The algorithm maintains an average question–answer response time of within 400 milliseconds, meeting the requirements for engineering applications.

The improvements made to the BERTserini algorithm proposed in this paper are versatile, with the expectation that they can be widely applied in the research and construction of intelligent question-answering systems for regulatory texts across various industries. The limitations of this study lie in the current engineering practices, which are currently confined to the power industry. There is a lack of engineering cases for the construction of intelligent question-answering systems in industries such as petroleum, steel, transportation, and others where regulatory knowledge is prevalent. The generalizability of the algorithmic process across multiple domains needs further validation.

The next research direction involves applying this algorithm to construct intelligent question-answering systems for regulatory texts in other industry sectors. Additionally, by incorporating algorithmic iterations and leveraging advancements in technology, particularly with large language models, there is a continuous effort to optimize and enhance the effectiveness of the question-answering system.

Author Contributions: Writing—original draft preparation, M.G.; validation, M.G. and T.J.; formal analysis, M.L.; investigation, Q.W.; writing—review and editing, N.W. and G.L.; visualization, T.J.; supervision, M.G.; M.G. and M.L. conceived the idea and provided resources. N.W. and G.L. wrote the manuscript. T.J. and M.G. provided guidelines. M.G., M.L., T.J., N.W., G.L. and Q.W. designed the study and participated in the experiment. All authors have read and agreed to the published version of the manuscript.

Funding: This research received no external funding.

Data Availability Statement: "Chinese Machine Reading Comprehension 2018 (CMRC2018)" at https://ymcui.com/cmrc2018 (accessed on 4 January 2018). "Delta Reading Comprehension Dataset (DRCD)" at https://github.com/DRCKnowledgeTeam/DRCD (accessed on 30 July 2018).

Conflicts of Interest: The authors declare no conflicts of interest.

References

1. Shum, H.Y.; He, X.; Li, D. From Eliza to XiaoIce: Challenges and Opportunities with Social Chatbots. *Front. Inf. Technol. Electron. Eng.* **2018**, *19*, 10–26. [CrossRef]
2. Xavier, A. Intelligence artificielle et psychiatrie: Noces d'or entre Eliza et Parry. *Linform. Psychiatr.* **2017**, *93*, 51–56.
3. Hb, B.G.; Kumar, M.A.; Kp, S. Vector Space Model as Cognitive Space for Text Classification. *arXiv* **2017**, arXiv:1708.06068. [CrossRef]
4. Ma, P.; Jiang, B.; Lu, Z.; Li, N.; Jiang, Z. Cybersecurity Named Entity Recognition Using Bidirectional Long Short-Term Memory with Conditional Random Fields. *Tsinghua Sci. Technol.* **2021**, *26*, 259–265. [CrossRef]
5. Mittal, A.; Agrawal, A.; Chouksey, A.; Shriwas, R. A Comparative Study of Chatbots and Humans. *Int. J. Adv. Res. Comput. Commun. Eng.* **2016**, *5*, 1055–1057.
6. Jabberwacky Website. Available online: http://www.jabberwacky.com (accessed on 30 August 2023).
7. Bohouta, G.; Kpuska, V.Z. Next-Generation of Virtual Personal Assistants (Microsoft Cortana, Apple Siri, Amazon Alexa and Google Home). In Proceedings of the IEEE CCWC 2018, the 8th IEEE Annual Computing and Communication Workshop and Conference, Las Vegas, NV, USA, 8–10 January 2018.
8. Poushneh, A. Impact of auditory sense on trust and brand affect through auditory social interaction and control. *J. Retail. Consum. Serv.* **2021**, *58*, 102281. [CrossRef]
9. Lei, X.; Tu, G.H.; Liu, A.X.; Ali, K.; Li, C.Y.; Xie, T. The Insecurity of Home Digital Voice Assistants—Amazon Alexa as a Case Study. *arXiv* **2017**, arXiv:1712.03327. [CrossRef]
10. Straga, D. Facebook Messenger as a Tool for Building Relationships with Customers: Bachelor Thesis. Ph.D. Thesis, Univerza v Ljubljani, Fakulteta za Družbene Vede, Ljubljana, Slovenia, 2017.
11. Berdasco, A.; López, G.; Diaz, I.; Quesada, L.; Guerrero, L.A. User Experience Comparison of Intelligent Personal Assistants: Alexa, Google Assistant, Siri and Cortana. *Proceedings* **2019**, *31*, 51. [CrossRef]
12. Devlin, J.; Chang, M.W.; Lee, K.; Toutanova, K. Bert: Pre-training of deep bidirectional transformers for language understanding. *arXiv* **2018**, arXiv:1810.04805.
13. Dehouche, N. Plagiarism in the age of massive Generative Pre-trained Transformers (GPT-3): "The best time to act was yesterday. The next best time is now". *Ethics Sci. Environ. Polit.* **2021**, *21*, 17–23. [CrossRef]
14. Jain, N.; Jain, P.; Kayal, P.; Sahit, J.; Pachpande, S.; Choudhari, J. AgriBot: Agriculture-specific question answer system. *Preprint* **2019**. [CrossRef]
15. Gao, Z.; Feng, A.; Song, X.; Wu, X. Target-dependent sentiment classification with BERT. *IEEE Access* **2019**, *7*, 154290–154299. [CrossRef]
16. Yang, W.; Xie, Y.; Lin, A.; Li, X.; Tan, L.; Xiong, K.; Li, M.; Lin, J. End-to-End Open-Domain Question Answering with Bertserini. *arXiv* **2019**, arXiv:1902.01718.
17. Abo Khamis, M.; Ngo, H.Q.; Rudra, A. FAQ: Questions Asked Frequently. In Proceedings of the 35th ACM SIGMOD-SIGACT-SIGAI Symposium on Principles of Database Systems, San Francisco, CA, USA, 26 June–1 July 2016. [CrossRef]
18. Kadhim, A.I. Term Weighting for Feature Extraction on Twitter: A Comparison between BM25 and TF-IDF. In Proceedings of the 2019 International Conference on Advanced Science and Engineering (ICOASE), Zakho, Iraq, 2–4 April 2019; pp. 124–128. [CrossRef]
19. Robertson, S.E.; Walker, S.; Jones, S.; Hancock-Beaulieu, M.M.; Gatford, M. Okapi at TREC-3. *Nist Spec. Publ. SP* **1995**, *109*, 109.
20. Yang, P.; Fang, H.; Lin, J. Anserini: Reproducible Ranking Baselines Using Lucene. *J. Data Inf. Qual. (JDIQ)* **2018**, *10*, 1–20. [CrossRef]
21. Cui, Y.; Che, W.; Liu, T.; Qin, B.; Yang, Z. Pre-Training with Whole Word Masking for Chinese Bert. *IEEE/ACM Trans. Audio Speech Lang. Process.* **2021**, *29*, 3504–3514. [CrossRef]
22. Cui, Y.; Liu, T.; Che, W.; Xiao, L.; Chen, Z.; Ma, W.; Wang, S.; Hu, G. A span-extraction dataset for Chinese machine reading comprehension. *arXiv* **2018**, arXiv:1810.07366.
23. Shao, C.C.; Liu, T.; Lai, Y.; Tseng, Y.; Tsai, S. DRCD: A Chinese machine reading comprehension dataset. *arXiv* **2018**, arXiv:1806.00920.
24. Rajpurkar, P.; Zhang, J.; Lopyrev, K.; Liang, P. Squad: 100,000+ questions for machine comprehension of text. *arXiv* **2016**, arXiv:1606.05250.
25. Bert Website. Available online: https://github.com/google-research/bert (accessed on 11 March 2020).

energies MDPI

Article

Hierarchical Blocking Control for Mitigating Cascading Failures in Power Systems with Wind Power Integration

Lun Cheng [1], Tao Wang [2,*], Yuhang Wu [2], Zeming Gao [1] and Ning Ji [1]

1 State Grid Hebei Electric Co., Ltd., Shijiazhuang 050000, China; chengl@he.sgcc.com.cn (L.C.); gzm@he.sgcc.com.cn (Z.G.); jin@he.sgcc.com.cn (N.J.)
2 Department of Mathematics and Physics, North China Electric Power University, Baoding 071000, China; 52151147@ncepu.edu.cn
* Correspondence: wangtao@ncepu.edu.cn

Abstract: The increasing uncertainty of wind power brings greater challenges to the control for mitigation of cascading failures. In order to minimize the risk of cascading failures in large-scale wind power systems at a lower economic cost, a multi-stage blocking control model is proposed based on sensitivity analysis. Firstly, the propagation mechanism of cascading failures in power systems with wind power integration is analyzed, and the propagation path of such failures is predicted. Subsequently, sensitive lines that are prone to failure are identified using the power sensitivity matrix, taking into account the effects of blocking control on the propagation path. By constraining the power flow of these sensitive lines, a multi-stage blocking control model for the predicted cascading failure path is proposed with the objective of minimizing the control cost and cascading failure probability. Based on probabilistic optimal power flow calculations, the constraints related to wind power uncertainty are transformed into opportunity constraints. To validate the effectiveness of the proposed model, the IEEE 39-node system is used as an example, and the results show that the obtained control method is able to balance economy and safety. In addition, the control costs for the same initial failure are higher as the wind power penetration rates and confidence levels increase.

Keywords: power systems; cascading failure; multi-stage blocking control; sensitive lines

Citation: Cheng, L.; Wang, T.; Wu, Y.; Gao, Z.; Ji, N. Hierarchical Blocking Control for Mitigating Cascading Failures in Power Systems with Wind Power Integration. *Energies* **2024**, *17*, 442. https://doi.org/10.3390/en17020442

Academic Editors: Lin Jiang, Bo Yang and Zhijian Liu

Received: 15 December 2023
Revised: 8 January 2024
Accepted: 12 January 2024
Published: 16 January 2024

1. Introduction

In recent years, global blackouts have become increasingly frequent, predominantly attributed to random failures that trigger abnormal operational states within the system, subsequently leading to cascading failures across various components [1–3]. Cascading failures are a serious threat to the safe operation of the power system.

According to data released by the Global Wind Energy Commission (GWEC), in 2022, the grid-connected capacity of global wind power increased by 77.6 GW, with an additional offshore installed capacity of 8 GW. The top five countries in terms of newly added wind power installations are China, the United States, Brazil, Germany, and Sweden. In China, wind power constitutes 13% of the total installed power generation capacity, providing 7.5% of the total electricity consumption. With the increasing integration of wind power sources into the grid, there is heightened uncertainty in source-side output and new dynamic behavioral properties of the grid, leading to an escalated risk of cascading failures in the power grid [4]. Therefore, it is very crucial to control cascading failures under the uncertain nodal power injection of wind power.

Currently, significant progress has been achieved in researching the prevention and control of cascading failures. Modern digital systems [5], complex systems theory [6,7], and complex network theory [8,9] are employed to investigate early warning, fault situation identification, and prevention control of cascading failures. However, modern digital systems are highly dependent on data, and monitoring systems within power systems involve numerous agents with varying data structures. Complex network theory and complex system theory primarily analyze the macroscopic characteristics of the power grid. Consequently, these limitations impede the guidance for formulating control schemes to prevent power grid cascading failures. Subsequently, adopting control methods based on deterministic safety criteria becomes crucial. This involves considering safety constraints during grid operation and employing sensitivity analysis [10], mathematical programming [11,12], and other methodologies to devise optimal control schemes. The objective is to eliminate abnormal operational states of the power grid and prevent cascading failures. In a specific emergency control model proposed in [10], sensitivity analysis is used to identify power adjustment nodes and prioritize the adjustment sequence of node sets based on their impact on line power adjustment. This approach successfully addresses the cascading overload tripping problem caused by line overload through multiple rounds of system power adjustment. Meanwhile, Dvorkin et al. [11] introduced a safety-constrained optimal power flow model into cascading failure preventive control, ensuring the lowest-cost dispatch of controllable generators while adhering to all operational constraints in both pre- and post-accident states. To address the high control cost or practical infeasibility associated with preventive control alone, Zhai et al. [12] established a cascading failure defense model based on preventive-emergency coordinated control. This model optimizes the initial operating state of the system using preventive control, considers the power grid's constraints on the tolerability of cascading failure consequences, and maximizes the dispatchable potential of generators during an accident, effectively suppressing cascading failures under economically optimal conditions. While these control models are generally effective for selected failure modes, they often overlook the influence of power grid operational status on component failure probability, leading to challenges in accurately quantifying the power grid's safety degree and resulting in conservative control schemes. Consequently, several studies have incorporated risk assessment theory into cascading failure control, considering uncertainties in system operation to develop control schemes that balance safety and economy. For instance, Wang et al. [13] proposed a risk assessment-based preventive-blocking coordinated control model to implement control measures at the link with the highest risk in the cascading failure path. This approach simplifies the optimization objective. Rui et al. [14] established a coordinated control model with system risk and control cost as indicators, and employed a multi-objective particle swarm optimization algorithm accounting for preferences to determine the optimal control scheme.

In fact, most studies focus on fault voltage ride-through control of wind turbines, security-constrained optimal power flow considering wind power uncertainty, and prevention control of cascading failures. However, hierarchical blocking control for mitigating cascading failures is rarely addressed. The predominant focus in existing research is on power grid operation scenarios where source output is predetermined. These studies deduce the propagation path of cascading failure based on the law of power flow transfer to formulate control schemes. However, in systems incorporating wind power, the substantial uncertainty in their power output heightens the complexity of controlling cascading failures. On one hand, when the proportion of wind power integrated into the grid is substantial, the fluctuating nature of its output introduces greater uncertainty in line power flow. Consequently, the challenge of addressing line overload becomes more severe and intricate to prevent. On the other hand, the uncertainty associated with wind power output may disrupt the power balance of the system. During such instances, the number and adjustment amplitude of conventional units participating in the adjustment process increase. This inevitably alters the flow transfer dynamics and may even lead to changes in the cascading failure path, accentuating the occurrence of new failures during the control process. Hence, formulating a cascading failure control scheme that considers both safety and economy is crucial. This approach ensures the safe and stable operation of the system by accounting for the impact of wind power output uncertainty on cascading failure.

Building upon the aforementioned analysis, this study presents a multi-stage blocking control model for cascading failure that explicitly considers the impact of wind power output uncertainty on sensitive lines within the framework of power systems with wind power integration. The key contributions of this article are outlined as follows:

(1) A probabilistic power flow is employed to delineate the uncertainties associated with wind power output, serving as the computational foundation for forecasting cascading failure pathways. We propose a cascading failure path prediction method tailored for power systems incorporating wind power integration.
(2) We analyze the mutual influences among the uncertainty of wind power output, the propagation path of cascading failure, and the control measures implemented to mitigate these effects. We propose the utilization of a power sensitivity matrix to identify susceptible lines prone to failure due to control measures. In the control model, distinct safety constraints are established for these sensitive lines to diminish the likelihood of failure and avoid causing new line failures during the control process, particularly in the context of cascading failures.
(3) Taking into account the uncertainties in control variables arising from wind power, constraints involving random variables are expressed as expected value constraints using probability optimal power flow. Additionally, the boundary constraints of output variables are portrayed as chance constraints. We construct a multi-stage blocking control model for cascading failures in wind power systems, ensuring the economic and safe implementation of the derived control scheme.
(4) The efficacy of the presented control model is validated through simulation analysis. The impact of large-scale wind power on cascading failure blocking control is examined by considering the wind power penetration rate and confidence level as key factors in the analysis.

The rest of this paper is organized as follows: Section 2 introduces the propagation path of cascading failure in power systems with wind power integration; Section 3 describes the sensitive lines in power systems with wind power integration. The blocking control model of cascading failure in power systems with wind power integration is proposed in Section 4; the simulation and results on IEEE test systems are given in Section 5, and Section 6 concludes the paper.

2. Propagation Path of Cascading Failure in Power Systems with Wind Power Integration

2.1. Probabilistic Power Flow Calculation

With the ongoing development of new energy sources, the consideration of uncertainty in new energy output at the source side is becoming increasingly crucial in probabilistic power flow calculations [15]. The analytical approach, exemplified by the cumulant method, linearizes the mathematical relationship between input and output random variables. It employs simple algebraic operations between cumulant variables instead of complex convolution operations to calculate the cumulant variables of each order of output variables. The probabilistic characteristics of output variables are then derived through the level expansion method, offering the advantages of reduced computational effort and higher computational efficiency [16]. Consequently, this paper develops a probabilistic power flow model that takes into account wind power uncertainty based on the cumulant method. Uncertainties in the wind power system primarily arise from the variability in user-side load demand and the fluctuations in the active output of the wind turbine due to changes in wind speed. In this paper, we assume that the power at the load node approximately follows a normal distribution, and the power output model is expressed by Equation (1):

$$\begin{cases} f(P_{\mathrm{L}}) = \frac{1}{\sqrt{2\pi}\sigma_{P_{\mathrm{L}}}} \exp[-\frac{(P_{\mathrm{L}}-\mu_{P_{\mathrm{L}}})^2}{2\sigma_{P_{\mathrm{L}}}{}^2}] \\ f(Q_{\mathrm{L}}) = \frac{1}{\sqrt{2\pi}\sigma_{Q_{\mathrm{L}}}} \exp[-\frac{(Q_{\mathrm{L}}-\mu_{Q_{\mathrm{L}}})^2}{2\sigma_{Q_{\mathrm{L}}}{}^2}] \end{cases} \quad (1)$$

where $f(P_L)$ is the probability density function of active power. $f(Q_L)$ is the probability density function of reactive power. $\mu_{P_L}, \sigma_{P_L}{}^2$ are the expectation and variance of load active power, respectively. $\mu_{Q_L}, \sigma_{Q_L}{}^2$ are the expectation and variance of the load reactive power, respectively.

In this paper, it is assumed that the wind speed follows the Weibull distribution, and its probability density function is represented by Equation (2):

$$f(v) = \frac{k}{c}\left(\frac{v}{c}\right)^{k-1} e^{-(v/c)^k} \tag{2}$$

where k is the shape parameter of the Weibull distribution, c is the scale parameter of the Weibull distribution. According to Reference [17], the relationship between wind power active output P_w and wind speed v satisfies can be described as:

$$P_w = \begin{cases} 0, & v < v_{ci} \ or \ v \geq v_{co} \\ \frac{v^3 - v_{ci}{}^3}{v_r{}^3 - v_{ci}{}^3} P_{wn} & v_{ci} \leq v \leq v_n \\ P_{wn} & v \geq v_n \end{cases} \tag{3}$$

where P_{wn} and v_n are the rated power and rated wind speed of the wind turbine, respectively. v_{ci} and v_{co} are the cut-in wind speed and cut-out wind speed of the wind turbine, respectively.

A probabilistic power flow model is employed in this work, which is based on t the AC power flow. In the AC power flow model, the nodal power equation is expressed in matrix form. Subsequently, a Taylor series expansion is conducted around the base run operating point, wherein higher order terms are disregarded, resulting in the following simplified expression:

$$\begin{cases} \boldsymbol{X} = \boldsymbol{X}_0 + \Delta\boldsymbol{X} = \boldsymbol{X}_0 + \boldsymbol{S}_0 \Delta\boldsymbol{W} \\ \boldsymbol{Z} = \boldsymbol{Z}_0 + \Delta\boldsymbol{Z} = \boldsymbol{Z}_0 + \boldsymbol{T}_0 \Delta\boldsymbol{W} \end{cases} \tag{4}$$

where $\boldsymbol{X}$ and $\Delta\boldsymbol{X}$ are the node state variables (including node voltage amplitude and phase angle) and their changes. $\boldsymbol{Z}$ and $\Delta\boldsymbol{Z}$ are the branch power flow variable (including branch active power flow and reactive power flow) and its variation. The subscript 0 indicates the value at the base run point. $\boldsymbol{S}_0$ and $\boldsymbol{T}_0$ are the sensitivity matrix. $\boldsymbol{S}_0 = \boldsymbol{J}_0{}^{-1}$; $\boldsymbol{T}_0 = \boldsymbol{G}_0 \boldsymbol{J}_0{}^{-1}$; $\boldsymbol{G}_0 = (\partial\boldsymbol{Z}/\partial\boldsymbol{X})|_{\boldsymbol{X}=\boldsymbol{X}_0}$; $\boldsymbol{J}_0$ is the Jacobi matrix; $\Delta\boldsymbol{W}$ is the node injection power perturbation.

According to the cumulant property, the random perturbation of the injected power at the node can be mathematically represented as:

$$\Delta w_i{}^{(k)} = \Delta w_{Gi}{}^{(k)} + \Delta w_{Li}{}^{(k)} \tag{5}$$

where $\Delta w_i{}^{(k)}$ add the k-order and half invariance of the power to the node i. $\Delta w_{Gi}{}^{(k)}$ and $\Delta w_{Li}{}^{(k)}$ add the k-order and half invariance of the power of the generator and the power of the load to the node i, respectively.

Thus, the k-order cumulant of the output variable can be obtained as:

$$\begin{cases} \Delta\boldsymbol{X}^k = \boldsymbol{S}_0^k \Delta\boldsymbol{W}^k \\ \Delta\boldsymbol{Z}^k = \boldsymbol{T}_0^k \Delta\boldsymbol{W}^k \end{cases} \tag{6}$$

where $\boldsymbol{S}_0{}^k$ and $\boldsymbol{T}_0{}^k$ are the matrices formed by the matrix $\boldsymbol{S}_0$ and the powers $\boldsymbol{T}_0$ of the k elements in the matrix, respectively.

After applying Equations (5) and (6) to calculate the cumulant of the output variable, the probability distribution function of the output variable, specifically the node voltage and the line power flow, can be derived using the Gram–Charlier progression expansion method.

2.2. Prediction of Cascading Failure

Cascading failure is a phenomenon characterized by the simultaneous occurrence of multiple factors, a short propagation time, and severe consequences. According to the literature [18], it is noted that the primary driving factor behind cascading failure in new energy power systems is still predominantly overload. Therefore, the present study aims to analyze and mitigate the occurrence of overload-dominated cascading failure. Overload-dominated cascading failure is a phenomenon in which a system experiences a failure that alters the optimal power flow transmission path. This leads to a redistribution of power flow from the faulty line to the non-faulty line, causing the non-faulty line to become overloaded. As a result, protection measures are triggered to remove the non-faulty line, leading to a chain reaction that ultimately results in the collapse of the entire system [19]. The probability of line tripping due to overload is commonly represented using a piecewise function [20].

$$c = \begin{cases} c_0,\ P_k \leq {P_k}^{\mathrm{r}} \\ c_0 + \frac{1-c_0}{{P_k}^{\mathrm{m}} - {P_k}^{\mathrm{r}}}(P_k - {P_k}^{\mathrm{m}}),\ {P_k}^{\mathrm{r}} < P_k \leq {P_k}^{\mathrm{m}} \\ 1,\ P_k > {P_k}^{\mathrm{m}} \end{cases} \tag{7}$$

where c_0 is the protection recessive fault probability, which is 0.01. P_k is the active power on the line k. ${P_k}^{\mathrm{r}}$ and ${P_k}^{\mathrm{m}}$ are the rated power and limit transmission power of the line k, respectively.

When the proportion of wind power is high, the fluctuating nature of its power output results in line power flow that exhibits uncertain characteristics. The probabilistic power flow model can be employed to determine the probability distribution state of power flow in power systems that incorporate wind power. However, the power flow along the line is no longer considered a constant value, but rather a probability density function. As a result, it is not possible to directly substitute this function into Equation (7) in order to determine the line outage probability. Therefore, Equation (8) combines the probability density function of the line power flow and the line outage probability to determine the line outage probability in power systems with wind power integration.

$$c_{\mathrm{w}} = \int_0^{+\infty} c \cdot f(P_k) dP_k \tag{8}$$

where $f(P_k)$ is the probability density function of line k active power. c is the probability of line tripping due to overload.

In the context of cascading failure propagation, the subsequent faulty lines are primarily impacted by the disconnection of the line in the previous stage if no mitigation measures are implemented. Therefore, the prediction of the propagation path of cascading failures in power systems with wind power integration can be achieved by continuously selecting the line with a higher probability of outage as the subsequent faulty line, as determined by Equation (8).

3. Sensitive Lines in Power Systems with Wind Power Integration

3.1. Impact of Blocking Control on the Propagation of Cascading Failure

Blocking control refers to the application of control measures in the early stages of the projected propagation trajectory of cascading failure, with the objective of hindering the further amplification of the failure. The control strategy for mitigating overload-dominated cascading failure primarily focuses on altering the power flow distribution state of the system by making adjustments to generator output and reducing the load. This measure is implemented in order to reduce the impact of excessive line load and prevent the occurrence of cascading failures. The redistribution of power flow within a system can result in alterations to the load rate of additional transmission lines and the outage probability of lines. This, in turn, has the potential to induce overload occurrences and trigger subsequent cascading reactions through the tripping of overloaded lines. Figure 1 depicts the dynamic relationship between the uncertainty of wind power output, the propagation of cascading failures, and the implementation of blocking control measures in power systems that incorporate the integration of wind power. The equilibrium of the system is disrupted due to the modification in wind power generation and the tripped lines caused by overload. The redistribution of power flow on tripped lines has a prominent influence on the optimal control measures. Currently, there is a rise in both the quantity and magnitude of conventional units participating in the process of adjustment. This increase can potentially exert a substantial influence on the operational status, resulting in a modification of the propagation path of cascading failures. Therefore, subsequent to the implementation of control measures, the propagation trajectory of cascading failure is influenced by both line disconnection and injection power alteration. The outcome will inevitably deviate from the initial predicted result, resulting in the failure of the original blocking control and hastening the malignant propagation of failure. In actuality, when there is a significant change in the power output of the generator, only a portion of the transmission lines are significantly affected and experience overloading. If the specific lines responsible for the cascading failure can be identified and appropriate defensive measures are implemented during the design of a blocking control scheme, it can effectively prevent the control measure from modifying the transmission path of the cascading failure. This ensures that the blocking control measure successfully achieves the intended outcome.

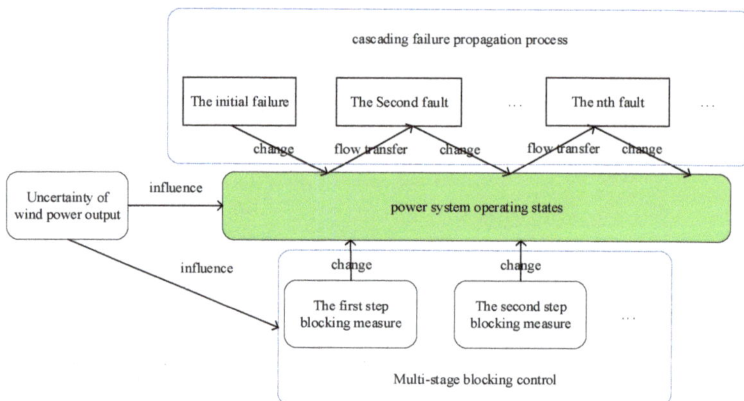

Figure 1. Interactive impact diagram of wind power, control measures, and cascading failure propagation process.

3.2. Determination of Sensitive Lines Based on Power Sensitivity Matrix

Based on the provided power grid topology and parameters, as well as the current correlation coefficient matrix, it is possible to derive a power sensitivity matrix that represents the relationship between the injected power variable of a node and the power variable of the line. From this, it can be concluded that the power sensitivity β_{k-i} between the power flow variables of the branch B_k in the power grid and the injected power variables of the node N_i is:

$$\beta_{k-i} = \frac{\lambda_{k-i} U_{k,B}}{U_{i,N}} \cos(\varphi_{k,B} - \varphi_{i,N}) \tag{9}$$

where λ_{k-i} is the current correlation coefficient between the current phasor of branch B_k and the injection current phasor of node N_i; $U_{k,B}$ and $\varphi_{k,B}$ are the first terminal voltage modulus and phase angle of branch B_k, respectively; $U_{i,N}$ and $\varphi_{i,N}$ are the voltage mode values and phase angles of node N_i, respectively.

Considering that the voltage amplitude and phase angle of each node in the power flow calculation of the transmission network are not significantly different, $\frac{U_{k,B}}{U_{i,N}}$ and $\cos(\varphi_{k,B} - \varphi_{i,N})$ can be approximately equal to 1. Therefore, Equation (9) can be equivalent to:

$$\beta_{k-i} = \lambda_{k-i} \tag{10}$$

The current correlation coefficient λ_{k-i} can be calculated by Equation (11):

$$\boldsymbol{C}(\lambda) = \boldsymbol{Y_B}\boldsymbol{A}^{\mathrm{T}}\boldsymbol{Y_N}^{-1} \tag{11}$$

where, for a power grid with n nodes and b branches, $\boldsymbol{C}(\lambda)$ is its current correlation coefficient matrix, $\boldsymbol{Y_B}$ is its branch admittance matrix, $\boldsymbol{A}$ is its node incidence matrix, $\boldsymbol{Y_N}$ is the node admittance matrix.

From Equation (10), it is evident that power sensitivity β_{k-i} is solely dependent on the parameters and topology of the power grid, and remains unaffected by the operational state of the system. The influence of node injection power change on the flow power on the branch is directly proportional to the power sensitivity β_{k-i} between each line and node.

The blocking control of cascading failure is primarily accomplished through the adjustment of the generator's output. Hence, the power sensitivity matrix for each controllable generator node and all lines is computed using Equation (10) prior to implementing the blocking control strategy. The lines that experience significant changes in their adjustment are identified as sensitive lines based on the magnitude of their power sensitivity. When implementing control measures, it is advisable to prioritize the prevention and control of these factors. This approach not only reduces the likelihood of new failures caused by control blockages, but also mitigates the issue of high control costs.

4. Blocking Control Model of Cascading Failure in Power Systems with Wind Power Integration

Based on the aforementioned analysis, it is evident that the incorporation of control constraints is crucial in order to effectively manage the power flow in sensitive transmission lines. This measure is imperative in order to guarantee that the implementation of blocking control measures does not modify the propagation path of cascading failure following the integration of wind power. After the integration of wind power, the control variables, such as node injection power adjustment, and the output state variables, including power flow and voltage, may exhibit uncertain behavior. Traditional control models that rely on deterministic optimal power flow are no longer applicable for blocking cascading failures. The application of probability optimal power flow with opportunity constraints and the integration of probability information of random variables have gained popularity in the optimization of new energy power systems. This approach entails the manipulation of different system parameters to account for the potential scenario in which the formulated scheme may not meet the constraint conditions with a certain level of confidence. Hence, the objective of this study is to mitigate the control cost associated with each link in the transmission path of cascading failure. This can be achieved through the adjustment of output from conventional units, discontinuation of wind power, and load reduction. The aforementioned objective is accomplished by employing the calculation of probabilistic optimal power flow. Constraints involving stochastic control variables are commonly formulated as expected value constraints, while boundary constraints on output state variables are expressed as opportunity constraints. The proposed multi-stage blocking control model aims to effectively mitigate the occurrence and propagation of cascading failure.

4.1. Objective Function

Considering that the purpose of cascading failure blocking control is to reduce the probability of cascading failure in the system with the most economical control cost possible, the objective function is:

$$f = \sum_{k=1}^{l} (\alpha(\sum_{i=1}^{N_G} \Delta\overline{P}_{uGi}^{k} + \sum_{i=1}^{N_G} \Delta\overline{P}_{dGi}^{k}) + \beta\sum_{i=1}^{N_W} \Delta\overline{P}_{dWi}^{k} + \gamma\sum_{i=1}^{N_D} \Delta\overline{P}_{dDi}^{k}) \tag{12}$$

where the superscript "—" indicates the expected value of the control variable. l is the number of links passed by the cascading failure propagation paths. $\sum_{i=1}^{N_G} \Delta\overline{P}_{uGi}^{k}$ and $\sum_{i=1}^{N_G} \Delta\overline{P}_{dGi}^{k}$ are the upward and downward power adjustment of the k link adjustable generator node i. $\sum_{i=1}^{N_W} \Delta\overline{P}_{dWi}^{k}$ is the downward power adjustment of the k link wind turbine node i. $\sum_{i=1}^{N_L} \Delta\overline{P}_{dDi}^{k}$ is the load reduction amount of the first link load node i. α, β and γ are the cost coefficients of adjusting generator output, wind abandonment penalty and load shedding; considering that grid dispatchers do not want to lose load on the grid, they will adjust generator output first and then consider wind abandonment afterwards, so the cost of load shedding is much higher than the generator dispatching cost and wind abandonment penalty [21]. N_G, N_W and N_D are the adjustable generators, the number of wind turbine nodes and the number of load nodes, respectively.

4.2. Expectation Constraints

The expectation constraints of the cascading failure blocking control model, which is based on probabilistic optimal power flow, primarily encompass the power balance and generator output range constraints.

(1) Power balance constraint:

$$\begin{cases} (\widetilde{P}_{Gi} + \sum\limits_{m=1}^{k} \Delta\widetilde{P}_{uGi}^{m} - \sum\limits_{m=1}^{k} \Delta\widetilde{P}_{dGi}^{m}) + (\widetilde{P}_{Wi} - \sum\limits_{m=1}^{k} \Delta\widetilde{P}_{dWi}^{m}) - \\ (\widetilde{P}_{Di} - \sum\limits_{m=1}^{k} \Delta\widetilde{P}_{dDi}^{m}) - \widetilde{V}_i \sum\limits_{j\in i} \widetilde{V}_j (G_{ij}\cos\widetilde{\theta}_{ij} + B_{ij}\sin\widetilde{\theta}_{ij}) = 0 \\ (\widetilde{Q}_{Gi} + \sum\limits_{m=1}^{k} \Delta\widetilde{Q}_{uGi}^{m} - \sum\limits_{m=1}^{k} \Delta\widetilde{Q}_{dGi}^{m}) + (\widetilde{Q}_{Wi} - \sum\limits_{m=1}^{k} \Delta\widetilde{Q}_{dWi}^{m}) - \\ (\widetilde{Q}_{Di} - \sum\limits_{m=1}^{k} \Delta\widetilde{Q}_{dDi}^{m}) - \widetilde{V}_i \sum\limits_{j\in i} \widetilde{V}_j (G_{ij}\sin\widetilde{\theta}_{ij} - B_{ij}\cos\widetilde{\theta}_{ij}) = 0 \end{cases} \tag{13}$$

(2) Upper and lower bound constraints on control variables:

$$\begin{cases} P_{Gi}^{\min} \le \overline{P}_{Gi} + \sum\limits_{m=1}^{k} \Delta\overline{P}_{uGi}^{m} - \sum\limits_{m=1}^{k} \Delta\overline{P}_{dGi}^{m} \le P_{Gi}^{\max} \\ Q_{Gi}^{\min} \le \overline{Q}_{Gi} + \sum\limits_{m=1}^{k} \Delta\overline{Q}_{uGi}^{m} - \sum\limits_{m=1}^{k} \Delta\overline{Q}_{dGi}^{m} \le Q_{Gi}^{\max} \\ P_{Wi}^{\min} \le \overline{P}_{Wi} - \sum\limits_{m=1}^{k} \Delta\overline{P}_{dWi}^{m} \le P_{Wi}^{\max} \\ Q_{Wi}^{\min} \le \overline{Q}_{Wi} - \sum\limits_{m=1}^{k} \Delta\overline{Q}_{dWi}^{m} \le Q_{Wi}^{\max} \end{cases} \tag{14}$$

$$\begin{cases} 0 \leq \Delta \overline{P}^k_{\mathrm{uG}i} \leq \Delta P^{\max}_{\mathrm{uG}i} \\ 0 \leq \Delta \overline{P}^k_{\mathrm{dG}i} \leq \Delta P^{\max}_{\mathrm{dG}i} \\ 0 \leq \Delta \overline{P}^k_{\mathrm{dW}i} \leq \Delta P^{\max}_{\mathrm{dG}i} \\ 0 \leq \Delta \overline{Q}^k_{\mathrm{uG}i} \leq \Delta Q^{\max}_{\mathrm{uG}i} \\ 0 \leq \Delta \overline{Q}^k_{\mathrm{dG}i} \leq \Delta Q^{\max}_{\mathrm{dG}i} \\ 0 \leq \Delta \overline{Q}^k_{\mathrm{dW}i} \leq \Delta Q^{\max}_{\mathrm{dG}i} \end{cases} \tag{15}$$

$$\begin{cases} 0 \leq \Delta \overline{P}^k_{\mathrm{dD}i} \leq P^{\max}_{\mathrm{D}i} \\ 0 \leq \Delta \overline{Q}^k_{\mathrm{dD}i} \leq Q^{\max}_{\mathrm{D}i} \\ P^{\min}_{\mathrm{D}i} \leq \sum\limits_{m=1}^{k} \Delta \overline{P}^m_{\mathrm{dD}i} \leq P^{\max}_{\mathrm{D}i} \\ Q^{\min}_{\mathrm{D}i} \leq \sum\limits_{m=1}^{k} \Delta \overline{Q}^m_{\mathrm{dD}i} \leq Q^{\max}_{\mathrm{D}i} \end{cases} \tag{16}$$

where the superscript "~" represents random variables. The superscript "—" represents the expected value of random variables. $P_{\mathrm{W}i}$ and $Q_{\mathrm{W}i}$ respectively represent the active and reactive power of the wind power at the feed node i. $P_{\mathrm{G}i}$ and $Q_{\mathrm{G}i}$ are the active power and reactive power of the generator i, respectively. P_{Di} and Q_{Di} respectively represent the active load and reactive load of the node i. $\Delta P^k_{\mathrm{uG}i}$, $\Delta P^k_{\mathrm{dG}i}$, $\Delta Q^k_{\mathrm{uG}i}$, $\Delta Q^k_{\mathrm{dG}i}$, $\Delta P^k_{\mathrm{dD}i}$, $\Delta Q^k_{\mathrm{dD}i}$, $\Delta P^k_{\mathrm{dW}i}$ and $\Delta Q^k_{\mathrm{dW}i}$ are the active power up-regulation, active power down-regulation, reactive power up-regulation, reactive power down-regulation, active load reduction, reactive load reduction and curtailment volume of the adjustable generator in the k link of the cascading failure, respectively. G_{ij} and B_{ij} are the real and imaginary parts of the elements of the row i and column j of the node derivative matrix, respectively, indicating that the node j is connected to the node i. V_{ij} and θ_{ij} are the voltage magnitude and phase angle between the node i and node j, respectively. The superscripts max and min correspond to the upper and lower limits of each variable, respectively. Also, assuming that the power balance can always be satisfied during the control process, Equation (10) can be transformed to take the desired power balance equation [22]:

$$\begin{cases} (\overline{P}_{\mathrm{G}i} + \sum\limits_{m=1}^{k} \Delta \overline{P}^m_{\mathrm{uG}i} - \sum\limits_{m=1}^{k} \Delta \overline{P}^m_{\mathrm{dG}i}) + (\overline{P}_{\mathrm{W}i} - \sum\limits_{m=1}^{k} \Delta \overline{P}^m_{\mathrm{dW}i}) - \\ (\overline{P}_{\mathrm{D}i} - \sum\limits_{m=1}^{k} \Delta \overline{P}^m_{\mathrm{dD}i}) - \overline{V}_i \sum\limits_{j \in i} \overline{V}_j (G_{ij} \cos \overline{\theta}_{ij} + B_{ij} \sin \overline{\theta}_{ij}) = 0 \\ (\overline{Q}_{\mathrm{G}i} + \sum\limits_{m=1}^{k} \Delta \overline{Q}^m_{\mathrm{uG}i} - \sum\limits_{m=1}^{k} \Delta \overline{Q}^m_{\mathrm{dG}i}) + (\overline{Q}_{\mathrm{W}i} - \sum\limits_{m=1}^{k} \Delta \overline{Q}^m_{\mathrm{dW}i}) - \\ (\overline{Q}_{\mathrm{D}i} - \sum\limits_{m=1}^{k} \Delta \overline{Q}^m_{\mathrm{dD}i}) - \overline{V}_i \sum\limits_{j \in i} \overline{V}_j (G_{ij} \sin \overline{\theta}_{ij} - B_{ij} \cos \overline{\theta}_{ij}) = 0 \end{cases} \tag{17}$$

4.3. Chance Constraints

In the context of cascading failure blocking control, the active power of the transmission lines and the amplitude of node voltages are crucial output state variables. The uncertainty associated with wind power introduces a probability distribution for these state variables. However, imposing strict security constraints to ensure compliance can result in high control costs that may not align with the practical scenario. Therefore, an alternative approach is to express these constraints as opportunity constraints, allowing for a certain level of violation of the security constraints with a specified confidence level. Furthermore, it is imperative to implement essential measures for preventing and controlling failures of sensitive lines, particularly those greatly affected by adjustments in generator output. The limitations on setting output state variables are as follows:

$$\begin{cases} \Pr\{|P_k| \le \mu P_k^{\mathrm{r}}\} \ge \rho \; \forall k \notin S \\ \Pr\{|P_k| \le P_k^{\mathrm{r}}\} \ge \rho \quad \forall k \in S \end{cases} \tag{18}$$

$$\Pr\left\{V_i^{\min} \le V_i \le V_i^{\max}\right\} \ge \rho \tag{19}$$

where $\Pr\{*\}$ is the probability that the inequality constraint is established. ρ is the preset confidence level. P_k and V_i are the active power flow on the line k and the voltage amplitude of the node i, respectively. μ is the artificially set control coefficient, S is a sensitive line set.

The approach used in this paper to handle chance constraints is based on the deterministic transformation method proposed in the literature [22,23]. Equations (18) and (19) can be expressed uniformly as follows.

$$\begin{cases} \Pr\{x \le x_{\max}\} \ge \rho \\ \Pr\{x \ge x_{\min}\} \ge \rho \end{cases} \tag{20}$$

where $x_{\max}$ and $x_{\min}$ generalize the upper and lower limits of the variable taken. From the probabilistic power flow calculation, the probability distribution state of the output variable can be obtained, and then the probability density function and quantile function of the random variable $Z(\rho)$ can be fitted. Suppose that the quantile of the random variable x at the confidence level is ρ, then Equation (20) can be transformed into:

$$\begin{cases} Z(\rho) \le x_{\max} \\ Z(1-\rho) \ge x_{\min} \end{cases} \tag{21}$$

The chance constraint is consequently converted into a deterministic inequality constraint, and the probabilistic power flow calculation, which relies on the chance constraint, is resolved through the subsequent steps.

First, without considering the uncertainty, $\rho = 1$ is transformed into a deterministic optimal power flow problem to solve the optimal control scheme.

Subsequently, the control scheme is utilized to solve the probabilistic power flow, taking into account the uncertainty in wind power output. This analysis yields the probability distribution and quantile function of line power flow and node voltages.

Finally, the quantile of the output variable is compared with the boundary of the variable using Equation (21) at a specified confidence level for the chance constraint. The success of the solution is determined by the satisfaction of the constraint. However, in the event that the constraint is breached, the computational boundary of the variable is modified to rectify the deterministic optimal trend. The boundary adjustment strategy is denoted by Equation (22).

$$\begin{cases} x_{\max} = x_{\max} * \max\left\{1 - \frac{Z(\rho) - x_{\max}}{x_{\max}}, 1 - \alpha\right\} \\ x_{\min} = x_{\min} * \min\left\{1 + \frac{x_{\min} - Z(\rho)}{x_{\min}}, 1 + \alpha\right\} \end{cases} \tag{22}$$

where a is the adjustment parameter to prevent unreasonable adjustment of the upper and lower limits of the random variables; 5% is desirable, and if $Z(\rho) > x_{\max}$, then the upper limit $x_{\max}$ of the calculation is adjusted. If $Z(1-\rho) < x_{\min}$, the lower limit $x_{\min}$ of the calculation is adjusted.

Furthermore, Equation (18) demonstrates that altering the control coefficient has an impact on the permissible power flow limit. A higher value of the control coefficient indicates a lower level of control effort, which in turn increases the likelihood of a line failure. However, this reduction in control effort is accompanied by a corresponding decrease in control costs. The objective of implementing cascading failure blocking control is to mitigate the likelihood of cascading failure while minimizing the associated control expenses. During the initial phases of cascading failure, the system exhibits a robust resistance to interference, which allows for the possibility of adjusting the control coefficient of non-sensitive lines. This adjustment permits a slight increase in power flow beyond the limit, thereby reducing control costs. However, it is crucial to prevent the occurrence of cascading failure before the system collapses.

4.4. Model Solution

The development of the proposed cascading failure blocking control model is based on the probabilistic optimal power flow, which involves solving a nonlinear programming problem using AC power flow. The direct solution to this problem is computationally intensive, and ensuring convergence is challenging. Hence, this study employs the concept of linearizing the AC power flow in order to estimate the nonlinear component of the constraints through the utilization of the Taylor series expansion method [24]. This approach transforms the problem into a linear programming problem. The proposed model and the optimization program are constructed using MATLAB programming tools, and the CPLEX solver is employed for solving. The flowchart of the proposed method is shown in Figure 2.

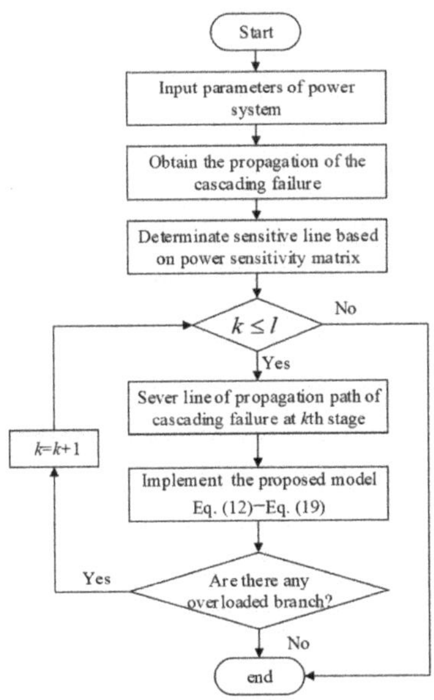

Figure 2. Flowchart of the proposed method.

5. Simulation and Results

In this paper, the IEEE 39-node system is used as a test arithmetic example; it contains 10 generators, 39 nodes and 46 lines with a total load of 6254.23 MW. The system section line diagram is shown in Figure 3. Among them, the generators on node 33 and node 37 are replaced by wind farms with installed capacities of 630 MW and 540 MW, respectively, where the total wind power penetration rate is about 20%. Assuming that the local wind speed approximately obeys the Weibull distribution and the wind farm active output is only related to the wind speed, the shape parameter and scale parameter of the Weibull distribution are 4 and 10.35, respectively, and the wind turbine models installed in the wind farm are WD5000 units, whose cut-in wind speed is 3 m/s, cut-out wind speed is 25 m/s, and rated wind speed is 11.3 m/s. α, β and γ will be taken as USD $1/MW, $10/MW and $100/MW, respectively. μ is 1.2 in the $l-1$ link of the cascading failure propagation path, and 1 in the l link. The generator adjustment parameters are shown in Table 1. The data, shown in Figure 3 and Table 1, come from the MATPOWER package in MATLAB.

Table 1. Generator adjustment parameters.

Generator No.	Initial Active Output (MW)	Upper/Lower Limit of Active Output (MW)	Initial Reactive Output (Mvar)	Upper/Lower Limit of Reactive Output (Mvar)
G1	250	1040/0	161.76	400/140
G2	677.87	976/0	221.57	300/−100
G3	650	725/0	206.97	300/150
WG4	630	630/0	108.29	250/0
G5	508	508/0	166.69	167/0
G6	650	687/0	210.66	300/−100
G7	560	580/0	100.17	240/0
WG8	540	540/0	0	250/0
G9	830	865/0	21.73	300/−150
G10	1000	1100/0	78.47	300/−100

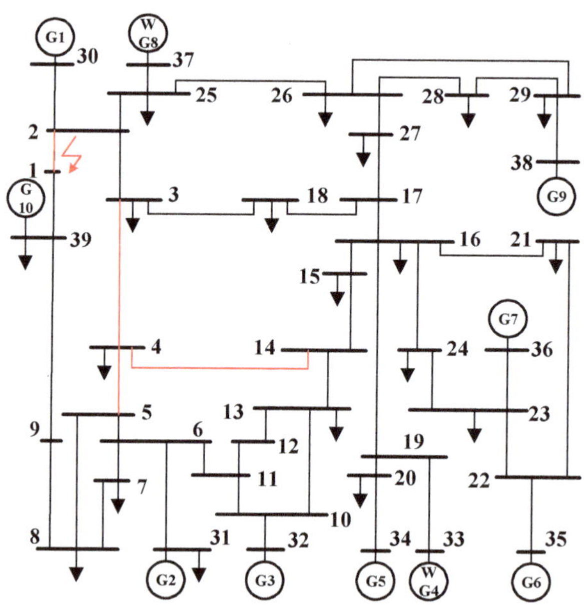

Figure 3. Diagram of IEEE 39 system.

5.1. Propagation Path of Cascading Failure

Line 1–2 disconnection is considered as a random initial fault, as shown by the red arrow in Figure 3. The first step is to utilize probability power flow to determine the probability distribution of power flow in each branch after the failure. Figure 4 illustrates the probability distribution curve of power flow through Line 4–5 after Line 1–2 is disconnected. Subsequently, Equation (2) is employed to calculate the probability of overload in other lines and the probability of failure. The largest probability of outage for each link is then selected to generate a cascading failure path, as presented in Table 2. This path is used as an example for blocking control.

Table 2. Propagation path of cascading failure without blocking control.

Failure Links	Failure Lines	Failure Probability
1	L1–2	1
2	L4–5	0.37
3	L4–14	0.66
4	L3–4	0.8

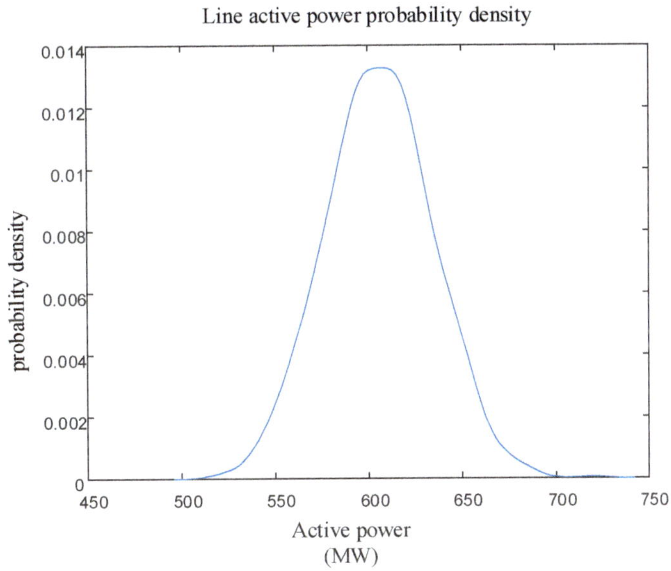

Figure 4. Probability distribution of power flow on line 4–14.

As evident in Table 2, Line 1–2 can be interpreted as representing the initial fault with a probability of failure set at 1. After the disconnection of Line 1–2, a significant shift in tides occurs in a large area, resulting in successive trips of Line 4–5 and Line 4–14 due to overload. Until the disconnection of Line 3–4 occurs, leading to system decoupling, there is a resultant load loss of approximately 500 MW.

5.2. Sensitive Line Set

According to power sensitivity analysis, it is possible to identify the lines that exhibit higher sensitivity to changes in generator output, as illustrated in Figure 5.

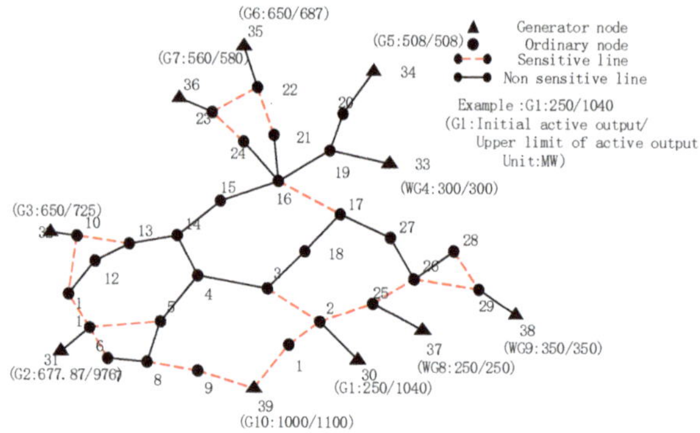

Figure 5. Sensitive lines of IEEE 39 system.

According to the findings presented in Figure 5, it is evident that making significant adjustments to the generator output results in notable fluctuations in power flow. However, these fluctuations primarily affect a limited number of sensitive transmission lines located in close proximity to the generator outlet. Consequently, these lines are more susceptible to experiencing overload conditions. After the generator G3 at node 32 experiences a substantial increase in its output, it has the potential to result in overload failure in the sensitive Line1–10 and Line10–13. Considering the varying adjustable capacities of each generator, it is expected that generators with smaller adjustable capacities, such as G7, will not cause significant fluctuations in the power flow of the corresponding sensitive line. Hence, to minimize control expenses and enhance solution efficiency, this study exclusively concentrates on the prevention and management of sensitive line failures associated with higher adjustable capacity and wind turbines.

5.3. Control Effect Analysis

Taking the propagation path of cascading failure in Section 5.1 as an example, the control model established in this article is used for blocking control:

Model 1: without considering multi-stage blocking control of sensitive lines.

Model 2: considering multi-stage blocking control of sensitive lines.

The control effect and cost are shown in Figures 6 and 7, respectively.

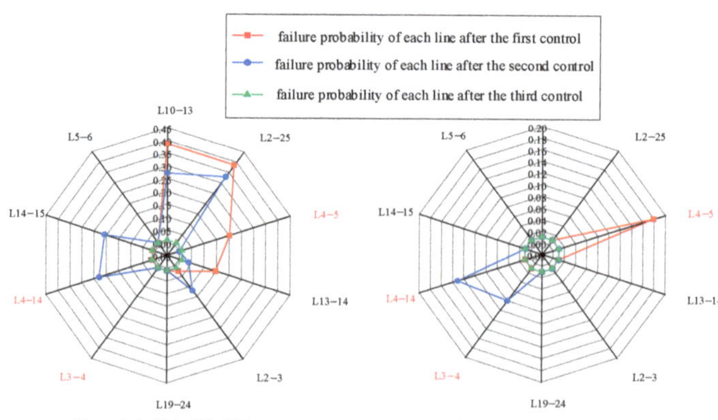

Figure 6. Control effect of Model 1 and Model 2.

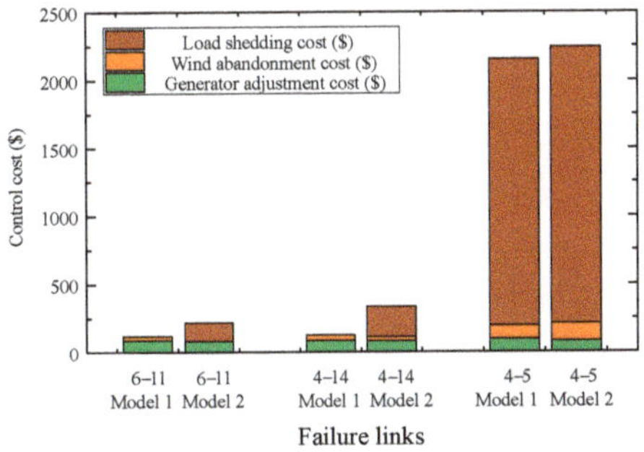

Figure 7. Control cost of Model 1 and Model 2.

From the analysis of Figure 6, it is evident that both models exhibit a significant reduction in the probability of failure at each stage when compared to the failure link without blocking control, as indicated in Table 2. However, Model 1 does not take into account the impact of blocking control measures. Upon implementing control measures for Line 1–2, the failure probability of the subsequent Line 4–5 decreased from 0.39 to 0.20. Nevertheless, as a result of the unpredictability of wind power generation and its consequential effects, a significant number of instances of active power shortages were observed subsequent to the occurrence of failures. The likelihood of failure in Line 2–25, located near the wind farm, and Line 10–13, situated near the generator with significant output adjustment, is greater compared to that of Line 4–5. Therefore, the subsequent propagation of cascading failure may deviate from the anticipated trajectory, resulting in the failure of the initial control scheme. Similarly, the occurrence of new line failures may also arise with the implementation of subsequent control measures. Model 2 effectively mitigates the deviation of the propagation path of cascading failure from the predicted outcomes through the proactive implementation of preventive control measures on vulnerable transmission lines. This ensures the effectiveness of control measures for each individual link.

From Figure 7, it is evident that during the initial phase of cascading failure, the implemented control measures involve modifying the generator output and implementing a limited amount of wind power abandonment. These measures aim to decrease the likelihood of subsequent failures. When the cascading failure extends to the point where Line 4–14 becomes disconnected, the two subnets within the system are only linked by Line 3–4. In order to avoid system disconnection, it is no longer feasible to restore the system to a safe operating state by solely making adjustments to conventional units and abandoning wind power. Therefore, the implementation of load shedding is necessary in order to regulate the power flow within the entire network line, ensuring that it remains below the rated value. In contrast to Model 1, Model 2 requires a slight reduction in load during the initial stages of cascading failure when generator adjustment is limited. This is necessary to maintain the power flow constraints of vulnerable transmission lines and ensure that the propagation path of cascading failure remains unchanged under limited generator adjustment conditions. Therefore, the initial control cost of Model 2 is comparatively high.

If multi-stage blocking control measures are not implemented, the enforcement of stringent control measures will be necessary during the initial stages of cascading failure. Specifically, the control coefficient will be consistently set to 1, while the other parameters will remain unaltered, as denoted by Model 3. The cost and effectiveness of control measures in Model 3 are outlined in Table 3.

Table 3. Model 3 control cost and control effectiveness.

Failure Links	Generator Adjustment Cost ($)	Wind Abandonment Cost ($)	Load Shedding Cost ($)	Failure Probability of Lines
1 (L1–2)	620.13	596.80	34,572.36	0.01

Based on the findings outlined in Table 3, it is apparent that disconnecting Line 1–2 leads to the successful mitigation of all potential line failures and greatly diminishes the probability of cascading failure. The achievement of this outcome is facilitated through the implementation of the single-stage control in model 3. Despite its advantages, the system requires a substantial amount of load shedding and adjustment of generator output, resulting in increased control costs and impeding the economic efficiency of the system.

5.4. *Influence of Wind Power on Blocking Control*

Then, the impact of wind power penetration rates or confidence levels on control measures is analyzed. Under the condition of maintaining a confidence level of 90% and keeping other parameter settings unchanged, the wind power penetration rate was increased to 30% by replacing generator G5 with a wind farm having an installed capacity of 500 MW. Subsequently, generator G9 was replaced by a wind farm with an installed capacity of 830 MW to achieve a wind power penetration rate of 40%. The cascading failure was successfully blocked and controlled under both operating conditions. The resulting control cost is presented in Figure 8.

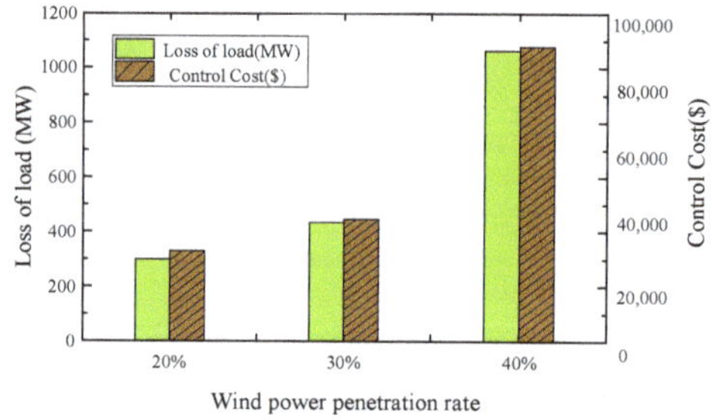

Figure 8. Control costs under different wind power penetration rates.

From the analysis presented in Figure 8, it is evident that the augmentation of wind power penetration leads to an exponential rise in control costs. When wind farms constitute a larger proportion of the power system and blocking control is implemented, there is a need to compensate for the active power deficit caused by insufficient wind power output. However, this compensation is constrained by the upper limit of the adjustable power generation capacity of conventional units. Consequently, significant load reduction is required to ensure the safety and stability of the system.

Furthermore, when applying the deterministic control method, it is necessary to ensure that the system state variables strictly adhere to the imposed constraints during the control process. This implies that the confidence levels of the variable constraints should all be 100%. However, this approach may lead to the issue of increased control costs. In this paper, the confidence levels of 90%, 95%, and 100% are employed for the comparison test. The remaining parameters are kept consistent with Model 1. The resulting control cost is presented in Figure 9.

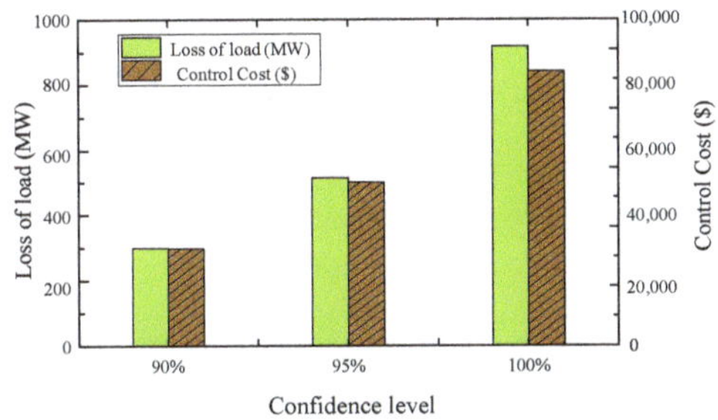

Figure 9. Control costs under different confidence levels.

From the analysis of Figure 9, it is evident that the control scheme is influenced by the uncertainty of wind power output. The confidence level plays a significant role in determining the level of conservatism in the control scheme. As the confidence level increases, the control scheme becomes more conservative. Consequently, there is an increase in both the amount of load shedding and control cost. Without the inclusion of a chance constraint, the control scheme will lead to a substantial amount of load shedding, even when the confidence level is set at 100%. However, the confidence level does not show a significant improvement in this scenario. Hence, the control model presented in this paper compromises a certain level of confidence in order to develop a scheme that may not fully satisfy the constraint with a low probability but significantly minimizes the control cost.

In conclusion, as the scale of wind power increases, it is important to consider the impact of uncertain wind power output and the limited adjustable power generation capacity of conventional units. In the event of a cascading failure in the system, various load shedding methods will be implemented to ensure the system operates within safe operating state constraints. However, these measures also result in increased control costs.

6. Conclusions

This article presents a novel approach to constructing a propagation path prediction model for cascading failure in power systems with wind power integration. The model is based on probabilistic power flow analysis. Based on the anticipated outcome, an analysis is conducted on the importance of incorporating sensitive lines into blocking control. Subsequently, a multi-stage blocking control model that takes into account sensitive lines is developed using probability optimal power flow. Through conducting simulation analysis on the IEEE 39-node system, several conclusions can be derived.

(1) The obtained blocking control measures can effectively reduce the risks of cascading failure. Sensitivity analysis is employed to reduce the solving dimension of the model by identifying the lines that exert a substantial influence on the control measures.
(2) The proposed control method can offer a viable solution for mitigating the impact of wind power while simultaneously minimizing control costs. This approach enables the development of a control scheme that effectively balances the objectives of safety and economy.
(3) The escalation of wind power penetration rate and confidence level will result in an increase in the expenses associated with cascading failure blocking control.

This paper exclusively focuses on the impact of cascading failure caused by overload, which is a steady-state problem. However, the outage of transmission lines may trigger transient issues, including voltage disturbances and frequency disturbances. These can lead to power system protection actions, including the disconnection of wind turbines from the grid. Furthermore, this paper does not address the subsequent effects of these disturbances on the propagation of cascading failures after the disconnection. Therefore, future work should focus on the transient cascading effects.

Author Contributions: L.C. and Y.W. formulated the conceptual framework presented in this work. L.C. authored the manuscript with the support of Y.W. and T.W. Z.G. and N.J. executed the simulations and conducted the data analysis. T.W. provided supervision for the project. All authors contributed critical insights, offered constructive feedback, and played integral roles in refining the research, analysis, and manuscript. All authors have read and agreed to the published version of the manuscript.

Funding: This research received no external funding.

Data Availability Statement: The data presented in this study are available on request from the corresponding author. The data are not publicly available due to commercial sensitivity.

Conflicts of Interest: Authors Lun Cheng, Zeming Gao and Ning Ji were employed by the company State Grid Hebei Electric Co., Ltd. The remaining authors declare that the research was conducted in the absence of any commercial or financial relationships that could be construed as a potential conflict of interest.

References

1. Choi, E.; Song, J. Cost-effective retrofits of power grids based on critical cascading failure scenarios identified by multi-group non-dominated sorting genetic algorithm. *Int. J. Disaster Risk Reduct.* **2020**, *49*, 101640. [CrossRef]
2. Busby, J.W.; Baker, K.; Bazilian, M.D.; Gilbert, A.Q.; Grubert, E.; Rai, V.; Rhodes, J.D.; Shidore, S.; Smith, C.A.; Webber, M.E. Cascading risks: Understanding the 2021 winter blackout in Texas. *Energy Res. Soc. Sci.* **2021**, *77*, 102106. [CrossRef]
3. Hines, P.; Apt, J.; Talukdar, S. Large blackouts in North America: Historical trends and policy implications. *Energy Policy* **2009**, *37*, 5249–5259. [CrossRef]
4. Athari, M.H.; Wang, Z. Impacts of Wind Power Uncertainty on Grid Vulnerability to Cascading Overload Failures. *IEEE Trans. Sustain. Energy* **2018**, *9*, 128–137. [CrossRef]
5. Yin, Y.K.; Chen, H.H.; Meng, X.P. Digital twin-driven identification of fault situation in distribution networks connected to distributed wind power. *Int. J. Electr. Power Energy Syst.* **2024**, *155*, 109415. [CrossRef]
6. Kendal, W.S. Self-organized criticality attributed to a central limit-like convergence effect. *Phys. A Stat. Mech. Its Appl.* **2015**, *421*, 141–150. [CrossRef]
7. Zhu, Z.Q.; Wu, L.J.; Xia, Z.P. An accurate subdomain model for magnetic field computation in slotted surface- mounted permanent-magnet machines. *IEEE Trans. Magn.* **2010**, *46*, 1100–1115. [CrossRef]
8. Zhou, B.; Meng, X.; Stanley, H.E. Power-law distribution of degree-degree distance: A better representation of the scale-free property of complex networks. *Proc. Natl. Acad. Sci. USA* **2020**, *117*, 14812–14818. [CrossRef] [PubMed]
9. Ju, Z.C.; Ma, J.L.; Xie, J.J.; Wang, Y.P.; Cui, H.M.; Duan, C.W. Cascading failure model for the mitigating edge failure of scale-free networks. *Pramana J. Phys.* **2019**, *92*, 62. [CrossRef]
10. Monteiro Pereira, R.M.; Machado Ferreira, C.M.; Maciel Barbosa, F.P. Influence of load shedding in the voltage stability of an electric power system using trajectory sensitivity analysis. In Proceedings of the 45th International Universities Power Engineering Conference, Cardiff, UK, 31 August–3 September 2010; pp. 1–6.
11. Dvorkin, Y.; Henneaux, P.; Kirschen, D.S.; Pandzic, H. Optimizing Primary response in preventive security-constrained optimal power flow. *IEEE Syst. J.* **2018**, *12*, 414–423. [CrossRef]
12. Zhai, C.; Zhang, H.; Xiao, G.; Pan, T.-C. A model predictive approach to protect power systems against cascading blackouts. *Int. J. Electr. Power Energy Syst.* **2019**, *113*, 310–321. [CrossRef]
13. Yingying, W.; Feng, L.; Jun, X. Prevention control of blackouts in power system based on fault chain and risk theory. In Proceedings of the 2014 International Conference on Power System Technology (POWERCON), Chengdu, China, 20–22 October 2014; IEEE: Piscataway, NJ, USA, 2014.
14. Yao, R.; Zhang, X.; Huang, S.; Mei, S.; Zhang, Z.; Li, X.; Zhu, Q. Cascading outage preventive control for large-scale AC-DC interconnected power grid. In Proceedings of the IEEE Power and Energy Society General Meeting, National Harbor, MA, USA, 27–31 July 2014; pp. 1–5.
15. Usaola, J. Probabilistic load flow with correlated wind power injections. *Electr. Power Syst. Res.* **2010**, *80*, 528–536. [CrossRef]
16. Zhang, P.; Lee, S.T. Probabilistic load flow computation using the method of combined cumulants and Gram-Charlie expansion. *IEEE Trans. Power Syst.* **2004**, *19*, 676–682. [CrossRef]
17. Yan, X.W.; Xu, Y.; Li, R.J. Multi-Time scale reactive power optimization of distribution grid based on model predictive control and including RDG regulation. *Trans. China Electrotech. Soc.* **2019**, *34*, 2022–2037.
18. Lu, Y.; Wang, Y.; Yong, P.; Zhang, N.; Kang, C.; Lu, D. Fast Power System Cascading Failure Path Searching with High Wind Power Penetration. *IEEE Trans. Sustain. Energy* **2020**, *11*, 2274–2283. [CrossRef]
19. Kirschen, D.; Allan, R.; Strbac, G. Contributions of individual generators to loads and flow. *IEEE Trans. Power Syst.* **1997**, *12*, 52–60. [CrossRef] [PubMed]
20. Zima, M.; Andersson, G. On security criteria in power systems operation. In Proceedings of the IEEE Power Engineering Society General Meeting, San Francisco, CA, USA, 16 June 2005; pp. 3089–3093.

21. Carreras, B.A.; Lynch, V.E.; Sachtjen, M.L.; Dobson, I.; Newman, D.E. Modeling blackout dynamics in power transmission networks with simple structure. In Proceedings of the 34th IEEE Hawaii International Conference on System Sciences, Maui, HI, USA, 3–6 January 2001; pp. 719–727.
22. Yu, H.; Chung, C.Y.; Wong, K.P.; Zhang, J.H. A chance constrained transmission network expansion planning method with consideration of load and wind farm uncertainties. *IEEE Trans. Power Syst.* **2009**, *24*, 1568–1576. [CrossRef]
23. Zou, B.; Xiao, Q. Solving probabilistic optimal power flow problem using quasi-Monte Carlo method and ninth-order polynomial normal transformation. *IEEE Trans. Power Syst.* **2014**, *29*, 300–307. [CrossRef]
24. Li, G.; Zhang, X.P. Stochastic optimal power flow approach considering correlated probabilistic load and wind farm generation. In Proceedings of the IET Conference on Reliability of Transmission and Distribution Networks, London, UK, 22–24 November 2011; pp. 1–7.

energies

Article

A Layered Parallel Equaliser Based on Flyback Transformer Multiplexed for Lithium-Ion Battery System

Hongrui Liu *, Xiangyang Wei, Junjie Ai and Xudong Yang

Faculty of Electrical Engineering, Kunming University of Science and Technology, Kunming 650500, China; 20212205024@stu.kust.edu.cn (X.W.); 201910501203@stu.kust.edu.cn (J.A.); 20212205061@stu.kust.edu.cn (X.Y.)
* Correspondence: lhr168@kust.edu.cn; Tel.: +86-156-8710-2699

Abstract: An effective equaliser is crucial for eliminating inconsistencies in the connected serial batteries and extending the life of the battery system. The current equalisers generally have the problems of low equalisation efficiency, slow equalisation speed, and complex switching control. A layered parallel equaliser based on a flyback transformer multiplexed for a lithium-ion battery system is proposed. The equaliser employs both hierarchical and parallel equalisation techniques, allowing for simultaneous processing of multiple objectives. This enhances both the efficiency and speed of the equalisation process. The efficiency of equalisation can be further improved by implementing PWM control with deadband complement. Additionally, the flyback transformer serves as an energy storage component for both layers of the equalisation module, resulting in a significant reduction in the size and cost of the equaliser. The circuit topology of the equaliser is presented, and its operational principle, switching control, and equalisation control strategy are analysed in detail. Finally, an experimental platform consisting of six lithium-ion batteries is constructed, and equalisation experiments are conducted to verify the advantages of the proposed equaliser in terms of equalisation speed, efficiency, and cost.

Keywords: equaliser; multiplexed flyback transformer; layered parallel equalisation; ithium-ion battery

Citation: Liu, H.; Wei, X.; Ai, J.; Yang, X. A Layered Parallel Equaliser Based on Flyback Transformer Multiplexed for Lithium-Ion Battery System. *Energies* **2024**, *17*, 754. https://doi.org/10.3390/en17030754

Academic Editor: Carlos Miguel Costa

Received: 16 December 2023
Revised: 18 January 2024
Accepted: 31 January 2024
Published: 5 February 2024

1. Introduction

Lithium-ion batteries are commonly used in grid energy storage and electric vehicles due to their high energy density, lack of memory effect, and long cycle life [1]. However, individual lithium-ion batteries have a low nominal voltage that does not meet the requirements for applications such as electric vehicles and energy storage. To achieve the necessary power levels, a large number of individual batteries must be connected in series and parallel [2]. However, lithium-ion batteries inevitably vary in terms of individual capacity and internal resistance during the manufacturing process, leading to inconsistency between batteries. This inconsistency gradually increases during the use cycle, making the battery pack prone to a single overcharge or overdischarge, which seriously affects the overall performance of the battery system. Thus, it is imperative to address the issue of inconsistency among battery system monomers. Battery equalisation technology proves to be an effective solution [3].

There are two main types of equalisation methods: passive equalisation and active equalisation [4,5]. The passive consumes the excess energy through a parallel resistor. Despite its simplicity, the passive equalisation structure results in high energy loss and slow equalisation speed [6].

The active equalisation employs an inductor, capacitor, and transformer as the energy storage elements to transfer energy between batteries. Capacitor-based equalisation is a method of transferring battery energy using capacitance. It controls the on-off of switching devices to transfer energy between adjacent batteries. This method avoids the need for direct electrical connections between batteries [7–9]. Inductor-based equalisation is a method of transferring battery energy using inductance, allowing for highly controllable equalisation currents. However, the process is time-consuming and can negatively impact efficiency [10–14]. LC-based equalisation utilises an inductor (L) and capacitor (C) to form a resonant tank for achieving energy equalisation. However, this method can only work at a specific frequency and duty cycle range, leading to a more complex design and control of LC-based equalisation circuits [15–19].

According to the type of the transformer, transformer-based equalisation can be divided into four types: forward transformer-based equalisation [20], flyback transformer-based equalisation [21,22], forward-flyback transformer-based equalisation [23–25], and multi-winding transformer-based equalisation [16,26–29]. Shang et al. [20] proposed a forward transformer equaliser that uses the voltage difference to automatically transfer energy between batteries. However, as the voltage difference decreases, equalisation becomes more difficult, resulting in poor equalisation results. Guo et al. [21] proposed a flyback transformer equaliser, which has the advantages of easy electrical isolation and simple control. However, the equalisation speed decreases, and the cost and volume increase as the number of series-connected batteries increases. Shang et al. [23] proposed a forward-flyback transformer-based equaliser that reduces the voltage difference between batteries by combining the characteristics of forward and flyback transformers. The equalisation speed is improved because the forward transformer is used before the flyback transformer. However, the equalisation effect of the forward transformer is poor when the voltage difference is small. Liu et al. [26] proposed a multi-winding transformer-based equaliser using forward conversion, which is easy to control, but the equalisation speed is slow, and the equalisation effect is poor when the voltage difference is low.

This paper proposes a layered parallel equaliser based on a flyback transformer multiplexed for a lithium-ion battery system to address the design issues mentioned above.

(1) The equaliser uses a layered equalisation mode to double the input voltage and improve equalisation efficiency. Additionally, the use of complementary PWM control with a dead band reduces switching loss, further improving equalisation efficiency.
(2) The parallel equalisation mode is utilised to enable simultaneous charging or discharging of multiple batteries or battery units, resulting in a significant improvement in equalisation speed.
(3) The equaliser employs flyback transformers as multiplexed energy storage elements, reducing both cost and size. The primary winding of the flyback transformer acts as an inductor in the first-layer equalisation, facilitating energy transfer between the two batteries. In the second-layer equalisation, the flyback transformer functions as an energy storage element, facilitating energy transfer between battery units through an energy transfer unit.
(4) The equaliser has a modular design, allowing for the simple addition of equalisation modules as the number of batteries connected in series increases while maintaining the parameters within the equalisation module.

The paper is structured into six sections. Section 2 provides a detailed explanation of the layered equalisation working principle and MOSFET control. Section 3 introduces the equalisation control strategy. Section 4 presents the experimental platform parameters, procedure, and results. Section 5 presents a comparative analysis of the proposed equaliser with various types of equalisers. Section 6 provides a general description of the main features of the proposed equaliser and identifies areas for improvement. Finally, Section 7 analyses the conclusion.

2. Proposed Equaliser

2.1. Structure of the Proposed Equaliser

The architecture of the proposed equaliser is illustrated in Figure 1 and consists of two equalisation modules: the first-layer equalisation module (FMi) and the second-layer equalisation module (SMi). The battery pack is divided into a single battery (Bi) and a battery unit (BUi) according to the equalisation process. The first-layer equalisation module primarily achieves energy equalisation between two single batteries within the battery unit. The second-layer equalisation module aims to achieve energy balance between the battery units. Once the first-layer equalisation is complete, the second-layer equalisation begins.

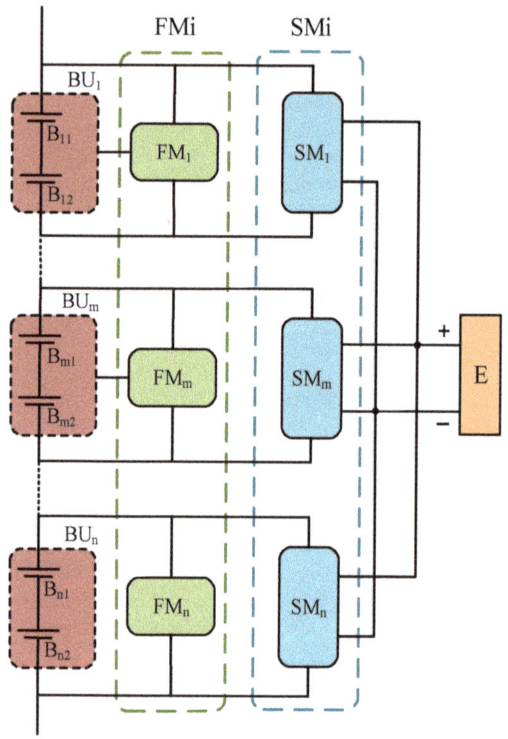

Figure 1. The architecture of the proposed equaliser.

The circuit topology of the proposed equaliser is shown in Figure 2, which consists of a battery system of n single batteries connected in series, n/2 flyback transformers, 3n MOSFETs, and an energy transfer unit E. Six MOSFET switches with anti-parallel diodes (Mi1 to Mi6) and a flyback transformer Ti form an equalisation module. Each battery unit BUi consists of two single batteries, which correspond to an equalisation module. The energy transit unit E is composed of two single batteries connected in series.

2.2. Operating Principle of the Proposed Equaliser

2.2.1. Operating Principle of the First-Layer Equalisation

To address the energy equalisation issue between two single batteries in the battery unit, the primary winding Li1 of the transformer is utilised as the energy storage component. The first-layer equalisation works in DCM mode in order to avoid the situation of magnetic saturation of the inductor. The corresponding switches adopt complementary PWM with a headband in order to further improve equalisation efficiency.

SOCBij represents the state of charge of the battery Bij. Taking batteries B11 and B12 as an example, assuming SOCB11 > SOCB12. Figure 3 shows the key waveforms of the first-layer equalisation. The first-layer equalisation stage can be divided into four steps.

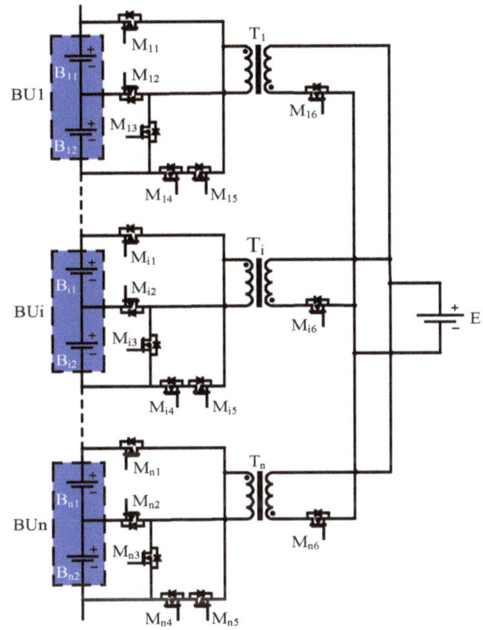

Figure 2. The circuit topology of the proposed equaliser.

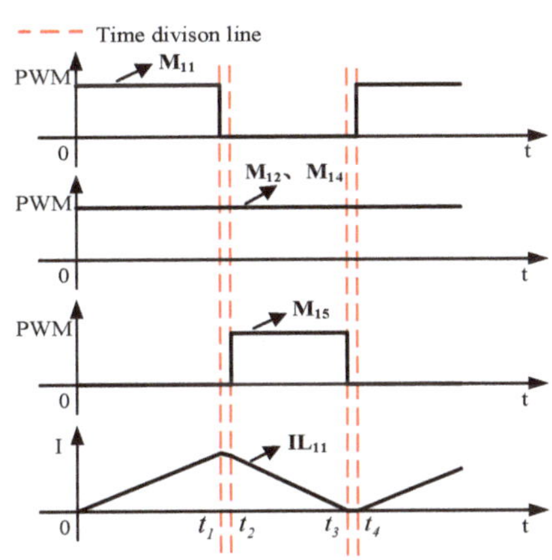

Figure 3. The key waveforms in the first-layer equalisation.

State I [$0-t_1$]: As shown in Figure 4a, M11 is turned on, M12 and M14 and M14 are stayed on, and the energy of B11 is transferred to the primary winding L11 of transformer T1. In this state, the current of the inductor L11 can be expressed as

$$i_{L11} = \frac{V_{B11} - 2V_{mos}}{L_{11}}t \tag{1}$$

where i_{L11} and L_{11} are the current and the inductance of L11, respectively. V_{B11} is the input voltage, V_{mos} is the conduction voltage drop of MOSFET.

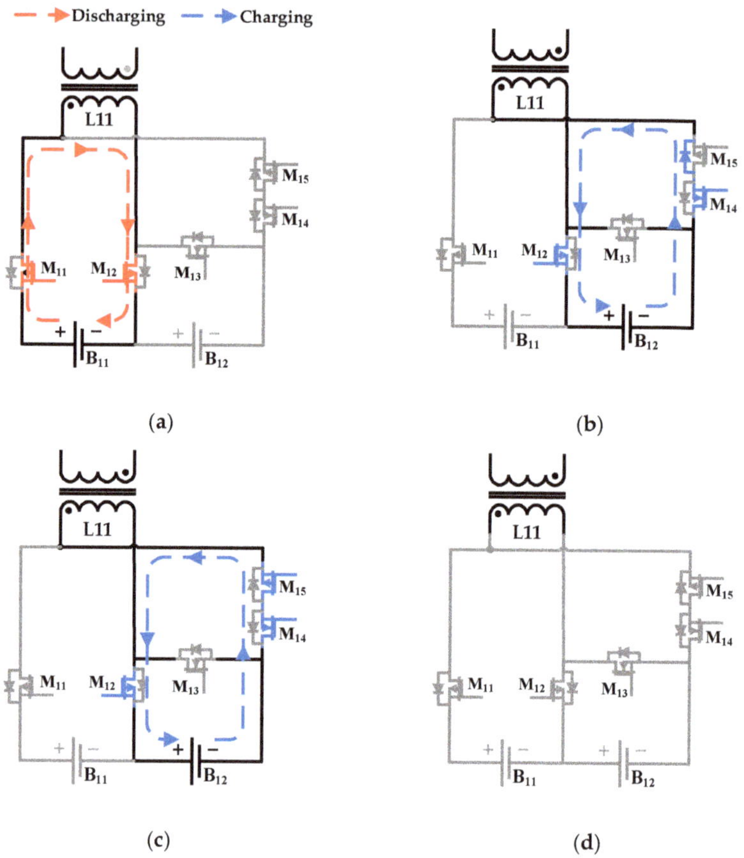

Figure 4. Operational states of the first−layer equalisation. (**a**) State I. (**b**) State II. (**c**) State III. (**d**) State IV.

State II [t_1-t_2]: As shown in Figure 4b, M11 is turned off, and then the energy in L11 flows to the battery B12 through the reverse parallel diode of M15. In this state, the current of L11 can be expressed as

$$i_{L11} = i_{L11}(t_1) - \frac{V_{out}+2V_{mos} + V_D}{L_{11}}(t - t_1) \quad (2)$$

where t_1 is the turn-on time of M11, t_1 is the output voltage, V_D is the forward voltage of the reverse parallel diodes of MOSFETs.

State III [t_2-t_3]: As shown in Figure 4c, M15 is turned on, and then the energy of L11 is transferred to battery B12. In this state, the current of L11 can be expressed as

$$i_{L11} = i_{L11}(t_2) - \frac{V_{out}+3V_{mos}}{L_{11}}(t - t_2) \quad (3)$$

State IV [t_3-t_4]: As shown in Figure 4d, M11, M13, M15 are turned off, and the current of L11 is zero. This indicates that the first-layer equalisation has been completed.

Considering the conduction losses of diodes and MOSFETs, the efficiency of the first-layer equalisation is:

$$\eta_1 = 1 - \frac{2\int_0^{t_1} R_{ds} i_{L11}{}^2 dt + \int_{t_1}^{t_2} i_{B12}(V_D + 2R_{ds} i_{L11})dt + 3\int_{t_2}^{t_3} R_{ds} i_{L11}{}^2 dt}{\int_0^{t_1} V_{B11} i_{L11} dt} \quad (4)$$

where R_{ds} is the on-resistance of MOSFETs, i_{L11} is the RMS current of L11.

2.2.2. Operating Principle of the Second-Layer Equalisation

The second-layer equalisation utilises the flyback transformer to transfer energy between different battery units via the energy transfer unit E. This process ultimately achieves energy equalisation among all battery units within the entire battery system. The equalisation process can be carried out simultaneously on multiple battery units during discharging or charging. High-energy battery units transfer energy to the energy transfer unit E during discharging, and the energy of E is transferred to low-energy battery units simultaneously. Additionally, the equalisation speed increases with the number of parallel equalisation targets.

To determine which battery unit is discharged or charged in the second-layer equalisation, it is necessary to compare the average value of all battery units, SOCBUi, with the average value of the SOC of the battery pack, SOCavg. As the first-layer equalisation is complete, the SOC values of B11 and B12 are equal, the SOC values of B21 and B22 are equal, and the SOC values of B31 and B32 are equal, respectively: SOCBU1 = SOCB11 = SOCB12, SOCBU2 = SOCB21 = SOCB22, and SOCBU3 = SOCB31 = SOCB32. For instance, let us consider the battery units BU1, BU2, and BU3. It is assumed that SOCBU1 > SOCBU2 > SOCavg > SOCBU3. The equalisation process can be divided into two states.

State I: As shown in Figure 5a, M11 and M21 are turned on, while M13 and M23 remain on. L11 absorbs and stores the energy released by battery units BU1 and BU2. The current on L11 can be expressed as

$$i_{L11} = \frac{V_{BU1} - 2V_{mos}}{L_{11}} t \quad (5)$$

where V_{BU1} is the voltage of BU1 and L_{11} is the inductance of the primary winding of T1.

The current i_{L21} on the primary winding L21 of T2 is as stated above.

When M11 and M21 are turned off, the energy induced by the secondary winding flows to E through the anti-parallel diodes of switches M16 and M26. At this stage, the current on the secondary winding L12 of T1 can be expressed as

$$i_E = \frac{V_{BU1} - 2V_{mos}}{L_{11}} DT + \frac{V_{BU2} - 2V_{mos}}{L_{21}} DT - \frac{V_{out} + V_D}{L_{12}} (t - DT) - \frac{V_{out} + V_D}{L_{22}} (t - DT) \quad (6)$$

where L_{12} and L_{22} are the inductance of the secondary winding of T1 and T2, respectively. V_{out} is the out voltage, D is the duty cycle of MOSFETs, T is the cycle-time.

State II: As shown in Figure 5b, M36 is turned on, and then E releases energy to the secondary winding L32 of T3. The current of E can be expressed as

$$i_E = \frac{V_E - V_{mos}}{L_{32}} t \quad (7)$$

where V_E is the voltage of E.

When M36 is turned off, M13 remained on. The energy stored in L31 is transferred to the battery unit BU3 through the anti-parallel diode of M31. The current of E can be expressed as

$$i_{L31} = \frac{V_E - V_{mos}}{L_{32}} DT - \frac{V_{out} + V_{mos} + V_D}{L_{31}} (t - DT) \quad (8)$$

Using the BU1 discharging as an example and considering the conduction losses of diodes and MOSFETs, the efficiency η_2 of the second-layer equalisation is:

$$\eta_2 = 1 - \frac{2\int_0^{DT} R_{ds} i_{L11}{}^2 dt + \int_{DT}^{(D+\alpha)T} V_D i_{L12} dt}{\int_0^{DT} V_{BU1} i_{L11} dt} \quad (9)$$

where α is the duty cycle required to release energy from the secondary winding L12 of T1.

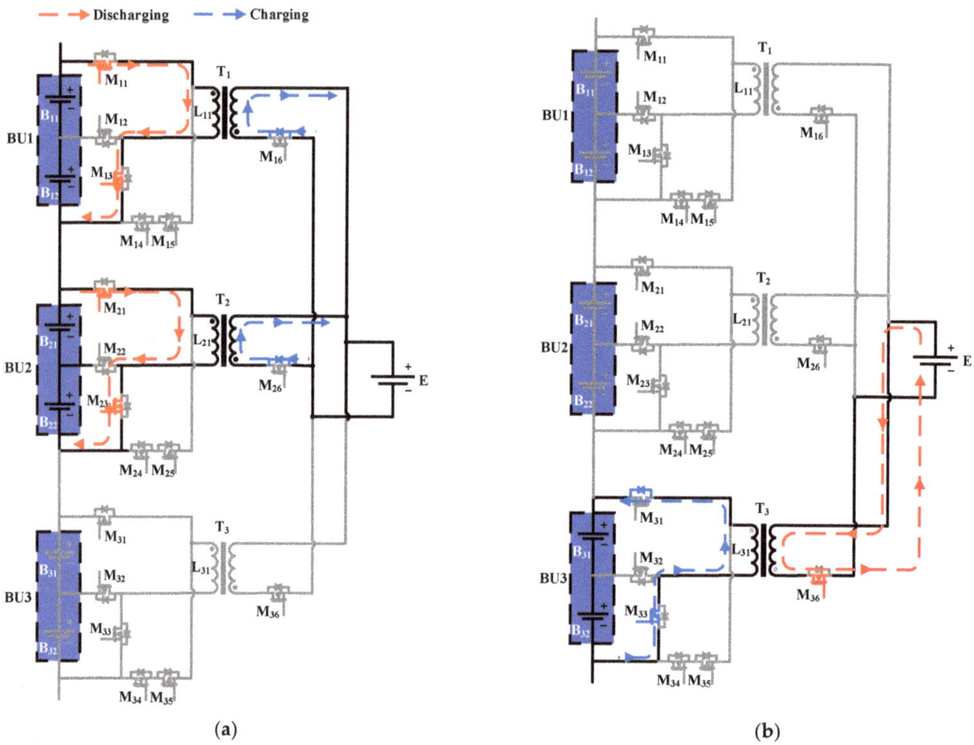

Figure 5. Operational states of the second-layer equalisation. (**a**) State I. (**b**) State II.

3. Equalisation Strategy Design

Figure 6 shows the equalisation control strategy of a layered parallel equaliser based on a flyback transformer multiplexed for a lithium-ion battery system. The control procedure for MOSFET equalisation is divided into two sequential layers. Before beginning the first-layer equalisation, it is necessary to estimate the quiescent state of charge (SOC) for all batteries.

The first-layer equalisation occurs when the difference in SOC between two batteries within the battery units exceeds the threshold value ε. The battery with the higher SOC value discharges, while the battery with the lower SOC value charges. The second-layer equalisation can only occur after all battery units have undergone internal equalisation. If any battery units remain unequalised, the first-layer equalisation will be repeated. The second-layer equalisation begins when the average SOC value SOCBUi of each battery unit differs from the average SOC value SOCavg of the battery pack by ε. Battery units larger than SOCavg discharge to the energy transfer unit E, while battery units smaller than SOCavg absorb energy from E. When the difference between the average SOC value SOCBUi of all the battery units and the average SOC value SOCavg of the battery pack is within the threshold value ε, the equalisation process ends.

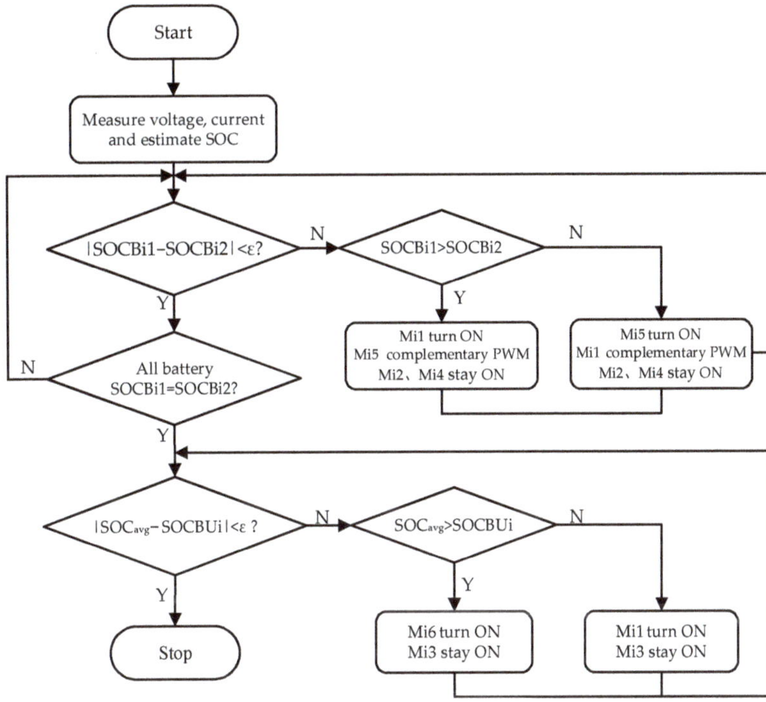

Figure 6. The equalisation strategy of the proposed equaliser.

4. Equalisation Experiment

4.1. Experimental Platform and Parameters

To demonstrate the effectiveness of this equaliser, an equalisation experimental platform consisting of six series-connected batteries was built, as shown in Figure 7. The rated voltage of the battery is 3.6 V, and the rated capacity is 2600 mAh. The equalisation experimental platform mainly consists of Labview, Battery String, Energy Transfer Battery, Measurement Module, Isolated Drive Supplies, Equaliser, Controller, DC Power Source and Oscilloscope. Table 1 shows the important parameters of the equalisation circuit.

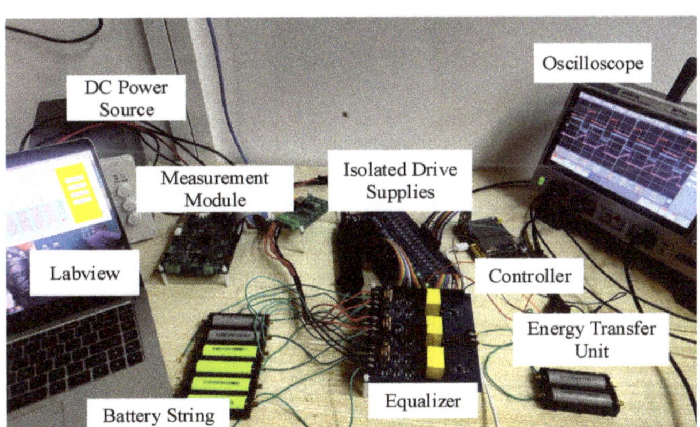

Figure 7. Experiment platform.

Table 1. Experimental parameters.

Parameters		Value
Battery	Model Nominal capacity Nominal voltage	LS18650-10A 2600 mAh 3.6 V
MOSFET	M_{ij}	TTD85N03AT (V_{DS} = 10 V, I_D = 85 A, R_{DS} = 4.5 mΩ)
Transformer	N_1:N_2 L_m L_k	1:1 21 µH 0.1 µH
	Equalisation start threshold (ε)	0.1%
	Switching frequency of the first-layer equalisation (f_1)	50 kHz
	Switching frequency of the second-layer equalisation (f_2)	50 kHz

The parameters for the equalisation circuit above are also applicable to a battery system composed of 2n (where n = 1, 2. . ., ∞) batteries connected in series.

4.2. The First-Layer Equalisation Experiment

Table 2 displays the initial SOC values of each battery. During the first-layer equalisation process, batteries B11, B21, and B31 are discharged simultaneously, while batteries B12, B22, and B32 are charged simultaneously. This allows for internal equalisation of the three battery units to occur simultaneously.

Table 2. Initial SOC value of each single battery.

Battery Number	SOC (%)	Voltage (V)	Battery Number	SOC (%)	Voltage (V)
B11	82.39	3.898	B22	67.34	3.891
B12	76.88	3.896	B31	62.75	3.888
B21	72.68	3.894	B32	57.28	3.884

Figure 8a shows the waveform of the current in the battery unit BU2 when the battery B21 is discharged and the waveforms of the complementary PWM driving signals of the switches M11 and M14 in the course of the experiment. Figure 8b shows the waveform of the current in the battery unit BU2 when the battery B22 is charged and the waveforms of the complementary PWM driving signals of the switches M11 and M14 in the course of the experiment. The equalisation current is about 2 A, the duty cycle for M11 is about 31%, the conduction duty cycle of M14 is about 67%, and the driving signals for both switches have a dead time of about 2%.

4.3. The Second-Layer Equalisation Experiment

The second-layer equalisation experiment of the battery system contains three battery units, where the battery unit whose average SOC value $SOCBUi_{avg}$ is higher than the battery system average SOC value SOCavg releases energy, and the battery unit whose average SOC value SOCBUiavg is lower than the battery system average SOC value SOCavg absorbs energy.

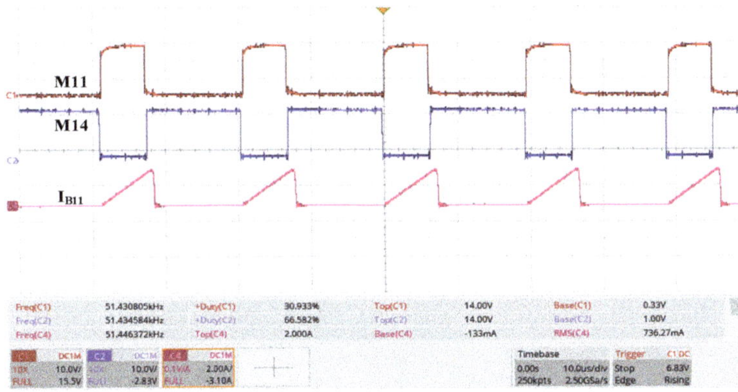

(**a**)

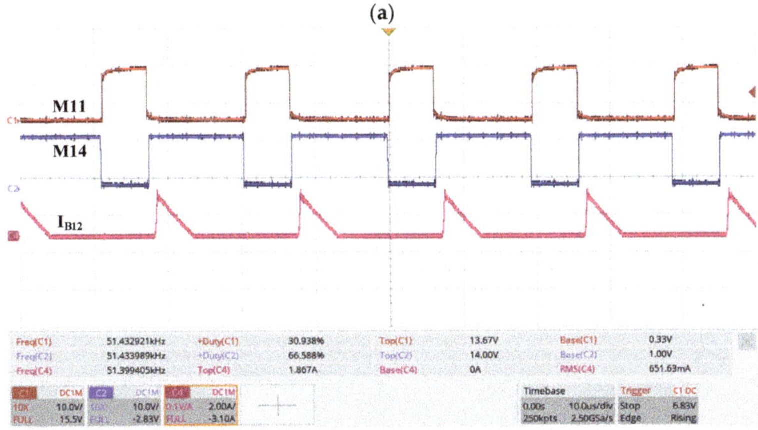

(**b**)

Figure 8. Experimental waveforms of the first-layer equalisation. (**a**) PWM signal waveforms of M11, M14 and discharging current waveform of B11. (**b**) PWM signal waveforms of M11, M14 and charging current waveform of B12.

Figure 9a shows the discharging current waveform of battery unit BU1 and the driving signal waveform of switching device M11 intercepted during the experiment. Figure 9b shows the charge current waveform of the energy transit battery E and the driving signal waveform of the switching device M11. The equalisation current is 2.467 A, and the conduction duty cycle of M11 is about 35%.

4.4. Equalisation experiment results

Figure 10 shows that the battery with a high SOC value in each battery unit releases energy, the battery with a low SOC value absorbs energy, and finally the SOC values between the two batteries reach the same. The equalisation times for the three battery units are 20 min, 21 min and 21 min, respectively, so the time required for the first layer of equalisation is 21 min. After equalisation, the SOCs of the three battery units are 79.44%, 70.11% and 59.89%, respectively.

Figure 10 also shows that the battery unit with a high SOC value in the battery system releases energy, and the battery unit with a low SOC value absorbs the energy. This process continues until the SOC values of all three battery units in the system are equalised. The second-layer equalisation process takes 77 min. Following equalisation, the SOC values of all three battery units are 69.85%.

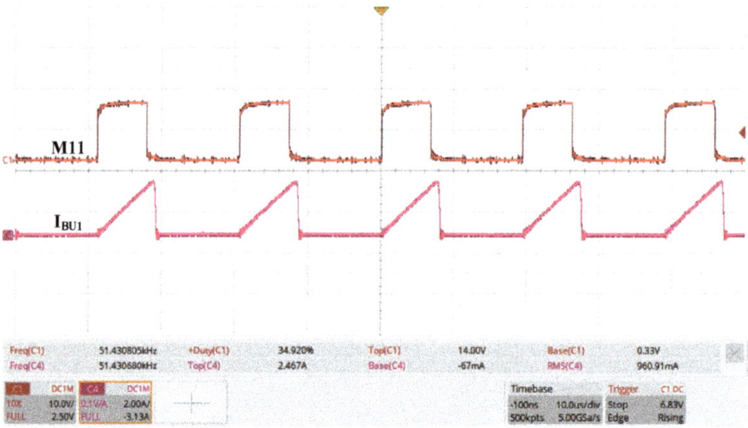

(**a**)

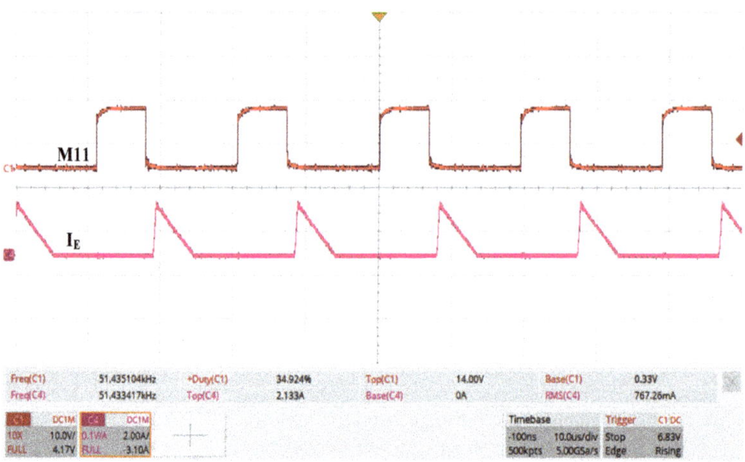

(**b**)

Figure 9. Experimental waveforms of the second-layer equalisation. (**a**) PWM signal waveforms of and discharge current waveform of BU1. (**b**) PWM signal waveforms of M11 and charge current waveform of E.

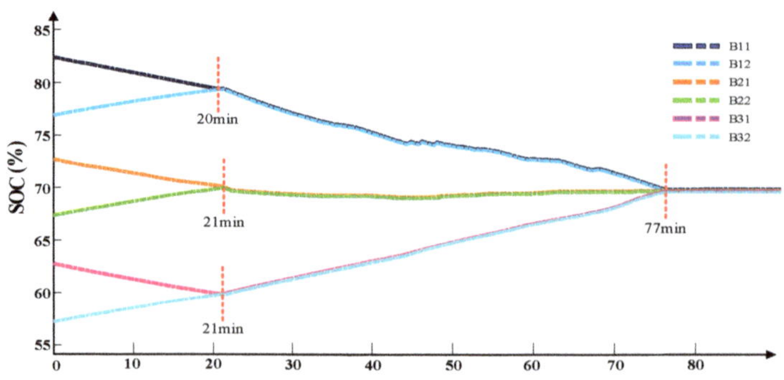

Figure 10. SOC curves of the battery string during the equalisation experiment.

5. Analysis of Experimental Results

5.1. Calculation of Equalisation Efficiency

5.1.1. Equalisation Efficiency of the First-Layer Equalisation

The first-layer equalisation utilises PWM with deadband complementary to operate the MOSFETs, resulting in reduced switching losses and improved equalisation efficiency. For instance, when considering batteries B11 and B12, the equalisation efficiency of the first-layer can be calculated using Equation (4) according to the relevant parameters and graphs from the equalisation experiment. The calculation process is shown as follows:

$$\begin{aligned}\eta_1 &= 1 - \frac{2\int_0^{0.31T} 4.5\times10^{-3}\times0.736^2 dt + \int_{0.31T}^{0.32T} 0.651\times\left(0.6+2\times4.5\times10^{-3}\times0.651\right)dt + 3\int_{0.32T}^{0.65T} 4.5\times10^{-3}\times0.651^2 dt}{\int_0^{0.31T} 3.898\times0.736 dt} \\ &\approx 98.8\%\end{aligned} \quad (10)$$

5.1.2. Equalisation Efficiency of the Second-Layer Equalisation

The second-layer equalisation process doubles the voltage of the battery unit by using it as the equalisation object. For instance, in the case of battery unit BU2, the equalisation efficiency of the second layer is calculated using Equation (9) and the relevant parameters and graphs from the equalisation experiment. The calculation process is shown as:

$$\eta_2 = 1 - \frac{2\int_0^{0.35T} 4.5\times10^{-3}\times0.961^2 dt + \int_{0.35T}^{0.67T} 0.6\times0.767 dt}{\int_0^{0.35T} (3.894+3.891)\times0.961} \approx 94.3\% \quad (11)$$

5.2. Comparison with Other Methods

To demonstrate the advantages of the proposed layered parallel equaliser based on a flyback transformer multiplexed for a lithium-ion battery system. Table 3 lists the equalisers that are compared with different types of equalisers in terms of the number of devices and equaliser performance. The number of devices included in the equaliser determines the size and cost of the equaliser, which is mainly compared to the number of MOSFETs (M), diodes (D), inductors (L), capacitors (C), and transformers (T) included in different types of equalisers. This paper classifies the comparison of equalisers into three categories: low (L), medium (M), and high (H), based on their cost, complexity, speed, efficiency, and size.

Table 3. Comparison of existing equalisers.

Equalisers	Components					Equalisation Performance				
	M	D	L	C	T	Cost	Efficiency	Speed	Complexity	Volume
[9]	2N	0	0	N − 1	0	L	93.1%	L	M	L
[10]	4N + 2	0	N − 1	0	0	H	80%	M	H	M
[15]	4N	0	N	N	0	H	95.2%	L	H	L
[16]	2N	0	2N	N	N	H	92%	M	H	H
[18]	4N + 1	0	1	1	0	M	96.05%	L	H	L
[19]	N/2	0	N/2	0	0	M	/	M	H	M
[20]	N	1	N/2	0	N/4	L	95.6%	M	L	H
[22]	2N + 6	0	1	0	1	M	80.4%	L	M	L
[25]	5/4N	0	5/4N	0	N/4	M	93.15%	M	L	M
[28]	2N	0	N	0	1	M	70.14%	L	M	L
Proposed equaliser	3N	0	0	0	N/2	M	98.8%/94.3%	H	M	M

The proposed equaliser uses a flyback transformer as an equalisation element. To provide a more intuitive comparison, it will be compared with a design based on transformer equalisation. In [16], half-bridge LC transformers are used as an equalisation topology. This topology uses more transformers and inductors, which are larger in size. Due to the fact that equalisation circuit is more intricate, the MOSFET control method is complex. The equaliser described in [22] only requires a flyback transformer, and the volume is small. Equalisation can only be conducted sequentially between batteries, and the equalisation speed is slow. The equalisation path contains more MOSFETs for energy flow, resulting in high switching losses and low equalisation efficiency. The topology in [25] employs multi-winding transformers as the equalisation element, resulting in faster equalisation with fewer MOSFETs and higher efficiency. However, it is unable to selectively charge during the charging process. In [28], the topology also employs multi-winding transformers as equalisation elements. However, as the number of batteries in series increases, the transformer has too many windings, resulting in poor equalisation efficiency.

Table 3 compares the proposed equaliser with various types of equalisers. It is evident that the proposed equaliser outperforms other equalisers in terms of efficiency and speed. Additionally, the proposed equaliser requires fewer components, resulting in a smaller size and lower cost.

6. Discussion

This paper proposes a layered parallel equaliser based on flyback transformer multiplexing for a lithium-ion battery system that has higher equalisation efficiency, faster equalisation speed, and greater scalability than other transformer-based equalisation designs. The equaliser employs layered and parallel equalisation to improve efficiency and speed. Additionally, the size of the equaliser is reduced by using a flyback transformer as a two-layer energy storage element. The efficiency of the equaliser is further improved by the short energy path and the small number of MOSFETs passing through the equalisation loop. The equaliser has a modular design and is highly scalable, with component parameters remaining constant as the number of series-connected batteries increases, making it suitable for large-scale energy storage battery systems. However, there are some areas for improvement in this proposed equaliser. For instance, the suggested equaliser uses a greater number of MOSFETs. By decreasing the number of MOSFETs, the cost of the equaliser can be further reduced. Additionally, MOSFET losses can be minimised through the implementation of resonant soft-switching techniques. It is anticipated that this can be improved in future research.

7. Conclusions

Aiming at the problems of slow equalisation speed, low equalisation efficiency and weak modularity in the large-scale energy storage lithium-ion battery system, this paper proposes a layered parallel equaliser based on a flyback transformer multiplexed for a lithium-ion battery system. The equaliser has the following characteristics:

1. High equalisation efficiency. The equaliser adopts a layered equalisation strategy. The first-layer equalisation targets a single battery and utilises a Buck-Boost circuit and complementary PWM driving method to prevent energy reflux. This approach results in low equalisation loss and high efficiency. The second-layer equalisation focuses on the battery unit and doubles the voltage to improve equalisation efficiency. This is achieved through a multi-winding transformer equalisation circuit. The first and second layer equalisation experiments resulted in an efficiency of 98.8% and 94.3%, respectively.
2. Fast equalisation speed. The multi-objective parallel equalisation method is utilised by the equaliser. The first-layer equalisation ensures that equalisation within each battery cell does not interfere with each other. The more parallel equalisation targets there are, the faster the entire battery system can be equalised. The second-layer equalisation module uses flyback transformers to parallel discharge or charge multiple battery units. It also enables simultaneous charging and discharging of multiple units, significantly increasing the speed of equalisation.
3. Strong modularity. With an increase in the number of series-connected batteries, the hardware parameters of the equaliser remain unchanged. It is only necessary to adjust the number of corresponding equalisation modules according to needs, which

greatly improves the equaliser's scalability. The equaliser is suitable for a large-scale lithium-ion battery energy storage system.

Author Contributions: H.L. proposed the equalisation ideal, X.W. designed the equaliser. X.W., X.Y. and J.A. conceived the research methods and control strategies. X.W. and J.A. designed and performed the experiments. X.W. completed the initial draft, H.L. completed the proofreading of the manuscript. All authors have read and agreed to the published version of the manuscript.

Funding: This work is supported by the National Natural Science Foundation of China (51967009).

Data Availability Statement: Data are contained within the article.

Conflicts of Interest: The authors declare no conflict of interest.

References

1. Deng, J.; Bae, C.; Marcicki, J.; Masias, A.; Miller, T. Safety modelling and testing of lithium-ion batteries in electrified vehicles. *Nat. Energy* **2018**, *3*, 261–266. [CrossRef]
2. Hannan, M.A.; Lipu, M.S.H.; Hussain, A.; Mohamed, A. A review of lithium-ion battery state of charge estimation and management system in electric vehicle applications: Challenges and recommendations (Review). *Renew. Sustain. Energy Rev.* **2017**, *78*, 834–854. [CrossRef]
3. Zhao, Z.Y.; Hu, H.T.; He, Z.Y.; Iu, H.H.C.; Davari, P.; Blaabjerg, F. Power Electronics-Based Safety Enhancement Technologies for Lithium-Ion Batteries: An Overview from Battery Management Perspective. *IEEE Trans. Power Electron.* **2023**, *38*, 8922–8955. [CrossRef]
4. Ghaeminezhad, N.; Ouyang, Q.; Hu, X.S.; Xu, G.T.; Wang, Z.S. Active Cell Equalization Topologies Analysis for Battery Systems: A Systematic Review. *IEEE Trans. Power Electron.* **2021**, *36*, 9119–9135. [CrossRef]
5. Izadi, Y.; Beiranvand, R. A Comprehensive Review of Battery and Supercapacitor Cells Voltage-Equalizer Circuits. *IEEE Trans. Power Electron.* **2023**, *38*, 15671–15692. [CrossRef]
6. Shang, Y.L.; Zhu, C.; Fu, Y.H.; Mi, C.C. An Integrated Heater Equalizer for Lithium-Ion Batteries of Electric Vehicles. *IEEE Trans. Ind. Electron.* **2019**, *66*, 4398–4405. [CrossRef]
7. Ye, Y.M.; Cheng, K.W.E.; Fong, Y.C.; Xue, X.D.; Lin, J.J. Topology, Modeling, and Design of Switched-Capacitor-Based Cell Balancing Systems and Their Balancing Exploration. *IEEE Trans. Power Electron.* **2017**, *32*, 4444–4454. [CrossRef]
8. Lee, S.; Noh, G.; Ha, J.I. Reconfigurable Power Circuits to Series or Parallel for Energy-Balanced Multicell Battery Pack. *IEEE Trans. Ind. Electron.* **2023**, *70*, 3641–3651. [CrossRef]
9. Shang, Y.L.; Cui, N.X.; Duan, B.; Zhang, C.H. Analysis and Optimization of Star-Structured Switched-Capacitor Equalizers for Series-Connected Battery Strings. *IEEE Trans. Power Electron.* **2018**, *33*, 9631–9646. [CrossRef]
10. Wang, S.; Yang, S.; Yang, W.; Wang, Y. A New Kind of Balancing Circuit with Multiple Equalization Modes for Serially Connected Battery Pack. *IEEE Trans. Ind. Electron.* **2021**, *68*, 2142–2150. [CrossRef]
11. Ding, X.F.; Zhang, D.H.; Cheng, J.W.; Wang, B.B.; Chai, Y.M.; Zhao, Z.H.; Xiong, R.; Luk, P.C.K. A Novel Active Equalization Topology for Series-Connected Lithium-ion Battery Systems. *IEEE Trans. Ind. Appl.* **2020**, *56*, 6892–6903. [CrossRef]
12. Manjunath, K.; Kalpana, R.; Singh, B.; Kiran, R. A Two-Stage Module Based Cell-to-Cell Active Balancing Circuit for Series Connected Lithium-Ion Battery Packs. *IEEE Trans. Energy Convers.* **2023**, *38*, 2282–2297. [CrossRef]
13. Wang, L.J.; Ke, J.Y.; Zhan, M.; Tian, A.N.; Jiang, J.C. Efficient and Fast Active Equalization Method for Retired Battery Pack Using Wide Voltage Range Bidirectional Converter and DBSCAN Clustering Algorithm. *IEEE Trans. Power Electron.* **2022**, *37*, 13824–13833.
14. Peng, F.; Wang, H.; Yu, L. Analysis and Design Considerations of Efficiency Enhanced Hierarchical Battery Equalizer Based on Bipolar CCM Buck–Boost Units. *IEEE Trans. Ind. Appl.* **2019**, *55*, 4053–4063. [CrossRef]
15. Liu, L.; Mai, R.; Xu, B.; Sun, W.; Zhou, W.; He, Z. Design of Parallel Resonant Switched-Capacitor Equalizer for Series-Connected Battery Strings. *IEEE Trans. Power Electron.* **2021**, *36*, 9160–9169. [CrossRef]
16. Liu, F.; Zou, R.; Liu, Y. An Any-Cell-to-Any-Cell Battery Equalizer Based on Half-Bridge LC Converter. *IEEE Trans. Power Electron.* **2023**, *38*, 4218–4223. [CrossRef]
17. Wei, Z.; Wang, H.; Lu, Y.; Shu, D.; Ning, G.; Fu, M. Bidirectional Constant Current String-to-Cell Battery Equalizer Based on L2C3 Resonant Topology. *IEEE Trans. Power Electron.* **2022**, *38*, 666–677. [CrossRef]
18. Noh, G.; Lee, J.; Ha, J.-I. Design and Analysis of Single-Inductor Power Converter for Both Battery Balancing and Voltage Regulation. *IEEE Trans. Ind. Electron.* **2022**, *69*, 2874–2884. [CrossRef]
19. Reema, N.; Jagadanand, G.; Sasidharan, N.; Shreelakshmi, M.P. An Enhanced Coupled Inductor Based Voltage Equalizer for Matrix Connected Ultracapacitor Bank. *IEEE Trans. Transp. Electrif.* **2023**, *9*, 3328–3337. [CrossRef]
20. Shang, Y.; Cui, N.; Duan, B.; Zhang, C. A Global Modular Equalizer Based on Forward Conversion for Series-Connected Battery Strings. *IEEE J. Emerg. Sel. Top. Power Electron.* **2018**, *6*, 1456–1469. [CrossRef]
21. Guo, X.; Geng, J.; Liu, Z.; Xu, X.; Cao, W. A Flyback Converter-Based Hybrid Balancing Method for Series-Connected Battery System in Electric Vehicles. *IEEE Trans. Veh. Technol.* **2021**, *70*, 6626–6635. [CrossRef]

22. Lee, K.M.; Lee, S.W.; Choi, Y.G.; Kang, B. Active Balancing of Li-Ion Battery Cells Using Transformer as Energy Carrier. *IEEE Trans. Ind. Electron.* **2017**, *64*, 1251–1257. [CrossRef]
23. Shang, Y.; Xia, B.; Zhang, C.; Cui, N.; Yang, J.; Mi, C.C. An Automatic Equalizer Based on Forward–Flyback Converter for Series-Connected Battery Strings. *IEEE Trans. Ind. Electron.* **2017**, *64*, 5380–5391. [CrossRef]
24. Narayanan, R.; Gangadharan, J.; Sasidharan, N.; Meleetil, P.S. A Novel Modularization Method for Voltage Equalization of Ultracapacitor Bank Using Coupled Inductor. *IEEE Trans. Ind. Electron.* **2024**, *71*, 3548–3558. [CrossRef]
25. Wang, S.; Wang, Y.; Chen, G.; Wei, D.; Shang, Y. An Efficient and Compact Equalizer Based on Forward-Flyback Conversion for Large-Scale Energy Storage Systems. *IEEE Trans. Transp. Electrif.* **2023**, *early accessed*. [CrossRef]
26. Liu, L.; Xu, B.; Yan, Z.; Zhou, W.; Li, Y.; Mai, R.; He, Z. A Low-Cost Multiwinding Transformer Balancing Topology for Retired Series-Connected Battery String. *IEEE Trans. Power Electron.* **2021**, *36*, 4931–4936. [CrossRef]
27. Fan, S.; Duan, J.; Sun, L.; Zhang, K. A Fast Modularized Multiwinding Transformer Balancing Topology for Series-Connected Super Capacitors. *IEEE Trans. Power Electron.* **2019**, *34*, 3255–3268. [CrossRef]
28. Chen, Y.; Liu, X.; Cui, Y.; Zou, J.; Yang, S. A MultiWinding Transformer Cell-to-Cell Active Equalization Method for Lithium-Ion Batteries with Reduced Number of Driving Circuits. *IEEE Trans. Power Electron.* **2016**, *31*, 4916–4929.
29. Nie, J.; Fu, R.; Cai, C.; Ma, J.; Shu, Z.; Ma, L. A High Efficiency Battery Equalizing Circuit Based on Half Bridge Topology with Multiport Transformer. *IEEE Trans. Ind. Electron.* **2024**, *71*, 2522–2532. [CrossRef]

energies

MDPI

Article

Interval State Estimation of Electricity-Gas Systems Considering Measurement Correlations

Yan Huang [1] and Lin Feng [1,2,*]

[1] School of Control Science and Engineering, Dalian University of Technology, Dalian 116024, China; huangyan139@mail.dlut.edu.cn
[2] School of Innovation and Entrepreneurship, Dalian University of Technology, Dalian 116024, China
* Correspondence: fenglin@dlut.edu.cn

Abstract: The popularization of electricity-gas systems leads to increasing demand for state management of systems. However, the existence of neglected measurement correlations brings uncertainties to the electricity-gas systems state estimation. In this paper, an interval state estimation method that considers measurement correlations existing in the electricity-gas systems is presented. We derive the linear measurement model for the electricity-gas systems through Taylor series expansion and estimate the measurement variance-covariance matrix with measurement correlations. The system parameter matrix and the measurement variance-covariance matrix containing measurement correlations are combined into an interval, and the interval state matrix considering measurement correlations is constructed. Then, the linear equations for the state estimation interval considering measurement correlations are established based on the measurement containing correlations and interval state matrix; as a result, the electricity-gas system state estimation model containing measurement correlations is established. In addition, a method for determining the range of state estimation intervals is proposed. Numerical tests on an integrated electricity-gas system comprising a 10-node natural gas network and IEEE 30-bus system indicate that the proposed approach has more advantages over the UT+KO approach in computation accuracy and computation efficiency.

Keywords: electricity-gas systems; measurement correlations; interval state estimation

Citation: Huang, Y.; Feng, L. Interval State Estimation of Electricity-Gas Systems Considering Measurement Correlations. *Energies* **2024**, *17*, 755. https://doi.org/10.3390/en17030755

Academic Editor: Marco Merlo

Received: 10 January 2024
Revised: 26 January 2024
Accepted: 3 February 2024
Published: 5 February 2024

1. Introduction

The popularization of electricity-gas systems leads to increasing demand for system state supervision, and electricity-gas system state estimation is becoming an important role of effective monitoring and control systems [1]. Most studies of state estimation assumed that the measurements follow the independent Gaussian distribution and ignored the correlations of the measurements. Nevertheless, this is not applicable in practice. Since measurement data is usually obtained by the data acquisition system and the same measurement equipment, every step from data acquisition may be affected by the same errors—after the superposition and propagation of these errors, there will be a certain correlation between the measurements [2]. The research shows that the long-term existence of measurement correlations may have adverse effects on the results of electricity-gas systems state estimation [3–5], such as the impact on the precision of state estimation in distribution systems [3], as well as the impact of bad data with measurement correlations on the measurement sets [4,5]. Thus, it is necessary to consider the existing measurement correlations in the state estimation of electricity-gas systems. However, in practice, the impact of these correlations on the system is hard to quantify. Therefore, we need to provide the uncertainty range of state variables induced by measurement correlations, and ensure that the state variables always exist in this range, for the sake of reducing the influence of measurement correlations on system safety and ensuring the stability and reliability of energy system.

Currently, there are relatively few studies considering the measurement correlations of the electricity-gas system; most focus on the study of the electricity system, and few involve both the measurement correlation of electricity system and natural gas system. As for the measurement correlations of electricity systems, the unscented transformation (UT) is regarded as an effective solution. For instance, a state estimation framework of electricity system based on UT for measurement correlations and measurements with incomplete time synchronization was developed in [6]. In [7–9], UT was employed to calculate the correlations between the measurements of the electricity system and the correlations between the voltage phasor, active and reactive power of the bus. These methods based on UT considered the measurement correlations of electricity systems, obtained the correlation between the measured values through symmetric sampling strategy, and then utilized the covariance to transfer the correlation. Although they had a certain tolerance for measurement correlations, they need to calculate the measurements independently and symmetrically, and are therefore mainly aimed at small-scale electricity systems. As the system size increases, the calculation efficiency will gradually decrease.

In addition to UT, some methods based on the Kalman filter were used to solve measurement correlations, including the Kalman filter (to predict the error covariance matrix) [10,11], and the point estimation method based on an extended Kalman filter [12]. In addition, the combination of untracked transform and Kalman filter has been used to detect bad data with measurement correlations [13]. These Kalman filter methods reduce the measurement correlations by filtering the measurement values, and the state estimation is therefore still based on the filtering and depends on the precise measurement values to a certain extent.

Furthermore, the measured data was modeled as multivariable time series to simulate the spatial correlation in [14,15]. Nevertheless, time series is only suitable for simulating the system state in the short term, and, because it is difficult to accurately describe long-term and fluctuating energy systems, such studies have not delved into the calculation of correlations. Apart from time series, some literatures combined least squares with point estimation to calculate the measurement correlations of electricity systems [16–18]. However, least squares is mainly used for linear calculations and is not very suitable for calculating nonlinear relationships in power systems. Although the correlations of load and input variables were considered in other studies, they mainly focus on pseudo-measurements and stochastic power flow calculation [19,20].

The above methods for considering the measurement correlations existing in the state estimation are mainly aimed at the electricity system. Of these methods considering the measurement correlations, the method based on UT obtained the correlation between the measured values through the symmetric sampling strategy, which is inefficient to solve. Although the Kalman filter method reduces the measurement correlation through filtering, the state estimation still depends on the accurate measurement value to a certain extent, and has low tolerance for measurement correlations. The time series is only suitable for simulating short-term system states, and the least squares method is not very suitable for nonlinear energy systems. Individual studies that considered correlation mainly focus on pseudo-measurements and stochastic power flow calculations. Therefore, these existing methods that considered measurement correlations in electricity systems still have some limitations, and current research rarely involves both electricity system and gas system.

To further effectively solve the state estimation of electricity-gas systems considering measurement correlations, an interval state estimation method is developed in this paper. Firstly, we derive the linear measurement model for the electricity-gas systems through Taylor series expansion and estimate the measurement variance-covariance matrix with measurement correlations. Subsequently, we combine the system parameter matrix with the measurement variance-covariance matrix containing measurement correlations into an interval to construct the interval state matrix considering measurement correlations. Then, the linear equations for the state estimation interval considering measurement correlations are established based on the measurement containing correlations and interval state matrix;

as a result, the electricity-gas system state estimation model with measurement correlations is constructed. Finally, a method for determining the range of state estimation intervals is presented. By comparing the state estimation interval with the safe operating range of system, we can determine if the safe operating range of the system fully covers the state estimation interval, which helps the system administrators enhance situational awareness capabilities, and guides administrators to make effective adjustments as well as to control the energy system when needed. Compared with stochastic methods, the interval state estimation method does not require iterative calculations, and the obtained interval results are more stable, avoiding the randomness of the results. Nevertheless, stochastic methods rely on random selection and have randomness and uncertainty. For multiple calculations of the same case, the results may differ significantly, making it difficult to ensure the reliability of the obtained results range. An integrated electricity-gas system comprised of the 10-node natural gas network and IEEE 30-bus system is taken for numerical tests, and the estimated results of gas demand, pressure at nodes, voltage amplitude, and voltage angle in the integrated electricity-gas system are validated under different levels of measurement correlation, to attest the effectiveness of the proposed approach.

The main contributions of this paper are summarized as follows:

(1) The derived linear model for measurements of electricity-gas systems transfers the nonlinear electricity-gas system model into the measurements-based linear model, describing the statistical characteristics of state variables in the nonlinear system through linear equations and converting them into measurements.
(2) The constructed interval state matrix and the linear equations of state estimation interval consider the correlation between measurements in the electricity-gas system (including the correlation between pressure at node and gas mass flow in the gas network, the correlation between active power and reactive power in the electricity system), and establish the electricity-gas system state estimation model containing these correlations.
(3) The proposed method for determining the range of state estimation interval allows the existence of measurement correlations, has a certain tolerance for measurement correlations, and provides the ideal distribution range of state variables under various measurement correlations.

The remainder of this paper is organized as follows. Section 2 gives the basic model and the linear measurement model for the electricity-gas systems. Section 3 gives the measurement variance-covariance matrix containing measurement correlations, and also establishes the interval state matrix and linear equations of state estimation interval. Section 4 presents a method for determining the range of state estimation intervals. Section 5 validates the effectiveness of the proposed method on an integrated electricity-gas system. Finally, some conclusions are drawn in Section 6.

2. The Linearized Model for Measuring Electricity-Gas Systems

This part mainly introduces the basic model of electricity-gas system, as well as the linear measurement model for the electricity-gas systems deduced by Taylor series expansion.

2.1. The Gas Pipeline System Model

The transmission of natural gas in a pipe exhibits slow dynamic characteristics, and the influence of temperature on the gas pipeline system can be ignored. It is generally considered that the temperature of natural gas in pipe is close to the ambient temperature, and the driving force of pressure causes natural gas to flow axially along a pipe, as shown in Figure 1.

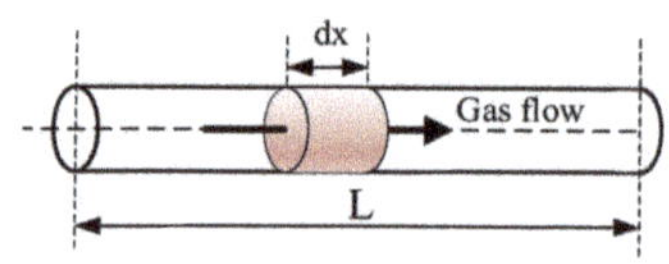

Figure 1. Schematic diagram of natural gas flow inside the pipeline.

The dynamic behavior of gas system is depicted via partial differential equations [21,22], and the state variables of natural gas flow mainly include pipeline pressure and gas flow. The dynamic behavior of natural gas system can be expressed as follows:

$$\frac{\partial(\pi)}{\partial t} + \frac{ZRT}{S} \cdot \frac{\partial\left(\dot{G}\right)}{\partial L} = 0 \tag{1a}$$

$$\frac{\partial(\pi)}{\partial L} + \frac{f \cdot ZRT \cdot \dot{G}\left|\dot{G}\right|}{2d \cdot S^2 \cdot \pi} = 0 \tag{1b}$$

where $\dot{G}$ denotes the natural gas mass flow, t denotes the time, $t = 1, \ldots, N$, π refers to the pressure, Z is the gas compressibility factor, R refers to ideal gas constant, f is the friction coefficient, T represents the average temperature of natural gas in the pipeline, L is the pipeline length, d represents the pipe inner diameter, S refers to the pipeline cross-section area.

Since the natural gas system model is a set of partial differential equations, direct calculation is more complex. Therefore, we linearize the above partial differential equations. It is assumed that gas flows unidirectionally in pipelines, so the time step is set as $\Delta t = T_n/N$, the spatial step is set as $\Delta L = L/M$, then Equation (1) is rewritten as follows:

$$\frac{\Delta\pi_{i+1}^{t} - \Delta\pi_{i+1}^{t-1}}{\Delta t} + \frac{ZRT}{S} \cdot \frac{\Delta\dot{G}_{i+1}^{t} - \Delta\dot{G}_{i}^{t}}{\Delta L} = 0 \tag{2a}$$

$$\frac{\Delta\pi_{i+1}^{t} - \Delta\pi_{i}^{t}}{\Delta L} + \frac{f \cdot ZRT \cdot \dot{G}_{st}}{2d \cdot S^2 \pi_{st}} \cdot \frac{\Delta\dot{G}_{i+1}^{t} + \Delta\dot{G}_{i}^{t}}{2} = 0,\ t = 1, \ldots, N,\ i = 1, \ldots, M-1 \tag{2b}$$

where $\Delta\pi_i^t$ and $\Delta\dot{G}_i^t$ are the pressure and gas flow in the pipe; in length of i at time t, $\Delta\pi_{i+1}^t$ denotes the pressure and $\Delta\dot{G}_{i+1}^t$ denotes gas flow in the pipe in length of $i+1$ at time t. $\dot{G}_{st}$ and π_{st} represent gas flow and pressure of steady state operation, respectively.

Then, we take a matrix to express the relationship between the gas flow and pressure at the inlet of the pipe and the pressure and gas flow at the pipe outlet, and convert Equation (2) into the matrix form in Equation (3).

$$\begin{bmatrix} \Delta\pi_{i+1}^{t} \\ \Delta\dot{G}_{i+1}^{t} \end{bmatrix} = A_g \begin{bmatrix} \frac{\Delta\pi_{i+1}^{t-1}}{\Delta t} + \frac{ZRT \cdot \Delta\dot{G}_{i}^{t}}{S \cdot \Delta L} \\ \frac{\Delta\pi_{i}^{t}}{\Delta L} - \frac{f \cdot ZRT \cdot \dot{G}_{st} \cdot \Delta\dot{G}_{i}^{t}}{4d \cdot S^2 \pi_{st}} \end{bmatrix} \tag{3}$$

where input variables consist of pressure $\Delta\pi_{i+1}^{t-1}$ in the pipe in length of $i+1$ at time $t-1$, gas flow $\Delta\dot{G}_i^t$ in the pipe in length of i at time t, pressure $\Delta\pi_i^t$ in the pipeline in length of i at time t. The output variables are composed of $\Delta\pi_{i+1}^t$ and $\Delta\dot{G}_{i+1}^t$. A_g denotes network parameter matrix.

For a natural gas network, taking the length of the pipeline as the step, the change of pressure and the mass flow are expressed as:

$$\begin{bmatrix} \Delta\pi_j^t \\ \Delta\dot{G}_j^t \end{bmatrix} = A_g \begin{bmatrix} \frac{\Delta\pi_j^{t-1}}{\Delta t} + \frac{ZRT\cdot\Delta\dot{G}_i^t}{S\cdot L_{i,j}} \\ \frac{\Delta\pi_i^t}{L_{i,j}} - \frac{f\cdot ZRT\cdot\dot{G}_{st}\cdot\Delta\dot{G}_i^t}{4dS^2\pi_{st}} \end{bmatrix} \quad i,j \in \{1,\ldots,N_n\} \tag{4}$$

where N_n is the number of node, $L_{i,j}$ is the pipeline length, $\Delta\pi_j^{t-1}$ is the pipeline outlet pressure at node j at time $t-1$, $\Delta\pi_j^t$ is pipeline outlet pressure and $\Delta\dot{G}_j^t$ is outlet gas flow. $\Delta\pi_i^t$ is pipeline inlet pressure and $\Delta\dot{G}_i^t$ is inlet gas flow. The network parameter matrix is expressed as Equation (5). The elements A_{11}, A_{12}, A_{21}, A_{22} of A are represented as below:

$$A_g = \begin{bmatrix} A_{11} & A_{12} \\ A_{21} & A_{22} \end{bmatrix} \tag{5a}$$

$$\begin{cases} A_{11} = f\cdot\dot{G}_{st}\cdot\Delta t\cdot L_{i,j}^2/A_0 \\ A_{12} = -4d\cdot S\cdot\pi_{st}\cdot\Delta t\cdot L_{i,j}/A_0 \\ A_{21} = -4d\cdot S^2\cdot\pi_{st}\cdot\Delta t\cdot L_{i,j}/(ZRT\cdot A_0) \\ A_{22} = 4d\cdot S^2\cdot\pi_{st}\cdot L_{i,j}^2/(ZRT\cdot A_0) \end{cases} \tag{5b}$$

$$A_0 = f\cdot\dot{G}_{st}L_{i,j}^2 - 4d\cdot S\cdot\pi_{st}\cdot\Delta t \tag{5c}$$

In a natural gas system, the gas flow at a node should satisfy the mass conservation constraint as:

$$\sum_{i\to}\dot{G}_{i,j} - \sum_{\to i}\dot{G}_{k,i} + \dot{G}_i^{load} - \dot{G}_i^{inject} = 0 \tag{6}$$

where $\sum_{i\to}\dot{G}_{i,j}$ is the sum of gas flows from node i, $\sum_{\to i}\dot{G}_{k,i}$ means that the gas flows into i from other nodes, $\dot{G}_i^{inject}$ represents the gas injection at node i, $\dot{G}_i^{load}$ is the gas load at node i.

2.2. The Electricity System Model

The electricity system model is represented as follows [23]:

$$P_i = \left(V_i^2 - V_iV_j\cos\theta_{ij}\right)g_{ij} - V_iV_j\sin\theta_{ij}b_{ij} \tag{7}$$

$$Q_i = -\left(V_i^2 - V_iV_j\cos\theta_{ij}\right)b_{ij} - V_iV_j\sin\theta_{ij}g_{ij} \tag{8}$$

where Q_i represents the reactive power at bus i, P_i denotes the active power at bus i, θ_{ij} denotes the voltage angle between bus i and bus j, V_i refers to the voltage magnitude at bus i, b_{ij} refers to the susceptance, g_{ij} represents the conductance.

In the integrated electricity-gas system, the general efficiency of gas-fired generators in converting natural gas into electric energy is stated as [24]:

$$H_i^g = \alpha_i^g + \beta_i^g\cdot\Delta P_i^g + \gamma_i^g\left(\Delta P_i^g\right)^2 \tag{9}$$

$$\Delta\dot{G}_i^g = \frac{H_i^g}{GHV} \tag{10}$$

where ΔP_i^g represents the electric energy generated by gas-fired generator, H_i^g denotes the heat value of gas, $\Delta\dot{G}_i^g$ represents the gas flow demand, GHV denotes the gross heat value of gas, α_i^g, β_i^g, and γ_i^g refer to fuel coefficients.

2.3. The Derivation of Linear Measurement Model for the Electricity-Gas Systems

In this section, we use Taylor series expansion to derive the linear measurement model of electricity-gas systems, thereby further linearizing the nonlinear gas system model, enabling us to then obtain a model for measurement. Firstly, we write the independent variables in the natural gas system, namely, $\Delta\pi_i^t$ and $\Delta\dot{G}_i^t$ at node i in the following function form:

$$\Delta\dot{G}_i^t = \varphi_1^{-1}\left[\varphi_1\left(\Delta\dot{G}_i^t\right)\right] \tag{11a}$$

$$\Delta\pi_i^t = \varphi_2^{-1}\left[\varphi_2\left(\Delta\pi_i^t\right)\right] \tag{11b}$$

Then, the outlet pressure $\Delta\pi_j^t$ and outlet gas flow $\Delta\dot{G}_j^t$ of the pipeline at node j in Equation (4) are respectively written in the following form:

$$\Delta\pi_j^t = A_{11}\left(\frac{\Delta\pi_j^{t-1}}{\Delta t} + \frac{ZRT\cdot\Delta\dot{G}_i^t}{S\cdot L_{i,j}}\right) + A_{12}\left(\frac{\Delta\pi_i^t}{L_{i,j}} - \frac{f\cdot ZRT\cdot\dot{G}_{st}\cdot\Delta\dot{G}_i^t}{4dS^2\pi_{st}}\right) \tag{12a}$$

$$\Delta\dot{G}_j^t = A_{21}\left(\frac{\Delta\pi_j^{t-1}}{\Delta t} + \frac{ZRT\cdot\Delta\dot{G}_i^t}{S\cdot L_{i,j}}\right) + A_{22}\left(\frac{\Delta\pi_i^t}{L_{i,j}} - \frac{f\cdot ZRT\cdot\dot{G}_{st}\cdot\Delta\dot{G}_i^t}{4dS^2\pi_{st}}\right) \tag{12b}$$

Subsequently, we linearize $\Delta\pi_j^t$, $\Delta\dot{G}_i^t$, and $\Delta\pi_i^t$ by Taylor series expansion, and obtain:

$$\alpha_g = \left.\frac{\partial\Delta\pi_j^t}{\partial\Delta\dot{G}_i^t}\right|_{\substack{\Delta\dot{G}_i^t = \Delta\dot{G}_{i,0}^t \\ \Delta\pi_i^t = \Delta\pi_{i,0}^t}} = A_{11}\cdot\frac{ZRT}{S\cdot L_{i,j}} - A_{12}\cdot\frac{f\cdot ZRT\cdot\dot{G}_{st}}{4d\cdot S^2\cdot\pi_{st}} \tag{13}$$

$$\beta_g = \left.\frac{\partial\Delta\pi_j^t}{\partial\Delta\pi_i^t}\right|_{\substack{\Delta\dot{G}_i^t = \Delta\dot{G}_{i,0}^t \\ \Delta\pi_i^t = \Delta\pi_{i,0}^t}} = A_{12}\cdot\frac{1}{L_{i,j}} \tag{14}$$

$$\Delta\dot{G}_i^t - \Delta\dot{G}_{i,0}^t \approx \left.\frac{\partial\Delta\dot{G}_i^t}{\partial\varphi_1\left(\Delta\dot{G}_i^t\right)}\right|_{\Delta\dot{G}_i^t = \Delta\dot{G}_{i,0}^t}\left\{\varphi_1\left(\Delta\dot{G}_i^t\right) - \varphi_1\left(\Delta\dot{G}_{i,0}^t\right)\right\} \tag{15}$$

$$\Delta\pi_i^t - \Delta\pi_{i,0}^t \approx \left.\frac{\partial\Delta\pi_i^t}{\partial\varphi_2\left(\Delta\pi_i^t\right)}\right|_{\Delta\pi_i^t = \Delta\pi_{i,0}^t}\left\{\varphi_2\left(\Delta\pi_i^t\right) - \varphi_2\left(\Delta\pi_{i,0}^t\right)\right\} \tag{16}$$

Then, we replace $\varphi_1\left(\Delta\dot{G}_i^t\right)$ with $\Delta\dot{G}_i^t$ and $\varphi_2\left(\Delta\pi_i^t\right)$ with $\Delta\pi_i^t$ to obtain the linearized model for measurement $\Delta\pi_j^t$ with respect to variables $\Delta\dot{G}_i^t$ and $\Delta\pi_i^t$:

$$\Delta\pi_j^t \approx \alpha_g\left(\Delta\dot{G}_i^t - \Delta\dot{G}_{i,0}^t\right) + \beta_g\left(\Delta\pi_i^t - \Delta\pi_{i,0}^t\right) + \Delta\pi_{j,0}^t \tag{17}$$

where, we take $\dot{G}_{st}$ and π_{st} in steady state as the initial values, expressed as $\Delta\pi_{i,0}^t$ and $\Delta\dot{G}_{i,0}^t$, respectively. $\Delta\pi_{j,0}^t$ can be derived from Equations (12a), (13)–(15).

For the electricity system, the voltage amplitude V and voltage angle θ_{ij} are represented in the following function form:

$$V = \phi_1^{-1}\left[\phi_1(V)\right] \tag{18a}$$

$$\theta_{ij} = \phi_2^{-1}[\phi_2(\theta_{ij})] \tag{18b}$$

Then, we linearize the active power P_i injected at bus i via Taylor series expansion to obtain the linearized model for measurement P_i, with respect to voltage amplitude V and voltage angle θ_{ij}:

$$\begin{aligned}
P_i - P_{i,0} &\approx \left.\frac{\partial P_i}{\partial V_i V_j}\right|_{\substack{V=V_0\\ \theta_{i,j}=\theta_{ij,0}}} (V_iV_j - V_{i,0}V_{j,0}) + \left.\frac{\partial P_i}{\partial \theta_{ij}}\right|_{\substack{V=V_0\\ \theta_{i,j}=\theta_{ij,0}}} (\theta_{ij} - \theta_{ij,0}) \\
&= \left.\frac{\partial P_i}{\partial V_i V_j}\right|_{\substack{V=V_0\\ \theta_{i,j}=\theta_{ij,0}}} \left\{ \left.\frac{\partial V_iV_j}{\partial \phi_1(V_i)}\right|_{V=V_0} \{\phi_1(V_i) - \phi_1(V_{i,0})\} + \left.\frac{\partial V_iV_j}{\partial \phi_1(V_j)}\right|_{V=V_0} \{\phi_1(V_j) - \phi_1(V_{j,0})\} \right\} \\
&+ \left.\frac{\partial P_i}{\partial \theta_{ij}}\right|_{\substack{V=V_0\\ \theta_{i,j}=\theta_{ij,0}}} \left\{ \left.\frac{\partial \theta_{ij}}{\partial \phi_2(\theta_{ij})}\right|_{\theta_{ij}=\theta_{ij,0}} \{\phi_2(\theta_{ij}) - \phi_2(\theta_{ij,0})\} \right\}
\end{aligned} \tag{19}$$

where we take voltage amplitude, as well as voltage angle at the balance node in electricity systems, as the initial values, denoted as $V_{i,0}$, $V_{j,0}$, and $\theta_{ij,0}$.

We define α_e and β_e as:

$$\alpha_e = \left.\frac{\partial P_i}{\partial V_i V_j}\right|_{\substack{V=V_0\\ \theta_{i,j}=\theta_{ij,0}}} \tag{20a}$$

$$\beta_e = \left.\frac{\partial P_i}{\partial \theta_{ij}}\right|_{\substack{V=V_0\\ \theta_{i,j}=\theta_{ij,0}}} \tag{20b}$$

Assuming that there is only one branch connection between bus i and bus j, we replace $\phi_1(V)$ with V and $\phi_2(\theta_{ij})$ with θ_{ij} to obtain a linearized model for measurement P_i with respect to voltage amplitude V and voltage angle θ_{ij}:

$$P_i \approx \alpha_e(V_iV_j - V_{i,0}V_{j,0}) + \beta_e(\theta_{ij} - \theta_{ij,0}) + P_{i,0} \tag{21}$$

It should be noted that Equations (19)–(21) represent the case where there is only one branch between bus i and bus j. For other special cases with multiple branches, the flux increment needs to be summed.

3. The Construction of State Matrix and Linear Equations of State Estimation Interval Considering Measurement Correlations

In this section, we calculate the measurement variance-covariance matrix containing measurement correlations, and combine the system parameter matrix and measurement variance-covariance matrix into a unified framework to construct a state matrix considering measurement correlations, and construct the state estimation model of the electricity-gas systems containing measurement correlations.

3.1. The Calculation of Measurement Variance-Covariance Matrix with Measurement Correlations

We extend the gas flow $\Delta \dot{G}_i^t$ and pressure $\Delta \pi_i^t$ at pipeline inlet of the natural gas network to $\Delta \pi_{i,k}^t$ and $\Delta \dot{G}_{i,k}^t$ with the same mean and variance, as follows:

$$\Delta \pi_{i,k}^t = \begin{cases} \Delta \overline{\pi}_i^t + \delta_{\Delta \pi_i^t}, \; if\; k = 1 \\ \Delta \overline{\pi}_i^t - \delta_{\Delta \pi_i^t}, \; if\; k = 2 \end{cases} \tag{22}$$

$$\dot{\Delta G}^t_{i,k} = \begin{cases} \overline{\dot{\Delta G}}^t_i + \delta_{\dot{\Delta G}^t_i}, \; if\; k = 1 \\ \overline{\dot{\Delta G}}^t_i - \delta_{\dot{\Delta G}^t_i}, \; if\; k = 2 \end{cases} \tag{23}$$

where $\Delta\overline{\pi}^t_i$ and $\overline{\dot{\Delta G}}^t_i$ are the means of $\Delta\pi^t_i$ and $\dot{\Delta G}^t_i$, respectively, and $\delta_{\Delta\pi^t_i}$ and $\delta_{\dot{\Delta G}^t_i}$ are the standard deviations of $\Delta\pi^t_i$ and $\dot{\Delta G}^t_i$, respectively.

Then, we replace $\dot{\Delta G}^t_i$ with $\dot{\Delta G}^t_{i,k}$ and $\Delta\pi^t_i$ with $\Delta\pi^t_{i,k}$, and rewrite (15) as:

$$\Delta\pi^t_{j,k} = \alpha_g\left(\dot{\Delta G}^t_{i,k} - \dot{\Delta G}^t_{i,0}\right) + \beta_g\left(\Delta\pi^t_{i,k} - \Delta\pi^t_{i,0}\right) + \Delta\pi^t_{j,0} \tag{24}$$

where $\Delta\pi^t_{j,k} = \left\{\Delta\pi^t_{j,1}, \Delta\pi^t_{j,2}\right\}$; similarly, we also need to calculate the expanded $\dot{\Delta G}^t_{j,k}$, and finally obtain $\dot{\Delta G}^t_{j,k} = \left\{\dot{\Delta G}^t_{j,1}, \dot{\Delta G}^t_{j,2}\right\}$.

We select N sets of experimental data and calculate the measurement variance-covariance matrix on the basis of an extended matrix:

$$\Delta\overline{\pi}^t_j = E\left[\Delta\pi^t_{j,k}\right] = \frac{1}{2N}\sum_{n=1}^{N}\sum_{k=1}^{2}\Delta\pi^{t,(n)}_{j,k} \tag{25}$$

$$\overline{\dot{\Delta G}}^t_j = E\left[\dot{\Delta G}^t_{j,k}\right] = \frac{1}{2N}\sum_{n=1}^{N}\sum_{k=1}^{2}\dot{\Delta G}^{t,(n)}_{j,k} \tag{26}$$

$$\sigma^2_{\Delta\pi^t_j} = \frac{1}{2N}\sum_{n=1}^{N}\sum_{k=1}^{2}\left[\left(\Delta\pi^{t,(n)}_{j,k} - \Delta\overline{\pi}^t_j\right)\left(\Delta\pi^{t,(n)}_{j,k} - \Delta\overline{\pi}^t_j\right)^T\right] \tag{27}$$

$$\sigma^2_{\dot{\Delta G}^t_j} = \frac{1}{2N}\sum_{n=1}^{N}\sum_{k=1}^{2}\left[\left(\dot{\Delta G}^{t,(n)}_{j,k} - \overline{\dot{\Delta G}}^t_{j,k}\right)\left(\dot{\Delta G}^{t,(n)}_{j,k} - \overline{\dot{\Delta G}}^t_{j,k}\right)^T\right] \tag{28}$$

where $\Delta\overline{\pi}^t_j$ and $\overline{\dot{\Delta G}}^t_j$ are the mean values of pressure and gas flow, respectively. $\sigma^2_{\Delta\pi^t_j}$, $\sigma^2_{\dot{\Delta G}^t_j}$ are variances of pressure and gas flow, respectively.

We further calculate the correlation parameters of gas network measurements. For the measurement variance-covariance matrix, the non-diagonal term corresponds to the value of the measurement correlations. We take the product of standard deviation of the measurements and the correlation coefficient between the measurements as measurement correlation parameters, as follows:

$$Corr_{\Delta\pi^t_j,\dot{\Delta G}^t_j} = \sigma_{\Delta\pi^t_j}\cdot\sigma_{\dot{\Delta G}^t_j}\cdot\rho_{\Delta\pi^t_j,\dot{\Delta G}^t_j} \tag{29}$$

$$\sum\nolimits_g = \begin{bmatrix} \sigma^2_{\Delta\pi^t_j} & Corr_{\Delta\pi^t_j,\dot{\Delta G}^t_j} \\ Corr_{\dot{\Delta G}^t_j,\Delta\pi^t_j} & \sigma^2_{\dot{\Delta G}^t_j} \end{bmatrix} \tag{30}$$

where $\sigma_{\Delta\pi^t_j}$ is the standard deviation of $\Delta\pi^t_j$, $\sigma_{\dot{\Delta G}^t_j}$ is standard deviation of the outlet gas flow $\dot{\Delta G}^t_j$, $\rho_{\Delta\pi^t_j,\dot{\Delta G}^t_j}$ is the correlation coefficient between the measurement $\Delta\pi^t_j$ and $\dot{\Delta G}^t_j$, $\sum_g$ is the measurement variance-covariance matrix in the gas system considering the measurement correlations.

Similar to the gas system, we extend voltage amplitude V as well as voltage angle θ_{ij} in the electricity system to $V_{i,k}$, $V_{j,k}$, $\theta_{ij,k}$ with the same mean and variance, as follows:

$$V_{i,k} = \begin{cases} \overline{V}_i + \delta_{V_i},\ if\ k = 1 \\ \overline{V}_i - \delta_{V_i},\ if\ k = 2 \end{cases} \tag{31}$$

$$V_{j,k} = \begin{cases} \overline{V}_j + \delta_{V_j},\ if\ k = 1 \\ \overline{V}_j - \delta_{V_j},\ if\ k = 2 \end{cases} \tag{32}$$

$$\theta_{ij,k} = \begin{cases} \overline{\theta}_{ij} + \delta_{\theta_{ij}},\ if\ k = 1 \\ \overline{\theta}_{ij} - \delta_{\theta_{ij}},\ if\ k = 2 \end{cases} \tag{33}$$

Furthermore, we assume that there is only one branch connection between bus i and bus j; we replace V_i with $V_{i,k}$, V_j with $V_{j,k}$, and θ_{ij} with $\theta_{ij,k}$, and Equation (21) is rewritten as:

$$P_{i,k} = \alpha_e \left(V_{i,k} V_{j,k} - V_{i,0} V_{j,0} \right) + \beta_e \left(\theta_{ij,k} - \theta_{ij,0} \right) + P_{i,0} \tag{34}$$

where $P_{i,k} = \{P_{i,1}, P_{i,2}\}$, Equation (34) represents the case where there is only one branch between bus i and bus j; for other special cases where there are multiple branches, the flux increment needs to be summed. Similarly, we also need to calculate the expanded $Q_{i,k}$, and finally obtain $Q_{i,k} = \{Q_{i,1}, Q_{i,2}\}$.

We calculate a measurement variance-covariance matrix of the electricity system as follows:

$$\overline{P}_i = E[P_{i,k}] = \frac{1}{2N} \sum_{n=1}^{N} \sum_{k=1}^{2} P_{i,k}^{(n)} \tag{35}$$

$$\overline{Q}_i = E[Q_{i,k}] = \frac{1}{2N} \sum_{n=1}^{N} \sum_{k=1}^{2} Q_{i,k}^{(n)} \tag{36}$$

$$\sigma_{P_i}^2 = \frac{1}{2N} \sum_{n=1}^{N} \sum_{k=1}^{2} \left[\left(P_{i,k}^{(n)} - \overline{P} \right) \left(P_{i,k}^{(n)} - \overline{P}_i \right)^T \right] \tag{37}$$

$$\sigma_{Q_i}^2 = \frac{1}{2N} \sum_{n=1}^{N} \sum_{k=1}^{2} \left[\left(Q_{i,k}^{(n)} - \overline{Q}_i \right) \left(Q_{i,k}^{(n)} - \overline{Q}_i \right)^T \right] \tag{38}$$

where $\overline{P}_i$ denotes mean value of active power and $\overline{Q}_i$ denotes mean value of reactive power, respectively; $\sigma_{P_i}^2$ is variance of active power, $\sigma_{Q_i}^2$ is variance of reactive power.

We then calculate measurement correlation parameters of electricity systems. For the electricity system measurement variance-covariance matrix, we take the product of standard deviation of measurement and correlation coefficient between measurements as the measurement correlation parameter, which is expressed as a non-diagonal term, as shown below:

$$Corr_{P_i,Q_i} = \sigma_{P_i} \cdot \sigma_{Q_i} \cdot \rho_{P_i,Q_i} \tag{39}$$

$$\sum\nolimits_e = \begin{bmatrix} \sigma_{P_i}^2 & Corr_{P_i,Q_i} \\ Corr_{Q_i,P_i} & \sigma_{Q_i}^2 \end{bmatrix} \tag{40}$$

where σ_{P_i} is standard deviation of active power, σ_{Q_i} is standard deviation of reactive power, ρ_{P_i,Q_i} represents the correlation coefficient between measurements P_i and Q_i, and $\sum_e$ represent the measurement variance-covariance matrix considering measurement correlations in electricity systems.

3.2. Constructing the State Matrix and Linear Equations of State Estimation Interval Considering Measurement Correlations

Considering the existence of measurement correlations, we combine the natural gas network parameter matrix A_g in Equation (5) with the measurement variance-covariance matrix $\sum_g$ in Equation (30) into a framework through interval to further construct the state matrix $\widetilde{A}_g$ considering the measurement correlations of natural gas network.

Firstly, we extend the natural gas network parameter matrix A_g in Equation (5), as follows:

$$A_g^* = \left[\begin{array}{c|c} A_g & \begin{pmatrix} 1 & 0 \\ 0 & 1 \end{pmatrix} \\ \hline \begin{pmatrix} 0 & 0 \\ 0 & 0 \end{pmatrix} & \begin{pmatrix} 1 & 0 \\ 0 & 1 \end{pmatrix} \end{array}\right] \tag{41}$$

where A_g^* is the expanded natural gas network parameter matrix.

We represent the natural gas network state variables $\Delta\pi_i^t$ and $\Delta\dot{G}_i^t$ in Equation (4) as X_g, and include the measurement noise $e_{\Delta\pi_j^t}$ and $e_{\Delta\dot{G}_j^t}$ in X_g, as follows:

$$X_g = \begin{bmatrix} \frac{\Delta\pi_{i+1}^{t-1}}{\Delta t} + \frac{ZRT\cdot\Delta\dot{G}_i^t}{S\cdot\Delta L} \\ \frac{\Delta\pi_i^t}{\Delta L} - \frac{f\cdot ZRT\cdot\dot{G}_{st}\cdot\Delta\dot{G}_i^t}{4dS^2\pi_{st}} \\ e_{\Delta\pi_j^t} \\ e_{\Delta\dot{G}_j^t} \end{bmatrix} \tag{42}$$

where $e_{\Delta\pi_j^t}$ and $e_{\Delta\dot{G}_j^t}$ take the standard deviation of measurements $\Delta\pi_j^t$ and $\Delta\dot{G}_j^t$.

Then, we use B_g^* to denote the measurements $\Delta\pi_j^t$ and $\Delta\dot{G}_j^t$, and extend the matrix as follows:

$$B_g^* = \begin{bmatrix} \Delta\pi_j^t \\ \Delta\dot{G}_j^t \\ e_{\Delta\pi_j^t} \\ e_{\Delta\dot{G}_j^t} \end{bmatrix} \tag{43}$$

The relationship between extended gas network parameter matrix A_g^*, natural gas network state variable X_g, and measurement B_g^* is established as follows:

$$\left[\begin{array}{c|c} A_g & \begin{pmatrix} 1 & 0 \\ 0 & 1 \end{pmatrix} \\ \hline \begin{pmatrix} 0 & 0 \\ 0 & 0 \end{pmatrix} & \begin{pmatrix} 1 & 0 \\ 0 & 1 \end{pmatrix} \end{array}\right] \begin{bmatrix} \frac{\Delta\pi_{i+1}^{t-1}}{\Delta t} + \frac{ZRT\cdot\Delta\dot{G}_i^t}{S\cdot\Delta L} \\ \frac{\Delta\pi_i^t}{\Delta L} - \frac{f\cdot ZRT\cdot\dot{G}_{st}\cdot\Delta\dot{G}_i^t}{4dS^2\pi_{st}} \\ e_{\Delta\pi_j^t} \\ e_{\Delta\dot{G}_j^t} \end{bmatrix} = \begin{bmatrix} \Delta\pi_j^t \\ \Delta\dot{G}_j^t \\ e_{\Delta\pi_j^t} \\ e_{\Delta\dot{G}_j^t} \end{bmatrix} \tag{44}$$

According to Equations (41)–(43), Equation (44) is further rewritten as:

$$A_g^* X_g = B_g^* \tag{45}$$

Furthermore, considering the computational dimension of the matrix and taking $\sum_g^{-1}\cdot\left[e_{\Delta\pi_j^t}; e_{\Delta\dot{G}_j^t}\right] = [0;0]$ as the target for calculation, we extend the measurement variance-covariance matrix $\sum_g$ considering measurement correlations in Equation (30) as follows:

$$\delta_g = \left[\begin{array}{c|c} \begin{pmatrix} 1 & 0 \\ 0 & 1 \end{pmatrix} & \begin{pmatrix} 1 & 0 \\ 0 & 1 \end{pmatrix} \\ \hline \begin{pmatrix} 0 & 0 \\ 0 & 0 \end{pmatrix} & \sum_g^{-1} \end{array}\right] \tag{46}$$

where δ_g is the extended matrix with respect to $\sum_g$.

We take $s = [s_1, s_2, \ldots, s_n]$ to represent correlation factor, and $s_i \in \{s_i \in \mathbb{R} | -1 \le s_i \le 1\}$, $i = 1, 2, \ldots, n$. We combine expanded matrix A_g^* and expanded measurement variance-covariance matrix δ_g considering measurement correlation into the interval, as follows:

$$\widetilde{A}_g = A_g^* + s\delta_g,\ \widetilde{A}_g \in \left[\widetilde{A}_g^* - \delta_g, \widetilde{A}_g^* + \delta_g\right] \tag{47}$$

where $\widetilde{A}_g$ denotes the natural gas system state matrix with measurement correlation.

We represent $\widetilde{A}_g$ in Equation (47) as the interval $\left[\widetilde{A}_g\right]$, where $\widetilde{A}_g \in \left[\underline{\widetilde{A}}_g, \overline{\widetilde{A}}_g\right]$, we define $\left[\widetilde{A}_g\right]$ as:

$$\left[\widetilde{A}_g\right] = \left[\underline{\widetilde{A}}_g, \overline{\widetilde{A}}_g\right] = \left\{\widetilde{A}_g \in \mathbb{R} \middle| \underline{\widetilde{A}}_g \le \widetilde{A}_g \le \overline{\widetilde{A}}_g\right\} \tag{48}$$

where $\underline{\widetilde{A}}_g$ is lower bound of interval $\left[\widetilde{A}_g\right]$, $\overline{\widetilde{A}}_g$ is upper bound of interval $\left[\widetilde{A}_g\right]$.

Similarly, we rewrite the measurements B_g^* containing the correlations of natural gas network measurements as interval $[B_g]$ and introduce ΔB_g as follows:

$$B_g = B_g^* + s\Delta B_g,\ B_g \in \left[B_g^* - \Delta B_g, B_g^* + \Delta B_g\right] \tag{49}$$

$$\Delta B_g = \begin{bmatrix} e_{\Delta\pi_j^t} \\ e_{\Delta\dot{G}_j^t} \\ 0 \\ 0 \end{bmatrix} \tag{50}$$

where $e_{\Delta\pi_j^t}$ and $e_{\Delta\dot{G}_j^t}$ take standard deviation of $\Delta\pi_j^t$, $\Delta\dot{G}_j^t$, respectively.

Then, the X_g of the natural gas network is represented as interval $[X_g]$. Based on Equations (45)–(50), combined with Equations (45), (47) and (49), we can obtain the approximate linear equation of state estimation interval that considers measurement correlations of the natural gas network.

$$\left[A_g^* + s\delta_g\right][X_g] \approx \left[B_g^* + s\Delta B_g\right] \tag{51}$$

For electricity systems, the transformation of system model into equation form is given by [25]; active power P_i, reactive power Q_i, voltage amplitude V, and voltage angle θ are classified according to different bus types:

$$\begin{bmatrix} P_L \\ P_S \\ P_R \\ Q_L \\ Q_S \\ Q_R \end{bmatrix} = \begin{bmatrix} A_{11} & A_{12} & A_{13} & A_{14} & A_{15} & A_{16} \\ A_{21} & A_{22} & A_{23} & A_{24} & A_{25} & A_{26} \\ A_{31} & A_{32} & A_{33} & A_{34} & A_{35} & A_{36} \\ A_{41} & A_{42} & A_{43} & A_{44} & A_{45} & A_{46} \\ A_{51} & A_{52} & A_{53} & A_{54} & A_{55} & A_{56} \\ A_{61} & A_{62} & A_{63} & A_{64} & A_{65} & A_{66} \end{bmatrix} \begin{bmatrix} \theta_L \\ \theta_S \\ \theta_R \\ V_L \\ V_S \\ V_R \end{bmatrix} + \begin{bmatrix} C_1 \\ C_2 \\ C_3 \\ C_4 \\ C_5 \\ C_6 \end{bmatrix} \tag{52}$$

$$\begin{cases} P = \begin{bmatrix} P_L^T & P_S^T & P_R^T \end{bmatrix}^T \\ Q = \begin{bmatrix} Q_L^T & Q_S^T & Q_R^T \end{bmatrix}^T \\ V = \begin{bmatrix} V_L^T & V_S^T & V_R^T \end{bmatrix}^T \\ \theta = \begin{bmatrix} \theta_L^T & \theta_S^T & \theta_R^T \end{bmatrix}^T \end{cases} \tag{53}$$

where subscript L denotes PQ bus, subscript S denotes PV bus, subscript R denotes $V\theta$ bus. A_{ij} is the electricity system parameter matrix, C_i is the constant term.

Considering the electricity system measurement correlations, the parameter matrix A_{ij} is represented as A_e and it is extended as:

$$A_e^* = \left[\begin{array}{c:c} A_e & \begin{pmatrix} 1 & 0 \\ 0 & 1 \end{pmatrix} \\ \hdashline \begin{pmatrix} 0 & 0 \\ 0 & 0 \end{pmatrix} & \begin{pmatrix} 1 & 0 \\ 0 & 1 \end{pmatrix} \end{array}\right] \tag{54}$$

where A_e^* refers to the expanded electricity system parameter matrix.

The X_e is employed to represent the V and θ of electricity system in Equation (52), and measurement noise e_{P_i} and e_{Q_i} are included as follows:

$$X_e = \begin{bmatrix} \theta \\ V \\ e_{P_i} \\ e_{Q_i} \end{bmatrix} \tag{55}$$

where e_{P_i} and e_{Q_i} take the standard deviation of P and Q, respectively.

Then, we use B_e^* to represent the measurements P and Q in the electricity system, and extend the matrix as follows:

$$B_e^* = \begin{bmatrix} P \\ Q \\ e_{P_i} \\ e_{Q_i} \end{bmatrix} \tag{56}$$

The relationship between the extended electricity system parameter matrix A_e^*, electricity system state variables X_e, and measurement B_e^* is established as follows:

$$\left[\begin{array}{c:c} A_e & \begin{pmatrix} 1 & 0 \\ 0 & 1 \end{pmatrix} \\ \hdashline \begin{pmatrix} 0 & 0 \\ 0 & 0 \end{pmatrix} & \begin{pmatrix} 1 & 0 \\ 0 & 1 \end{pmatrix} \end{array}\right] \begin{bmatrix} \theta \\ V \\ e_{P_i} \\ e_{Q_i} \end{bmatrix} = \begin{bmatrix} P \\ Q \\ e_{P_i} \\ e_{Q_i} \end{bmatrix} \tag{57}$$

We denote Equation (57) as:

$$A_e^* X_e = B_e^* \tag{58}$$

Then, considering the computational dimension of the matrix and taking $\sum_e^{-1} \cdot [e_{P_i}; e_{Q_i}] = [0; 0]$ as the target for calculation, we extend the measurement variance-covariance matrix $\sum_e$ considering measurement correlations in Equation (40) as follows:

$$\delta_e = \left[\begin{array}{c:c} \begin{pmatrix} 1 & 0 \\ 0 & 1 \end{pmatrix} & \begin{pmatrix} 1 & 0 \\ 0 & 1 \end{pmatrix} \\ \hdashline \begin{pmatrix} 0 & 0 \\ 0 & 0 \end{pmatrix} & \sum_e^{-1} \end{array}\right] \tag{59}$$

where δ_e is the expanded measurement variance-covariance matrix with regard to $\sum_e$.

Similar to gas systems, we combine expanded electricity system parameter matrix A_e^* and the expanded measurement variance-covariance matrix δ_e considering measurement correlations in the interval, as follows:

$$\widetilde{A}_e = A_e^* + s\delta_e,\ \widetilde{A}_e \in \left[\widetilde{A}_e^* - \delta_e, \widetilde{A}_e^* + \delta_e\right] \tag{60}$$

where $\widetilde{A}_e$ is the electricity system state matrix containing measurement correlations.

We rewrite the measurement B_e^* containing the measurement correlations of electricity system, and further express it via interval $[B_e]$, and introduce ΔB_e, as follows:

$$B_e = B_e^* + s\Delta B_e,\ B_e \in [B_e^* - \Delta B_e, B_e^* + \Delta B_e] \tag{61}$$

$$\Delta B_e = \begin{bmatrix} e_{P_i} \\ e_{Q_i} \\ 0 \\ 0 \end{bmatrix} \tag{62}$$

where e_{P_i} and e_{Q_i} take the standard deviation of P and Q, respectively.

We represent X_e of electricity system as the interval $[X_e]$. According to Equations (58)–(61), the state estimation interval approximates linear equation; in considering measurement correlations of electricity systems, we can obtain:

$$[A_e^* + s\delta_e][X_e] \approx [B_e^* + s\Delta B_e] \tag{63}$$

4. Determining the Range of State Estimation Intervals with Measurement Correlations

In this section, we put forward an approach to determine the range of state estimation intervals containing measurement correlations. We represent the natural gas network state variables X_g and the electricity system state variables X_e as interval $[X]$, the gas system measurement B_g as well as the electricity system measurement B_e as electricity-gas system measurement $[B]$; and A_g^* and A_e^* as A^*, respectively. Combining Equations (51) and (63), the linear equation of state estimation interval considering electricity-gas system measurement correlations is stated as:

$$[A^* + s\delta][X] \approx [B^* + s\Delta B] \tag{64}$$

We define $U = (A^*)^{-1} s\delta$ and multiply Equation (64) by $(A^*)^{-1}$ to obtain:

$$[X] + U[X] = (A^*)^{-1}B^* + (A^*)^{-1}s\Delta B \tag{65}$$

Then, we replace $(A^*)^{-1}B^*$ with $\widetilde{X}$, and further define $[X] = \widetilde{X} + s\Delta X$, and Equation (65) is stated as:

$$\widetilde{X} + s\Delta X + U\left(\widetilde{X} + s\Delta X\right) = \widetilde{X} + (A^*)^{-1}s\Delta B \tag{66}$$

After simplification, we obtain:

$$s\Delta X^{(k+1)} = \left(\widetilde{A}^*\right)^{-1} \cdot s\Delta B - U\widetilde{X} - Us\Delta X^{(k)} \tag{67}$$

The termination condition for iteration is set to:

$$\left\|\Delta X^{(k+1)}\right\|_\infty - \left\|\Delta X^{(k)}\right\|_\infty < \varepsilon,\ \varepsilon > 0 \tag{68}$$

5. Case Studies

In the current experimental conditions, our method is validated on an integrated electricity-gas system shown in Figure 2, which is comprised of a 10-node natural gas network and the IEEE 30-bus system. In Figure 2, there are six generators in the electricity system, where G2, G5 are gas-fired generators. In the natural gas network, nodes five and ten are load nodes. Node six is connected to G5, and node nine is connected to G2. Table 1 provides standard parameters of gas in the pipeline network, Table 2 gives the gas network parameters, Table 3 shows the node parameters. The IEEE 30-bus system parameters are given by Matpower 4.0 [26]. We also simulate the integrated electricity-gas system to calculate the established state estimation model in MATLAB/Simulink (2021 version). MATLAB is an effective tool for simulating hybrid energy systems [27], which can, in combination with Matpower [28,29], be used to test and calculate the power flow of the power system. The widespread use of MATLAB in various energy system situations sufficiently demonstrates its effectiveness.

Table 1. Natural gas operating parameters.

Term	Value	Term	Value	Term	Value
Z	0.9	R	500 J/(kg·K^{-1})	T	278 K

Table 2. Node parameters of 10-node natural gas network.

Node	Gas Injection (kg/s)	Pressure (MPa)
5	−16.3	3.01
6	−11.8	3.04
9	−13.2	2.86
10	−18.7	2.93

Table 3. 10-node natural gas network parameters.

Number	From	To	L (km)	f	d (m)	Mass Flow (kg/s)
1	1	2	10	0.01	0.5	60
2	2	3	20	0.012	0.4	34.6
3	2	4	15	0.011	0.45	25.4
4	3	5	10	0.01	0.5	16.3
5	3	7	15	0.011	0.45	18.3
6	4	6	10	0.01	0.5	11.8
7	4	7	20	0.012	0.4	13.6
8	7	8	5	0.01	0.4	31.9
9	8	9	5	0.01	0.4	13.2
10	8	10	5	0.01	0.4	18.7

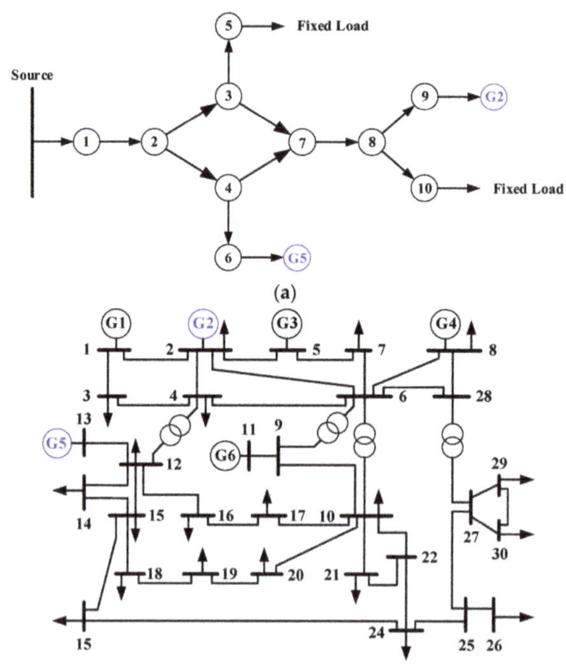

Figure 2. The integrated electricity-gas system comprising 10-node natural gas network and IEEE 30-bus system. (**a**) Topology of 10-node natural gas network; (**b**) Topology of IEEE 30-bus system.

5.1. Case 1: The Correlation Coefficient $\rho_{\Delta\pi_j^t,\Delta\dot{G}_j^t}$ between Measurements $\Delta\pi_j^t$ and $\Delta\dot{G}_j^t$ and the Correlation Coefficient ρ_{P_i,Q_i} between Measurements P_i and Q_i Are Set to 0.15

In this section, considering the existence of measurement correlations, our approach is used to estimate integrated electricity-gas system state. As shown in Figure 2, the natural gas network is connected to the electricity system through gas-fired generators G5 and G2 respectively, and nodes six and nine are supplied with gas by nodes four and eight respectively. Therefore, changes in gas flow at node six and node nine will result in changes in pressure of nodes four and eight. When gas demand of gas-fired generator G2 changes constantly, we track inlet pressure dynamic response of node 8. We set measurement correlation coefficients $\rho_{\Delta\pi_j^t,\Delta\dot{G}_j^t}$ and ρ_{P_i,Q_i} in Equations (29) and (39) to 0.15, and set termination threshold ε in Equation (68) as $\varepsilon = 10^{-4}$. The gas flow of G2 at node 9 and pressure at node 8 are estimated with our method, and the results are compared with those of the Krawczyk operator (KO) interval method based on unscented transformation (UT). Krawczyk Operator (KO) is an iterative method that takes the interval vector obtained by the interval Gaussian elimination method as the initial value, and obtains the solution set of the interval linear equation by iterating the interval vector. The Krawczyk operator (KO) interval method based on unscented transformation (UT) mainly obtains the correlation between measurements through the symmetric sampling strategy of unscented transformation (UT), and describes and calculates the interval range where the correlation exists by using the Krawczyk operator (KO) interval method.

Figure 3 provides state estimation uncertainty ranges caused by measurement correlations obtained through the proposed method and UT+KO method. This paper mainly studies the slow change of natural gas flow. It can be intuitively seen that, compared with the UT+KO method, the interval boundary of the state variable of gas flow variation and pressure change given by the proposed method is more compact; while the interval upper boundary and lower boundary provided by the UT+KO method is far from real values, the interval range is larger. This is because UT requires independent calculation of measurements, and the KO interval method requires measurements to participate in multiple iterative operation. Multiple iterative operation of measurement interval at the same time will lead to the superposition and transmission of measurement correlations, which inevitably expands the interval range. These will easily exceed the operating limit of the system, thus losing the reference value.

Then, the voltage magnitude and voltage angle of electricity system is estimated. Figures 4 and 5 give voltage amplitude interval, voltage angle interval, and corresponding interval widths when the measurement correlation coefficient $\rho_{P_i,Q_i} = 0.15$. In Figures 4a and 5a, we use purple vertical lines to represent the voltage amplitude range obtained through the proposed method, green vertical lines to represent the results provided by the UT+KO method, and blue horizontal lines to represent the real value. In Figures 4b and 5b, we highlight the voltage amplitude interval width and voltage angle interval width given by our method and UT+KO method in red and blue, respectively. Obviously, the upper boundary and lower boundary of interval obtained by our method are closer to real values, while interval range provided by the UT+KO method is larger, and the estimation results are somewhat conservative. This is mainly due to the superposition and transfer of measurement correlations of each measurement interval in the process of iterative calculation of multiple measurement intervals by the UT+KO method, which enlarges the interval and makes estimation results more conservative.

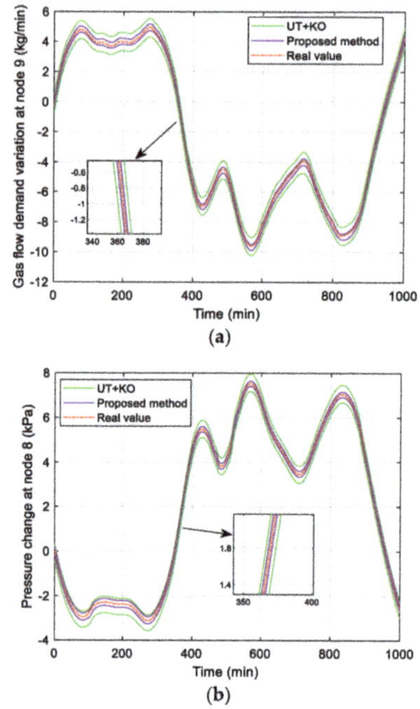

Figure 3. State estimation interval bound of the natural gas system, considering the measurement correlation coefficient $\rho_{\Delta\pi_j^t,\Delta\dot{G}_j^t} = 0.15$. (**a**) Gas flow demand variation; (**b**) Pressure change.

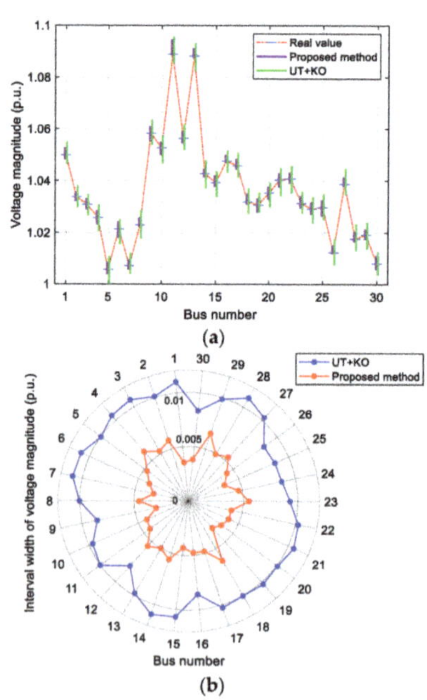

Figure 4. State estimation interval bound of IEEE 30-bus system with measurement correlation coefficient $\rho_{P_i,Q_i} = 0.15$. (**a**) Voltage magnitude; (**b**) Interval width of voltage magnitude.

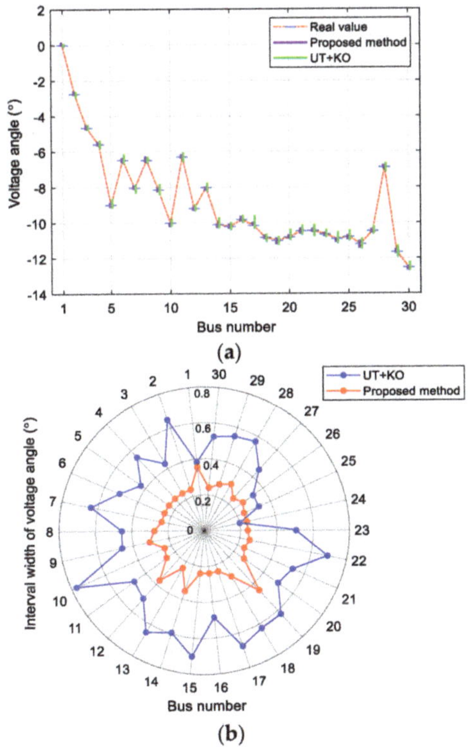

Figure 5. State estimation interval bound of IEEE 30-bus system with measurement correlation coefficient $\rho_{P_i,Q_i} = 0.15$. (**a**) Voltage angle; (**b**) Interval width of voltage angle.

5.2. Case 2: The Correlation Coefficient $\rho_{\Delta\pi_j^t,\Delta\dot{G}_j^t}$ between Measurements $\Delta\pi_j^t$ and $\Delta\dot{G}_j^t$ and the Correlation Coefficient ρ_{P_i,Q_i} between Measurements P_i and Q_i Are Set to 0.3

To further verify the effectiveness of our method at different measurement correlation levels, the measurement correlation coefficients $\rho_{\Delta\pi_j^t,\Delta\dot{G}_j^t}$ and ρ_{P_i,Q_i} are set to 0.3, and the changes in gas flow demand and pressure at nodes are estimated, as shown in Figure 6.

In Figure 6, the estimated range of gas flow demand variation and pressure change at $\rho_{\Delta\pi_j^t,\Delta\dot{G}_j^t} = 0.3$ is greater than that at $\rho_{\Delta\pi_j^t,\Delta\dot{G}_j^t} = 0.15$ in Figure 3. The measurement correlation level therefore affects the estimation accuracy. When the measurement correlation is large, the upper bound and lower bound of estimation interval will deviate from the ideal state interval. Although state estimation interval range will increase as the measurement correlations increase, our method is superior to the UT+KO method.

Furthermore, we estimate the electricity system state when the measurement correlation coefficient $\rho_{P_i,Q_i} = 0.3$; the results are shown in Figures 7 and 8.

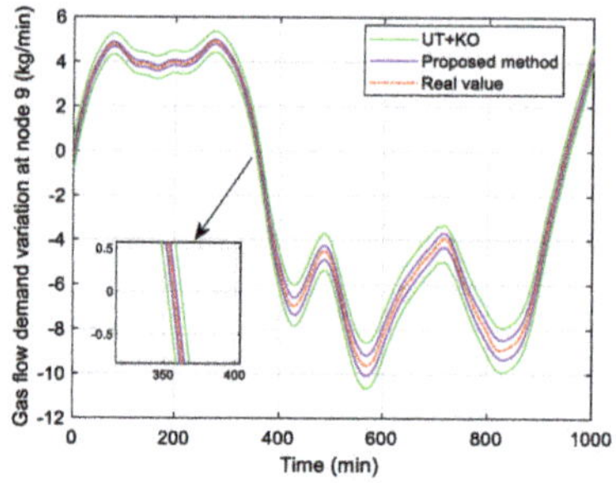

(a)

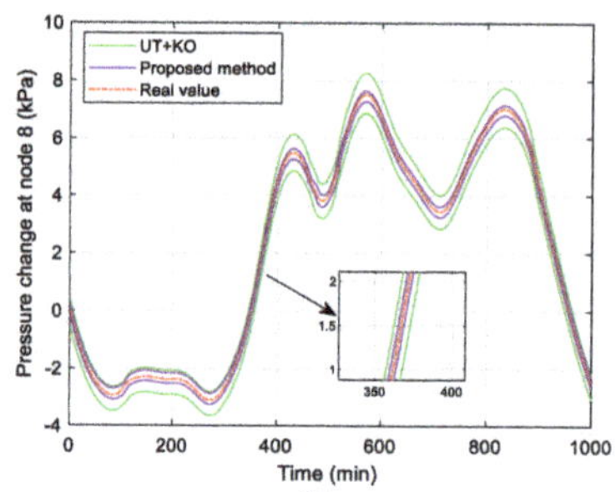

(b)

Figure 6. State estimation interval bound of natural gas system with measurement correlation coefficient $\rho_{\Delta\pi_j^t,\Delta\dot{G}_j^t} = 0.3$. (**a**) Gas flow demand variation; (**b**) Pressure change.

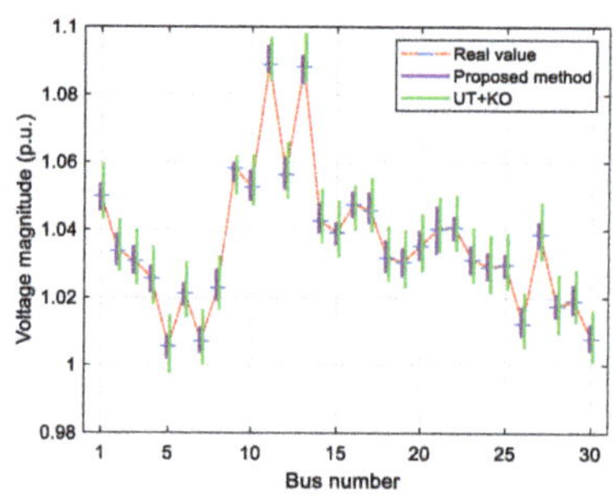

(a)

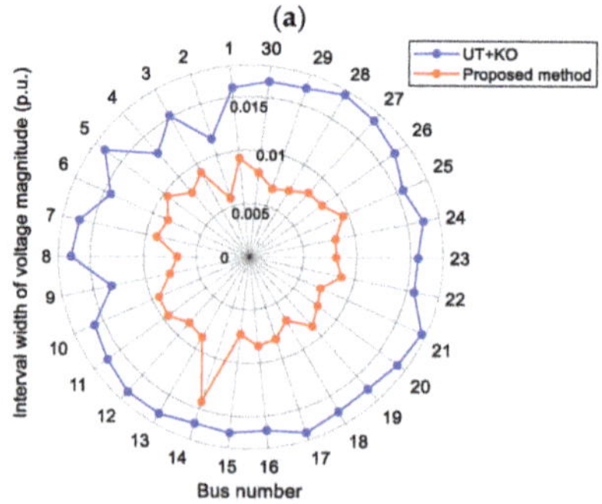

(b)

Figure 7. State estimation interval bound of IEEE 30-bus system with measurement correlation coefficient $\rho_{P_i,Q_i} = 0.3$. (**a**) Voltage magnitude; (**b**) Interval width of voltage magnitude.

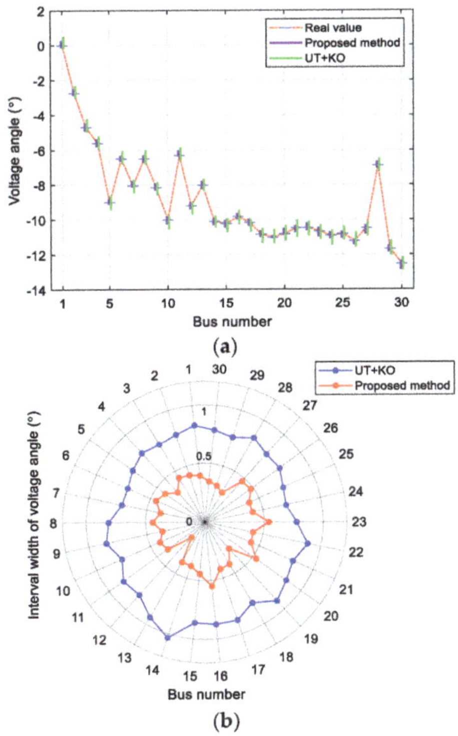

Figure 8. State estimation interval bound of IEEE 30-bus system with measurement correlation coefficient $\rho_{P_i,Q_i} = 0.3$. (**a**) Voltage angle; (**b**) Interval width of voltage angle.

In Figures 7 and 8, when measurement correlation coefficient ρ_{P_i,Q_i} changes from 0.15 to 0.3, the estimated ranges of voltage magnitude and voltage angle expand. Both our method and the UT+KO method are affected to a certain extent, but the overall variation of our method is relatively small. This is because our method does not need to perform the iterative calculation of measurement interval at the same time, which avoids the accumulation and transmission of measurement correlations.

Furthermore, we evaluate the accuracy of results by using the following two indicators:

$$M_1 = \frac{1}{n}\sum\nolimits_{i=1}^{n}(\overline{x}_i - \underline{x}_i) \tag{69}$$

$$M_2 = \max(\overline{x}_i - \underline{x}_i) \tag{70}$$

where $\overline{x}_i$ represents the upper boundary, and $\underline{x}_i$ represents the lower boundary of the interval variable, respectively. M_1 is the interval width average value, M_2 is the interval width maximum value. When M_1 and M_2 are smaller, accuracy is higher.

Table 4 provides statistics for the estimation results of integrated electricity-gas system given by two methods, and provides the corresponding estimation accuracy indicators when the measurement correlation coefficient is 0.15 and 0.3, respectively. The results indicate that although estimation results accuracy decreases as the measurement correlation increases, the M_1 and M_2 indicators corresponding to the proposed method are smaller. This is mainly due to the accumulation of the measurement correlation of UT+KO method during the multiple iterative calculation of various measurement intervals, which enlarges interval range and reduces estimation results accuracy. In addition, Table 5 provides the calculation time of our method and UT+KO method. It can be seen that when the measurement correlation coefficients are smaller, the average execution time is shorter. The UT+KO method uses multiple measurement intervals iterative calculation, so the calculation time is longer.

Table 4. Estimation accuracy (of the IEGS) of different measurement correlation coefficients.

Accuracy Indices		The Measurement Correlation Coefficients $\rho_{\Delta\pi_j^t,\Delta\dot{G}_j^t}$=0.15, ρ_{P_i,Q_i}=0.15		The Measurement Correlation Coefficients $\rho_{\Delta\pi_j^t,\Delta\dot{G}_j^t}$=0.3, ρ_{P_i,Q_i}=0.3	
		M_1	M_2	M_1	M_2
Proposed method UT+KO	Gas flow demand variation at node 9: $\Delta\dot{G}_i^t$	0.3964 0.8070	0.4541 0.9681	0.5391 1.2772	0.9308 2.0476
Proposed method UT+KO	Pressure change at node 8: $\Delta\pi_i^t$	0.2749 0.7523	0.3516 0.7846	0.3704 0.8961	0.6096 1.3971
Proposed method UT+KO	Voltage magnitude of IEEE 30-bus system: V	0.0046 0.0099	0.0065 0.0110	0.0084 0.0161	0.0143 0.0176
Proposed method UT+KO	Voltage angle of IEEE 30-bus: θ_{ij}	0.371 0.527	0.4486 0.7703	0.4055 0.8143	0.5495 1.0329

Table 5. IEGS computation time with different measurement correlation coefficients.

Method	Average Execution Time (s)	
	The Measurement Correlation Coefficients $\rho_{\Delta\pi_j^t,\Delta\dot{G}_j^t}$=0.15, ρ_{P_i,Q_i}=0.15	The Measurement Correlation Coefficients $\rho_{\Delta\pi_j^t,\Delta\dot{G}_j^t}$=0.3, ρ_{P_i,Q_i}=0.3
Proposed method UT+KO	2.68 43.59	3.41 54.37

6. Conclusions

An interval state estimation method that considers the measurement correlations of electricity-gas systems is presented in this paper. The linear measurement model of electricity-gas systems is derived via Taylor series expansion, and the measurement variance-covariance matrix with measurement correlations is estimated. Then, the system parameter matrix and the measurement variance-covariance matrix with measurement correlations are combined into an interval, and the interval state matrix and linear equations of state estimation interval considering measurement correlations are constructed; as a result, the state estimation method for the electricity-gas system with measurement correlations is proposed. Finally, a method for determining state estimation interval range is presented. Comparing the state estimation interval with the safe operating limit range of the system will help system administrators to make effective judgments and effectively adjust electricity-gas systems. The numerical tests on an integrated electricity-gas system illustrate that our method outperforms the UT+KO method in terms of calculation accuracy and efficiency.

This paper quantifies and describes the measurement correlations of electricity-gas systems, but does not deeply consider the effects of measurement correlations on electricity-gas systems. In future work, we will further analyze the impacts of measurement correlations on electricity-gas systems, and will develop an adaptive interval state estimation method that dynamically adjusts the estimation process on the basis of real-time measurement data. This will improve the robustness and adaptability of interval state estimation, helping it to adapt to different working conditions and system changes.

Author Contributions: Conceptualization, Y.H.; methodology, Y.H.; software, L.F.; validation, Y.H. and L.F.; formal analysis, Y.H.; investigation, Y.H.; resources, Y.H.; data curation, L.F.; writing—original draft preparation, Y.H.; writing—review and editing, L.F.; visualization, Y.H.; supervision, L.F.; project administration, L.F.; funding acquisition, L.F. All authors have read and agreed to the published version of the manuscript.

Funding: This research was funded by the National Natural Science Foundation of PR China (61972064) the Liaoning Revitalization Talents Program (XLYC1806006); and the Fundamental Research Funds for the Central Universities (DUT19RC(3)012).

Data Availability Statement: All data are referenced in the article.

Conflicts of Interest: The authors declare no conflicts of interest.

References

1. Jiang, Y.; Ren, Z.; Yang, X. A steady-state energy flow analysis method for integrated natural gas and power systems based on topology decoupling. *Appl. Energy* **2022**, *306*, 118007. [CrossRef]
2. Aminifar, F.; Shahidehpour, M.; Fotuhi-Firuzabad, M. Power system dynamic state estimation with synchronized phasor measurements. *IEEE Trans. Instrum. Meas.* **2013**, *63*, 352–363. [CrossRef]
3. Muscas, C.; Pau, M.; Pegoraro, P.A. Effects of measurements and pseudomeasurements correlation in distribution system state estimation. *IEEE Trans. Instrum. Meas.* **2014**, *63*, 2813–2823. [CrossRef]
4. Santhosh, K.C.; Rajawat, K.; Chakrabarti, S. Robust distribution system state estimation with hybrid measurements. *IET Gener. Transm. Distrib.* **2020**, *14*, 3250–3259. [CrossRef]
5. Caro, E.; Conejo, A.J.; Minguez, R. Multiple bad data identification considering measurement dependencies. *IEEE Trans. Power Syst.* **2011**, *26*, 1953–1961. [CrossRef]
6. Zhao, J.; Wang, S.; Mili, L. A robust state estimation framework considering measurement correlations and imperfect synchronization. *IEEE Trans. Power Syst.* **2018**, *33*, 4604–4613. [CrossRef]
7. Singh, A.K.; Pal, B.C. Decentralized dynamic state estimation in power systems using unscented transformation. *IEEE Trans. Power Syst.* **2013**, *29*, 794–804. [CrossRef]
8. Caro, E.; Valverde, G. Impact of transformer correlations in state estimation using the unscented transformation. *IEEE Trans. Power Syst.* **2013**, *29*, 368–376. [CrossRef]
9. Zhao, J. Power system dynamic state estimation considering measurement correlations. *IEEE Trans. Energy Conver.* **2017**, *32*, 1630–1632. [CrossRef]
10. Zanni, L.; Le Boudec, J.Y.; Cherkaoui, R. A prediction-error covariance estimator for adaptive Kalman filtering in step-varying processes: Application to power-system state estimation. *IEEE Trans. Contr. Syst. Technol.* **2016**, *25*, 1683–1697. [CrossRef]
11. Lu, Z.; Gong, D.; Sun, Y. A New State Updating Approach in Power System Dynamic State Estimation Considering Correlated Measurements. In Proceedings of the 2017 Eighth International Conference on Intelligent Control and Information Processing (ICICIP), Hangzhou, China, 3–5 November 2017.
12. Lu, Z.; Wei, Z.; Sun, Y. Power system dynamic state estimation considering correlation of measurement error from PMU and SCADA. *Concurr. Comp.-Pract. Exp.* **2019**, *31*, e4726. [CrossRef]
13. Yang, Y.; Hu, W.; Min, Y. Projected unscented Kalman filter for dynamic state estimation and bad data detection in power system. In Proceedings of the 2017 Iet International Conference on Developments in Power System Protection, Copenhagen, Denmark, 31 March–3 April 2014.
14. Chakhchoukh, Y.; Vittal, V.; Heydt, G.T. PMU based state estimation by integrating correlation. *IEEE Trans. Power Syst.* **2013**, *29*, 617–626. [CrossRef]
15. Zhao, J.; Zhang, G.; Dong, Z.Y. Robust forecasting aided power system state estimation considering state correlations. *IEEE Trans. Smart Grid.* **2016**, *9*, 2658–2666. [CrossRef]
16. Chakhchoukh, Y.; Vittal, V.; Heydt, G.T. LTS-based robust hybrid SE integrating correlation. *IEEE Trans. Power Syst.* **2016**, *32*, 3127–3135. [CrossRef]
17. Lu, Z.; Yang, S.H.; Yang, S. Mixed measurement-based power system state estimation with measurement correlation. In *2016 Advances in Power and Energy Engineering: Proceedings of the 8th Asia-Pacific Power and Energy Engineering Conference*; CRC Press: Boca Raton, FL, USA, 2016.
18. Caro, E.; Conejo, A.J.; Minguez, R. Power system state estimation considering measurement dependencies. *IEEE Trans. Power Syst.* **2009**, *24*, 1875–1885. [CrossRef]
19. Abedi, B.; Ghadimi, A.A.; Abolmasoumi, A.H. An improved TPM-based distribution network state estimation considering loads/DERs correlations. *Electr. Eng.* **2021**, *103*, 1541–1553. [CrossRef]
20. Valverde, G.; Saric, A.T.; Terzija, V. Stochastic monitoring of distribution networks including correlated input variables. *IEEE Trans. Power Syst.* **2012**, *28*, 246–255. [CrossRef]
21. Thorley, A.; Tiley, C.H.; Terzija, V. Unsteady and transient flow of compressible fluids in pipelines—A review of theoretical and some experimental studies. *Int. J. Heat Fluid Flow.* **1987**, *8*, 3–15. [CrossRef]
22. Osiadacz, A. Simulation of transient gas flows in networks. *Int. J. Numer. Methods Fluids* **1984**, *4*, 13–24. [CrossRef]
23. Abur, A.; Exposito, A.G.; Sun, Y. *Power System State Estimation: Theory and Implementation*; CRC Press: Boca Raton, FL, USA, 2004; pp. 79–93.
24. Sheikhi, A.; Bahrami, S.; Ranjbar, A.M. An autonomous demand response program for electricity and natural gas networks in smart energy hubs. *Energy* **2015**, *89*, 490–499. [CrossRef]

25. Liu, Y.; Wang, Y.; Zhang, N. A data-driven approach to linearize power flow equations considering measurement noise. *IEEE Trans. Smart Grid.* **2019**, *11*, 2576–2587. [CrossRef]
26. Zimmerman, R.D.; Murillo-Sánchez, C.E.; Thomas, R.J. MATPOWER: Steady-state operations, planning, and analysis tools for power systems research and education. *IEEE Trans. Power Syst.* **2010**, *26*, 12–19. [CrossRef]
27. Xu, X.; Jia, H.; Chiang, H.D. Dynamic modeling and interaction of hybrid natural gas and electricity supply system in microgrid. *IEEE Trans. Power Syst.* **2014**, *30*, 1212–1221. [CrossRef]
28. Xu, J.; Wu, Z.; Dou, X. An interval arithmetic-based state estimation for unbalanced active distribution networks. In Proceedings of the 2017 IEEE Power & Energy Society General Meeting, Chicago, IL, USA, 16–20 July 2017.
29. D'Antona, G. Power system static-state estimation with uncertain network parameters as input data. *IEEE Trans. Instrum. Meas.* **2016**, *65*, 2485–2494. [CrossRef]

electronics

Article

Control Strategies of Thrust Ripple Suppression for Electromagnetic Microgravity Facility

Yuman Li [1,2], Wenbo Dong [1,2,3,*], Congmin Lv [1,2,3], Zhe Wang [1,3] and Yongkang Zhang [1,3]

1 Technology and Engineering Center for Space Utilization, Chinese Academy of Sciences, Beijing 100094, China; lvcongmin@csu.ac.cn (C.L.)
2 University of Chinese Academy of Sciences, Beijing 100049, China
3 Key Laboratory of Space Utilization, Chinese Academy of Sciences, Beijing 100094, China
* Correspondence: wbdong@csu.ac.cn

Abstract: This paper presents an innovative solution that is able to suppress the thrust ripple in a high-power asynchronous linear induction motor (LIM) used in a microgravity experiment facility electromagnetic launch (MEFEL) system. By addressing the crucial need for low levels of thrust ripple in MEFEL applications, we propose a dynamic model-based adaptive controller (MAC) and an enhanced quasi-proportional-resonant (PR) controller. The MAC is designed to compensate for the inherent impedance asymmetry of the linear motor. The PR controller minimizes thrust ripple by eliminating harmonics within the current loop. A comparative analysis indicates that both MAC and PR control are effective in reducing harmonics, suppressing the thrust ripple, and maintaining system stability. Computer simulations show a noteworthy 75% reduction in the thrust ripple and a decrease in the negative current. Partial tests on the MEFEL device validate the practical efficacy of the proposed control methods, emphasizing the method's ability to enhance the quality of microgravity in real-world scenarios significantly.

Keywords: linear induction motor; control system simulation; model-based adaptive control; thrust ripple suppression; microgravity facility; electromagnetic launch

Citation: Li, Y.; Dong, W.; Lv, C.; Wang, Z.; Zhang, Y. Control Strategies of Thrust Ripple Suppression for Electromagnetic Microgravity Facility. *Electronics* **2024**, *13*, 1247. https://doi.org/10.3390/electronics13071247

Academic Editor: Antonio Moreno-Munoz

Received: 25 January 2024
Revised: 15 March 2024
Accepted: 16 March 2024
Published: 27 March 2024

1. Introduction

The electromagnetic microgravity experimental facility is a device used to simulate space environments on Earth [1]. Microgravity drop towers usually simply drop the experimental capsule from the top of the tower to create a microgravity environment. The improved drop towers are equipped with catapult systems to launch the experimental capsule so that the experimental capsule can perform the round-way free-fall motion. The ability to achieve microgravity time could be doubled by using catapult methods. There are two main ways to catapult the experimental capsule in drop towers, hydraulic catapult and linear motor catapult. Compared with hydraulic catapult, the method using linear motors has a better overload character which benefits the microgravity experimental objects.

This paper researches an electromagnetic microgravity facility system as shown in Figure 1, which utilizes linear induction motors to drive the experimental capsule [2]. The microgravity environment within the experimental capsule is generated by vertically catapulting the experimental capsule using a LIM. During the free-fall motion of the experimental capsule, microgravity experiments can be conducted within the capsule [3]. To achieve the precision necessary for microgravity experiments [4], the LIM's natural thrust ripple must be controlled through advanced methods.

Apart from their application in microgravity facilities, LIMs are widely used in high-power and high-speed applications [5,6], ranging from aircraft carrier electromagnetic catapults to the field of rail transportation. Thrust ripple suppression is a common challenge faced in the study of LIMs. The structure of the LIM is shown in Figure 2, where the finiteness of both stator and mover generates end effects, thereby causing the thrust ripple.

In order to investigate the exact mechanisms of the thrust ripple, scholars have meticulously analyzed the model of the LIM through various approaches. Scholarly research on thrust ripples in LIMs is primarily focused on three aspects. The first involves analyzing the magnetic field changes in the structure of linear motors from a magnetic field perspective to understand the characteristics of flux variations and the resulting thrust ripples [7,8]. However, this approach often relies on simulation for theoretical analysis and is challenging to integrate with practical control applications. The second aspect involves analyzing the impedance asymmetry of LIMs from a mathematical model perspective [9,10]. One study [9] mainly analyzes impedance asymmetry in the ABC coordinate system, while another [10] provides a dq coordinate system's inductance matrix model, which is more convenient for application in field-oriented control (FOC) systems. The third aspect involves analyzing the equivalent circuit perspective, introducing a coefficient $f(Q)$ at the excitation inductance to reflect the end effects of a LIM [11–13].

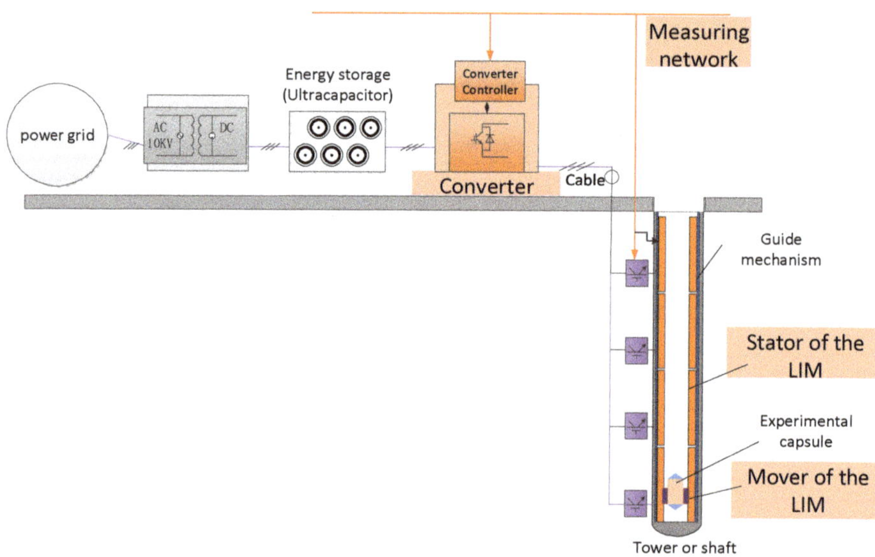

Figure 1. Microgravity electromagnetic catapult experimental system.

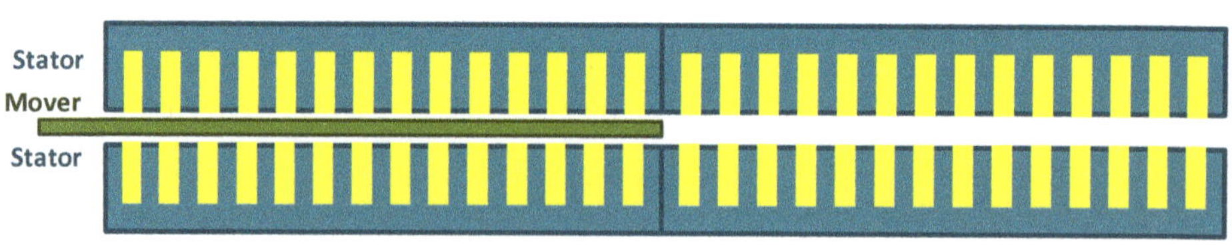

Figure 2. Linear induction motor.

This study focuses on analyzing the thrust ripples of LIMs based on a mathematical model, where the parameters can be obtained through experimental measurements. Research indicates that the impedance asymmetry in LIMs manifests as an imbalance in currents in the control system [9,14,15]. Therefore, reducing thrust ripples can be achieved by eliminating current imbalances. Study [16] indicates that when the stator currents in a LIM exhibit three-phase balance, the three-phase voltages must be unbalanced. Thus, improvements to traditional three-phase voltage sources are necessary to eliminate thrust ripples. One proposed enhancement involves injecting negative sequence voltage harmonics [17–19]. However, this method faces difficulties in practical engineering applications due to complex calculations and the need to parallel two power sources, which is detrimental to the stability of the control system. Some scholars have addressed the issue of current imbalance by eliminating harmonics in the *dq*-axis currents in FOC, such as applying dynamic adjustments of the reference values for *dq*-axis currents to eliminate thrust ripples [20], or adding PR controllers to *dq*-axis currents [21,22]. However, *dq*-axis currents are obtained through Clark–Park transformation, and the parameters in Clark–Park are based on a three-phase symmetrical sinusoidal system. For LIMs, at least one of the currents or voltages is imbalanced, making direct improvements to *dq*-axis currents in FOC less precise.

In summary, previous research has either focused on the establishment of models or on model-free control methods. This research investigates whether the control effect can be further improved by utilizing model-based adaptive control based on known motor parameters. This is achieved by introducing a model-based controller for adjusting the voltage phase, thereby reducing current imbalance and suppressing the thrust ripple. In the current controller, the harmonics of i_d and i_q are analyzed and a quasi-PR controller is employed to reduce the harmonics which also contributes to the suppression of the thrust ripple. The results and discussions of the simulation demonstrate that both the model-based controller and the quasi-PR controller can reduce the current imbalance and suppress the thrust ripple.

2. LIM Models and Design of MAC

This section discusses the differences between LIM models and rotary motor models. By addressing the uniqueness of the LIM, a model-based adaptive controller is proposed for eliminating the imbalance of current phase.

The representation of the linear induction motor in the equivalent circuit can manifest in various forms [11–13], involving the introduction of end effect coefficients at the rotor leakage inductance and excitation inductance to capture the asymmetry inherent in the LIM model. This study researched the relationship between excitation inductance in the equivalent model and the mutual inductance in the flux model, providing the feasibility of fixing the current imbalance through adjusting the voltage phase.

The steady-state model of a rotary induction motor is commonly expressed as the following equivalent circuit shown in Figure 3, where R_S and $L_{s\sigma}$ are the stator resistance and stator leakage inductance. R_r/s and $L_{r\sigma}$ are the rotor resistance with slip rate and rotor leakage inductance referred to the stator side. L_m is the excitation inductance. U_s is the voltage of stator winding.

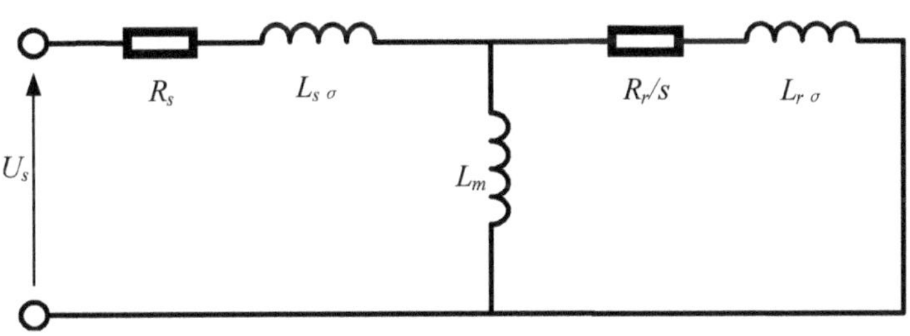

Figure 3. T-type equivalent circuit.

The excitation inductance L_m in the equivalent circuit satisfies the following:

$$L_m = \frac{3}{2} L_{m1} \tag{1}$$

where L_m is the excitation inductance and L_{m1} is the main self-inductance of stator windings. The circuit itself reflects the voltage–current relationship of a single-phase winding. In rotary induction motors, the three-phase parameters are equal so the equivalent circuit can represent three-phase windings.

However, in LIMs the excitation inductances of three-phase windings are not equal due to the imbalance of stator inductance matrix. When the currents are balanced, the excitation inductances can be expressed as (The detailed derivation process can be found in Appendix A)

$$\begin{aligned} L_{mA} &= L_{AA} - \tfrac{1}{2}L_{AB} - \tfrac{1}{2}L_{AC} \\ L_{mB} &= L_{BB} - \tfrac{1}{2}L_{BA} - \tfrac{1}{2}L_{BC} \\ L_{mC} &= L_{CC} - \tfrac{1}{2}L_{CA} - \tfrac{1}{2}L_{CB} \end{aligned} \quad (2)$$

$$L_{ss} = \begin{bmatrix} L_{AA} & L_{AB} & L_{AC} \\ L_{BA} & L_{BB} & L_{BC} \\ L_{CA} & L_{CB} & L_{CC} \end{bmatrix} \quad (3)$$

where L_{mA}, L_{mB} and L_{mC} are the excitation inductances of phase A, phase B, and phase C. L_{AA}, L_{BB} and L_{CC} are the self-inductances of phase A, phase B, and phase C. $L_{AB} = L_{BA}$ is the mutual inductance of phase A and phase B. $L_{AC} = L_{CA}$ is the mutual inductance of phase A and phase C. $L_{BC} = L_{CB}$ is the mutual inductance of phase A and phase C. L_{ss} is the stator inductance matrix. In a rotary motor, the stator flux caused by stator current can be expressed as follows:

$$\psi_{sA} = L_{AA}i_A + L_{AB}i_B + L_{AC}i_C = (L_{s\sigma} + \frac{3}{2}L_{m1})i_A \quad (4)$$

where ψ_{sA} represents the stator–stator flux in phase A. i_A, i_B, and i_C are the phase currents of the stator. In LIM, the stator–stator flux in phase A can be expressed as follows:

$$\begin{aligned} \psi'_{sA} &= L_{AA}i_A + L_{AB}i_B + L_{AC}i_C \\ &= [L_{s\sigma} + L_{AA} + L_{AB}\cos(-120^\circ) + L_{AC}\cos(120^\circ)]i_A \\ &= \left(L_{s\sigma} + L_{AA} - \tfrac{1}{2}L_{AB} - \tfrac{1}{2}L_{AC}\right)i_A \end{aligned} \quad (5)$$

The calculation process for phases B and C is the same as that for phase A.

The asymmetry of L_{ss} results in the imbalance of three-phase currents because the voltage references from the control system are standard sinusoidal three-phase waves. Hence, the motivation of this research is to find a controller which can fix the current imbalance problem by adjusting the voltages.

Sine waves of voltage and current can both be represented using vectors as shown in Figure 4. The impedance is a complex value that can also be represented using vector notation. Based on the Equation (2), the impedances of each phase winding are expressed as $\dot{Z}_A$, $\dot{Z}_B$, and $\dot{Z}_C$. The phase difference of voltage can be calculated through the impedance of the three-phase windings when current is balanced.

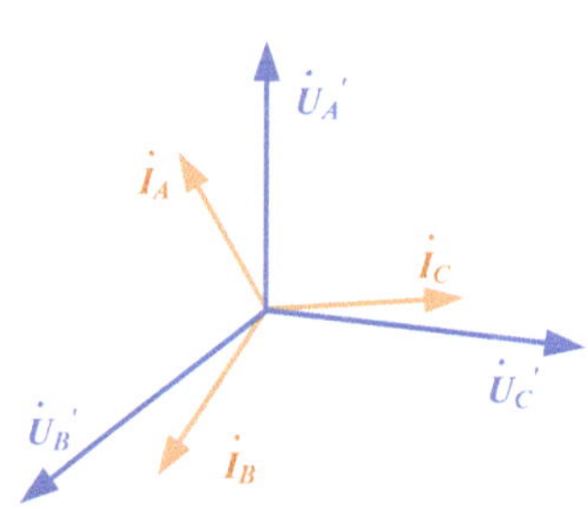

Figure 4. Vector diagram of voltages and currents.

Define φ_{AB} and φ_{AC} which satisfy the following:

$$\varphi_{AB} = 120^{\circ} + tan^{-1}\frac{Img\left(\frac{\dot{Z_A}}{\dot{Z_B}}\right)}{Real\left(\frac{\dot{Z_A}}{\dot{Z_B}}\right)} \tag{6}$$

$$\varphi_{AC} = 120^{\circ} + tan^{-1}\frac{Img\left(\frac{\dot{Z_A}}{\dot{Z_C}}\right)}{Real\left(\frac{\dot{Z_A}}{\dot{Z_C}}\right)} \tag{7}$$

Hence, the model-based adaptive controller can be designed to adjust the phase differences of the voltages by tuning the angles in the Clark transformation, thereby suppressing current imbalance.

$$\begin{bmatrix} u'_A \\ u'_B \\ u'_C \end{bmatrix} = \begin{bmatrix} 1 & 0 \\ \cos\varphi_{AB} & \sin\varphi_{AB} \\ \cos\varphi_{AC} & \sin\varphi_{AC} \end{bmatrix} \frac{2}{3} \begin{bmatrix} 1 & \cos\left(\frac{2\pi}{3}\right) & \cos\left(-\frac{2\pi}{3}\right) \\ 0 & \sin\left(\frac{2\pi}{3}\right) & \sin\left(-\frac{2\pi}{3}\right) \end{bmatrix} \begin{bmatrix} u_A \\ u_B \\ u_C \end{bmatrix} \tag{8}$$

where u_A, u_B, and u_C represent standard sinusoidal voltage reference values before adjustment and u'_A, u'_B, and u'_C represent the post-adjustment voltage reference values (We can prove that the current phase is balanced after the adjustment in Appendix B).

Based on the provided equations, the controller can optimize the voltage signals of the three phases, thereby reducing the current imbalance and suppressing the thrust ripple. In the commonly used field-oriented control (FOC) system for LIMs, the Equation (9) can be employed as a replacement of the $dq - abc$ transformation module.

$$\begin{bmatrix} u'_A \\ u'_B \\ u'_C \end{bmatrix} = \begin{bmatrix} \cos\theta & -\sin\theta \\ \cos(\theta + \varphi_{AB}) & -\sin(\theta + \varphi_{AB}) \\ \cos(\theta + \varphi_{AC}) & -\sin(\theta + \varphi_{AC}) \end{bmatrix} \begin{bmatrix} u_d \\ u_q \end{bmatrix} \tag{9}$$

where u_d and u_q are the voltages of d axis and q axis in FOC, and θ is the electrical angle. Here, we define this control strategy as a MAC, since the control matrix is adaptive to the model and changes over time.

3. Thrust Ripple Analysis and Design of Quasi-PR Controller

The introduction of the MAC eliminated the "phase imbalance" in current phase by adjusting the voltage. However, the issue of current "amplitude imbalance" has not been addressed. In the field-oriented control (FOC) system, the controlled variables i_d and i_q are both in direct current. The imbalance in current amplitude manifests as fluctuations in the d-axis and q-axis currents. Therefore, this paper considers adopting an appropriate dynamic control method to mitigate the fluctuations in the d-axis and q-axis currents.

The unbalanced three-phase sinusoidal waves can be decomposed into positive-sequence, negative-sequence, and zero-sequence components using the orthogonal component method. In LIMs, the negative-sequence component of the current introduces thrust ripple and fluctuations in the d-axis and q-axis currents. There are two types of thrust ripple that exist in LIMs [10]. One is the thrust ripple at a frequency of $2f$ (where f represents the supply frequency) generated by the stator negative-sequence current, and the other is the thrust fluctuation at a frequency of $2sf$ (where s represents slip) generated by the rotor negative-sequence current. In FOC, the thrust ripple at a frequency of $2f$ can be suppressed by reducing the stator negative-sequence current.

In the control system under FOC, the thrust ripple caused by negative sequences of stator currents results in harmonics of i_d and i_q in current loop. The proportional-resonant

(PR) controller exhibits excellent control performance in eliminating harmonics at specific frequencies. The transfer function of the ideal PR controller is given by

$$G_{PR}(s) = K_p + \frac{K_r s}{s^2 + 2\omega_c s + \omega_0^2} \tag{10}$$

Due to the high precision requirements of the ideal PR controller for the controlled object's parameters, its practical application may lead to less desirable control results. Therefore, a quasi-PR controller is utilized to address this issue [20,21].

$$G_{PR}(s) = K_p + \frac{2K_r \omega_c s}{s^2 + 2\omega_c s + \omega_0^2} \tag{11}$$

Typically, a quasi-PR controller is used to eliminate specific harmonics with fixed value of ω_c. In the context of this study, the harmonic frequency ω_c is $4\pi f$ where f is the frequency of stator currents. Thus, the transfer function for the quasi-PR controller used in the LIM can be expressed as follows:

$$G_{PR}(s) = K_p + \frac{8\pi K_r f s}{s^2 + 8\pi f s + \omega_0^2} \tag{12}$$

The PI controller can be expressed as:

$$G_{PI}(s) = K_p + \frac{K_i}{s} \tag{13}$$

Both the quasi-PR controller and the PI controller include a proportional component. In this study, the value of K_p for the quasi-PR controller is set to 0. In the subsequent simulations and experiments, the PR controller employed is the modified quasi-PR controller, as illustrated in Equation (12). The current loop is represented as shown in Figure 5, in which the asterisk (*) denotes the reference value.

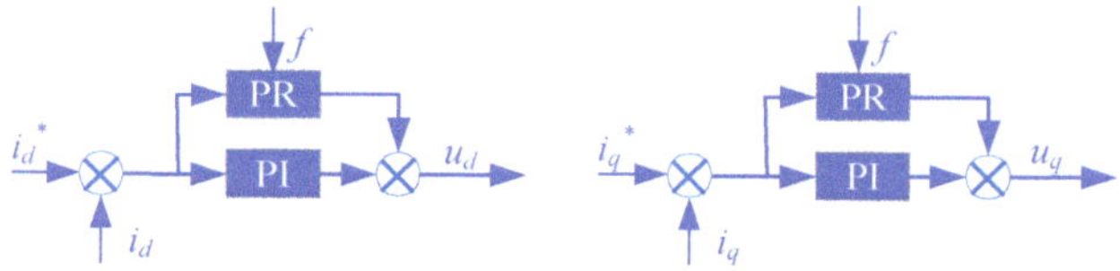

Figure 5. Current controllers for the adaptation of the quasi-PR controller.

The control system diagram of the research is illustrated as shown in Figure 6, in which the asterisk (*) denotes the reference value. Two control methods are employed to suppress the thrust ripple caused by stator current imbalance. The first method involves introducing the MAC to correct the phase imbalance of the stator current, and the second method incorporates the PR controller to correct the amplitude imbalance of the stator current.

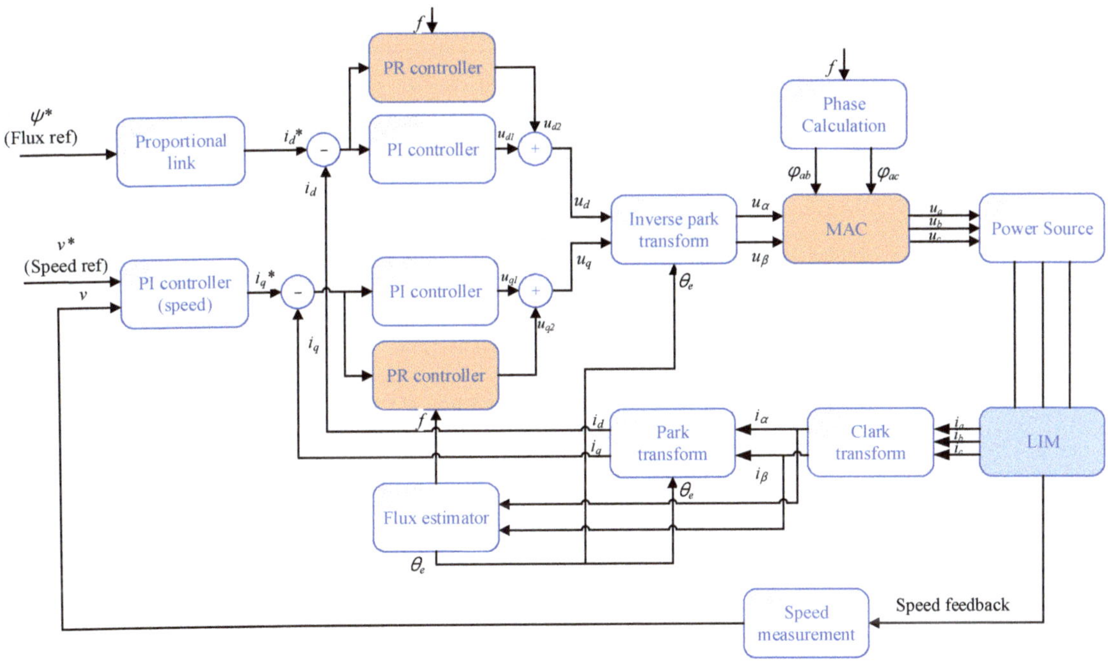

Figure 6. Framework of the control system.

4. Simulation Results and Discussions

4.1. Simulation Parameters

In this section, simulation results validate the optimization results of the MAC and the PR controller in the FOC system. The simulation system is built using Simulink (Matlab R2022b), and the LIM is implemented through an S-function based on the magnetic flux linkage equation. The designed parameters of the LIM investigated in this study are presented as shown in Table 1.

Table 1. Designed parameters of the LIM for simulation.

Symbol	Definition	Value
L_s	Leakage inductance of the stator	7.8093×10^{-2} mH
L_r	Leakage inductance of mover	9.0628×10^{-2} mH
L_m	Magnetizing inductance	1.3344 mH
R_s	Stator resistance	0.0385 Ω
R_r	Mover resistance	0.01763 Ω
τ	Electrical Cycle Length	0.213 m
m	Load mass	500 kg

In the simulated linear motor model, the inductance matrix is configured with the L_{AA} set at 1.28×10^{-3} mH, whereas both L_{BB} and L_{CC} are established at 0.97×10^{-3} mH. This deliberate parameterization serves to emulate the impedance asymmetry characteristics of practical motor systems. This paper models the motor control system using Simulink (Matlab R2022b), where the motor model is designed as a flux linkage model based on the motor impedance parameters and implemented as an S-function block. The studied control system is a FOC system. In the current loop section, the PI controller and quasi-PR controller are connected in parallel. The parameters for the *d*-axis current PI controller are $K_p = 0.5$, $K_i = 100$, and those for the quasi-PR controller are $K_r = 1000$, with a bandwidth of $b = 10$ Hz. The parameters for the q-axis current PI controller are $K_p = 0.7$, $K_i = 80$, and those for the quasi-PR controller are $K_r = 1000$, with a bandwidth of $b = 10$ Hz. The control parameters are shown in Table 2.

Table 2. Control parameters.

Parameters	PI Controller (Current Loop)	PR Controller (Current Loop)	PI Controller (Velocity Loop)
Switching Frequency	8 kHz	8 kHz	/
Sampling time	10^{-4} s	10^{-4} s	10^{-4} s
Rising time	0.001 s	/	0.1 s
Setting time	0.05 s	/	0.2 s
Bandwidth	8 kHz	10 Hz	500 Hz

4.2. Simulation Curves and Data

4.2.1. Mechanical Characteristics Simulation Results

The LIM underwent a 5 s acceleration, with the introduction of the MAC at 2.5 s and the incorporation of a PR controller at 3.5 s. The velocity curve in Figure 7 indicates that the control system effectively tracked the speed signal.

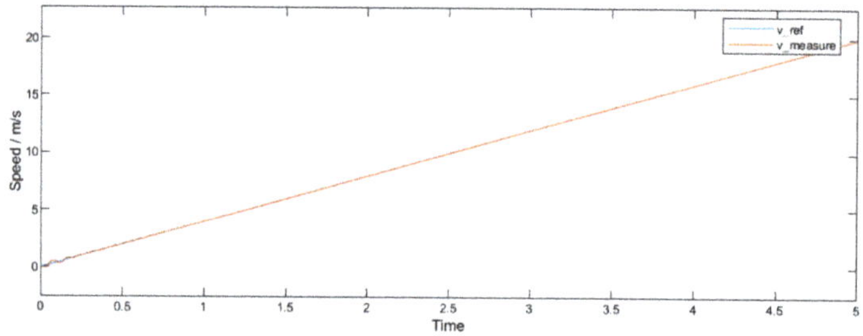

Figure 7. Simulation result of reference and measured speed.

This paper primarily focuses on the motor acceleration process; hence, the corresponding speed reference values continuously increase. It can be observed that the tracking performance of the speed is quite satisfactory.

The thrust simulation results shown in Figure 8 are as follows: (a) provides an overview of the thrust results, the MAC is introduced at 2.5 s, and the PR control is added at 3.5 s; (b) presents the magnified thrust results under the original FOC control, showing thrust ripple fluctuating between 4400 N and 4600 N; (c) displays the magnified results of FOC+MAC, demonstrating a reduction in peak-to-peak thrust value to around 100 N; and (d) exhibits the magnified results of FOC+MAC+PR, where the peak-to-peak thrust value is further reduced to 50 N. Additionally, it is evident that the system experiences oscillation around 3.5 s, attributed to the alteration of the PI control's steady-state behavior upon the introduction of the PR controller. The simulation in this study involves adding the PR control when the PI controller reaches steady state to facilitate the observation of performance enhancement with the PR control. However, in practical applications, controllers are not typically changed during operation; instead, the decision to adopt the PR control is made at the outset.

4.2.2. Current Characteristics Simulation Results

The simulation results of the stator current, including its positive-sequence component, negative-sequence component, and zero-sequence component, are illustrated in Figure 9.

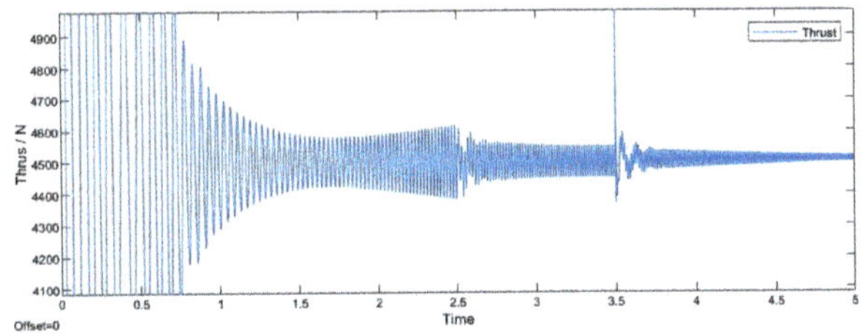

(a)

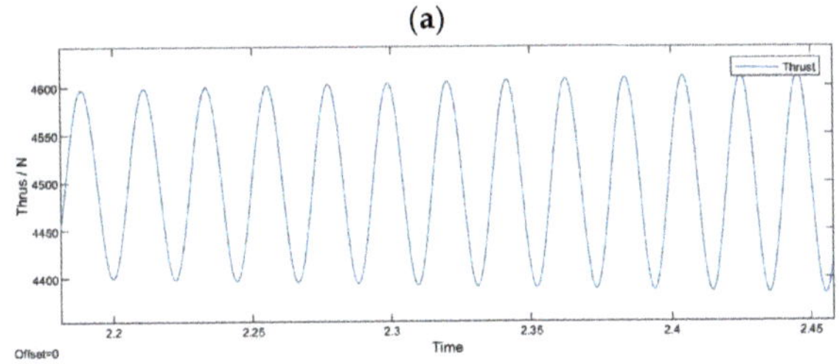

(b)

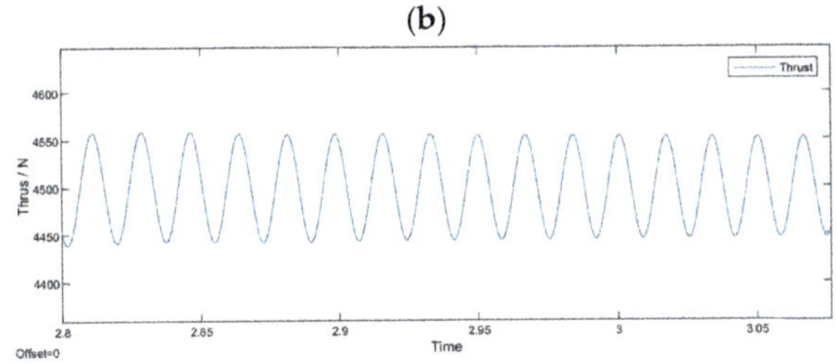

(c)

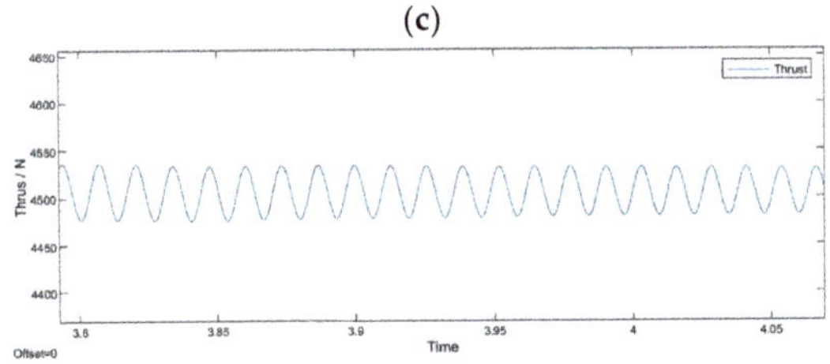

(d)

Figure 8. Thrust simulation results: (**a**) thrust results overview; (**b**) thrust results under FOC; (**c**) thrust results under FOC+MAC; and (**d**) Thrust results under FOC+MAC+PR.

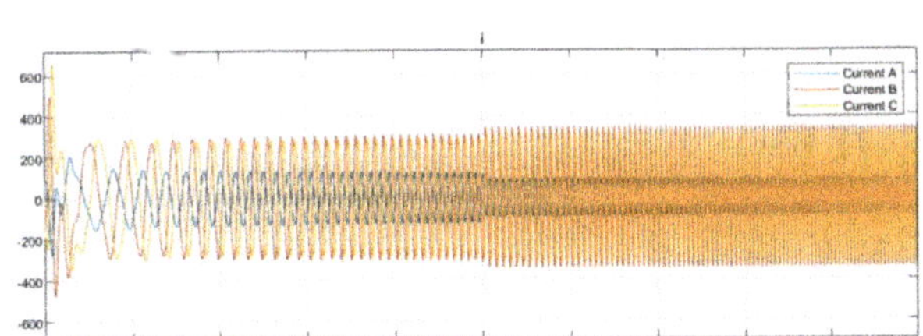

(a)

Figure 9. *Cont.*

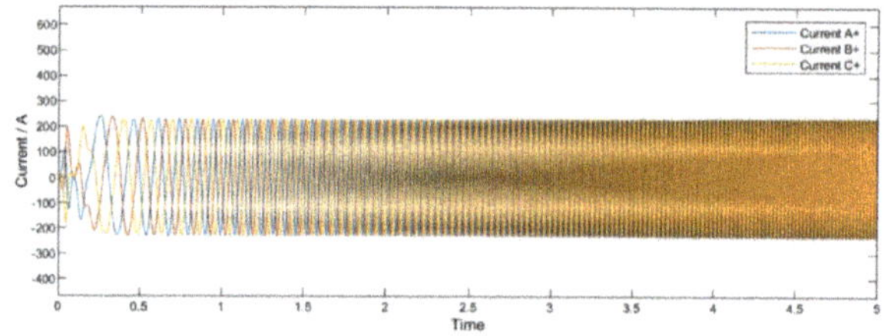

(b)

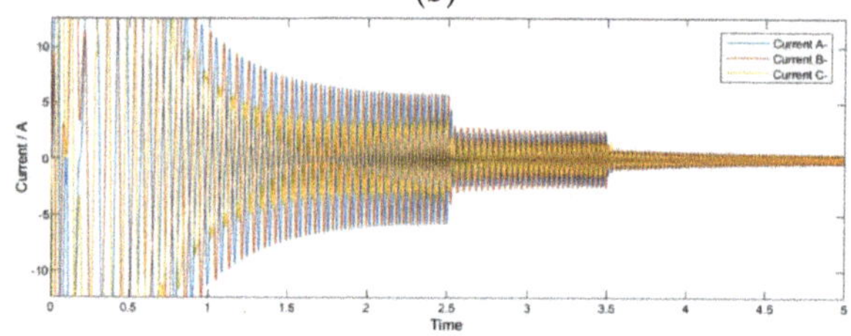

(c)

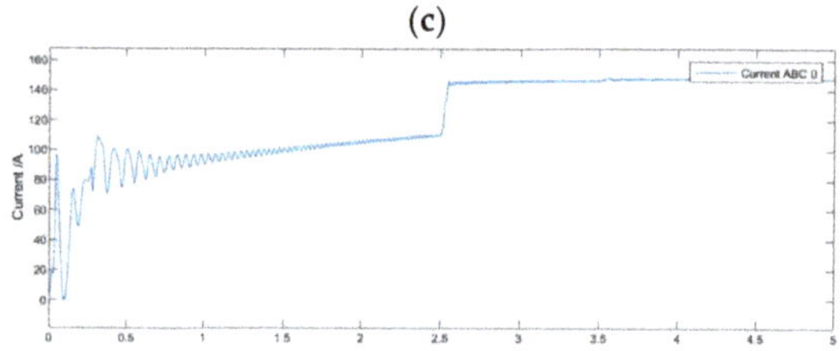

(d)

Figure 9. Current simulation results overview: (**a**) stator currents; (**b**) positive-sequence stator currents; (**c**) negative-sequence stator currents; and (**d**) zero-sequence stator currents.

In Figure 9, subgraph (a) represents the stator current, while subgraphs (b), (c), and (d) represent the positive-sequence component, negative-sequence component, and zero-sequence component decomposed from the stator current using the orthogonal component method. It can be observed that the positive-sequence current component shows no significant change. It is evident that the negative-sequence current decreases at 2.5 s with the introduction of the MAC and further diminishes at 3.5 s with the implementation of the PR control. Additionally, an increment in the zero-sequence current is observed. However, in the context of this study, no specific analysis has been undertaken for the zero-sequence current due to its negligible impact on thrust ripple.

It can be observed from the simulation results that Figure 10 has shown that, after the MAC is introduced at 2.5 s, there is a reduction in negative-sequence current from 6 A to 2 A. At 3.5 s, with the addition of PR control, there is a further reduction in negative-sequence current to approximately 1 A. Based on the depicted graph, it is evident that the waveform of the negative-sequence current is not a conventional sine wave before the integration of the PR controller. We attribute this deviation to the presence of harmonics. The PR controller proves adept at reducing these harmonics. The simulation results for the amplitude of the negative-sequence current are illustrated in Figure 11. The results indicate that both the MAC and PR controllers can effectively reduce the amplitude of the negative sequence current, but that the PR controller excels in eliminating harmonics.

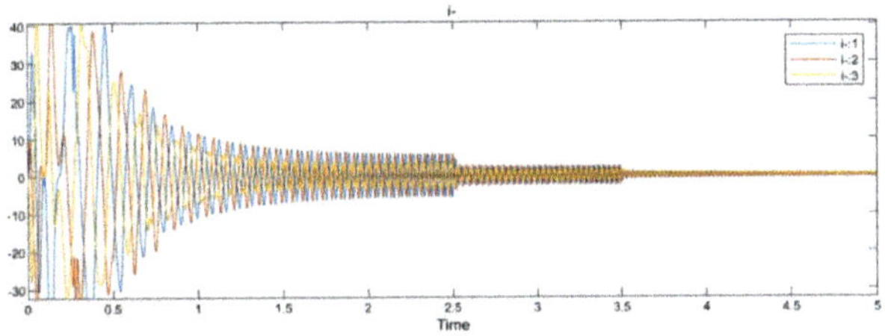

(a)

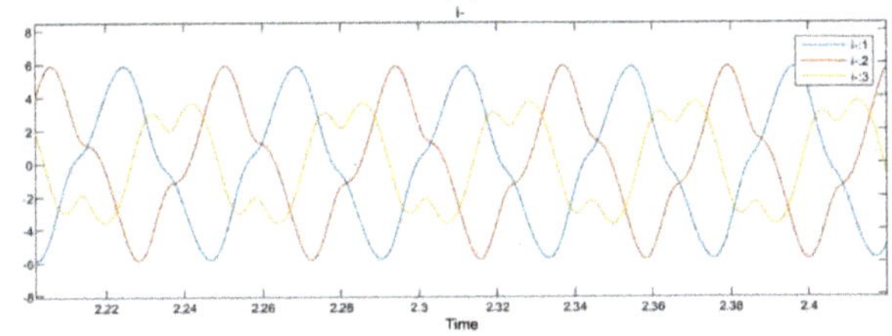

(b)

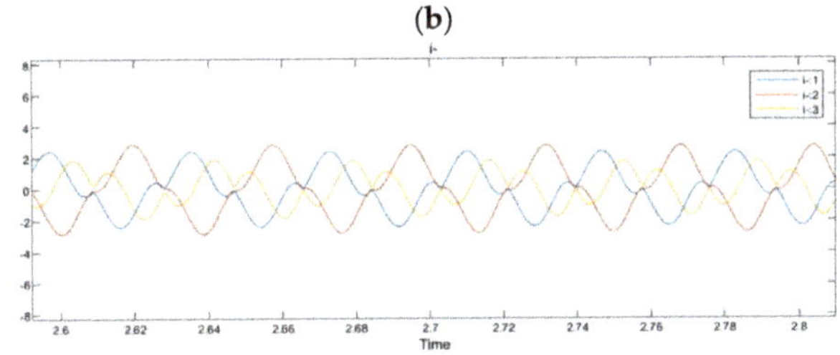

(c)

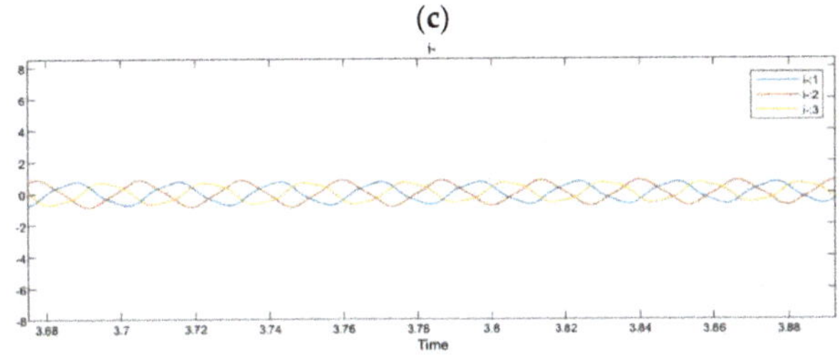

(d)

Figure 10. Negative-sequence current: (**a**) negative-sequence currents overview; (**b**) negative-sequence currents under FOC; (**c**) negative-sequence currents under FOC+MAC; and (**d**) negative-sequence currents under FOC+MAC+PR.

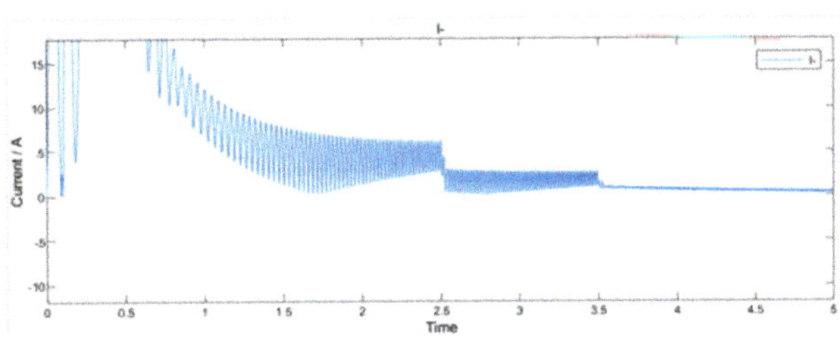

Figure 11. Amplitude of negative sequence current.

The simulation results for i_d and i_q are shown in Figure 12. In the FOC control system, the controlled variables are i_d and i_q, and the introduction of the PR controller directly aims to mitigate the fluctuations in i_d and i_q.

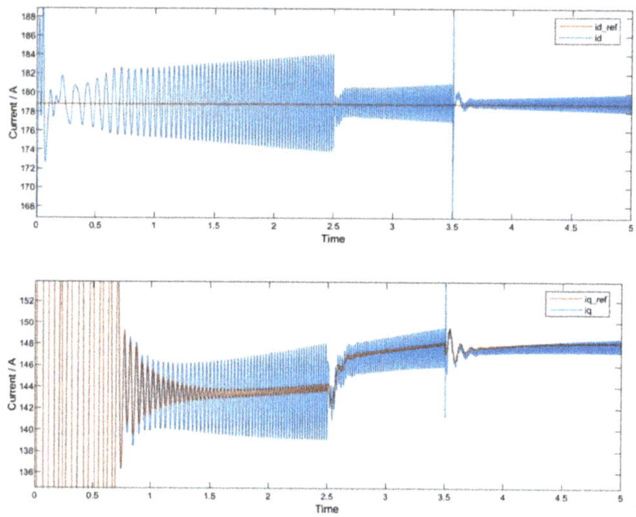

Figure 12. Simulation results for i_d and i_q.

From Figure 12, it can be observed that the addition of both the MAC and PR significantly reduces the harmonics in both i_d and i_q. It can be observed that at 3.5 s, the system experienced some instability, which was caused by the introduction of the PR controller. It is important to emphasize that the simulation involved adding the PR controller during the runtime to compare the effects introduced by the PR controller. However, in actual operation, the PR controller should be added at the beginning of the runtime.

Figure 13 shows that the difference in the voltage phase calculated by the MAC module is approximately 2.2 rad, i.e., 126.5°; however, when the MAC is not used, the difference in the voltage phase is 120°.

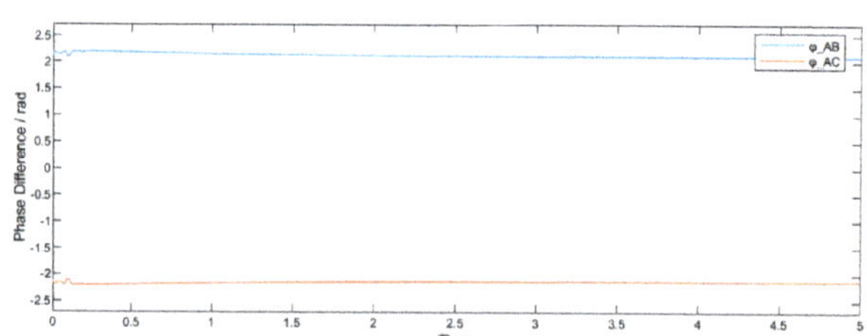

Figure 13. Phase calculation results.

In conclusion, based on the simulation and experimental results, the proposed MAC can effectively reduce the thrust ripple by altering the voltage phase. Furthermore, with the incorporation of the quasi-PR controller, the suppression of the thrust ripple is further enhanced. Simulation results also indicate that the suppression of the thrust ripple is related to the reduction in both the current harmonics in the dq axis and the negative-sequence component of the stator current.

5. Experimental Physical Test and Validation

To validate the effectiveness of the simulation results, physical experiments were conducted on the platform shown in Figure 14. The experimental platform consists of a double-sided primary induction linear motor, with the stator and mover depicted in the figure. The LIM depicted in Figure 14 is a long-stator LIM that operates using segmented power supply.

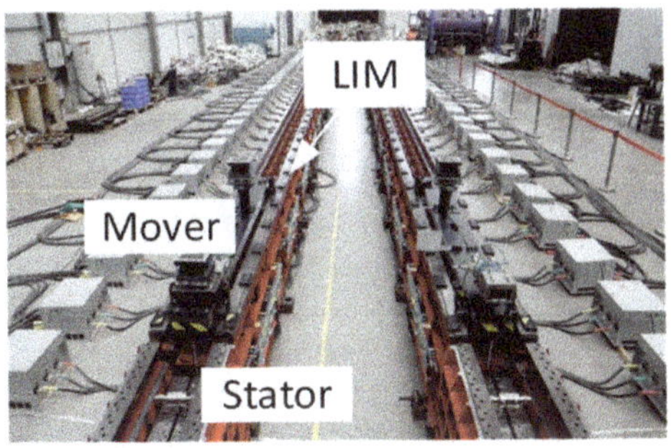

Figure 14. LIM experimental platform.

We also conducted experiments using the MEFEL device as Figure 15 shows, which utilizes a motor with the same parameters as those employed in the horizontal experimental platform. The difference is that the motor in this facility operates vertically. The experimental capsule is driven by two LIMs, with the mover installed on the outer side of the experimental capsule and the stator mounted on the inner wall of the tower. Additionally, rails are installed to ensure the directional movement of the experimental capsule. Each operation involves launching the experimental capsule upward using the LIM, allowing the experimental capsule to undergo free-fall motion, and then retrieving it. This study primarily focuses on the launch phase, but in the actual testing environment, the launch, free-fall, and recovery phases are conducted together.

Figure 15. MEFEL experimental platform with vertical LIM.

The design parameters of the LIM used in both of these experimental platforms are the same, but improvements are made to the manufacturing process; this results in the LIM exhibiting a slightly better performance in MEFEL. The design parameters of the LIM are shown in Table 3.

Table 3. LIM parameters.

Parameter	Value	Unit
Nominal Power	82	kW
DC Bus Voltage	600	V
Control Period	2	ms
Sample Time	150	μs
Primary Inductance	0.17	mH
Nominal Speed	20	m/s

5.1. Experimental Results in the Horizontal Experimental Platform

Due to the limitations that existed in the experimental environment, we only conducted tests under blocked conditions on the horizontal platform and obtained the following current results. The experimental results of the stator current before and after the utilization of the MAC and PR controller are shown in Figure 16.

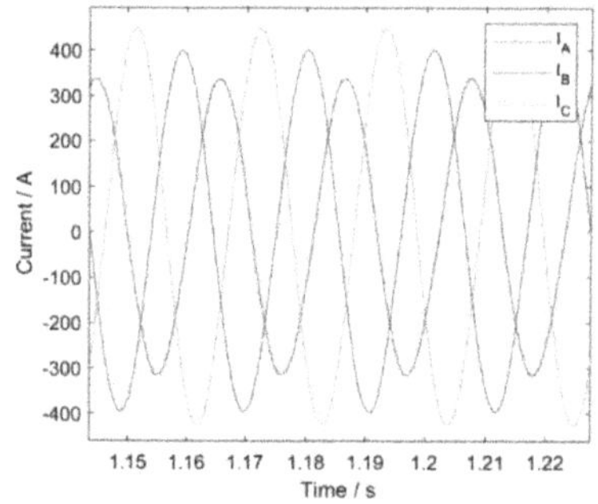

(a)

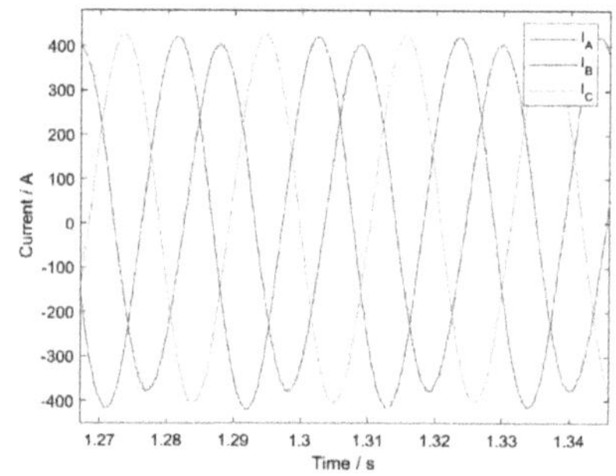

(b)

Figure 16. Stator current: (**a**) no compensation method; (**b**) MAC and PR employed.

The experimental results indicate that the peak stator currents for the three phases before the use of MAC and quasi-PR controllers were 392.5 A, 320.1 A, and 451.6 A, respectively. After applying the MAC and quasi-PR controllers, the peak stator currents for the three phases were 410.7 A, 395.4 A, and 417.7 A, respectively. The three-phase current imbalance decreased from 29.1% to 5.28%, indicating a significant reduction in current imbalance. The experimental results demonstrate that the adoption of the MAC and quasi-PR controllers can effectively reduce the current imbalance in the stator, leading to the LIM exhibiting an improved performance.

5.2. Experimental Results in the MEFEL Platform

In the MEFEL platform, we conducted three sets of comparative dynamic experiments. The first group utilized the FOC control system, the second group employed the FOC+MAC system, and the third group utilized the FOC+MAC+PR system. Our objective was to observe the improvements in thrust brought about by the proposed methods; however, direct measurement of the thrust of the linear motor was challenging to achieve. We could only calculate the acceleration by processing the velocity signals, thereby indirectly observing the optimization of thrust fluctuations. The experimental results regarding the velocity are shown in Figure 17.

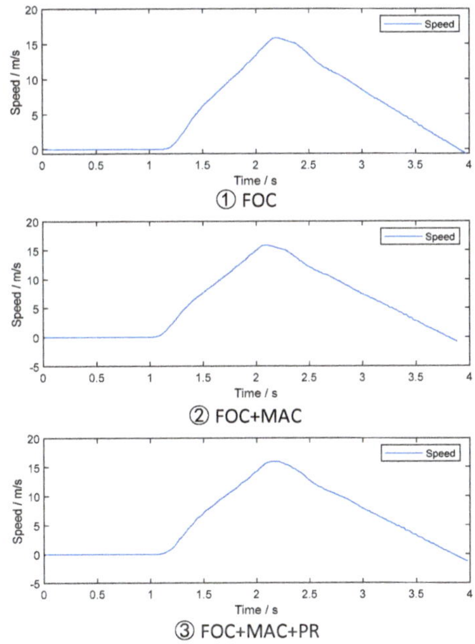

Figure 17. Speed results of MEFEL.

The operation requirements of the electromagnetic microgravity facility studied in this paper dictate a specific speed profile, meaning that regardless of changes in the motor-controlled current loop, we aim for the speed loop to remain consistent. The results of the speed operation are shown in Figure 17, where it can be observed that under different control methods, the tracking of the speed loop is similar, ultimately meeting the system's performance requirements. Specifically, the performance of the current loop can be indirectly observed by observing the acceleration. Direct measurement of thrust in linear motors is challenging; hence, this study indirectly observes thrust and thrust ripples by measuring the acceleration of the experimental capsule using an accelerometer. The differentiation of speed in Figure 17 is roughly equivalent to the acceleration in Figure 18, but the signals in Figures 17 and 18 are measured by different sensors. The speed results in Figure 17 are derived from the differentiation of the position signal obtained by the grating displacement sensor, while acceleration is directly measured using an accelerometer.

It can be observed from the acceleration results that, under FOC, the acceleration fluctuates at approximately 5 m/s^2. Building upon this, the utilization of the MAC reduces the acceleration fluctuation to around 3.5 m/s^2, and further implementing the PR control narrows the acceleration fluctuation to approximately 2 m/s^2. It is important to note that the system also encounters other disruptive forces, such as mechanical friction, air resistance, and mechanical vibrations. The vibration reduction method proposed in this paper solely focuses on suppressing the fluctuations caused by LIM thrust. Vibrations originating from the mechanical structure within the experimental environment require mechanical damping solutions, which can be achieved via the modification of the construction design. The processed acceleration data can provide insights into the thrust fluctuations observed in the linear motor. Utilizing the MAC and PR controls individually can partially suppress the thrust ripple, but the most effective suppression of thrust fluctuation is achieved when both the MAC and PR controls are employed simultaneously. It is worth noting that the accuracy of the acceleration data is constrained, as it is derived from velocity signals through differentiation and filtering. Therefore, the extent of acceleration fluctuation is closely related to the level of filtering applied.

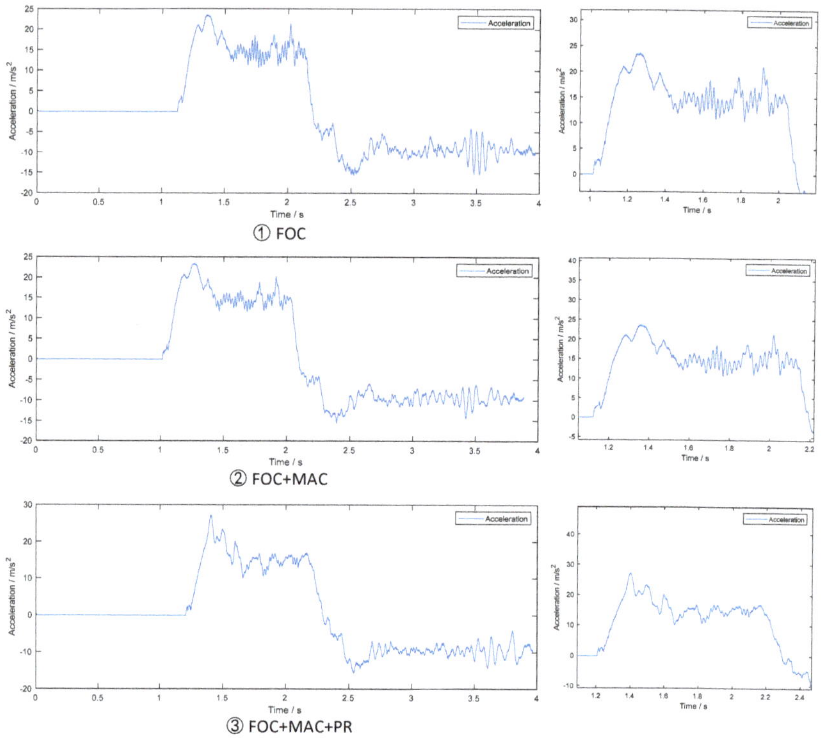

Figure 18. Acceleration results of MEFEL.

The motor in the actual test environment belongs to a catapult system, and the operation process is divided into catapult stage, free fall stage, and recovery stage. The research content of this paper mainly focuses on the launch section, but the experimental results show the free fall stage and the recovery stage as well. In simulation we can simulate only the catapult stage. In the simulation results, the two control methods proposed in this paper are added in the operation process, while in the actual test environment, it is impossible to add the controller in the operation process to compare the control effect, and many experiments are needed.

Through the comparison of the acceleration results of the actual test and the thrust results of the simulation, it can be seen that the actual acceleration has a very low frequency and a high amplitude fluctuation, which is caused by the segmented stator. This part of the fluctuation can also be improved by optimizing the segmented control strategy, but it is not the focus of this study. This paper mainly studies the elimination of the thrust ripple caused by the physical characteristics of the linear motor itself. Additionally, it can be observed that there are many irregular interferences in the thrust of the actual operation, which may be from the mechanical vibration in the environment or the friction of the guide rail, etc., while the thrust ripple in the simulation is only caused by the physical characteristics of the linear motor. Although there are many differences between physical testing and simulation testing, the effectiveness of the method proposed in this article can still be seen from the actual test results.

6. Conclusions

This study addresses the challenge of the thrust ripple in a high-power LIM for the microgravity facility. It introduces a model-based adaptive controller (MAC) and an enhanced quasi-proportional-resonant (PR) controller for current control, taking into account the impedance asymmetry inherent in LIMs. The MAC adjusts the voltage phase to counteract the current imbalance that results from the LIM's impedance asymmetry. The improved quasi-PR controller, in tandem with the PI controller, works to eliminate harmonics in the current *d*-*q* axis. The comprehensive theoretical derivations and calculation methods employed for both control strategies are thoroughly examined. Simulation results indicate that after adopting the MAC to correct the current phase, the thrust ripple decreases to 50%. A further reduction to 25% is achieved by implementing the PR controller to correct the current amplitude. The experimental results demonstrate the effectiveness of the method proposed in this paper in practical engineering applications.

Author Contributions: Conceptualization, Y.L. and W.D.; Methodology, Y.L.; Software, Z.W.; Validation, Y.L.; Investigation, Y.L. and W.D.; Resources, W.D., C.L. and Y.Z.; Data curation, Z.W.; Writing—original draft, Y.L.; Writing—review & editing, Y.L., W.D. and Z.W.; Visualization, Z.W.; Supervision, C.L. and Y.Z.; Project administration, C.L. and Y.Z.; Funding acquisition, Y.Z. All authors have read and agreed to the published version of the manuscript.

Funding: This research was funded by the Chinese Manned Space Program—Space Utilization System (412021000031), Scientific Instruments and Equipment Development Program of the Chinese Academy of Sciences (YJKYYQ20180017) (YJKYYQ20210028), and the Youth Innovation Promotion Association CAS.

Data Availability Statement: The data presented in this study are available in this article.

Conflicts of Interest: The authors declare no conflict of interest.

Appendix A

Inductance Equivalent Expression of LIMs

The model of a rotary motor can be expressed as follows:

$$\begin{array}{c} U = RI + \frac{d}{dt}\psi \\ \psi = LI \end{array}$$

in which L satisfies

$$L = \begin{bmatrix} L_{ss} & L_{sr} \\ L_{rs} & L_{rr} \end{bmatrix}$$
$$L_{ss} = \begin{bmatrix} L_{AA} & L_{AB} & L_{AC} \\ L_{BA} & L_{BB} & L_{BC} \\ L_{CA} & L_{CB} & L_{CC} \end{bmatrix}$$
$$L_{sr} = \begin{bmatrix} L_{Aa} & L_{Ab} & L_{Ac} \\ L_{Ba} & L_{Bb} & L_{Bc} \\ L_{Ca} & L_{Cb} & L_{Cc} \end{bmatrix}$$
$$L_{rs} = \begin{bmatrix} L_{aA} & L_{aB} & L_{aC} \\ L_{bA} & L_{bB} & L_{bC} \\ L_{cA} & L_{cB} & L_{cC} \end{bmatrix}$$
$$L_{rr} - \begin{bmatrix} L_{aa} & L_{ab} & L_{ac} \\ L_{ba} & L_{bb} & L_{bc} \\ L_{ca} & L_{cb} & L_{cc} \end{bmatrix}$$

In the matrix of L_{ss}, L_{AA}, L_{BB} and L_{CC} represent the self-inductances of the stator windings. L_{AB}, L_{AC} and other inductances in the matrix represent the mutual inductance of the stator windings.

The self-inductance consists of leakage inductance and main self-inductance.

$$\begin{array}{c} L_{AA} = L_{BB} = L_{CC} \\ = L_{s\sigma} + L_{m1} \end{array}$$

Due to the intentional design of a $\frac{2\pi}{3}$ electrical angular difference between windings, the mutual inductances satisfy the following relationship:

$$\begin{aligned} L_{AB} &= L_{AC} = L_{BC} = L_{BA} = L_{CA} = L_{CB} \\ &= \cos\left(\pm\tfrac{2\pi}{3}\right)L_{m1} \\ &= = -\tfrac{1}{2}L_{m1} \end{aligned}$$

The steady-state model of motor is commonly analyzed using the T-type equivalent circuit model, as shown in Figure A1.

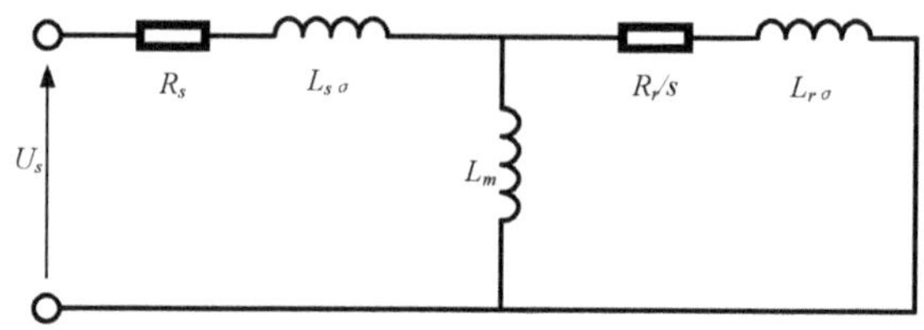

Figure A1. T-type equivalent circuit.

It reflects the voltage–current relationship on the single-phase winding, where the excitation inductance L_m is equal to $\frac{3}{2}L_{m1}$.

Up to this point, we have observed an interesting phenomenon: in the magnetic flux linkage equation, the voltage of the single-phase winding is related to the currents of other windings, while in the T-type equivalent circuit, it seems to be only related to the current of the winding itself. This is because in three-phase motors, sinusoidal currents have a 120° phase difference, allowing the inductance matrix L_{ss} to be transformed into a diagonal matrix through calculation. (A reminder: the calculation of excitation inductance is independent of the rotor, as the motor inherently functions as a transformer model. After winding transformation calculation, the excitation inductance of the rotor is equal to that of the stator. In other words, the circuit model parameters on the rotor side are not equivalent to the magnetic flux model parameters, so we do not analyze the rotor-side magnetic flux.)

In phase A, the stator flux caused by stator current can be expressed as follows:

$$\begin{aligned} \psi_{sA} &= L_{AA}i_A + L_{AB}i_B + L_{AC}i_C \\ &= (L_{s\sigma} + L_{m1})i_A + \left(-\tfrac{1}{2}L_{m1}\right)i_B + \left(-\tfrac{1}{2}L_{m1}\right)i_C \\ &= (L_{s\sigma} + L_{m1})i_A - \tfrac{1}{2}L_{m1}[cos(120°) + cos(-120°)]i_A \\ &= (L_{s\sigma} + \tfrac{3}{2}L_{m1})i_A \end{aligned}$$

The calculation process for phases B and C is the same as that for phase A. In this case the L_{ss} can be expressed as

$$L_{ss} = \begin{bmatrix} L_{s\sigma} + \frac{3}{2}L_{m1} & 0 & 0 \\ 0 & L_{s\sigma} + \frac{3}{2}L_{m1} & 0 \\ 0 & 0 & L_{s\sigma} + \frac{3}{2}L_{m1} \end{bmatrix}$$

However, in a LIM the impedance matrix cannot be transformed into a diagonal matrix due to the asymmetry of the impedance. Even when the currents are balanced, ψ_{sA} can only be expressed as follows:

$$\begin{aligned} \psi'_{sA} &= L_{AA}i_A + L_{AB}i_B + L_{AC}i_C \\ &= [L_{s\sigma} + L_{AA} + L_{AB}\cos(-120°) + L_{AC}\cos(120°)]i_A \\ &= \left(L_{s\sigma} + L_{AA} - \tfrac{1}{2}L_{AB} - \tfrac{1}{2}L_{AC}\right)i_A \end{aligned}$$

The calculation process for phases B and C is the same as that for phase A. Thus L_m of the three-phase windings in the equivalent circuit cannot be expressed as one parameter. The equations for the three-phase excitation inductances are as follows:

$$\begin{aligned} L_{mA} &= L_{AA} - \tfrac{1}{2}L_{AB} - \tfrac{1}{2}L_{AC} \\ L_{mB} &= L_{BB} - \tfrac{1}{2}L_{BA} - \tfrac{1}{2}L_{BC} \\ L_{mC} &= L_{CC} - \tfrac{1}{2}L_{CA} - \tfrac{1}{2}L_{CB} \end{aligned}$$

In this case, the L_{ss} can be expressed as

$$\boldsymbol{L_{ss}} = \begin{bmatrix} L_{AA} - \frac{1}{2}L_{AB} - \frac{1}{2}L_{AC} & 0 & 0 \\ 0 & L_{BB} - \frac{1}{2}L_{BA} - \frac{1}{2}L_{BC} & 0 \\ 0 & 0 & L_{CC} - \frac{1}{2}L_{CA} - \frac{1}{2}L_{CB} \end{bmatrix}$$

It should be noted that this expression is based on the steady-state model of motor, assuming the motor speed changes smoothly.

Appendix B

Demonstration of the Effect of MAC for Current Imbalance Correction

The MAC can be represented as follows:

$$\begin{bmatrix} u'_A \\ u'_B \\ u'_C \end{bmatrix} = \begin{bmatrix} 1 & 0 \\ \cos\varphi_{AB} & \sin\varphi_{AB} \\ \cos\varphi_{AC} & \sin\varphi_{AC} \end{bmatrix} \frac{2}{3} \begin{bmatrix} 1 & \cos\left(\frac{2\pi}{3}\right) & \cos\left(-\frac{2\pi}{3}\right) \\ 0 & \sin\left(\frac{2\pi}{3}\right) & \sin\left(-\frac{2\pi}{3}\right) \end{bmatrix} \begin{bmatrix} u_A \\ u_B \\ u_C \end{bmatrix}$$

$$= \frac{2}{3} \begin{bmatrix} 1 & \cos\left(\frac{2\pi}{3}\right) & \cos\left(-\frac{2\pi}{3}\right) \\ \cos\varphi_{AB} & \cos\left(\varphi_{AB-} - \frac{2\pi}{3}\right) & \cos\left(\varphi_{AB} + \frac{2\pi}{3}\right) \\ \cos\varphi_{AC} & \cos\left(\varphi_{AC} - \frac{2\pi}{3}\right) & \cos\left(\varphi_{AC} - \frac{2\pi}{3}\right) \end{bmatrix} \begin{bmatrix} u_A \\ u_B \\ u_C \end{bmatrix}$$

As u_A, u_B, and u_C are sinusoidal three-phase waves which can be represented as

$$\begin{aligned} u_A &= u_m \cos\omega t \\ u_B &= u_m \cos\left(\omega t - \tfrac{2\pi}{3}\right) \\ u_C &= u_m \cos\left(\omega t + \tfrac{2\pi}{3}\right) \end{aligned}$$

where u_m is amplitude, ω is angular velocity and t is time.

Based on the above equations, it can be calculated that:

$$\begin{aligned} u'_A &= u_m \cos\omega t \\ u'_B &= u_m \cos(\omega t - \varphi_{AB}) \\ u'_C &= u_m \cos(\omega t - \varphi_{AC}) \end{aligned}$$

Hence, when $\varphi_{AB} = \frac{2\pi}{3}$ and $\varphi_{AC} = -\frac{2\pi}{3}$, $[u'_A; u'_B; u'_C] = [u_A; u_B; u_C]$.

The AC waveforms u'_A, u'_B and u'_C can be represented using vectors:

$$\begin{aligned} \dot{U}'_A &= U_m \angle 0 \\ \dot{U}'_B &= U_m \angle -\varphi_{AB} \\ \dot{U}'_C &= U_m \angle -\varphi_{AC} \end{aligned}$$

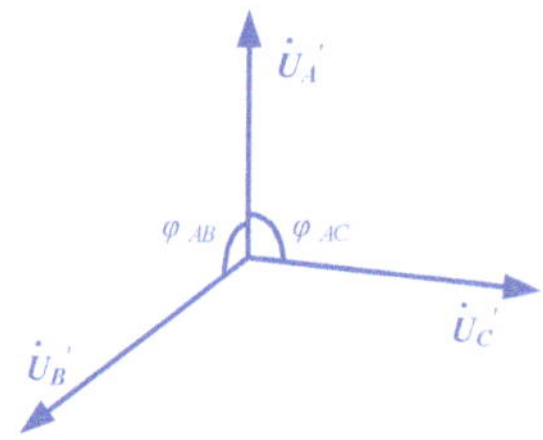

Figure A2. Voltage vectors.

Hence

$$\begin{bmatrix} \dot{I}_A' \\ \dot{I}_B' \\ \dot{I}_C' \end{bmatrix} = \begin{bmatrix} \dot{Z}_A & 0 & 0 \\ 0 & \dot{Z}_B & 0 \\ 0 & 0 & \dot{Z}_C \end{bmatrix}^{-1} \begin{bmatrix} \dot{U}_A' \\ \dot{U}_B' \\ \dot{U}_C' \end{bmatrix}$$

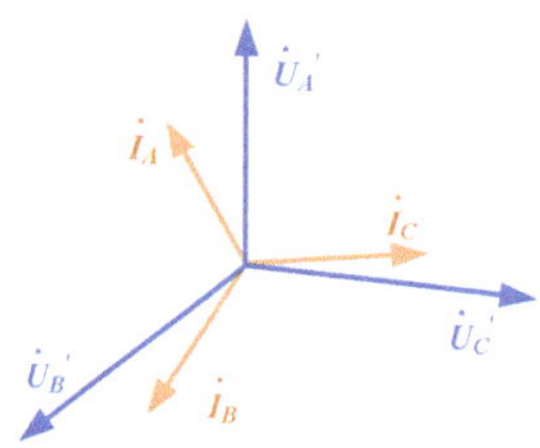

Figure A3. Voltage vectors and current vectors.

define $\varphi_{Z_A Z_B}$ which satisfies the following relationship:

$$\frac{\dot{Z}_A}{\dot{Z}_B} = \left|\frac{\dot{Z}_A}{\dot{Z}_B}\right| \angle \varphi_{Z_A Z_B}$$

The voltage phase difference of phase A and phase B satisfies the following:

$$\varphi_{AB} = \frac{2\pi}{3} + \varphi_{Z_A Z_B}$$

Then, calculate the phase difference of currents phase A and phase B:

$$\begin{aligned} \frac{\dot{I}_A'}{\dot{I}_B'} &= \frac{\dot{Z}_B}{\dot{Z}_A}\frac{\dot{U}_A'}{\dot{U}_B'} = \left|\frac{\dot{Z}_B}{\dot{Z}_A}\right| \frac{U_m}{U_m} \angle\left(-\varphi_{Z_A Z_B} + 0 - (-\varphi_{AB})\right) \\ &= \left|\frac{\dot{Z}_B}{\dot{Z}_A}\right| \angle\left(\varphi_{AB} - \varphi_{Z_A Z_B}\right) \\ &= \left|\frac{\dot{Z}_B}{\dot{Z}_A}\right| \angle \frac{2\pi}{3} \end{aligned}$$

Hence, the phase difference is $\frac{2\pi}{3}$. By using the same method, the phase difference between $I_A{\prime}$ and $I_C{\prime}$ can be calculated to be $-\frac{2\pi}{3}$.

Thereby, we can conclude that the phase difference of I_A', I_B', and I_C' is $\frac{2\pi}{3}$, after balancing correction.

Additionally, it can be observed that the current imbalance has only been fixed in terms of phase deviation, while the amplitude deviation still exists. In the control system under FOC, the currents i_d and i_q can reflect the problem of unequal current amplitudes, and dynamic control can be applied for adjustment. This is described in Section 3 of this paper, where it is achieved through a PI+PR controller.

References

1. Könemann, T.; Kaczmarczik, U.; Gierse, A.; Greif, A.; Lutz, T.; Mawn, S.; Siemer, J.; Eigenbrod, C.; von Kampen, P.; Lämmerzahl, C. Concept for a next-generation drop tower system. *Adv. Space Res.* **2015**, *55*, 1728–1733. [CrossRef]
2. Zhang, J.; Dong, W.; Wang, Z.; Zhang, Y.; Zhang, X.; Cai, Z.; Ma, W.; Li, Y.; Li, X.; Cui, X. Development of a New Microgravity Experiment Facility with Electromagnetic Launch. *Microgravity Sci. Technol.* **2021**, *33*, 1–18. [CrossRef]
3. Li, Y.; Dong, W.; Zhang, S.; Zhang, Y. A Prototype of Microgravity Facility Operated by Linear Motors: Motion Plan and Control. In Proceedings of the IAF Microgravity Sciences and Processes Symposium, International Astronautical Congress 2020, Washington, DC, USA, 12–14 October 2020.
4. Zhao, J.; Wang, S.; Liu, Q.; He, Z.; Zhang, W.; Li, K.; Zhou, Z.; Luo, X.; Miao, J.; Zheng, H.; et al. Retrospect and Perspective on Microgravity Science in China. *Chin. J. Space Sci.* **2021**, *41*, 34–45. [CrossRef]
5. Nasar, S.A.; Boldea, I. *Linear Motion Electric Machine*; John Wiley & Sons Inc.: Hoboken, NJ, USA, 1976; pp. W43–W56.
6. Long, X. *Theory and Electromagnetic Design for Linear Induction Motor*; Press of Science: Beijing, China, 2006; pp. 67–72.
7. Nie, S.; Fu, L.; Xu, J.; Li, W.; Sun, Z. Asymmetrical Model and Parameter Calculation of Segment-powered Linear Inductive Motor Mover. *J. Electr. Mach. Control.* **2017**, *21*, 10–17.
8. Lu, J.; Ma, W.; Li, L. Research on Longitudinal End Effect of High-speed Long Primary Double-sided Linear Induction Motor. *Proc. CSEE* **2008**, *28*, 73–78.
9. Sun, Z.; Liu, D.; Ma, W.; Lu, J.; Xu, J. Research of Static Longitudinal End Effect and Impedance Matrix for Long Primary Double-sided Linear Induction Motors. *Proc. Chin. Soc. Electr. Eng.* **2010**, *30*, 1–6.
10. Lv, G.; Cui, L.; Zhi, R. Inductance Analysis of Transverse Flux Linear Synchronous Motor for Maglev Trains Considering Three-Dimensional Operating Conditions. *IEEE Trans. Ind. Electron.* **2024**, *71*, 769–776. [CrossRef]
11. Duncan, J. Linear induction motor—Equivalent-circuit model. *Proc. Inst. Electr. Eng.* **1983**, *130*, 51–57.
12. Xu, W.; Zhu, J.G.; Zhang, Y.; Li, Z.; Li, Y.; Wang, Y.; Guo, Y.; Li, Y. Equivalent circuits for single-sided linear induction motors. *IEEE Trans. Ind. Appl.* **2010**, *46*, 2410–2423. [CrossRef]
13. Xu, W.; Zhu, J.G.; Zhang, Y.; Li, Y.; Wang, Y.; Guo, Y. An improved equivalent circuits of a single-sided linear induction motor. *IEEE Trans. Veh. Technol.* **2010**, *59*, 2277–2289. [CrossRef]
14. Sun, Z.; Gao, J.; Ma, W.; Lu, J.; Xu, J. Impedance Matrix and Parameters Measurement Research for Long Primary Double-Sided Linear Induction Motor. *IEEE Trans. Plasma Sci.* **2019**, *47*, 2703–2709. [CrossRef]
15. Lu, J.; Ma, W. Investigation of Phase Unbalance Characteristics in the Linear Induction Coil Launcher. *IEEE Trans. Plasma Sci.* **2011**, *39*, 110–115. [CrossRef]
16. Han, Z.; Xu, J.; Rui, W.; Zhang, Y.; Li, M. Research on Thrust Ripple of Aperiodic Transient Linear Induction Motors. In Proceedings of the 2021 13th International Symposium on Linear Drives for Industry Applications (LDIA), Wuhan, China, 1–3 July 2021; pp. 1–5.
17. Liu, H.; Wang, Y.; Zhang, Q.; Zhang, Z.; Song, T. Optimization of Thrust Ripple of Long Primary Double Sided Linear Induction Motor Injected with Harmonic Current. *J. Electr. Mach. Control* **2023**, *27*, 49–60.
18. Zhou, W.; Sun, Z.; Zeng, R.; Mao, Y.; Ding, A. A Thrust Ripple Suppression Method for Long-Primary Double-Sided Linear Induction Motors Based on Harmonic Current Injection. *IEEE Trans. Transp. Electrif.* **2023**, *9*, 2348–2355. [CrossRef]
19. Sun, X.; Xu, J.; Zhu, J.; Xu, W. Thrust Ripple Suppression Based on Negative Current Control for Short-Primary Low-Speed Large LIM Under Transient Operation. *IEEE Trans. Energy Convers.* **2023**, *38*, 1566–1575. [CrossRef]
20. Wang, K.; Li, Y.; Ge, Q.; Shi, L. An Improved Indirect Field-Oriented Control Scheme for Linear Induction Motor Traction Drives. *IEEE Trans. Ind. Electron.* **2018**, *65*, 9928–9937. [CrossRef]
21. Gu, H.; Cai, Z.; Wang, Z.; Zhang, Y.; Dong, W.; Guo, R. Three-Phase Current Unbalance Suppression Method of Linear Induction Motor Based on PR Controller. In Proceedings of the 2022 4th International Conference on Smart Power & Internet Energy Systems (SPIES), Beijing, China, 27–30 October 2022; pp. 883–888.
22. Cui, D.; Wei, R.; Liu, J.; Ge, Q. Linear synchronous motor drive system based on proportion resonant current controller in maglev transportation. In Proceedings of the 2014 17th International Conference on Electrical Machines and Systems (ICEMS), Hangzhou, China, 22–25 October 2014; pp. 1289–1292.

energies

MDPI

Article

Optimization Operation Strategy for Shared Energy Storage and Regional Integrated Energy Systems Based on Multi-Level Game

Yulong Yang, Tao Chen *, Han Yan, Jiaqi Wang, Zhongwen Yan and Weiyang Liu

School of Electrical Engineering, Northeast Electric Power University, Jilin 132000, China; yangyulong@neepu.edu.cn (Y.Y.); 2202200346@neepu.edu.cn (H.Y.); 2202200180@neepu.edu.cn (J.W.); 2202300130@neepu.edu.cn (Z.Y.); 2202300068@neepu.edu.cn (W.L.)
* Correspondence: 2202100008@neepu.edu.cn

Abstract: Regional Integrated Energy Systems (RIESs) and Shared Energy Storage Systems (SESSs) have significant advantages in improving energy utilization efficiency. However, establishing a coordinated optimization strategy between RIESs and SESSs is an urgent problem to be solved. This paper constructs an operational framework for RIESs considering the participation of SESSs. It analyzes the game relationships between various entities based on the dual role of energy storage stations as both energy consumers and suppliers, and it establishes optimization models for each stakeholder. Finally, the improved Differential Evolution Algorithm (JADE) combined with the Gurobi solver is employed on the MATLAB 2021a platform to solve the cases, verifying that the proposed strategy can enhance the investment willingness of energy storage developers, balance the interests among the Integrated Energy Operator (IEO), Energy Storage Operator (ESO) and the user, and improve the overall economic efficiency of RIESs.

Keywords: Regional Integrated Energy System; shared energy storage; optimized operation; multi-level game; integrated demand response

Citation: Yang, Y.; Chen, T.; Yan, H.; Wang, J.; Yan, Z.; Liu, W. Optimization Operation Strategy for Shared Energy Storage and Regional Integrated Energy Systems Based on Multi-Level Game. *Energies* **2024**, *17*, 1770. https://doi.org/10.3390/en17071770

Academic Editor: Abdul-Ghani Olabi

Received: 7 March 2024
Revised: 2 April 2024
Accepted: 6 April 2024
Published: 8 April 2024

1. Introduction

Against the backdrop of the "Dual Carbon" goals, improving the current energy supply structure and promoting sustainable economic development have become mainstream trends [1]. The development of clean energy and the enhancement of the overall energy utilization efficiency are essential aspects of this trend [2]. This places higher demands on the stability and flexibility of electricity dispatch. In this context, the Regional Integrated Energy System (RIES) has garnered widespread attention due to its features of multi-energy coupling and reliable energy supply [3,4].

In the design and evaluation of the RIES, it is crucial to allocate energy storage systems appropriately in order to ensure the sustainability of the energy supply, as well as the economic efficiency and reliability of the RIES. By the end of 2023, China had completed the construction and operation of new energy storage projects, with a cumulative installed capacity of 3.139 GW/6.687 GWh. However, the current independent construction of energy storage systems involves a large capital investment and a long payback period, which limits the development of energy storage systems [5,6]. The new generation of Shared Energy Storage Systems (SESSs) that have emerged in the context of the energy internet provide an effective solution to these issues [7,8]. A SESS integrates independently dispersed energy storage resources on the source, grid, and load sides for unified collaborative scheduling, which improves the utilization rate of energy storage resources and the reliability of energy supply and use in RIES. Reference [9] applies a community energy storage system, taking into account the flexible resources on the demand side, such as heating, ventilation, and air conditioning systems, and proposes a coordinated optimization strategy that satisfies user comfort while reducing energy consumption costs. Reference [10] applies a combination of long- and short-term storage to an RIES, reducing the load forecasting errors through rolling optimization of long-term energy balance models and short-term power balance models, fully utilizing energy storage resources, and meeting the autonomous operation requirements of multi-energy microgrids. Reference [11] constructs a mechanism considering energy interaction between multiple IESs in neighboring regions, using shared storage to balance the energy flow and reduce the overall operating costs of the system. However, the above studies did not consider the energy storage system as an independent stakeholder, did not consider multi-party conflicts of interest, and the interests of energy storage investors were not guaranteed, which is not conducive to the promotion and construction of energy storage systems. To address these issues, references [12,13] establish a multi-party energy interaction mechanism based on a carbon trading incentive mechanism considering the system's low-carbon characteristics, solving the multi-party interest conflict in the RIES while reducing environmental pollution. References [14,15] analyze various operation modes of shared energy storage systems and propose a model for leasing energy storage capacity, focusing on the scientific configuration of energy storage system parameters, which is more in line with the actual operation of RIESs and better leverages the flexibility of energy storage systems to regulate loads.

Game theory is one of the standard tools for analyzing the economics of complex sequential strategy models. By using game theory, the cooperation and competition among multiple stakeholders in RIESs can be clarified, and the enthusiasm of each stakeholder to participate in energy transactions can be fully utilized [16–18]. References [19,20] consider the multiple uncertainties caused by the emergence of large-scale producer–consumer clusters. Reference [20] utilizes a non-cooperative game based on stochastic dynamic programming to develop a universal framework for both day-ahead and real-time electricity markets, significantly reducing the operational costs of microgrids and the charging costs of electric vehicles. Reference [21] examines the interests among multiple stakeholders in RIESs with shared energy storage, constructs a "one-leader-multiple-follower" game model, and develops a dynamic pricing mechanism for electric and thermal energy, achieving a win–win situation for all involved parties. Reference [22] establishes a game model based on Stackelberg game theory, with the distributed energy storage system as the leader and the load side as the follower, considering both the economic benefits and the load-side energy satisfaction index. Reference [23], considering the competitive and cooperative relationships among various stakeholders, constructs both a cooperative game model and a non-cooperative game model to achieve optimal energy distribution. However, these studies are all based on the traditional master–slave game structure of the participating subjects in the game analysis, the Energy Storage Operator (ESO) is fixed in the conditions of the leader or follower, which does not take into account the duality of the participating subjects as both the follower and the leader, as well as the simultaneous master–slave relationship and competitive relationship that exists between the two subjects under the circumstance, and the lack of mobilization of the subjects to participate in the optimization of the enthusiasm.

Aiming at the above problems, this paper proposes an optimal operation strategy for SESSs and RIESs based on a multi-level game. The strategy analyzes the master–slave relationship between the two when the ESO chooses to charge from the Integrated Energy Operator (IEO) and the competitive relationship between the two when meeting the load demand on the user side, and it integrates the user-side electric demand response and electric heat production. A multilevel master–slave game model is constructed with the IEO as the head leader, the ESO as the secondary leader, and the user as the follower. The model can stimulate the enthusiasm of each subject to participate in the coordinated operation, thus enhancing the overall economic efficiency of RIESs. In order to solve the model, the adaptive differential evolution algorithm with external archiving (JADE) combined with the Gurobi solver is used, and the results of the case analysis verify the effectiveness of the optimized operation strategy proposed in this paper.

2. Structure and Operation Mode of the RIES

The RIES architecture proposed in this paper consists of three entities: the IEO, the ESO, and the user. The energy coupling relationships are illustrated in Figure 1.

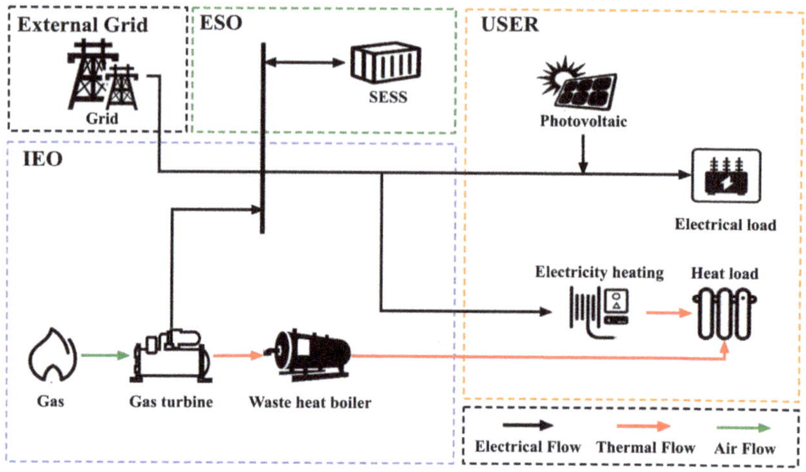

Figure 1. The RIES system structure with an SESS.

The IEO serves as an intermediary between the external grid, the user, and the ESO within the RIES, being capable of bidirectional energy transmission. The IEO is equipped with a micro gas turbine, which purchases gas from the external gas network for power generation. Simultaneously, it generates waste heat that enters a waste heat boiler and provides thermal energy to the user side through a heat exchange device.

The SESS is configured on the ESO side, and its deployment reduces the load volatility of the RIES while also increasing the diversity of energy choices for the user. The SESS chooses to charge from the IEO side and then provides discharge services to the user side. The ESO adjusts its charging and discharging operations based on price signals and user energy demands, setting reasonable selling prices and selecting appropriate time periods to enhance its own revenue.

The user side includes both electrical and thermal loads, equipped with photovoltaic power generation units and electric heating devices. The user can optimize their internal operating status and flexibly choose various energy sources to meet their energy needs based on their own interests and load requirements. For electrical loads, when the self-generated electricity meets the demand, the surplus electricity can be sold to the external grid; when the self-generated electricity is insufficient to meet the demand, the user can choose to purchase electricity from the IEO or ESO based on the time-of-use electricity prices. For thermal loads, in addition to the heat exchange units provided by the IEO, the thermal loads can also be satisfied by the user-side electric heating devices.

3. Game Relationships among Stakeholders

In the RIES, different entities have their own interests and conflicts, leading to a situation of mutual game theory. The IEO formulates a set of strategies for the daily electricity and heat selling prices. The ESO formulates charging and discharging strategies and selling prices for the user based on the IEO's pricing strategy, considering its own profit and the SESS capacity, and it adjusts the charging and discharging strategies and selling prices reasonably. Upon receiving the selling price strategies for energy from the IEO and ESO, the user side adjusts the electricity and heat load for each time period in real time based on the price signals and energy demand, plans the energy purchasing strategy, and uploads the strategy to the IEO and ESO.

When the selling energy price from the IEO is high, the ESO will adjust the selling energy price to increase profit, and the user side will prioritize cheaper energy sources for purchase. If the user side purchases a large amount of energy, the IEO and ESO will re-formulate the selling energy price strategy, and the ESO will adjust the charging and discharging strategies accordingly.

Due to the sequential decision-making process of the three parties, the entire process can be viewed as a multi-level game model with the IEO as the head leader, the ESO and user as followers, and the ESO as the secondary leader of the user in the RIES.

The IEO, the ESO and the user objective functions are F_{IEO}, F_{ESO} and F_{user}, respectively, all of which are optimized to maximize the operational benefits, as shown in the game relationship diagram in Figure 2.

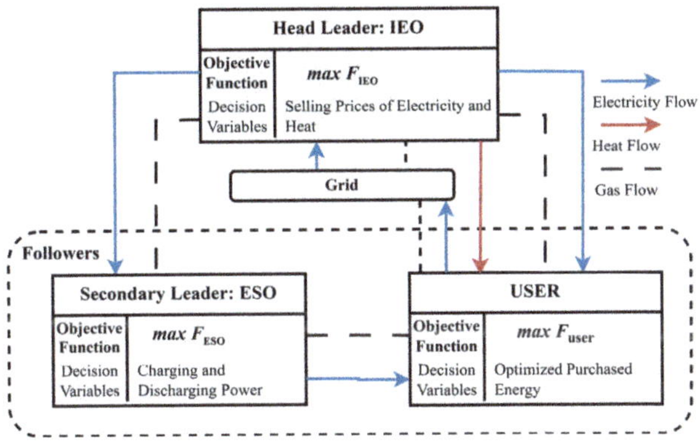

Figure 2. Multi-stakeholder leader–follower game framework.

4. Decision-Making Model of Each Stakeholder

4.1. Subsection Operation Model of the IEO

The IEO is responsible for integrating and coordinating various energy resources, such as electricity, natural gas, and thermal energy. Additionally, as the head leader within the RIES, the IEO aims to maximize its operational efficiency by employing dynamic time-of-use electricity pricing and heating prices as decision variables. The optimization objective of the IEO is to maximize its operational benefits, and its objective function is defined as follows:

$$\max F_{\text{IEO}} = \sum_{t=1}^{T}(C_{\text{ESO,sell}}^{t} + C_{\text{user,sell}}^{t} - C_{\text{grid}}^{t} - C_{\text{mt}}^{t}) \tag{1}$$

In the formula, divide one day into T time slots, where t represents the t-th time slot. In the t-th time slot, $C_{\text{ESO,sell}}^{t}$ and $C_{\text{user,sell}}^{t}$, respectively, represent the revenue for the IEO after selling energy to the ESO and the user, C_{grid}^{t} represents the interaction cost between the IEO and the grid (IEO acquiring/selling electricity, with C_{grid}^{t} being positive/negative), and C_{mt}^{t} represents the gas cost used by IEO. The expressions for the above are as follows:

$$C_{\text{grid}}^{t} = [\max(L_{\text{u}}^{t} + P_{\text{char}}^{t} - P_{\text{mt,e}}^{t}, 0)p_{\text{e,s}}^{t} + \min(L_{\text{u}}^{t} + P_{\text{chr}}^{t} - P_{\text{mt,e}}^{t}, 0)p_{\text{e,b}}^{t}]\Delta t \tag{2}$$

$$C_{\text{ESO,sell}}^{t} = p_{\text{e,s}}^{t} P_{\text{chr}}^{t} \Delta t \tag{3}$$

$$C_{\text{user,sell}}^{t} = (L_{\text{h}}^{t} p_{\text{h,s}}^{t} + L_{\text{e}}^{t} p_{\text{e,s}}^{t})\Delta t \tag{4}$$

In the formula, P_{chr}^{t}, L_{h}^{t}, L_{e}^{t} represent the electric power purchased by the ESO from the IEO and the electric and thermal power purchased by the user from the IEO during the t-th time slot; and $p_{\text{e,s}}^{t}$, $p_{\text{h,s}}^{t}$ represent the selling prices of electricity and heat by the IEO.

The IEO is equipped with a micro gas turbine, responsible for providing electricity and heat, with its main fuel being natural gas. The relationship between its operating cost and electric-thermal output during the t-th time slot can be expressed as:

$$C_{\mathrm{mt}}^{t} = (\lambda_{\mathrm{gas}} \frac{P_{\mathrm{mt,e}}^{t}}{\mathrm{H}_{\mathrm{gas}} \eta_{\mathrm{mt,e}}}) \Delta t \tag{5}$$

$$P_{\mathrm{mt,h}}^{t} = \frac{1 - \eta_{\mathrm{mt,e}} - \eta_{\mathrm{mt,loss}}}{\eta_{\mathrm{mt,e}}} \eta_{\mathrm{mt,h}} P_{\mathrm{mt,e}}^{t} \tag{6}$$

In the formula, λ_{gas} is the unit price of natural gas; $P_{\mathrm{mt,e}}^{t}$, $P_{\mathrm{mt,h}}^{t}$ are the electric and thermal power outputs during the t-th time slot; $\mathrm{H}_{\mathrm{gas}}$ is the low calorific value of gas; and $\eta_{\mathrm{mt,e}}$, $\eta_{\mathrm{mt,h}}$, $\eta_{\mathrm{mt,loss}}$ are, respectively, the power generation efficiency, heating coefficient, and heat dissipation loss rate of the micro gas turbine.

When the IEO interacts with the external grid for energy exchange, it needs to adhere to the power constraints of the grid to ensure that the energy purchased or sold does not exceed the carrying capacity of the interconnection lines, thereby maintaining the stable operation of the grid. At this point, the following constraints should be met:

$$\begin{cases} 0 \leqslant P_{\mathrm{eg,s}}^{t} \leqslant \zeta_{\mathrm{eg,s}}^{t} P_{\mathrm{eg,s}}^{\max} \\ 0 \leqslant P_{\mathrm{eg,b}}^{t} \leqslant \zeta_{\mathrm{eg,b}}^{t} P_{\mathrm{eg,b}}^{\max} \\ \zeta_{\mathrm{eg,b}}^{t} + \zeta_{\mathrm{eg,s}}^{t} \leqslant 1 \end{cases} \tag{7}$$

In the equation, $P_{\mathrm{eg,s}}^{t} / P_{\mathrm{eg,b}}^{t}$ represent the selling/purchasing electric power from the grid to the IEO; $P_{\mathrm{eg,b}}^{\max} / P_{\mathrm{eg,s}}^{\max}$ represent the maximum selling/purchasing electric power from the grid to the IEO; and $\zeta_{\mathrm{eg,s}}^{t}$ and $\zeta_{\mathrm{eg,b}}^{t}$ are Boolean variables, indicating the flag for selling/purchasing electric power from the grid to the IEO, meaning that only one action of selling/purchasing electricity can be conducted at a time

Additionally, in order to ensure the interests of various stakeholders, the energy purchase/sale prices of the IEO need to satisfy the following constraints:

$$p_{\mathrm{eg,b}}^{t} < p_{\mathrm{e,s}}^{t} < p_{\mathrm{eg,s}}^{t} \tag{8}$$

$$p_{\mathrm{h,min}}^{t} < p_{\mathrm{h,s}}^{t} < p_{\mathrm{h,max}}^{t} \tag{9}$$

$$\sum_{i=1}^{T} p_{\mathrm{e,s}}^{t} \leqslant T \overline{p_{\mathrm{e}}^{t}} \tag{10}$$

$$\sum_{i=1}^{T} p_{\mathrm{h,s}}^{t} \leqslant T \overline{p_{\mathrm{h}}^{t}} \tag{11}$$

In the equations, $p_{\mathrm{eg,b}}^{t}$, $p_{\mathrm{eg,s}}^{t}$ represent the purchasing and selling electricity prices from the grid during the t-th time slot of the day; $p_{\mathrm{h,min}}^{t}$, $p_{\mathrm{h,max}}^{t}$ represent the upper and lower limits of the selling price of heat specified by the IEO; and $\overline{p_{\mathrm{e}}^{t}}$, $\overline{p_{\mathrm{h}}^{t}}$ represent the average selling electricity and heat prices.

4.2. Operation Model of the ESO

The ESO references the selling energy prices of the IEO and the energy consumption situation on the user side, reasonably formulating its electricity price strategy and charging/discharging plans to maximize its total revenue while meeting the energy needs of the user. The objective function can be expressed as:

$$\max F_{\mathrm{ESO}} = \sum_{t=1}^{T} (C_{\mathrm{ESO,dis}}^{t} - C_{\mathrm{ESO,sell}}^{t} - C_{\mathrm{ESO,op}}^{t}) \tag{12}$$

$$C^t_{\mathrm{ESO,dis}} = \sum_{t=1}^{T} p^t_{\mathrm{dis}} P^t_{\mathrm{dis}} \Delta t \tag{13}$$

$$C^t_{\mathrm{ESO,op}} = k_c C^t_{\mathrm{ESO,sell}} \tag{14}$$

In the equation, P^t_{chr}, P^t_{dis} represent the charging and discharging power of the ESO in the t-th time slot; p^t_{dis} represents the price at which the ESO sells electricity in the t-th time slot; and k_c represents the operational and maintenance costs of the ESO.

In order to ensure the ESO's healthy participation in market competition, the following constraints are imposed on its selling electricity prices:

$$0.8 p^t_{\mathrm{e,s}} \le p^t_{\mathrm{dis}} \le 1.4 p^t_{\mathrm{e,s}} \tag{15}$$

In order to ensure that the shared energy storage system can continuously provide services in consecutive time slots within the same period and across different periods, that is, to maintain energy continuity and ensure the conservation of charging and discharging power within a single period, the following conditions must be satisfied:

$$E^t_{\mathrm{eso}} = E^{t-1}_{\mathrm{eso}}(1-\eta_{\mathrm{o}}) + (P^t_{\mathrm{chr}}\eta^+ - \frac{P^t_{\mathrm{dis}}}{\eta^-})\Delta t \tag{16}$$

$$\sum_{t=1}^{T} P^t_{\mathrm{chr}}\eta^+ - \sum_{t=1}^{T} \frac{P^t_{\mathrm{dis}}}{\eta^-} = 0 \tag{17}$$

In the equation, η_{o}, η^+, η^- represent the energy self-loss rate when the ESO is operating, the efficiency when the ESO is charging, and the efficiency when the ESO is discharging, respectively.

At any given moment, the charging or discharging power is subject to the ESO's allowed maximum power limit, while the total charging or discharging power in any time period is subject to the ESO's allowed maximum capacity limit. These constraints should be satisfied as follows:

$$\begin{cases} \left| \frac{E^{t+1}_{\mathrm{eso}} - E^t_{\mathrm{eso}}}{\Delta t} \right| \le \zeta^t_{\mathrm{chr}} P^{\max}_{\mathrm{chr}} \\ \left| \frac{E^{t+1}_{\mathrm{eso}} - E^t_{\mathrm{eso}}}{\Delta t} \right| \le \zeta^t_{\mathrm{dis}} P^{\max}_{\mathrm{dis}} \\ \zeta^t_{\mathrm{chr}} + \zeta^t_{\mathrm{dis}} \le 1 \end{cases} \tag{18}$$

In the equation, $P^{\max}_{\mathrm{chr}}$, $P^{\max}_{\mathrm{dis}}$ represent the allowed maximum charging and discharging power by the ESO, while ζ^t_{chr}, ζ^t_{dis} are Boolean variables representing the charging/discharging state of the energy storage device at the t-th time slot, restricting the energy storage device to perform only one charging or discharging action at a time.

In order to ensure the stability and lifespan of the shared energy storage system, the constraints on the total number of charging and discharging cycles and its maximum capacity within one period for the ESO are as follows:

$$\sum_{t=1}^{T} (\zeta^t_{\mathrm{chr}} + \zeta^t_{\mathrm{dis}}) \le \chi^{\max} \tag{19}$$

$$E^{\min}_{\mathrm{eso}} \le E^t_{\mathrm{eso}} \le E^{\max}_{\mathrm{eso}} \tag{20}$$

In the equation, $\chi^{\max}$ represents the maximum number of charging and discharging cycles allowed by the ESO within period the t-th time slot, while $E^{\min}_{\mathrm{eso}}$, $E^{\max}_{\mathrm{eso}}$, respectively, represent the minimum and maximum values of the capacity specified for the SESS.

4.3. Operation Model of the User

The user side adjusts the energy load based on its energy demand, selects energy sources reasonably, and considers the user load satisfaction index, aiming to maximize the operational efficiency of the user side. The objective function is:

$$\max F_{\text{user}} = \sum_{t=1}^{T} (U_{\text{user}}^{t} - C_{\text{user}}^{t} - C_{\text{q}}^{t}) \tag{21}$$

In the equation, U_{user}^{t} represents the electricity utility function; C_{user}^{t} represents the total cost of purchasing and selling electricity and heat by the user side; and C_{q}^{t} represents the penalty cost for reducing the heat load and the decrease in user load satisfaction due to the interrupted heat load.

The above equations can be further expressed as:

$$U_{\text{user}}^{t} = \left(a{L_{\text{u}}^{t}}^{2} + bL_{\text{u}}^{t}\right)\Delta t \tag{22}$$

$$C_{\text{user}}^{t} = (p_{\text{h,s}}^{t} L_{\text{h}}^{t} + p_{\text{e,s}}^{t}(L_{\text{e}}^{t} + L_{\text{eh}}^{t}) + p_{\text{dis}}^{t} L_{\text{e,dis}}^{t} - p_{\text{l,s}}^{t} L_{\text{l,s}}^{t})\Delta t \tag{23}$$

$$C_{\text{q}}^{t} = \beta({L_{\text{w,h}}^{t}}^{2})\Delta t \tag{24}$$

$$L_{\text{u}}^{t} = L_{\text{e}}^{t} + L_{\text{e,dis}}^{t} - L_{\text{pv}}^{t} \tag{25}$$

In the equations, L_{u}^{t}, $L_{\text{e,dis}}^{t}$, L_{eh}^{t}, $L_{\text{l,s}}^{t}$, $L_{\text{w,h}}^{t}$ and L_{pv}^{t} represent the net electrical load after adjusting the energy consumption load, the electrical energy obtained by the user side from the ESO, the electrical energy consumed by the user side's electric heating equipment for heat production, the electrical energy sold by the user side to the external grid, the actual heat load reduction on the user side, and the forecasted output of photovoltaic equipment, respectively, within the t-th time slot. a and b represent the parameters of the quadratic function characterizing the electricity utility function; $p_{\text{l,s}}^{t}$ represents the selling electricity price to the external grid by the user side in the t-th time slot; and β represents the penalty coefficient for reducing and interrupting the heat load.

The user-side electricity load can be divided into two parts: the rigid electricity load with high reliability requirements and the interruptible flexible electricity load. In this paper, the former is not adjusted, while the flexible electricity load is moved to periods with lower electricity prices based on the price signals received by the user side, aiming to reduce electricity costs. Meanwhile, the user-side heat load can be partially reduced to lower electricity costs. The constraints for shifting and reducing the load on the user side should be satisfied as follows:

$$L_{\text{u}}^{t} = L_{\text{e,g}}^{t} + L_{\text{e,s}}^{t} + L_{\text{eh}}^{t} \tag{26}$$

$$L_{\text{u,h}}^{t} = L_{\text{v,h}}^{t} - L_{\text{w,h}}^{t} \tag{27}$$

$$L_{\text{e,s,min}}^{t} \le L_{\text{e,s}}^{t} \le L_{\text{e,s,max}}^{t} \tag{28}$$

$$0 \le L_{\text{w,h}}^{t} \le L_{\text{w,h,max}}^{t} \tag{29}$$

$$\sum_{t}^{T} L_{\text{e,s}}^{t}\Delta t = 0 \tag{30}$$

In the equations, $L_{\text{e,g}}^{t}$, $L_{\text{e,s}}^{t}$ represent the rigid electricity load and the flexible electricity load; $L_{\text{v,h}}^{t}$ represents the heat load before adjustment; and $L_{\text{e,s,min}}^{t}/L_{\text{e,s,max}}^{t}$ and $L_{\text{w,h,max}}^{t}$ represent the minimum/maximum values for shifting the flexible electricity load and the maximum reduction amount for the heat load.

The electric heating equipment on the user side in this paper can meet all forms of the user's heat demand. The user can consider selecting the source of heat energy based on price signals to maximize the benefits for the user side. The constraints for the operation of the electric heating equipment are as follows:

$$L^{t}_{\mathrm{e,h}} = \eta^{l}_{\mathrm{h}} L^{t}_{\mathrm{eh}} \tag{31}$$

$$0 \leq L^{t}_{\mathrm{e,h}} \leq L^{t}_{\mathrm{e,h,max}} \tag{32}$$

In the equations, $L^{t}_{\mathrm{e,h}}$ and $L^{t}_{\mathrm{e,h,max}}$ represent the heat generated by the electric heating equipment and the upper limit of heat generation in the t-th time slot of the day; and η^{l}_{h} represents the conversion efficiency of the user-side electric heating equipment.

For the methods of obtaining heat energy on the user side, the following options are provided:

$$L^{t}_{\mathrm{u,h}} = \begin{cases} \zeta_{\mathrm{h}} L^{t}_{\mathrm{w,h}} + (1-\zeta_{\mathrm{h}}) L^{t}_{\mathrm{e,h}} & (L^{t}_{\mathrm{u,h}} \leq L^{t}_{\mathrm{w,h,max}}) \\ \zeta_{\mathrm{h}} L^{t}_{\mathrm{w,h,max}} + L^{t}_{\mathrm{e,h}} & (L^{t}_{\mathrm{u,h}} > L^{t}_{\mathrm{w,h,max}}) \end{cases} \tag{33}$$

$$\zeta_{\mathrm{h}} = \begin{cases} 0 & (\frac{p^{t}_{\mathrm{e,s}}}{\eta^{l}_{\mathrm{h}}} \leq p^{t}_{\mathrm{h,s}}) \\ 1 & (\frac{p^{t}_{\mathrm{e,s}}}{\eta^{l}_{\mathrm{h}}} > p^{t}_{\mathrm{h,s}}) \end{cases} \tag{34}$$

5. Decision-Making Model of Each Stakeholder

The multi-level leader–follower game model is solved using the adaptive differential evolution algorithm with external archiving (JADE) combined with the Gurobi solver on the MATLAB 2021a platform.

In the JADE algorithm, the crossover and scaling factors of the IEO are dynamically maintained based on the historical performance of the objective function in each iteration. The strategy is as follows: if the current individual performs better than its neighbors in the previous generation, then the original crossover and scaling factors remain unchanged; if the current individual performs worse in the previous generation, an exponential transformation is adopted to gradually adjust the values of the crossover and scaling factors. This allows the better-performing individuals to pass on their factors to the offspring, increasing the efficiency of the offspring search, thus finding the global optimum faster. This makes the JADE algorithm suitable for solving high-dimensional nonlinear optimization problems such as maximizing the total revenue of the IEO. The steps and flowchart of the game solution are shown in Figure 3.

The specific implementation process of the multi-level leader–follower game is as follows:

(1) Initialize the parameters of the IEO, the ESO, and the user side, set the initial population number in the JADE algorithm to 40, with mutation probability and crossover probability of 10% and 80%; $K = 0$, $K_{\max} = 120$.
(2) The leader IEO will transmit the pricing strategies of the initial electricity and heat prices generated to the followers, the ESO and the user.
(3) The ESO responds first, formulating the charging strategy to transmit to the IEO while also determining the selling price to transmit to the user side.
(4) The user side accepts the selling price strategies from the IEO and ESO, optimizes its energy consumption structure to maximize its benefits, formulates its energy purchase strategy, and reports it to the IEO and ESO.
(5) Based on the genetic crossover selection and mutation, priority is given to ensuring the highest overall system efficiency. Subsequently, the IEO ensures the maximization of its own operational efficiency, guiding the ESO and the user in dynamic optimization, updating the ESO's charging strategy and the user side's energy purchasing strategy. Where F^{k*}_{IEO} is the maximum value of the objective function of the IEO in the previous iteration process.
(6) Repeat steps (3) to (5) until the selling price strategy of the IEO, the charging strategy and selling price strategy of the ESO, and the energy purchasing strategy of the user

side reach a game equilibrium solution or the maximum number of iterations is reached, then output the optimization results.

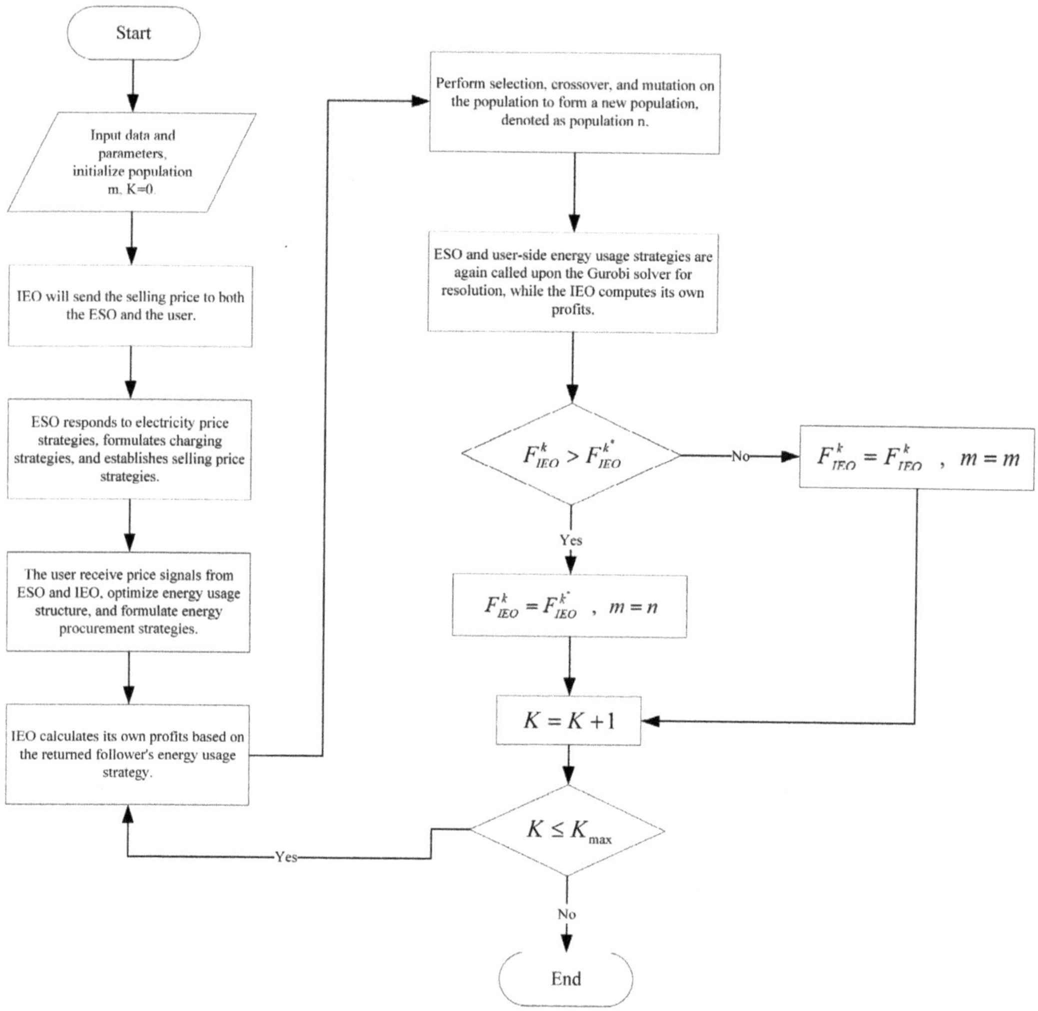

Figure 3. Multi-level game solving process.

6. Case Study Analysis

The actual example described in this paper is a community model including the IEO and ESO supplying energy to several residential buildings. It is assumed that a day can be divided into $T = 24$ time periods, with each residence equipped with electric heating equipment and solar power generation devices on the roof. The IEO side is equipped with a micro gas turbine. The electricity and heat energy demand of the user side, as well as the photovoltaic output within one cycle, are shown in Figure 4.

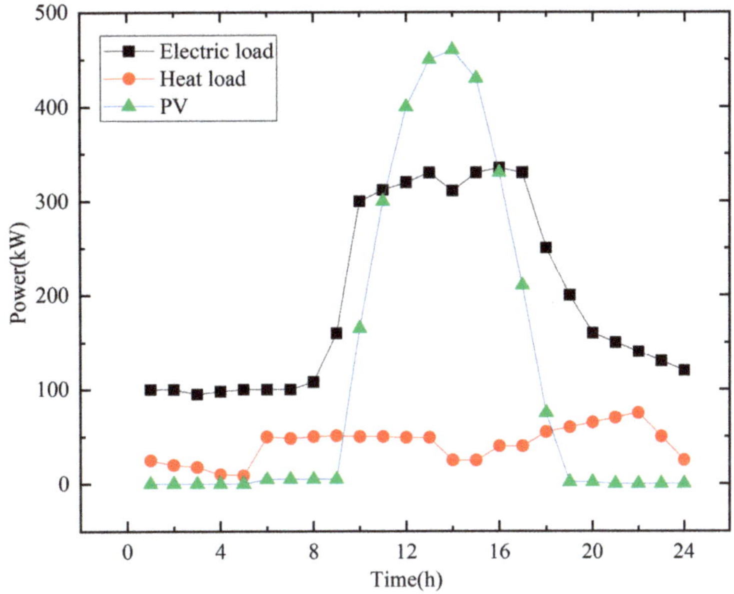

Figure 4. Predicted curves of the electricity and heat loads on the user side, along with the photovoltaic output.

6.1. Analysis of the Economic Results of RIES Operation

Using the operational strategies described in this paper, after each party reaches a game equilibrium, the electricity selling price curves of the IEO and ESO are as shown in Figure 5a, where the overall trends of the IEO's and ESO's electricity selling prices follow the same fluctuations as the grid's electricity selling prices.

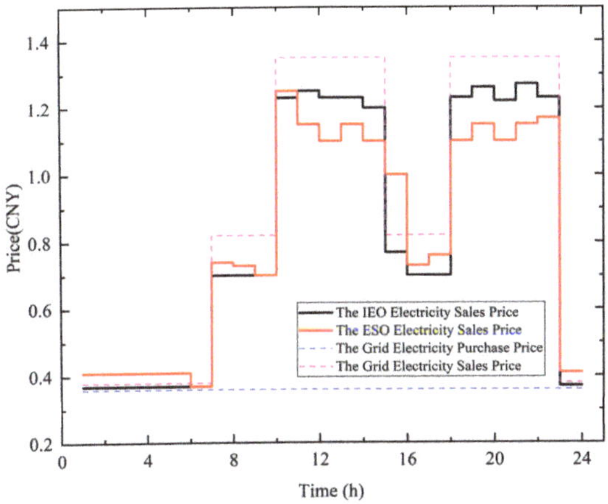

(**a**) Optimized electricity sales price by IEO and ESO.

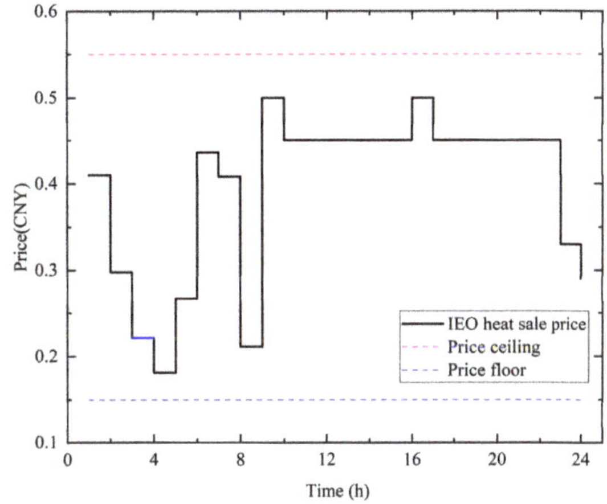

(**b**) Optimized heat sales price by IEO.

Figure 5. Optimization results of the selling prices by IEO and ESO.

During the off-peak hours of the grid's time-of-use electricity prices from 0:00~6:00 and 22:00~24:00, the ESO's selling price is generally higher than the IEO's, averaging 9.5% higher. This is because the ESO cannot perform charging and discharging activities simultaneously, so it should choose to charge during the periods when the IEO's electricity selling prices are relatively low throughout the entire cycle. At these times, setting a high selling price by the ESO can encourage the user to purchase electricity from the IEO. On the other hand, this is also due to the efficiency losses during charging and discharging, making the IEO's selling cost higher than the ESO's. During the peak hours of the grid's time-of-use electricity prices from 9:00~14:00 and 17:00~22:00, the ESO's selling price is generally 19.1% lower compared to the IEO's, making the user more inclined to obtain electricity from the ESO. During the normal hours of the grid's time-of-use electricity prices from 6:00~9:00 and 14:00~17:00, the ESO's selling price is generally 9.2% higher than IEO's, at which times the user is more inclined to obtain electricity from the IEO.

From Figure 6, it can be observed that the SESS charges during 87.5% of the off-peak periods of the electricity price (0:00~5:00 and 22:00~24:00) and discharges during 80% of the peak periods (9:00~11:00, 12:00~14:00 and 17:00~22:00). The ESO follows the principle of "charge low, discharge high", but it does not charge to the full capacity during charging, nor does it discharge in every peak period. This is because, in the game process, this charging and discharging mode effectively reduces the decrease in the IEO's profits caused by the addition of the ESO, and it also reduces the user side's electricity purchase costs during peak price periods. During the flat-rate electricity price periods (6:00~9:00 and 14:00~17:00), the capacity of the SESS remains unchanged. This ensures that the energy storage equipment reduces the number of charge and discharge cycles, reduces the operating costs, and extends its service life.

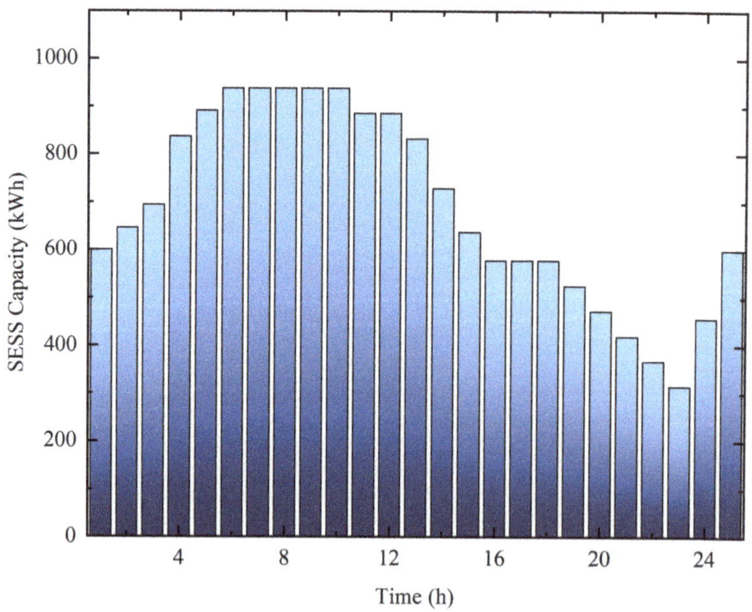

Figure 6. Variation in the SESS's capacity.

Figure 7 shows the supply–demand balance of the user-side electricity load after the game optimization. The flexible electricity load has been shifted from the continuous high-price period of 9:00~18:00 to the two continuous periods of 0:00~9:00 and 18:00~24:00, which has improved the user's benefits while achieving the goal of "peak shaving and valley filling". However, the period 18:00~22:00, which receives the transferred flexible electricity load, belongs to the high-price period. This is mainly due to three reasons. Firstly, this method of electricity load transfer ensures that electricity loads with long continuous operation times can also participate in the transfer, increasing the proportion of flexible electricity load in the total load and making the adjustment more flexible. Secondly, there is a proportional limit to the flexible electricity load that can be absorbed in each unit time period of the entire cycle. Finally, the micro gas turbine used by the IEO in this paper operates according to a heat-to-electricity strategy, and the electricity generated by the micro gas turbine can be thermally absorbed by the transferrable flexible electricity load.

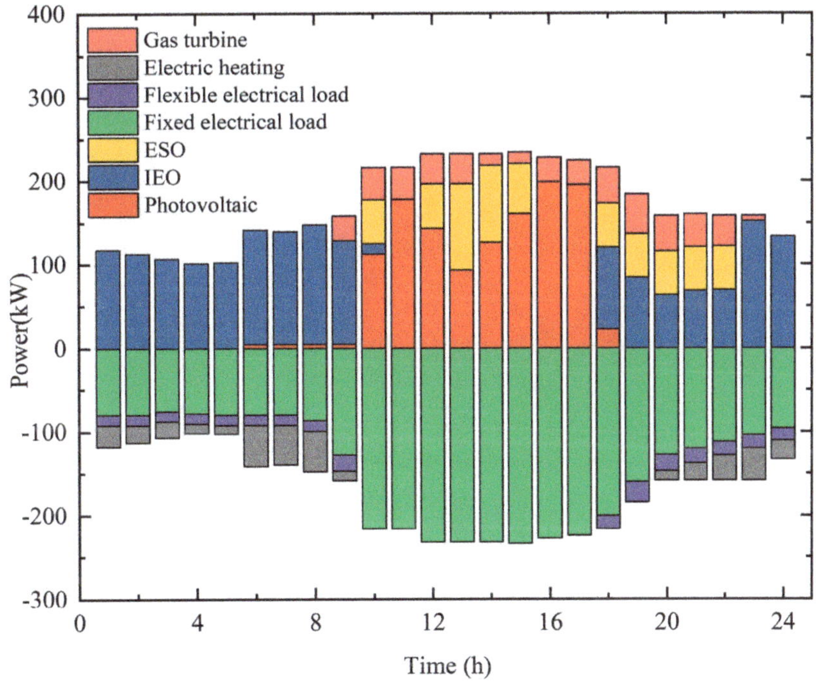

Figure 7. Supply-demand balance of the electricity load.

The IEO's thermal energy selling price curve is shown in Figure 5b, combined with the supply–demand balance of the user's thermal load after game optimization shown in Figure 8. During the periods of low or relatively low electricity prices from 0:00~8:00 and 23:00~24:00, electric heating is used as the source of thermal energy supply, effectively reducing the user's heating costs. During the period of higher electricity prices from 9:00~22:00, the user tends to use the IEO as the source of thermal energy supply. During some time periods, the thermal energy supply of the user side comes from both electric heating and the IEO. This is because, on the one hand, thermal energy regulation has inertia, and on the other hand, for example, the thermal load surges during the period of 19:00~22:00, thus a combined heating supply mode is adopted. The optimized user thermal load is reduced throughout the entire cycle. Observing the IEO's thermal selling price curve, it can be seen that the selling price of thermal energy is lower during 0:00~9:00 and 22:00~24:00, which coincides with the use of the IEO as the source of thermal energy during 8:00~9:00 and 22:00~23:00. In other time periods, even if the selling price of thermal energy is high, it still maintains a price advantage as electricity prices increase. Therefore, there is not much fluctuation in prices, and the user also tends to use the IEO as the source of thermal energy.

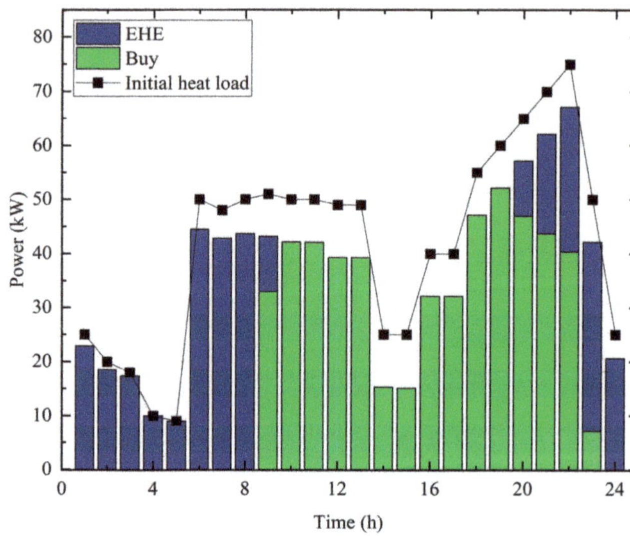

Figure 8. Before and after optimization of the heat load.

6.2. Analysis of the Economic Results of RIES Operation

To illustrate the effectiveness of the multi-level game optimization scheduling proposed in this paper in terms of economy and system stability, the following three scenarios are set for comparative analysis:

- Scenario 1: The economic operation strategy proposed in this paper, where the IEO's electricity and heat selling prices, as well as the ESO's electricity selling prices, are all non-fixed values.
- Scenario 2: The IEO's electricity and heat selling prices are non-fixed values, while the ESO's electricity selling prices are fixed.
- Scenario 3: The IEO's electricity and heat selling prices are non-fixed values, with no ESO participation in the system operation.

Table 1 reflects the returns of each subject and the overall returns of RIES under different scenarios, Comparing the revenues of each entity in Scenario 1 and Scenario 3, it can be observed that in Scenario 1 compared to Scenario 3, where the ESO did not participate in the energy system operation, the user-side revenue increased by CNY 152.81, a 2.8% increase. The ESO's participation in the system operation resulted in an increase of CNY 382.51 in revenue. The overall revenue of the RIES increased by CNY 486.63, a 7.8% increase. It is evident that the ESO's participation in the RIES's operation reduces the load fluctuation of the RIES, improves the overall economic efficiency, and enhances the overall performance of the energy system. However, the IEO's revenue decreased by CNY 48.69, a 7.2% reduction. This decrease is attributed to the competitive relationship between the IEO and the ESO in supplying energy to the user side. After optimization, the total energy consumption on the user side remains constant, leading to a decrease in the IEO's revenue.

Table 1. Revenues of various entities and overall revenues in different scenarios.

Revenue (CNY) / Scenario	IEO	ESO	User	RIES
1	628.01	382.51	5705.11	6715.63
2	543.54	207.85	5874.52	6625.91
3	676.70	0.00	5552.30	6229.00

To verify the impact of the ESO's electricity price strategy on the economic efficiency of the system after the ESO participates in the energy system operation, a comparison of the revenues of each entity in Scenario 1 and Scenario 2 was conducted. Relative to Scenario 2, in Scenario 1, the IEO's revenue increased by CNY 84.47, a 15.5% increase; the ESO's revenue increased by CNY 174.66, an 84.03% increase; the user revenue decreased by CNY 169.41, a 2.9% reduction; and the overall system revenue increased by CNY 89.72, a 1.4% increase. It is evident that the non-fixed electricity price strategy of the ESO led to an overall increase in revenue. At the same time, the ESO's revenue increased significantly, enhancing the enthusiasm for the ESO's participation in the investment.

The graph in Figure 9 shows the capacity variation of the ESO's energy storage system in Scenarios 1 and 2 during different time periods within one cycle. It can be seen that the charging and discharging periods of the ESO are almost the same in both scenarios. However, in Scenario 2, where the ESO's selling price is fixed, the ESO cannot adjust its price appropriately during peak periods of the IEO's electricity price and the user's energy demand to increase its revenue, thus not fully utilizing the flexibility of the ESO. Additionally, due to the lower ESO electricity price in this scenario, the user is more inclined to choose the ESO as their energy source, leading to higher user-side profits in Scenario 2 compared to Scenario 1, but further reducing the ESO's benefits.

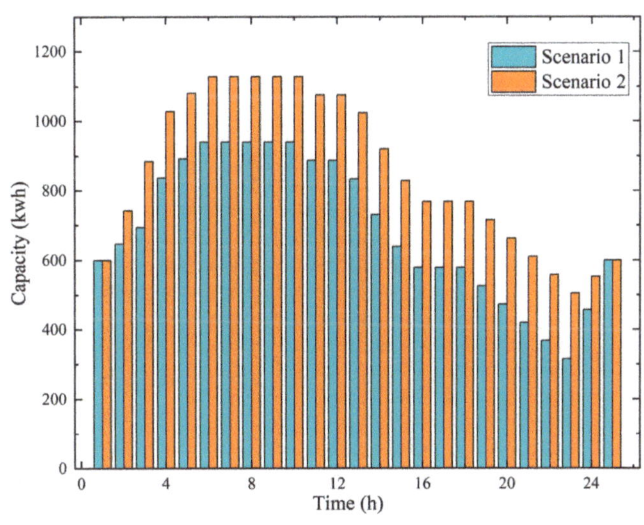

Figure 9. Comparison of the SESS capacity changes between Scenarios 1 and 2.

The graph in Figure 10 shows the user-side electricity load curve for the entire cycle under the three scenarios. It can be seen that in all three scenarios, compared to the original scenario, the electricity load achieves the goal of "peak shaving and valley filling", resulting in a smoother load curve. With the participation of the ESO in the energy system operation, the entire load curve becomes smoother, reducing the peak and valley loads and improving the overall system stability. However, compared to Scenario 1, in Scenario 2, the ESO's selling price is generally lower during the 9:00~16:00 period, leading the user side to prefer electric heating to meet the heating demand, resulting in a higher electricity load in Scenario 2 compared to Scenario 1.

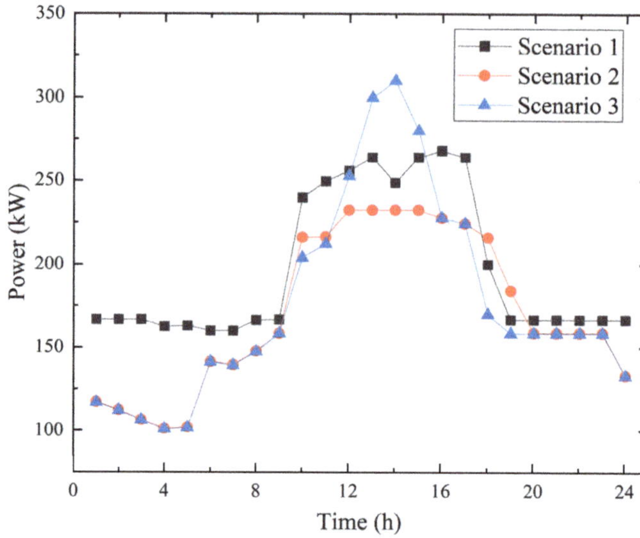

Figure 10. Comparison of the user-side electrical load curves in the three scenarios.

7. Conclusions

(1) Existing studies mostly utilize the leader–follower mode of master–slave game models, where each entity in the model can only choose to be a leader or a follower. This paper considers the dual role of the ESO as a buyer/seller of energy, constructing a multi-level game model that fully exploits the tripartite flexibility of source–load–storage, achieving multi-entity and multi-level coordinated optimization scheduling.

(2) With the development of shared energy storage services and the increasing variety of user-side energy-consuming devices, this paper considers the coupling of the IEO-side electricity and heat, as well as the participation of the user-side electric heating devices. It also takes into account the competitive relationship between the IEO and ESO in energy supply, making the constructed model more comprehensive and better suited to the actual operational scenarios of energy systems.

(3) Compared to centralized algorithms, the method of using the heuristic JADE algorithm combined with the Gurobi solver for model solving achieves the distributed solving of the model, better safeguarding the data privacy of the IEO, the ESO, and the user.

(4) The comparison of three scenarios validates that the energy system economic operation strategy proposed in this paper, based on multi-level games with shared energy storage, can maximize the overall benefits of the system, achieving mutually beneficial cooperation between the IEO, the ESO, and the user.

The optimization operating strategy for the RIES presented in this paper is designed based on a scenario where all the stakeholders provide complete information. However, in actual trading markets, it cannot be guaranteed that all the parties involved in the game are absolutely rational, and the phenomenon of incomplete information is inevitable, leading to unreliable operating strategies. Therefore, how to achieve the optimal resource allocation strategy in scenarios of incomplete information merits attention.

Author Contributions: Conceptualization, Y.Y. and T.C.; methodology, Y.Y., T.C. and H.Y.; software, Y.Y., T.C. and H.Y.; validation, Y.Y., T.C. and J.W.; formal analysis, Y.Y. and T.C.; investigation, Z.Y. and W.L.; resources, Y.Y.; data curation, T.C., H.Y. and J.W.; writing—original draft preparation, T.C.; writing—review and editing, Y.Y., T.C. and H.Y.; supervision, T.C; project administration, Y.Y.; funding acquisition, Y.Y. All authors have read and agreed to the published version of the manuscript.

Funding: This research was funded by the Science and Technology Planning Project of Xinjiang Uygur Autonomous Region (2022A01007-3).

Data Availability Statement: All data are referenced in the article.

Conflicts of Interest: The authors declare no conflicts of interest.

References

1. Zhang, S.; Wang, D.; Cheng, H.; Song, Y.; Yuan, K. Key technologies and challenges of low-carbon integrated energy system planning for carbon emission peak and carbon neutrality. *Autom. Electr. Power Syst.* **2022**, *46*, 189–207.
2. Chen, H.; Mao, W.; Zhang, R.; Yu, W. Low-carbon optimal scheduling of a power system source-load considering coordination based on carbon emission flow theory. *Power Syst. Prot. Control* **2021**, *49*, 1–11.
3. Shuai, X.; Ma, Z.; Wang, X.; Guo, H.; Zhang, H. Optimal operation of shared energy storage and integrated energy microgrid based on leader-follower game theory. *Power Syst. Technol.* **2023**, *47*, 679–690.
4. Cao, Y.; Yu, J.; Li, Y.; Guo, J.Q.; Zhu, S.Y.; Yong, W.K. Evolutionary game based competition strategy selection method for retail side of regional integrated energy market. *Autom. Electr. Power Syst.* **2023**, *47*, 104–113.
5. Yan, D.; Chen, Y. Review on business model and pricing mechanism for shared energy storage. *Autom. Electr. Power Syst.* **2022**, *46*, 178–191.
6. Li, Y.; Zhang, N.; Zhang, S.; Liu, Y.; Du, E.; Wang, P.; Xiao, W.; Ci, S. Multienergy cloud energy storage for power systems: Basic concepts and research prospects. *Proc. CSEE* **2023**, *43*, 2179–2190.
7. Lai, S.Y.; Qiu, J.; Tao, Y.C. Individualized pricing of energy storage sharing based on discount sensitivity. *IEEE Trans. Ind. Inform.* **2022**, *18*, 4642–4653. [CrossRef]
8. Kalathil, D.; Wu, C.; Poolla, K.; Varaiya, P. The sharing economy for the electricity storage. *IEEE Trans. Smart Grid* **2019**, *10*, 556–567. [CrossRef]
9. Hosseini, S.M.; Carli, R.; Jantzen, J.; Dotoli, M. Multi-Block ADMM Approach for Decentralized Demand Response of Energy Communities with Flexible Loads and Shared Energy Storage System. In Proceedings of the 2022 30th Mediterranean Conference on Control and Automation (MED), Athens, Greece, 28 June–1 July 2022; pp. 67–72.
10. Zhao, B.; Wang, W.; Chen, Z.; Luo, B.; Li, Z.; Wang, X. Energy-Power Distributionally Robust Optimization for Multi-energy Microgrid Considering Long Short-term Hybrid Energy Storage. *Autom. Electr. Power Syst.* **2023**, *47*, 22–33.
11. Hu, X.; Tian, W.; Song, G.; Yu, H. Bi-objective Rolling Operation Optimization Based on Surrogate Model Acceleration of Community-Level Integrated Energy Systems. *Electr. Power* **2023**, *56*, 138–145.
12. Wang, R.; Cheng, S.; Wang, Y.; Dai, J.; Zuo, X. Low-carbon and economic optimization of a regional integrated energy system based on a master-slave game with multiple stakeholders. *Power Syst. Prot. Control* **2022**, *50*, 12–21.
13. Cui, Y.; Zeng, P.; Zhong, W.; Cui, W.; Zhao, Y. Low-carbon economic dispatch of electricity-gas-heat integrated energy system based on ladder-type carbon trading. *Electr. Power Autom. Equip.* **2021**, *41*, 10–17.
14. Xi, J.; Sun, L.; Ge, P.; Zhang, R. Source-charge Coordinated Economic Dispatching of Photothermal-electrothermal Integrated Energy System Considering Electric Heating Demand Response. *J. Northeast Electr. Power Univ.* **2023**, *43*, 61–71.
15. Zhang, W.; Lu, X. Sharing and self-operating multimode trading model of energy storage aggregators with peer-to-peer trade. *Autom. Electr. Power Syst.* **2022**, *46*, 15–23.
16. Li, Z.; Chen, L.; Liu, D.; Chen, F.; Zheng, T.; Mei, S. Subsidy pricing method for stackelberg-game-based energy storage system. *High Volt. Eng.* **2020**, *46*, 519–526.
17. Srinivasan, D.; Rajgarhia, S.; Radhakrishnan, B.M.; Sharma, A.; Khincha, H.P. Game-theory based dynamic pricing strategies for demand side management in smart grids. *Energy* **2017**, *126*, 132–143. [CrossRef]
18. Taylor, P.D.; Jonker, L.B. Evolutionary stable strategies and game dynamics. *Math. Biosci.* **1978**, *40*, 145–156. [CrossRef]
19. Cheng, Y.; Guo, Q. Bi-level Optimal Scheduling Strategy of Multi-agent Integrated Energy System With Dynamic Energy Prices and Shared Energy Storage Power Station. *Mod. Electr. Power* **2024**, *41*, 10–20.
20. Hosseini, S.M.; Carli, R.; Parisio, A.; Dotoli, M. Robust Decentralized Charge Control of Electric Vehicles under Uncertainty on Inelastic Demand and Energy Pricing. In Proceedings of the 2020 IEEE International Conference on Systems, Man, and Cybernetics (SMC), Toronto, ON, Canada, 11–14 October 2020; pp. 1834–1839.
21. Rathor, S.K.; Saxena, D. Decentralized Energy Management System for LV Microgrid Using Stochastic Dynamic Programming With Game Theory Approach Under Stochastic Environment. *IEEE Trans. Ind. Appl.* **2021**, *57*, 3990–4000. [CrossRef]
22. Tian, X.; Chen, L.; Li, X.; Yuan, W. Optimal Scheduling for Energy Storage Sharing Among Communities With Photovoltaic Resource Based on Stackelberg Game and Improved Shapley Value. *Power Syst. Technol.* **2023**, *47*, 2252–2261.
23. Karavas, C.-S.; Arvanitis, K.; Papadakis, G. A Game Theory Approach to Multi-Agent Decentralized Energy Management of Autonomous Polygeneration Microgrids. *Energies* **2017**, *10*, 1756. [CrossRef]

energies

MDPI

Technical Note

An Intelligent Controller of LED Street Light Based on Discrete Devices

Zhan Wang *, Dehua Zhang, Jishen Li and Wei Zhang

College of Electrical Engineering, Zhejiang University, Hangzhou 310027, China
* Correspondence: wzwang@zju.edu.cn

Abstract: To combat global environmental deterioration and energy scarcities, it is crucial to implement energy-saving upgrades for urban road lighting. Comparatively, LEDs have emerged as an advanced and eco-friendly lighting option due to their low energy consumption, excellent performance, high color rendering index, and prolonged lifespan. By incorporating solar cell technology, a smart LED street light controller based on small-scale integrated circuits was developed to enable intelligent control for various lighting needs such as dimming, timing, automatic detection, and sound and light control. Through circuit simulations and experimental outcomes, it has been validated that the controller's structure and performance parameters align with the design specifications. This design encompasses knowledge from diverse fields, including fundamentals of circuit and electronic technology, photovoltaic cell technology, power electronics, and sensor technology, showcasing robust engineering and practicality. Its utilization in the experimental course for second-year college students majoring in electrical engineering contributes to the grooming of professionals and expands the perspectives of future talents, enriching their application of knowledge and practical innovation capabilities.

Keywords: controller of LED light; DC-DC converter; photovoltaic cell; switch power supply

Citation: Wang, Z.; Zhang, D.; Li, J.; Zhang, W. An Intelligent Controller of LED Street Light Based on Discrete Devices. *Energies* **2024**, *17*, 1838. https://doi.org/10.3390/en17081838

Academic Editor: Alon Kuperman

Received: 4 February 2024
Revised: 9 April 2024
Accepted: 10 April 2024
Published: 11 April 2024

1. Introduction

To reduce greenhouse gas emissions and address global energy shortages, the application and development of renewable energy technologies have received widespread attention. In the process of new energy conversion, power electronics technology provides reliable technical support and has become an important component, and related technological research continues to make new progress. Elrefaey et al. introduced an enhanced topology for a DC–DC converter suitable for PV applications, with the ability to be powered by multiple DC sources and to output to multiple channels [1]. Kulasekaran and Dasarathan focused on integrating rooftop solar with the DC microgrid and proposed a high-gain DC–DC converter for photovoltaic systems (HGBC-PVS) to connect to a higher-voltage network [2]. A smart controller designed by A. Derrouazin based on fuzzy control was developed to oversee a sustainable hybrid power system, aiming to sustain a typical residential environment [3]. Bodele and Kulkarni employed the bidirectional modular PV battery system (BMPBS), which utilizes non-isolated buck and boost converter combinations, to mitigate module mismatch losses due to factors such as partial shading (PS), dissimilar aging effects, different power ratings, etc. [4].

The advancement and implementation of new energy technologies have resulted in a growing need for engineering and technical expertise in relevant industries. Within this context, creating design projects centered on power electronics technology and new energy technology holds practical significance and promotional value for enhancing the knowledge and skills of second-year college students majoring in electrical engineering. The LED street light controller based on small-scale integrated circuits integrates fundamentals of circuit and electronic technology, photovoltaic cell technology, sensing technology, and switch power supply technology. The design's tasks and objectives encompass the following three components. The design of an energy-efficient street light intelligent controller utilizes simulator devices.

1.1. Power Module

Utilizing a lithium-ion battery, model 18650, with main parameters of 4800 mWh and 3.7 V, as the power source, the controller circuit is supplied with ±5 V DC power. When the ambient light conditions are met, the lithium battery is charged using a solar panel, selected with parameters of 5 V and 200 mA in this design.

1.2. Controller

When light is detected, the LED street light is turned off. When no light is detected, the LED street light is illuminated with low brightness, and the intensity of the low-brightness illumination can be adjusted. When someone approaches, a sound signal is detected, or a manual switch is triggered, the LED street light will illuminate at high brightness for a certain period of time, and the duration of high-brightness illumination can be adjusted.

1.3. LED Driver

The street light utilizes LED light source and requires a corresponding LED constant current driving circuit.

The remainder of the paper is organized as follows. Section 2 will introduce the whole implement of the controller. Section 3 will describe a simulation based on OrCAD used to verify the whole implement. Section 4 will provide an implementation of the intelligent controller and verify availability.

2. Design of Intelligent Controller

The intelligent controller's structural diagram is depicted in Figure 1. It primarily encompasses (1) the DC power supply unit, which furnishes DC power for detection and control circuits via DC-DC conversion. This unit comprises three modules: a boost circuit, a negative voltage generation circuit, and a battery-charging circuit [5]. (2) The sensor detection unit comprises four modules: the light detection circuit, human infrared detection circuit, sound detection circuit, and manual switch circuit, which yield multiple logic levels. (3) The logic control unit transforms these multiple logic levels into control signals to regulate the operation, shutdown, and brightness of LED street lights. The brightness of the LED light is adjusted by the PWM modulation circuit, and the duration of high-brightness lighting is determined by the timing circuit. (4) The LED driver unit accepts the control signals and drives the LED street lights to operate with a constant current.

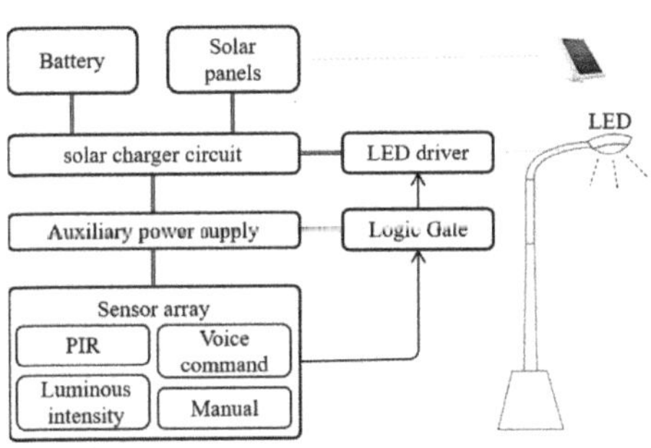

Figure 1. Structure of intelligent controller.

The organizational structure of the main circuit and the design of parameters are outlined as follows.

2.1. Passive Infra-Red Sensing Signal Processing Circuit

The passive infra-red (PIR) sensor, specifically the KB-500B model (Shenzhen Haiwang Sensor Co., Ltd., Shenzhen, China) chosen for this design, integrates an N-channel junction field-effect transistor in its configuration, connected using a common drain setup for impedance transformation. This converts the charge output from the pyroelectric detection element into a voltage output. Since the infrared center wavelength of human radiation falls within the 9–10 μm range, the detection element's wavelength sensitivity spans from 0.2 to 20 μm. Typically, a window with a filter lens is positioned atop the sensor, allowing the passage of light within the 7 to 10 μm wavelength range. This facilitates the creation of an infrared sensor tailored for human radiation detection. To enhance detector sensitivity and expand the detection range, a Fresnel lens is commonly installed in front of the detector. This lens, composed of transparent plastic with equidistant tooth patterns engraved on one side, limits the incident light's peak wavelength to approximately 10 μm. Alongside an amplification circuit, it can amplify the detection signal by over 70 dB and detect human body movement within a 20 m range [6]. Furthermore, sensor sensitivity is influenced by the direction of human movement. Radial movement has low sensitivity, while transverse movement (i.e., perpendicular to the radius) is highly sensitive. Therefore, strategic installation positions must be selected to achieve optimal detection sensitivity.

The sensing signal is directed to a dedicated integrated circuit, with the BISS0001 [7] selected for processing in this design. The BISS0001 comprises an operational amplifier, voltage comparator, state controller, delay time timer, blocking time timer, and reference voltage source, primarily utilized for signal amplification, control, and timing functions. As the sensor's detection element encounters varying infrared radiation, it undergoes impedance conversion and transmits the voltage signal to the integrated circuit's input terminal. Subsequently, it undergoes multi-stage amplification and bidirectional amplitude discrimination to identify the effective triggering signal. This then triggers a delay time timer to produce high-level pulses of a specific duration (T_x). Subsequently, the blocking time timer is initiated to obstruct the input end for a defined period (T_i) to suppress various interferences arising during the load switching process. Consequently, the detection of human infrared radiation is converted into a high-level pulse signal output.

The schematic diagram of the infrared pyroelectric sensing signal processing circuit is shown in Figure 2. The main parameters include the following.

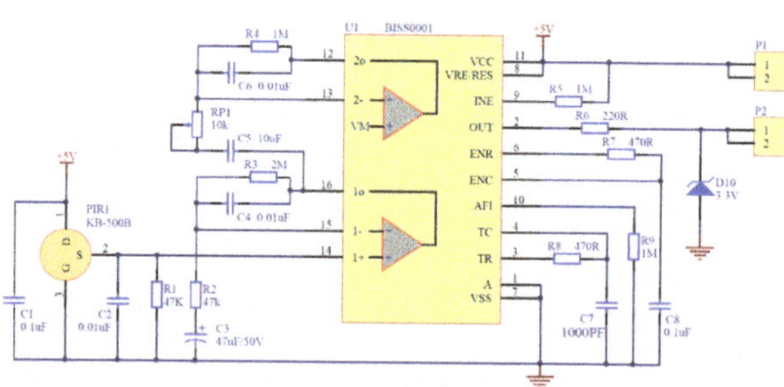

Figure 2. PIR sensing signal processing circuit.

The first stage of voltage magnification is:

$$\dot{A}_{v1} = 1 + \frac{R_3}{R_2} = 43.6$$

The second stage of voltage magnification is:

$$\dot{A}_{v2} = -\frac{R_4}{R_{P1}} > -100$$

The total magnification can achieve 73 dB by adjusting potentiometer RP1.

The delay time of the output high level is:

$$T_x = 4.9 \times 10^5 R_8 C_7 = 230 \text{ ms}$$

The input hold-off time is:

$$T_i = 24 R_7 C_8 = 1.1 \text{ ms}$$

2.2. Sound Detection and Processing Circuit

The schematic diagram depicts the operation of the sound detection and processing circuit, as shown in Figure 3. The circuit first captures the sound signal through an electret microphone and converts it into an electrical signal. This signal is then amplified through a two-stage amplification circuit to produce a voltage signal corresponding to the sound signal.

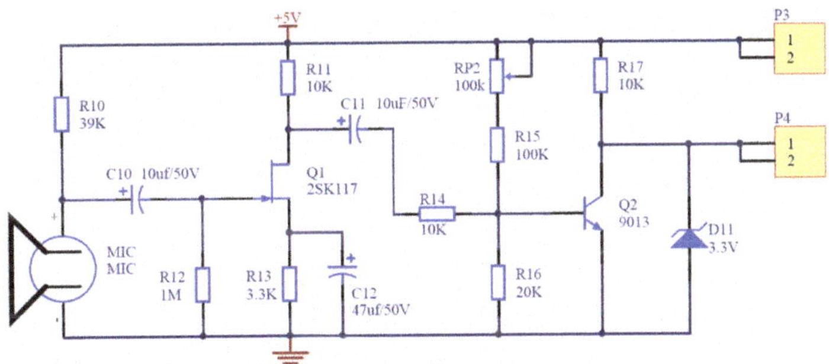

Figure 3. Sound detection and processing circuit.

The design incorporates an NPN transistor (model 9013 [8]), which switches between the on/off states. Initially, it is in the cutoff region, generating a high-level output from the collector. As the microphone's sound signal is converted into a weak electrical signal, it undergoes initial amplification through a common source amplifier circuit composed of N-channel field-effect transistors (model 2SK117 [9]). The signal then passes through a common emitter amplifier circuit, causing the transistors to enter the saturation zone and resulting in the collector outputting a low level.

The designed parameters provide flexibility. According to the parameter calculation in the diagram, when the microphone's output signal amplitude is 10 mV, the transistor enters the saturation zone, leading to a low-level output from the collector. During actual debugging, the potentiometer RP2 can be adjusted to set the transistor's initial state to "cut-off", ensuring that the collector initially outputs a high level.

2.3. Light Detection and Processing Circuit

The diagram depicted in Figure 4 illustrates the light detection and processing circuit. The main circuit consists of an in-phase voltage comparator made up of an integrated operational amplifier (specifically, model LM358 [10] was chosen for this design). When light is present, the resistance of the photoresistor (model 5537 was used in this design) decreases, with a bright resistance of about 18–50 kΩ. This results in the input voltage of the operational amplifier dropping below the reference voltage, leading to negative saturation of the operational amplifier output. In the absence of light, the resistance value of the photoresistor increases, with a dark resistance of about 2 MΩ, causing the input voltage of the operational amplifier to surpass the reference voltage and resulting in positive saturation of the operational amplifier output. The sensitivity of the circuit to light intensity can be adjusted by manipulating potentiometer RP3.

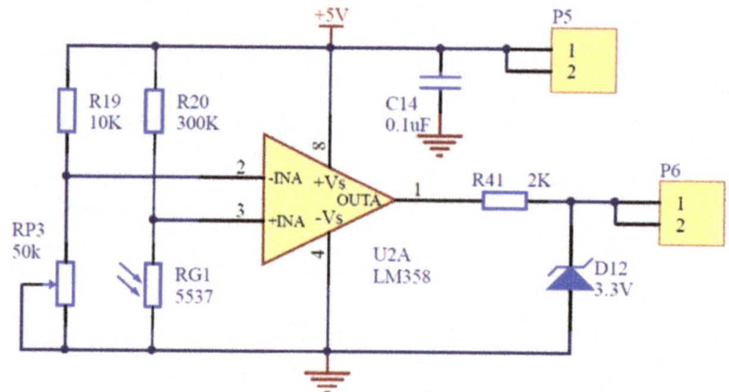

Figure 4. Light detection and processing circuit.

2.4. PWM Generation Circuit

The circuit of PWM generation is shown in Figure 5. The circuit is composed of a triangular wave generator and comparator module. The triangular wave generator adopts an integral circuit structure of constant current charging and discharging to generate triangular waves with excellent linearity. The period of the triangular wave can be obtained through the following equation.

$$T = 4R_{26}C_{21}\frac{R_{25}}{R_{24}} = 96.8\ \mu\text{s}$$

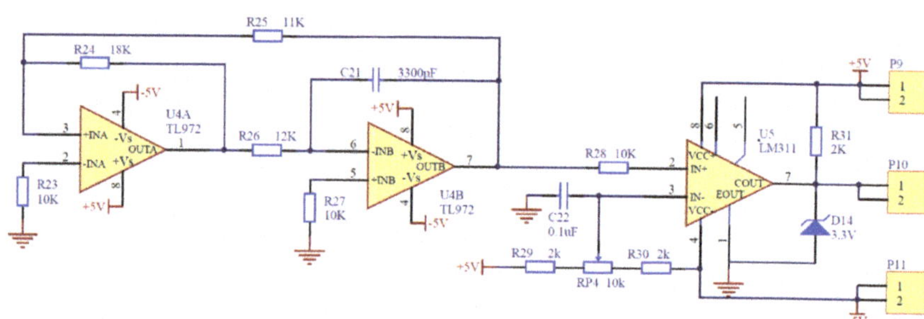

Figure 5. PWM generation circuit.

And the frequency is given by:

$$f = \frac{1}{T} = 10.3\ \text{kHz}$$

The amplitude of the triangular wave is:

$$V_{\text{om}} = \frac{R_{25}}{R_{24}}V_{\text{Z}} = 3.06\ \text{V}$$

The duty of output PWM is:

$$D = \frac{V_{\text{r}}}{V_{\text{om}}}$$

where V_{r} is the reference level of the same phase input terminal of the comparator. Adjusting the potentiometer RP4 can change the size of this value. In this scheme, the integrated operational amplifier uses TL972 [11] and the voltage comparator uses LM311 [12].

2.5. Timing Circuit

The timing circuit is utilized to produce intense illumination for a set period, employing the standard monostable circuit configuration of 555 [13] integrated circuits. The circuit diagram is depicted in Figure 6. The input terminal experiences a falling edge and is non-retriggerable.

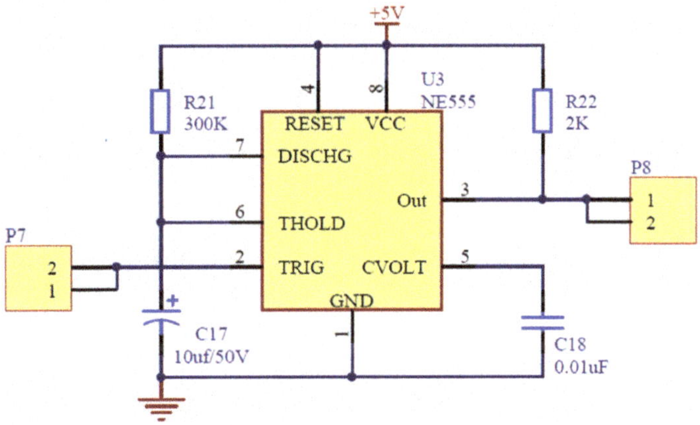

Figure 6. Timing circuit.

The delay time is given by the following equation:

$$t_W = R_{21}C_{17}\ln\frac{V_{DD}-0}{V_{DD}-\frac{2}{3}V_{DD}} = R_{21}C_{17}\ln 3 = 3.6\text{ s}$$

2.6. Logic Circuit

The logical circuit is the kernel of implementing intelligent control, which converts the logic level output by the multi-channel detection circuits into the control signal of the LED driver. The logical relationship is shown in Table 1.

Table 1. Port output status description.

Port Label	Port Source	Description
$J3$	Sound detection circuit	If not sounds, output H level; otherwise, output L level
$J4$	PIR sensing circuit	If no one detected, output L level; otherwise, H level
$J9$	Manual switch	An H level is output when manual is switched, or keep L level
$J8$	Light detection circuit	If no light, output H level, or output L level
$J10$	PWM generation circuit	Brightness control by duty cycle
$J6$	LED driver circuit	If $O_{555} = 1$, J_6 obtains H level; otherwise, J_6 obtains L level

Logical expressions are designed according to the following requirements:

$$In_{555} = \overline{\overline{J_3} + J_4 + J_9}$$

$$J_6 = \overline{\overline{O_{555}} J_{10}} \cdot J_8$$

The circuit schematic diagram of this scheme using logic gate circuits 74LS00 [14], 74LS08 [15], and 74LS32 [16] is shown in Figure 7.

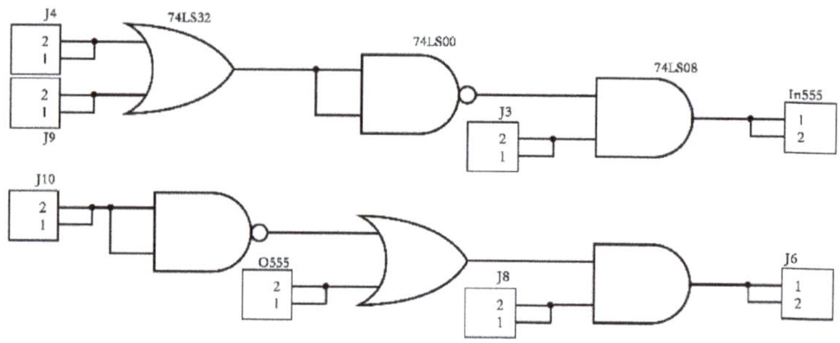

Figure 7. Implementation of intelligent logic based on 74 Series IC.

2.7. LED Driver Circuit

The schematic diagram of the LED driver circuit is shown in Figure 8, mainly composed of a TPS61165 boost converter [17]. By changing the input voltage of the VIN pin and the control signal of the CTRL pin, LED on/off and brightness control can be achieved.

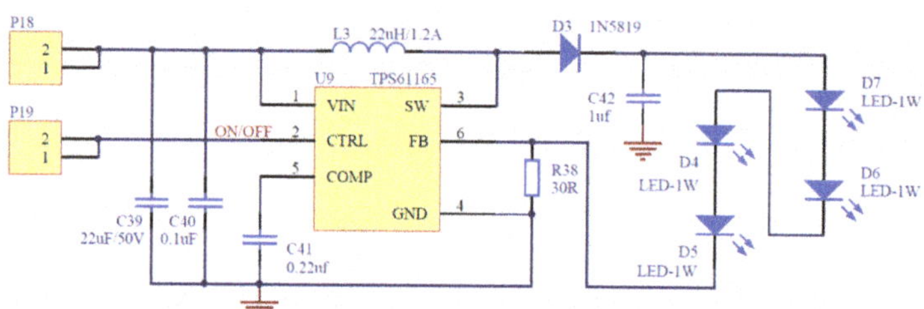

Figure 8. LED driver circuit.

This design load comprises a series connection of four white 1 W high-luminance (80–90 lm) LEDs, with a maximum operating current of 350 mA and a voltage of 3.0–3.4 V per LED. When the CTRL pin is held high, the reference voltage from the FB pin is 200 mV, with a 2% accuracy. The minimum value of sampling resistance is:

$$R_{38} = \frac{V_{FB}}{I_{LED}} = \frac{0.2}{0.35} = 0.57\ \Omega$$

The driving circuit operates at a fixed switching frequency of 1.2 MHz, with a switching current limit of 1.2 A. According to

$$\Delta i_L = 20\% \cdot I_{omax} = 70\text{ mA},\ V_{IN} = 5\text{ V},\ V_{OUT} = 12\text{ V},\ D = 0.58$$

the minimum inductor is $L_3 = 17.3\ \mu H$. This design case selects an inductance value of 22 μH. Appropriately increasing the R_{38} resistance value can decrease the output current and reduce LED brightness.

2.8. Battery Charging Circuit

The schematic diagram of the charging circuit is shown in Figure 9. This design uses a BQ21040 integrated charger [18] to charge lithium-ion batteries using solar panels. According to the typical parameter K_{ISET} in the BQ21040 data manual $K_{ISET} = 540\text{A}\Omega$, taking $R_{ISET} = R_{33} = 2\text{ k}\Omega$, the output fast charge current can be obtained as:

$$I_{out} = \frac{K_{ISET}}{R_{33}} = 0.27\text{ A}$$

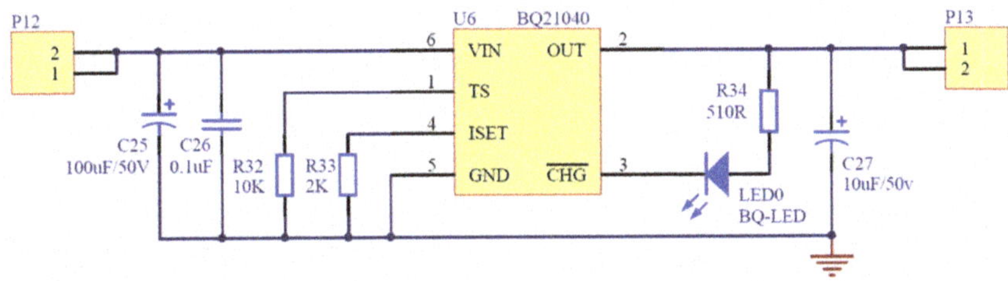

Figure 9. Battery charging circuit.

Reducing the R_{ISET} resistance value can increase the charging current, but cannot exceed the maximum continuous charging current range of lithium batteries.

2.9. Boost Circuit

The boost circuit's schematic diagram is depicted in Figure 10. This design employs the LM2735 integrated circuit [19] to elevate the DC voltage from the 3.7–4.2 V output of lithium batteries to 5 V, and accomplishes voltage stabilization. Utilizing a boost topology based on dedicated integrated circuits enables boost conversion to be achieved under voltage closed-loop feedback control.

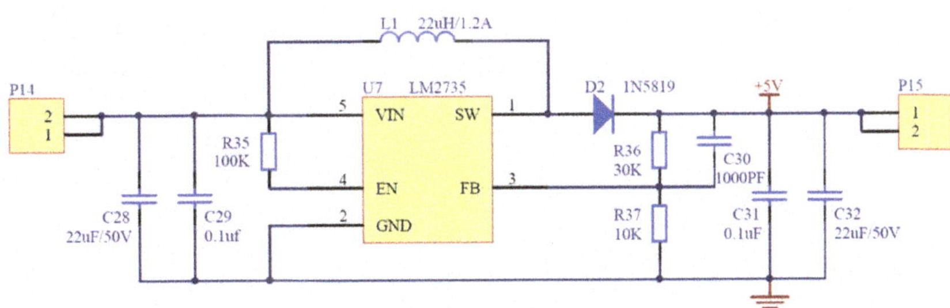

Figure 10. Boost circuit.

According to the reference voltage $V_{REF} = 1.255$ V provided in the LM2735 data manual, the sampling voltage divider resistor can be designed using the following equation.

$$\frac{R_{36}}{R_{37}} = \frac{V_{OUT}}{V_{REF}} - 1 = \frac{5}{1.225} - 1 \approx 3$$

Take $R_{36} = 30$ kΩ, $R_{37} = 10$ kΩ, at which point the duty cycle of the boost circuit is 0.26. Based on the LM2735 data manual, the output inductance is selected as 22 µH in order to keep the boost circuit operating in CCM mode.

2.10. Negative Voltage Generation Circuit

The diagram in Figure 11 illustrates the schematic of the negative voltage generation circuit. This design utilizes the TPS60401 integrated charge pump voltage converter [20] to accomplish polarity reversal from +5 V to −5 V. Following the passage through series inductors and parallel capacitors, a stable −5 V DC voltage is produced. According to the typical parameters outlined in the TPS60401 data manual, the capacitance C_{36} is chosen as 10 µF. The fixed switching frequency is 20 kHz, and the inductance L_2 is selected as 22 µH. The capacitance C_{38} is also chosen as 22 µF.

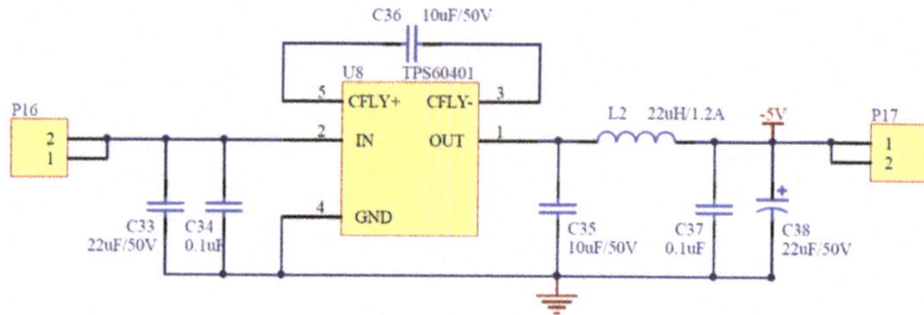

Figure 11. Negative voltage generation circuit.

3. Simulation Based on OrCAD

3.1. PIR Sensing Signal Processing Circuit

The parameters for the PIR sensing circuit model using LM358 and 555 are to be constructed, and both quiescent point and transient analysis are to be completed. The simulation results are displayed in Figure 12a,b. Based on the simulation results, the PIR sensor triggers the monostable trigger inside the chip when it receives a sine signal with an offset.

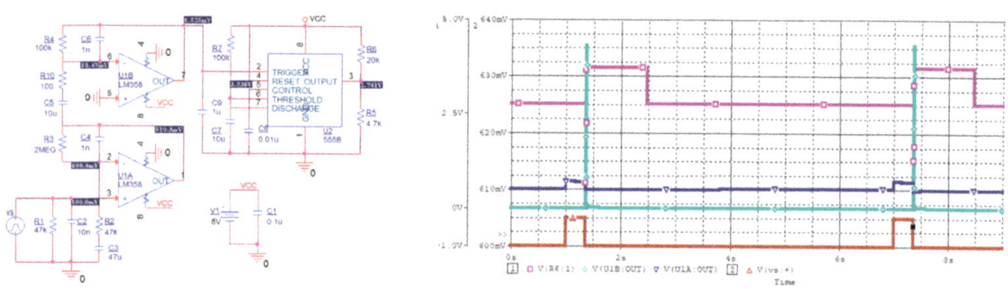

(a) (b)

Figure 12. (**a**) Quiescent point analysis of PIR sensing signal processing circuit; (**b**) transient analysis of PIR sensing signal processing circuit.

3.2. Sound Detection and Processing Circuit

To create a sound detection and processing circuit based on the provided design schematic, completing both quiescent point and transient analysis is necessary. The simulation results are illustrated in Figure 13a,b. According to the transient analysis, when the circuit receives a 10 mV, 1 kHz sine signal, the output remains in a saturated state, aligning with the design principle.

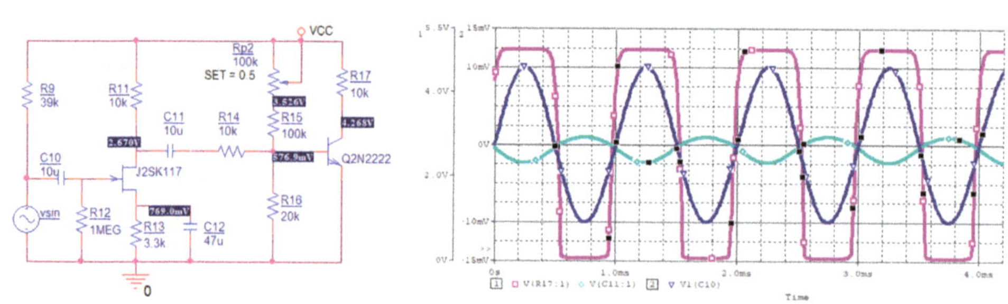

(a) (b)

Figure 13. (**a**) Quiescent point analysis of sound detection and processing circuit; (**b**) transient analysis of sound detection and processing circuit.

3.3. Light Detection and Processing Circuit

To construct a light detection and processing circuit based on the design schematic, we replaced the photosensitive resistor with an adjustable resistor and completed the quiescent point analysis. The simulation results are shown in Figure 14. The analysis indicated that the circuit produces low and high output levels when the adjustable resistance values are 50 kΩ (representing illuminated) and 500 kΩ (representing non-illuminated), respectively, aligning with the design principle.

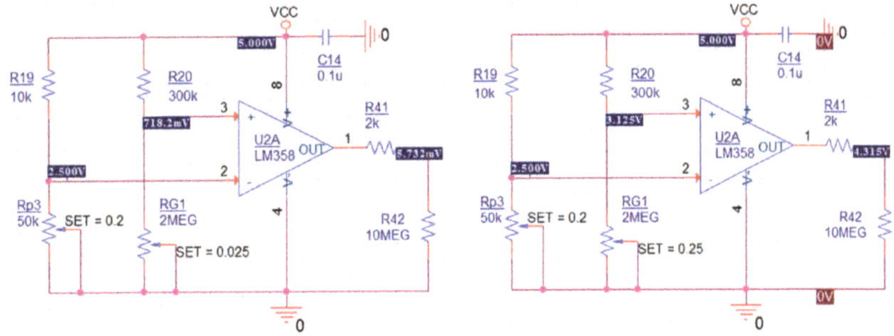

Figure 14. Quiescent point analysis of light detection and processing circuit.

3.4. PWM Generation Circuit

To construct a PWM generation circuit based on the design schematic, we completed quiescent point analysis and transient analysis, and the simulation results are shown in Figure 15a,b. The transient analysis revealed that the triangular wave generation circuit produces positive and negative alternating triangular waves, subsequently generating a square wave with an adjustable duty cycle amplitude of 5 V through the LM311 comparator.

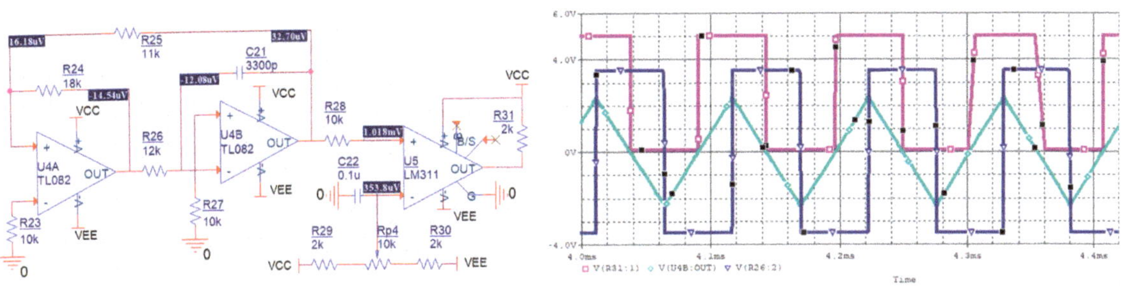

(a) (b)

Figure 15. (**a**) Quiescent point analysis of PWM generation circuit; (**b**) transient analysis of PWM generation circuit.

3.5. Timing Circuit

To construct a timing circuit based on the design schematic, we completed quiescent point analysis and transient analysis, and the simulation results are shown in Figure 16a,b. The transient analysis results indicate that the falling edge of the input signal triggers the timing circuit. The capacitor begins to charge, and the circuit outputs a high level with a pulse width of approximately 3.3 s. When the voltage of the capacitor reaches 3.3 V, the capacitor begins to discharge, and the circuit outputs a low level, which is consistent with the design principle.

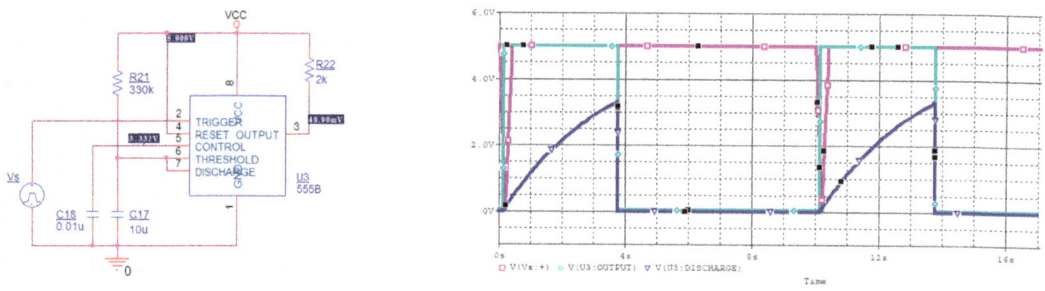

Figure 16. (**a**) Quiescent point analysis of timing circuit; (**b**) transient analysis of timing circuit.

3.6. Logic Circuit

The logic circuit, as per the design schematic, utilizes the Intel MAX 10 series FPGA (model 10M50DAF484C7G [21]). The triangular wave generator, Triangle: u2, and the ADC [7:0], as a comparator input, are employed to produce PWM waves. The simulation was conducted using ModelSim (Quartus-lite-17.1.0.590-windows), and the results are depicted in Figure 17.

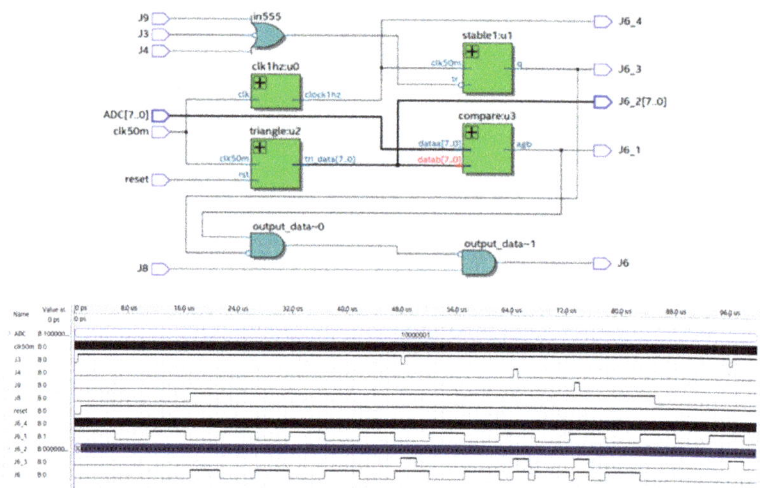

Figure 17. ModelSim simulation based on Intel MAX 10 Series FPGA.

The simulation results revealed that when the light detection circuit produces a low level, the J6 output remains at a low level. When the light detection circuit outputs a high level, but the timing circuit is not triggered, J6 outputs a PWM wave, and the duty cycle can be adjusted by the ADC output, indicating adjustable brightness of the LED at low intensity. When the light detection circuit outputs a high level and triggers the timing circuit, J6 produces a high level for a specific duration determined by the timing circuit, indicating the LED's operation at high brightness for a certain period.

4. Implementation of Controller Circuit

We implemented a controller testing platform as shown in Figure 18, and based on this platform, we completed the functional and characteristic testing of the following main modules.

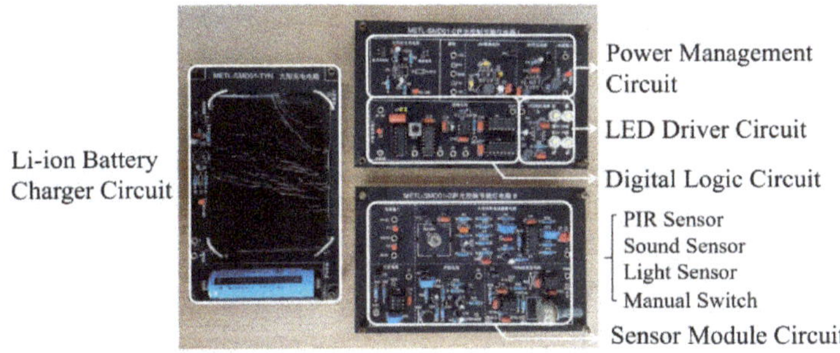

Figure 18. Testing platform of intelligent controller.

4.1. LED Driver Circuit

When the LED illuminated and the intensity was high, oscilloscope channel 1 was connected to the SW pin of the TPS61165 driver chip, and the waveform was observed as depicted in Figure 19. Based on the waveform measurement, the operational frequency of the driving circuit was 1.22 MHz, and the average output voltage was 11.32 V, which is largely consistent with the intended value. During practical measurement, the sampling resistance can be set to 2 Ω, resulting in an output current of approximately 100 mA, aiming to reduce LED brightness and output power for ease of debugging.

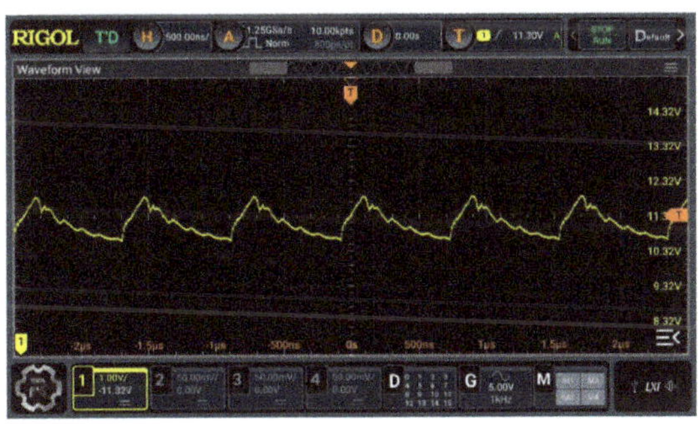

Figure 19. SW pin of LED driver circuit.

4.2. Timing Circuit

Masking the photoresistor to replicate a light-free environment causes the LED to function in a state of low-brightness illumination. The manual switch is activated, triggering the timing circuit. The process of the LED transitioning from low to high brightness, then returning to low brightness, can then be observed. The oscilloscope's channel 1 is connected to the SW pin of the driver chip TPS61165, and the waveform is depicted in Figure 20.

Based on waveform measurements, the average output voltage of the LED under low-brightness illumination is approximately 9.03 V, while under high-brightness illumination, the average output voltage is around 11.25 V. The duration of maintaining this value is approximately 4.05 s, with a theoretical calculation of 3.6 s, indicating an error of approximately 12%. This is primarily influenced by the precision of the selected electrolytic capacitor and resistor, which is within an acceptable range.

4.3. Sound Detection and Processing Circuit

The microphone is lightly tapped to activate the sound detection circuit, causing the LED lighting brightness to transition from low to high. The output of the sound detection circuit is used as the trigger source for the oscilloscope, employing a single trigger capture. A digital logic probe is utilized to observe the output waveforms of each circuit, as illustrated in Figure 21. In the figure, probes D0~D5, respectively, indicate the output signals of light detection, sound detection, infrared sensing, manual control, and PWM circuit, as well as the logical status of LED drive control signals.

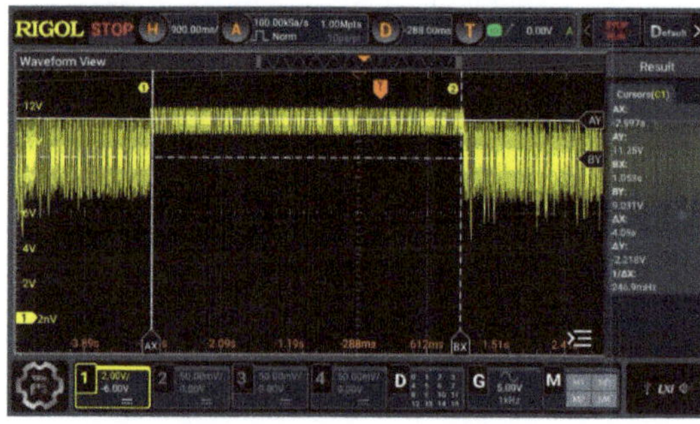

Figure 20. SW pin after starting the timing circuit.

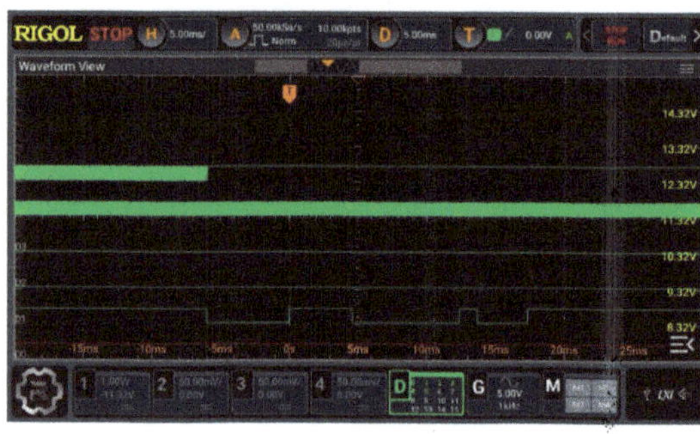

Figure 21. LED control triggered by sound detection circuit.

Upon waveform observation, it is evident that when the output of the sound detection circuit shifts from a high level (silent) to a low level (audible), the LED drive control signal switches from PWM mode to high-level output mode, indicating the brightness of the LED lighting transitioning from weak to strong. As the sound is received by the microphone in wave form, it can be observed from the graph that the D1 waveform generates multiple pulses within the 25 ms range.

4.4. PIR Sensing Signal Processing Circuit

The oscilloscope logic channel connection method in Section 4.3 is maintained. Moving the human body to effectively trigger the infrared sensing circuit, using the output of the infrared sensing circuit as the oscilloscope trigger source, and capturing the waveform with a single trigger as shown in Figure 22 should be performed. Upon waveform observation, it is evident that when the PIR sensing circuit is triggered, a high-level pulse is generated and maintained for a period of time before returning to the low level. Using the oscilloscope cursor measurement, it can be seen that the duration of the high level is approximately 227 ms, which is consistent with the design value of 230 ms. The use of the dedicated integrated circuit BISS0001 allows for clear visibility of rising and falling edges without jitter.

4.5. Manual Switch Circuit

The connection method for the logic channel of the oscilloscope is maintained as described in Section 4.3. The LED lighting brightness change is triggered manually using a switch. The switch output is used as the trigger source for the oscilloscope, and the waveform is depicted in Figure 23.

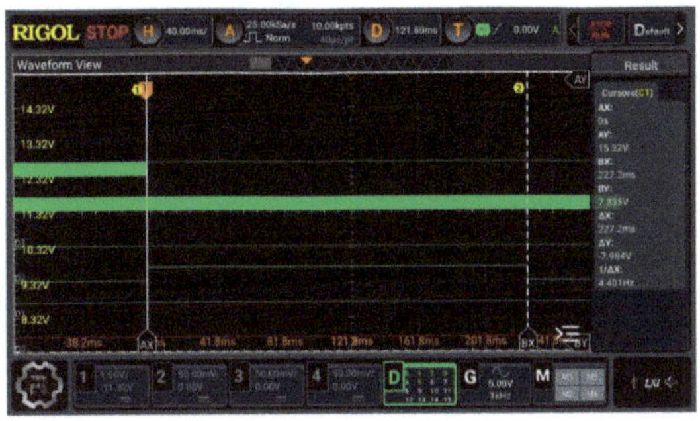

Figure 22. LED control triggered by PIR detection circuit.

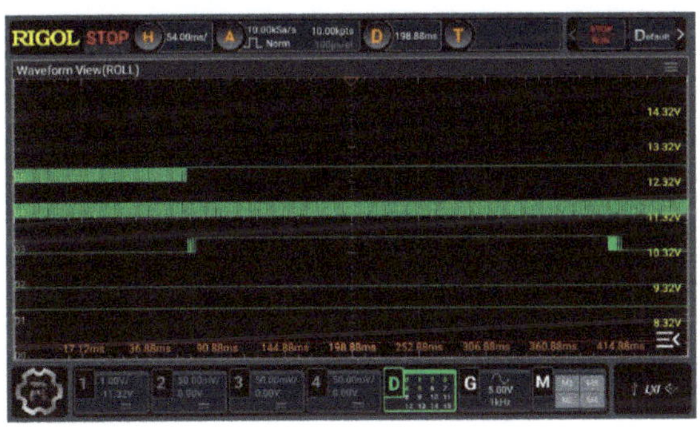

Figure 23. LED control triggered by manual switch.

Upon waveform observation, it is evident that when the switch is closed and opened, there will be short-term rapid fluctuations. When the first jitter occurs, the LED driver control signal is triggered to switch from PWM mode to high-level output mode, representing the brightness of the LED lighting transitioning from low to high. According to the cursor measurement results, the switch action lasted for approximately 325 ms.

4.6. Light Detection and Processing Circuit

The methodology for maintaining the connection of the oscilloscope logic channel in Section 4.3 and linking the oscilloscope analog channel CH1 to the SW pin of the LED driver chip TPS61165 should be upheld. The lighting simulation involves using a flashlight and using the output of the light control circuit as the trigger source for the oscilloscope, capturing the waveform with a single trigger, as illustrated in Figure 24. Upon observation of the waveform, it is evident that the transition of the lighting detection circuit output from a high level (representing no illumination) to a low level (representing illumination) causes the LED driver control signal to shift from PWM mode to low-level output mode, indicating the change in the LED light's operational state from low-brightness illumination to off. The output waveform of the oscilloscope analog channel demonstrates a gradual decrease in the output voltage of the driving circuit to 5 V as the output of the optical control circuit changes. As per the cursor measurement, the output voltage drop process in the figure takes approximately 10 ms.

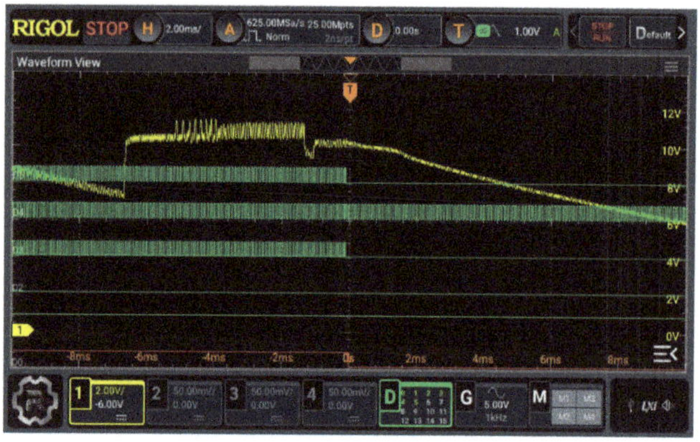

Figure 24. LED control triggered by light detection circuit.

The simulation and actual measurement results of the primary unit circuits affirm that the intelligent controller's circuit functions and main performance indicators fulfill the design requirements. It can intelligently regulate the operational status of LED street light through light intensity, human infrared, sound, and manual switches, thereby achieving energy-saving effects.

5. Conclusions

This paper presents an intelligent controller of LED street light, leveraging simulator devices to enable intelligent perception, dimming, and control of LED street light across multiple channels. The design encompasses fundamental circuit principles, digital and analog electronic technology, as well as core professional courses like power electronic technology and sensor technology. The training focuses on theoretical design, EDA simulation, and experimental debugging. It also integrates the design and implementation of basic unit circuits such as digital logic circuits, BJT and MOS transistor circuits, operational amplifiers, and comparator circuits. Furthermore, it includes targeted expansion in typical applications of 555 chips and DC-DC power supply chips, with a focus on practical teaching value. By substituting and combining local unit circuits, advanced designs for this controller can be developed. For example, using FPGA technology to achieve PWM generation and control is expected to improve controller performance and development efficiency. Due to the focus of this solution on the circuit implementation of analog devices and the lack of in-depth research on FPGA applications, this will be the main research area of the team in the next stage.

Author Contributions: Conceptualization, Z.W., D.Z., and W.Z.; methodology, Z.W., D.Z., and W.Z.; software, Z.W. and J.L.; validation, J.L. and D.Z.; formal analysis, D.Z. and W.Z.; investigation, Z.W. and D.Z.; resources, W.Z.; data curation, Z.W.; writing—original draft preparation, Z.W. and J.L.; writing—review and editing, Z.W. and J.L.; visualization, Z.W. and J.L.; supervision, Z.W.; project administration, Z.W., D.Z., and W.Z. All authors have read and agreed to the published version of the manuscript.

Funding: This research received no external funding.

Data Availability Statement: The data presented in this study are all available in the article.

Conflicts of Interest: The authors declare no conflicts of interest.

References

1. Elrefaey, M.S.; Ibrahim, M.E.; Eldin, E.T.; Hegazy, H.Y.; El-Kholy, E.E.; Abdalfatah, S. Multiple-source single-output buck-boost DC–DC converter with increased reliability for photovoltaic (PV) applications. *Energies* **2023**, *16*, 216. [CrossRef]
2. Kulasekaran, P.S.; Dasarathan, S. Design and analysis of interleaved high-gain bi-directional DC–DC converter for microgrid application integrated with photovoltaic systems. *Energies* **2023**, *16*, 5135. [CrossRef]
3. Derrouazin, A.; Aillerie, M.; Mekkakia-Maaza, N.; Charles, J.-P. Multi input-output fuzzy logic smart controller for a residential hybrid solar-wind-storage energy system. *Energy Convers. Manag.* **2017**, *148*, 238–250. [CrossRef]
4. Bodele, N.J.; Kulkarni, P.S. Modular battery-integrated bidirectional single-stage DC–DC converter for solar PV based DC nano-grid application. *Sol. Energy* **2023**, *259*, 1–14. [CrossRef]
5. Patel, M.; Zhou, Z. An interleaved battery charger circuit for a switched capacitor inverter-based standalone single-phase photovoltaic energy management system. *Energies* **2023**, *16*, 7155. [CrossRef]
6. Urfaliglu, O.; Soyer, E.B.; Toreyin, B.U.; Cetin, A.E. PIR-sensor based human motion event classification. In Proceedings of the 2008 IEEE 16th Signal Processing, Communication and Applications Conference, Aydin, Türkiye, 20–22 April 2008; pp. 1–4.
7. Micro Power PIR Motion Detector, IC. Available online: http://www.ladyada.net/media/sensors/BISS0001.pdf (accessed on 14 January 2024).
8. 2N9013 Transistor. Available online: http://www.unisonic.com.tw/datasheet/9013.pdf (accessed on 14 January 2024).
9. Field Effect Transistor Silicon N-Channel Junction Type 2SK117 Low Noise Audio Amplifier Applications. Available online: https://www.futurlec.com/Datasheet//Transistors/2SK117pdf (accessed on 30 January 2024).
10. Industry-Standard Dual Operational Amplifiers. Available online: https://www.ti.com/lit/gpn/lm358 (accessed on 14 January 2024).
11. TL97x Output Rail-To-Rail Very-Low-Noise Operational Amplifiers. Available online: https://www.ti.com/lit/gpn/tl972 (accessed on 14 January 2024).
12. LM111, LM211, LM311 Differential Comparators. Available online: https://www.ti.com/lit/gpn/lm311 (accessed on 14 January 2024).
13. LM555 Timer. Available online: https://www.ti.com/lit/gpn/lm555 (accessed on 14 January 2024).
14. SNx400, SNx4LS00, and SNx4S00 Quadruple 2-Input Positive-NAND Gates. Available online: https://www.ti.com/lit/gpn/sn74ls00 (accessed on 14 January 2024).
15. Quadruple 2-Input Positive-and Gates. Available online: https://www.ti.com/lit/gpn/sn74ls08 (accessed on 14 January 2024).
16. Quadruple 2-Input Positive-or Gates. Available online: https://www.ti.com/lit/gpn/sn74ls32 (accessed on 14 January 2024).
17. TPS61165 High-Brightness, White LED Driver in WSON and SOT-23 Packages. Available online: https://www.ti.com/lit/gpn/tps61165 (accessed on 14 January 2024).
18. BQ21040 0.8-A, Single-Input, Single Cell Li-Ion and Li-Pol Battery Charger. Available online: https://www.ti.com/lit/gpn/bq21040 (accessed on 14 January 2024).
19. LM2735 520-kHz and 1.6-MHz Space-Efficient Boost and SEPIC DC/DC Regulator. Available online: https://www.ti.com/lit/gpn/lm2735 (accessed on 14 January 2024).
20. TPS6040x Unregulated 60-mA Charge Pump Voltage Inverter. Available online: https://www.ti.com/lit/gpn/tps60401 (accessed on 14 January 2024).
21. Intel®MAX®10 FPGA Device Datasheet. Available online: https://cdrdv2-public.intel.com/666319/m10_datasheet-683794-666319.pdf (accessed on 14 January 2024).

energies

MDPI

Article

Collaborative Optimization Scheduling of Multi-Microgrids Incorporating Hydrogen-Doped Natural Gas and P2G–CCS Coupling under Carbon Trading and Carbon Emission Constraints

Yuzhe Zhao * and Jingwen Chen

School of Electrical and Control Engineering, Shaanxi University of Science and Technology, Xi'an 710021, China
* Correspondence: 210612082@sust.edu.cn

Abstract: In the context of "dual carbon", restrictions on carbon emissions have attracted widespread attention from researchers. In order to solve the issue of the insufficient exploration of the synergistic emission reduction effects of various low-carbon policies and technologies applied to multiple microgrids, we propose a multi-microgrid electricity cooperation optimization scheduling strategy based on stepped carbon trading, a hydrogen-doped natural gas system and P2G–CCS coupled operation. Firstly, a multi-energy microgrid model is developed, coupled with hydrogen-doped natural gas system and P2G–CCS, and then carbon trading and a carbon emission restriction mechanism are introduced. Based on this, a model for multi-microgrid electricity cooperation is established. Secondly, design optimization strategies for solving the model are divided into the day-ahead stage and the intraday stage. In the day-ahead stage, an improved alternating direction multiplier method is used to distribute the model to minimize the cooperative costs of multiple microgrids. In the intraday stage, based on the day-ahead scheduling results, an intraday scheduling model is established and a rolling optimization strategy to adjust the output of microgrid equipment and energy purchases is adopted, which reduces the impact of uncertainties in new energy output and load forecasting and improves the economic and low-carbon operation of multiple microgrids. Setting up different scenarios for experimental validation demonstrates the effectiveness of the introduced low-carbon policies and technologies as well as the effectiveness of their synergistic interaction.

Keywords: multi-microgrids; low-carbon; collaborative optimization; hydrogen-doped natural gas; P2G–CCS; carbon trading; carbon emission constraints

Citation: Zhao, Y.; Chen, J. Collaborative Optimization Scheduling of Multi-Microgrids Incorporating Hydrogen-Doped Natural Gas and P2G–CCS Coupling under Carbon Trading and Carbon Emission Constraints. *Energies* **2024**, *17*, 1954. https://doi.org/10.3390/en17081954

Academic Editors: Alberto Pettinau and Gaetano Zizzo

Received: 24 March 2024
Revised: 13 April 2024
Accepted: 17 April 2024
Published: 19 April 2024

1. Introduction

Dual carbon goals have prompted existing research to focus on the low-carbon optimization of multi-microgrids. There are several ways to achieve this: either through the implementation of low-carbon policies, such as carbon trading or carbon trading markets [1–5], or by employing low-carbon technologies, such as the individual or combined use of carbon capture technology, electrolytic hydrogen production technology, and other low-carbon technologies [6–10]. These measures have effectively reduced the carbon emissions of multi-microgrids and have enhanced their environmental friendliness.

Regarding low-carbon policies, Reference [11] established a model focusing on carbon emissions. Based on the carbon trading mechanism, a cooperative model was proposed to determine optimal power trading [12], and Reference [13] established a low-carbon transaction mechanism based on the carbon trading mechanism and the Stackelberg game theory, and Reference [14] developed a trading strategy between multiple entities in the system using the double Stackelberg game framework, focusing on the uncertainty of the carbon trading mechanism. Reference [15] established an integrated energy microgrid (IEM) model focusing on carbon trading. On this basis, a multi-IEM collaborative operation model was constructed. Reference [16] established a multi-microgrid system with a comprehensive and flexible operation mode based on the carbon trading mechanism, and the carbon trading mechanism was also introduced by Reference [17] to establish a low-carbon optimization scheduling model of multi-microgrids. In Reference [18], carbon emission and green certificate trading were introduced to establish a multi-microgrid cooperative transaction model, focusing on the coupling of energy and carbon. A double-layer electric heating and sharing model of multi-microgrids was constructed, and the carbon trading mechanism was introduced by Reference [19]. Reference [20] proposed a local power- and carbon-trading method for interconnecting multi-energy microgrids, introducing a carbon trading mechanism, and Reference [21] established a multi-microgrids collaborative operation model, in which the ladder carbon trading mechanism was considered.

Regarding low-carbon technology, Reference [22] established a coupling operation framework for a combined heat and power generation and a power-togas and carbon capture system, and integrated this with a carbon trading mechanism to reduce carbon emissions. Reference [23] established a multi-agent cooperative operation model based on Nash bargaining theory, adding carbon capture systems and power-to-gas devices to the sub-model under the constraints of comprehensive demand response and carbon trading. Reference [24] constructed a microgrid model integrating electricity-to-gas and carbon capture systems, and a multi-energy coordinated microgrid model and an optimal scheduling scheme for a regional integrated energy system cluster (RIESG), which combined power-to-gas and inter-park power assistance, was proposed by Reference [25]. In Reference [26], a cooperative model of multi-microgrids was established, and a power-to-hydrogen device was added to the model.

In conclusion, current research has demonstrated that low-carbon policies, represented by carbon trading and low-carbon technologies as well as carbon capture systems, have effectively reduced emissions in industry. It is evident that the combined applications of various emission reduction policies and technologies will become a research focus in the future for the power system research field. Therefore, the synergistic application of multiple low-carbon policies and technologies, within the cooperative optimization scheduling of multi-microgrids, holds significant research significance. It provides a feasible approach for exploring the low-carbon operation of the power system.

This paper proposes a multi-microgrids electric energy cooperation optimization scheduling strategy based on carbon trading and carbon emission constraints, including P2G–CCS coupling and hydrogen-doped natural gas system. Firstly, it establishes a multi-energy microgrid model incorporating hydrogen-doped natural gas and a P2G–CCS coupling system, and it introduces a tiered carbon trading and carbon emission constraint mechanism. Based on this, a model for multi-microgrid electric energy cooperation is established. Secondly, the optimization strategy for solving the model is designed with the two following stages: the day-ahead and intraday stages. In the day-ahead stage, an alternating direction multiplier method is used for the distributed solving of the model, thereby minimizing the cooperative cost of multi-microgrids; in the intraday stage, an intraday scheduling model is established, based on the day-ahead scheduling results. A rolling optimization strategy is adopted to adjust the output of the microgrid devices and the amount of energy purchases, reducing the impact of uncertainty in the renewable energy and load forecasting and improving the economic and low-carbon operation of multiple microgrids. Finally, different scenarios are set, to demonstrate the proposed method's in-depth analysis of the cost and carbon emissions impact on multiple microgrids under the coordination of low-carbon policies and technologies.

2. Multi-Microgrids Model

Figure 1 is the model of the multi-microgrids.

As shown in Figure 1, a multi-microgrid (MMG) is a highly intelligent and flexible energy system composed of multiple microgrids. Each microgrid functions as a small-scale energy system, and energy sharing among different subsystems within the multi-microgrids is achieved through the flow of electrical energy. This enables the efficient utilization and flexible distribution of energy within the system. In addition to energy sharing among subsystems, connections to the main power grid and gas networks provide backup energy sources and flexibility for the multi-microgrid system.

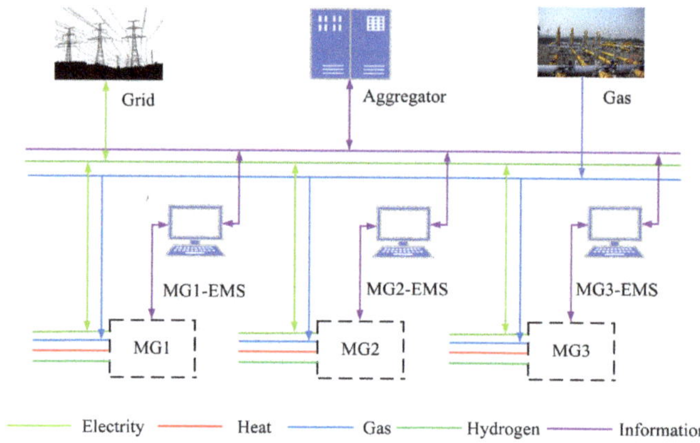

Figure 1. Structure of multi-microgrids.

In order to analyze the microgrid energy flow, Figure 2 shows the multi-energy microgrid model incorporating P2G–CCS coupling and a hydrogen-doped natural gas system.

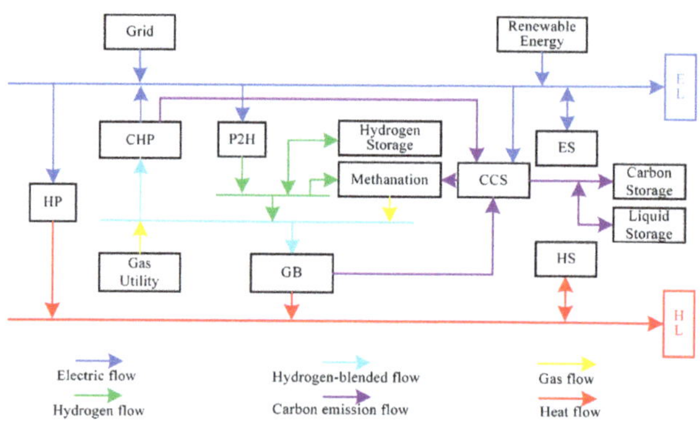

Figure 2. Multi-energy microgrid model construction diagram.

As shown in Figure 2, the model includes renewable energy generation, a combined heat and power (CHP) unit, a gas boiler (GB), a two-stage power-to-gas (P2G) system (the system includes a power-to-hydrogen (P2H) and a methanation unit), a heat pump (HP), and a carbon capture system (CCS; this system includes carbon storage and liquid storage), as well as electric energy storage (ES), heat storage (HS), and a hydrogen storage system (HSS).

3. Modeling of Hydrogen-Doped Natural Gas and P2G–CCS Coupling

Figure 3 is the subsystem structure of P2G–CCS coupling and hydrogen-doped natural gas.

In Figure 3, the multi-microgrid system generates a certain amount of carbon dioxide during its operation. There are various pathways and methods for handling these carbon emissions. Some of them are directly emitted into the atmosphere through flue gas diversion, while another portion is sent to the CCS facilities for treatment. However, as carbon capture technology is not 100% efficient, a portion of the carbon dioxide treated by the CCS facility is indirectly emitted into the atmosphere. Within the carbon capture and storage facility, a portion of the carbon dioxide is used as feedstock for methanation and sent to the P2G facility. In this process, carbon dioxide reacts with hydrogen to form methane, thus enabling the coupled operation of P2G and CCS. This coupled operation not only helps reduce the carbon dioxide content in the atmosphere but also generates renewable natural gas resources, enabling carbon recycling. Another portion of the carbon dioxide treated by the carbon capture and carbon storage facility undergoes carbon sequestration. The P2G process is refined into two stages: P2H and methane production. Firstly, there is the electrolysis process for hydrogen production. In this stage, electrical energy is consumed to generate hydrogen gas. A portion of the hydrogen is supplied to the CHP and the GB units, while the remaining part undergoes methane production. In the methane production process, methane is produced by the reaction of hydrogen with captured carbon dioxide, which is then injected into the natural gas supply.

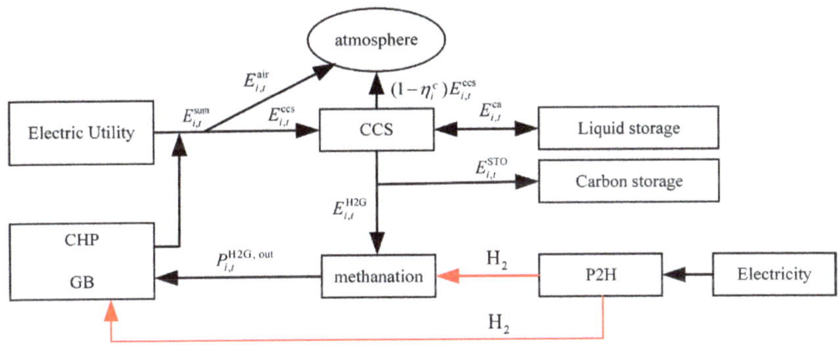

Figure 3. P2G–CCS coupling and hydrogen-doped natural gas subsystem structure.

3.1. Modeling of P2G–CCS Coupling

(1) Carbon Capture System model

In order to exploit the low-carbon potential, a CCS facility with a liquid storage unit is employed. $P_{i,t}^{\text{ccssum}}$ consists of fixed energy consumption $P_{i,t}^{\text{b}}$ and operational energy consumption $P_{i,t}^{\text{ccs}}$, and the expression is as follows:

$$P_{i,t}^{\text{ccs}} = e_{\text{c}} E_{i,t}^{\text{ccs}} \tag{1}$$

$$P_{i,t}^{\text{ccssum}} = P_{i,t}^{\text{b}} + P_{i,t}^{\text{ccs}} \tag{2}$$

$$0 \leq P_{i,t}^{\text{ccs}} \leq P_{i,\max}^{\text{ccs}} \tag{3}$$

$$P_{i,\min}^{\text{ccssum}} \leq P_{i,t}^{\text{ccssum}} \leq P_{i,\max}^{\text{ccssum}} \tag{4}$$

In Equations (1)–(4), Equation (1) calculates the energy conversion between the operating power consumed by the CCS and the mass of CO_2 treated, Equation (2) calculates the total power consumption of the CCS, and Equations (3) and (4) calculate the CCS operating power and total power constraints, respectively. e_{c} represents the unit energy consumption coefficient for processing carbon dioxide, which is 0.269; $E_{i,t}^{\text{ccs}}$ denotes the amount of CO_2 absorbed by the CCS unit in microgrid i at time t; $P_{i,\max}^{\text{ccs}}$ denotes the maximum operating power of the CCS unit in microgrid i; and $P_{i,t}^{\text{ccssum}}$ is the total power of CCS unit in microgrid i at time t, and it ranges from its minimum value, $P_{i,\min}^{\text{ccssum}}$, to its maximum value, $P_{i,\max}^{\text{ccssum}}$.

The CCS unit employed in this paper introduces a flue gas bypass system and a liquid storage unit. The specific expressions are as follows:

$$\begin{cases} E_{i,t}^{\text{ccs}} = E_{i,t}^{\text{sum}} - E_{i,t}^{\text{air}} \\ E_{i,t}^{\text{ccs}} = \frac{1}{\eta_i^{\text{C}}} (E_{i,t}^{\text{H2G}} + E_{i,t}^{\text{STO}} + E_{i,t}^{\text{ca}}) \end{cases} \tag{5}$$

Equation (5) calculates the mass of CO_2 treated by the CCS. In Equation (5), $E_{i,t}^{\text{sum}}$ is the total amount of carbon emissions from the units in microgrid i at time t, $E_{i,t}^{\text{air}}$ is the amount of CO_2 emitted into the atmosphere by the units in microgrid i at time t, $E_{i,t}^{\text{ca}}$ is the amount of CO_2 provided by the liquid storage unit in microgrid i at time t, $E_{i,t}^{\text{H2G}}$ represents the amount of CO_2 utilized for methanation from the CCS in microgrid i at time t, $E_{i,t}^{\text{STO}}$ represents the amount of CO_2 stored from the CCS in microgrid i at time t, and η_i^{C} represents the efficiency of the CCS unit in microgrid i.

The constraint for the emissions into the atmosphere from the CCS unit is as follows:

$$\begin{cases} 0 \leq E_{i,t}^{\text{air}} \leq E_{i,\max}^{\text{air}} \\ E_{i,\max}^{\text{air}} = \sigma_{\text{c}} E_{i,t}^{\text{sum}} \end{cases} \tag{6}$$

In Equation (6), $E_{i,\max}^{\text{air}}$ is the maximum amount of carbon emissions released into the atmosphere by the units in microgrid i at time t, and σ_{c} denotes the gas partition coefficient for the CCS unit.

The liquid storage unit is an essential component of the CCS unit. Following the method proposed, carbon dioxide in the liquid storage unit exists in the form of compounds in amine solution. The expressions for the process are as follows [27]:

$$V_{i,t}^{\text{ca}} = \frac{E_{i,t}^{\text{ca}} M_{\text{MEA}} \theta_i}{M_{\text{CO}_2} C_{\text{R}} \rho_{\text{R}}} \tag{7}$$

$$E_{i,\min}^{\text{ca}} \leq E_{i,t}^{\text{ca}} \leq E_{i,\max}^{\text{ca}} \tag{8}$$

In Equations (7) and (8), Equation (7) is the conversion equation for the liquid storage unit, Equation (8) is the constraint of the liquid storage unit, $V_{i,t}^{\text{ca}}$ denotes the volume of CO_2 provided by the liquid storage unit installed in microgrid i at time t, $E_{i,\min}^{\text{ca}}$ and $E_{i,\max}^{\text{ca}}$ are the minimum and maximum volume of CO_2 provided by the liquid storage unit in microgrid i, M_{MEA} is the molar mass of monoethanolamine, θ_i is the conversion coefficient of the liquid storage unit in microgrid i, C_{R} is the concentration of monoethanolamine solution, ρ_{R} is the density of the monoethanolamine solution, and M_{CO_2} is the molar mass of CO_2.

The constraint expression for the liquid storage unit is given in Equation (9), as follows:

$$\begin{cases} V_{i,t}^{\text{F}} = V_{i,t-1}^{\text{F}} - V_{i,t}^{\text{ca}} \\ V_{i,t}^{\text{P}} = V_{i,t-1}^{\text{P}} + V_{i,t}^{\text{ca}} \\ 0 \leq V_{i,t}^{\text{F}} \leq V_i^{\text{CR}} \\ 0 \leq V_{i,t}^{\text{P}} \leq V_i^{\text{CR}} \\ V_{i,0}^{\text{F}} = V_{i,24}^{\text{F}} \\ V_{i,0}^{\text{P}} = V_{i,24}^{\text{P}} \end{cases} \tag{9}$$

In Equation (9), $V_{i,t}^{\text{F}} = V_{i,t-1}^{\text{F}} - V_{i,t}^{\text{ca}}$ and $V_{i,t}^{\text{P}} = V_{i,t-1}^{\text{P}} + V_{i,t}^{\text{ca}}$ are the expressions for the change in volume of the liquid-rich and liquid-poor units, respectively; $0 \leq V_{i,t}^{\text{F}} \leq V_i^{\text{CR}}$ and $0 \leq V_{i,t}^{\text{P}} \leq V_i^{\text{CR}}$ are the liquid-rich and liquid-poor unit constraints, respectively; and $V_{i,0}^{\text{F}} = V_{i,24}^{\text{F}}$ and $V_{i,0}^{\text{P}} = V_{i,24}^{\text{P}}$ are expressed as no change in the liquid-rich and liquid-poor unit reserves at the end of the dispatch cycle, respectively. V_i^{CR} is the capacity of the liquid storage unit in microgrid i; $V_{i,t}^{\text{F}}$ and $V_{i,t}^{\text{P}}$ are the reserves of the liquid storage units that store rich and lean liquid in microgrid i, respectively; $V_{i,0}^{\text{F}}$ and $V_{i,0}^{\text{P}}$ are the initial reserves of the liquid storage units storing liquid-rich and liquid-poor in microgrid i, respectively; and $V_{i,24}^{\text{F}}$ and $V_{i,24}^{\text{P}}$ are the final reserves of the liquid storage units storing liquid-rich and liquid-poor in microgrid i, respectively.

(2) Two-stage Power-to-Gas model

This model, based on P2G, is subdivided into P2H and methanation steps. The model expression is as follows:

$$\begin{cases} P_{i,t}^{\mathrm{EL,H_2}} = \eta_i^{\mathrm{EL}} P_{i,t}^{\mathrm{EL}} \\ P_{i,t}^{\mathrm{H2G,\ out}} = \eta_i^{\mathrm{H2G}} P_{i,t}^{\mathrm{H2G,\ in}} \end{cases} \tag{10}$$

In Equation (10), $P_{i,t}^{\mathrm{EL,H_2}} = \eta_i^{\mathrm{EL}} P_{i,t}^{\mathrm{EL}}$ is the expression for P2H and $P_{i,t}^{\mathrm{H2G,\ out}} = \eta_i^{\mathrm{H2G}} P_{i,t}^{\mathrm{H2G,\ in}}$ is the expression for methanation. $P_{i,t}^{\mathrm{EL}}$ represents the electric power consumed by electrolysis in microgrid i at time t. $P_{i,t}^{\mathrm{EL,H_2}}$ represents the hydrogen production power consumed by electrolysis in microgrid i at time t. $P_{i,t}^{\mathrm{H2G,\ out}}$ denotes the methane production power of the methanation equipment in microgrid i at time t. $P_{i,t}^{\mathrm{H2G,\ in}}$ denotes the hydrogen consumption power of the methanation equipment in microgrid i at time t. η_i^{EL} and η_i^{H2G} are the efficiencies of the P2H and methanation processes in microgrid i, respectively.

The constraints for the P2H process are as follows:

$$\begin{cases} P_{i,\min}^{\mathrm{EL}} \le P_{i,t}^{\mathrm{EL}} \le P_{i,\max}^{\mathrm{EL}} \\ P_{i,\mathrm{d}}^{\mathrm{EL}} \le P_{i,t}^{\mathrm{EL}} - P_{i,t-1}^{\mathrm{EL}} \le P_{i,\mathrm{up}}^{\mathrm{EL}} \end{cases} \tag{11}$$

In Equation (11), $P_{i,\min}^{\mathrm{EL}} \le P_{i,t}^{\mathrm{EL}} \le P_{i,\max}^{\mathrm{EL}}$ is the constraint on the P2H consumption of electrical energy, and $P_{i,\mathrm{d}}^{\mathrm{EL}} \le P_{i,t}^{\mathrm{EL}} - P_{i,t-1}^{\mathrm{EL}} \le P_{i,\mathrm{up}}^{\mathrm{EL}}$ is the constraint on the P2H consumption of the electrical energy climbing rate. $P_{i,\max}^{\mathrm{EL}}$ and $P_{i,\min}^{\mathrm{EL}}$ are the maximum and minimum electric power consumption for electrolysis in the P2H process in microgrid i at time t, respectively; $P_{i,\mathrm{up}}^{\mathrm{EL}}$ and $P_{i,\mathrm{d}}^{\mathrm{EL}}$ are the maximum and minimum climbing power in the P2H process in microgrid i at time t, respectively.

The expression for the consumption of carbon dioxide by the P2G unit can be represented as follows [28]:

$$E_{i,t}^{\mathrm{H2G}} = \frac{3600 P_{i,t}^{\mathrm{H2G,\ out}} \rho_{\mathrm{CO_2}}}{Q_{\mathrm{CH_4}}} \tag{12}$$

$$\begin{cases} E_{i,t}^{\mathrm{H2G,SUM}} = E_{i,t}^{\mathrm{H2G}} + E_{i,t}^{\mathrm{STO}} \\ E_{i,t}^{\mathrm{H2G,SUM}} = \eta_i^{\mathrm{C}} E_{i,t}^{\mathrm{ccs}} - E_{i,t}^{\mathrm{ca}} \end{cases} \tag{13}$$

The constraint for carbon storage is as follows:

$$0 \le E_{i,t}^{\mathrm{STO}} \le E_{i,\max}^{\mathrm{STO}} \tag{14}$$

In Equations (12) and (13), Equation (12) is the conversion equation for CO_2 required for the methanation process and Equation (13) calculates the total mass of CO_2 consumed by the P2G unit. $\rho_{\mathrm{CO_2}}$ is the density of carbon dioxide, $E_{i,\max}^{\mathrm{STO}}$ is the maximum amount of carbon sequestration in microgrid i, and $E_{i,t}^{\mathrm{H2G,SUM}}$ is the amount of CO_2 consumed by the P2G unit in microgrid i at time t.

As the volume of carbon dioxide consumed during the methanation process is consistent with the volume of methane generated, determining the required mass of carbon dioxide becomes a crucial step. This mass can be calculated using Equation (12). Equation (13) indicates that all the carbon dioxide required for methanation in the P2G unit comes from the CCS, thereby achieving the coupling of P2G–CCS and enhancing the economic and low-carbon operation of the unit.

3.2. Modeling of Hydrogen Blending in Combined Heat and Power Units and Gas Boilers

(1) Hydrogen blending in combined heat and power units

When using CHP units with a certain proportion of hydrogen blended into natural gas, it is safe to burn natural gas with a hydrogen blending ratio of 10–20% [29]. The expression of the model is as follows:

$$\begin{cases} P_{i,t}^{\text{e,chp}} = \eta_i^{\text{e,chp}}(P_{i,t}^{\text{g,chp}} + P_{i,t}^{\text{h}_2\text{,chp}}) \\ P_{i,t}^{\text{h,chp}} = \eta_i^{\text{h,chp}}(P_{i,t}^{\text{g,chp}} + P_{i,t}^{\text{h}_2\text{,chp}}) \\ P_{i,t}^{\text{g,chp}} = V_{i,t}^{\text{g,chp}} Q_{\text{CH}_4} \\ P_{i,t}^{\text{h}_2\text{,chp}} = V_{i,t}^{\text{h}_2\text{,chp}} Q_{\text{H}_2} \\ Y_{i,t}^{\text{h}_2\text{,chp}} = V_{i,t}^{\text{h}_2\text{,chp}} / (V_{i,t}^{\text{g,chp}} + V_{i,t}^{\text{h}_2\text{,chp}}) \\ 0 \leq P_{i,t}^{\text{e,chp}} \leq P_{i,\max}^{\text{e,chp}} \\ 0 \leq P_{i,t}^{\text{h,chp}} \leq P_{i,\max}^{\text{h,chp}} \\ \lambda_{i,\min}^{\text{e,chp}} \leq P_{i,t}^{\text{e,chp}} - P_{i,t-1}^{\text{e,chp}} \leq \lambda_{i,\max}^{\text{e,chp}} \end{cases} \tag{15}$$

In Equation (15), $P_{i,t}^{\text{e,chp}} = \eta_i^{\text{e,chp}}(P_{i,t}^{\text{g,chp}} + P_{i,t}^{\text{h}_2\text{,chp}})$ and $P_{i,t}^{\text{h,chp}} = \eta_i^{\text{h,chp}}(P_{i,t}^{\text{g,chp}} + P_{i,t}^{\text{h}_2\text{,chp}})$ are the conversion equations for the relationship between the electricity and heat production of the CHP unit, respectively; $P_{i,t}^{\text{g,chp}} = V_{i,t}^{\text{g,chp}} Q_{\text{CH}_4}$ and $P_{i,t}^{\text{h}_2\text{,chp}} = V_{i,t}^{\text{h}_2\text{,chp}} Q_{\text{H}_2}$ are the natural gas and hydrogen power conversion equations for CHP consumption, respectively; $Y_{i,t}^{\text{h2,chp}} = V_{i,t}^{\text{h}_2\text{,chp}} / (V_{i,t}^{\text{g,chp}} + V_{i,t}^{\text{h}_2\text{,chp}})$ is the hydrogen blending ratio for the CHP unit; and $0 \leq P_{i,t}^{\text{e,chp}} \leq P_{i,\max}^{\text{e,chp}}$ and $0 \leq P_{i,t}^{\text{h,chp}} \leq P_{i,\max}^{\text{h,chp}}$ are the constraints on the electrical and heat power produced by the CHP unit, respectively. $\lambda_{i,\min}^{\text{e,chp}} \leq P_{i,t}^{\text{e,chp}} - P_{i,t-1}^{\text{e,chp}} \leq \lambda_{i,\max}^{\text{e,chp}}$ is the ramp rate constraint on CHP unit electrical power. $P_{i,t}^{\text{e,chp}}$ and $P_{i,t}^{\text{h,chp}}$, respectively, denote the power generated by the CHP unit for electricity and heat in microgrid i at time t, respectively. $P_{i,t}^{\text{g,chp}}$ and $P_{i,t}^{\text{h}_2\text{,chp}}$ are the power consumption of natural gas and hydrogen by the CHP unit in microgrid i at time t, respectively. $V_{i,t}^{\text{g,chp}}$ and $V_{i,t}^{\text{h}_2\text{,chp}}$ are the volume of natural gas and hydrogen consumed by the CHP unit in microgrid i at time t, respectively; $Y_{i,t}^{\text{h}_2\text{,chp}}$ is the hydrogen blending ratio in microgrid i at time t; $P_{i,\max}^{\text{e,chp}}$ and $P_{i,\max}^{\text{h,chp}}$ are the maximum electrical and heat power output of the hydrogen-blended CHP unit in microgrid i, respectively; $\lambda_{i,\min}^{\text{e,chp}}$ and $\lambda_{i,\max}^{\text{e,chp}}$ are the minimum and maximum ramp rate of the hydrogen-blended CHP unit in microgrid i, respectively; and Q_{CH_4} and Q_{H_2} are the heating values of CH_4 and H_2, respectively.

(2) Hydrogen blending in gas boilers

The blending ratio of hydrogen with natural gas is within the range of 10–20% by molar mass. The expression of the model is as follows:

$$\begin{cases} P_{i,t}^{\text{h,gb}} = (P_{i,t}^{\text{g,gb}} + P_{i,t}^{\text{h2,gb}})\eta_i^{\text{gb}} \\ P_{i,t}^{\text{g,gb}} = V_{i,t}^{\text{g,gb}} Q_{\text{CH}_4} \\ P_{i,t}^{\text{h}_2\text{,gb}} = V_{i,t}^{\text{h}_2\text{,gb}} Q_{\text{H}_2} \\ Y_{i,t}^{\text{h}_2\text{,gb}} = \frac{V_{i,t}^{\text{h}_2\text{,gb}} \rho_{\text{H}_2}}{M_{\text{H}_2}} / (\frac{V_{i,t}^{\text{g,gb}} \rho_{\text{CH}_4}}{M_{\text{CH}_4}} + \frac{V_{i,t}^{\text{h}_2\text{,gb}} \rho_{\text{H}_2}}{M_{\text{H}_2}}) \\ 0 \leq P_{i,t}^{\text{h,gb}} \leq P_{i,\max}^{\text{h,gb}} \\ \lambda_{i,\min}^{\text{h,gb}} \leq P_{i,t}^{\text{h,gb}} - P_{i,t-1}^{\text{h,gb}} \leq \lambda_{i,\max}^{\text{h,gb}} \end{cases} \tag{16}$$

In Equation (16), $P_{i,t}^{\text{h,gb}} = (P_{i,t}^{\text{g,gb}} + P_{i,t}^{\text{h2,gb}})\eta_i^{\text{gb}}$ is the heat power conversion equation for the GB unit; $P_{i,t}^{\text{g,gb}} = V_{i,t}^{\text{g,gb}} Q_{\text{CH}_4}$ and $P_{i,t}^{\text{h}_2\text{,gb}} = V_{i,t}^{\text{h}_2\text{,gb}} Q_{\text{H}_2}$ are the natural gas and hydrogen consumption equations for the GB unit, respectively; $Y_{i,t}^{\text{h}_2\text{,gb}} = \frac{V_{i,t}^{\text{h}_2\text{,gb}} \rho_{\text{H}_2}}{M_{\text{H}_2}} / (\frac{V_{i,t}^{\text{g,gb}} \rho_{\text{CH}_4}}{M_{\text{CH}_4}} + \frac{V_{i,t}^{\text{h}_2\text{,gb}} \rho_{\text{H}_2}}{M_{\text{H}_2}})$ is the hydrogen blending ratio for the GB; $0 \leq P_{i,t}^{\text{h,gb}} \leq P_{i,\max}^{\text{h,gb}}$ is the heat power constraint for the GB unit; and $\lambda_{i,\min}^{\text{h,gb}} \leq P_{i,t}^{\text{h,gb}} - P_{i,t-1}^{\text{h,gb}} \leq \lambda_{i,\max}^{\text{h,gb}}$ is the ramp rate constraint on the GB's heat power. $P_{i,t}^{\text{h,gb}}$ is the power generated by the GB unit for heat in microgrid i at time t. $P_{i,t}^{\text{g,gb}}$ and $P_{i,t}^{\text{h}_2\text{,gb}}$ are the power consumption of natural gas and hydrogen by the GB unit in microgrid i at time t, respectively. $V_{i,t}^{\text{g,gb}}$ and $V_{i,t}^{\text{h}_2\text{,gb}}$ are the volume of natural gas and hydrogen consumed by the GB unit in microgrid i at time t, respectively; $P_{i,\max}^{\mathrm{h,gb}}$ is the maximum heat power output of the GB unit in microgrid i; $\lambda_{i,\max}^{\mathrm{h,gb}}$ and $\lambda_{i,\min}^{\mathrm{h,gb}}$ are the maximum and minimum ramp rate of the GB unit in microgrid i, respectively; $Y_{i,t}^{\mathrm{h_2,gb}}$ is the hydrogen blending ratio (by molar mass) in microgrid i at time t; $\rho_{\mathrm{H_2}}$ and $\rho_{\mathrm{CH_4}}$ are the density of H_2 and CH_4, respectively; and $M_{\mathrm{H_2}}$ and $M_{\mathrm{CH_4}}$ are the molar mass of H_2 and CH_4, respectively.

4. Staircase Carbon Trading Mechanism and Carbon Emission Constraints

4.1. Carbon Trading Costs

The carbon emission quota is as follows:

$$E_{i,t}^{0} = D^{\mathrm{chp}}(P_{i,t}^{\mathrm{e,chp}} + P_{i,t}^{\mathrm{h,chp}}) + D^{\mathrm{gb}} P_{i,t}^{\mathrm{h,gb}} + D^{\mathrm{res}} P_{i,t}^{\mathrm{res}} \tag{17}$$

where D^{chp}, D^{gb} and D^{res} are the carbon quota coefficients for the CHP, the GB and the renewable energy unit, respectively, and $E_{i,t}^{0}$ is the carbon emission quota in microgrid i at time t.

The equation for calculating the carbon emissions of the microgrid is as follows:

$$\begin{cases} E_{i,t}^{\mathrm{CO_2}} = a^{\mathrm{CO_2}} V_{i,t}^{\mathrm{g,chp}} Q_{\mathrm{CH4}} + b^{\mathrm{CO_2}} V_{i,t}^{\mathrm{g,gb}} Q_{\mathrm{CH4}} + c^{\mathrm{CO_2}} + \lambda_{\mathrm{e}} P_{i,t}^{\mathrm{BUY}} - \eta_i^{\mathrm{C}} E_{i,t}^{\mathrm{ccs}} \\ E_{i,t}^{\mathrm{sum}} = a^{\mathrm{CO_2}} V_{i,t}^{\mathrm{g,chp}} Q_{\mathrm{CH4}} + b^{\mathrm{CO_2}} V_{i,t}^{\mathrm{g,gb}} Q_{\mathrm{CH4}} + c^{\mathrm{CO_2}} + \lambda_{\mathrm{e}} P_{i,t}^{\mathrm{BUY}} \end{cases} \tag{18}$$

where $a^{\mathrm{CO_2}}$ and $b^{\mathrm{CO_2}}$ are the carbon emission coefficients for the CHP and GB units, respectively; $c^{\mathrm{CO_2}}$ is the carbon emission constant; and λ_{e} is the carbon emission conversion coefficient for purchased electricity. $E_{i,t}^{\mathrm{CO_2}}$ is the total amount of carbon emissions of the microgrids in microgrid i at time t.

The quantity of CO_2 involved in carbon trading is set to $C_{i,t}^{\mathrm{CO_2}}$, which is calculated using the following Equation (19):

$$C_{i,t}^{\mathrm{CO_2}} = E_{i,t}^{\mathrm{CO_2}} - E_{i,t}^{0} \tag{19}$$

The cost of ladder-type carbon trading is calculated using the following [30]:

$$CE_{i,t}^{\mathrm{CO_2}} = \begin{cases} -\chi(2+3\alpha)L + \chi(1+3\alpha)\left(C_{i,t}^{\mathrm{CO_2}} + 2L\right) \\ C_{i,t}^{\mathrm{CO_2}} \le -2L \\ -\chi(1+\alpha)L + \chi(1+2\alpha)\left(C_{i,t}^{\mathrm{CO_2}} + L\right) \\ -2L < C_{i,t}^{\mathrm{CO_2}} \le -L \\ \chi(1+\alpha)C_{i,t}^{\mathrm{CO_2}},\ -L < C_{i,t}^{\mathrm{CO_2}} \le 0 \\ \chi E_{i}^{\mathrm{CO_2}},\ 0 < C_{i,t}^{\mathrm{CO_2}} \le L \\ \chi L + \chi(1+\alpha)\left(C_{i,t}^{\mathrm{CO_2}} - L\right) \\ L < C_{i,t}^{\mathrm{CO_2}} \le 2L \\ \chi(2+\alpha)L + \chi(1+2\alpha)\left(C_{i,t}^{\mathrm{CO_2}} - 2L\right) \\ 2L \le C_{i,t}^{\mathrm{CO_2}} \end{cases} \tag{20}$$

where $CE_{i,t}^{\mathrm{CO_2}}$ is the carbon trading cost in microgrid i at time t; χ is the base carbon emission price; L is the carbon emission interval; and α is the carbon emission price growth rate.

4.2. Carbon Emission Constraints

To conduct an in-depth study of the effects of carbon emission constraints in the context of the collaborative operation and scheduling of multi-microgrids, φ is introduced as a carbon emission constraint coefficient and the constraint is set as follows in Equation (21):

$$\begin{cases} E_i^{CO_2} \leq (1-\varphi)E_{i,\max}^{CO_2} \\ E_i^{CO_2} = \sum_{t=1}^{T} E_{i,t}^{CO_2} \end{cases} \tag{21}$$

where $E_{i,\max}^{CO_2}$ is the maximum carbon emissions in the day-ahead stage without considering carbon emission constraints in microgrid i, and $E_i^{CO_2}$ is the cumulative carbon emissions over the entire scheduling period in microgrid i.

5. The Optimization Strategy for Multi-Microgrids

The strategy for multi-microgrids comprises the day-ahead scheduling stage and the intraday scheduling stage. In the day-ahead scheduling stage, an optimization model is established and solved based on the forecasted values of renewable energy generation and load, taking into account microgrid constraints and carbon emission limitations. This process produces the optimal energy dispatch plan for the next day. In the intraday scheduling stage, an intraday scheduling model is developed, derived from the outcomes of the day-ahead scheduling. This model considers current data on renewable energy generation, the current load, and forecasted information for other time periods, as well as the sharing of electricity between microgrids. Additionally, it incorporates the actual output values of P2G and CCS devices during the optimization period. A rolling optimization strategy is employed to optimize the output of energy devices and the amount of energy purchased, aiming to reduce errors in renewable energy generation and load forecasts. Furthermore, it aims to minimize the penalty costs associated with deviations from the day-ahead plan, ultimately enhancing the economic performance of multi-microgrid operations.

The process of the optimization strategy is illustrated in Figure 4.

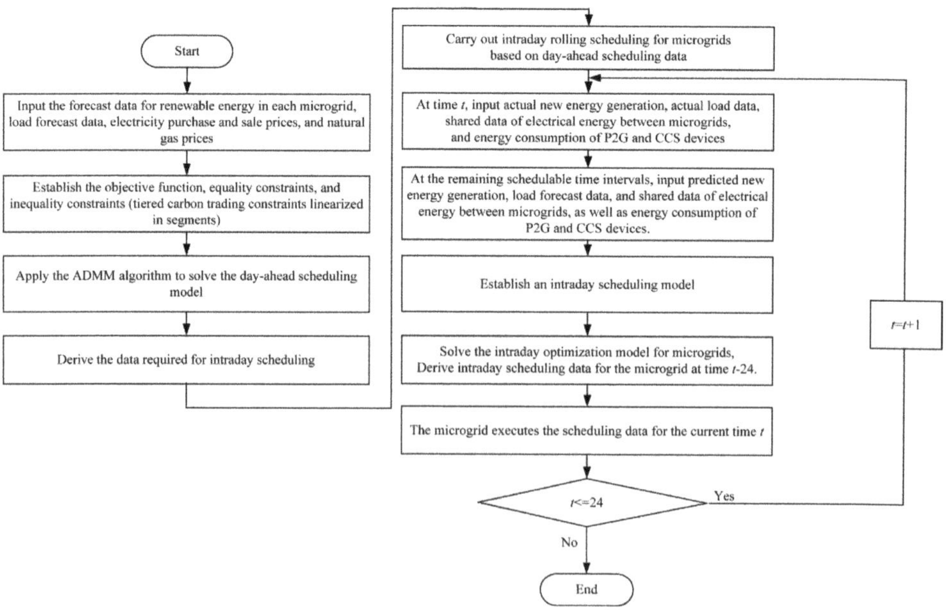

Figure 4. Flowchart of the model solution.

5.1. Day-Ahead Optimization Scheduling

5.1.1. The Objective Function

Day-ahead scheduling is aimed at minimizing the total cost of operating the multi-microgrids, with the objective function as follows:

$$\begin{cases} \min \prod_{i=1}^{\Theta} [C_i^{\text{MG}}] \\ C_i^{\text{MG}} = C_i^{\text{out}} + C_i^{\text{ESS}} + C_i^{\text{CO}_2} + C_i^{\text{load}} + C_i^{\text{cur}} \end{cases} \tag{22}$$

where Θ is the set of microgrids and C_i^{MG} is the operating cost of microgrid i.

(1) The cost of reducing renewable energy output

The cost of reducing renewable energy output in microgrid i, denoted by C_i^{cur}, is calculated using the following Equation (23):

$$C_i^{\text{cur}} = \sum_{t=1}^{T} (\lambda^{\text{cur}} P_{i,t}^{\text{cur}}) \tag{23}$$

where λ^{cur} is the unit cost of reducing renewable energy output power and $P_{i,t}^{\text{cur}}$ is the reduction power of renewable energy in microgrid i at time t.

(2) Carbon emission cost

The cost of carbon emission in microgrid i, denoted by $C_i^{\text{CO}_2}$, is calculated using the following Equation (24):

$$C_i^{\text{CO}_2} = \sum_{t=1}^{T} CE_i^{\text{CO}_2} \tag{24}$$

(3) Energy storage cost

The energy storage cost of microgrid i, denoted by C_i^{ESS}, is given by Equation (25), as follows:

$$C_i^{\text{ESS}} = \sum_{t=1}^{T} [\alpha^{\text{ES}}(P_{i,t}^{\text{ES,c}} + P_{i,t}^{\text{ES,d}}) + \alpha^{\text{HS}}(P_{i,t}^{\text{HS,c}} + P_{i,t}^{\text{HS,d}}) + \alpha^{\text{H}_2}(P_{i,t}^{\text{H,c}} + P_{i,t}^{\text{H,d}})] \tag{25}$$

where α^{ES}, α^{HS} and α^{H_2} are the scheduling cost coefficients for electrical energy, heat energy and hydrogen energy storage, respectively; $P_{i,t}^{\text{ES,c}}$ and $P_{i,t}^{\text{ES,d}}$ are the charging and discharging power of ES in microgrid i during time period t, respectively; $P_{i,t}^{\text{HS,c}}$ and $P_{i,t}^{\text{HS,d}}$ are the charging and discharging power of HS in microgrid i during time period t, respectively; and $P_{i,t}^{\text{H,c}}$ and $P_{i,t}^{\text{H,d}}$ are the charging and discharging power of HSS in microgrid i during time period t, respectively.

(4) Cost of the load demand

The cost of the load demand in microgrid i, denoted by C_i^{load}, is calculated using the following Equation (26):

$$C_i^{\text{load}} = \sum_{t=1}^{T} \left[\lambda^{\text{cut,e}} P_{i,t}^{\text{e,cut}} + \lambda^{\text{trans,e}} P_{i,t}^{\text{e,tr}} + \lambda^{\text{cut,h}} P_{i,t}^{\text{h,cut}} + \lambda^{\text{trans,h}} P_{i,t}^{\text{h,tr}}\right] \tag{26}$$

where $\lambda^{\text{cut,e}}$ is the compensation coefficient for electric load reduction, $\lambda^{\text{cut,h}}$ is the compensation coefficient for heat load reduction, $\lambda^{\text{trans,e}}$ is the compensation coefficient for electric load transfer, and $\lambda^{\text{trans,h}}$ is the compensation coefficient for heat load transfer.

(5) External interaction costs

The external interaction costs of microgrid i, denoted by C_i^{out}, mainly consist of two parts; namely, electricity purchase/sale costs and gas purchase costs, which are calculated using the following Equation (27):

$$C_i^{\text{out}} = \sum_{t=1}^{T} \left[(M_t^{\text{CH}_4} V_{i,t}^{\text{BUY}}) + (M_t^{\text{BUY}} P_{i,t}^{\text{BUY}} - M_t^{\text{SELL}} P_{i,t}^{\text{SELL}}) \right] \tag{27}$$

where $M_t^{\text{CH}_4}$ is the natural gas price in period t; $V_{i,t}^{\text{BUY}}$ is the natural gas purchase quantity of microgrid i in period t; $P_{i,t}^{\text{BUY}}$ and $P_{i,t}^{\text{SELL}}$ are the purchased and sold power from the grid by microgrid i in the t-th time period, respectively; and M_t^{BUY} and M_t^{SELL} are the electricity purchase price and electricity selling price from the grid in period t, respectively.

5.1.2. Constraints

(1) Electrical power balance constraint

The constraint of electrical power balance is:

$$P_{i,t}^{\text{re}} + P_{i,t}^{\text{e,chp}} + P_{i,t}^{\text{BUY}} + P_{i,t}^{\text{ES,d}} = P_{i,t}^{\text{e,hp}} + P_{i,t}^{\text{e}} + P_{i,t}^{\text{ES,c}} + P_{i,t}^{\text{SELL}} + \sum_{j \neq i}^{\Theta} P_{i-j,t} + P_{i,t}^{\text{ccssum}} + P_{i,t}^{\text{EL}} \tag{28}$$

where $P_{i,t}^{\text{re}}$ is the output power of renewable energy in microgrid i at time t; $P_{i-j,t}$ is the amount of electrical energy exchanged between microgrid i and microgrid j in time period t; $i \in \Theta, j \in \Theta$; $P_{i,t}^{\text{e}}$ is the electrical load of microgrid i in the t-th time period, $P_{i,t}^{\text{ES,c}}$ and $P_{i,t}^{\text{ES,d}}$ are the charging and discharging power of the electrical energy storage system in microgrid i in the t-th time period, respectively; and $P_{i,t}^{\text{e,hp}}$ is the electric power consumed by the heat pump in the i-th time period of the microgrid i.

(2) Heat power balance constraint

The constraint of heat power balance is as follows:

$$P_{i,t}^{\text{h,hp}} + P_{i,t}^{\text{h,chp}} + P_{i,t}^{\text{h,gb}} + P_{i,t}^{\text{HS,d}} = P_{i,t}^{\text{h}} + P_{i,t}^{\text{HS,c}} \tag{29}$$

where $P_{i,t}^{\text{h}}$ is the heat load of microgrid i at time t; $P_{i,t}^{\text{h,hp}}$ is the heat power generated by the HP in microgrid i at time t; and $P_{i,t}^{\text{HS,c}}$ and $P_{i,t}^{\text{HS,d}}$ are the charging and discharging power of the HS in microgrid i at time t, respectively.

(3) Gas power balance constraint

The constraint of gas power balance is as follows:

$$\begin{cases} P_{i,t}^{\text{g,BUY}} = P_{i,t}^{\text{g,chp}} + P_{i,t}^{\text{g,gb}} - P_{i,t}^{\text{H2G, out}} \\ P_{i,t}^{\text{g,BUY}} = V_{i,t}^{\text{BUY}} Q_{\text{CH}_4} \end{cases} \tag{30}$$

where $P_{i,t}^{\text{g,BUY}}$ is the power of gas purchased externally in microgrid i at time t.

(4) Hydrogen power balance constraint

The constraint of hydrogen power balance is as follows:

$$P_{i,t}^{\text{EL,H}_2} + P_{i,t}^{\text{H,d}} = P_{i,t}^{\text{h}_2\text{,chp}} + P_{i,t}^{\text{h}_2\text{,gb}} + P_{i,t}^{\text{H,c}} + P_{i,t}^{\text{H2G, in}} \tag{31}$$

(5) Renewable energy supply constraint

The constraint of renewable energy is as follows:

$$\begin{cases} P^{e,r}_{i,t} = P^{cur}_{i,t} + P^{re}_{i,t} \\ 0 \leq P^{re}_{i,t} \leq P^{e,r}_{i,t} \end{cases} \tag{32}$$

The actual renewable energy output is seen as the aggregate of the predicted value and the stochastic forecasting error. The prediction error of renewable energy output follows a normal distribution, and it is expressed in Equation (33), as follows [31]:

$$P^{e,r}_{i,t} \sim \overline{P}^{e,r}_{i,t} + N(0, \sigma^2_{i,t}) \tag{33}$$

where $P^{e,r}_{i,t}$ is the actual renewable energy generation power; $\overline{P}^{e,r}_{i,t}$ is the predicted renewable energy generation power; and $\sigma^2_{i,t}$ is the variance of renewable energy generation, =0.1.

(6) Hydrogen energy storage system constraint

The constraint of the HSS is as follows:

$$\begin{cases} E^{H_2}_{i,t} = E^{H_2}_{i,t-1} + \eta^{H,c}_i P^{H,c}_{i,t} \Delta t - \frac{P^{H,d}_{i,t} \Delta t}{\eta^{H,d}_i} \\ E^{H_2}_{i,0} = E^{H_2}_{i,T} \\ E^{H_2}_{i,\min} \leq E^{H_2}_{i,t} \leq E^{H_2}_{i,\max} \\ 0 \leq P^{H,c}_{i,t} \leq u^{H,c}_t P^{H,c}_{i,\max} \\ 0 \leq P^{H,d}_{i,t} \leq u^{H,d}_t P^{H,d}_{i,\max} \\ 0 \leq u^{H,d}_t + u^{H,c}_t \leq 1 \end{cases} \tag{34}$$

In Equation (34), $E^{H_2}_{i,t} = E^{H_2}_{i,t-1} + \eta^{H,c}_i P^{H,c}_{i,t} \Delta t - \frac{P^{H,d}_{i,t} \Delta t}{\eta^{H,d}_i}$ is the conversion formula for hydrogen energy storage capacity and $E^{H_2}_{i,0} = E^{H_2}_{i,T}$ indicates that the stored energy of the HSS is the same after the optimization period ends. $E^{H_2}_{i,\min} \leq E^{H_2}_{i,t} \leq E^{H_2}_{i,\max}$ is the constraint on hydrogen energy storage capacity. $0 \leq P^{H,c}_{i,t} \leq u^{H,c}_t P^{H,c}_{i,\max}$ and $0 \leq P^{H,d}_{i,t} \leq u^{H,d}_t P^{H,d}_{i,\max}$ are the constraints on the power of energy storage and discharge in hydrogen storage, respectively. $0 \leq u^{H,d}_t + u^{H,c}_t \leq 1$ indicates that energy storage and discharge cannot occur simultaneously. $E^{H_2}_{i,t}$ is the hydrogen storage of the HSS in microgrid i at time t; $\eta^{H,c}_i$ and $\eta^{H,d}_i$ are the charging and the discharging efficiency of the hydrogen energy storage device in the microgrid i; and $E^{H_2}_{i,\min}$ and $E^{H_2}_{i,\max}$ are the minimum and maximum storage capacity of the HSS in the microgrid i, respectively. The binary variables $u^{H,d}_t$ and $u^{H,c}_t$ indicate that hydrogen storage and hydrogen discharge cannot occur simultaneously.

(7) Electrical energy storage system constraint

The ES in microgrid i during the t-th time period is formulated as follows:

$$\begin{cases} E^{ES}_{i,t} = E^{ES}_{i,t-1} + \eta^{ES,c}_i P^{ES,c}_{i,t} \Delta t - \frac{P^{ES,d}_{i,t} \Delta t}{\eta^{ES,d}_i} \\ E^{ES}_{i,0} = E^{ES}_{i,T} \\ E^{ES}_{i,\min} \leq E^{ES}_{i,t} \leq E^{ES}_{i,\max} \\ 0 \leq P^{ES,c}_{i,t} \leq u^{ES,c}_t P^{ES,c}_{i,\max} \\ 0 \leq P^{ES,d}_{i,t} \leq u^{ES,d}_t P^{ES,d}_{i,\max} \\ 0 \leq u^{ES,c}_t + u^{ES,d}_t \leq 1 \end{cases} \tag{35}$$

In Equation (35), $E^{ES}_{i,t} = E^{ES}_{i,t-1} + \eta^{ES,c}_i P^{ES,c}_{i,t} \Delta t - \frac{P^{ES,d}_{i,t} \Delta t}{\eta^{ES,d}_i}$ is the conversion relationship formula for electrical energy storage capacity and $E^{ES}_{i,0} = E^{ES}_{i,T}$ indicates that the energy stored in the ES is the same after the entire optimization scheduling period. $E_{i,\min}^{\mathrm{ES}} \leq E_{i,t}^{\mathrm{ES}} \leq E_{i,\max}^{\mathrm{ES}}$ is the constraint on electrical energy storage capacity. $0 \leq P_{i,t}^{\mathrm{ES,c}} \leq u_t^{\mathrm{ES,c}} P_{i,\max}^{\mathrm{ES,c}}$ and $0 \leq P_{i,t}^{\mathrm{ES,d}} \leq u_t^{\mathrm{ES,d}} P_{i,\max}^{\mathrm{ES,d}}$ are the constraints on the power of energy storage and discharge in ES, respectively. $0 \leq u_t^{\mathrm{ES,c}} + u_t^{\mathrm{ES,d}} \leq 1$ indicates that energy storage and discharge cannot occur simultaneously. $E_{i,t}^{\mathrm{ES}}$ is the electrical energy storage capacity in the ES of microgrid i in the t-th time period; $\eta_i^{\mathrm{ES,c}}$ and $\eta_i^{\mathrm{ES,d}}$ are the charging and discharging efficiency of the ES in microgrid i, respectively; $E_{i,\min}^{\mathrm{ES}}$ and $E_{i,\max}^{\mathrm{ES}}$ are the minimum and maximum storage capacity of the ES in microgrid i, respectively; and $P_{i,\max}^{\mathrm{ES,c}}$ and $P_{i,\max}^{\mathrm{ES,d}}$ are the maximum charging and discharging power of the ES in microgrid i, respectively. The two binary variables $u_t^{\mathrm{ES,c}}$ and $u_t^{\mathrm{ES,d}}$ indicate that charging and discharging cannot occur simultaneously.

(8) Heat energy storage constraint

The constraint of HS is as follows:

$$\begin{cases} E_{i,t}^{\mathrm{HS}} = E_{i,t-1}^{\mathrm{HS}} + \eta_i^{\mathrm{HS,c}} P_{i,t}^{\mathrm{HS,c}} \Delta t - \frac{P_{i,t}^{\mathrm{HS,d}} \Delta t}{\eta_i^{\mathrm{HS,d}}} \\ E_{i,0}^{\mathrm{HS}} = E_{i,T}^{\mathrm{HS}} \\ E_{i,\min}^{\mathrm{HS}} \leq E_{i,t}^{\mathrm{HS}} \leq E_{i,\max}^{\mathrm{HS}} \\ 0 \leq P_{i,t}^{\mathrm{HS,c}} \leq u_t^{\mathrm{HS,c}} P_{i,\max}^{\mathrm{HS,c}} \\ 0 \leq P_{i,t}^{\mathrm{HS,d}} \leq u_t^{\mathrm{HS,d}} P_{i,\max}^{\mathrm{HS,d}} \\ 0 \leq u_t^{\mathrm{HS,d}} + u_t^{\mathrm{HS,c}} \leq 1 \end{cases} \tag{36}$$

In Equation (36), $E_{i,t}^{\mathrm{HS}} = E_{i,t-1}^{\mathrm{HS}} + \eta_i^{\mathrm{HS,c}} P_{i,t}^{\mathrm{HS,c}} \Delta t - \frac{P_{i,t}^{\mathrm{HS,d}} \Delta t}{\eta_i^{\mathrm{HS,d}}}$ is the conversion relationship formula for heat energy storage capacity and $E_{i,0}^{\mathrm{HS}} = E_{i,T}^{\mathrm{HS}}$ indicates that the heat energy storage capacity of the HS is the same after the optimization period ends. $E_{i,\min}^{\mathrm{HS}} \leq E_{i,t}^{\mathrm{HS}} \leq E_{i,\max}^{\mathrm{HS}}$ is the constraint on heat energy storage capacity. $0 \leq P_{i,t}^{\mathrm{HS,c}} \leq u_t^{\mathrm{HS,c}} P_{i,\max}^{\mathrm{HS,c}}$ and $0 \leq P_{i,t}^{\mathrm{HS,d}} \leq u_t^{\mathrm{HS,d}} P_{i,\max}^{\mathrm{HS,d}}$ are the constraints on the power of energy storage and discharge in heat storage, respectively. $0 \leq u_t^{\mathrm{HS,d}} + u_t^{\mathrm{HS,c}} \leq 1$ indicates that energy storage and discharge cannot occur simultaneously. $E_{i,t}^{\mathrm{HS}}$ is the energy stored in the HS of microgrid i in the t-th time period; $\eta_i^{\mathrm{HS,c}}$ and $\eta_i^{\mathrm{HS,d}}$ are the charging and discharging efficiency of the HS in microgrid i, respectively; and $P_{i,\max}^{\mathrm{HS,c}}$ and $P_{i,\max}^{\mathrm{HS,d}}$ are the maximum charging and discharging power of the HS in microgrid i, respectively.

(9) Heat pump constraint

The constraint of HP is as follows:

$$\begin{cases} P_{i,t}^{\mathrm{h,hp}} = \eta_i^{\mathrm{hp}} P_{i,t}^{\mathrm{e,hp}} \\ P_{i,\min}^{\mathrm{e,hp}} \leq P_{i,t}^{\mathrm{e,hp}} \leq P_{i,\max}^{\mathrm{e,hp}} \end{cases} \tag{37}$$

In Equation (37), $P_{i,t}^{\mathrm{h,hp}} = \eta_i^{\mathrm{hp}} P_{i,t}^{\mathrm{e,hp}}$ is the power conversion relationship between electricity and heat in the HP and $P_{i,\min}^{\mathrm{e,hp}} \leq P_{i,t}^{\mathrm{e,hp}} \leq P_{i,\max}^{\mathrm{e,hp}}$ is the electrical power constraint relationship in the HP. η_i^{hp} is the electrical-to-heat conversion efficiency of the HP in the microgrid i. $P_{i,\min}^{\mathrm{e,hp}}$ and $P_{i,\max}^{\mathrm{e,hp}}$ are the minimum and maximum electric power consumption of the HP in microgrid i, respectively.

(10) The constraints between the microgrid and power grid

The constraint between the microgrid and power grid is as follows:

$$\begin{cases} 0 \le P_{i,t}^{\mathrm{BUY}} \le P_{i,\max}^{\mathrm{BUY}} \\ 0 \le P_{i,t}^{\mathrm{SELL}} \le P_{i,\max}^{\mathrm{SELL}} \end{cases} \tag{38}$$

In Equation (38), $0 \le P_{i,t}^{\mathrm{BUY}} \le P_{i,\max}^{\mathrm{BUY}}$ and $0 \le P_{i,t}^{\mathrm{SELL}} \le P_{i,\max}^{\mathrm{SELL}}$ are the constraints on purchasing electricity from the grid and selling electricity to the grid, respectively. $P_{i,\max}^{\mathrm{BUY}}$ and $P_{i,\max}^{\mathrm{SELL}}$ are the upper limits of power bought from and sold to the grid in microgrid i.

(11) Electrical and heat load constraints

The electrical load of microgrid i at time t, denoted by $P_{i,t}^{\mathrm{e}}$ is composed of the following: the fixed electrical load, $P_{i,t}^{\mathrm{e,g}}$; the transferable electrical load, $P_{i,t}^{\mathrm{e,tr}}$; and the reducible electrical load, $P_{i,t}^{\mathrm{e,cut}}$. The relationship is modeled as follows:

$$\begin{cases} P_{i,t}^{\mathrm{e}} = P_{i,t}^{\mathrm{e,g}} + P_{i,t}^{\mathrm{e,tr}} - P_{i,t}^{\mathrm{e,cut}} \\ -a^{\mathrm{tr}} P_{i,t}^{\mathrm{e,g}} \le P_{i,t}^{\mathrm{e,tr}} \le a^{\mathrm{tr}} P_{i,t}^{\mathrm{e,g}} \\ 0 \le P_{i,t}^{\mathrm{e,cut}} \le b^{\mathrm{cut}} P_{i,t}^{\mathrm{e,f}} \\ \sum\limits_{t=1}^{T} P_{i,t}^{\mathrm{e,tr}} = 0 \end{cases} \tag{39}$$

In Equation (39), $-a^{\mathrm{tr}} P_{i,t}^{\mathrm{e,g}} \le P_{i,t}^{\mathrm{e,tr}} \le a^{\mathrm{tr}} P_{i,t}^{\mathrm{e,g}}$ and $0 \le P_{i,t}^{\mathrm{e,cut}} \le b^{\mathrm{cut}} P_{i,t}^{\mathrm{e,f}}$ are the constraints on transferred and curtailable electrical loads. $\sum\limits_{t=1}^{T} P_{i,t}^{\mathrm{e,tr}} = 0$ indicates no loss of transferable load during the dispatch cycle. a^{tr} is the transfer load coefficient and b^{cut} is the reducible load coefficient.

The heat load of microgrid i at time t, denoted by $P_{i,t}^{\mathrm{h}}$, is composed of the following: the fixed heat load, $P_{i,t}^{\mathrm{h,g}}$; the transferable heat load, $P_{i,t}^{\mathrm{h,tr}}$; and the reducible heat load, $P_{i,t}^{\mathrm{h,cut}}$. The relationship is modeled as follows:

$$\begin{cases} P_{i,t}^{\mathrm{h}} = P_{i,t}^{\mathrm{h,g}} + P_{i,t}^{\mathrm{h,tr}} - P_{i,t}^{\mathrm{h,cut}} \\ -a^{\mathrm{tr}} P_{i,t}^{\mathrm{h,g}} \le P_{i,t}^{\mathrm{h,tr}} \le a^{\mathrm{tr}} P_{i,t}^{\mathrm{h,g}} \\ 0 \le P_{i,t}^{\mathrm{h,cut}} \le b^{\mathrm{cut}} P_{i,t}^{\mathrm{h,f}} \\ \sum\limits_{t=1}^{T} P_{i,t}^{\mathrm{h,tr}} = 0 \end{cases} \tag{40}$$

In Equation (40), $-a^{\mathrm{tr}} P_{i,t}^{\mathrm{h,g}} \le P_{i,t}^{\mathrm{h,tr}} \le a^{\mathrm{tr}} P_{i,t}^{\mathrm{h,g}}$ and $0 \le P_{i,t}^{\mathrm{h,cut}} \le b^{\mathrm{cut}} P_{i,t}^{\mathrm{h,f}}$ are the constraints on transferred and curtailable heat loads. $\sum\limits_{t=1}^{T} P_{i,t}^{\mathrm{h,tr}} = 0$ indicates no loss of transferable load during the dispatch cycle.

(12) Constraint of energy transfer between microgrids

The constraint of energy transfer between microgrids is as follows:

$$|P_{i-j,t}| \le P_{i-j,\max} \tag{41}$$

where $P_{i-j,\max}$ is the limitation on power transfer between microgrid i and microgrid j. The other constraints are described by Equations (1)–(21).

5.1.3. Model Linearization and Solution

The day-ahead model, integrating tiered carbon trading mechanisms, P2G–CCS coupling technology and gas hydrogenation technology, require linearization due to their mixed-integer non-linear nature.

Equation (20) is linearized by introducing 0–1 variables, and the linearization of the piecewise function is expressed as follows [28]:

$$y = \begin{cases} a_1x + b_1, & c_1 \leq x \leq c_2 \\ a_2x + b_2, & c_2 \leq x \leq c_3 \\ a_3x + b_3, & c_3 \leq x \leq c_4 \end{cases} \tag{42}$$

By introducing binary variables, denoted by d_i, and continuous variables, denoted by z_i, the original Equation (42) is transformed into Equation (43), to impose interval constraints on the values of x and b. This step ensures that the values of variables x and b are limited to specific single intervals, enhancing the accuracy and applicability of the mathematical model.

$$\begin{cases} y = \sum\limits_{i}^{3} a_i z_i + b_i d_i \\ x = z_1 + z_2 + z_3 \\ d_1 + d_2 + d_3 = 1 \\ d_1 c_1 \leq a_1 z_1 \leq d_1 c_2 \\ d_2 c_2 \leq a_2 z_2 \leq d_2 c_3 \\ d_3 c_3 \leq a_3 z_3 \leq d_3 c_4 \end{cases} \tag{43}$$

5.1.4. Model Solution

The alternating direction method of multipliers (ADMM) is applied to solve the day-ahead scheduling model by taking the following steps:

(1) Introduce auxiliary variables, $P_{j-i,t}$, to construct auxiliary expressions, as follows:

$$P_{i-j,t} + P_{j-i,t} = 0, \forall i \tag{44}$$

where $P_{j-i,t}$ is the amount of electrical energy exchanged between microgrid j and microgrid i in time period t.

(2) Formulate the augmented Lagrangian function expression for the day-ahead issue, as follows:

$$\begin{cases} L_i = \min[C_i^{\mathrm{MG}}] + \sum\limits_{j}^{\Theta} \sum\limits_{t=1}^{T} \lambda_{i-j}(P_{i-j,t} + P_{j-i,t}) + \sum\limits_{j}^{\Theta} \frac{\rho_i}{2} \sum\limits_{t=1}^{T} ||P_{i-j,t} + P_{j-i,t}||_2^2 \\ \text{s.t.}(1)\text{–}(21),\ (28)\text{–}(43) \end{cases} \tag{45}$$

where λ_{i-j} is the Lagrange multiplier and $\rho_i(k)$ is the penalty parameter at the k-th iteration.

(3) Initialize the iteration number $k = 1$ and iterate through the following steps:

Update the decision variables for the microgrid, $P_{i-j,t}(k+1)$, as follows:

$$P_{i-j,t}(k+1) = \arg\min L_i(\lambda_{i-j}(k), P_{i-j,t}(k), P_{j-i,t}(k)) \tag{46}$$

Based on the variable $P_{i-j,t}(k+1)$, microgrid j updates its decision variables $P_{j-i,t}(k+1)$ through Equation (47), as follows:

$$P_{j-i,t}(k+1) = \arg\min L_j(\lambda_{j-i}(k), P_{i-j,t}(k+1), P_{j-i,t}(k)) \tag{47}$$

Each microgrid updates its variables in each iteration process.

(4) Update the Lagrange multiplier parameters after each iteration, as follows:

$$\lambda_{i-j}(k+1) = \lambda_{i-j}(k) + \rho_i(k)(P_{i-j,t} + P_{j-i,t}) \tag{48}$$

(5) Determine the convergence of the algorithm, as follows:

$$\begin{cases} \sum_{t=1}^{T}\sum_{i=1}^{\Theta}\left\|P_{i-j,t}(k+1)+P_{j-i,t}(k+1)\right\|_2 \le \varepsilon^{\text{pri}} \\ \sum_{t=1}^{T}\sum_{i=1}^{\Theta}\left\|P_{i-j,t}(k+1)-P_{i-j,t}(k)\right\|_2 \le \varepsilon^{\text{dual}} \end{cases} \tag{49}$$

where ε^{pri} denotes the primal residuals and $\varepsilon^{\text{dual}}$ denotes the dual residuals.

If the convergence condition of Equation (49) or $k > k_{\max}$ is satisfied, the iteration terminates; otherwise, update the iteration number to $k+1$ and update the penalty parameter, repeating steps (3)–(5).

5.2. Intraday Optimization Scheduling

The objective expression for the intraday dispatch of microgrids is as follows:

$$\min\left\{\overline{C}_i^{\text{cur}}+\overline{C}_i^{\text{outer}}+\overline{C}_i^{\text{ESS}}+\overline{C}_i^{\text{CO}_2}+\overline{C}_i^{\text{pun}}\right\} \tag{50}$$

where:

$$\overline{C}_i^{\text{pun}} = \sum_{\tau\in T_S}\lambda^{\text{P}}\left[\begin{array}{l}(\overline{P}_{i,\tau}^{\text{e,chp}}-\hat{P}_{i,\tau}^{\text{e,chp}})^2+(\overline{P}_{i,\tau}^{\text{h,gb}}-\hat{P}_{i,\tau}^{\text{h,gb}})^2+(\overline{P}_{i,\tau}^{\text{ES,c}}-\hat{P}_{i,\tau}^{\text{ES,c}})^2+(\overline{P}_{i,\tau}^{\text{ES,d}}-\hat{P}_{i,\tau}^{\text{ES,d}})^2\\+(\overline{P}_{i,\tau}^{\text{HS,c}}-\hat{P}_{i,\tau}^{\text{HS,c}})^2+(\overline{P}_{i,\tau}^{\text{HS,d}}-\hat{P}_{i,\tau}^{\text{HS,d}})^2+(\overline{P}_{i,\tau}^{\text{SELL}}-\hat{P}_{i,\tau}^{\text{SELL}})^2+(\overline{P}_{i,\tau}^{\text{BUY}}-\hat{P}_{i,\tau}^{\text{BUY}})^2\\+(\overline{V}_{i,\tau}^{\text{BUY}}-\hat{V}_{i,\tau}^{\text{BUY}})^2+(\overline{P}_{i,\tau}^{\text{hp}}-\hat{P}_{i,\tau}^{\text{hp}})^2+(\overline{P}_{i,t}^{\text{H,d}}-\hat{P}_{i,t}^{\text{H,d}})^2+(\overline{P}_{i,t}^{\text{H,c}}-\hat{P}_{i,t}^{\text{H,c}})^2\end{array}\right] \tag{51}$$

In Equation (51), $(\overline{\bullet})$ and $(\hat{\bullet})$ are the day-ahead and intraday variables and λ^{P} is the unit penalty cost coefficient. T_S is the intraday optimization scheduling horizon at time t, $T_S = \{t, t+1, \ldots, T\}$.

During the intraday dispatch phase, the rolling optimization process proceeds as follows: Based on the actual values of renewable energy output and load at time t and their predicted values for the remaining time, the rolling optimization model within the intraday time domain is solved to obtain the optimal solution. The microgrid executes the optimization solution in time domain t, and the process is repeated for the next scheduling cycle at time $t+1$. The intraday dispatch adjusts the solution derived from the outcomes of the day-ahead optimization dispatch, utilizing the electricity power sharing data between microgrids obtained from the day-ahead stage, as well as the actual load values forecasted beforehand.

Some constraints in the intraday stage are derived from the day-ahead stage, such as constraints (1)–(21), (32)–(38) and (42)–(43); different constraints in the intraday stage are calculated as follows:

$$\begin{array}{l}\overline{P}_{i,\tau}^{\text{re}}+\overline{P}_{i,\tau}^{\text{e,chp}}+\overline{P}_{i,\tau}^{\text{BUY}}+\overline{P}_{i,\tau}^{\text{ES,d}} = \\ \overline{P}_{i,\tau}^{\text{e,hp}}+\overline{P}_{i,\tau}^{\text{SELL}}+\overline{P}_{i,\tau}^{\text{ES,c}}+\overline{P}_{i,\tau}^{\text{e}}+\Delta\overline{P}_{i,\tau}^{\text{e}}+\sum_{j\neq i}^{\Theta}\hat{P}_{i-j,\tau}+\hat{P}_{i,t}^{\text{ccssum}}+\hat{P}_{i,t}^{\text{EL}}, \forall\tau\in T_S, \forall i\in\Theta\end{array} \tag{52}$$

$$\overline{P}_{i,\tau}^{\text{h,hp}}+\overline{P}_{i,\tau}^{\text{h,chp}}+\overline{P}_{i,\tau}^{\text{h,gb}}+\overline{P}_{i,\tau}^{\text{HS,d}} = \overline{P}_{i,\tau}^{\text{HS,c}}+\overline{P}_{i,\tau}^{\text{h}}+\Delta\overline{P}_{i,\tau}^{\text{h}}, \forall\tau\in T_S, \forall i\in\Theta \tag{53}$$

$$\overline{P}_{i,t}^{\text{g,BUY}} = \overline{P}_{i,t}^{\text{g,chp}}+\overline{P}_{i,t}^{\text{g,gb}}-\overline{P}_{i,t}^{\text{H2G, out}}, \forall\tau\in T_S, \forall i\in\Theta \tag{54}$$

$$\overline{P}_{i,t}^{\text{EL,H}_2}+\overline{P}_{i,t}^{\text{H,d}} = \overline{P}_{i,t}^{\text{h}_2\text{,chp}}+\overline{P}_{i,t}^{\text{h}_2\text{,gb}}+\overline{P}_{i,t}^{\text{H,c}}+\overline{P}_{i,t}^{\text{H2G, in}}, \forall\tau\in T_S, \forall i\in\Theta \tag{55}$$

In Equations (52) and (53), $\Delta\overline{P}_{i,\tau}^{\text{e}}$ is the discrepancy between the actual and forecasted electrical load values of microgrid i at time t, and $\Delta\overline{P}_{i,\tau}^{\text{h}}$ is the discrepancy between the actual and forecasted heat load values of microgrid i at time t.

6. Experimental Verification

6.1. Parameter Settings

The multi-microgrids model in this chapter consists of three microgrids, denoted as microgrid 1, 2, and 3 (MG1, MG2, and MG3), each equipped with hydrogen blending units, carbon capture, and a two-stage P2G unit. The pricing for electricity purchase and sale transactions between the microgrids and the grid is displayed in Table 1. The microgrid system parameters are detailed in Table 2. The forecasted power output of renewable energy generation for microgrids is shown in Figure 5. The predicted data for electricity and heat load are presented in Figures 6 and 7 [32]. The parameters for ladder-type carbon trading are set as follows: $\chi = 250$ (CNY)/t, $L = 100$ kg, and $a = 25\%$. The natural gas price is 3.5 (CNY)/m^3, and both the day-ahead and intraday stages have the same carbon emission restriction coefficient: $\varphi = 0.05$.

Table 1. Electricity purchase and sale price chart.

Category	Time Period	Price ((CNY)/kWh)
Electricity price	23:00–7:00	0.40
	08:00–11:00, 15:00–18:00	0.75
	12:00–14:00, 19:00–22.00	1.20
Electricity sale	0:00–24:00	0.20

Table 2. Microgrid parameters.

Parameter	Value	Parameter	Value	Parameter	Value
$P_{i,\min}^{e,chp}$ (kW)	0	$P_{i,\min}^{HS,c}$ (kW)	0	$\eta_i^{e,chp}$	0.3
$P_{i,\max}^{e,chp}$ (kW)	2000	$P_{i,\max}^{HS,c}$ (kW)	300	Q_{CH_4}	9.7
$P_{i,\min}^{h,chp}$ (kW)	0	$P_{i,\max}^{HS,d}$ (kW)	300	Q_{H_2}	3.55
$P_{i,\max}^{h,chp}$ (kW)	2000	$P_{i,\min}^{HS,d}$ (kW)	0	D^{chp}	0.01
$P_{i,\max}^{ccs}$ (kW)	1700	$E_{i,\min}^{HS}$ (kWh)	200	D^{gb}	0.01
$P_{i,t}^{b}$ (kW)	300	$E_{i,\max}^{HS}$ (kWh)	1200	D^{res}	0.01
$P_{i,\min}^{ccssum}$ (kW)	0	$E_{i,\min}^{H_2}$ (kWh)	200	a^{co_2}	0.5
$P_{i,\min}^{ccssum}$ (kW)	2000	$E_{i,\max}^{H_2}$ (kWh)	1200	b^{co_2}	0.65
$P_{i,\max}^{h,gb}$ (kW)	2000	$\eta_i^{H_2(c)}$	0.95	c^{co_2}	18.20
$\lambda_{i,\min}^{e,chp}$ (kW)	−1000	$\eta_i^{H_2,d}$	0.95	λ_e	0.2
$\lambda_{i,\max}^{e,chp}$ (kW)	1000	$P_{i,\max}^{H_2,c}$ (kW)	300	η_i^{C}	0.9
$\lambda_{i,\min}^{h,gb}$ (kW)	−1000	$P_{i,\max}^{H_2,d}$ (kW)	300	σ_c	0.1
$\lambda_{i,\max}^{h,gb}$ (kW)	1000	$P_{i,\min}^{EL}$ (kW)	0	η_i^{hp}	0.35
$P_{i,\min}^{e,hp}$ (kW)	0	$P_{i,\max}^{EL}$ (kW)	1500	η_i^{gb}	0.9
$P_{i,\max}^{e,hp}$ (kW)	1000	$P_{i,d}^{EL}$ (kW)	−300	$\eta_{\min}^{soc}$	0.2
$P_{i,\min}^{ES,c}$ (kW)	0	$P_{i,up}^{EL}$ (kW)	300	$\eta_{\max}^{soc}$	0.9
$P_{i,\max}^{ES,c}$ (kW)	300	$P_{i,\max}^{buy}$ (kWh)	1800	C_i^{rate}	2000
$P_{i,\min}^{ES,d}$ (kW)	0	V_i^{CR} (m^3)	100,000	η_i^{EL}	0.85
$P_{i,\max}^{ES,d}$ (kW)	300	M_{MEA} (g/mol)	61.08	η_i^{H2G}	0.70
M_{CO_2} (g/mol)	44	α^{ES}	0.016	λ^{cur}	0.03
$\lambda^{cut,e}$	0.3	α^{HS}	0.016	$\lambda^{cut,h}$	0.1
$\lambda^{trans,e}$	0.3	α^{H2}	0.016	$\lambda^{trans,h}$	0.1
θ_i	3.3	C_R (%)	30	ρ_R (g/mL)	1.01

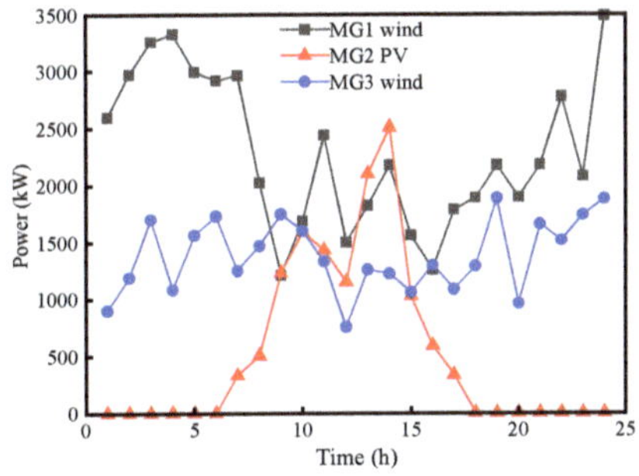

Figure 5. Forecast of renewable energy.

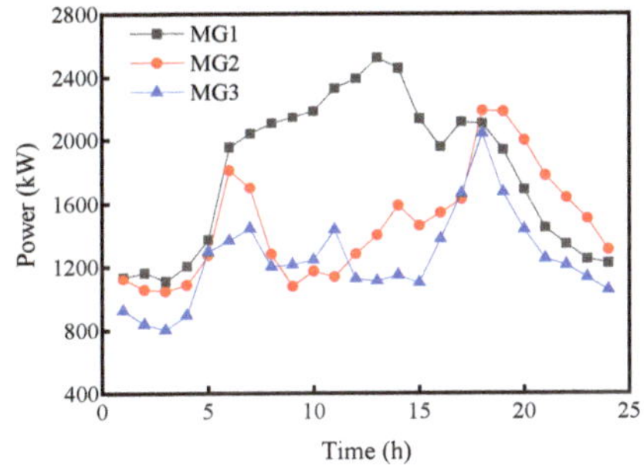

Figure 6. Forecast of electricity load.

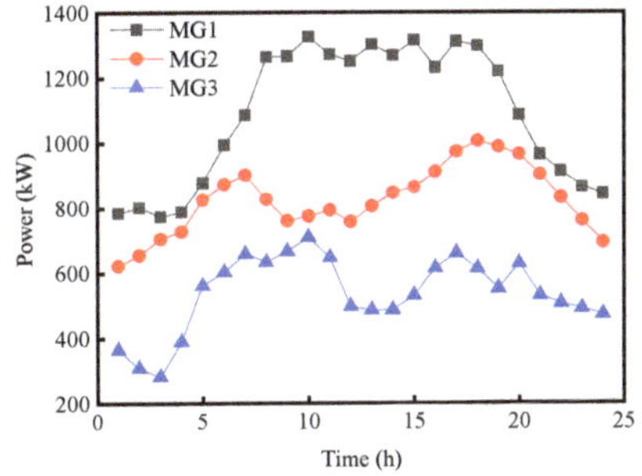

Figure 7. Forecast of heat load.

6.2. Analysis of Day-Ahead Results

$\rho_i(1) = 3 \times 10^{-4}$ and $k_{\max} = 100$; the iteration and residual convergence results of the operating costs for each microgrid are shown in Figure 8 in the day-ahead stage, and the inter-microgrid power exchange is depicted in Figure 9.

In Figure 8a–c, it can be observed that the costs of MG 1, 2, and 3 in the day-ahead stage are 16,126.41 (CNY), 27,865.84 (CNY), and 9699.53 (CNY), respectively. The calculated carbon emissions are 16,823.12 kg, 18,542.12 kg, and 7682.12 kg. From Figure 8d, it can be inferred that convergence is achieved after 37 iterations, with both the primal and dual residuals being less than 10^{-3}, ensuring the accuracy of the results of the algorithm.

Figure 9 shows the power exchange between microgrids, indicating the presence of power interaction among them. This demonstrates that the scheduling model proposed in this chapter enables cooperative operation among microgrids in the day-ahead stage, facilitating shared electricity operation.

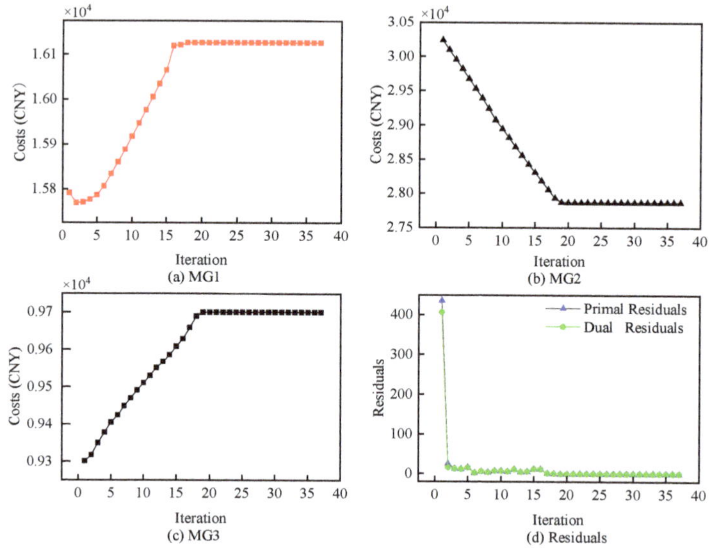

Figure 8. Multi-microgrid cost iteration and residual iteration case diagrams.

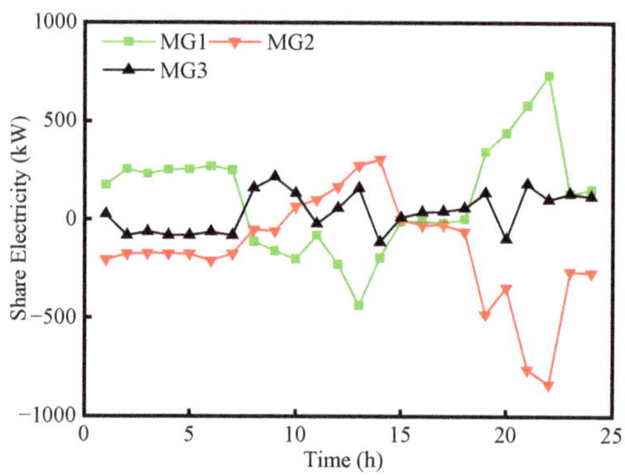

Figure 9. Interaction power between microgrids.

6.3. Analysis of Carbon Trading Mechanism

Three scenarios have been established for comparing and analyzing the ladder-type carbon trading mechanism, where each microgrid contains fixed hydrogen blending units and P2G–CCS coupling units.

Scenario 1 involves an MMG without the carbon trading mechanism; scenario 2 involves an MMG with the traditional fixed carbon trading mechanism; scenario 3 involves an MMG with the ladder-type carbon trading mechanism.

Table 3 shows the intraday results for the three scenarios.

Table 3. Results of scheduling under different scenarios.

Scenario	Carbon Emissions (kg)	Energy Cost (CNY)	Total Cost (CNY)
1	45,755.23	47,495.76	58,104.66
2	40,357.65	41,342.16	57,004.18
3	37,330.62	38,377.12	52,116.13

Table 3 illustrates that, compared to scenario 2, the carbon emissions and the total costs of scenario 3 decrease by 7.5% and 8.57%, respectively; furthermore, compared to scenario 1, the carbon emissions and the total costs of scenario 2 decrease by 10.9% and 11.7%, respectively.

In scenario 1, because of the lack of constraints from a carbon trading mechanism, the microgrid prioritizes minimizing operating costs, heavily relying on low-cost but high-carbon-emitting external energy resources. This choice results in extensive use of carbon-emitting units, leading to a high level of carbon emissions in scenario 1 microgrids. Moreover, the over-reliance on external energy resources increases the overall operating costs of microgrids.

In scenario 2, a single fixed carbon trading mechanism incorporates carbon trading costs into the objective function of operating costs. This mechanism imposes constraints on carbon emissions, requiring microgrids to balance economics and low-carbon considerations during optimization dispatch. Therefore, microgrids prioritize the use of renewable energy or other low-carbon energy sources to enhance the utilization efficiency of renewable energy. Compared to scenario 1, the carbon emissions and external energy procurement costs of microgrids decrease in scenario 2. However, the carbon trading mechanism limits the potential for further reducing carbon emissions in the microgrid in scenario 2.

In scenario 3, a carbon trading mechanism is implemented, where the carbon trading price increases with the increase in carbon emissions. This mechanism effectively guides the flexibility output of various units in the microgrid, reduces reliance on external resource purchases, and decreases the system's energy procurement costs. Compared to scenario 2, the ladder-type carbon trading mechanism has advantages in achieving low-carbon operation and maintaining the economic operation of microgrids in scenario 3. Therefore, the ladder-type carbon trading mechanism has greater potential in promoting low-carbon and economically efficient operation of microgrids.

In summary, through the comparison of scenarios 1, 2, and 3, the ladder-type carbon trading mechanism demonstrates significant economic and environmental advantages in the optimization dispatch of multi-microgrids.

6.4. Analysis of Low-Carbon Technologies

To validate the effectiveness, economic viability, and low-carbon nature of the introduced hydrogen-doped natural gas units and two-stage P2G–CCS coupling model in multi-microgrids, four scenarios are compared and analyzed, as illustrated in Table 4. In scenario 2, the required hydrogen gas is purchased externally, with the price of externally purchased H_2 being 2.8 (CNY)/m^3. The ladder-type carbon trading mechanism is applied to four scenarios.

Table 4. Comparison of four operation schemes.

Scenarios	Hydrogen-Doped Natural Gas Units	P2G–CCS Coupling Devices
1	NO	NO
2	YES	NO
3	NO	YES
4	YES	YES

Tables 5 and 6 show the intraday results for carbon emissions and the equipment output of microgrid devices in the multi-microgrids under the four scenarios.

Compared to scenario 3, the costs of each sub-microgrid in scenario 4 decrease by 538.59 (CNY), 240.37 (CNY), and 31.17 (CNY), respectively, with carbon emissions also decreasing by 268.11 kg, 156.75 kg, and 16.44 kg, respectively. Due to the decrease in external gas purchase costs and carbon trading prices for the multi-microgrids being greater than the increase in purchased electricity costs, the overall expenses of the multi-microgrids are further diminished.

Table 5. Carbon emissions and costs of multi-microgrids under different scenarios.

Scenarios	MG	Cost (CNY)	Carbon Emission (kg)
1	1	16,746.75	17,317.44
	2	31,284.36	18,584.86
	3	9347.55	9199.38
2	1	18,088.01	16,272.90
	2	26,561.42	17,488.37
	3	10,169.73	9329.31
3	1	16,109.25	14,384.31
	2	27,186.89	15,974.36
	3	9630.12	7413.25
4	1	15,570.66	14,116.20
	2	26,946.52	15,817.61
	3	9598.95	7396.81

Table 6. The output of each microgrid device in four scenarios.

Scenarios	MG	Grid (kWh)	Gas Energy (m^3)	Electricity of CHP Unit (kWh)	Heat of CHP Unit (kWh)	Heat of GB (kWh)	Power of CCS (kWh)	Power of P2G Unit (kWh)
1	1	3072.02	2818.85	945.49	1181.87	23,642.32	0	0
	2	26,510.23	2218.38	1339.41	1674.26	16,912.08	0	0
	3	2203.27	1453.68	1131.93	1414.92	10,536.27	0	0
2	1	3716.26	2465.27	907.92	1134.90	22,534.67	0	0
	2	20,841.08	2091.13	1425.01	1781.26	16,792.99	0	0
	3	2359.30	1379.50	1252.43	1565.54	10,616.53	0	0
3	1	3308.81	2251.50	942.35	1177.94	23,421.58	8011.19	5912.95
	2	23,139.46	2245.10	1524.98	1906.23	16,436.41	7829.59	121.04
	3	2865.38	1469.66	1341.41	1676.76	10,392.36	7811.19	169.80
4	1	3328.50	2195.46	1066.61	1318.27	22,892.97	7954.69	6412.61
	2	24,793.59	2103.59	1719.49	2149.36	16,427.22	7801.07	1654.26
	3	3679.05	1378.86	1417.49	1771.86	10,197.53	7803.01	1131.21

Compared to scenario 2, the costs of the microgrid in scenario 4 decrease by 2517.35 (CNY), −385.1 (CNY), and 570.78 (CNY), respectively, with carbon emissions also decreasing by 2156.7 kg, 1670.76 kg, and 1932.5 kg, respectively. This is due to the CCS, which reduces the carbon emissions; additionally, the P2G device provides the hydrogen gas required for hydrogen blending and the natural gas required for system operation, reducing the cost of purchasing gas and hydrogen for the operation of the multi-microgrids more than the increase in the cost of purchasing electricity; this consequently leads to a further decrease in the overall cost of the multi-microgrids.

Compared to scenario 1, the costs of the microgrid in scenario 4 decrease by 1176.09 (CNY), 4337.84 (CNY), and −251.4 (CNY), respectively, with carbon emissions also decreasing by 3201.24 kg, 2767.25 kg, and 1802.57 kg, respectively. Due to the increase in output of the hydrogen-doped natural gas units in scenario 4, the heat coupling effect of the cogeneration unit increases its heat output, reducing the heat output of the gas boiler with higher carbon emissions per unit, and the CCS device's carbon capture function results in lower carbon emissions for the multi-microgrids. In addition, the operation of the P2G device generates a certain amount of hydrogen gas and natural gas to supply the system, and the decrease of the carbon trading cost and gas purchase cost leads to a reduction in the total operating cost of the multi-microgrids, compared to the increase in the cost of purchasing electricity.

In conclusion, the hydrogen-doped natural gas and the P2G–CCS coupling units improve the economic and environmental performance of the multi-microgrids.

6.5. Synergistic Optimization Results of Intraday Low-Carbon Technologies and Low-Carbon Policies

Section 6.3 validated the effectiveness of the low-carbon policy of tiered carbon trading, while Section 6.4 validated the effectiveness of low-carbon technologies such as hydrogen-doped natural gas and P2G–CCS coupled operation.

In order to verify the collaborative optimization of low-carbon technologies and policies, the following scenarios were established: scenario 1, which considers hydrogen-doped natural gas and P2G–CCS coupled operation and ladder-type carbon trading without considering carbon emission constraints; and scenario 2, which comprehensively considers hydrogen-doped natural gas and P2G–CCS coupled operation, carbon trading, and carbon emission constraints.

In scenario 2, the carbon emission restriction is 5%. The intraday results of costs and carbon emissions and the output of equipment for microgrid in two scenarios are shown in Tables 7 and 8.

Table 7. The cost and carbon emissions of microgrids under four scenarios.

Scenario	MG	Cost (CNY)	Carbon Emissions (kg)
1	1	15,570.66	14,116.20
	2	26,946.52	15,817.61
	3	9598.95	7396.81
2	1	13,942.42	13,410.39
	2	27,385.01	15,026.72
	3	9503.18	7026.96

Table 8. The output of microgrid equipment in different scenarios.

Scenario	MG	Grid (kWh)	Gas Energy (m^3)	Electricity of CHP Unit (kWh)	Heat of CHP Unit (kWh)	Heat of GB (kWh)	Power of CCS (kWh)	Power of P2G Unit (kWh)	Power of HP (kW)
1	1	3328.50	2195.46	1066.61	1318.27	22,892.97	7954.69	6412.61	0
	2	24,793.59	2103.59	1719.49	2149.36	16,427.22	7801.07	1654.26	0
	3	3679.05	1378.86	1417.49	1771.86	10,197.53	7803.01	1131.21	0
2	1	3891.60	2207.89	899.00	1123.75	21,570.35	7981.64	5939.68	5150.82
	2	26,281.50	1935.89	231.59	289.49	18,181.55	7855.92	1498.15	0
	3	5042.38	1312.23	797.20	996.50	10,939.39	7855.65	1038.58	99.56

Tables 7 and 8 indicate that the cost in scenario 2 decreases by 1285.52 (CNY), compared to scenario 1, indicating a decrease in total operating costs of 2.46%. This is because, with the addition of carbon emission constraints, there is an increase in the purchased power from the grid, a decrease in the output of the cogeneration unit within the microgrid system, a reduction in the quantity of gas procured from external suppliers, and a decrease in carbon trading costs, due to the decrease in carbon emissions. Therefore, the increase in the cost of purchased power is less than the decrease in the cost of gas purchased and the reduction in carbon trading costs.

In scenario 2, the total carbon emissions of the multi-microgrid decrease by 1866.55 kg, indicating a decrease of 5.00% in the total carbon emissions of the multi-microgrids in scenario 2 compared to scenario 1. Because scenario 2 builds upon scenario 1 by incorporating carbon emission constraints, the limitation on carbon emissions reduces the forced output of the hydrogen-doped natural gas unit, leading to a decrease in its output. Microgrids choose to use carbon-free heat pumps or increase the output of gas boilers to compensate for the heat deficit. Moreover, they increase the power of purchased electricity to fulfill the electricity and HP requirements. Therefore, since the increase in carbon emissions from purchased electricity and the increase in output of gas boilers are less than the decrease in output of cogeneration and the decrease in carbon emissions due to the use of HP, the collaborate optimization of low-carbon technologies and policies in reducing emissions in microgrid operation are verified.

In scenario 2, the total costs of intraday stage scheduling under this strategy decrease by 5.32%, and carbon emissions decrease by 17.61% compared to the day-ahead stage, verifying the effectiveness of the strategy.

In order to deeply analyze the effect of collaborative optimization of low-carbon technologies and policies, Figures 10–12 show the electricity, heat, and hydrogen power balance of the microgrid in scenario 2.

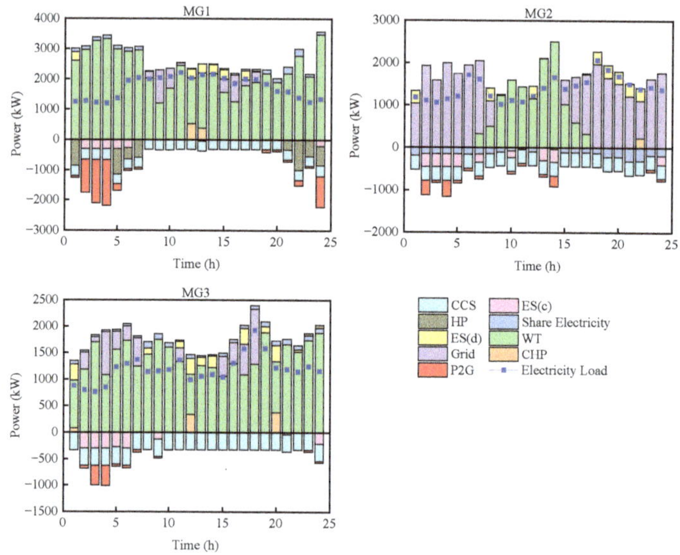

Figure 10. Electric power balance of each microgrid under scenario 2.

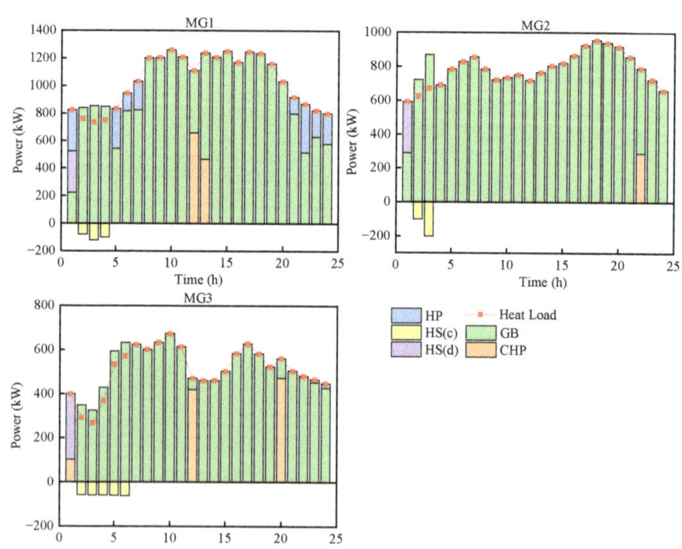

Figure 11. Heat power balance of each microgrid in scenario 2.

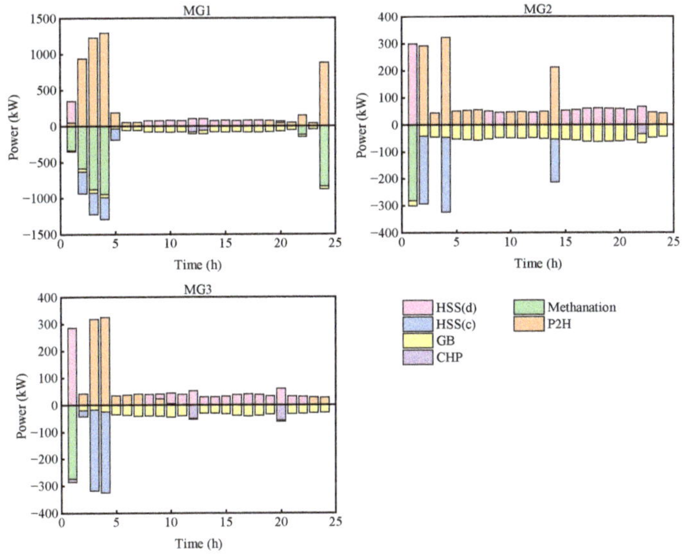

Figure 12. Hydrogen power balance of each microgrid in scenario 2.

Figure 10 indicates that the electrical load is mainly provided by renewable energy units, external purchased electricity, the electric energy storage unit, and the CHP unit, among which renewable energy generation provides most of the electricity demand between the energy-consuming unit. When the output of renewable energy exceeds the demand, P2G consumes a large amount of renewable energy output to improve the utilization rate of renewable energy. The electric storage unit charges energy during off-peak electricity pricing periods (23:00–7:00) and discharges stored energy output during peak electricity pricing periods (12:00–14:00 and 19:00–22:00). MG1 and MG3 share electricity, which reduces the power of external purchased electricity for MG2 and improves the economic efficiency of the multi-microgrids operation.

Figure 11 indicates that the heat load is mainly met by hydrogen-blended GB, with the shortfall mainly supplemented by the hydrogen-doped natural gas unit and heat pumps.

Figure 12 indicates that more hydrogen is utilized during the hydrogen blending process, effectively leveraging the energy advantages of hydrogen and increasing its utilization rate. The utilization of hydrogen energy does not produce carbon emissions during combustion; therefore, the microgrid can mitigate the system's carbon emissions by hydrogen-doped natural gas.

In conclusion, the hydrogen-doped natural gas and two-stage P2G–CCS coupled operation strategy, which considers carbon emission constraints and carbon trading, as depicted in this chapter, enhances the economic and environmental performance of multi-microgrid operation.

7. Conclusions

This article proposes a cooperative optimization scheduling strategy for multi-microgrid systems under the coupling operation of natural gas–hydrogen blending systems and P2G–CCS, considering carbon trading and carbon emission constraints. Through setting up simulation experiments, the following conclusions are drawn:

1. The introduced carbon trading mechanism is validated. Compared to fixed carbon trading, carbon emissions decrease by 7.51% and total costs decrease by 8.57%. Compared to no carbon trading mechanism, carbon emissions decrease by 18.41% and total costs decrease by 10.32%. Regarding the establishment of P2G–CCS coupling and hydrogen-doped natural gas units, carbon emissions decrease by 17.23% and

total costs decrease by 9.17%, compared to scenarios without low-carbon equipment, effectively reducing carbon emissions and operating costs.

2. The introduced P2G–CCS coupling operation and hydrogen-doped natural gas unit models are validated. Compared to conventional models, the costs of the microgrid decrease by 1176.09 (CNY), 4337.84 (CNY), and −251.4 (CNY), respectively, with carbon emissions also decreasing by 3201.24 kg, 2767.25 kg, and 1802.57 kg, respectively, reducing the costs and carbon emissions of multi-microgrids.
3. This article proposes an optimization strategy, incorporating carbon emission constraints on the above basis. Through day-ahead and intraday scheduling, the experimental results show that, compared to the day-ahead stage, the total operating costs of intraday stage scheduling under this strategy decrease by 5.32%, and carbon emissions decrease by 17.61%, verifying the effectiveness of the strategy. Compared to scenarios that do not consider carbon emission constraints, the total operating costs of intraday stage scheduling under this strategy decrease by 2.46% and carbon emissions decrease by 5.00%. Therefore, the synergistic effect of low-carbon policies and low-carbon technologies on multi-microgrid optimization scheduling can further tap into the emission reduction potential of microgrids, reducing the costs and carbon emissions of multi-microgrid systems, and providing a pathway for the low-carbon transformation of the power system.

Author Contributions: Conceptualization, Y.Z.; Methodology, Y.Z.; Writing—original draft, Y.Z.; Writing—review & editing, J.C. All authors have read and agreed to the published version of the manuscript.

Funding: This research received no external funding.

Data Availability Statement: The original contributions presented in the study are included in the article, further inquiries can be directed to the corresponding author.

Conflicts of Interest: The authors declare no conflict of interest.

Nomenclature

Indices

t	index of time slots.
i	index of multi-microgrids.
T	set of time slots.
Θ	set of microgrids.
T_S	intraday optimization scheduling horizon.
τ	index of intraday time slots.
k	index of iterations of ADMM.

Parameters

$P_{i,\max}^{\text{ccs}}$	maximum operating power of the CCS in microgrid i.
$P_{i,\min}^{\text{ccssum}}/P_{i,\max}^{\text{ccssum}}$	minimum/maximum total power of the CCS in microgrid i.
η_i^{C}	efficiency of the carbon capture equipment in microgrid i.
$E_{i,\max}^{\text{air}}$	maximum carbon emissions to the atmosphere by the units in microgrid i.
σ_{c}	flue gas partition coefficients for CCS unit.
$E_{i,\min}^{\text{ca}}/E_{i,\max}^{\text{ca}}$	minimum/maximum volume of carbon dioxide provided by the liquid storage unit in microgrid i.
M_{MEA}	molar mass of monoethanolamine.
θ_i	conversion coefficient of the liquid storage unit in microgrid i.
C_{R}	concentration of monoethanolamine solution.
ρ_{R}	density of the monoethanolamine solution.

M_{CO_2}	molar mass of carbon dioxide.
η_i^{EL}	conversion efficiencies of P2H processes in microgrid i.
η_i^{H2G}	conversion efficiencies of methanation processes in microgrid i.
$P_{i,\min}^{EL}/P_{i,\max}^{EL}$	minimum/maximum electric power for electrolysis in the P2H process in microgrid i.
$P_{i,d}^{EL}/P_{i,up}^{EL}$	minimum/maximum climbing power in the P2H process in microgrid i.
$E_{i,\max}^{STO}$	maximum amount of carbon sequestration in microgrid i.
$P_{i,\max}^{e,chp}/P_{i,\max}^{h,chp}$	maximum electrical/heat power output of the hydrogen-blended CHP in microgrid i.
$\lambda_{i,\min}^{e,chp}/\lambda_{i,\max}^{e,chp}$	minimum/maximum ramp rate of the hydrogen-blended CHP in microgrid i.
Q_{CH_4}/Q_{H_2}	heating value of CH_4/H_2.
$P_{i,\max}^{h,gb}$	maximum heat power output of the GB in microgrid i.
$\lambda_{i,\min}^{h,gb}/\lambda_{i,\max}^{h,gb}$	minimum/maximum ramp rate of the GB in microgrid i.
ρ_{H_2}/ρ_{CH_4}	the density of H_2/CH_4.
M_{H_2}/M_{CH_4}	the molar mass of H_2/CH_4.
$D^{chp}/D^{gb}/D^{res}$	carbon quota coefficient for CHP/GB/renewable energy.
a^{CO_2}/b^{CO_2}	carbon emission coefficient for CHP/GB.
c^{CO_2}	carbon emission constant.
λ_e	carbon emission conversion coefficient for purchased electricity.
χ	base carbon emission price.
L/α	carbon emission interval/carbon emission price growth rate.
φ	carbon emission constraint coefficient.
$E_{i,\max}^{CO_2}$	maximum carbon emissions in the day-ahead stage in microgrid i, without considering carbon emission constraints.
λ^{cur}	reduction in the cost of renewable energy.
$M_t^{CH_4}$	natural gas unit price in period t.
M_t^{BUY}/M_t^{SELL}	electricity purchase price/electricity selling price from the grid in period t.
η_i^{hp}	the electrical-to-heat conversion efficiency of the HP in the microgrid.
$P_{i,\min}^{e,hp}/P_{i,\max}^{e,hp}$	minimum/maximum electric power consumption of the HP in the microgrid i.
$P_{i,\max}^{BUY}/P_{i,\max}^{SELL}$	maximum power purchase/sold from the main grid in microgrid i.
a^{tr}/b^{cut}	transfer load coefficient/reducible load coefficient.
$P_{i-j,\max}$	maximum power transfer between microgrid i and j.
λ^{P}	penalty cost coefficient.
ρ_{CO_2}	density of carbon dioxide.
$\eta_i^{H,c}/\eta_i^{H,d}$	charging/discharging efficiency of the HSS in the microgrid i.
$E_{i,\min}^{H_2}/E_{i,\max}^{H_2}$	minimum/maximum storage capacity of the HSS in the microgrid i.
$E_{i,\min}^{ES}/E_{i,\max}^{ES}$	minimum/maximum storage capacity of the ES in microgrid i.
$P_{i,\max}^{ES,c}/P_{i,\max}^{ES,d}$	maximum charging/discharging power of the ES in microgrid i.
$P_{i,\max}^{HS,c}/P_{i,\max}^{HS,d}$	maximum charging/discharging power of the HS in microgrid i.
$\eta_i^{HS,c}/\eta_i^{HS,d}$	charging/discharging efficiency of the HS in microgrid i.
$\varepsilon^{pri}/\varepsilon^{dual}$	primal/dual residuals.
λ_{i-j}	Lagrange multiplier.
$(\bar{\bullet})/(\hat{\bullet})$	day-ahead/intraday variables.

Variables

$P_{i,t}^{\text{ccs}}$	operating power of the CCS unit in microgrid i at time t.
$P_{i,t}^{\text{b}}$	fixed power of the CCS unit in microgrid i at time t.
$P_{i,t}^{\text{ccssum}}$	total power of the CCS unit in microgrid i at time t.
$E_{i,t}^{\text{ccs}}$	amount of carbon dioxide absorbed by the CCS unit in microgrid i at time t.
$E_{i,t}^{\text{sum}}$	total carbon emissions from the units in microgrid i at time t.
$E_{i,t}^{\text{air}}$	amount of carbon dioxide emitted into the atmosphere by the units in microgrid i at time t.
$E_{i,t}^{\text{ca}}$	amount of carbon dioxide provided by the liquid storage unit in microgrid i at time t.
$E_{i,t}^{\text{H2G}}$	amount of carbon dioxide utilized for methanation from the CCS in microgrid i at time t.
$E_{i,t}^{\text{STO}}$	amount of carbon dioxide storage from the CCS in microgrid i at time t.
$V_{i,t}^{\text{ca}}$	volume of carbon dioxide provided by the liquid storage unit installed in microgrid i at time t.
V_{i}^{CR}	liquid storage unit capacity in microgrid i.
$V_{i,t}^{\text{F}}/V_{i,t}^{\text{P}}$	reserves of the liquid storage units for storing rich/lean liquid in microgrid i.
$V_{i,0}^{\text{F}}/V_{i,0}^{\text{P}}$	initial reserves of the liquid storage units for storing rich liquid/lean liquid in microgrid i.
$V_{i,24}^{\text{F}}/V_{i,24}^{\text{P}}$	final reserves of the liquid storage units storing rich liquid/lean liquid in microgrid i.
$P_{i,t}^{\text{EL}}$	electric power consumed by electrolysis in microgrid i at time t.
$P_{i,t}^{\text{EL,H}_2}$	hydrogen production power consumed by electrolysis in microgrid i at time t.
$P_{i,t}^{\text{H2G, out}}$	methane production power of the methanation equipment in microgrid i at time t.
$P_{i,t}^{\text{H2G, in}}$	hydrogen consumption power of the methanation equipment in microgrid i at time t.
$E_{i,t}^{\text{H2G,SUM}}$	amount of carbon dioxide consumed by P2G in microgrid i at time t.
$P_{i,t}^{\text{e,chp}}/P_{i,t}^{\text{h,chp}}$	power generated by the CHP unit for electricity/heat in microgrid i at time t.
$P_{i,t}^{\text{g,chp}}/P_{i,t}^{\text{h}_2\text{,chp}}$	power consumption of natural gas/hydrogen by the CHP unit in microgrid i at time t.
$V_{i,t}^{\text{g,chp}}/V_{i,t}^{\text{h}_2\text{,chp}}$	volume of natural gas/hydrogen consumed by the CHP unit in microgrid i at time t.
$Y_{i,t}^{\text{h2,chp}}$	hydrogen blending ratio in microgrid i at time t.
$P_{i,t}^{\text{h,gb}}$	power generated by the GB unit for heat in microgrid i at time t.
$P_{i,t}^{\text{g,gb}}/P_{i,t}^{\text{h}_2\text{,gb}}$	power consumption of natural gas/hydrogen by the GB unit in microgrid i at time t.
$V_{i,t}^{\text{g,gb}}/V_{i,t}^{\text{h2,gb}}$	volume of natural gas/hydrogen consumed by the GB in microgrid i at time t.
$Y_{i,t}^{\text{h}_2\text{,gb}}$	hydrogen blending ratio (by molar mass) in microgrid i at time t.
$C_{i,t}^{\text{CO}_2}$	amount of carbon dioxide involved in carbon trading in microgrid i at time t.
$E_{i,t}^{0}$	carbon emission quota in microgrid i at time t.
$E_{i,t}^{\text{CO}_2}$	carbon emissions of the microgrids in microgrid i at time t.
$CE_{i,t}^{\text{CO}_2}$	carbon trading cost in microgrid i at time t.
$E_{i}^{\text{CO}_2}$	total carbon emissions of microgrid i.
C_{i}^{MG}	operating cost of microgrid i.
C_{i}^{cur}	reduction in the cost of renewable energy generation in microgrid i.

$C_i^{CO_2}$	carbon emission cost of microgrid i.
C_i^{ESS}	energy storage cost of microgrid i.
C_i^{load}	load demand of cost microgrid i.
C_i^{out}	external interaction costs of microgrid i.
$P_{i,t}^{cur}$	reduction power of renewable energy in microgrid i at time t.
$V_{i,t}^{BUY}$	natural gas purchase quantity in microgrid i at time t.
$V_{i,t}^{BUY}$	natural gas purchase quantity of microgrid i in period t.
$P_{i,t}^{BUY}$ / $P_{i,t}^{SELL}$	purchased/sold power from the grid in microgrid i at time t.
$P_{i,t}^{re}$	output power of renewable energy.
$P_{i-j,t}/P_{j-i,t}$	amount of electrical energy exchanged between microgrid i and j/microgrid j and i in time period t.
$P_{i,t}^{e}$	electrical load of microgrid i in the t-th time period.
$P_{i,t}^{ES,c}$ / $P_{i,t}^{ES,d}$	charging/discharging power of the electrical energy storage system in microgrid i in the t-th time period.
$P_{i,t}^{e,hp}$	electric power consumed by the heat pump in the i-th time period of the microgrid i.
$P_{i,t}^{h}$	heat load of microgrid i at time t.
$P_{i,t}^{h,hp}$	heat power generated by the heat pump in microgrid i at time t.
$P_{i,t}^{HS,c}$ / $P_{i,t}^{HS,d}$	charging/discharging power of the heat energy storage system in microgrid i at time t.
$P_{i,t}^{g,BUY}$	power of gas purchased externally in microgrid i at time t.
$P_{i,t}^{e,r}$	actual renewable energy generation power.
$\overline{P}_{i,t}^{e,r}$	predicted renewable energy generation power.
$\sigma_{i,t}^2$	variance of renewable energy generation.
$E_{i,t}^{H_2}$	energy storage capacity of the HSS in the i-th microgrid in the t-th time period.
$u_t^{H,d}$ / $u_t^{H,c}$	binary variables in HSS.
$E_{i,t}^{ES}$	energy storage capacity in the ES of microgrid i in the t-th time period.
$\eta_i^{ES,c}$ / $\eta_i^{ES,d}$	charging/discharging efficiency of the ES in microgrid i.
$u_t^{ES,c}$ / $u_t^{ES,d}$	binary variables in ES.
$E_{i,t}^{HS}$	energy storage capacity in the HS of microgrid i in the t-th time period.
$P_{i,t}^{e,g}$ / $P_{i,t}^{e,tr}$ / $P_{i,t}^{e,cut}$	fixed/transferable/reducible electrical load in microgrid i at time t.
$P_{i,t}^{h,g}$ / $P_{i,t}^{h,tr}$ / $P_{i,t}^{h,cut}$	fixed/transferable/reducible heat load in microgrid i at time t.
$\Delta\overline{P}_{i,\tau}^{e}$	deviation between the actual and predicted electrical load values of microgrid i at time t.
$\Delta\overline{P}_{i,\tau}^{h}$	deviation between the actual and predicted heat load values of microgrid i at time t.
$\rho_i(k)$	penalty parameter at the k-th iteration.
Acronyms	
ADMM	alternating direction method of multipliers.
CHP	combined heat and power.
GB	gas boiler.
HP	heat pump.
ES/HS	electric/heat energy storage.
HSS	hydrogen energy storage.
MMG	multi-microgrids.

P2G	power-to-gas.
CCS	carbon capture system.
P2H	power-to-hydrogen.

References

1. Zhang, J.; Liu, Z. Low Carbon Economic Scheduling Model for a Park Integrated Energy System Considering Integrated Demand Response, Ladder-Type Carbon Trading and Fine Utilization of Hydrogen. *Energy* **2024**, *290*, 130311. [CrossRef]
2. Wang, R.; Wen, X.; Wang, X.; Fu, Y.; Zhang, Y. Low Carbon Optimal Operation of Integrated Energy System Based on Carbon Capture Technology, LCA Carbon Emissions and Ladder-Type Carbon Trading. *Appl. Energy* **2022**, *311*, 118664. [CrossRef]
3. Gao, L.; Fei, F.; Jia, Y.; Wen, P.; Zhao, X.; Shao, H.; Feng, T.; Huo, L. Optimal Dispatching of Integrated Agricultural Energy System Considering Ladder-Type Carbon Trading Mechanism and Demand Response. *Int. J. Electr. Power Energy Syst.* **2024**, *156*, 109693. [CrossRef]
4. Wang, L.; Shi, Z.; Dai, W.; Zhu, L.; Wang, X.; Cong, H.; Shi, T.; Liu, Q. Two-Stage Stochastic Planning for Integrated Energy Systems Accounting for Carbon Trading Price Uncertainty. *Int. J. Electr. Power Energy Syst.* **2022**, *143*, 108452. [CrossRef]
5. Liu, R.; Bao, Z.; Yu, Z.; Zhang, C. Distributed Interactive Optimization of Integrated Electricity-Heat Energy Systems Considering Hierarchical Energy-Carbon Pricing in Carbon Markets. *Int. J. Electr. Power Energy Syst.* **2024**, *155*, 109628. [CrossRef]
6. Shi, Z.; Yang, Y.; Xu, Q.; Wu, C.; Hua, K. A Low-Carbon Economic Dispatch for Integrated Energy Systems with CCUS Considering Multi-Time-Scale Allocation of Carbon Allowance. *Appl. Energy* **2023**, *351*, 121841. [CrossRef]
7. Lei, D.; Zhang, Z.; Wang, Z.; Zhang, L.; Liao, W. Long-Term, Multi-Stage Low-Carbon Planning Model of Electricity-Gas-Heat Integrated Energy System Considering Ladder-Type Carbon Trading Mechanism and CCS. *Energy* **2023**, *280*, 128113. [CrossRef]
8. Dong, W.; Lu, Z.; He, L.; Geng, L.; Guo, X.; Zhang, J. Low-Carbon Optimal Planning of an Integrated Energy Station Considering Combined Power-to-Gas and Gas-Fired Units Equipped with Carbon Capture Systems. *Int. J. Electr. Power Energy Syst.* **2022**, *138*, 107966. [CrossRef]
9. Wang, S.; Wang, S.; Zhao, Q.; Dong, S.; Li, H. Optimal Dispatch of Integrated Energy Station Considering Carbon Capture and Hydrogen Demand. *Energy* **2023**, *269*, 126981. [CrossRef]
10. Gao, J.; Meng, Q.; Liu, J.; Wang, Z. Thermoelectric Optimization of Integrated Energy System Considering Wind-Photovoltaic Uncertainty, Two-Stage Power-to-Gas and Ladder-Type Carbon Trading. *Renew. Energy* **2024**, *221*, 119806. [CrossRef]
11. Bao, H.; Sun, Y.; Zheng, S. A Collaborative Training Approach for Multi Energy Systems in Low-Carbon Parks Accounting for Response Characteristics. *IET Renew. Power Gener.* **2024**, *18*, 456–475. [CrossRef]
12. Chen, J.; Tang, Z.; Huang, Y.; Qiao, A.; Liu, J. Asymmetric Nash Bargaining-Based Cooperative Energy Trading of Multi-Park Integrated Energy System under Carbon Trading Mechanism. *Electr. Power Syst. Res.* **2024**, *228*, 110033.
13. Wang, R.; Cheng, S.; Zuo, X.; Liu, Y. Optimal Management of Multi Stakeholder Integrated Energy System Considering Dual Incentive Demand Response and Carbon Trading Mechanism. *Int. J. Energy Res.* **2022**, *46*, 6246–6263. [CrossRef]
14. Zhang, M.; Yang, J.; Yu, P.; Tinajero, G.D.A.; Guan, Y.; Yan, Q.; Zhang, X.; Guo, H. Dual-Stackelberg Game-Based Trading in Community Integrated Energy System Considering Uncertain Demand Response and Carbon Trading. *Sustain. Cities Soc.* **2024**, *101*, 105088.
15. Li, Y.; Zhang, X.; Wang, Y.; Qiao, X.; Jiao, S.; Cao, Y.; Xu, Y.; Shahidehpour, M.; Shan, Z. Carbon-Oriented Optimal Operation Strategy Based on Stackelberg Game for Multiple Integrated Energy Microgrids. *Electr. Power Syst. Res.* **2023**, *224*, 109778. [CrossRef]
16. Lyu, Z.; Lai, Y.; Yi, J.; Liu, Q. Low Carbon and Economic Dispatch of the Multi-microgrids Integrated Energy System Using CCS-P2G Integrated Flexible Operation Method. *Energy Sources Part Recovery Util. Environ. Eff.* **2023**, *45*, 3617–3638. [CrossRef]
17. Liu, Y.; Li, X.; Liu, Y. A Low-Carbon and Economic Dispatch Strategy for a Multi-microgrids Based on a Meteorological Classification to Handle the Uncertainty of Wind Power. *Sensors* **2023**, *23*, 5350. [CrossRef]
18. Chen, P.; Qian, C.; Lan, L.; Guo, M.; Wu, Q.; Ren, H.; Zhang, Y. Shared Trading Strategy of Multiple Microgrids Considering Joint Carbon and Green Certificate Mechanism. *Sustainability* **2023**, *15*, 10287. [CrossRef]
19. Wang, H.; Wang, C.; Zhao, L.; Ji, X.; Yang, C.; Wang, J. Multi-Micro-Grid Main Body Electric Heating Double-Layer Sharing Strategy Based on Nash Game. *Electronics* **2023**, *12*, 214. [CrossRef]
20. Zhong, X.; Liu, Y.; Xie, K.; Xie, S. A Local Electricity and Carbon Trading Method for Multi-Energy Microgrids Considering Cross-Chain Interaction. *Sensors* **2022**, *22*, 6935. [CrossRef]
21. Zhang, Z.; Du, J.; Fedorovich, K.S.; Li, M.; Guo, J.; Xu, Z. Optimization Strategy for Power Sharing and Low-Carbon Operation of Multi-microgrids IES Based on Asymmetric Nash Bargaining. *Energy Strategy Rev.* **2022**, *44*, 100981. [CrossRef]
22. Zhang, K.; Gao, C.; Zhang, G.; Xie, T.; Li, H. Electricity and Heat Sharing Strategy of Regional Comprehensive Energy Multi-microgrids Based on Double-Layer Game. *Energy* **2024**, *293*, 130655. [CrossRef]
23. Duan, P.; Zhao, B.; Zhang, X.; Fen, M. A Day-Ahead Optimal Operation Strategy for Integrated Energy Systems in Multi-Public Buildings Based on Cooperative Game. *Energy* **2023**, *275*, 127395. [CrossRef]
24. Xu, J.; Yi, Y. Multi-microgrids Low-Carbon Economy Operation Strategy Considering Both Source and Load Uncertainty: A Nash Bargaining Approach. *Energy* **2023**, *263*, 125712. [CrossRef]

25. Lyu, Z.; Liu, Q.; Liu, B.; Zheng, L.; Yi, J.; Lai, Y. Optimal Dispatch of Regional Integrated Energy System Group Including Power to Gas Based on Energy Hub. *Energies* **2022**, *15*, 9401. [CrossRef]
26. Hu, Q.; Zhou, Y.; Ding, H.; Qin, P.; Long, Y. Optimal Scheduling of Multi-microgrids with Power to Hydrogen Considering Federated Demand Response. *Front. Energy Res.* **2022**, *10*, 1002045. [CrossRef]
27. Cui, Y.; Deng, G.; Zeng, P.; Zhong, W.; Zhao, Y.; Liu, X. Multi-Time Scale Source-Load Dispatch Method of Power System with Wind Power Considering Low-Carbon Characteristics of Carbon Capture Power Plant. *Proc. CSEE* **2022**, *42*, 5869–5886.
28. Chen, D.; Liu, F.; Liu, S. Optimization of Virtual Power Plant Scheduling Coupling with P2G-CCS and Doped with Gas Hydrogen Based on Stepped Carbon Trading. *Power Syst. Technol.* **2022**, *46*, 2042–2054.
29. Wang, K.; Liang, Y.; Jia, R.; Wu, X.; Wang, X.; Dang, P. Two-Stage Stochastic Optimal Scheduling for Multi-microgrids Networks with Natural Gas Blending with Hydrogen and Low Carbon Incentive under Uncertain Envinronments. *J. Energy Storage* **2023**, *72*, 108319. [CrossRef]
30. Guo, R.; Ye, H.; Zhao, Y. Low Carbon Dispatch of Electricity-Gas-Thermal-Storage Integrated Energy System Based on Stepped Carbon Trading. *Energy Rep.* **2022**, *8*, 449–455. [CrossRef]
31. Wang, Y.; Wang, X.; Shao, C.; Gong, N. Distributed Energy Trading for an Integrated Energy System and Electric Vehicle Charging Stations: A Nash Bargaining Game Approach. *Renew. Energy* **2020**, *155*, 513–530. [CrossRef]
32. Du, J.; Han, X.; Wang, J. Distributed Cooperation Optimization of Multi-microgrids under Grid Tariff Uncertainty: A Nash Bargaining Game Approach with Cheating Behaviors. *Int. J. Electr. Power Energy Syst.* **2024**, *155*, 109644. [CrossRef]

electronics

MDPI

Article

A Novel Improved Variational Mode Decomposition-Temporal Convolutional Network-Gated Recurrent Unit with Multi-Head Attention Mechanism for Enhanced Photovoltaic Power Forecasting

Hua Fu [1], Junnan Zhang [1,*] and Sen Xie [2]

[1] Faculty of Electrical and Control Engineering, Liaoning Technical University, Huludao 125105, China; fuhua@lntu.edu.cn
[2] Institute of Intelligence Science and Engineering, Shenzhen Polytechnic University, Shenzhen 518055, China; xiesen@szpu.edu.cn
* Correspondence: 472010030@stu.lntu.edu.cn

Abstract: Photovoltaic (PV) power forecasting plays a crucial role in optimizing renewable energy integration into the grid, necessitating accurate predictions to mitigate the inherent variability of solar energy generation. We propose a novel forecasting model that combines improved variational mode decomposition (IVMD) with the temporal convolutional network-gated recurrent unit (TCN-GRU) architecture, enriched with a multi-head attention mechanism. By focusing on four key environmental factors influencing PV output, the proposed IVMD-TCN-GRU framework targets a significant research gap in renewable energy forecasting methodologies. Initially, leveraging the sparrow search algorithm (SSA), we optimize the parameters of VMD, including the mode component K-value and penalty factor, based on the minimum envelope entropy principle. The optimized VMD then decomposes PV power, while the TCN-GRU model harnesses TCN's proficiency in learning local temporal features and GRU's capability in rapidly modeling sequence data, while leveraging multi-head attention to better utilize the global correlation information within sequence data. Through this design, the model adeptly captures the correlations within time series data, demonstrating superior performance in prediction tasks. Subsequently, the SSA is employed to optimize GRU parameters, and the decomposed PV power mode components and environmental feature attributes are inputted into the TCN-GRU neural network. This facilitates dynamic temporal modeling of multivariate feature sequences. Finally, the predicted values of each component are summed to realize PV power forecasting. Validation using real data from a PV station corroborates that the novel model demonstrates a substantial reduction in RMSE and MAE of up to 55.1% and 54.5%, respectively, particularly evident in instances of pronounced photovoltaic power fluctuations during inclement weather conditions. The proposed method exhibits marked improvements in accuracy compared to traditional PV power prediction methods, underscoring its significance in enhancing forecasting precision and ensuring the secure scheduling and stable operation of power systems.

Keywords: photovoltaic power forecasting; gated recurrent units; minimum envelope entropy; VMD decomposition; TCN

Citation: Fu, H.; Zhang, J.; Xie, S. A Novel Improved Variational Mode Decomposition-Temporal Convolutional Network-Gated Recurrent Unit with Multi-Head Attention Mechanism for Enhanced Photovoltaic Power Forecasting. *Electronics* **2024**, *13*, 1837. https://doi.org/10.3390/electronics13101837

Academic Editor: Enrique Romero-Cadaval

Received: 8 April 2024
Revised: 2 May 2024
Accepted: 7 May 2024
Published: 9 May 2024

1. Introduction

As fossil energy is restricted by resource reserves and environmental problems, it has become a consensus of global development to vigorously develop and efficiently utilize renewable energy [1,2]. Because of data released by the national energy administration, for solar power generation in China, the installed capacity is about 650 million kilowatts as of the end of February 2024—a year-on-year increase of 56.9%. In the future, the proportion of new energy installations will continue to increase, and the photovoltaic and other new energy industries will need to develop rapidly. It not only provides strong guarantees for energy security but also injects new impetus into economic growth and the achievement of carbon peaking and carbon neutrality goals.

However, photovoltaic power is distinguished by its unpredictability and instability, posing significant disruptions to the regular functioning of extensive grid-connected solar photovoltaic systems and presenting considerable hurdles to the power grid's quality and stability [3–5]. This fluctuation or intermittency is caused by several factors, i.e., humidity, air pressure, irradiance, and temperature. When the meteorological factors change, large power fluctuations at the power supply side are produced in the power system, bringing operation risks to the active power balance and frequency regulation and affecting the economy of the power system [6–8]. Thus, to alleviate the problem, accurate prediction of PV power as the key technology is present. Meanwhile, it supplies guidance for unit commitment, thereby reducing the power generation cost and strengthening the competitiveness of photovoltaic energy in the electricity market. Hence, reliable photovoltaic forecasting technology of the power system is crucial for the economical and safe operation and photovoltaic field management.

Currently, numerous studied on photovoltaic power forecasting have been carried out, and the forecasting approaches have been presented as time series [9–11], neural networks [12–14], support vector machines [15,16], Markov chain [17,18] and a combination of corresponding methods [19–21]. As the scale of power plants continues to expand, the amount of data produced by power plants has also exploded. In fact, due to the quantity and quality of the source data of power plants, the traditional neural network photovoltaic power forecasting model is restricted by not considering environmental factors [22], thereby lacking reasonable utilization of complex sequence information. In addition, considering the nonlinear change in photovoltaic power and multiple environment sequence information, the convergence rate of the model slows down and overfitting appears with the increase in network input variables [23–25]. At the same time, the accuracy of photovoltaic power forecasting is also affected by time-varying factors [26,27]. Therefore, to guarantee the feasibility of photovoltaic power forecasting, it is beneficial to fully analyze the impact of environmental factors on the modeling of photovoltaic power forecasting. Moreover, the long short-term memory (LSTM) network, as referenced in the literature [28,29], represents a type of deep neural network. Within the framework of deep learning models, the LSTM network stands out for its exceptional proficiency in addressing issues related to time series forecasting, attributable to its distinctive architectural design. But, the LSTM architecture is characterized by a higher parameter count compared to the GRU architecture. Specifically, LSTM incorporates a greater number of gating units and parameters, resulting in increased model complexity and computational demands. This elevated parameterization in LSTM models may consequently lead to escalated training and inference costs. GRU has been identified as a well-suited solution for managing and predicting the challenges associated with extended time intervals and temporal delays within time series data. Its efficacy in addressing such issues has led to widespread adoption across various industrial processes. In addition, time series modeling has been studied and explored in terms of photovoltaic power forecasting [30]. However, GRU network performance is greatly affected by parameters, and whether the model parameters are reasonable has a great influence on the forecasting results [31,32].

Moreover, due to some random factors such as weather, there are many uncertainties in the actual photovoltaic sequence. In addition, the photovoltaic power has non-stationary and nonlinear characteristics, and a single forecasting model is insufficient to satisfy forecasting accuracy requirements. The hybrid forecasting model with a decomposition algorithm effectively reduces the original photovoltaic sequence characteristics and has better forecasting performance. There are familiar decomposition approaches which include the variational mode, empirical mode [33], and wavelet decomposition [34]. Nevertheless, the selection of base functions and thresholds is depended on the WD effect. EMD and its derived methods lack a mathematical theoretical foundation due to endpoint effects.

VMD can effectively suppress noise and is regarded as the most effective decomposition technique. Nevertheless, intrinsic mode functions (IMFs) and the number of modes of VMD have a remarkable effect on the decomposition effect.

The above methods provide inspiration and motivation for the proposed forecasting strategy in this paper. However, the actual photovoltaic power is greatly interfered by the external environment and has the characteristics of instability and obvious intermittent fluctuation [35,36]. In addition, there are differences in photovoltaic power forecasting models under different environments, and a single model cannot meet the actual production needs. Moreover, the forecasting accuracy is directly affected by whether the selection of forecasting model parameter is reasonable. Therefore, a novel hierarchical VMD-TCN-GRU multi-head attention mechanism for photovoltaic power forecasting is present in this paper. The main innovation points of this article are as follows:

(1) To decompose photovoltaic power, the variational mode decomposition method is used. Meanwhile, the optimal mode and penalty factor are searched based on the minimum envelope entropy to enhance the adaptability of the variational mode decomposition algorithm.
(2) Different TCN-GRU models are constructed for different PV modal components decomposed via the improved variational mode decomposition algorithm, and the main environmental factors, for example, atmospheric pressure, air temperature, solar irradiance, and component temperature, are considered as TCN-GRU model inputs.
(3) According to the SSA, the hidden layer neural element number, training frequency, and learning rate parameters that have a significant impact on network performance were optimized. The forecasting results under multiple photovoltaic modes are integrated to obtain better photovoltaic power forecasting. Finally, for the proposed forecasting strategy, the photovoltaic power of the actual power plant is applied to illustrate the feasibility.

The remaining parts of this article include Section 2, which is dedicated to the detailed exposition of the methodology employed in our study, elucidating the theoretical framework and computational techniques utilized in our research investigation. Section 3 comprehensively describes the simulation results derived from the application of the proposed methodology, presenting empirical data and analysis to support our research findings. In Section 4, a rigorous discussion is conducted to evaluate and interpret the significance of the obtained results, thereby validating the rationality of our research approach and its implications for the field of study. Finally, the conclusions drawn from our research endeavor are summarized in Section 5, encapsulating the key findings, implications, and potential avenues for future research exploration.

2. Materials and Methods

2.1. TCN Network

The TCN represents a convolutional neural network architecture tailored for addressing time-series problems, integrating dilated causal convolution (DCC) and residual connection (RC) mechanisms. This architecture effectively captures the interdependencies between data points, facilitating subsequent predictions.

Dilated convolution, a key component of TCN, expands the receptive field by selectively skipping portions of the input. By adjusting the dilation factor, dilated convolution modulates the size of the receptive field, enabling the network to flexibly control the historical information incorporated into the output. In the context of one-dimensional sequential data $x \in R^n$ and filters $f : \{0, 1, \cdots, k-1\} \to R$, the convolutional kernel, characterized by filter coefficients k and dilation factor d, extends the receptive field. The operation of dilated convolution is expressed as follows:

$$F(x) = \sum_{i=0}^{k-1} f(i) \cdot x_{s-d\cdot i} \quad (1)$$

where d denotes the dilation factor; $s - d \cdot i$ represents historical data in the input sequence; and k stands for the filter coefficient [37].

The dilated causal convolution, as illustrated in Figure 1, reveals that the receptive field size of a point Y_t in the output sequence is modulated via k and d. Importantly, the output at a given point is influenced solely by the preceding historical data. The TCN architecture employed in this study utilizes dilated causal convolutions with dilation factors d set to 1, 2, 4, and 8 and a filter coefficient k of 3, as depicted in Figure 2. By flexibly adjusting the receptive field, the model comprehensively considers temporal features within the power data. Tailoring the memory length of output nodes based on varying input time scales effectively addresses the issue of historical data neglect observed in traditional methods. This adaptability proves advantageous, particularly in the context of short-term photovoltaic forecasting.

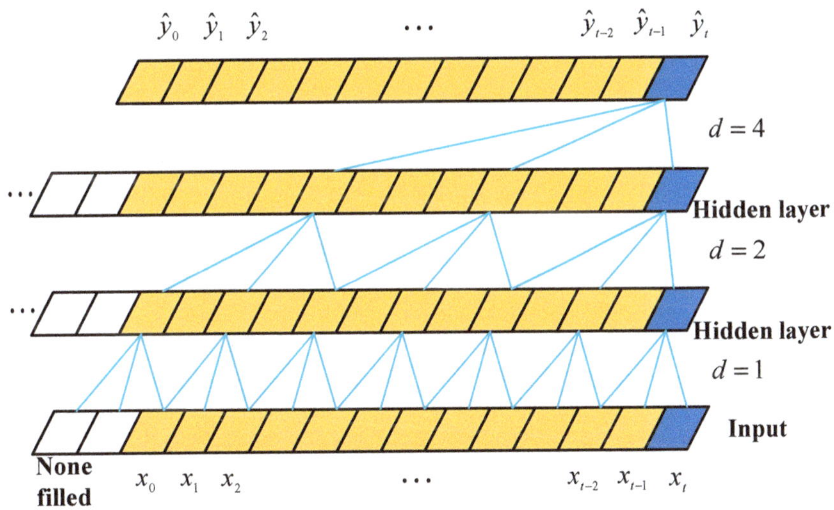

Figure 1. Structure diagram of dilated causal convolution network.

The residual connection has been demonstrated as an effective approach in training deep neural networks, enabling the network to propagate information across layers. The construction of a residual block to replace a single convolutional layer is depicted in the following Figure 2. A residual block comprises two convolutional layers and non-linear mapping, with WeightNorm and Dropout incorporated at each layer for network regularization.

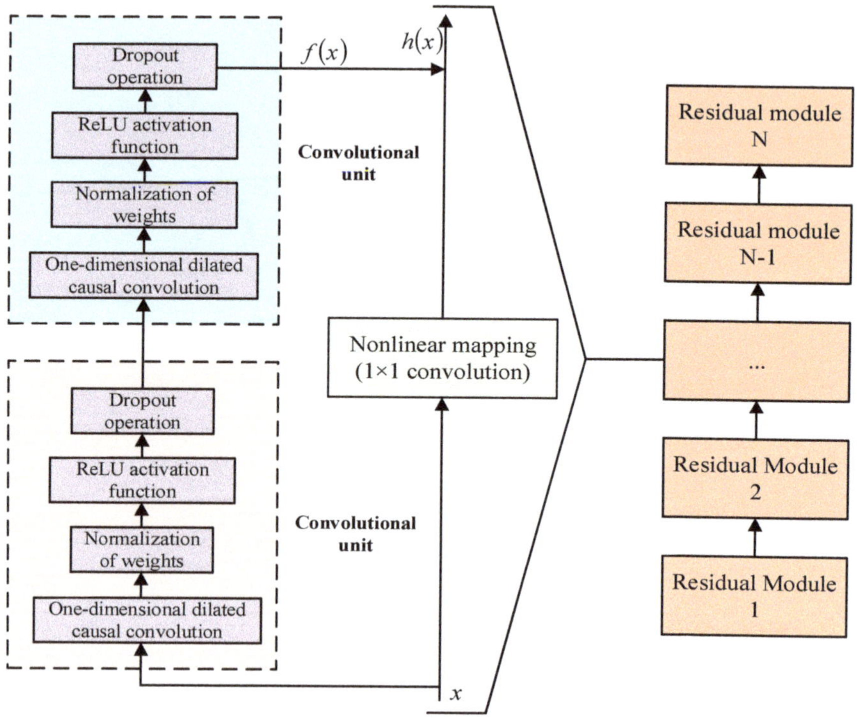

Figure 2. TCN residual unit structure diagram.

2.2. TCN-GRU

The traditional GRU network architecture represents a specialized form of recurrent neural network (RNN) models, offering an optimized alternative based on the LSTM network structure. While maintaining comparable network accuracy, the GRU achieves simplification of unit complexity by adjusting the structures of the input gate, forget gate, and output gate within the LSTM computational framework. Notably, the integration of the LSTM forget gate and input gate into a unified update gate contributes to a reduction in the model's learning and training duration, enhancing overall computational efficiency. In the GRU unit structure, the reset gate regulates the proportion of historical moment memory values entering the output gate, while the update gate determines the quantity of retained historical moment memory information, thereby governing the state update of the hidden layer. The unit structure is illustrated in Figure 3. This optimization in unit architecture exemplifies the GRU's ability to streamline computational complexity while preserving the essential memory dynamics, showcasing its potential for expedited learning and improved operational efficiency compared to conventional LSTM models.

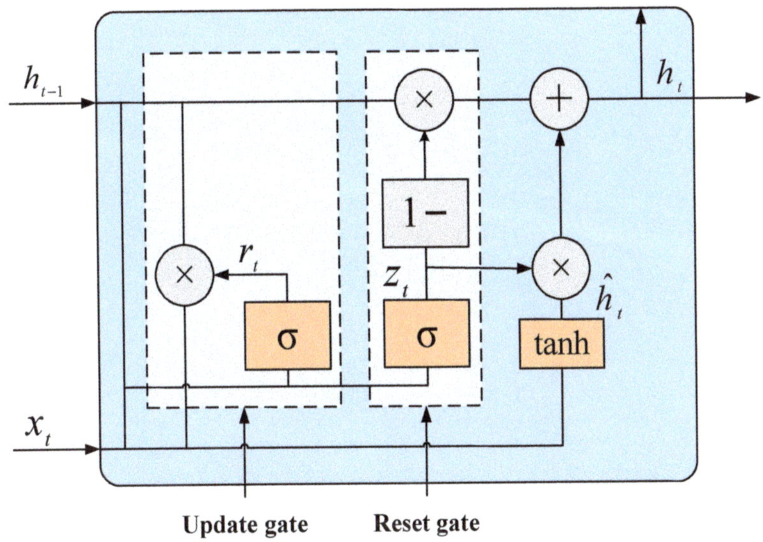

Figure 3. Structure diagram of GRU.

The expressions for each gate function are as follows:

$$r_t = \sigma(U_r h_{t-1} + W_r x_t + b_r), \tag{2}$$

$$z_t = \sigma(U_z h_{t-1} + W_z x_t + b_z), \tag{3}$$

$$\hat{h}_t = \tanh[W_h x_t + U_h(r_t \odot h_{t-1}) + b_h], \tag{4}$$

$$h_t = (1 - z_t) \odot h_{t-1} + z_t \odot \hat{h}_{t-1} \tag{5}$$

where h_{t-1} represents the model's output at the previous time step; x_t denotes the input at time t; $\sigma(\cdot)$ signifies the activation function, typically modeled as the sigmoid function; r_t and z_t denote the reset gate and update gate, respectively; h_t signifies the state output of the model at time t; W and U represent weight matrices; tanh denotes the hyperbolic tangent function; $\odot$ represents the Hadamard product between two matrices; and b signifies the bias terms for each input [38].

In the temporal convolutional network, the application of DCC allows TCN to possess a larger receptive field with fewer layers, enabling it to process longer historical data. The DCC utilizes an activation function and undergoes weight normalization and regularization operations. Meanwhile, the RC ensures stability in TCN by employing skip connections from the input to the output, especially in deeper TCN architectures.

The gated recurrent unit, distinct from traditional RNNs, introduces changes to the hidden layer architecture, incorporating memory cells, an update gate, and related gates. GRU determines when to update memory cells with candidate values through the update gate. Compared to one-dimensional convolutional neural networks (CNNs), TCN, due to its use of DCC and RC, can process longer historical data with increased stability. Therefore, TCN is selected for high-dimensional feature extraction from input data. For time series prediction, GRU demonstrates performance almost equivalent to LSTM but with faster training speeds, justifying its selection for sequence prediction tasks. In summary, this paper proposes a PV power prediction framework, VMD-TCN-GRU, based on VMD and incorporating TCN and GRU components. The workflow involves initial data preprocessing steps such as data cleaning and standardization, followed by the application of VMD decomposition to the processed data. Subsequently, the individual VMD modes are input into TCN residual blocks for high-dimensional feature extraction. Finally, the output from the TCN residual blocks is fed into the GRU network for prediction, yielding the final forecast results.

2.3. Multi Head Attention Mechanism

While the GRU demonstrates excellent performance in sequence prediction tasks, it is not immune to the issue of error accumulation. Photovoltaic power data, being continuous over time, are subject to considerable uncertainty due to external environmental factors and unforeseen events. The impact of abrupt data changes amplifies errors over multiple time steps during training, leading to suboptimal prediction outcomes. Attention mechanisms, capable of adaptively capturing global dependencies within the data, offer the advantage of focusing not only on the information at the current position in the sequence but also on information at other positions. However, attention mechanisms necessitate the computation of weight relationships between each sequence, leading to significant computational resource requirements, especially when dealing with long sequences.

To address these challenges, this paper introduces a novel structure called a multi-head attention gated recurrent unit (MAGRU), combining the strengths of GRU and multi-head attention mechanisms. The MAGRU structure is presented in Figure 4. In this approach, a sliding window is introduced at the position of the GRU output hidden state h_t, aggregating the information from the preceding m time steps into a new sequence. Here, h denotes the dimensionality of the hidden state h_t. Subsequently, multiple sets of learnable weight matrices W_i^Q, W_i^K, and $W_i^V \in R^{h \times d}$ are introduced for each head, serving as Query, Key, and Value matrices, respectively, where i represents the group index, and d is the dimensionality of the attention mechanism. The calculations for the Query, Key, and Value matrices are formulated as follows [39]:

$$Q_i = HW_q, K_i = HW_k, V_i = HW_v \tag{6}$$

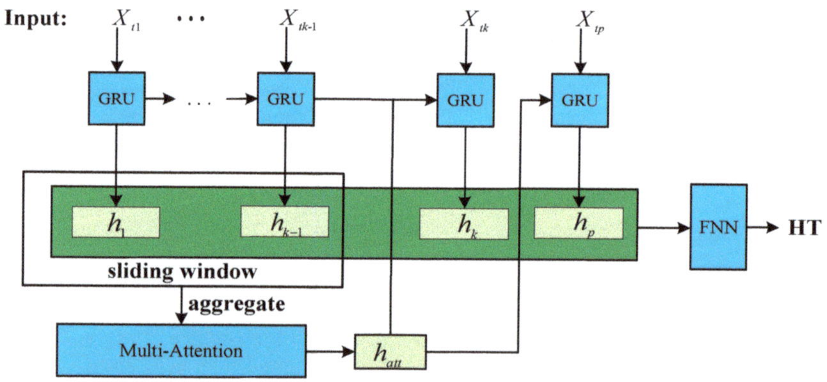

Figure 4. Diagram of multi-head attention GRU.

For each hidden state at time step t, use the Query, Key, and Value matrices to calculate its attention weight. The formula is as follows:

$$Attention_i(Q_i, K_i, V_i) = \text{softmax}(\frac{Q_i K_i^T}{\sqrt{d_a}})V_i \tag{7}$$

where $\sqrt{d_a}$ is used to scale the size of the inner product and avoid the input of the softmax function being too large or too small [40]. Concatenate the output vectors of multiple attention heads to obtain matrix Z:

$$Z = concat(Attention_1, \ldots, Attention_i). \tag{8}$$

Perform a linear transformation on matrix Z to obtain attention score output h_{att} as follows:

$$h_{att} = Z \cdot W \in R^{m \times h} \tag{9}$$

where W is a learnable weight matrix. h_{att} will participate in training as the hidden state of the input for the next time step GRU [41].

Finally, the obtained output is passed on to a feedforward neural network FNN to adjust its dimension and obtain the temporal feature $HT \in R^{Q \times N \times D}$ of PV power data.

2.4. Forecasting Modeling Based on Improved TCN-GRU

Since the actual photovoltaic power system is characterized by instability and intermittent fluctuations, to guarantee the precision and feasibility for photovoltaic power forecasting, an improved TCN-GRU network forecasting framework is proposed, as shown in Figure 5. Firstly, the photovoltaic power is decomposed using VMD, and the penalty factor and decomposition mode number of VMD are determined according to the minimum envelope entropy principle. Subsequently, daily environmental data and photovoltaic power data are utilized as TCN-GRU network inputs. In addition, the TCN-GRU is established for photovoltaic power forecasting under different modal components, and the network parameters are optimized using the SSA algorithm. Finally, to obtain final photovoltaic power forecasting, the forecasting results of the SSA-TCN-GRU model under different modal components are reconstructed.

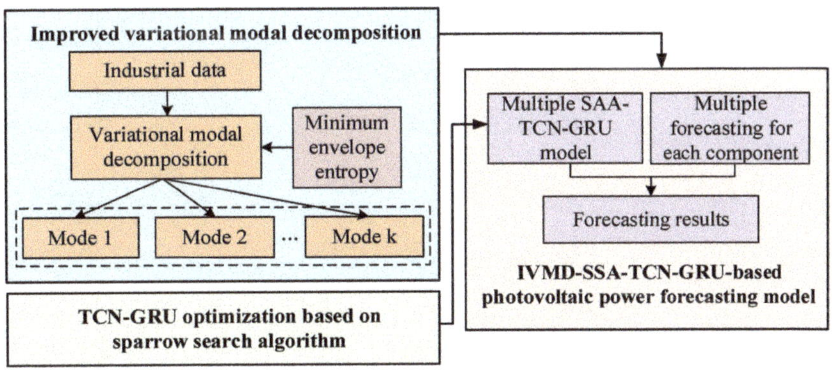

Figure 5. The framework of improved TCN-GRU network forecasting.

2.5. Improved Variational Modal Decomposition

2.5.1. Preliminary of VMD

The original photovoltaic power sequence is decomposed to obtain the separation of stationary series and non-stationary series, reduce the randomness and non-stationary of photovoltaic power, and decrease the interference in the forecasting process. The VMD represents a variational approach rooted in the amalgamation of frequency mixing, Hilbert transform, and classical Wiener filter methodologies. Distinguished by its non-recursive nature and adaptive signal processing capabilities, VMD emerges as a robust method for addressing signal decomposition challenges. Thus, original photovoltaic power signal $f(t)$ is decomposed through the VMD algorithm into k discrete photovoltaic power mode $u_k(t)$, that is, the signal is decomposed into a limited number of mode components with different IMFs. Compared to empirical mode decomposition, the endpoint effects and modal aliasing problems are overcome using the VMD method.

The specific construction steps are present as follows:

1. For each mode function, through Hilbert transform, the analytic signal $u_k(t)$ is obtained to acquire its unilateral spectrum $[\delta(t) + \frac{j}{\pi t}] * u_k(t)$;

2. The frequency spectrum for each mode $u_k(t)$ is modulated to the corresponding base band $[(\delta(t)+\frac{j}{\pi t})*u_k(t)]e^{-j\omega_k t}$ by mixing the exponential terms $e^{-j\omega_k t}$ of its corresponding center frequency ω_k;
3. Through the Gaussian smoothness of the demodulation signal, for each mode signal, the bandwidth is estimated, and the constrained variational problem is obtained.

The extended Lagrange expression is as follows [42]:

$$\begin{cases} \min\limits_{\{u_k\},\{\omega_k\}} \left\{ \sum_{k=1}^{k} \|\partial_t\left[\left(\delta(t)+\frac{1}{\pi t}\right)*u_k(t)\right]e^{-j\omega_k t}\|_2^2 \right\} \\ s.t.\sum\limits_k u_k = f \end{cases}. \tag{10}$$

By introducing Lagrange multiplication operator λ and quadratic penalty factor α, an augmented Lagrange expression is present whereby the constrained variation is converted to the unconstrained one, as follows:

$$\begin{aligned} L(\{u_k\},\{\omega_k\},\lambda) &= \alpha\sum_k \|\partial_t\left[\left(\delta(t)+\frac{j}{\pi t}\right)*u_k(t)\right]e^{-j\omega_k t}\|_2^2 \\ &+\|f(t)-\sum_k u_u(t)\|_2^2+\left\langle \lambda(t), f(t)-\sum_k u_k(t)\right\rangle \end{aligned}. \tag{11}$$

The solution of each mode function is as follows:

$$\hat{u}_k^{n+1}(\omega) = \frac{\hat{f}(\omega)-\sum\limits_{i\neq k}\hat{u}_i(\omega)+\frac{\hat{\lambda}(\omega)}{2}}{1+2\alpha(\omega-\omega_k)^2}. \tag{12}$$

The solution of the center frequency for each mode is as follows:

$$\omega_k^{n+1} = \frac{\int_0^\infty \omega|\hat{u}_k(\omega)|^2 d\omega}{\int_0^\infty |\hat{u}_k(\omega)|^2 d\omega}. \tag{13}$$

λ is updated as follows:

$$\hat{\lambda}_k^{n+1}(\omega) \leftarrow \hat{\lambda}^n(\omega)+\tau\left(\hat{f}(\omega)-\sum_k \hat{u}_k^{n+1}(\omega)\right). \tag{14}$$

Determine whether the termination condition is met.

$$\sum_k \|\hat{u}_k^{n+1}-\hat{u}_k^n\|_2^2 \Big/ \|\hat{u}_k^n\|_2^2 < \varepsilon. \tag{15}$$

The specific flow of the VMD algorithm is shown in Algorithm 1. Compared to EMD, VMD has a strict mathematical model and often has better robustness in dealing with noise. VMD not only effectively decomposes various harmonics but also does not consider the relative amplitude between harmonics and the distance between their respective center frequencies during mode separation. Moreover, the VMD method has high decomposition accuracy, fewer decomposition layers and no mode aliasing [43]. For photovoltaic power, it can be used as an effective means to accurately decompose the various frequency components, which is conducive to photovoltaic power forecasting under unstable and intermittent fluctuation conditions.

Algorithm 1 The process of the VMD algorithm.

Complete Optimization of VMD
Initialize $\{\hat{u}_k^1\}, \{\omega_k^1\}, \{\hat{\lambda}_k^1\}, n \leftarrow 0$
Repeat
$n \leftarrow n+1$
for k = 1: K do
Update $\hat{u}_k$ for all $\omega \geq 0$:

$$\hat{u}_k^{n+1}(\omega) \leftarrow \frac{\hat{f}(\omega) - \sum_{i<k} \hat{u}_i^{n+1}(\omega) - \sum_{i>k} \hat{u}_i^{n}(\omega) + \frac{\hat{\lambda}^n(\omega)}{2}}{1+2\alpha(\omega-\omega_k^n)^2}$$

Update ω_k:

$$\omega_k^{n+1} = \frac{\int_0^\infty \omega |\hat{u}_k^{n+1}(\omega)|^2 d\omega}{\int_0^\infty |\hat{u}_k^{n+1}(\omega)|^2 d\omega}$$

end for
Dual ascent for all $\omega \geq 0$:

$$\hat{\lambda}_k^{n+1}(\omega) \leftarrow \hat{\lambda}^n(\omega) + \tau\left(\hat{f}(\omega) - \sum_k \hat{u}_k^{n+1}(\omega)\right)$$

Until convergence: $\sum_k \|\hat{u}_k^{n+1} - \hat{u}_k^n\|_2^2 / \|\hat{u}_k^n\|_2^2 < \varepsilon$

2.5.2. VMD with Minimum Envelope Entropy

In fact, decomposition mode number k has a certain degree of impact on the decomposition effect of VMD. Moreover, in multiple IMFs, mode overlap occurs when the same component appears if the k is big. On the contrary, in the same IMF, there are multiple components. So far, the empirical value has been used by most mode classification studies. To solve this problem, envelope entropy is introduced to search for the optimal mode number k and equilibrium factor, and an improved VMD is proposed. The detailed steps for IVMD are presented:

Step 1: Input the signal $x(i)$ $(i = 1, 2, \ldots, N)$ into IVMD model.

Step 2: Initialize parameters; let k = 2. In addition, IVMD is performed.

Step 3: Calculate the envelope entropy, which is represented as Equation (16), and determine whether it meets the condition of minimum envelope entropy under the condition of meeting the error limit.

$$\begin{cases} E_p = -\sum_{i=1}^{N} p_i \log p_i \\ p_i = a(i) / \sum_{i=1}^{N} a(i) \end{cases} \tag{16}$$

where the entropy value of p_i is the envelope entropy E_p, and N is the number of sampling points. Decomposed using VMD after Hilbert demodulation, $a(i)$ is the envelope signal of k mode components; p_i is the probability distribution sequence [44].

Step 4: Stop decomposition and obtain k when the condition is reached. Otherwise, $k = k + 1$, until the condition is satisfied.

2.6. GRU Optimization Using SSA

The forecasting performance according to the GRU network is greatly affected by its parameters. To fully demonstrate the effectiveness of GRU, a GRU model is established for each mode component, respectively. In addition, the number of hidden layer neurons is optimized through the SSA, training times, and the learning rate for GRU network model parameters. The SSA originates from the anti-capture and foraging behavior.

(1) The position $X_{i,j}^{t+1}$ of the founder is updated as follows:

$$X_{i,j}^{t+1} = \begin{cases} X_{i,j}^t \exp(-i/\alpha \cdot iter_{\max}) & , R_2 < \mathrm{ST} \\ X_{i,j}^t + Q \cdot L & , R_2 \geq \mathrm{ST} \end{cases} \tag{17}$$

where $iter_{\max}$ represent the maximum number of iterations and t denotes the current iteration number. The position of the ith sparrow in dimension j is represented by $X_{i,j}^t$; the range of j is $\{1, 2, \cdots, d\}$. Additionally, $ST \in [0.5, 1]$ and $R_2 \in [0, 1]$ refer to the security threshold and alarm value, respectively. A random number $\alpha \in (0, 1]$ is utilized in the algorithm, and $L(1 \times d)$ is a complete matrix with all elements equal to 1. A normally distributed random number, denoted as Q, signifies the absence of predators, prompting the finder to engage in wide-area search mode when the condition $R_2 < ST$ is met. Conversely, if there is impending danger $R_2 > ST$, all sparrows swiftly relocate to alternative safe locations [45].

(2) The position of the joiner sparrow is expressed as follows:

$$X_{i,j}^{t+1} = \begin{cases} Q \cdot \exp\left((X_{\text{worst}} - X_{i,j}^t)/i^2\right), i > n/2 \\ X_p^{t+1} + \left|X_{i,j}^t - X_p^{t+1}\right| \cdot A^+ \cdot L, \text{others} \end{cases} \tag{18}$$

where X_{worst} denotes the global worst position, X_p represents the optimal position of the current discoverer, and a 1xd matrix A is defined with elements randomly assigned 1 or -1: $A^+ = A^{\mathrm{T}}\left(AA^{\mathrm{T}}\right)^{-1}$. The condition where $i > n/2$ indicates that the participant with poor fitness is at a heightened risk of starvation [46].

(3) Assuming that 10%–20% of the sparrow population perceives danger and promptly relocates to a safe area, the guard position $X_{i,j}^{t+1}$ is determined as follows:

$$X_{i,j}^{t+1} = \begin{cases} X_{\text{best}}^t + \beta \cdot \left|X_{i,j}^t - X_{\text{best}}^t\right|, f_i > f_g \\ X_{i,j}^t + K \cdot \left(\frac{\left|X_{i,j}^t - X_{\text{worst}}^t\right|}{(f_i - f_w) + \varepsilon}\right), f_i = f_g \end{cases} \tag{19}$$

where β follows a normal distribution with mean 0 and variance 1, representing a random number for the step control parameter. Other variables include X_{best} as the global optimal position; $K \in [-1, 1]$ as the step control parameter, denoting the direction of sparrow movement; f_g as the global optimal fitness value; f_i as the fitness value of individual sparrows at the current step; and f_w as the global worst fitness value. To prevent division by zero, ε is introduced as a minimum constant. In addition, sparrows located at the edges of the population face increased vulnerability to predators when $f_i > f_g$. Conversely, sparrows positioned in the middle of the population effectively communicate the awareness of danger to other sparrows, thereby reducing the risk of predation under the condition $f_i = f_g$ [47].

In the context of optimizing the parameters of a GRU using the SSA, a systematic procedure is outlined as follows:

Step 1: Initialization of parameters

The initial steps involve setting up the number of iterations, determining the ratio of predators within the population, and initializing the population size.

Step 2: Fitness evaluation and sorting

Subsequently, the fitness value of each individual sparrow is computed, and the population is sorted in descending order based on their fitness values.

Step 3: Update of discoverer's location

The position of the discoverer, representing the sparrow with the optimal fitness value, is then updated according to the SSA algorithm.

Step 4: Update of joiner's location

Similarly, the location of the joiner, denoting a sparrow seeking to improve its fitness by joining the discoverer, is adjusted based on the algorithm's principles.

Step 5: Update of vigilant position

The vigilant position, indicating a sparrow aware of potential danger, is updated in accordance with the algorithm's specifications.

Step 6: Fitness calculation and position update

The fitness value of the sparrows is recalculated, and their positions are updated iteratively to enhance the optimization process.

Step 7: If the stop conditions is met, the command output is displayed. Otherwise, repeat Step 2 to Step 6.

2.7. IVMD-SSA-TCN-GRU-Based Photovoltaic Power Forecasting Strategy

Since the intermittent fluctuation and instability characteristics for photovoltaic power, the photovoltaic power time series are firstly decomposed into different modes, and then a TCN-GRU model optimized using the SSA is established for each mode. Furthermore, the forecasting results for each mode are integrated to achieve power forecasting. Figure 6 presents the flowchart of the IVMD-SSA-TCN-GRU photovoltaic power forecast, and the specific forecasting steps are demonstrated as follows.

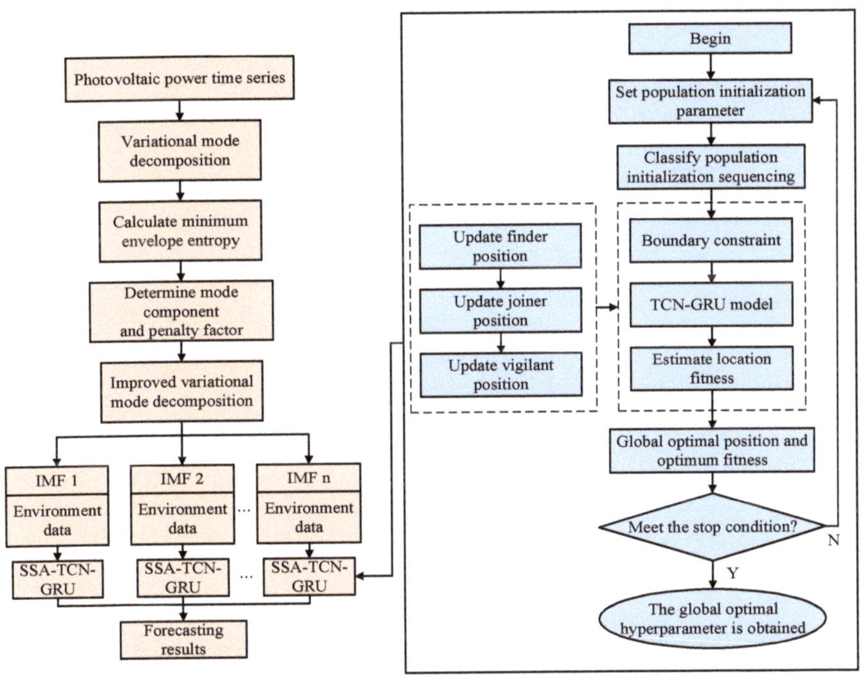

Figure 6. The flowchart of IVMD-SSA-TCN-GRU photovoltaic power forecasting.

Step 1: Select the environment information as the model input.

Step 2: Use the IVMD method to decompose the original photovoltaic power sequence and obtain the k components.

Step 3: Firstly, set the parameter range (number of learning rate η, training times E, and hidden layer neurons H), search range of sparrow population size N, and maximum number of iterations M. Moreover, set the mean square error as an objective function for the optimization algorithm. Furthermore, set up the coupling model of SSA-TCN-GRU.

Step 4: Establish SSA-TCN-GRU forecasting models for each component and obtain k forecasting models.

Step 5: Add the corresponding forecasted values of k forecasting models to obtain the forecasting result of photovoltaic power.

3. Results

To illustrate the feasibility of the photovoltaic power forecasting strategy proposed for a photovoltaic power station in China, the historical photovoltaic power data are used for case study analysis. Photovoltaic datasets with different weather conditions (sunny, cloudy, and rainy) were selected with 864 samples, of which the first 605 were used for training and the last 259 were used for forecasting. The sampling interval was 5min, and the installed capacity of the photovoltaic field was 603 MW.

3.1. Data Processing

Since PV power and meteorological data contain outliers and missing values, the data error interference affects the accuracy of the models. Therefore, data reconciliation was used to clean the data, eliminate outliers, and fill in missing values [48,49].

Due to the difference in dimensionality of data for different variables, the data were standardized to ensure normal calculation.

$$\widetilde{x}(i) = \frac{x(i) - x_{\min}}{x_{\max} - x_{\min}} \quad (20)$$

where $x(i)$ is the sample of the original photovoltaic sequence or meteorological sequence; $\widetilde{x}(i)$ represents the normalized processed sequences, which are in [0, 1]; and for the sample data, $x_{\max}$ and $x_{\min}$ are the maximum and minimum value of the sample data, respectively.

3.2. Evaluation Index of Forecasting Model

As forecasting evaluation indexes, MAE and RMSE are utilized to quantitatively analyze the forecasting performance and generalization ability of the model.

$$\text{RMSE} = \sqrt{\frac{1}{N}\sum_{i=1}^{N}(y_i - \hat{y}_i)^2} \quad (21)$$

$$\text{MAE} = \frac{1}{N}\sum_{i=1}^{N}|y_i - \hat{y}_i| \quad (22)$$

where N is the testing sample; y_i is the true data of photovoltaic power; $\hat{y}_i$ is the forecasting result of photovoltaic power; and for the actual photovoltaic power, $\overline{y}$ is the average value.

3.3. Simulation Analysis

In addition, air temperature, irradiance, air pressure, and module temperature as well as historical photovoltaic power are determined as the IVMD-TCN-GRU forecasting model inputs. For predetermined parameters K and α, to solve the problem in the traditional VMD algorithm, an adaptive IVMD algorithm is proposed based on envelope entropy. Moreover, the non-stationary and nonlinear characteristics are decomposed for photovoltaic power. Figure 7 illustrates the iterative process of envelope entropy values under different weather conditions, and the optimal parameter combination and corresponding minimum envelope entropy obtained from this are shown in Table 1. The decomposition results of photovoltaic power corresponding to the optimal K value under various weather conditions are shown in Figure 8. The IMF indicates that sub modes are obtained after photovoltaic power decomposition. The different modes after VMD decomposition not only have stronger stationarity, but also maintain the trend characteristics of original photovoltaic data well. Considering the chaotic nature of photovoltaic weather processes and the nonlinear relationship between photovoltaic weather and output photovoltaic power, TCN-GRU photovoltaic power forecasting models are established for each mode component decomposed using IVMD. In addition, to acquire the forecasted photovoltaic power, the forecasted output of each model is added. The five hyperparameters of the model are determined, that is, the number of hidden layers is 1, the dimension of the input layer is

24, the time step of the input layer is 10, the dimension of the output variable is 1, and the dimension of each hidden layer is 10. Finally, Figure 9 presents the experimental results of photovoltaic power prediction based on IVMD-TCN-GRU.

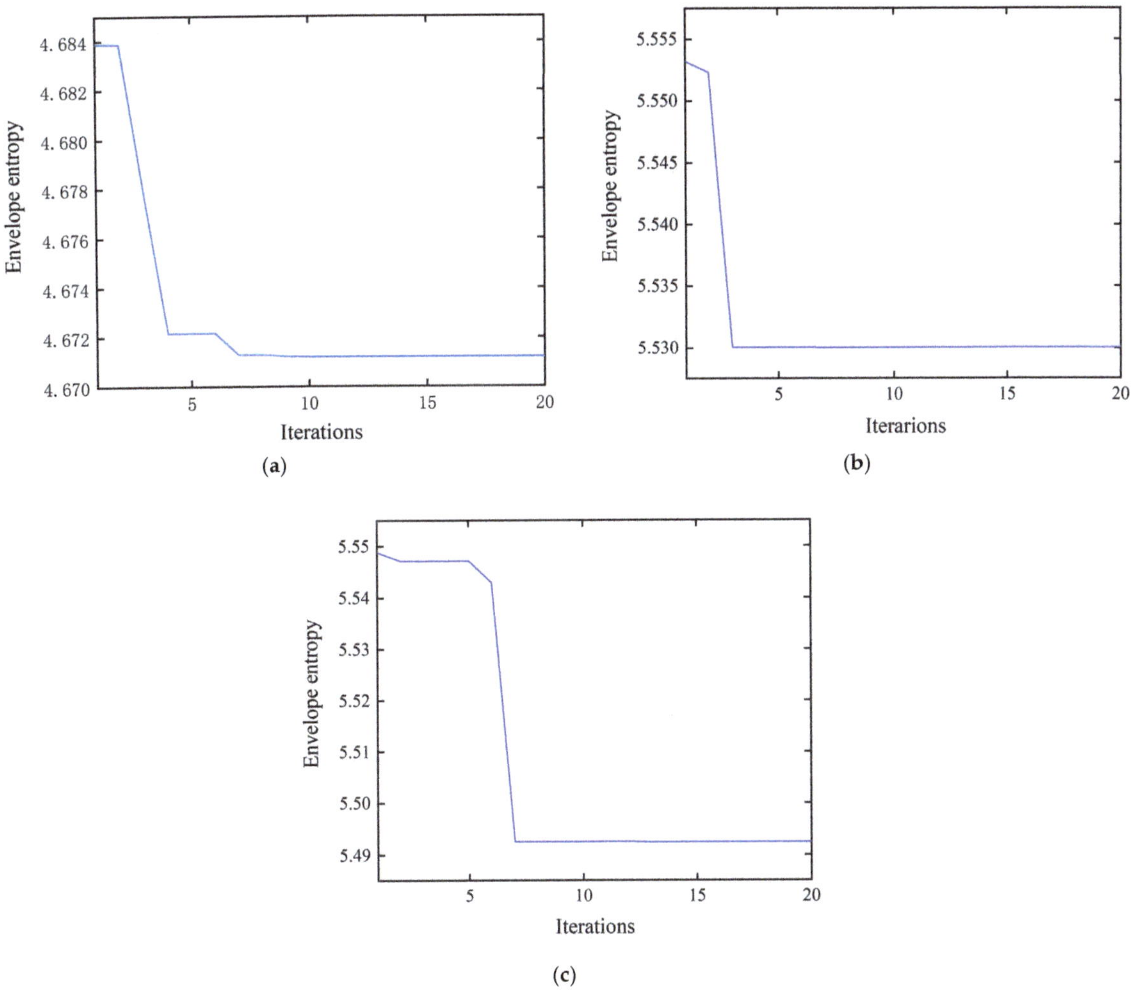

Figure 7. Envelope entropy iteration process. (**a**) Envelope entropy on sunny days. (**b**) Envelope entropy on cloudy days. (**c**) Envelope entropy on rainy days.

Table 1. The optimal parameters of IVMD.

Weather Types	Minimum Envelope Entropy	K	α
Sunny day	4.6712	3	925
Cloudy day	5.5301	6	59
Rainy day	5.4925	7	93

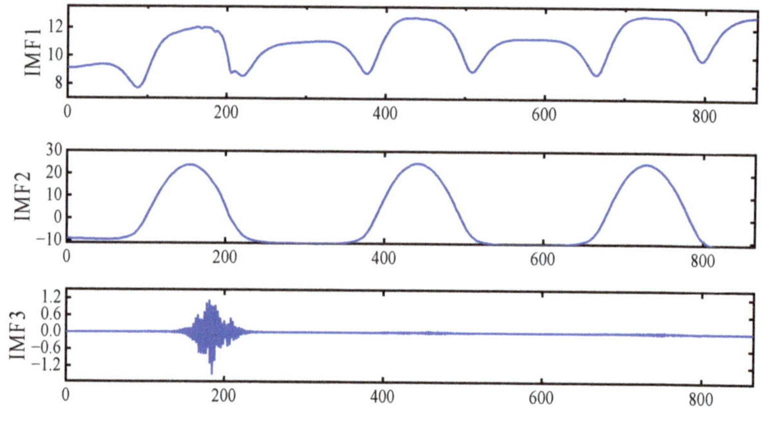

(a)

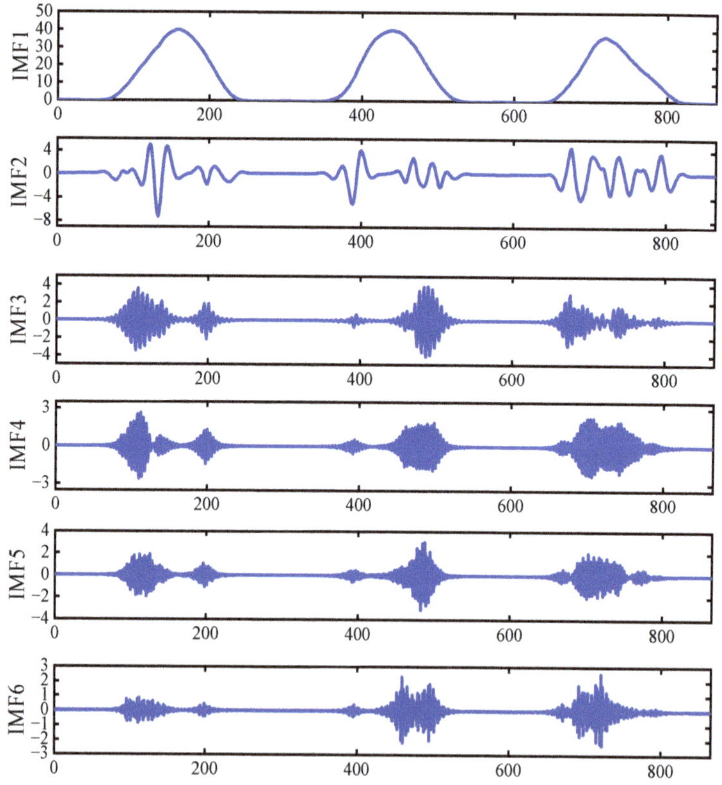

(b)

Figure 8. *Cont.*

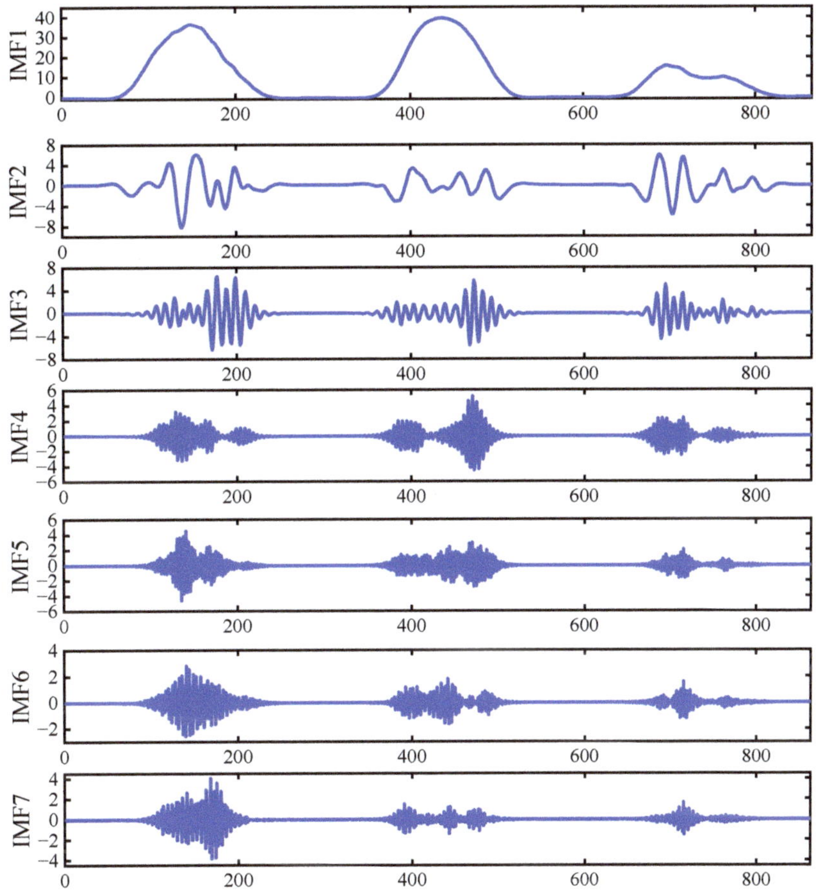

(c)

Figure 8. Photovoltaic power in IVMD decomposition. (**a**) Photovoltaic power IVMD decomposition on sunny days. (**b**) Photovoltaic power IVMD decomposition on cloudy days. (**c**) Photovoltaic power IVMD decomposition on rainy days.

Taking photovoltaic power in sunny days as an example, the comparative experimental results of TCN-GRU, IVMD-TCN-GRU, and IVMD-SSA-TCN-GRU are shown in Figure 10. In addition, in Table 2, RMSE, MAE, time consumption are evaluated to quantitatively compare the performance for each model. It can be observed that, in comparison with TCN-GRU, RMSE and MAE of the IVMD-SSA-TCN-GRU forecasting strategy are reduced by 34.1% and 36.3%, respectively. This is because non-stationary and nonlinear characteristics of photovoltaic power are weakened by photovoltaic power decomposition. Compared with the IVMD-TCN-GRU model, RMSE and MAE decrease by 16.7% and 5.2%, respectively. Because the optimal parameter combination is matched using the SSA for the TCN-GRU network, it can better preserve the original information when processing high-dimensional data, reduce the original data sequence complexity, alleviate the time delay characteristics and the fluctuation range. It can achieve higher accuracy than other single methods. In general, the IVMD-SSA-TCN-GRU forecasting model shows stronger forecasting ability and higher forecasting accuracy.

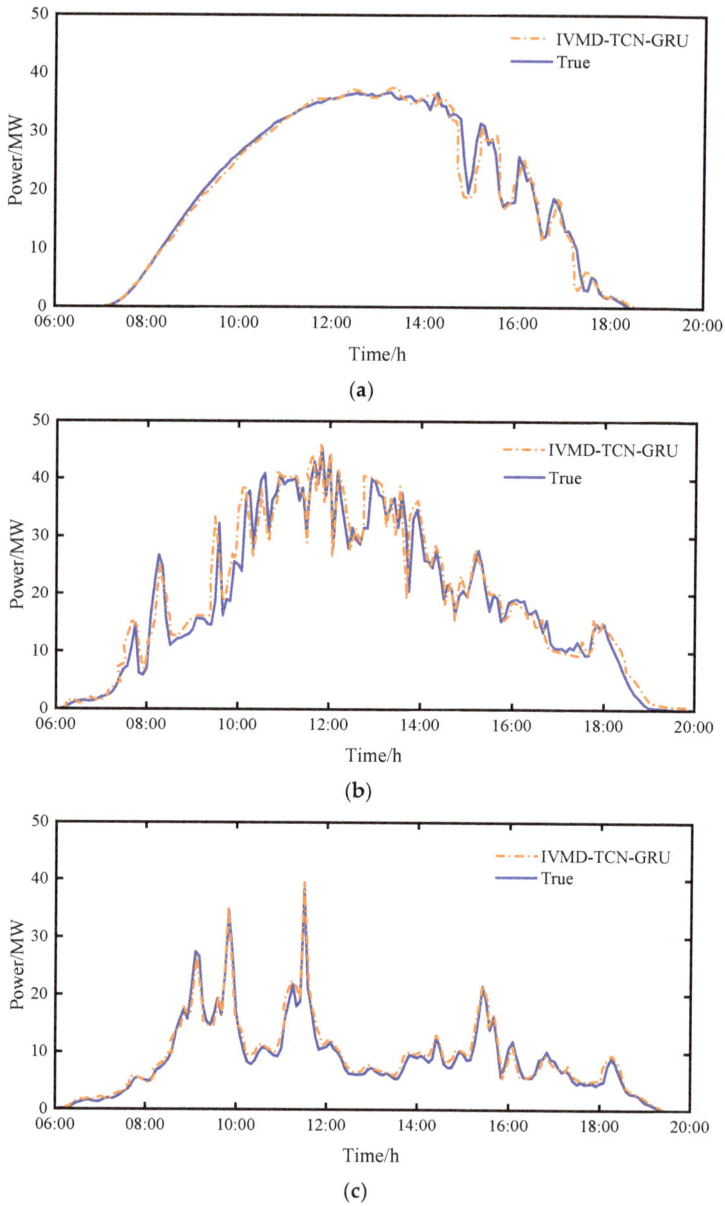

Figure 9. IVMD-TCN-GRU forecasting results. (**a**) Comparison diagram on sunny days. (**b**) Comparison diagram on cloudy days. (**c**) Comparison diagram on rainy days.

Table 2. Forecasting errors of the IVMD-SSA-TCN-GRU method.

Evaluation Indexes	TCN-GRU	IVMD-TCN-GRU	IVMD-SSA-TCN-GRU
RMSE	1.7479	1.3832	1.152
MAE	1.4819	0.9968	0.94461
Time	0.0141	0.0132	0.0047

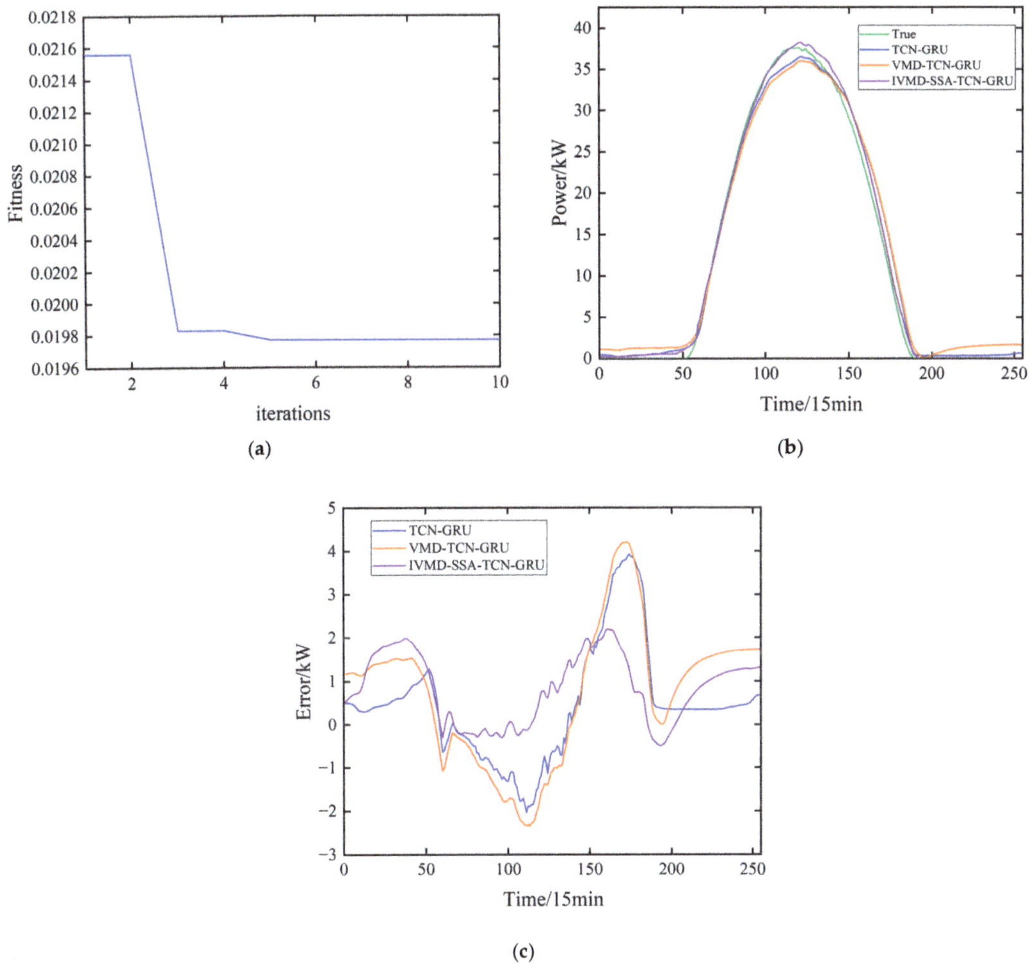

Figure 10. Comparison of photovoltaic power forecasting on sunny days. (**a**) The iteration of SSA-TCN-GRU on sunny days. (**b**) Photovoltaic power forecasting on sunny days. (**c**) The forecasting error on sunny days.

Moreover, the EMD-SSA-TCN-GRU and IVMD-SSA-Elman models are used as comparisons to verify the superiority under the conditions of cloudy and rainy days with large fluctuations in photovoltaic power. Figure 11 shows the forecasting results with cloudy conditions; under rainy conditions, the forecasting results are illustrated in Figure 12. The performance indexes of the different approaches are exhibited in Table 3. The proposed forecasting model has certain improvement in both RMSE and MAE, verifying the good effect of preserving environmental characteristics on photovoltaic weather type classification and model establishment. Meanwhile, VMD is applied to decompose photovoltaic power, and a forecasting model is set up for each mode before reconstruction, reducing amount of data and shortening the forecasting time. The proposed model has stronger forecasting ability and higher forecasting accuracy compared with other models. Moreover, compared with the EMD-SSA-TCN-GRU forecasting model, for the IVMD-SSA-TCN-GRU forecasting model, RMSE and MAE are reduced by 37.1% and 27.8%. The reason is that IVMD has better decomposition performance and is more suitable for decomposing and forecasting photovoltaic power. Compared with IVMD-SSA-Elman, RMSE and MAE declined by 55.1% and 54.5%, respectively; due to issues when dealing with time series problems, the TCN-GRU network has better performance. Table 4 presents the performance metrics of the novel method proposed in this study alongside the approaches WOA-BiLSTM-Attention [50], LSTM-TCN [51], and CNN-GRU [52] in scenarios characterized by rainy conditions and substantial fluctuations in photovoltaic power generation. Our findings reveal that the proposed method outperforms the existing techniques in terms of predictive accuracy and dependability, as evidenced by the lower MAE and RMSE values obtained by our model.

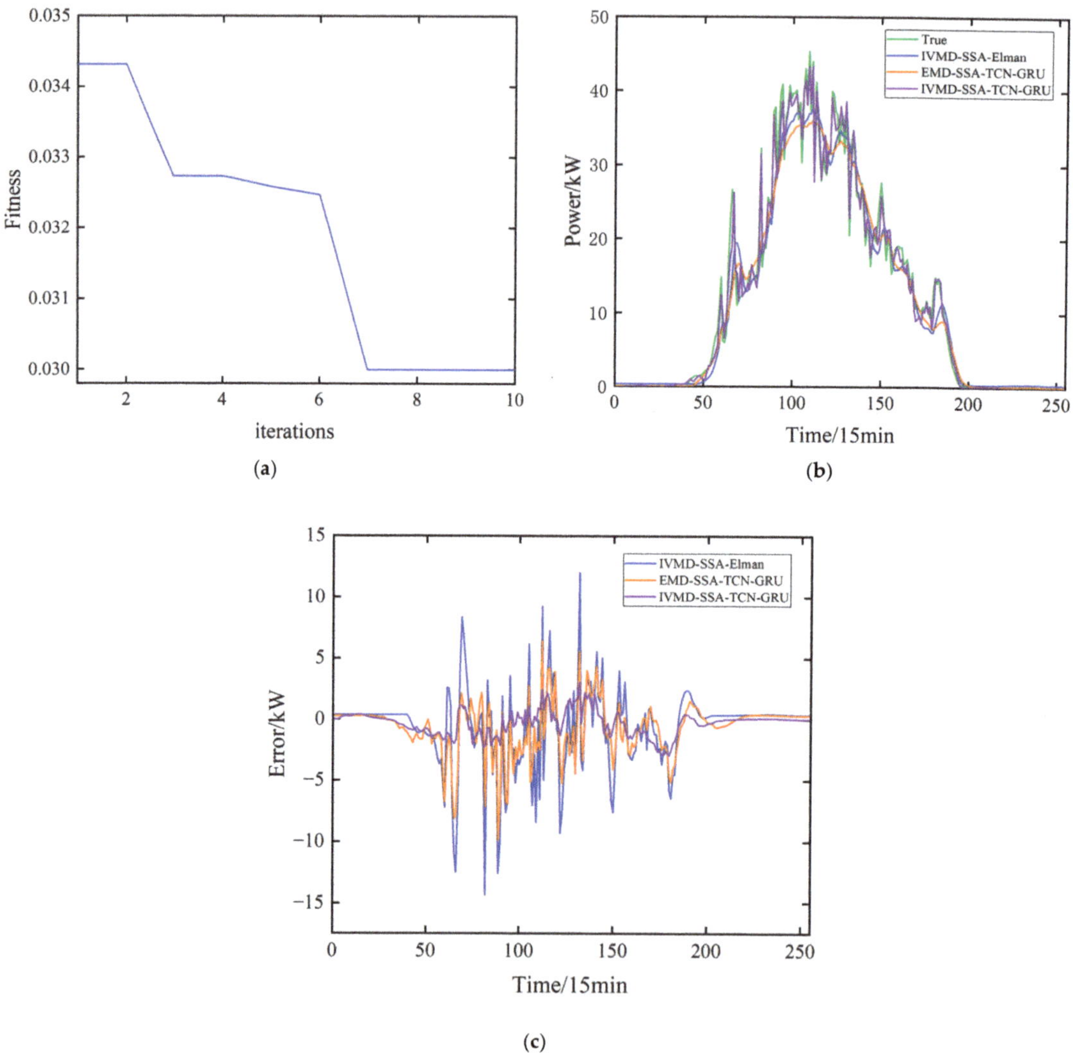

Figure 11. Comparison of photovoltaic power forecasting in cloudy days. (**a**) The iteration of SSA-TCN-GRU in cloudy days. (**b**) Photovoltaic power forecasting on cloudy days. (**c**) The forecasting error on cloudy days.

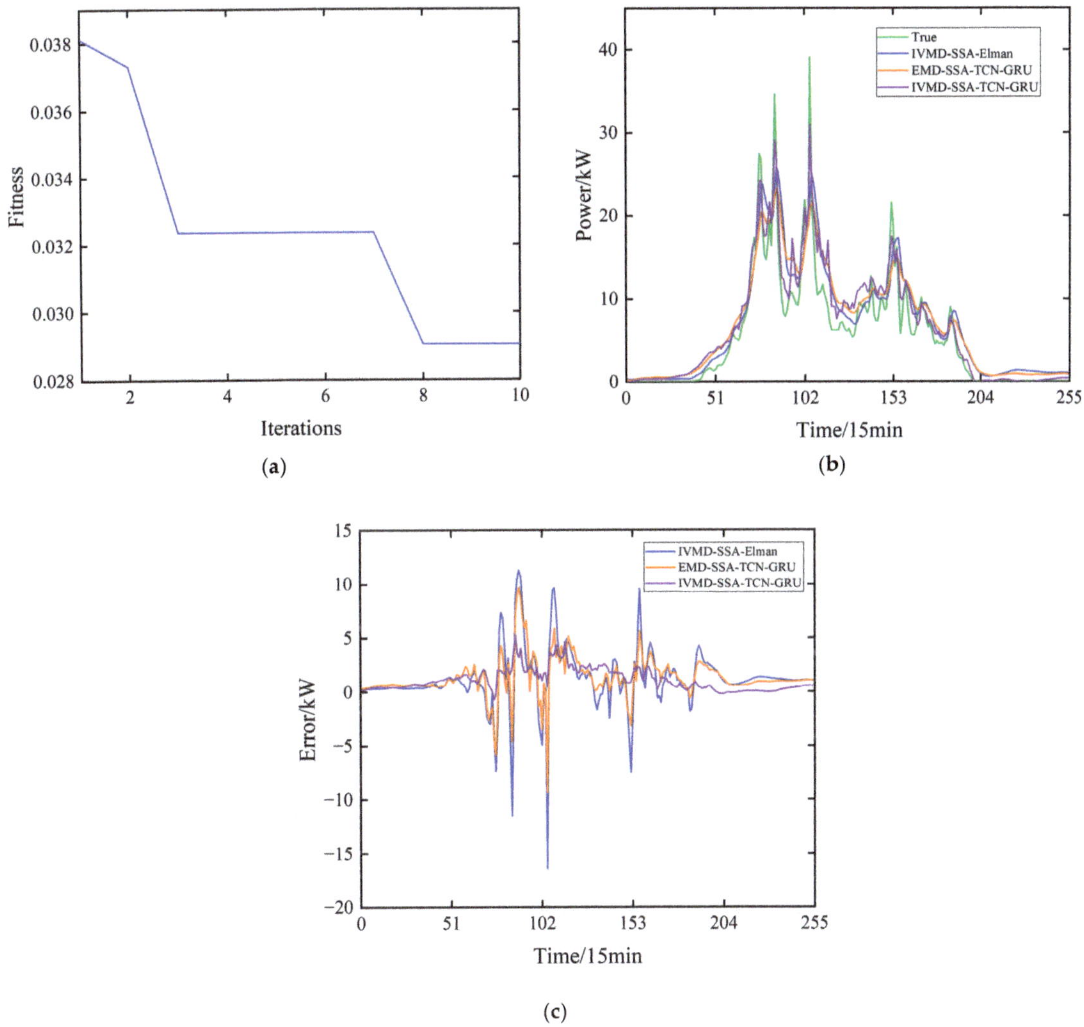

Figure 12. Comparison of photovoltaic power forecasting on rainy days. (**a**) The iteration of SSA-TCN-GRU on rainy days. (**b**) Photovoltaic power forecasting on rainy days. (**c**) The forecasting error in rainy days.

Table 3. The forecasting errors of different models.

Weather Type	Model	MAE	RMSE
Sunny day	IVMD-SSA-Elman	2.32	4.17
	EMD-SSA-TCN-GRU	2.17	3.76
	IVMD-SSA-TCN-GRU	1.98	3.41
Cloudy day	IVMD-SSA-Elman	3.07	4.84
	EMD-SSA-TCN-GRU	2.74	4.82
	IVMD-SSA-TCN-GRU	2.71	4.66
Rainy day	IVMD-SSA-Elman	5.01	8.15
	EMD-SSA-TCN-GRU	3.16	5.82
	IVMD-SSA-TCN-GRU	2.28	3.66

Table 4. The forecasting errors of different models on rainy days.

Method	MAE	RMSE
IVMD-SSA-TCN-GRU	2.28	3.66
WOA-BiLSTM-Attention	2.45	3.73
LSTM-TCN	2.56	3.91
CNN-GRU	2.71	4.14

4. Discussion

The present study introduces a novel hierarchical approach to photovoltaic power forecasting that integrates IVMD into the TCN-GRU framework, further enhanced by a multi-head attention mechanism. This integration aims to tackle the inherent complexities and variabilities in PV power generation, which are significantly influenced by environmental factors. Our findings underscore the effectiveness of combining advanced signal processing techniques with deep learning models to improve the accuracy of PV power forecasts, which are crucial for the efficient management and integration of solar energy into the power grid.

The use of IVMD, optimized using the SSA, for the decomposition of PV power data marks a significant advancement in the preprocessing stage of forecasting [53–55]. This methodological choice allows for a refined extraction of the intrinsic modes within the power generation data, facilitating a more detailed and accurate analysis of the power output fluctuations. The optimization of the modal components and penalty factors through the SSA not only enhances the decomposition process but also tailors it specifically to the characteristics of the PV power data, thereby maximizing the relevance and efficiency of the subsequent forecasting model.

The core of our forecasting model combines the strengths of TCN and GRU, augmented with a multi-head attention mechanism [56–58]. This design leverages the TCN's capability to extract local feature patterns within time series data and the GRU's proficiency in capturing long-term dependencies, addressing two critical aspects of time series forecasting. The addition of a multi-head attention mechanism further elevates the model's performance by enabling a dynamic focus on the most relevant features across the time series, thereby improving the accuracy and reliability of the forecasts. This integration not only harnesses the individual strengths of these components but also mitigates their limitations, illustrating the synergistic potential of hybrid modeling approaches in complex forecasting tasks.

Incorporating environmental factors into the model represents a holistic approach to forecasting, acknowledging the significant impact of external variables on PV power output. This inclusion ensures that the model captures not only the internal dynamics of the time series data but also the influence of external conditions, providing a comprehensive framework for forecasting. The empirical validation of the model using real-world data from a PV station demonstrates its superior performance compared to traditional forecasting methods, highlighting its practical significance and potential impact on the energy sector.

However, the sophistication and computational demands of our proposed model pose challenges for real-time application and scalability. Future research could focus on optimizing the model's efficiency and exploring the feasibility of real-time forecasting, potentially broadening its applicability and utility in operational settings. Moreover, investigating the model's performance across diverse geographical locations and under varying environmental conditions would be invaluable, further affirming its robustness and adaptability. Firstly, integrating multiple components, such as the IVMD, TCN, and GRU, requires careful design and optimization to ensure the seamless functioning of the overall architecture. Coordinating the interactions between these components and fine-tuning hyperparameters can be a non-trivial task that demands computational resources and expertise. Moreover, validating the performance of the proposed model involves addressing issues related to data quality, model interpretability, and generalizability. Ensuring the robustness of the model across different datasets, geographic locations, and weather conditions requires rigorous testing and validation procedures. Additionally, interpreting the results of the forecasting model and identifying the factors influencing its predictions can be challenging, especially when dealing with complex neural network architectures and attention mechanisms. Furthermore, the scalability and computational efficiency of the proposed model may present practical challenges, particularly when deploying the forecasting system in real-world applications with stringent latency and resource constraints. Optimizing the algorithm for efficient inference and deployment on various platforms while maintaining high forecasting accuracy is a critical consideration in operationalizing the proposed approach.

In conclusion, this study contributes a significant advancement to the field of PV power forecasting by proposing a comprehensive and integrative model that adeptly addresses the complexities of solar power generation. The innovative combination of IVMD, TCN-GRU, and a multi-head attention mechanism not only showcases the potential of hybrid models in enhancing forecast accuracy but also sets a foundation for future research aimed at optimizing and expanding the applicability of advanced forecasting techniques in the renewable energy sector.

5. Conclusions

In this study, we proposed a novel hierarchical forecasting model for PV power based on a multi-head attention mechanism integrated with VMD, TCN, and GRU. Through extensive experimentation and validation using real-world PV power data, we have drawn several important conclusions regarding the effectiveness and applicability of our proposed model.

(1) Our results demonstrate that the integration of VMD, TCN, GRU, and a multi-head attention mechanism significantly improves the accuracy and reliability of PV power forecasting compared to traditional methods. By leveraging VMD for signal decomposition and TCN-GRU for dynamic time series modeling, our model effectively captures both local temporal features and long-term dependencies in the data, leading to more precise predictions.

(2) The incorporation of a multi-head attention mechanism enables our model to exploit global contextual information in the time series data, further enhancing its forecasting performance. The attention mechanism allows the model to dynamically weigh the importance of different input features, thereby improving the utilization of relevant information for prediction.

(3) The optimization of VMD parameters using the SSA and the fine-tuning of GRU parameters contribute to the overall effectiveness of our proposed model. The optimization process ensures that the model is able to adapt to the specific characteristics of the input data, thereby improving its generalization capability and robustness.

Overall, our study highlights the importance of incorporating advanced machine learning techniques and considering environmental factors in PV power forecasting. The proposed hierarchical VMD-TCN-GRU multi-head attention mechanism offers a promising solution for accurately predicting PV power output, which is essential for optimizing the operation and management of solar energy systems. This research contributes to the advancement of PV power forecasting methodologies and provides valuable insights for researchers and practitioners in the field of renewable energy forecasting. The proposed model holds significant potential for facilitating the integration of solar energy into the power grid and supporting the transition towards a sustainable energy future.

Author Contributions: Conceptualization, H.F. and J.Z.; methodology, J.Z.; software, J.Z.; validation, J.Z. and H.F.; formal analysis, J.Z.; investigation, J.Z. and S.X.; resources, J.Z.; data curation, J.Z.; writing—original draft preparation, J.Z.; writing—review and editing, J.Z.; visualization, J.Z.; supervision, H.F.; project administration, H.F.; funding acquisition, H.F. All authors have read and agreed to the published version of the manuscript.

Funding: This research was funded by the National Natural Science Foundation of China, grant number 51974151.

Data Availability Statement: The data are unavailable due to privacy.

Acknowledgments: We are grateful for financial and logistical support from the H.F. Model Worker Innovation Laboratory and thank all of the original partners that supported data collection and analyses for the initial work on the Photovoltaic Power Prediction Project.

Conflicts of Interest: The authors declare no conflicts of interest.

References

1. Pillot, B.; Muselli, M.; Poggi, P.; Dias, J.B. Historical trends in global energy policy and renewable power system issues in Sub-Saharan Africa: The case of solar PV. *Energy Policy* **2019**, *127*, 113–124. [CrossRef]
2. Ospina, J.; Newaz, A.; Faruque, M.O. Forecasting of PV plant output using hybrid wavelet-based LSTM-DNN structure model. *IET Renew. Power Gen.* **2019**, *13*, 1087–1095. [CrossRef]
3. Sun, W.; Wang, Y. Short-term wind speed forecasting based on fast ensemble empirical mode decomposition, phase space reconstruction, sample entropy and improved back-propagation neural network. *Energy Convers. Manag.* **2018**, *157*, 1–12. [CrossRef]
4. Ssekulima, E.B.; Anwar, M.B.; Hinai, A.A.; Moursi, M.S.E. Wind speed and solar irradiance forecasting techniques for enhanced renewable energy integration with the grid: A review. *IET Renew. Power Gen.* **2016**, *10*, 885–898. [CrossRef]
5. Zhang, Q.; Ma, W.H.; Li, G.L.; Xie, M.; Shao, Q.Z. Partition fault diagnosis of power grids based on improved PNN and GRA. *IEEE J. Trans. Electr. Electr.* **2020**, *16*, 57–66. [CrossRef]
6. Liu, J.; Liu, M.W.; Wang, Z.M.; Yang, J.W.; Lou, S.H. Multi-flexibility resources planning for power system considering carbon trading. *Sustainability* **2022**, *14*, 13296. [CrossRef]
7. Chen, X.; Lou, S.H.; Liang, Y.J.; Wu, Y.W.; He, X.L. Optimal scheduling of a regional power system aiming at accommodating clean energy. *Sustainability* **2021**, *13*, 2169. [CrossRef]
8. Sen Biswas, R.; Pal, A.; Werho, T.; Vittal, V. A graph theoretic approach to power system vulnerability identification. *IEEE Trans. Power Syst.* **2021**, *36*, 923–935. [CrossRef]
9. Wu, T.; Wang, X.C.; Qiao, S.J.; Xian, X.P.; Liu, Y.B.; Zhang, L. Small perturbations are enough: Adversarial attacks on time series prediction. *Inform. Sci.* **2022**, *587*, 794–812. [CrossRef]
10. Huang, B.Q.; Zheng, H.A.; Guo, X.B.; Yang, Y.; Liu, X.M. A novel model based on DA-RNN network and skip gated recurrent neural network for periodic time series forecasting. *Sustainability* **2022**, *14*, 326. [CrossRef]
11. Salles, R.; Pacitti, E.; Bezerra, E.; Porto, F.; Ogasawara, E. TSPred: A framework for nonstationary time series prediction. *Neurocomputing* **2021**, *467*, 197–202. [CrossRef]
12. Guo, K.; Hu, Y.L.; Qian, Z.; Liu, H.; Zhang, K.; Sun, Y.F.; Gao, J.B.; Yin, B.C. Optimized graph convolution recurrent neural network for traffic prediction. *IEEE Trans. Intell. Transp.* **2021**, *22*, 1138–1149. [CrossRef]
13. Lou, Y.; Wu, R.Z.; Li, J.L.; Wang, L.; Li, X.; Chen, G.R. A learning convolutional neural network approach for network robustness prediction. *IEEE Trans. Cybern.* **2022**, *53*, 4531–4544. [CrossRef] [PubMed]
14. Topic, J.; Skugor, B.; Deur, J. Neural network-based prediction of vehicle fuel consumption based on driving cycle data. *Sustainability* **2022**, *14*, 744. [CrossRef]
15. Zhang, X.S.; He, B.A.; Sabri, M.M.S.; Al-Bahrani, M.; Ulrikh, D.V. Soil liquefaction prediction based on bayesian optimization and support vector machines. *Sustainability* **2022**, *14*, 11944. [CrossRef]
16. Huang, J.L.; Jin, T.; Liang, M.L.; Chen, H.L. Prediction of heat exchanger performance in cryogenic oscillating flow conditions by support vector machine. *Appl. Therm. Eng.* **2020**, *182*, 116053. [CrossRef]
17. Zhou, L.S.; Zhou, X.T.; Liang, H.; Huang, M.T.; Li, Y. Hybrid short-term wind power prediction based on Markov chain. *Front. Energy Res.* **2022**, *10*, 899692. [CrossRef]
18. Mao, C.Y.; Bao, L.W.; Yang, S.D.; Xu, W.J.; Wang, Q. Analysis and prediction of pedestrians' violation behavior at the intersection based on a Markov chain. *Sustainability* **2021**, *13*, 5690. [CrossRef]
19. Huang, Y.; Yu, J.H.; Dai, X.H.; Huang, Z.; Li, Y.Y. Air-quality prediction based on the EMD-IPSO-LSTM combination model. *Sustainability* **2022**, *14*, 4889. [CrossRef]
20. Wang, X.Q.; Xu, N.K.; Meng, X.R.; Chang, H.Q. Prediction of gas concentration based on LSTM-LightGBM variable weight combination model. *Energies* **2022**, *15*, 827. [CrossRef]
21. Li, P.D.; Gao, X.Q.; Li, Z.C.; Zhou, X.Y. Effect of the temperature difference between land and lake on photovoltaic power generation. *Renew. Energy* **2022**, *185*, 86–95. [CrossRef]
22. Nelega, R.; Greu, D.I.; Jecan, E.; Rednic, V.; Zamfirescu, C.; Puschita, E.; Turcu, R.V.F. Prediction of power generation of a photovoltaic power plant based on neural networks. *IEEE Access* **2023**, *11*, 20713–20724. [CrossRef]
23. Zhang, H.C.; Zhu, T.T. Stacking model for photovoltaic-power-generation prediction. *Sustainability* **2022**, *14*, 5669. [CrossRef]
24. Li, Y.L.; Yan, L.C.; He, H.; Zha, W.T. Regional ultra-short-term wind power combination prediction method based on fluctuant/smooth components division. *Front. Energy Res.* **2022**, *10*, 840519. [CrossRef]

25. Hui, L.; Ren, Z.Y.; Yan, X.; Li, W.Y.; Bo, H. A multi-data driven hybrid learning method for weekly photovoltaic power scenario forecast. *IEEE Trans. Sustain. Energy* **2021**, *13*, 91–100.
26. Zha, Y.X.; Lin, J.; Li, G.J.; Wang, Y.; Yi, Z. Analysis of inertia characteristics of photovoltaic power generation system based on generalized droop control. *IEEE Access* **2021**, *9*, 37834–37839. [CrossRef]
27. Qian, Z.; Pei, Y.; Zareipour, H.; Chen, N. A review and discussion of decomposition-based hybrid models for wind energy forecasting applications. *Appl. Energy* **2019**, *235*, 939–953. [CrossRef]
28. Lee, J.; Kim, H.; Kim, H. Commercial vacancy prediction using LSTM neural networks. *Sustainability* **2021**, *13*, 5400. [CrossRef]
29. Xiao, Z.X.; Tang, F.; Wang, M.Y. Wind power short-term forecasting method based on LSTM and multiple error correction. *Sustainability* **2023**, *15*, 3798. [CrossRef]
30. Xiang, X.; Li, X.; Zhang, Y.; Hu, J. A short-term forecasting method for photovoltaic power generation based on the tcn-ecanet-gru hybrid model. *Sci. Rep.* **2024**, *14*, 6744. [CrossRef]
31. Moradzadeh, A.; Zakeri, S.; Shoaran, M.; Mohammadi-Ivatloo, B.; Mohammadi, F. Short-term load forecasting of microgrid via hybrid support vector regression and long short-term memory algorithms. *Sustainability* **2020**, *12*, 7076. [CrossRef]
32. Sun, Z.; Zhao, S.; Zhang, J. Short-term wind power forecasting on multiple scales using VMD decomposition, K-means clustering and LSTM principal computing. *IEEE Access* **2019**, *18*, 17–29. [CrossRef]
33. Lang, X.; Rehman, N.U.; Zhang, Y.F.; Xie, L.; Su, H.Y. Median ensemble empirical mode decomposition. *Signal Process.* **2020**, *176*, 107686. [CrossRef]
34. Zhu, W.; Yang, Y.; Zhi, P.; Liang, Z. A control strategy of photovoltaic hybrid energy storage system based on adaptive wavelet packet decomposition. *Int. J. Electrochem. Sci.* **2022**, *17*, 221144. [CrossRef]
35. Khan, F.; Alshahrani, T.; Fareed, I.; Kim, J.H. A comprehensive degradation assessment of silicon photovoltaic modules installed on a concrete base under hot and low-humidity environments: Building applications. *Sustain. Energy Technol. Assess.* **2022**, *52*, 102314. [CrossRef]
36. Khan, F.; Rezgui, B.D.; Kim, J.H. Reliability study of c-si pv module mounted on a concrete slab by thermal cycling using electroluminescence scanning: Application in future solar roadways. *Materials* **2020**, *13*, 470. [CrossRef]
37. Perera, M.; De Hoog, J.; Bandara, K.; Senanayake, D.; Halgamuge, S. Day-ahead regional solar power forecasting with hierarchical temporal convolutional neural networks using historical power generation and weather data. *Appl. Energy* **2024**, *361*, 122971. [CrossRef]
38. Mahjoub, S.; Chrifi-Alaoui, L.; Marhic, B.; Delahoche, L. Predicting Energy Consumption Using LSTM, Multi-Layer GRU and Drop-GRU Neural Networks. *Sensors* **2022**, *22*, 4062. [CrossRef] [PubMed]
39. Yu, X.; Zhang, D.; Zhu, T.; Jiang, X. Novel hybrid multi-head self-attention and multifractal algorithm for non-stationary time series prediction. *Inf. Sci.* **2022**, *613*, 541–555. [CrossRef]
40. He, J.; Zhang, X.; Zhang, X.; Shen, J. Remaining useful life prediction for bearing based on automatic feature combination extraction and residual multi-head attention gru network. *Meas. Sci. Technol.* **2023**, *35*, 036003. [CrossRef]
41. Zhang, Y.-M.; Wang, H. Multi-head attention-based probabilistic cnn-bilstm for day-ahead wind speed forecasting. *Energy* **2023**, *278*, 127865. [CrossRef]
42. Zhang, Y.G.; Pan, G.F.; Chen, B.; Han, J.Y.; Zhao, Y.; Zhang, C.H. Short-term wind speed prediction model based on GA-ANN improved by VMD. *Renew. Energy* **2020**, *156*, 1373–1388. [CrossRef]
43. Wang, X.; Ma, W. A hybrid deep learning model with an optimal strategy based on improved vmd and transformer for short-term photovoltaic power forecasting. *Energy* **2024**, *295*, 131071. [CrossRef]
44. Yang, Y.; Liu, H.; Han, L.; Gao, P. A feature extraction method using vmd and improved envelope spectrum entropy for rolling bearing fault diagnosis. *IEEE Sens. J.* **2023**, *23*, 3848–3858. [CrossRef]
45. Xue, J.; Shen, B. A novel swarm intelligence optimization approach: Sparrow search algorithm. *Syst. Sci. Control Eng.* **2020**, *8*, 22–34. [CrossRef]
46. Gharehchopogh, F.S.; Namazi, M.; Ebrahimi, L.; Abdollahzadeh, B. Advances in sparrow search algorithm: A comprehensive survey. *Arch. Comput. Methods Eng.* **2023**, *30*, 427–455. [CrossRef] [PubMed]
47. Yue, Y.G.; Cao, L.; Lu, D.W.; Hu, Z.Y.; Xu, M.H.; Wang, S.X.; Li, B.; Ding, H.H. Review and empirical analysis of sparrow search algorithm. *Artif. Intell. Rev.* **2023**, *56*, 10867–10919. [CrossRef]
48. Xie, S.; Wang, H.Z.; Peng, J.C.; Liu, X.L.; Yuan, X.F. A hierarchical data reconciliation based on multiple time-delay interval estimation for industrial processes. *ISA Trans.* **2020**, *105*, 198–209. [CrossRef]
49. Xie, S.; Yang, C.H.; Yuan, X.F.; Wang, X.L.; Xie, Y.F. A novel robust data reconciliation method for industrial processes. *Contr. Eng. Pract.* **2019**, *83*, 203–212. [CrossRef]
50. Yu, M.; Niu, D.; Wang, K.; Du, R.; Yu, X.; Sun, L.; Wang, F. Short-term photovoltaic power point-interval forecasting based on double-layer decomposition and woa-bilstm-attention and considering weather classification. *Energy* **2023**, *275*, 127348. [CrossRef]
51. Limouni, T.; Yaagoubi, R.; Bouziane, K.; Guissi, K.; Baali, E.H. Accurate one step and multistep forecasting of very short-term pv power using lstm-tcn model. *Renew. Energy* **2023**, *205*, 1010–1024. [CrossRef]
52. Sabri, N.M.; El Hassouni, M. Accurate photovoltaic power prediction models based on deep convolutional neural networks and gated recurrent units. *Energy Sources Part A Recovery Util. Environ. Eff.* **2022**, *44*, 6303–6320. [CrossRef]

53. Cai, L.; Hu, D.; Zhang, C.; Yu, S.; Xie, J. Tool vibration feature extraction method based on ssa-vmd and svm. *Arab. J. Sci. Eng.* **2022**, *47*, 15429–15439. [CrossRef]
54. Zhou, S.; Yao, Y.; Luo, X.; Jiang, N.; Niu, S. Dynamic response evaluation for single-hole bench carbon dioxide blasting based on the novel ssa–vmd–pcc method. *Int. J. Geomech.* **2023**, *23*, 04022248. [CrossRef]
55. Gao, X.; Guo, W.; Mei, C.; Sha, J.; Guo, Y.; Sun, H. Short-term wind power forecasting based on ssa-vmd-lstm. *Energy Rep.* **2023**, *9*, 335–344. [CrossRef]
56. Cai, C.; Li, Y.; Su, Z.; Zhu, T.; He, Y. Short-term electrical load forecasting based on vmd and gru-tcn hybrid network. *Appl. Sci.* **2022**, *12*, 6647. [CrossRef]
57. Li, L.; Li, Y.; Mao, R.; Li, L.; Hua, W.; Zhang, J. Remaining useful life prediction for lithium-ion batteries with a hybrid model based on tcn-gru-dnn and dual attention mechanism. *IEEE Trans. Transp. Electrif.* **2023**, *9*, 4726–4740. [CrossRef]
58. Pu, X.; Xiao, H.; Wang, J.; Pei, W.; Yang, J.; Zhang, J. A novel gru-tcn network based interactive behavior learning of multi-energy microgrid under incomplete information. *Energy Rep.* **2023**, *9*, 608–616. [CrossRef]

energies

MDPI

Article

Exploring Motion Stability of a Novel Semi-Submersible Platform for Offshore Wind Turbines

Hongxu Zhao [1], Xiang Wu [1] and Zhou Zhou [1,2,*]

1 Department of Mechanical Engineering, Hunan Institute of Engineering, Xiangtan 411104, China
2 Hunan Province Engineering Laboratory of Wind Power Operation, Maintenance and Testing Technology, Xiangtan 411104, China
* Correspondence: zhouzhou@hnie.edu.cn

Abstract: The stability of offshore floating wind turbine foundation platforms is a fundamental requirement for the efficiency and safety of wind power generation systems. This paper proposes a novel small-diameter float-type semi-submersible platform to improve system stability. To evaluate the superior motion stability of the proposed floating platform, a comprehensive frequency–domain response analysis and experimental study were conducted in comparison with the OC4-DeepCwind platform developed by the National Renewable Energy Laboratory (NREL). The respective comparison of the frequency–domain response analysis and the experimental results demonstrated that the proposed floating wind turbine platform shows better hydrodynamic characteristics and resonance avoidance capability. This not only reduces the Response Amplitude Operators (RAOs), but also enhances the system stability, namely, effectively avoiding the regions of concentrated wave loading and low-frequency ranges. Furthermore, the proposed small-diameter semi-submersible platform has the potential to reduce manufacturing costs, providing valuable insights for the manufacturing of offshore floating wind turbine systems.

Keywords: offshore wind turbine; floating platform; seakeeping; hydrodynamic response

Citation: Zhao, H.; Wu, X.; Zhou, Z. Exploring Motion Stability of a Novel Semi-Submersible Platform for Offshore Wind Turbines. *Energies* **2024**, *17*, 2313. https://doi.org/10.3390/en17102313

Academic Editor: Frede Blaabjerg

Received: 14 March 2024
Revised: 1 May 2024
Accepted: 8 May 2024
Published: 10 May 2024

1. Introduction

In recent years, offshore floating wind energy technology has garnered increasing interest due to the richer and more stable wind resources available in deep-sea areas compared to coastal regions [1]. However, harnessing wind energy in deep-sea locations poses greater challenges compared to exploiting nearshore wind energy resources. Among the numerous challenges, one of the most critical issues is how to mitigate wave loads on offshore floating wind turbines (OFWTs) to enhance their motion stability [2].

Efforts have been made in recent years to improve the motion stability of OFWTs. For instance, improvements have been pursued through structural design modifications aimed at reducing platform motion and enhancing stability. Lemmer et al. [3] optimized design parameters such as column spacing, diameter, and height based on the OC4DeepCwind platform, demonstrating that the optimized structure effectively resists structural fatigue and ultimate loads. Zhang et al. [4] proposed a novel offshore wind turbine platform design and optimization method, introducing a tilted-column fully submerged OFWT platform with superior hydrodynamic performance and economic feasibility. Zhou et al. [5] improved the original OC4 model by integrating optimized multi-segment mooring lines and tilted columns, resulting in a new model capable of effectively reducing the heave and pitch motions of semi-submersible floating wind power platforms. Scicluna [6] developed a self-aligning single-point mooring buoy and conducted a stability analysis, revealing greater stability along the transverse axis compared to the pitch axis, meeting all requirements stipulated by DNV regulations. Rezanejad et al. [7] conducted experimental research on the hydrodynamic performance of a novel dual-chamber floating oscillating water column device, providing valuable insights for the design of such systems.

In addition, considerable scholarly efforts have been devoted to studying the hydrodynamic characteristics of these platforms. Chen et al. [8] developed a numerical model that fully couples air–water hydrodynamic analysis to evaluate OFWTs, offering precise forecasts of platform motions and dynamic loads critical for structural integrity assessments. Tabeshpour et al. [9] established a three-dimensional boundary element model of the Amirkabir platform using the boundary element method for hydrodynamic analysis, accurately obtaining RAOs. Yang et al. [10] formulated an aero-hydro-servo-elastic coupling framework for OFWT analysis, capturing the intricate interactions among aerodynamics, fluid dynamics, and structural components, and thus providing a nuanced representation of system dynamic behavior. Gao et al. [11–13], employing the OpenFOAM model, investigated the transient gap resonance between two solids induced by varying focusing wave amplitudes and spectral peak periods, yielding insights into mitigating vibration and motion response in such floating body systems. Zou et al. [14] introduced a novel short-term prediction method for OFWTs using the Kriging-MOGA algorithm, emphasizing the substantial influence of aerodynamic loads on floating wind turbine platform stability. Bashetty and Ozcelik [15] conducted a comprehensive review of OFWT dynamics, emphasizing the diverse challenges associated with operational aspects and underscoring the necessity for precise modeling and simulation techniques to ensure the safety and efficacy of these systems. Deng et al. [16] proposed a novel method employing neural networks for predicting semi-submersible platform motions, furnishing an accurate and efficient tool for platform dynamic evaluation. Rezanejad et al. [17], utilizing internally coupled tools to integrate far-field nonlinear waves, applied High-Order Spectral (HOS) methods to enhance the prediction of low-frequency motion responses of OFWTs.

Continuing from the aforementioned work, this paper presents a novel semi-submersible floating platform designed to support OFWTs, with an iterative optimization of its initial parameters to meet desired stability requirements. To facilitate this study, a hydrodynamic numerical model of the proposed semi-submersible floating wind turbine platform was first established, and its hydrodynamic performance, dynamic response, and wind–wave experimental performance were studied using ANSYS AQWA 2022 R2. Subsequently, scale models of both the proposed platform and the OC4-DeepCwind platform developed by the NREL were fabricated and tested in a multifunctional water tank. The comparison between these two platforms revealed that the proposed floating wind turbine platform has a superior natural period and is able to effectively avoid energy concentration areas and low-frequency ranges of waves, thereby reducing platform wave loads and actual motion response RAOs, and showing better hydrodynamic performance and stability. The specific computational theory will be introduced in Section 2. Section 3 will detail the main parameters of the platform model and mooring system. Section 4 will conduct the hydrodynamic analysis and the RAO comparisons for the two models, followed by Section 5, which will introduce the experimental equipment and validate the simulations through scaled model testing. Finally, a summary is given in Section 6.

2. Mathematical Theory

To provide a clear exposition of the research approach, Figure 1 displays a flowchart that depicts the study's content and outlined methodology. The diagram encompasses the investigation of numerical models, the theoretical framework utilized, and the data derived from hydrodynamic calculations. Finally, a comparative and validating analysis of the two models is conducted through experimental methods.

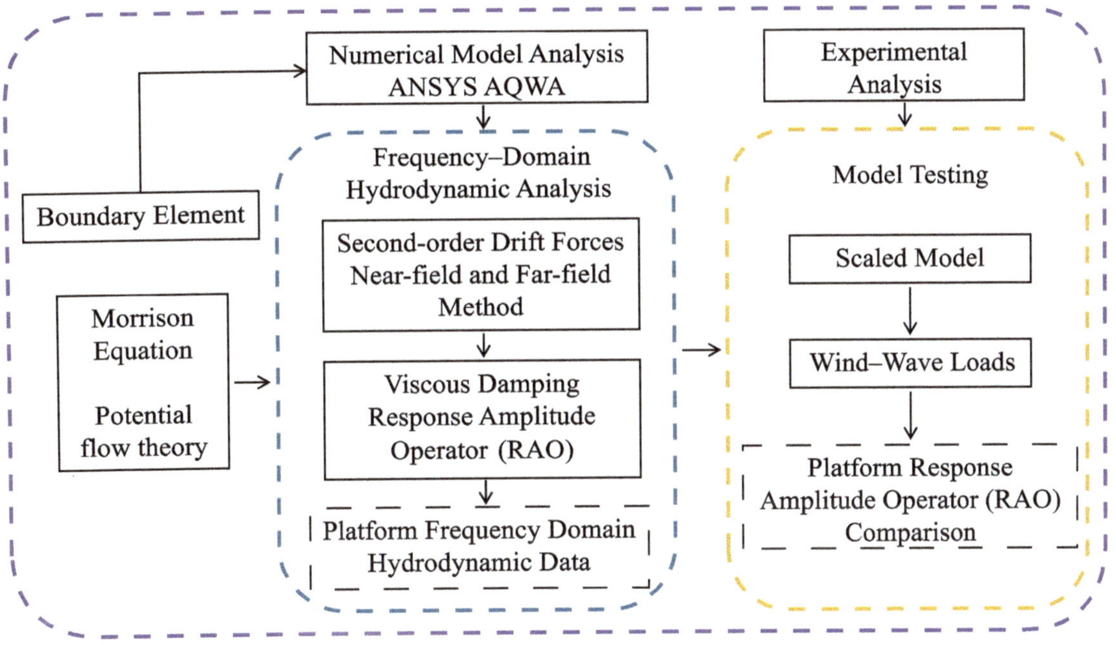

Figure 1. Flowchart of research approach and methodology.

2.1. Morrison Equation

For small-diameter components (where the ratio of component diameter D to wavelength L satisfies $L/D \leq 0.15$) [18], their presence has minimal impact on the surrounding flow field. When calculating wave forces, the Morrison equation method is employed. According to this method, wave forces can be divided into two parts: one part is the velocity force generated by the undisturbed velocity field, and the other part is the acceleration force generated by the wave acceleration field. For floating foundations, considering the motion of the foundation, the wave load on a unit height column can be expressed as follows:

$$\begin{aligned} F &= F_i + F_d \\ &= C_d \frac{\rho}{2} D u_r |u_r| \Delta Z + (1 + C_a) \rho \frac{\pi D^2}{4} \left(\frac{\partial u_r}{\partial_t} + \frac{\partial u_w}{\partial_t} \right) \Delta Z \end{aligned} \tag{1}$$

In the equation, F_i represents the inertial force acting on the unit height column; F_d denotes the velocity force exerted on the unit height column; D stands for the diameter of the column; ΔZ represents the unit height; ρ signifies the density of water; u_r denotes the relative velocity between the water particle and the floating foundation; u_W represents the velocity of the water particle; and C_u and C_d, respectively, represent the added mass coefficient and the drag coefficient. The value of C_d, C_a is typically set to 1.0.

2.2. Potential Flow Theory

For large-diameter components [18], the presence of structures will affect the surrounding flow field. In the theoretical calculation of wave loads, three-dimensional potential flow theory is commonly employed. This theory assumes that the water is an incompressible, inviscid, and irrotational ideal fluid, where waves are considered as potential motions, and linearized free surface boundary conditions are used for the solution. When the floating platform moves in still water, the first-order velocity potential $\varnothing(x, y, z, t)$ in the flow field satisfies the Laplace equation:

$$\nabla^2 \varnothing = 0 \tag{2}$$

Here, ∇^2 is the Laplace operator, representing the sum of the second-order spatial derivatives of the velocity potential. By solving this equation, the distribution of velocity potential in the flow field can be obtained, and thereby we can calculate the wave loads on the floating platform.

The first-order linear wave force is obtained through computation.

$$\begin{aligned}
&\vec{F} = \vec{F}_I + \vec{F}_D + \vec{F}_R \\
&\vec{F} + \vec{F}_D = -\int_S i\omega\rho(\varnothing_i + \varnothing_d)n_j ds \\
&\vec{F}_R = \ddot{x}_k \frac{\rho}{\omega} \int_S \varnothing_{ik,Im} n_j ds - \dot{x}_k \rho \int_S \varnothing_{ik}, Re\ n_j ds \\
&= A_{jk}\ddot{x}_k - B_{jk}\dot{x}_k
\end{aligned} \tag{3}$$

In the equation, $\vec{F}$ represents the wave load generated by incident potential; $\vec{F}_D$ denotes the diffraction force; $\vec{F}_R$ signifies the radiation force; ω stands for the wave frequency; $\vec{n}$ denotes the normal direction; S represents the wetted surface area; $\varnothing_i$ stands for the incident potential; $\varnothing_d$ stands for the diffracted potential; ρ stands for the density of water; n_j stands for the direction factors; $\varnothing_{ik,Im}$ and $\varnothing_{ik,Re}$ denote the imaginary and real parts of the incident potential, respectively; and A_{jk} and B_{jk} represent the added mass coefficient and radiation damping coefficient, where k, j represents the degrees of freedom of the structure.

2.3. Hydrodynamic Theory

Based on the potential flow theory approach, considering the Bernoulli equation and the velocity potential equation, hydrodynamics are calculated based on the distribution of the fluid velocity field and pressure field. Reducing the waterline area will alter the hydrodynamic characteristics of the floating body. The specific statement is as follows:

$$F_{water} = \int_S \rho\vec{v}\left(\vec{v} - \vec{v}_{flowfield}\right)dS \tag{4}$$

Here, ρ represents the density of the fluid, $\vec{v}$ is the velocity field on the floating body, $\vec{v}_{flowfield}$ denotes the velocity field of the fluid flow, and S represents the wetted surface area.

3. The Proposed Floating Platform

3.1. Numerical Model

The proposed floating platform features a distinctive inward concave design at the waterline of its three external columns, setting it apart from the conventional OC4-DeepCwind platform. This unique design leads to a noteworthy reduction in the waterplane area of the proposed floating platform. Consequently, the platform, in theory, would be subjected to decreased wave loads. For this reason, the proposed design would significantly benefit the motion stability of the OFWTs in their practical application. For ease of understanding, the conceptual design of the proposed floating platform and the OC4-DeepCwind platform are shown in Figure 2.

To enable the study of the proposed platform and the comparison with the OC4-DeepCwind platform, numerical models of both platforms were developed in ANSYS AQWA. By referencing the pertinent parameters of the OC4-DeepCwind platform [19], the initial dimensions of the proposed floating platform and the OFWT supported by it were determined. The structures of the platforms are depicted in Figure 3, and their numerical parameters are listed in Table 1.

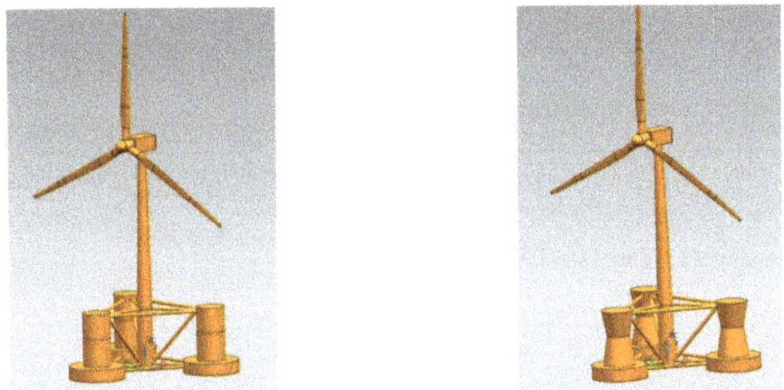

(a) OC4-DeepCwind. (b) The proposed floating platform.

Figure 2. The OC4-DeepCwind platform and the proposed platform.

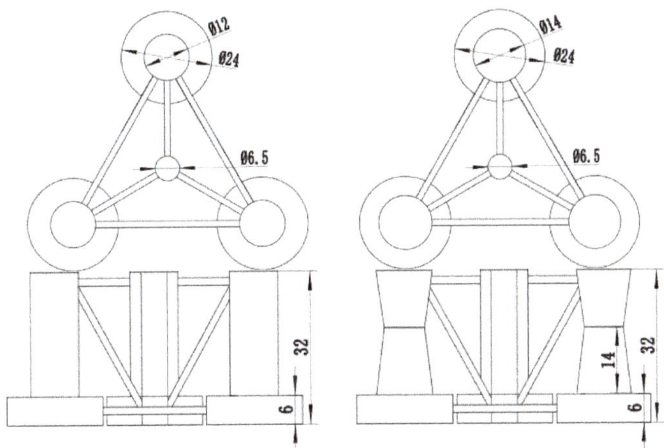

(a) OC4-DeepCwind platform. (b) The proposed floating platform.

Figure 3. Schematic diagram of the two floating platforms.

Table 1. Detailed design parameters of the floating platforms.

Parameters	Proposed Floating Platform	OC4-DeepCwind Platform
Draft (m)	20	20
Height above waterline of floater (m)	12	12
Outer bottom floater diameter (m)	24	24
Diameter at the outer draft line (m)	10	12
Diameter of outer floater (m)	12	12
Diameter of lower conical floater (m)	12	12
Height of lower conical float below Waterline (m)	14	14
Height of upper conical floater above Waterline (m)	12	12
Distance between outer-side pontoons (m)	50	50
Diameter of central pontoon (m)	6	6
Height of central pontoon (m)	30	30
Diameter of support column (m)	1.6	1.6
Total mass (kg)	1.335×10^7	1.347×10^7
Moment of inertia Ixx (kg/m^2)	7.370×10^9	6.827×10^9
Moment of inertia Iyy (kg/m^2)	7.371×10^9	6.827×10^9
Moment of inertia Izz (kg/m^2)	1.114×10^{10}	1.226×10^{10}

3.2. Mooring System

The mooring system of the proposed platform is composed of three steel catenary mooring chains and their associated cable ducts. The cable ducts are designed to be located 14 m below the waterline on the large-diameter hanging cylinder at the bottom of the construction. The three mooring anchors are located in the seabed, and the cable duct and anchor point are connected to both ends of the catenary line. This type of semi-submersible platform has a large self-weight, providing self-stability. Figure 4 shows the schematic diagram of the mooring system, and Table 2 provides the parameters of the mooring chains.

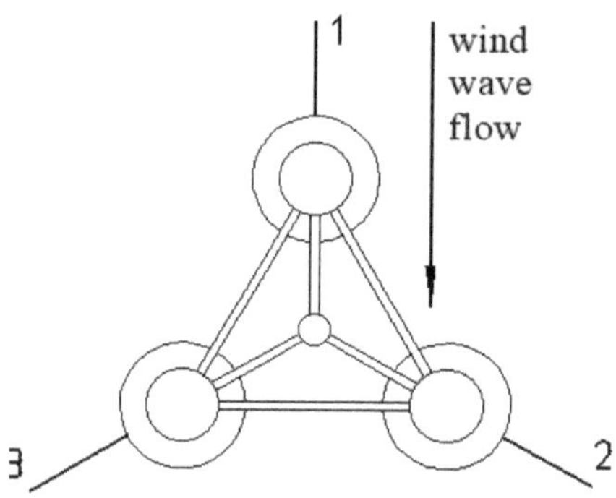

Figure 4. The schematic diagram of the mooring system.

Table 2. Mooring chain parameters.

Parameters	Numerical Value
Quantity	3
Length (m)	507
Mass per unit length (kg/m)	150
Equivalent interfacial area (m^2)	0.01
Axial stiffness (N/m)	753,600,000
Breaking force (N)	6,090,000
Added mass coefficient	1.1
Drag coefficient	1.0
Axial drag coefficient	0.1
Longitudinal damping coefficient	0.025

4. Frequency–Domain Analysis

In this section, the hydrodynamic performances of the OFWTs that are supported by the proposed floating platform and the OC4-DeepCwind platform are studied numerically with the aid of ANSYS AQWA. During the numerical calculations, by following a consistency check, the whole computational domain is meshed by 12,000 elements, about 8000 of which are diffraction elements. In the numerical models, the mass of the wind turbine is simplified to a point force acting on the top surface of the platform. The structural meshing is shown in Figure 5.

In this study, both the near-field and far-field methods described in [20–22] were applied to calculate the second-order wave forces for validating the effectiveness of the meshing results and the reliability of the hydrodynamic model. The second-order mean drift forces in the surge direction obtained for the OFWTs supported by the two types of floating platforms are shown in Figure 6.

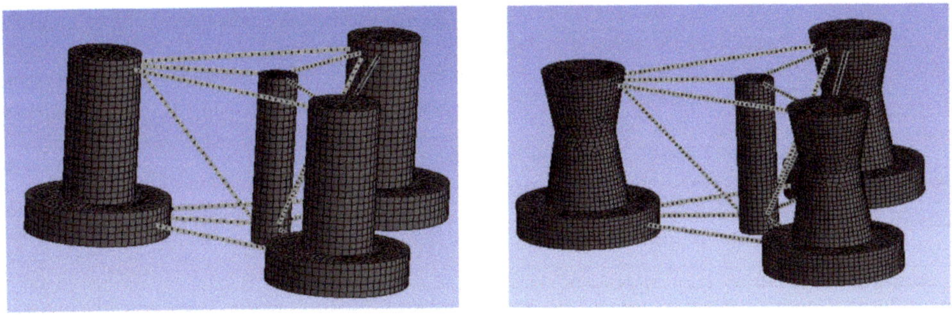

(a) The OC4-DeepCwind platform. (b) The proposed floating platform.

Figure 5. Meshing results of the two types of floating platforms.

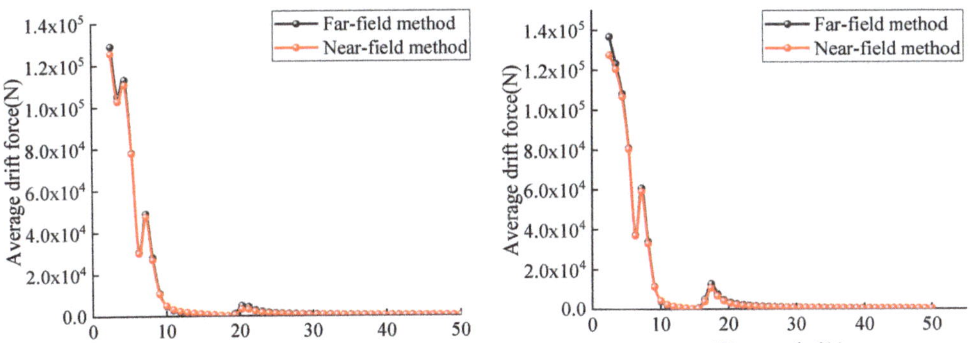

(a) Turbine supported by the proposed platform. (b) OC4-DeepCwind wind turbine.

Figure 6. Second-order mean drift forces in the surge direction.

From Figure 6, it can be seen that the results of the average drift force in the surge direction obtained using the near-field and far-field methods are consistent in both trend and magnitude. Therefore, it can be concluded that the model meshing method meets the hydrodynamic calculation accuracy requirements very well.

4.1. Hydrodynamic Viscous Damping

The OFWT supported by the proposed floating platform will be deployed in an offshore wind farm where the water is 200 m deep and the density of the seawater is 1025 kg/m^3. The incident wave period ranges from 2.6 to 50 s (i.e., 2.147 to 0.126 rad/s). Based on the potential flow theory for the hydrodynamic performance analysis of OFWTs [23–25], the issue of the overestimated hydrodynamic response of the OFWT is addressed by introducing additional viscous damping as a corrective measure in the calculation. The additional viscous damping effectively mitigates the overestimation, thereby leading to more accurate and reliable predictions.

Assuming the critical damping coefficient of the system in a particular Degree Of Freedom (DOF) is C_r, it can be mathematically expressed as

$$C_r = 2\sqrt{k(M + M_a)} \tag{5}$$

where M is the moment of inertia, M_a is the added mass of the platform in the DOF of interest, and k represents the hydrostatic restoring coefficient of the platform.

Take 5% of C_r in every DOF as the additional viscous damping in the corresponding direction to assess the hydrodynamic performance of the OFWT. The estimated results of the additional viscous damping for the OFWT supported by the two types of floating platforms are listed in Table 3.

Table 3. Additional viscous damping of the OFWT.

Platform	DOFs	Inertial Mass/kg	Added Mass/kg	Restoring Stiffness/N × m^{-1}	Critical Damping /N × (m/s)$^{-1}$ or /N × m × (°/s)$^{-1}$	Viscous Damping /N × (m/s)$^{-1}$ or /N × m × (°/s)$^{-1}$
Proposed platform	Heave	1.33×10^7	1.44×10^7	2.75×10^8	1.75×10^6	8.74×10^5
	Roll	7.37×10^9	1.08×10^8	1.67×10^7	7.07×10^8	3.53×10^7
	Pitch	7.37×10^9	1.08×10^8	1.67×10^7	7.07×10^8	3.53×10^7
OC4-Deep Cwind	Heave	1.34×10^7	1.42×10^7	3.79×10^6	2.05×10^7	1.02×10^6
	Roll	6.82×10^9	1.08×10^8	2.55×10^7	8.42×10^8	4.21×10^7
	Pitch	6.82×10^9	1.08×10^8	2.55×10^7	8.42×10^8	4.21×10^7

To verify the effect of the additional viscous damping on motion stability, the RAOs of the OFWT in the heave and roll directions were calculated before and after considering additional viscous damping, respectively. The results obtained when the wave incident angle is 0° are shown in Figure 7. Figure 7a depicts the calculated RAOs of the OFWT when it is supported by the proposed floating platform, while Figure 7b shows the RAOs of the OFWT when it is supported by the OC4-DeepCwind platform.

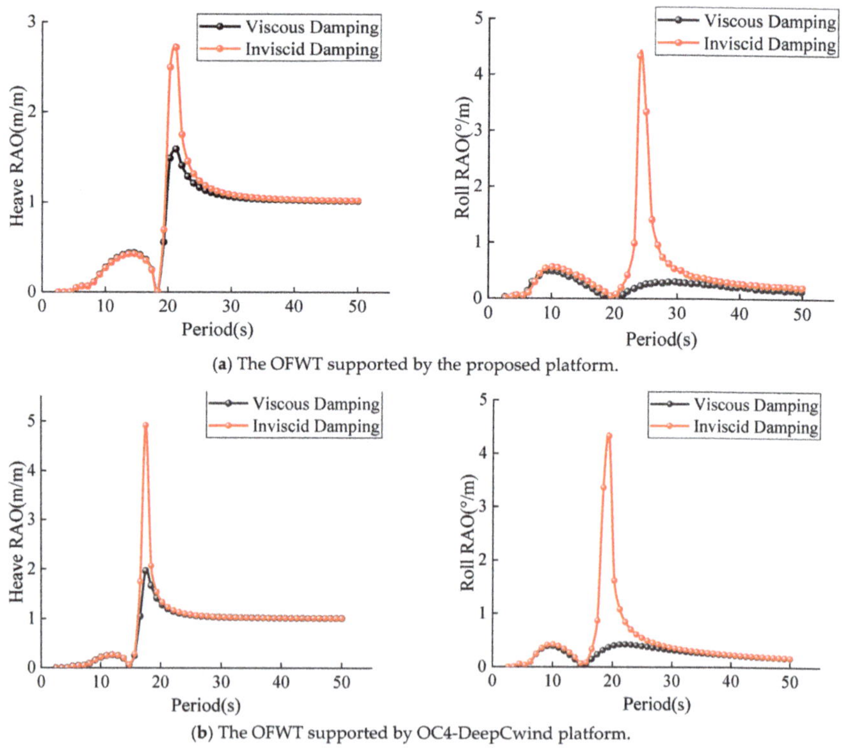

(**a**) The OFWT supported by the proposed platform.

(**b**) The OFWT supported by OC4-DeepCwind platform.

Figure 7. RAOs of the OFWT were obtained before and after considering additional viscous damping.

From Figure 7, it can be seen that regardless of the supporting platform, the hydrodynamic response of the OFWT is significantly biased before additional viscous damping is considered. By contrast, after considering additional viscous damping, the RAOs of the OFWT in both the heave and roll directions are diminished to varying extents across a wide range of wave periods. Therefore, it is imperative to consider additional viscous damping when employing ANSYS AQWA to evaluate the hydrodynamic responses of OFWTs.

4.2. Hydrodynamic Response

The RAO, i.e., the ratio of the motion amplitude of the structure in each DOF to the amplitude of sea waves, indicates the hydrodynamic response characteristics of a floating structure in waves. To investigate the hydrodynamic response of the OFWT under various wave conditions when it is supported by the proposed floating platform and to demonstrate the superiority of the proposed platform over the OC4-DeepCwind platform in ensuring the safety and reliability of the OFWT in extreme weather conditions, the RAOs of the OFWT supported by the proposed platform and the OC4-DeepCwind platform, respectively, are calculated with the incident waves approaching from various directions.

Firstly, Figure 8 shows the RAOs of the OFWT in all six DOFs when the OFWT is subjected to regular waves characterized by different wave periods and incident angles.

From Figure 8, it can be seen that, as expected, the dynamic response of the OFWT in the heave direction is insensitive to the change in the incident wave angle. In other words, the RAOs of the OFWT at the corresponding wave periods remain unchanged when the incident wave angle changes. By contrast, the RAOs of the OFWT at the corresponding wave periods in the other five DOFs show significant change as soon as the incident wave angle changes. This suggests that the hydrodynamic responses of the OFWT in the other five DOFs are sensitive to the change in the approaching direction of incident waves.

To demonstrate the superiority of the proposed platform over the conventional OC4-DeepCwind platform in ensuring the safety and reliability of the OFWT in extreme weather conditions, the RAOs of the OFWT supported by the proposed platform and the OC4-DeepCwind platform, respectively, are compared in this study. The comparison results in the heave and pitch directions are illustrated in Figure 9. Herein, the RAOs of the OFWT in the heave and pitch directions are of interest in the comparison because the safety and power generation efficiency of the OFWT can be significantly affected by the motions of the turbine in these two DOFs.

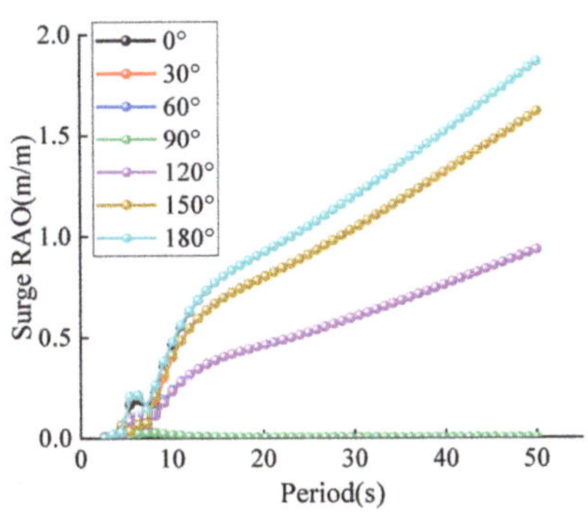

(a) Surge.

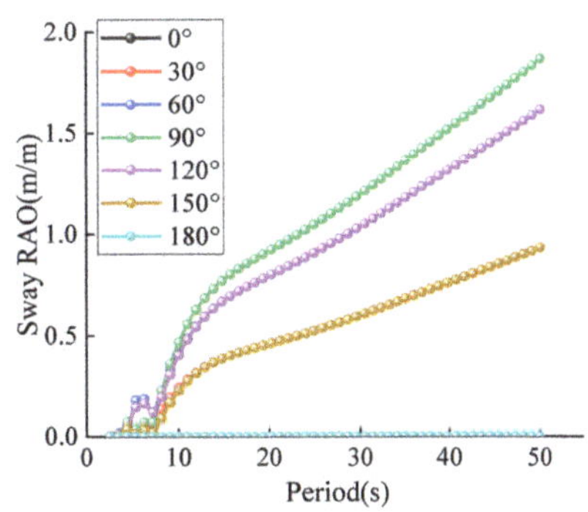

(b) Sway.

Figure 8. *Cont.*

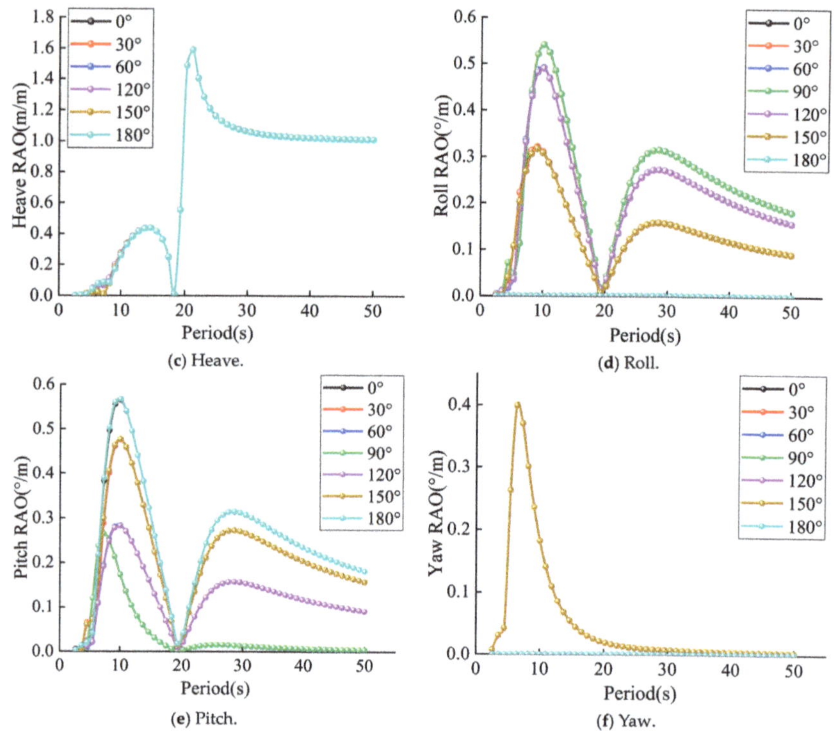

Figure 8. RAOs of the OFWT when supported by the proposed floating platform.

It is well known that the occurrence of resonant vibrations poses a big risk to the safe operation of OFWTs in waves. This is because resonant vibrations can lead to substantial fatigue loads on turbines' structures and components, potentially resulting in immediate damage or, in the most severe cases, the complete sinking of the turbine. From Figure 9, it is found that the OFWT supported by the proposed platform implies smaller RAO values in both the heave and pitch directions when resonant vibration occurs. This implies that, in comparison to the OC4-DeepCwind platform, the proposed platform provides safer and more reliable support to the OFWT operating in waves.

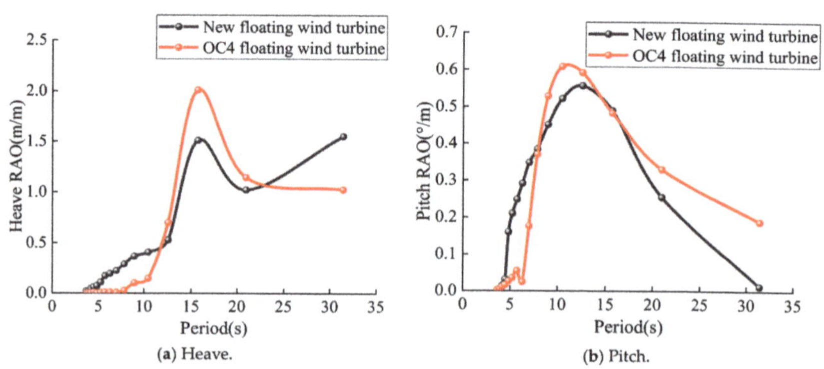

Figure 9. Comparison of the RAOs when the OFWT is supported by the two types of platform.

5. Verification Test

5.1. Experimental Setup

The water tank used for the model tests in this study had a length of 14.65 m, a width of 5.9 m, and a height of 1.8 m. At both ends of the tank, the outer diameter was 5.9 m and the inner diameter was 2.4 m. One side of the tank featured a glass window, which was 3 m long and 1.2 m high, as illustrated in Figure 10. The wind–wave generation system of the tank can generate waves with a maximum height of 0.2 m and a wave period ranging from 0.5 s to 5 s. Additionally, the system is capable of generating a maximum wind speed of 16 m/s in the testing section.

(a) Front view of the water tank.

(b) Side view of the water tank.

(c) The wind–wave control system.

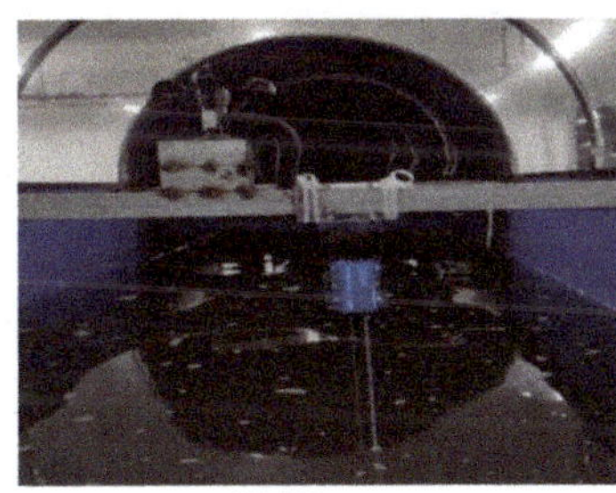

(d) Wave gauge.

Figure 10. Water tank used for the model tests.

The experiment used Smart sensor software(CBG03) for collecting wave probe data and the Qualisys motion tracking system for motion capture. The six DOF motions of the OFWT were measured using four Oqus motion capture cameras provided by Qualisys, as shown in Figure 11b. The scaled models of the OFWT supported by the two types of floating platform are depicted in Figure 12.

(a) Motion capture cameras.

(b) Locations of the cameras.

Figure 11. Qualisys motion capture system.

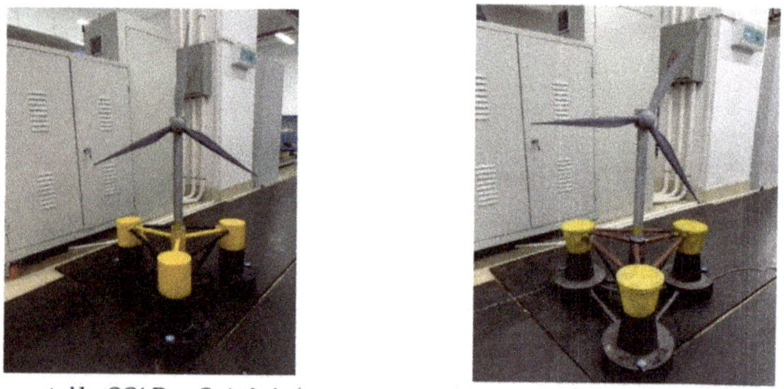

(**a**) Supported by OC4-DeepCwind platform. (**b**) Supported by the proposed platform.

Figure 12. The scaled models of the OFWT.

During the test, two distinct load condition categories were considered. The first category focused on assessing the seakeeping performance of the OFWT exclusively in pure waves. Meanwhile, the second category was designed to evaluate the seakeeping performance of the OFWT under combined wave and wind conditions, resembling the actual offshore environment in which the OFWT operates. Detailed information regarding the wave and wind parameters for each loading condition is given in Table 4.

Table 4. Matrix of testing conditions.

Loading Conditions	Wave Period (s)	Wave Height (m)	Wind Speed (m/s)
Pure wave	0.50	0.05	---
	0.75	0.05	---
	1.00	0.05	---
	1.25	0.05	---
	1.50	0.05	---
	1.75	0.05	---
	2.00	0.05	
Wind–wave combined	0.50	0.05	6.0
	0.75	0.05	6.0
	1.00	0.05	6.0
	1.25	0.05	6.0
	1.50	0.05	6.0
	1.75	0.05	6.0
	2.00	0.05	6.0

5.2. Results and Discussion

The proposed floating platform and the OC4-DeepCwind platform serve as the support structures for the OFWT during the test, respectively. The movements of the OFWT in all six DOFs under different external load conditions were measured using the Qualisys motion capture system. Given that the turbine's movements in the heave and pitch directions profoundly influence its safety and power generation efficiency, the test results in these two directions under various external load conditions are shown in Figures 13 and 14.

It is worth noting that, to minimize the influence of size on the evaluation results, Figures 13 and 14 employed RAOs, which represent the ratios of turbine motion in a specific direction to the amplitude of the waves, thereby describing the seakeeping performance of the OFWT under various external loading conditions. From the results depicted in Figures 13 and 14, when compared to the OC4 floating offshore wind turbine, it is obvious to see that the novel floating offshore wind turbine shows favorable heave and pitch motion performance in different wave periods, both under pure wave loading and combined loading conditions with varying wave periods and steady winds.

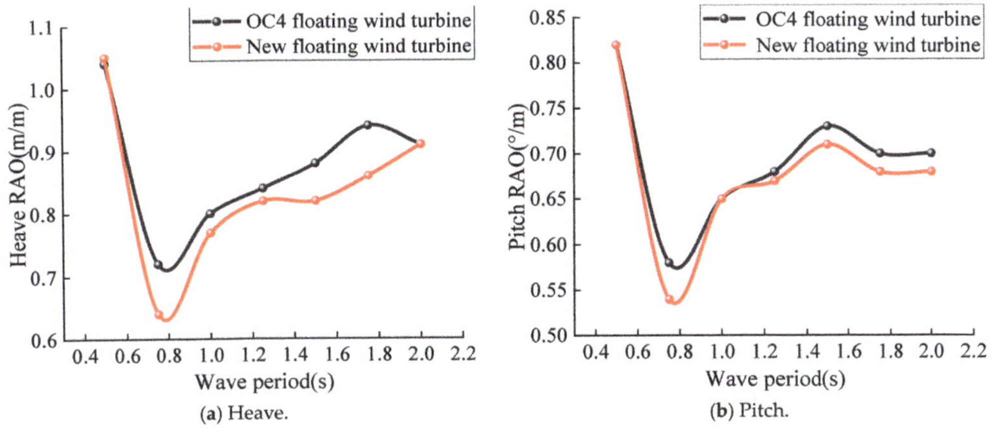

Figure 13. Test results were obtained under pure wave excitation.

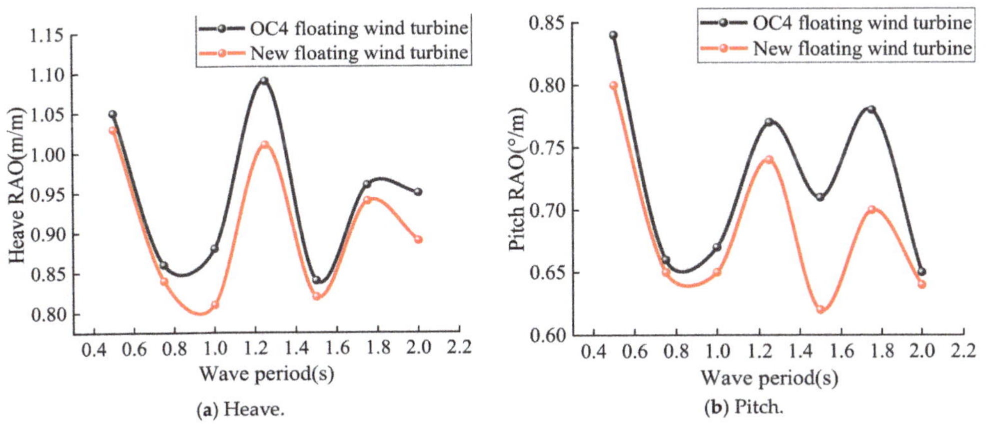

Figure 14. Test results obtained under wind–wave combined excitation.

6. Conclusions

A novel floating wind turbine foundation has been proposed based on the OC4-DeepCwind floating foundation. Models of the systems were created and frequency–domain response analyses were conducted using Ansys AQWA, along with wind–wave experiments. The research findings lead to the following conclusions.

1. Introducing viscous damping enables a more accurate determination of the corresponding natural period's RAO, thus providing a more realistic reflection of the hydrodynamic characteristics of floating wind turbines. This enhances the reliability of the simulation comparisons between the two semi-submersible wind turbine platforms. The results indicate a lower RAO for the new floating wind turbine platform.
2. The comparative frequency–domain analysis between the new design of the semi-submersible floating wind turbine and the OC4-DeepCwind platform revealed that a smaller diameter float contacting the water surface resulted in a better motion performance. However, the interaction between the floater and the water surface changes as the platform moves. Therefore, the next step could involve altering the height of the floater's small waterline area and conducting fully coupled simulation experiments for the better optimization of the new design of the semi-submersible floating platform's structure and performance.

3. In scale model experiments under wind–wave conditions, the comparative analysis between the experimental and numerical simulations revealed the significant impact of steady wind waves on OFWTs at different wave periods. Compared to the OC4-DeepCwind platform, the new floating offshore wind turbine could better avoid energy concentration areas and low-frequency wave ranges, thereby preventing platform resonance and improving motion stability.
4. The reduced diameter of the float at the waterline in the new design of the semi-submersible floating wind turbine could lower manufacturing costs as compared to the OC4-DeepCwind platform. However, the extent of diameter reduction must be considered in conjunction with overall strength. Future work will involve comprehensive strength testing of the new design of the semi-submersible floating platform.

Author Contributions: Methodology, X.W.; Software, X.W.; Formal analysis, Z.Z.; Data curation, H.Z. and Z.Z.; Writing—original draft, H.Z.; Writing—review & editing, Z.Z. All authors have read and agreed to the published version of the manuscript.

Funding: The work was supported by the Hunan Natural Science Foundation (Grant Nos. 2024JJ7110 and 2021JJ50105), the Xiangtan Science and Technology Plan Project (Grant Nos. ZX-YB20231004), and the Postgraduate Scientific Research Innovation Project of Hunan Province (CX20231280).

Data Availability Statement: The original contributions presented in the study are included in the article, further inquiries can be directed to the corresponding author.

Conflicts of Interest: The authors declare no conflict of interest.

References

1. Edwards, E.C.; Holcombe, A.; Brown, S.; Ransley, E.; Hann, M.; Greaves, D. Trends in floating offshore wind platforms: A review of early-stage devices. *Renew. Sustain. Energy Rev.* **2024**, *193*, 114271. [CrossRef]
2. Zhou, B.; Zhang, Z.; Li, G.; Yang, D.; Santos, M. Review of key technologies for offshore floating wind power generation. *Energies* **2023**, *16*, 710. [CrossRef]
3. Lemmer, F.; Yu, W.; Müller, K.; Cheng, P.W. Semi-submersible wind turbine hull shape design for a favorable system response behavior. *Mar. Struct.* **2020**, *71*, 102725. [CrossRef]
4. Zhang, H.; Wang, H.; Cai, X.; Xie, J.; Wang, Y.; Zhang, N. Novel method for designing and optimising the floating platforms of offshore wind turbines. *Ocean Eng.* **2022**, *266*, 112781. [CrossRef]
5. Liu, Z.; Zhou, Q.; Tu, Y.; Wang, W.; Hua, X. Proposal of a novel semi-submersible floating wind turbine platform composed of inclined columns and multi-segmented mooring lines. *Energies* **2019**, *12*, 1809. [CrossRef]
6. Scicluna, D.; De Marco Muscat-Fenech, C.; Sant, T.; Vernengo, G.; Tezdogan, T. Preliminary analysis on the hydrostatic stability of a self-aligning floating offshore wind turbine. *J. Mar. Sci. Eng.* **2022**, *10*, 2017. [CrossRef]
7. Rezanejad, K.; Gadelho, J.F.M.; Xu, S.; Soares, C.G. Experimental investigation on the hydrodynamic performance of a new type floating Oscillating Water Column device with dual-chambers. *Ocean Eng.* **2021**, *234*, 109307. [CrossRef]
8. Cheng, P.; Huang, Y.; Wan, D. A numerical model for fully coupled aero-hydrodynamic analysis of floating offshore wind turbine. *Ocean Eng.* **2019**, *173*, 183–196. [CrossRef]
9. Tabeshpour, M.R.; Hajnoruzi, F. Conceptual study on dynamic responses of semi-submersible platforms. In *Proceedings of the Institution of Civil Engineers-Maritime Engineering*; Thomas Telford Ltd.: London, UK, 2023; Volume 176, pp. 31–43.
10. Yang, Y.; Bashir, M.; Michailides, C.; Li, C.; Wang, J. Development and application of an aero-hydro-servo-elastic coupling framework for analysis of floating offshore wind turbines. *Renew. Energy* **2020**, *161*, 606–625. [CrossRef]
11. Gao, J.L.; Lyu, J.; Wang, J.H.; Zhang, J.; Liu, Q.; Zang, J.; Zou, T. Study on transient gap resonance with consideration of the motion of floating body. *China Ocean Eng.* **2022**, *36*, 994–1006. [CrossRef]
12. Gao, J.L.; Lyu, J.; Zhang, J.; Zang, J. Influences of floater motion on gap resonance triggered by focused wave groups. *China Ocean Eng.* **2023**, *37*, 685–697. [CrossRef]
13. Gong, S.K.; Gao, J.L.; Mao, H.F. Investigations on fluid resonance within a narrow gap formed by two fixed bodies with varying breadth ratios. *China Ocean Eng.* **2023**, *37*, 962–974. [CrossRef]
14. Zou, Q.; Lu, Z.; Shen, Y. Short-term prediction of hydrodynamic response of a novel semi-submersible FOWT platform under wind, current and wave loads. *Ocean Eng.* **2023**, *27*, 114471. [CrossRef]
15. Bashetty, S.; Ozcelik, S. Review on dynamics of offshore floating wind turbine platforms. *Energies* **2021**, *14*, 6026. [CrossRef]
16. Deng, Y.; Feng, W.; Xu, S.; Chen, X.; Wang, B. A novel approach for motion predictions of a semi-submersible platform with neural network. *J. Mar. Sci. Technol.* **2020**, *26*, 883–895. [CrossRef]
17. Zhang, Y.; Xu, H.; Law, Y.; Santo, H.; Magee, A. Hydrodynamic analysis and validation of the floating DeepCwind semi-submersible under 3-h irregular wave with the HOS and CFD coupling method. *Ocean Eng.* **2023**, *287*, 115701. [CrossRef]

18. Falnes, J.; Kurniawan, A. *Ocean Waves and Oscillating Systems: Linear Interactions Including Wave-Energy Extraction*; Cambridge University Press: Cambridge, UK, 2020.
19. Robertson, A.; Jonkman, J.; Masciola, M.; Song, H.; Goupee, A.; Coulling, A.; Luan, C. *Definition of the Semi-Submersible Floating System for Phase II of OC4*; National Renewable Energy Lab.(NREL): Golden, CO, USA, 2014.
20. Choi, Y.R.; Hong, S.Y.; Choi, H.S. An analysis of second-order wave forces on floating bodies by using a higher-order boundary element method. *Ocean Eng.* **2001**, *28*, 117–138. [CrossRef]
21. Tran, T.T.; Kim, D.H. The coupled dynamic response computation for a semi-submersible platform of floating offshore wind turbine. *J. Wind Eng. Ind. Aerodyn.* **2015**, *147*, 104–119. [CrossRef]
22. Liu, Z.; He, J.; Meng, Y.; Zhang, H.; Zhou, Y.; Tao, L. Numerical and experimental study on the influence of a moonpool on motion performance and stability of a drillship. *Ocean Eng.* **2022**, *262*, 112241. [CrossRef]
23. Lopez-Pavon, C.; Souto-Iglesias, A. Hydrodynamic coefficients and pressure loads on heave plates for semi-submersible floating offshore wind turbines: A comparative analysis using large scale models. *Renew. Energy* **2015**, *81*, 864–881. [CrossRef]
24. Joseph, D.D. Potential flow of viscous fluids: Historical notes. *Int. J. Multiph. Flow* **2006**, *32*, 285–310. [CrossRef]
25. Felder, S.; Chanson, H. Turbulence, dynamic similarity and scale effects in high-velocity free-surface flows above a stepped chute. *Exp. Fluids* **2009**, *47*, 1–18. [CrossRef]

electronics

MDPI

Article

Modeling and Suppression of Conducted Interference in Flyback Power Supplies Based on GaN Devices

Jichi Yan [1,2,*], Haoyuan Wu [2], Xueliang Fu [3], Mingtong Li [3] and Yannan Yu [1,2]

[1] Key Laboratory of Advanced Manufacturing and Automation Technology, Guilin University of Technology, Education Department of Guangxi Zhuang Autonomous Region, Guilin 541006, China; 2019174@glut.edu.cn
[2] College of Mechanical and Control Engineering, Guilin University of Technology, Guilin 541006, China; 2120221219@glut.edu.cn
[3] Greatwall Power Supply Technology Co., Ltd., Shenzhen 518000, China; fuxueliang@gwpst.com (X.F.); limingtong@gwpst.com (M.L.)
* Correspondence: 2019178@glut.edu.cn

Abstract: The application of GaN power devices has significantly increased the power density of flyback power supplies but has also caused severe electromagnetic interference (EMI) issues. To address the challenge of conducted interference in flyback power supplies, a comprehensive analysis of the transmission mechanism of conducted common-mode noise is undertaken. This analysis involves simplifying the equivalent model of conducted interference and leveraging the circuit characteristics of conducted noise to propose a solution for attenuating common-mode noise. Considering the constraints of external compensation capacitors, a balanced winding is further introduced to mitigate the impact of noise. To enhance the efficacy of conducted interference suppression, it is suggested to change the winding structure of the transformer and incorporate a shielding winding. This configuration aims to minimize the generation and propagation of common-mode noise within the transformer. Finally, experimental verification is carried out using a 150 W GaN flyback power supply prototype. The experimental results demonstrate that the proposed method effectively suppresses common-mode noise in the circuit.

Keywords: flyback converter; electromagnetic interference (EMI); distributed capacitance; winding structure; shielding winding

Citation: Yan, J.; Wu, H.; Fu, X.; Li, M.; Yu, Y. Modeling and Suppression of Conducted Interference in Flyback Power Supplies Based on GaN Devices. *Electronics* **2024**, *13*, 2360. https://doi.org/10.3390/electronics13122360

Academic Editors: Lin Jiang, Bo Yang and Zhijian Liu

Received: 15 May 2024
Revised: 4 June 2024
Accepted: 6 June 2024
Published: 16 June 2024

1. Introduction

Flyback converters are renowned for their simple structure, requiring fewer components while providing a stable output and high efficiency. They are commonly used in low- and medium-power supplies. As the demand for higher efficiency and power density in power supplies increases, wide-bandgap semiconductor devices such as Gallium Nitride (GaN) and Silicon Carbide (SiC) are being increasingly employed. These devices exhibit superior switching characteristics, enabling power supplies to operate at ultrahigh switching frequencies. Consequently, this significantly enhances power density and reduces power losses in power supplies [1–4]. Power converters based on GaN devices achieve higher switching frequencies but are also associated with higher dv/dt and di/dt conversion rates, generating electromagnetic interference (EMI) noise that can disrupt low-voltage electronics and sensitive digital circuits [5,6]. To meet electromagnetic compatibility requirements, EMI filters are widely used in flyback converters. However, the typical volume of EMI filters constitutes about one-third of the converter, which is not conducive to cost savings and power density improvement [7,8]. This study aims to investigate the noise sources and propagation paths of flyback power supplies based on GaN devices, establish a noise propagation model, and propose an effective common-mode noise suppression method. The goal is to reduce the inherent noise of the flyback power supply, simplify the design of

EMI filters, and thereby decrease the size and weight of these filters, ultimately improving the power density of the power supply.

Current research on the suppression of converter conduction interference primarily focuses on the following aspects: direct suppression at the noise source, effective cutoff or compensation of the noise propagation path, and offsetting of the displacement current. Reference [9] analyzed the conduction paths of the flyback common-mode interference current, measured the equivalent common-mode capacitance values on different paths, and suppressed the interference by balancing the primary and secondary common-mode currents to reduce the total common-mode displacement current to zero. Reference [10] proposed, for multi-output flyback converters, that the boost converter should take two isolated and one non-isolated output to prevent voltage spikes in the coupling inductors and reduce device stress, thereby reducing electromagnetic interference. Reference [11] suggested placing a switching tube between the two coils to equalize noise using the symmetry of the primary coils; however, this method complicates the circuit and is difficult to control, making its feasibility low. References [12–14] proposed altering the distributed capacitance between windings by changing the transformer structure and winding method, thereby improving the system's EMC characteristics. Using copper foil shielding for transformers to suppress common-mode noise is also common, but the overall design theory guidance is insufficient [15]. Reference [16] utilized core shielding to address the coupling problem caused by the transformer core and employed a converter to solve the issues of multi-Y capacitance and capacitive filtering, providing better EMI suppression compared to conventional pulse-width modulation. Reference [17] highlighted the significant influence of transformer common-mode coupling capacitance on noise suppression and demonstrated that the feasibility of the transformer can be verified without an EMI receiver, using only a signal generator and an oscilloscope. References [18–20] incorporated balanced winding techniques to achieve effective shielding and offset common-mode noise, and they provided a theoretical method for calculating the number of turns in a balanced winding, which is valuable for this study. The common-mode noise coupling mechanism proposed in references [11,21–23] is more challenging to understand, and although it performs detailed modeling reasoning, the introduced shielding layer in the transformer is overly complex. References [16,24–26] discuss the integration of shielded windings and the modeling of capacitance distribution in flyback converters, which serves as a good guide for transformer EMC design. Reference [27] analyzes the EMI characteristics of wide-bandwidth semiconductor power systems, facilitating a subsequent analysis of conducted interference propagation.

Based on recent studies on conducted interference suppression, several methods are available to directly cancel common-mode noise or suppress it within the noise loop; however, their general applicability is limited. This paper proposes three methods to suppress common-mode noise interference in GaN flyback power supplies and provides a detailed analysis of each method. An experimental platform is constructed using a 150 W GaN flyback power supply, and the experimental results validate the effectiveness and feasibility of the proposed methods.

The rest of the paper is organized as follows. Section 2 describes the common-mode noise transmission mechanism in GaN-based flyback power supplies. Section 3 details three methods for common-mode noise suppression. The experimental results and conclusions are presented in Sections 4 and 5, respectively.

2. Transmission Mechanisms of Conducted Common-Mode Noise

The conducted interference transmission model of the flyback converter is shown in Figure 1. The front dashed lines, L_1, L_2, C_1, C_2, R_1, and R_2, collectively form a Line Impedance Stabilization Network (LISN). This network ensures a stable impedance for accurately testing the conducted noise of the device under test and extracting clean power quality. C_{Y1}, C_{Y2}, C_{X1}, C_{X2}, and L constitute a simple primary EMI filter. C_{X1} and C_{Y2} filter out differential-mode interference between the grids. C_Y acts as a common-mode capacitor, and C_{Y1}, C_{Y2}, C_{Y3}, and C_{Y4} are connected between a high voltage and the ground to filter out common-mode interference. Typically, ceramic capacitors of 4.7 nF and 2.2 nF are selected for this purpose. A separator is used to separate the differential-mode and common-mode noise, while an EMI receiver is utilized to receive the noise signal from the separator. D_1, D_2, D_3, and D_4 are the four diodes of the rectifier bridge. C_{IN} is the input capacitance, C_{OUT} is the output capacitance, C_{PS} is the primary noise source on the secondary-side winding's equivalent capacitance, C_{SP} is the secondary-side noise source on the primary winding's equivalent capacitance, C_{PH} is the distribution capacitance between the switching tubes and heatsink, C_{SG} is the distribution capacitance of the heatsink with respect to the ground, and C_S is the distribution capacitance of the secondary side with respect to the ground. V_O is the output voltage, PE means the ground, PG means the transformer's primary-side ground, and SG is the transformer's secondary-side ground. For safety reasons, SG is generally directly connected to the ground.

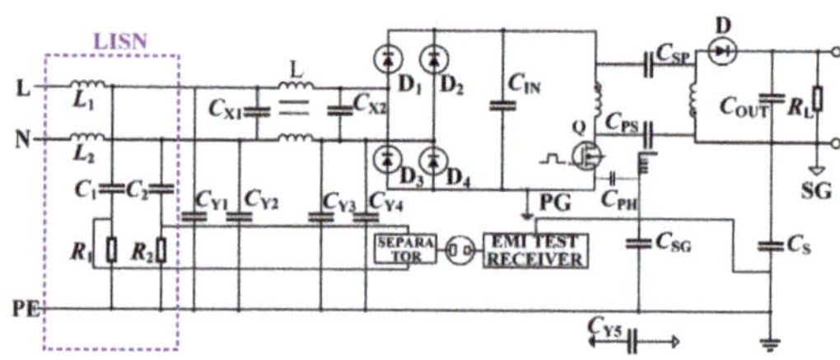

Figure 1. Conducted interference transmission model of the flyback converter.

Figure 2 shows the common-mode noise transfer path from the primary side of the transformer to the secondary side. In this scenario, noise originating from voltage jumps (V_P) in the primary-side switching tube generates a noise current. This current flows through the primary equivalent capacitance (C_{PS}) of the transformer, the secondary-side distributed capacitance relative to the ground (C_S), and the LISN, forming a complete loop for noise transfer from the primary side to the secondary side.

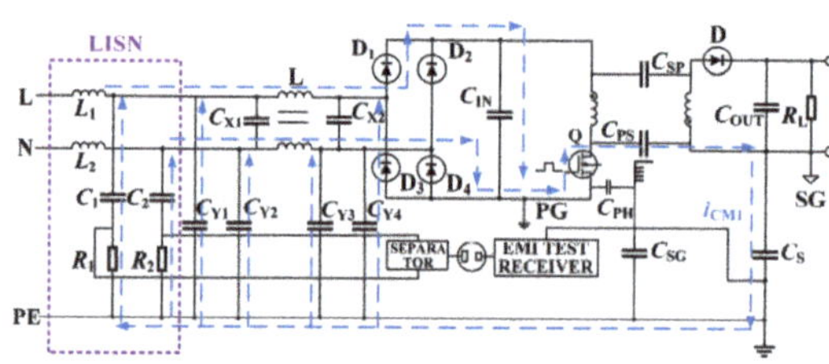

Figure 2. Common-mode noise transmitted from the primary side to the secondary side of the transformer.

Given that common-mode noise currents typically exhibit low magnitudes (in the μA range), the voltage drop across the LISN's equivalent resistor is minimal and negligible compared to the noise source potential. The raw common-mode noise intensity for this path (without filtering) can be described by the following equation:

$$i_{CM1} = \frac{(C_{PS}C_S)}{C_{PS} + C_S}\frac{du_P}{dt} \tag{1}$$

Figure 3 shows the common-mode noise transmission path from the secondary side of the transformer to the primary side. Here, noise originating from potential jumps in the secondary-side diode generates a noise current. This current traverses various components, including the secondary equivalent capacitance (C_{SP}) of the transformer, the output capacitance (C_{OUT}), the secondary-side distributed capacitance relative to the ground (C_S), and the LISN, forming a complete noise transmission loop from the secondary side to the primary side.

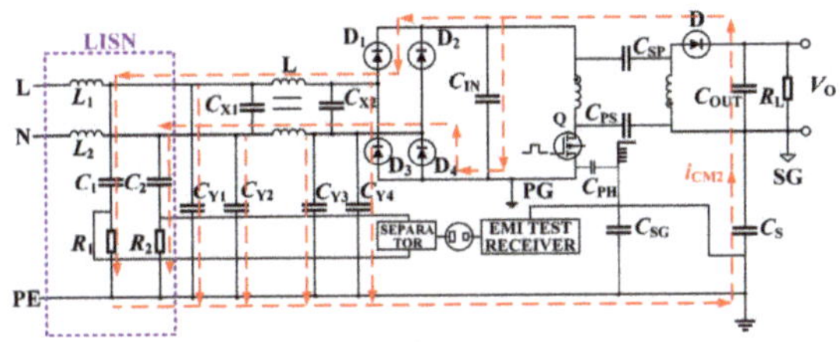

Figure 3. Common-mode noise transmitted from the secondary side to the primary side of the transformer.

The raw common-mode noise intensity, i_{CM2}, (without filtering) in this pathway can be calculated using the following equation:

$$i_{CM2} = \frac{(C_{SP}C_S)}{C_{SP} + C_S}\frac{du_D}{dt} \tag{2}$$

The effect of the output capacitance (C_{OUT}) is not considered in Equation (2) because the capacitive reactance of C_{OUT} is significantly lower than that of C_{SP} within the frequency range of 150 kHz to 30 MHz, which is used for evaluating conducted electromagnetic interference (EMI). Consequently, C_{OUT} has minimal impact on the common-mode noise in this pathway.

As shown in Figure 4, the transmission path of the common-mode noise, i_{CM3}, involves a heatsink. Here, noise arising from voltage jumps (V_P) in the primary-side switching tube generates a noise current. This current flows through the distributed capacitance of the heatsink (C_{PH}), the distributed capacitance of the heatsink relative to the ground (C_{SG}), and the LISN, forming a complete noise transmission loop.

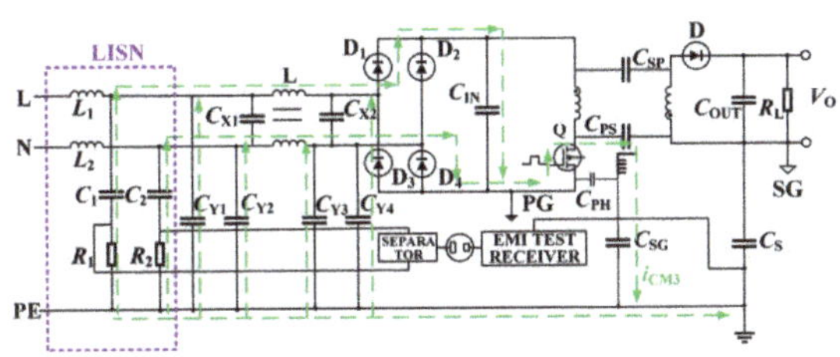

Figure 4. Common-mode noise flowing through the heatsink.

The raw common-mode noise intensity, i_{CM3}, (without filtering) in this method can be calculated using the following equation:

$$i_{CM3} = \frac{(C_{PH}C_{SG})}{C_{SP} + C_{SG}}\frac{du_P}{dt} \tag{3}$$

In an actual engineering environment, when the leakage current permitted by safety regulations is less than 0.35 mA, the capacitance value of the applied capacitor is adjusted. Simultaneously, based on the current value on the main line inductor (3A), the inductance is chosen to be 19 mH. The upper and lower stages of the common-mode inductance, L, function as a common-mode choke. When the common-mode current flows through them, the interference current aligns with the common-mode current, creating a high-impedance characteristic that effectively suppresses the common-mode current. This plays a significant role in reducing common-mode noise in the circuit.

To further clarify the impact of conducted interference, the interference propagation path is equivalently simplified below. The equivalent circuit is shown in Figure 5: R_1 is the 20 Ω equivalent impedance of the LISN, V_C is the voltage on the LISN, V_P is the equivalent voltage of the switching tube, V_D is the equivalent voltage of the diode, i_{PS} is the equivalent capacitance current of the primary-side noise source with respect to the secondary-side winding, and i_{SP} is the equivalent capacitance current of the secondary-side noise source with respect to the primary-side winding. Obviously, the common-mode displacement current will exist equally.

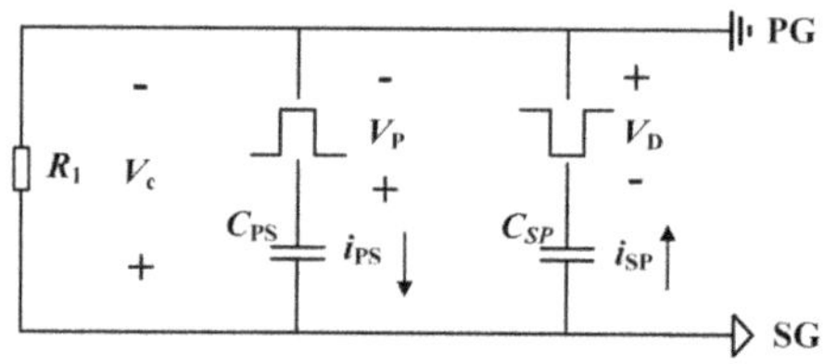

Figure 5. Conducted interference equivalent model.

3. Conducted Common-Mode Noise Improvement Measures

3.1. Compensation Capacitors Improve Noise Effects

To optimize the EMC performance of flyback power supplies, additional filter capacitors, especially Y capacitors for common-mode noise suppression, are usually introduced into the transformer's primary and secondary circuits to mitigate the effects of the conducted noise. This principle involves measuring the actual noise current, i_{PS}, on the primary side and the noise current, i_{SP}, on the secondary side, then adjusting the impedance of the noise path according to the measured values by repositioning the added capacitor compensation accordingly to suppress common-mode noise generation. If the primary noise, i_{PS}, increases, the capacitor, C_{Y5}, across the primary and secondary edges is placed on top of the transformer winding, and the secondary impedance is lowered by C_{Y5} to increase i_{SP}.

$$i_{SP} = \frac{(C_{SP} + C_{Y5})dV_P}{dt} \tag{4}$$

The secondary noise current increases with the increase in the secondary impedance. When the secondary noise is similar to the primary noise, the common-mode noise of the primary and secondary sides cancels each other out. As a result, the common-mode current flowing through the LISN impedance network decreases. Therefore, if the secondary noise, i_{SP}, is larger, the compensation capacitor, C_{Y6}, should be placed in the opposite direction. At this time, the primary-side impedance decreases, and the value of i_{PS} increases accordingly.

$$i_{PS} = \frac{(C_{PS} + C_{Y6})dV_D}{dt} \tag{5}$$

The total common-mode noise is $i_{CM} = i_{PS} - i_{SP}$. When the primary and secondary noises are equal, the conducted common-mode noise can be suppressed from the source. Based on the filtering theory, we can initially estimate the capacitance value of the Y capacitor. Under rational conditions, assuming that the maximum allowable ripple voltage is ΔV, the capacitor needs to provide a charge in one switching cycle equal to the charge change of the input current ripple, I_{rip}, in that cycle, and according to the basic capacitor charging and discharging principle, the capacitance calculation formula is as follows:

$$C = \frac{(I_{rip} \times T)}{\Delta V} \tag{6}$$

where I_{rip} is the desired common-mode noise current ripple, T is the operating period of the switching power supply, and ΔV is the maximum common-mode noise voltage fluctuation allowed on both sides of the capacitor. Experimental investigations on various values of compensation capacitance suggest an optimal range of 5 pF to 10 pF. If the value of the compensation capacitance is too small, the compensation effect may not be significant. Conversely, if the compensation capacitance is too large, it may result in overcompensation, leading to a deterioration in the common-mode noise suppression effect. The circuit model with a compensation capacitor added is shown in Figure 6.

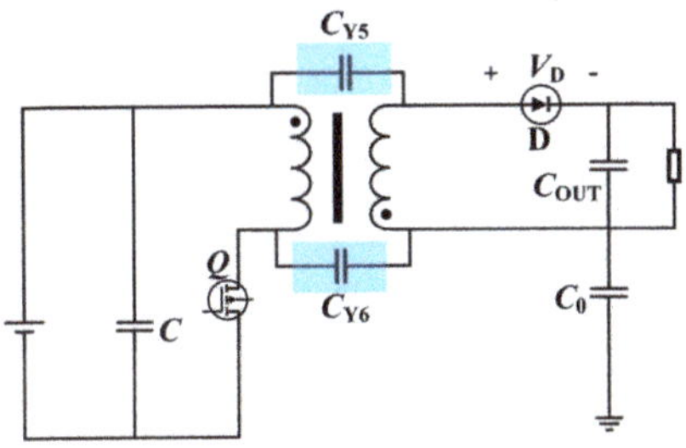

Figure 6. Circuit model with compensation capacitor added (C_{Y5} or C_{Y6}).

3.2. Balanced Winding Improves Noise Impact

The high-frequency switching characteristics of GaN devices lead to an asymmetric distribution of current in the circuit. The current conversion between the primary and secondary windings of the transformer, combined with the transformer's leakage inductance and other parasitic parameters, unavoidably generates common-mode noise. To mitigate these issues, balanced windings are added to the transformer's primary side to compensate for and suppress the noise. When the common-mode noise current passes through the power line, the balanced winding senses this noise and generates a reverse displacement current. This reverse displacement current creates a magnetic field in the core that cancels out the magnetic field of the original common-mode noise, significantly reducing the propagation of common-mode noise on the power line and greatly enhancing the EMI performance of the power system.

The transformer circuit model with a balanced winding added is shown in Figure 7. In this model, Pri denotes the primary winding, Sec denotes the secondary winding, and Bal denotes the balanced winding. V_{SW} represents the voltage across the original primary winding, and Pri1 and Pri2 refer to two single-layer primary windings connected in series with turn numbers N_{P1} and N_{P2}, respectively. In the actual design, the balanced winding is closely surrounded by the outer secondary winding, and its heterodyne end is connected to the primary ground (PG) of the transformer. This ensures that the transformer's original noise can be canceled out with the common-mode noise transmitted from the secondary winding to the balanced winding. The winding arrangement of the transformer with the balanced winding added is shown in Figure 8.

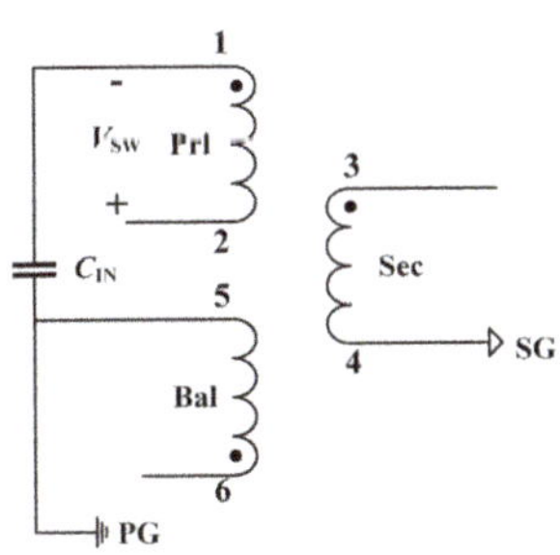

Figure 7. Transformer circuit model with balanced winding added.

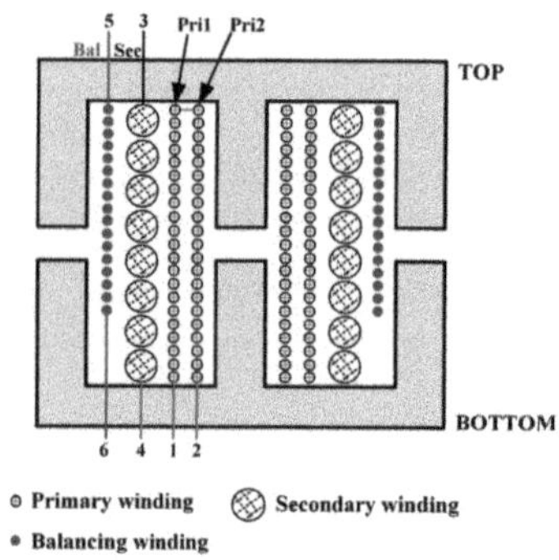

Figure 8. Winding arrangement of the transformer with balanced winding added.

By disregarding the leakage flux phenomenon and ignoring the AC impedance changes caused by the skin effect and proximity effect, we can assume that the potential is uniformly distributed in the winding, and the parasitic capacitance is also uniformly distributed along the winding. Referring to the layout of the winding arrangement and its endpoint connection relationship shown in Figure 8, we can plot the potential distribution of the winding at x = 0 along the y-axis, as shown in Figure 9. Here, h represents the height of the core window, l is the length of one turn of the winding, N_B is the number of turns in the balanced winding, and h_B is the width of the balanced winding.

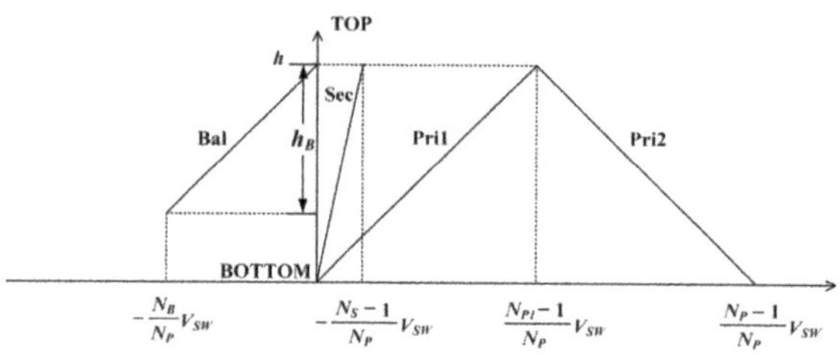

Figure 9. Winding potential distribution diagram (at x = 0).

From the winding potential distribution relationship shown in Figure 9, the potential expression for the primary winding, Pri1, is $L(x,y)$, and the potential expression for the secondary winding, Sec, is $Q(x,y)$:

$$Q(x,y) = \frac{N_S - 1}{N_P} \cdot \frac{y}{h} V_{SW} + \frac{1}{N_P} \cdot \frac{x}{l} V_{SW} \tag{7}$$

$$L(x,y) = \frac{N_{P1} - 1}{N_P} \cdot \frac{y}{h} V_{SW} + \frac{1}{N_P} \cdot \frac{x}{l} V_{SW} \tag{8}$$

The common-mode displacement current, $i_{\text{Pri1\_Sec}}$, from the primary winding, Pri1, to the secondary winding, Sec, can be derived using Equation (9), where $C_{\text{P1\_S}}$ is the parasitic capacitance between the primary winding, Pri1, and the secondary winding, Sec, and $d_{\text{P1\_S}}$ is the distance between the primary winding, Pri1, and the secondary winding, Sec.

$$i_{\text{Pri1\_Sec}} = \int_0^h \int_0^l \frac{C_{\text{P1\_S}}}{h \cdot l} \cdot \frac{d}{dt}[L(x,y) - Q(x,y)]dxdy \tag{9}$$

$$C_{\text{P1\_S}} \approx \varepsilon \frac{hl}{d_{\text{P1\_S}}} \tag{10}$$

The expression for the potential of the balanced winding can be calculated from Equations (7)–(10) as follows:

$$N(x,y) = \frac{N_B}{N_P} \cdot \frac{y}{h} V_{SW} - \frac{N_B}{N_P} \cdot \frac{h}{h} V_{SW} + \frac{1}{N_P} \cdot \frac{x}{l} V_{SW} \quad (11)$$

The common-mode displacement current, i_{Sec_Bal}, from the secondary winding to the balance winding can be determined using Equation (12), where C_{S_B} is the parasitic capacitance between the secondary winding and the balance winding, d_{S_B} is the distance between Sec and Bal, d_B is the diameter of the balance winding, and ε is the dielectric constant.

$$i_{Sec\_Bal} = \int_{h-h_B}^{h} \int_{0}^{l} \frac{C_{S\_B}}{h_B \cdot l} \cdot \frac{d}{dt}[Q(x,y) - N(x,y)]dxdy \quad (12)$$

$$C_{S\_B} \approx \varepsilon \frac{h_B l}{d_{S\_B}} \quad (13)$$

$$N_B \approx \frac{h_B}{d_B} \quad (14)$$

Based on Formulas (7)–(14), and making $i_{Pri1\_Sec} = i_{Sec\_Bal}$, a numerical solution can be found by determining the number of turns in the balanced winding.

3.3. Transformer Structure Improves Noise Impact

Transformers play a crucial role in the electromagnetic compatibility (EMC) performance of power supplies. For flyback power supplies based on GaN devices, the noise generated by the high-frequency switching characteristics of GaN devices is superimposed on the noise produced by the transformer, affecting the EMI characteristics of the power supply. Therefore, it is necessary to implement improvement measures for the transformer. Elements such as the winding structure, winding sequence, and shielding form of the transformer significantly impact the electromagnetic interference (EMI) performance of the entire system. The distributed capacitance of the transformer is influenced by its structural parameters, and the size of the distributed capacitance between windings can be controlled by altering the winding structure and adding shielding layers to enhance the EMI performance of the flyback converter.

Parasitic resonance phenomena caused by the transformer's interlayer capacitance pose a substantial challenge to the electromagnetic compatibility (EMC) performance of converters. To address these issues, the voltage and current stresses on switching tubes during switching are reduced by decreasing the interlayer capacitance, which in turn improves the transformer's operating performance. Figure 10 shows the voltage distribution on the winding under four different winding methods (A, B, C, D). For the same winding, the inter-turn voltages of the two layers of the winding exhibit a decreasing trend, and the equivalent interlayer capacitance decreases accordingly.

Figure 11 shows the winding arrangement structure of the transformer using the ordinary winding method and sandwich winding method, respectively. Compared to the normal winding method, the sandwich winding method features a larger spacing between the two layers of primary windings, resulting in reduced interlayer capacitance. Consequently, the sandwich winding can diminish parasitic resonance in the circuit and mitigate the adverse effects of parasitic oscillations on the EMC performance of the converter.

Referring to Figure 11b, although the sandwich winding method decreases the interlayer capacitance, the increased area between the primary and secondary windings leads to an elevated intergroup capacitance. This increase in the intergroup capacitance exacerbates the common-mode noise of the transformer. To address this issue, a shielding layer can be added between the adjacent primary and secondary windings to reduce direct magnetic coupling, thereby effectively suppressing the propagation of common-mode noise.

When the flyback power supply operates at high frequencies, the common-mode noise generated by the high-speed switching of the transistors creates a strong electromagnetic field within the transformer. A sandwich-type winding structure can mitigate most of this noise by utilizing a shielding layer, which effectively guides the noise current to the ground, thereby eliminating conduction through the power line. Consequently, it significantly reduces electromagnetic interference with the external environment and the power grid.

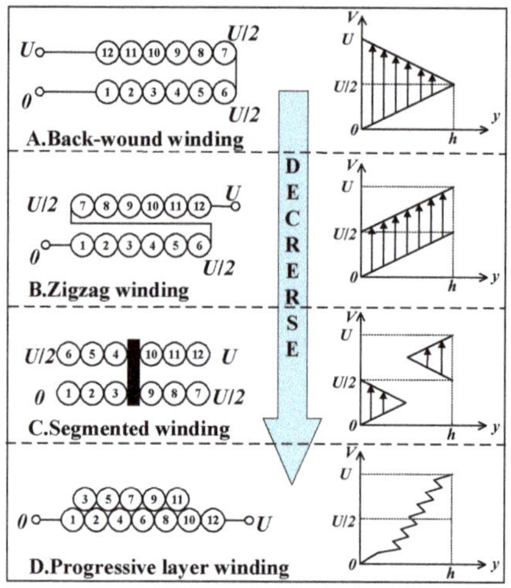

Figure 10. Different winding structures and their corresponding winding voltage differences.

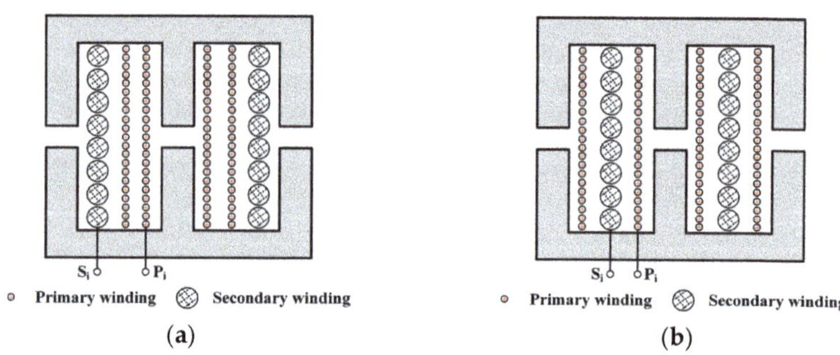

Figure 11. Winding arrangement structure corresponding to the two different winding methods for the transformer: (**a**) the ordinary winding method; (**b**) the sandwich winding method.

3.3.1. Transformer Distributed Capacitance Modeling

The common-mode noise transmission path with a shielding layer added is shown in Figure 12. This figure incorporates the path i_{SSH} into the original common-mode noise transmission path. The noise is generated by a voltage jump, V_P, in the primary-side switching tube, forming a noise current that flows through the distributed capacitance, C_{SSH}, between the secondary winding and the shielding layer. It then passes through the rectifier bridge and the LISN, eventually returning to the diode, D, from the ground. Figure 13 shows an equivalent model of conducted interference with a shielding layer added.

The shielding layer can be categorized into two forms: copper foil shielding and winding shielding. Figure 14 shows the two forms of transformer shielding.

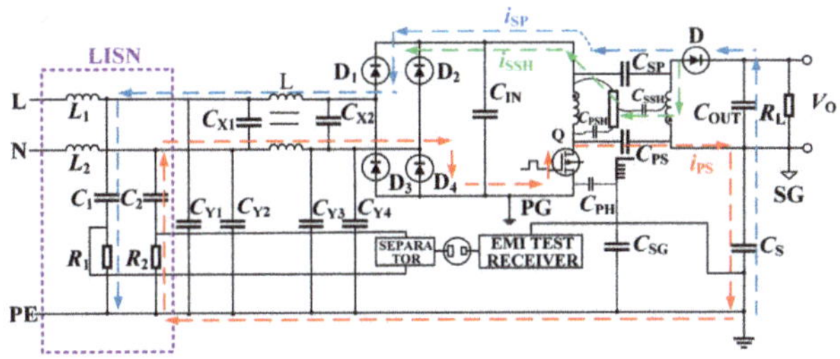

Figure 12. Common-mode noise transmission path with shielding layer added.

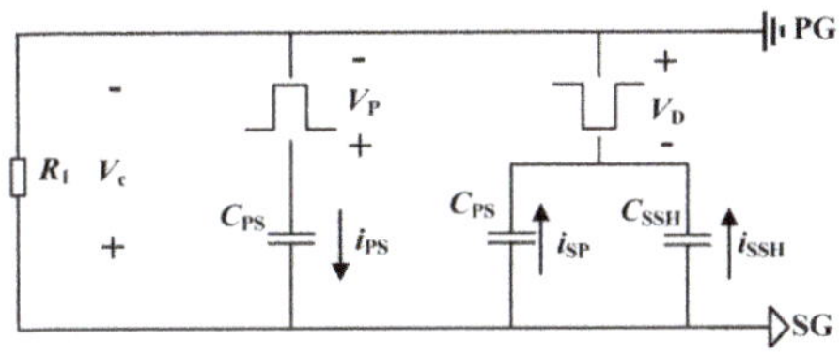

Figure 13. Equivalent model of conducted interference with shielding layer added.

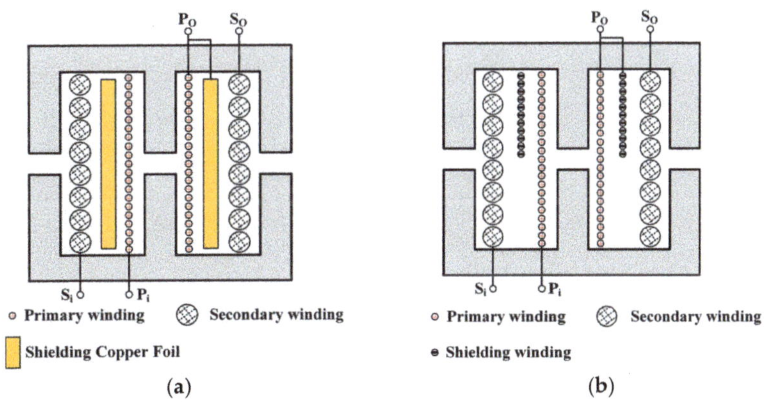

Figure 14. Two forms of transformer shielding layers: (**a**) transformer with copper foil shielding; (**b**) transformer with shielding winding.

During the production process, it is crucial to prevent short circuits caused by overlapping copper foil. This can be achieved by properly wrapping insulating tape at the beginning and end of the foil. Enameled wire with diameters ranging from 0.1 mm to 0.3 mm is commonly used for shielding windings, as it helps maintain consistency in wire diameter and simplifies the manufacturing process. The number of turns and layout of the shielding winding significantly impact the overall shielding efficiency. Therefore, during fine winding operations, it is essential to use insulating tape to ensure robust and effective insulation at the winding ends. In summary, compared to relying solely on copper foil for shielding, the shielding winding design offers more advantages in terms of production convenience and cost control.

When exploring the practical engineering applications of sandwich-structure transformers, the winding point of the shielded winding is a critical yet often overlooked detail. The arrangement order of the windings is as follows: primary winding I, secondary winding, shielding winding, and primary winding II. The starting position of the shielded windings directly affects the transformer's EMI characteristics. During the specific implementation process, the shielding winding should be wound from the transformer's reference ground potential (RGP), with its end connected to the magnetic core. This connection strategy is significant as it ensures that during the transformer's operation, the core and the winding's voltages change in opposite directions, achieving a mutual offset effect and maximizing the core's shielding effectiveness (the drain winding refers to the winding connected to the switching device). However, if the polarity of the shielding winding is misconfigured (i.e., the start and end are reversed), the voltage change direction will align with that of the drain windings, resulting in a loss in the shielding effect in the core. The winding order of the sandwich-structure transformer and the correct winding points for the shielded windings are shown in Figure 15.

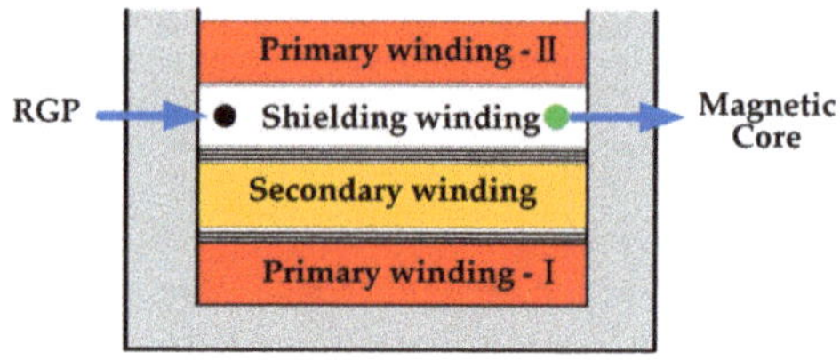

Figure 15. The order of the sandwich-structure transformer's windings and the correct winding points for the shielding winding.

3.3.2. Shielding Winding Design for Flyback Transformers

According to the common-mode noise canceling principle of the flyback transformer, the optimal number of shielding winding turns can be determined by calculating the common-mode current and the number of winding turns. The complete inference steps are described below.

Figure 16 shows the common-mode interference transmission path with a shielding winding added. Here, i_{SHS} represents the common-mode current transferred from the shielding winding to the secondary side, and i_{SSH} represents the common-mode current transferred from the secondary side to the shielding winding.

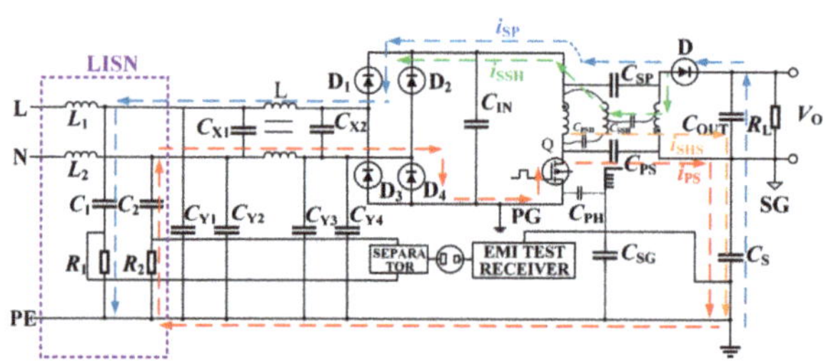

Figure 16. Common-mode interference transmission path with shielding winding added.

The total common-mode current can be obtained from Figure 16 as follows:

$$i_{CM} = (i_{PS} - i_{SP}) + (i_{SHS} - i_{SSH}) \quad (15)$$

where $(i_{PS} - i_{SP})$ represents the common-mode current between the primary and secondary windings, and $(i_{SHS} - i_{SSH})$ represents the common-mode current between the shielding and secondary windings.

The structure of the transformer windings significantly affects the magnitude of the common-mode current and the voltage distribution within the transformer. By utilizing the connections between the transformer windings, further calculations for the parameters $(i_{PS} - i_{SP})$ and $(i_{SHS} - i_{SSH})$ can be performed. Figure 17 shows the structural parameters between the primary and secondary windings of the transformer. Figure 18 shows the structural parameters between the shield winding and the secondary winding of the transformer.

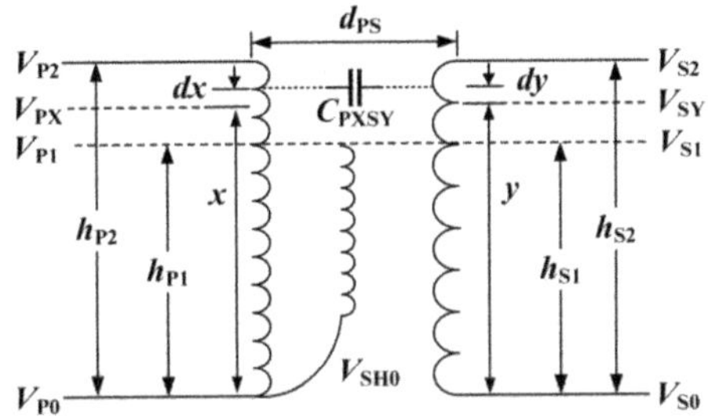

Figure 17. Structural parameters between the primary and secondary windings of the transformer.

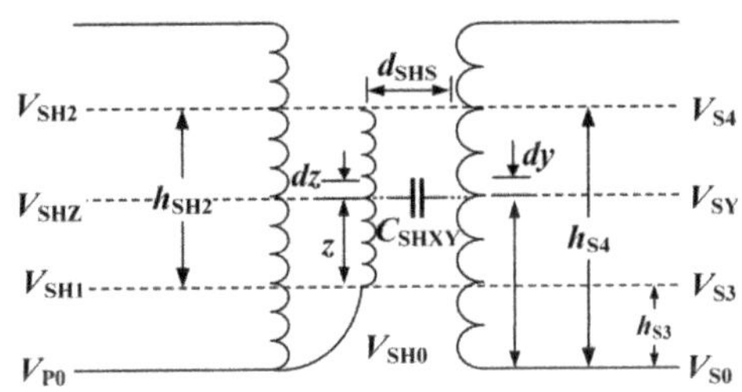

Figure 18. Structural parameters between the shield winding and the secondary winding of the transformer.

V_{P0}, V_{S0}, and V_{SH0} denote the low-potential reference points on the primary, secondary, and shielding windings, respectively. h_{P1}, h_{P1}, and x represent the distances from a point on the original edge to the low-potential point. V_{P1}, V_{P2}, and V_{PX} are the voltages of the primary winding at heights h_{P1}, h_{P2}, and x from the low-potential reference point. h_{S1}, h_{S2}, h_{S3}, h_{S4}, and y denote the distances from a point on the secondary side to the low-potential reference point. V_{S1}, V_{S2}, V_{S3}, V_{S4}, and V_{SY} represent the voltages on the secondary side at heights h_{S1}, h_{S2}, h_{S3}, h_{S4}, and y from the low-potential reference point. h_{SH1}, h_{SH2}, and z denote the distances from a point on the shielding winding to the low-potential reference point. V_{SH1}, V_{SH2}, and V_{SHZ} denote the voltages on the shielding winding at heights h_{SH1}, h_{SH2}, and z from the low-potential reference point. d_{PS} represents the distance from the primary winding to the secondary winding. d_{SHS} represents the distance between the shield and the secondary winding. C_{PXSY} represents the magnitude of the capacitance element between the primary winding and the secondary winding. C_{SHSY} represents the magnitude of the capacitance element between the shield winding and the secondary winding. Assuming that the distributed capacitance between adjacent ends of the winding is uniformly distributed, we obtain the following:

$$\begin{cases} C_{PXSY} = \frac{C_{PS0}}{(h_{P2}-h_{P1})\times(h_{S2}-h_{S1})} \\ C_{SHSY} = \frac{C_{shs0}}{(h_{SH2}-h_{SH1})\times(h_{S4}-h_{S3})} \end{cases} \quad (16)$$

V_{PN0} represents the voltage difference between the first and last terminals of the primary winding, V_{SN0} represents the voltage difference between the first and last terminals of the secondary winding, V_{SHN0} represents the voltage difference between the first and last terminals of the shielding winding, N_P represents the number of turns on the primary winding, N_S represents the number of turns on the secondary winding, and v represents the voltage increment for each turn on the winding. By considering these variables, the following relationship can be derived:

$$\begin{cases} V_{PN0} = N_P \times v \\ V_{SN0} = N_S \times v \\ V_{SHN0} = N_{SH} \times v \end{cases} \tag{17}$$

The heights of the primary winding, secondary winding, and shielding winding are denoted as H_P, H_S, and H_{SH}, respectively. The following relationship can then be established:

$$\begin{cases} \frac{dV_{PX}}{dt} = \frac{d\left(V_{P0}+\frac{x}{H_P}\times N_P\times v\right)}{dt} \\ \frac{dV_{SY}}{dt} = \frac{d\left(V_{S0}+\frac{y}{H_S}\times N_S\times v\right)}{dt} \\ \frac{dV_{SHZ}}{dt} = \frac{d\left(V_{SH0}+\frac{z}{H_{SH}}\times N_{SH}\times v\right)}{dt} \end{cases} \tag{18}$$

Based on the structural parameters labeled in Figures 17 and 18, and in conjunction with the current and voltage relationship in the capacitor, the following equation can be formulated:

$$\begin{cases} i_P = \int_{h_{S1}}^{h_{S2}} \int_{h_{P1}}^{h_{P2}} C_{PXSY} \frac{d(V_{PX} - V_{SY})}{dt} dxdy \\ i_{SH} = \int_{h_{S3}}^{h_{S4}} \int_{h_{SH1}}^{h_{SH2}} C_{SHSY} \frac{d(V_{SHZ} - V_{SY})}{dt} dzdy \end{cases} \tag{19}$$

By substituting Equation (18) into (19), we obtain the following:

$$\begin{cases} i_P = \int_{h_{S1}}^{h_{S2}} \int_{h_{P1}}^{h_{P2}} C_{PXSY} \left[\frac{d(V_{P0} - V_{S0})}{dt} + \frac{x}{H_P} \times N_P \frac{dv}{dt} - \frac{y}{H_S} \times N_S \frac{dv}{dt} \right] dxdy \\ i_{SH} = \int_{h_{S3}}^{h_{S4}} \int_{h_{SH1}}^{h_{SH2}} C_{SHSY} \left[\frac{d(V_{SH0} - V_{S0})}{dt} + \frac{z}{H_{SH}} \times N_{SH} \frac{dv}{dt} - \frac{y}{H_S} \times N_S \frac{dv}{dt} \right] dzdy \end{cases} \tag{20}$$

The values of $d(V_{P0} - V_{S0})/dt$, $d(V_{SH0} - V_{S0})/dt$, and dv/dt in Equation (20) are associated with the interference source. The relationship between these values can be expressed as shown in Equation (21):

$$\begin{cases} i_P = \int_{h_{S1}}^{h_{S2}} \int_{h_{P1}}^{h_{P2}} C_{PXSY} \frac{d(V_{PX} - V_{SY})}{dt} dxdy \\ i_{SH} = \int_{h_{S3}}^{h_{S4}} \int_{h_{SH1}}^{h_{SH2}} C_{SHSY} \frac{d(V_{SHZ} - V_{SY})}{dt} dzdy \end{cases} \tag{21}$$

By substituting Equation (21) into Equation (20), the following relationship is obtained:

$$\begin{cases} i_P = C_{PS0} \frac{dv}{dt} \times \left(N_0 + \frac{h_{P1}+h_{P2}}{2\times H_P} \times N_P - \frac{h_{S1}+h_{S2}}{2\times H_S} \times N_S \right) \\ i_{SH} = C_{SHS0} \frac{dv}{dt} \times \left(N_1 + \frac{h_{SH1}+h_{SH2}}{2\times H_{SH}} \times N_{SH} - \frac{h_{S3}+h_{S4}}{2\times H_S} \times N_S \right) \end{cases} \tag{22}$$

C_{PS0} represents the intergroup distributed capacitance between the primary and secondary windings, while C_{SHS0} represents the intergroup distributed capacitance between the secondary and shielding windings. C_{PS0} and C_{SHS0} can be represented by the following equations:

$$\begin{cases} C_{PS0} = k_{PS}\varepsilon_{PS} \frac{(h_{P2}-h_{P1})\times l_{avps}}{d_{PS}} \\ C_{SHS0} = k_{SHS}\varepsilon_{SHS} \frac{(h_{SH2}-h_{SH1})\times l_{avshs}}{d_{SHS}} \end{cases} \tag{23}$$

Here, k represents a constant, ε represents the dielectric constant, and l represents the average circumference of the winding.

3.3.3. Determination of Optimum Number of Turns for Shielding Winding

Given the overall architecture and materials of the transformer, the value of the structural capacitance depends solely on the number of turns on the shielding winding, N_{SH}. There is a functional relationship between the common-mode current and the shielding winding, which can be represented as $i_{CM} = g(N_{SH}) \times dv/dt$, where g($N_{SH}$) is a function of the number of turns on the shielding winding.

For any section of the transformer where the primary winding is adjacent to the secondary winding, the common-mode current, i_P, between that adjacent section can be represented as follows:

$$i_P = g_P(N_{SH}) \times \frac{dv}{dt} \tag{24}$$

$$g_P(N_{SH}) = k_{PS}\varepsilon_{PS}\frac{(h_{P2} - h_{P1}) \times l_{avps}}{d_{PS}}\left(N_0 + \frac{h_{P1} + h_{P2}}{2 \times H_P} \times N_P - \frac{h_{S1} + h_{S2}}{2 \times H_S} \times N_S\right) \tag{25}$$

Similarly, the common-mode current, i_{SH}, between any section of the shielding winding and the secondary winding in the transformer can be represented as follows:

$$i_{SH} = g_{SH}(N_{SH}) \times \frac{dv}{dt} \tag{26}$$

$$g_{SH}(N_{SH}) = k_{SHS}\varepsilon_{SHS}\frac{(h_{SH2} - h_{SH1}) \times l_{avshs}}{d_{SHS}}\left(N_1 + \frac{h_{SH1} + h_{SH2}}{2 \times H_{SH}} \times N_{SH} - \frac{h_{S3} + h_{S4}}{2 \times H_S} \times N_S\right) \tag{27}$$

H_{SH}, h_{P1}, h_{P2}, h_{S1}, h_{S2}, h_{S3}, and h_{S4} can be represented either by known transformer construction parameters or by a function, $g(N_{SH})$, that depends on the number of turns on the shielding winding. This is shown in Equation (27), where w_P represents the width of each turn on the primary winding, w_S is the width of each turn of the primary winding, w_{SH} is the width of each turn on the shielding winding, H_0 is the height between the bottom end of the shielding winding and the bottom end of the primary winding, N_{P2} represents the number of turns on the winding with a corresponding height of h_{P2} in the primary-side winding, and N_{S2} represents the number of turns on the winding with a corresponding height of h_{S2} in the secondary winding.

$$\begin{cases} H_{SH} = w_{SH} \times N_{SH} \\ h_{P1} = H_0 + w_{SH} \times N_{SH} \\ h_{P2} = w_P \times N_{P2} \\ h_{S1} = H_0 + w_{SH} \times N_{SH} \\ h_{S2} = w_S \times N_{S2} \\ h_{S4} = h_{S3} + w_{SH} \times N_{SH} \end{cases} \tag{28}$$

The total common-mode current in the transformer, as a function of the number of turns in the shielding winding, can be represented as follows:

$$i_{CM} = \sum_{j=1}^{n} i_{pj} + \sum_{k=1}^{m} i_{shk} = g(N_{SH}) \times \frac{dv}{dt} \tag{29}$$

i_{pj} represents the common-mode current in the region adjacent to the primary winding and the secondary winding in section j. n represents the number of segments in the transformer in which an adjacent region exists between the primary winding and the secondary winding. i_{shk} represents the common-mode current in the region adjacent to the shielding winding and the secondary winding in section k. m represents the number of segments in which an adjacent region exists between the shielding winding and the secondary winding.

Combined with Equation (29), when $g(N_{SH})$ is at its minimum, the common-mode current, i_{CM}, is also minimized. This can be solved by finding the root of $|g(N_{SH})| = 0$. The optimum number of turns in the shielding winding can be obtained by taking the integer part of the root of $|g(N_{SH})| = 0$.

4. Discussion

To validate the feasibility of the proposed conduction noise suppression method, a 150 W GaN flyback power supply was developed as the test subject, with its main parameters listed in Table 1. To ensure experimental accuracy, three noise reduction schemes (labeled as A, B, and C) were tested using a single experimental prototype. During these tests, only one component of the flyback power supply was altered at a time, such as adding a compensation capacitor, installing a balanced winding in the transformer, or installing a shielding winding in the transformer.

Table 1. Electrical parameters of the prototype.

Electrical Parameters	Rating
Input voltage, V_{in}/V	220
Output voltage, V_o/V	12
Output power, P_o/W	150
Switching frequency, f/Hz	1×10^5

To validate the common-mode interference suppression method proposed in this paper, a conducted interference test platform was constructed in accordance with the CISPR 22 standard. The electrical connection block diagram of the experimental platform is shown in Figure 19. The experimental environment and noise testing instruments are shown in Figure 20. The LISN model used is the Cybertek EM5040B (by Shenzhen Zhiyong Electronics Co., Ltd., Shenzhen, China.), the EMI receiver model is the RS ESCI7 (by Rohde & Schwarz, Munich, Germany.), and the test frequency range is from 150 kHz to 30 MHz. The model of the separator is the TBLM1 MATE (by Satake, Hiroshima, Japan). During the testing process, the LISN was employed to eliminate power grid fluctuations and isolate the required signal, which was then transmitted to the EMI receiver via the separator. The EMI receiver performed spectral analysis on the received signal to obtain the noise spectrogram, which was subsequently transmitted to the host computer for processing and storage.

The EMI test was conducted on the original flyback power supply prototype without implementing the proposed noise reduction method. The transformer in the original flyback power supply prototype employs a standard winding technique, utilizing an EFD20 core model(by TDG, Zhejiang, China), and the core material is TP4 from TDG (Zhejiang, China). The EMI spectrum of the original flyback power supply is shown in Figure 21.

Scheme A: The experimental power supply in this scheme is based on the original flyback design, with an external compensation capacitor added. Figure 22 illustrates the EMI spectrum of the flyback power supply with a 5 pF compensation capacitor, while Figure 23 shows the EMI spectrum with a 20 pF compensation capacitor.

Comparing Figure 22 with the original EMI spectrogram, it is evident that the noise profile decreases by approximately 15 dB·µV in the 150 kHz to 1 MHz range and by about 10 dB·µV in the 1 MHz to 10 MHz range. The addition of appropriately valued compensation capacitors effectively reduces the common-mode noise level, particularly in the low- and mid-frequency bands of the noise spectrum, demonstrating a significant compensatory effect. This solution employs a simple circuit and is quick and effective to implement.

By comparing the test results in Figure 23 with the noise profile in Figure 22, it is evident that increasing the compensation capacitance from 5 pF to 20 pF results in a rise of approximately 20 dB·µV in the noise profile within the 4.5 MHz to 5.5 MHz range, indicating a significant deterioration in the compensation effect. As the compensation capacitance increases, noise propagation is further exacerbated.

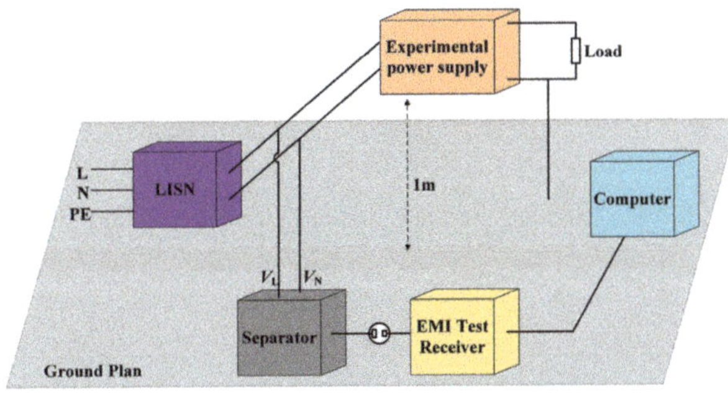

Figure 19. Electrical connection block diagram of the experimental platform.

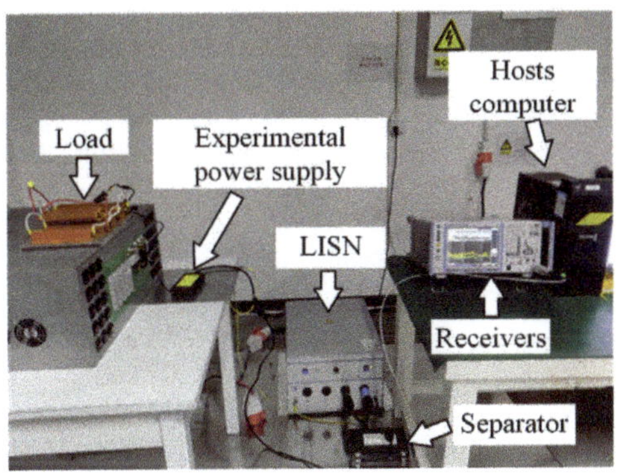

Figure 20. Experimental environment and noise testing instruments.

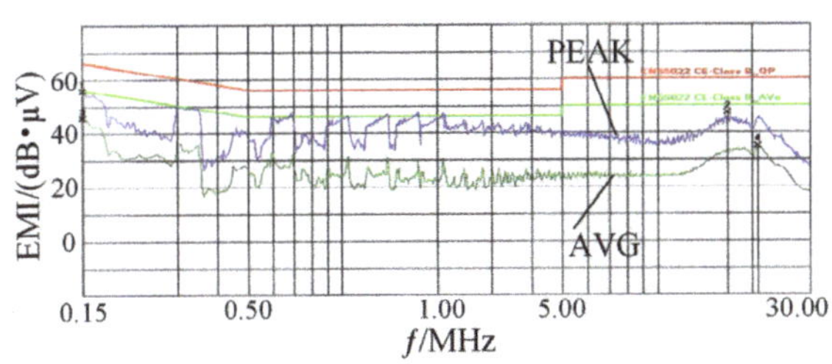

Figure 21. Original EMI spectrum.

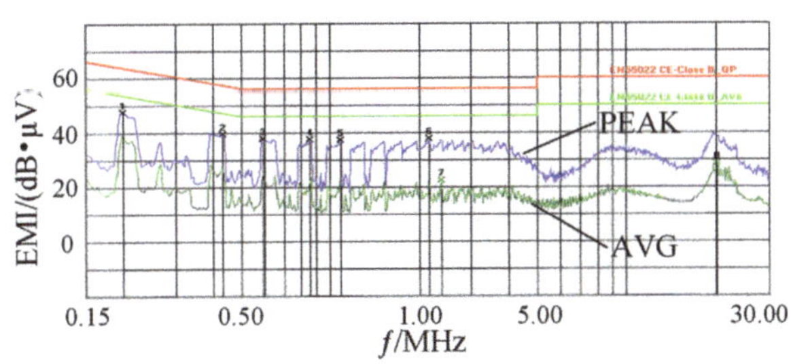

Figure 22. EMI spectrum with compensation capacitor added (5 pF).

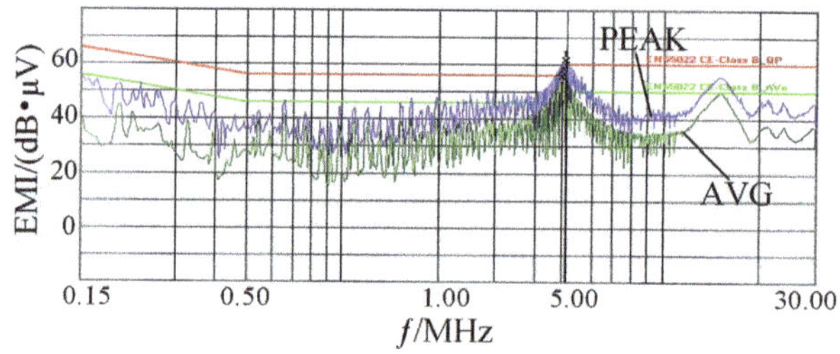

Figure 23. EMI spectrum with compensation capacitor added (20 pF).

Scheme B: This experimental power supply is based on the original flyback power supply, with the addition of a balanced winding in the transformer. The specific parameters of the transformer are detailed in Table 2. Figure 24 shows the EMI spectrum of the transformer with the balanced winding added. Comparing the noise curve in Figure 24 with the original spectrum, it can be observed that the noise level of the converter decreases by approximately 15 dB·μV in the range of 150 kHz to 1 MHz and by about 20 dB·μV in the range of 10 MHz to 20 MHz. This analysis indicates that the balanced winding significantly suppresses common-mode noise in both low- and high-frequency bands.

Table 2. Transformer parameters with balanced winding added.

Parameters (Units)	Rating
N_P (turns)	120
N_S (turns)	9
h (mm)	9.66
d_{P1_S} (mm)	0.325
d_{S_SH} (mm)	0.285
d_{P2_S} (mm)	0.452
d_{S_B} (mm)	0.325
d_B (mm)	0.200

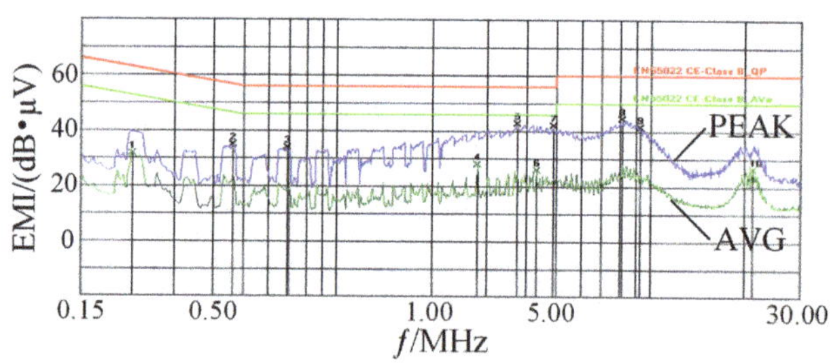

Figure 24. The EMI spectrum of the transformer with the balanced winding added.

Scheme C: The experimental power supply in this scheme is based on the original flyback power supply, with the transformer employing a sandwich winding method and an additional shielding winding. The specific parameters of this transformer are shown in Table 3. Figure 25 shows the EMI spectrum of the transformer with the sandwich structure and additional shielding winding. Comparing the noise curve in Figure 25 with the original noise spectrum, it is evident that the noise spectrum decreased by approximately 15 dB·μV in the range of 150 kHz to 1 MHz, 10 dB·μV in the range of 10 MHz to 20 MHz, and 20 dB·μV in the range of 10 MHz to 30 MHz. The high-frequency noise peaks are significantly attenuated. In summary, the addition of the shielding winding to the transformer significantly improved the full-frequency-band common-mode noise of the flyback converter. Compared with conventional solutions, this program can achieve full-band noise suppression to a certain extent.

Table 3. Transformer parameters with shielding winding added.

Parameters (Units)	Rating
N_P (turns)	120
N_S (turns)	9
N_{SH} (turns)	40
h_{P1} (mm)	8
h_{P2} (mm)	8
h_{P3} (mm)	6.28
w_{SH} (mm)	0.23

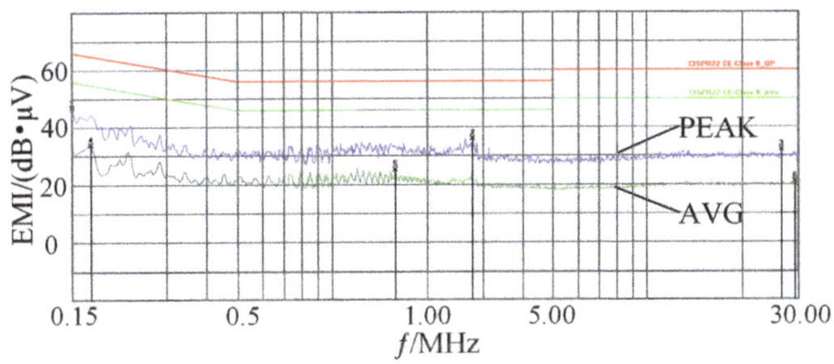

Figure 25. The EMI spectrum of the transformer with sandwich structure and additional shielding winding.

5. Conclusions

Flyback power supplies utilize GaN devices to increase efficiency and power density. However, the high-frequency switching characteristics of GaN devices cause EMI problems, jeopardizing the stable performance of the power supply. This paper provides a detailed analysis of the noise transmission mechanism in GaN-based flyback converters and proposes three improvement schemes to address electromagnetic interference. Scheme A: Based on the characteristics of the noise circuit, an external compensation capacitor is used to improve the impedance of the noise path, allowing primary and secondary noise to cancel each other out. Scheme B: Introducing balanced windings in the transformer effectively reduces common-mode noise caused by transformer leakage inductance and parasitic parameters. Scheme C: To suppress electromagnetic interference at the source, a method involving changes to the transformer structure and the use of shielded windings is proposed, significantly reducing common-mode noise levels and enhancing the electromagnetic compatibility of the GaN-based flyback power supply. Finally, a comparative experiment was conducted using the GaN-based flyback power supply, and the experimental results fully verify the feasibility of the proposed conducted interference suppression methods.

In this paper, three novel improvement schemes are proposed to enhance noise suppression in GaN-based flyback power supplies, providing significant practical guidance for improving their EMI performance.

Author Contributions: Conceptualization, J.Y. and Y.Y.; methodology, J.Y. and H.W.; software, J.Y. and H.W.; validation, J.Y., H.W. and X.F.; formal analysis, M.L.; investigation, Y.Y.; resources, Y.Y. and H.W.; data curation, Y.Y.; writing—original draft preparation, J.Y. and Y.Y.; writing—review and editing, J.Y. and H.W.; visualization, H.W. and X.F.; supervision, H.W.; project administration, J.Y.; funding acquisition, J.Y. and M.L. All authors have read and agreed to the published version of the manuscript.

Funding: This research was funded by the Natural Science Foundation of Guangxi Province of China, grant number 2021GXNSFAA220038.

Data Availability Statement: Data are contained within the article.

Conflicts of Interest: Author Xueliang Fu and Mingtong Li were employed by the company Greatwall Power Supply Technology Co., Ltd. The remaining authors declare that the research was conducted in the absence of any commercial or financial relationships that could be construed as a potential conflict of interest.

References

1. Kozak, J.P.; Barchowsky, A.; Hontz, M.R.; Koganti, N.B.; Stanchina, W.E.; Reed, G.F.; Mao, Z.H.; Khanna, R. An analytical model for predicting turn-ON overshoot in normally-OFF GaN HEMTs. *IEEE J. Emerg. Sel. Top. Power Electron.* **2020**, *8*, 99–110. [CrossRef]
2. Li, F.; Wang, R.; Huang, H.; Ren, Y.; Liang, Z.; Ren, G.; Tao, P.; Sun, Z.; Sun, N.; Zhao, C.; et al. Temperature-dependent hot electron effects and degradation mechanisms in 650-V GaN-based MIS-HEMT power devices under hard switching operations. *IEEE J. Emerg. Sel. Top. Power Electron.* **2021**, *9*, 6424–6431. [CrossRef]
3. Collings, W.; Nelson, T.; Sellers, A.; Khanna, R.; Courtay, A.; Jimenez, S.; Lemmon, A. Optimization algorithms for dynamic tuning of wide bandgap semiconductor device models. In Proceedings of the 2021 IEEE Applied Power Electronics Conference and Exposition (APEC), Phoenix, AZ, USA, 14–17 June 2021; pp. 2427–2433.
4. Sun, J.; Li, J.; Costinett, D.J.; Tolbert, L.M. A GaN-based CRM totem-pole PFC converter with fast dynamic response and noise immunity for a multi-receiver WPT system. In Proceedings of the IEEE Energy Conversion Congress and Exposition, Detroit, MI, USA, 11–15 October 2020; pp. 2555–2562.
5. Yao, J.; Li, Y.; Wang, S.; Huang, X.; Lyu, X. Modeling and reduction of radiated EMI in a GaN IC-based active clamp flyback adapter. *IEEE Trans. Power Electron.* **2021**, *36*, 5440–5449. [CrossRef]
6. Zhuolin, D.; Dong, Z.; Tao, F. Modeling and Prediction of Electromagnetic Interference in SiC Motor Drive Systems. *Trans. China Electrotech. Soc.* **2020**, *35*, 4726–4738.
7. Fei, C.; Yang, Y.; Li, Q.; Lee, F.C. Shielding technique for planar matrix transformers to suppress common-mode EMI noise and improve efficiency. *IEEE Trans. Ind. Electron.* **2018**, *65*, 1263–1272. [CrossRef]
8. Huang, X.; Feng, J.; Lee, F.C.; Li, Q.; Yang, Y. Conducted EMI analysis and filter design for MHz active clamp flyback front-end converter. In Proceedings of the 2016 IEEE Applied Power Electronics Conference and Exposition, Long Beach, CA, USA, 20–24 March 2016; pp. 1534–1540.
9. Fu, K.; Chen, W. Evaluation Method of Flyback Converter Behaviors on Common-Mode Noise. *IEEE Access* **2019**, *7*, 28019–28030. [CrossRef]
10. Li, Y.; Wang, S.; Sheng, H.; Lakshmikanthan, S. Investigate and Reduce Capacitive Couplings in a Flyback Adapter With a DC-Bus Filter to Reduce EMI. *IEEE Trans. Power Electron.* **2020**, *35*, 6963–6973. [CrossRef]
11. Bai, Y.; Yang, X.; Li, X.; Zhang, D.; Chen, W. A novel balanced winding topology to mitigate EMI without the need fora Y-capacitor. In Proceedings of the 2016 IEEE Applied Power Electronics Conference and Exposition(APEC), Long Beach, CA, USA, 20–24 March 2016; pp. 3623–3628.
12. Hsieh, H.; Shih, S. Effects of transformer structures on the noise balancing and cancellation mechanisms of switching power converters. In Proceedings of the 2014 International Power Electronics Conference (IPEC-Hiroshima 2014—ECCE ASIA), Hiroshima, Japan, 18–21 May 2014; pp. 2380–2384.
13. Hsieh, H.I.; Shih, S.F.; Wang, H.; Liao, Y.H. Novel techniques to reduce common-mode (CM) and mix-mode (MM) EMI noise emissions based on noise balance and cancellation. In Proceedings of the 2016 IEEE International Conference on Power and Energy (PECon), Melaka, Malaysia, 28–29 November 2016; pp. 145–149.
14. Pahlevaninezhad, M.; Hamza, D.; Servansing, A.; Bakhshai, A.; Jain, P. An integrated EMI suppression Y-caps applicable to high frequency planar transformers in high power DC-DC converters used for electric vehicle. In Proceedings of the IECON 2012—38th Annual Conference on IEEE Industrial Electronics Society, Montreal, QC, USA, 25–28 October 2012; pp. 4246–4251.
15. Chen, H.; Xiao, J. Determination of Transformer Shielding Foil Structure for Suppressing Common-Mode Noise in Flyback Converters. *IEEE Trans. Magn.* **2016**, *52*, 1–9. [CrossRef]
16. Li, Y.; Zhang, H.; Wang, S.; Sheng, H.; Chng, C.P.; Lakshmikanthan, S. Investigating Switching Transformers for Common Mode EMI Reduction to Remove Common Mode EMI Filters and Y-Capacitors in Flyback Converters. *IEEE J. Emerg. Sel. Top. Power Electron.* **2018**, *6*, 2287–2301. [CrossRef]
17. Fu, K.; Chen, W.; Lin, S. A General Transformer Evaluation Method for Common-Mode Noise Behavior. *Energies* **2019**, *12*, 1984. [CrossRef]
18. Guangdong, D.; Fanghua, Z. Research on Common Mode Conducted Noise Cancellation Method in Flyback Converter With Noise Balance Principle. *Proc. CSEE* **2019**, *39*, 3320–3327.
19. Liu, Y.; Dong, G.; Zhang, F. Common Mode Noise Suppression in Flyback Converters. *Trans. China Electrotech. Soc.* **2019**, *34*, 4795–4803.

20. Chan, Y.P.; Pong BM, H.; Poon, N.K.; Liu, J.C.P. Common-mode noise cancellation in switching-mode power supplies using an equipotential transformer modeling technique. *IEEE Trans. Electromagn. Compat.* **2012**, *54*, 594–602. [CrossRef]
21. Xie, L.; Ruan, X.; Ye, Z. Equivalent noise source: An effective method for analyzing common-mode noise in isolated power converters. *IEEE Trans. Ind. Electron.* **2016**, *63*, 2913–2924. [CrossRef]
22. Bai, Y.; Chen, W.; He, R.; Zhang, D.; Yang, X. EMI noise cancelation by optimizing transformer design without need for the traditional Y-capacitor. In Proceedings of the IEEE Applied Power Electronics Conference and Exposition, Long Beach, CA, USA, 20–24 March 2016; pp. 766–771.
23. Kong, P.; Lee, F.C. Transformer structure and its effects on common mode EMI noise in isolated power converters. In Proceedings of the IEEE Applied Power Electronics Conference and Exposition, Palm Springs, CA, USA, 21–25 February 2010; pp. 1424–1429.
24. Kang, B.G.; Chung, S.K.; Won, J.S.; Kim, H.S. EMI reduction technique of flyback converter based on capacitance model of transformer with wire shiel. In Proceedings of the 2015 9th International Conference on Power Electronics and ECCE Asia (ICPE-ECCE Asia), Seoul, Republic of Korea, 1–5 June 2015; pp. 163–169.
25. Biela, J.; Kolar, J.W. Using Transformer Parasitics for Resonant Converters—A Review of the Calculation of the Stray Capacitance of Transformers. *IEEE Trans. Ind. Appl.* **2008**, *44*, 223–233. [CrossRef]
26. Chen, H.; Zheng, Z.; Xiao, J. Determining the Number of Transformer Shielding Winding Turns for Suppressing Common-Mode Noise in Flyback Converters. *IEEE Trans. Electromagn. Compat.* **2018**, *60*, 1606–1609. [CrossRef]
27. Fan, J.W.T.; Chow, J.P.W.; Chan, W.T.; Zhang, K.; Relekar, A.; Ho, K.W.; Tung, C.P.; Wang, K.W.; Chung, H.S.H. Modeling and experimental assessment of the EMI characteristics of switching converters with power semiconductor filters. *IEEE Trans. Power Electron.* **2020**, *35*, 2519–2533. [CrossRef]

energies

MDPI

Article

Research on a Distributed Photovoltaic Two-Level Planning Method Based on the SCMPSO Algorithm

Ang Dong [1,2,*] and Seon-Keun Lee [1]

[1] Department of Energy and Electrical Engineering, Woosuk University, Jeonju 376-701, Republic of Korea; caiserrisk@woosuk.ac.kr

[2] Electrical Engineering and Automation, School of Physics and Electronic Engineering, Yan Cheng Teachers University, Yancheng 224002, China

* Correspondence: yixing0930@outlook.com; Tel.: +82-010-8351-7328

Abstract: In response to challenges such as voltage limit violations, excessive currents, and power imbalances caused by the integration of distributed photovoltaic (distributed PV) systems into the distribution network, this study proposes at two-level optimization configuration method. This method effectively balances the grid capacity and reduces the active power losses, thereby decreasing the operating costs. The upper-level optimization enhances the distribution network's capacity by determining the siting and sizing of distributed PV devices. The lower-level aims to reduce the active power losses, improve the voltage stability margins, and minimize the voltage deviations. The upper-level planning results, which include the siting and sizing of the distributed PV, are used as initial conditions for the lower level. Subsequently, the lower level feeds back its optimization results to further refine the configuration. The model is solved using an improved second-order oscillating chaotic map particle swarm optimization algorithm (SCMPSO) combined with a second-order relaxation method. The simulation experiments on an improved IEEE 33-bus test system show that the SCMPSO algorithm can effectively reduce the voltage deviations, decrease the voltage fluctuations, lower the active power losses in the distribution network, and significantly enhance the power quality.

Keywords: distributed photovoltaic (PV); distribution network capacity; siting and sizing; SCMPSO; two-level optimization

Citation: Dong, A.; Lee, S.-K. Research on a Distributed Photovoltaic Two-Level Planning Method Based on the SCMPSO Algorithm. *Energies* **2024**, *17*, 3251. https://doi.org/10.3390/en17133251

Academic Editor: Oscar Barambones

Received: 3 June 2024
Revised: 26 June 2024
Accepted: 28 June 2024
Published: 2 July 2024

1. Introduction

Solar energy, as one of the most abundant clean energy sources on Earth, is widely regarded as an effective solution to mitigate the current energy crisis [1]. Among the various ways to utilize solar energy, photovoltaic (PV) power generation is particularly seen as an efficient method. Currently, distributed PV power generation is a common mode of PV power generation, typically integrated into low-voltage distribution networks as a power source [2,3]. This generation method has advantages such as small size, flexible installation locations, and short construction periods, allowing for quick integration into the distribution network. However, the integration of PV power generation also alters the topology and power flow distribution of the distribution network. The uncertainty and variability of its power output may impact the stable operation of the distribution network [4–6]. Once connected to the grid, PV can potentially reduce the power quality and safety of the distribution network. Therefore, a reasonable layout of PV systems is key to addressing these issues.

To effectively assess the performance of PV grid integration, current standards primarily include power system interconnection standards, power quality standards, safety standards, and technical performance standards [7]. Among these standards, the distribution network's capacity to accommodate PV is a crucial indicator within the power system interconnection standards. This indicator is essential to ensure that PV systems can effectively connect to the distribution network, thereby helping to maintain the overall operational efficiency and safety of the distribution network. The capacity of a PV distribution network refers to the maximum PV capacity that can be integrated under the maximum power load or current level that the distribution network can withstand. However, this indicator does not reflect the operational status of the distribution network equipment, the effects of PV grid integration, and other influencing factors [8]. In [9], the siting and sizing planning of PV grid integration based on considering capacity are studied. In [10], the distribution network capacity is evaluated based upon three aspects: the equipment level, load supply capability, and distribution network structure. While considering the capacity of the distribution network, PV grid integration must also consider various constraints. The main constraints limiting the integration capacity of PV include the voltage thresholds at each node, PV output constraints, harmonic content, and grid distortion levels. These conditions collectively determine the safety and efficiency of the PV system's grid integration [11–13].

Currently, there is a substantial amount of international research on the planning of the distributed photovoltaic (PV) grid integration. Most studies have established optimization objective models based on a single factor, such as solely optimizing for node voltage thresholds, harmonics, or the generalized short-circuit ratio. However, these optimization models do not have universality and are not universally applicable to other constraints. In practical applications, distributed PV grid integration needs to consider multiple constraints to meet real-world requirements. For example, the model proposed in reference [14] performs distributed PV capacity analysis constrained by voltage quality and short-circuit capacity; reference [15] considers the economic perspective and proposes an optimal planning model for the investment, operation, and maintenance of distributed PV grid integration—while references [16,17] optimize based on PV output, node voltage thresholds, and the maximum carrying capacity of currents. However, PV grid integration planning is a nonlinear problem with multiple objective constraints, characterized by complexity and a large computational scale. This makes it challenging for traditional algorithms to effectively optimize and find the optimal solution.

Building on previous research, this paper proposes a new two-level PV grid integration planning method that comprehensively considers multiple aspects such as the distribution network capacity and active power loss. The upper-level planning aims to enhance the network capacity by the precise siting and sizing of PV equipment, while the lower-level planning focuses on reducing active power loss, lowering daily operating costs, and minimizing voltage deviations [18]. To solve this model, an improved second-order oscillating chaotic map particle swarm optimization algorithm (SCMPSO) combined with a dual relaxation method is employed. Case study analysis verifies the convergence and effectiveness of this algorithm.

2. Siting and Sizing Planning of PV Equipment

Figure 1 illustrates the structure of distributed PV integration into the distribution network. As shown in Figure 1, the distribution network feeder contains a total of n nodes. U_i represents the voltage at node i, measured in kilovolts (kv); i denotes any node. Ri and Xi are the equivalent resistance and reactance of branch i, measured in ohms (Ω). Pi and Qi represent the active and reactive loads at node i, measured in kilowatts (kw) and kilovars (kvar), respectively. P_{pv} denotes the power input from the PV system, measured in kilowatts (kw). In a power system, $R_i + jX_i$ represents the impedance, where R_i denotes the resistance between node i and node i+1, and Xi represents the reactance between node i and node i + 1. The j typically represents the imaginary unit. In Figure 1, S represents the power source. The red box displays a schematic diagram of the components of a distributed photovoltaic system.

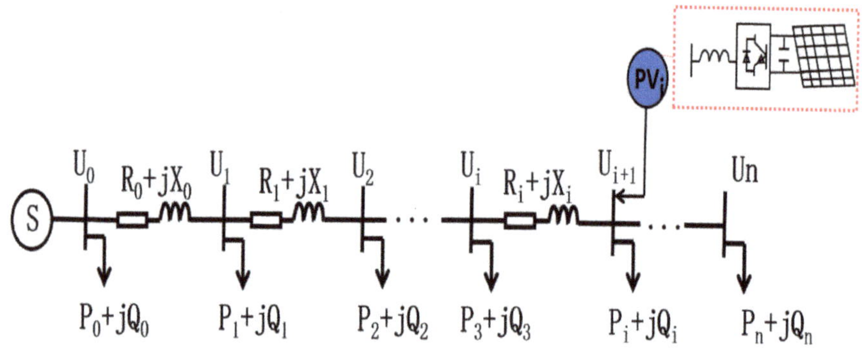

Figure 1. Distribution network PV integration structure.

After the integration of PV, the power flow and voltage distribution in the distribution network are altered. For example, at node i, the voltage difference between the adjacent nodes before PV integration is expressed as:

$$\Delta U_i = U_i - U_{i-1} = -\frac{R_i \sum_{m=i}^{n} P_m + X_i \sum_{m=1}^{n} Q_m}{U_{i-1}} \tag{1}$$

In Equation (1), P_m represents the real power consumed or generated by the system from the power source to node m. Q_m represents the reactive power consumed or generated by the system from the power source to node m.

The node voltage expression for node i is:

$$\begin{aligned} U_i = U_0 + \sum_{k=1}^{i} \Delta U_k &= U_0 - \sum_{k=1}^{i} \frac{R_k \sum_{m=k}^{n} P_m + X_k \sum_{m=k}^{n} Q_m}{U_{k-1}} \\ &= U_0 - \sum_{k=1}^{i} \frac{R_k \sum_{m=k}^{n} (P_{L,j} - P_{DG,j}) + X_k \sum_{m=k}^{n} (Q_{L,j} - Q_{DG,j})}{U_{k-1}} \end{aligned} \tag{2}$$

In Equation (2), U_0 represents the voltage at the node connecting the distribution network to the higher-level grid, i.e., the voltage at node $k-1$, where k takes values 1, 2, . . ., n. $P_{PV,j}$ and $Q_{PV,j}$ represent the active and reactive power of the distributed PV at node j, respectively, with j ranging from 1 to n. $P_{L,j}$ and $Q_{L,j}$ denote the active and reactive loads at node j, respectively. R_k and X_k are the resistance and reactance between nodes k and $k-1$, respectively.

When PV equipment is integrated into the distribution network, the integration point is represented by any node t. The voltage deviation between any two upstream nodes and the voltage expression for any upstream node iii are as follows:

$$\Delta U_i' = U_i' - U_{i-1}' = -\frac{R_i(\sum_{m=i}^{n} P_m - P_{pv}) + X_i \sum_{m=i}^{n} Q_m}{U_{i-1}'} \tag{3}$$

$$U_i' = U_0 - \sum_{k=1}^{i} \frac{R_k \times (\sum_{m=k}^{n} P_m - P_{pv}) + X_k \sum_{m=k}^{n} Q_m}{U_{k-1}'} \tag{4}$$

When the PV equipment is integrated into the distribution network, the integration point is represented by any node t. The voltage deviation between any two downstream nodes of the integration point, as well as the voltage expression for any downstream node i, are as follows:

$$\Delta U'_i = U'_i - U'_{i-1} = -\frac{R_i \sum_{m=i}^{n} P_m + X_i \sum_{m=i}^{n} Q_m}{U'_{i-1}} \tag{5}$$

$$\begin{aligned} U'_i = U_0 - \sum_{k=1}^{t} \frac{R_k \times \left(\sum_{m=k}^{n} P_m - P_{pv}\right) + X_k \times \sum_{m=k}^{n} Q_m}{U'_{k-1}} \\ - \sum_{k=t+1}^{i} \frac{R_k \times \sum_{m=k}^{n} P_m + X_k \times \sum_{m=k}^{n} Q_m}{U'_{k-1}} \end{aligned} \tag{6}$$

According to Equations (1) and (2), it can be observed that, in the absence of PV integration, the voltage gradually decreases along the transmission line. From Equation (3) to (6), it can be seen that, when the output power of the distributed PV at the integration node exceeds the load power, the power flow at the node will reverse, turning the node into an active node. In this scenario, the node will inject power into the distribution network. As the output power of the distributed PV increases, the node voltage rises accordingly, especially at locations close to the load end, where the voltage rise is more significant. This may cause the voltage at the distribution network nodes to exceed the normal threshold range.

Considering the impact of distributed PV integration on the carrying capacity of the distribution network, this study rationally plans the siting and sizing of PV equipment from the perspectives of power quality and system operational safety to ensure the safe and economical operation of the system. This study plans the PV integration capacity while ensuring the maximum carrying capacity of the distribution network and conducts the detailed planning and optimization of the integration location and capacity of the equipment [19].

3. Two-Level Planning for Distributed PV

The integration of PV into the distribution network affects its carrying capacity. To enhance the carrying capacity of the distribution network, reduce active power losses, increase economic benefits, and ensure the practical feasibility of PV integration, this paper proposes a new two-level model construction scheme [19]. This scheme aims to determine the siting and sizing of PV and energy storage equipment to improve the system stability and economic efficiency. The planning framework is as follows:

Figure 2 provides a detailed introduction to the two-level planning framework for distributed photovoltaics. The PV model planning mentioned in this paper is divided into two levels. The upper-level planning mainly focuses on maximizing the load-bearing capacity of the distribution network. It establishes a planning model by combining the investment costs of the distributed PV with the system active power loss to determine the siting and sizing of PV equipment. The lower-level planning, on the other hand, aims to reduce the system active power loss, improve the system stability margin, and minimize the voltage deviation. This is achieved by solving the model using an improved second-order oscillating chaotic mapping particle swarm optimization algorithm. The siting and sizing of equipment determined by the upper-level planning become the initial conditions for the lower-level optimization. The results of the lower-level planning are then fed back to the upper level, further optimizing the installation locations and capacities of the PV equipment.

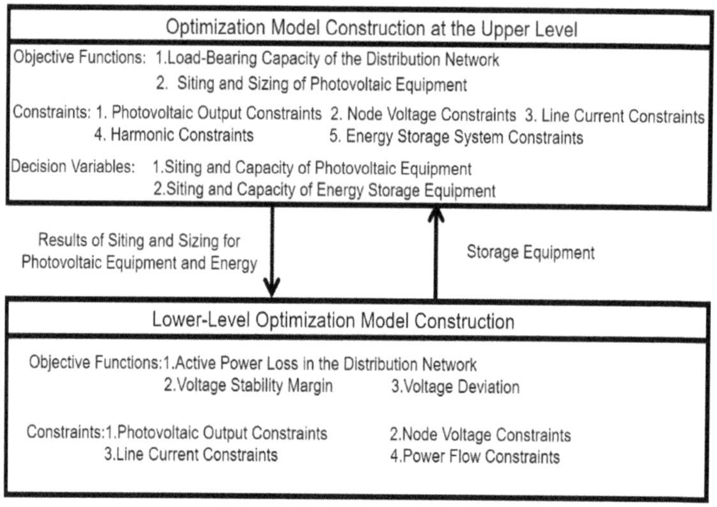

Figure 2. Two-level planning framework for distributed photovoltaics.

4. Upper-Level Model Construction

4.1. Distribution Network Maximum Load-Bearing Capacity Model

4.1.1. Establishment of the Objective Function

The establishment of an evaluation model for the load-bearing capacity of the distribution network under multiple constraints:

$$F_1 = \max\{\sum_{i \in \Psi DG} S_i\} \tag{7}$$

In Equation (7), F_1 represents the maximum load-bearing capacity that the system can withstand, and Si represents the maximum allowable capacity of PV at node i (KVA). Ψ_{DG} denotes the set of distributed photovoltaic points.

4.1.2. Constraints

1. Distributed Photovoltaic Output Constraints:

$$P_{DG,i,min} \leq P_{DG,i} \leq P_{DG,i,max} \tag{8}$$

$$Q_{DG,i,max} \leq Q_{DG,i} \leq Q_{DG,i,max} \tag{9}$$

In the above equations, $P_{DG,i}$ and $Q_{DG,i}$ represent the active and reactive power outputs of the distributed photovoltaic at node i, respectively. $P_{DG,i,min}$ and $P_{DG,i,max}$ denote the lower and upper limits of active power output at node i, respectively. Similarly, $Q_{DG,i,min}$ and $Q_{DG,i,max}$ denote the lower and upper limits of reactive power output at node i, respectively.

2. Node Voltage Constraints:

Translation: According to the "Electric Power Quality Supply Voltage Deviation" GB/T 12325-2008 [20], lines of different voltage levels are allowed different ranges of voltage deviation, and node voltages are constrained by these voltage deviations.

$$U_N(1-\varepsilon) \leq U_i \leq U_N(1+\varepsilon) \tag{10}$$

In Equation (10): U_N represents the rated voltage of the distribution network at that level (KV); ε represents the voltage deviation rate at that level.

3. Line Current Constraints:

$$I_{i-j,min} \le I_{i-j} \le I_{i-j,max} \tag{11}$$

In Equation (11), I_{i-j} represents the current value between branch i-j, $I_{i-j,min}$, and $I_{i-j,max}$, respectively, indicating the minimum and maximum currents that can pass through that branch.

4. Harmonic Constraints:

According to GB/T12325—2008 "Power Quality Public Network Harmonics" [20], the maximum allowable harmonic current values vary for systems with different voltage levels and reference short-circuit capacities. To ensure the effectiveness of harmonic current constraints, we determine the harmonic constraints under the maximum output of the photovoltaic equipment:

$$I_{i-j} = \frac{S_i}{\sqrt{3}U_n} \times \mu \le I_k \tag{12}$$

In Equation (12), K represents the k-th order harmonic current, and μ represents the maximum allowable harmonic current as per the national standard GB/T 1549-1993 "Power Quality Public Network Harmonics" [21].

5. Energy Storage Constraints:

$$P^t_{j,ESS} = P^t_{j,ch} - P^t_{j,dc} \tag{13}$$

$$|\mu^t_{j,ch}| + |\mu^t_{j,dch}| \le 1 \tag{14}$$

$$0 \le P^t_{j,ch} \le P^{max}_{j,ESS} \tag{15}$$

$$0 \le |P^t_{j,dch}| \le |\mu^t_{j.dch} P^{max}_{j,ESS}| \tag{16}$$

$P^t_{j,ch}$ and $P^t_{j,dc}$ represent the charging and discharging powers of the energy storage device j at time t, respectively. $\mu^t_{j,ch}$ and $\mu^t_{j,dch}$ are binary variables where E equals 1 during charging and 0 otherwise; μ_{dch} is the opposite. $P^{max}_{j,ESS}$ indicates the maximum charging and discharging power of the energy storage device at node j. For the ease of comparison in the charging and discharging diagram, it should be noted that, during charging, $P^t_{j,ch}$ is shown below the axis; during discharging, $P^t_{j,ch}$ is shown above the axis.

4.2. Site Selection and Capacity Determination Model

4.2.1. Development of the Objective Function for Site Selection and Capacity Determination

This paper introduces a model for site selection and capacity determination for PV systems, predicated on the maximum carrying capacity of the distribution network. The objective is to minimize investment costs and active power losses, thereby enhancing the economic efficiency of the system. After confirming the compliance with the carrying capacity requirements, a secondary model is formulated:

$$F_2 = w_1 \times a \times F_1 + w_2 \times b \times f_2 + w_3 \times c \times f_3 \tag{17}$$

In Equation (17), w_1, w_2, and w_3 represent the weights within the objective function, satisfying the condition $w_1 + w_2 + w_3 = 1$. F_1 denotes the carrying capacity value, and the site selection and capacity determination model is based on the maximum carrying capacity model of the distribution network; hence, w_1 is the predominant weight, such that $w_1 \geq w_2$ and $w_1 \geq w_3$. The specific values for the weights w_2 and w_3 are determined based on actual conditions. F_1 is elaborated upon in Equation (7), f_2 indicates investment costs, and f_3 represents active network losses. a, b, and c serve as penalty factors.

4.2.2. Investment Costs

The investment costs for a distributed PV primarily comprise the initial capital expenditures and ongoing operation and maintenance expenses. The cost model is described as follows:

$$f_2 = f(a,b) * \sum_{i}^{N_e} C_{I\_PV} P_{PV,i} + C_{OM\_PV} \sum_{i}^{N_e} P_{PV,i} \quad (18)$$

$$f_{(a,b)} = \frac{a \times (a+1)^b}{(a+1)^b - 1} \quad (19)$$

In Equation (18), $f_{(a,b)}$ represents the loss factor of the PV equipment, reflecting the ratio of its current value to its initial investment. a is the annual depreciation rate for the PV equipment, and b denotes its expected lifespan. Ne indicates the number of PV units invested, CI_PV is the cost per unit capacity for installing PV equipment, P_{PV_i} is the installed capacity for the i-th PV, and COM_PV represents the operational and maintenance costs per unit capacity of the PV equipment.

4.2.3. Active Network Loss Optimization Model

Strategically installing PV within the limits of the distribution network's maximum carrying capacity can effectively reduce the active network losses. To minimize these losses, a mathematical model is constructed:

$$f_3 = P_{loss} = \sum_{T=1}^{N} \sum_{i=1}^{n} I_i^2 R_i \quad (20)$$

In Equation (20), N signifies the system's operational duration, n denotes the number of system branches, Ii is the current in branch i, and R_i is the resistance of branch i.

5. Development of the Lower-Level Model

5.1. Lower-Level Objective Function Model Development

In PV planning, reducing the active network losses during system operation is crucial. Building on this, reactive power optimization becomes the central issue of the lower-level optimization. This process relies on the decisions made in the upper-level planning regarding the site selection and capacity determination of PV equipment. The primary goal is to minimize active network losses. Voltage stability margin and voltage deviation are also considered secondary objectives to effectively regulate the equipment. This approach integrates the goals of both the upper and lower-level planning to ensure the power system's efficient and stable operation. The model for the lower-level objective function is:

$$\min G = w_1^*\left(\sum_{t}^{T} \sum_{i=1}^{n} P_{loss,t}\right) + w_2^*\left(\sum_{t}^{T} \sum_{i=1}^{n} VSM_{i,t}\right) + w_3^*\left(\sum_{t}^{T} \sum_{i=1}^{n} \frac{U_{i,t} - U_N}{U_N}\right) \quad (21)$$

In Equation (21), w^*_1, w^*_2, and w^*_3 are the weights of the objective function, fulfilling the condition $w^*_1 + w^*_2 + w^*_3 = 1$. The primary optimization target in this model is the system's active network loss, and hence, w^*_1 has the greatest proportion, $w^*_1 \geq w^*_2$ and $w^*_1 \geq w^*_3$. $U_{i,t}$ represents the actual voltage at node i during time t.

5.2. Constraints

The lower-level planning constraints slightly differ from those at the upper level, mainly reflecting the temporal changes in state variables [22].

1. Load flow constraints:

$$P_{DG,i,t} - P_{L,i,t} - U_{i,t}\sum_{j=1}^{N} U_{j,t}(G_{i,j}\cos\theta_{i,j,t} + B_{i,j}\sin\theta_{i,j,t}) = 0 \quad (22)$$

$$Q_{DG,i,t} - Q_{L,i,t} + U_{i,t}\sum_{j=1}^{N} U_{j,t}(G_{i,j}\cos\theta_{i,j,t} - B_{i,j}\sin\theta_{i,j,t}) = 0 \quad (23)$$

$P_{DGi,t}$ and $Q_{DGi,t}$ are the active and reactive powers at node i at time t, $i \in N$; $\theta_{i,j,t}$ is the voltage angle difference between nodes i and j at time t, where $G_{i,j}$ and $B_{i,j}$ are the conductance and susceptance between these nodes.

2. Node voltage constraints

$$U_{imin} \leq U_{i,t} \leq U_{imax} \quad (24)$$

U_{imin} and U_{imax} represent the upper and lower voltage limits for node i, respectively.

3. Line current constraints, as per Equation (11).
4. Distributed photovoltaic output constraints:

$$(P_{i,t}^{PV})^2 + (Q_{i,t}^{PV})^2 = (S_{i,t}^{PV})^2 \quad (25)$$

$$\eta_i = \sum_{t=1}^{T} P_{i,t}^{PV} / \sum_{t=1}^{T} P_{i,t}^{PV_0} \quad (26)$$

$$\eta \geq \eta_{min} \quad (27)$$

5. SVC Reactive Power Output Constraints:

$S^{PV}{}_{Vi}$ is the capacity of the distributed photovoltaic at node i, and $P^{PV0}{}_{i,t}$ is the theoretical output of the distributed photovoltaic at node i. This represents the minimum utilization efficiency of the distributed photovoltaics.

$$q_{i,min}^{s} \leq q_{i,t}^{s} \leq q_{i,max}^{s} \quad (28)$$

In Equation (24), represent the lower and upper limits of the SVC output at node i at time t.

6. Solving the Photovoltaic Grid-Connected Model Using an Improved Second-Order Oscillatory Chaos Mapping Particle Swarm Optimization Algorithm

Using the improved second-order oscillatory chaos mapping particle swarm optimization algorithm, rapid solutions are achieved for both upper- and lower-level models. This algorithm facilitates the efficient transmission of the global optimum between the two levels through internal iterations, and the data transfer and feedback between models, ultimately determining the locations and capacities of the distributed photovoltaic systems [23,24].

6.1. Overview of PSO

The traditional particle swarm optimization (PSO) is an optimization algorithm developed to mimic the flight trajectories of birds during foraging [25]. At the start of the algorithm, a swarm of particles is generated, and the optimal solution is sought by simulating the movement of particles in space. The PSO algorithm takes related variables as target fitness and updates the state based on the current position, velocity, and historical best status of the particles. The algorithm terminates when the number of iterations reaches a preset maximum [26]. Different iteration methods of the particle swarm can affect the accuracy of the final result. In each iteration, the velocity and position of the particles are calculated according to specific formulas. Through multiple iterations, particles locate the position of the optimal solution. The iteration process continues until the set number of iterations is reached. The formulas for updating the individual particle velocity and position are as follows:

$$V_i^{k+1} = wV_i^k + r_1c_1(Xpbest_i^k - x_i^k) + r_2c_2(Xgbest_i^k - x_i^k) \tag{29}$$

$$X_i^{k+1} = X_i^{k+1} + V_i^{k+1} \tag{30}$$

W is the inertia weight, c_1 is the individual learning factor, c_2 is the social learning factor, r_1 and r_2 are random numbers between 0 and 1. ${V_i}^k$ is the current velocity of the particle, ${X_{pbesti}}^k$ is the current individual best position, and ${X_{gbesti}}^k$ is the global best position at iteration k.

Traditional PSO can easily fall into local optima, converge prematurely, and exhibit uneven particle distribution during iterations [18]. To avoid these shortcomings, the particle swarm algorithm has been improved.

6.2. *Improved Particle Swarm Optimization Algorithm*

6.2.1. Chaotic Mapping of Population

In traditional PSO, randomly generated initial positions and velocities can lead to uneven particle distribution in the iteration search space, affecting the algorithm's global search capability. To address this issue, chaotic mapping is introduced to initialize particles. This establishes chaotic mapping for initializing particle positions, enhancing the search range of the swarm and increasing the randomness of particle solutions [27]. In this paper, random numbers generated by the logistic map are used as initial positions to increase the diversity of particle solutions, accelerating the convergence speed and enhancing the performance of the algorithm.

The mathematical expression for the logistic map:

$$X_{i+1}^k = F(x, r) = r * X_i^k(1 - X_i^k) \tag{31}$$

In Equation (31), r is the control parameter, also known as the chaos mapping parameter. When r is between 0 and 4, the system exhibits periodic and chaotic behavior, and it may also converge to a stable state. At r = 4, the system exhibits the maximum chaotic behavior.

Steps for logistic population initialization:

(1) Define the logistic chaos function to generate a D-dimensional vector, with each component within the rand (0,1) range.
(2) Use the logistic function to map the population position.

6.2.2. Self-Adaptive Inertia Weight Iteration Improvement

The value of the inertia weight W significantly impacts the algorithm's global and local search capabilities. A high W value can accelerate the convergence of the swarm but reduces the search range, diminishing the diversity of the particle swarm and making, it prone to falling into local optima. Conversely, a low W value can lead to an excessive local search, slowing convergence, and losing global search capability, similarly making it easy to fall into local optima [28].

To balance local and global search capabilities, this paper adopts a self-adaptive inertia weight method, adjusting the inertia weight value appropriately as the number of iterations increases. In the early stages of iteration, a larger inertia weight is set to enhance the global search capability; in later stages, the inertia weight is gradually reduced in a nonlinear manner to enhance the algorithm's local search capability. The expression for self-adaptive inertia weight is:

$$b = \frac{fit_i - fit_{gbest}}{fit_0 - fit_{gbest}} \tag{32}$$

$$w = \begin{cases} nw_{max} + (w_{max} - w_{min})(1 - \frac{1}{b}), b \geq 1 \\ w_{min} + b(w_{max} - w_{min}), b < 1 \end{cases} \tag{33}$$

In Equation (32), fit_i is the current fitness of the particle, fit_{gbest} is the best fitness among the particle population; and fit_0 is the critical population fitness. In Equation (33), W_{max} is the maximum weight factor, and W_{min} is the minimum weight factor.

6.2.3. Iterative Algorithm Improvement

In the PSO algorithm, the updated particle velocity is determined by the individual's historical best position and the global historical best position, but the relative positions of particles are not fully considered. To improve the precision and coverage of the global search, this paper introduces oscillatory variables, which involve perturbations near the optimal solution to generate new solutions, thereby allowing the algorithm's particles not only to seek existing optimal solutions but also to enhance the diversity of particles [29,30]. Such improvements help prevent the algorithm from prematurely converging to local optima, while also enhancing the exploration capability of the algorithm. The update formula for the iterative algorithm is:

$$V_i^{k+1} = wv_i^k + u_1\left[Xpbest_i^k - (1+\lambda_1)Xgbest_i^k + \lambda_1 Xgbest_i^{k-1}\right] + u_2\left[Xpbest_i^k - (1+\lambda_2)Xgbest_i^k + \lambda_2 Xgbest_i^{k-1}\right] \tag{34}$$

In Equation (34), λ_1 and λ_2 represent the convergence rates for gradual and oscillatory convergence within the particle swarm, respectively:

$$u_1 = c_1 * r_1,\ c_1 = 1.496,\ r_1 = \text{random} \tag{35}$$

$$u_2 = c_2 * r_2,\ c_2 = 1.496,\ r_2 = \text{random} \tag{36}$$

When the iteration count $t \leq T/2$, the algorithm exhibits an oscillatory convergence:

$$\lambda_1 \geq \frac{2\sqrt{c_1 r_1} - 1}{c_1 r_1}\ \lambda_2 \geq \frac{2\sqrt{c_2 r_2} - 1}{c_2 r_2} \tag{37}$$

When the iteration count $t > T/2$, the algorithm undergoes progressive convergence:

$$\lambda_1 \leq \frac{2\sqrt{c_1 r_1} - 1}{c_1 r_1}\ \lambda_2 \leq \frac{2\sqrt{c_2 r_2} - 1}{c_2 r_2} \tag{38}$$

Figures 3 and 4 respectively show the convergence curves at different numbers of population iterations. The PSO algorithm was validated using benchmark functions. Five common benchmark functions were selected to test the algorithm. The sphere function was used to evaluate the convergence efficiency of the algorithm; the Rosenbrock function was used to assess the evaluation accuracy of the algorithm; the Levy function and Griewank function, both multimodal benchmark functions, were used to test the algorithm's performance in the presence of multiple local optima; the Goldstein–Price function was employed to examine the algorithm's global search capability. Table 1 provides relevant data for these benchmark functions.

Table 1. Parameters of benchmark functions.

Function Name	Dimension	Search Domain	Acceptance	Optimum
Sphere	50	[−100, 100] D	0.01	0
Rosenbrock	50	[−100, 100] D	0.01	0
Levy	50	[−10, 10] D	0.01	0
Griewank	50	[−50, 50] D	0.01	0
Goldstein–Price	50	[−5, 5] D	0.01	0

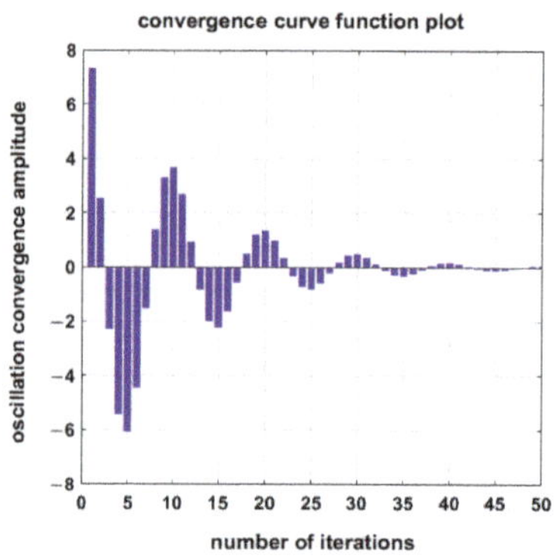

Figure 3. Oscillatory convergence curve.

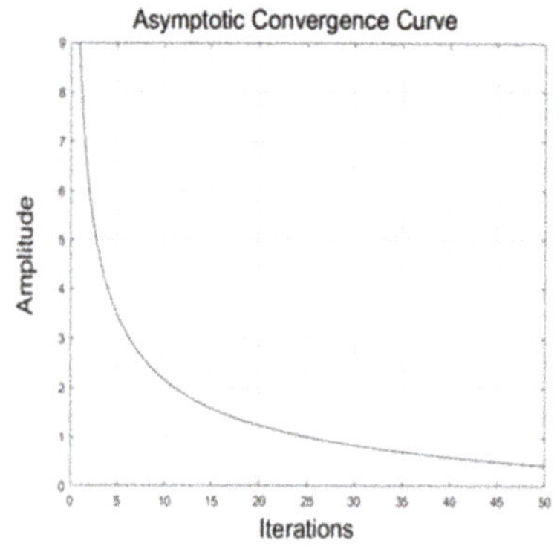

Figure 4. Progressive convergence curve.

To verify that the improved PSO algorithm's convergence is superior to the original PSO algorithm and to other improved PSO algorithms, comparative experiments were conducted using the same benchmark parameters. The specific settings were as follows: the particle swarm size was 50, the dimensionality was 50, and the number of iterations was 2000. The benchmark function test results and the comparison of the convergence precision for each function are shown in Figures 5 and 6. Table 1 shows the relevant parameters of the benchmark functions.

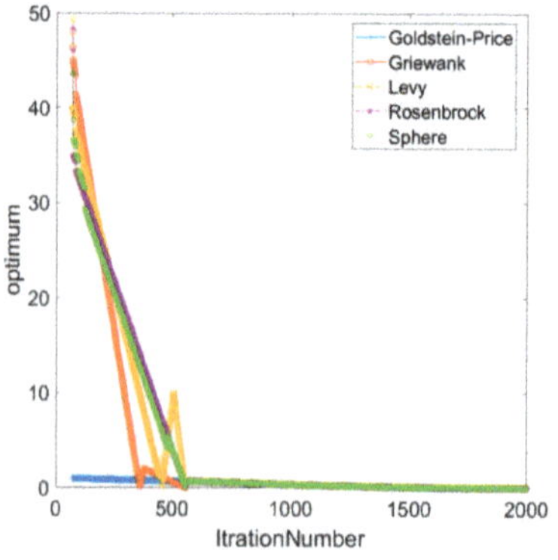

Figure 5. Test results of various benchmark functions.

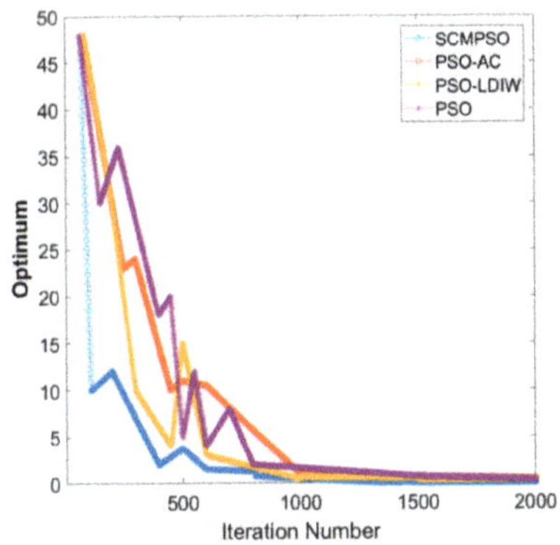

Figure 6. Iteration comparison of different particle swarm optimization algorithms.

Figure 6 shows that the SCMPSO algorithm demonstrates significant advantages early in the iteration process. When the iterations reach around 500, the results of the four particle swarm optimization algorithms converge to approximately zero. To ensure the reliability of the algorithm, the number of iterations should not be less than 500. In this study, 2000 iterations were selected to avoid the potential issue of insufficient iterations compromising the validation of the algorithm's effectiveness.

From Figures 5 and 6, it is evident that the proposed SCMPSO algorithm outperforms other PSO algorithms in terms of convergence speed, search range, and convergence precision. Additionally, when addressing complex problems, this algorithm exhibits the characteristics of rapid convergence, high efficiency, and high stability.

6.3. Model Solution

The flowchart for solving the two-layer distributed photovoltaic planning model in the distribution network using the SCMPSO algorithm is shown in Figure 7.

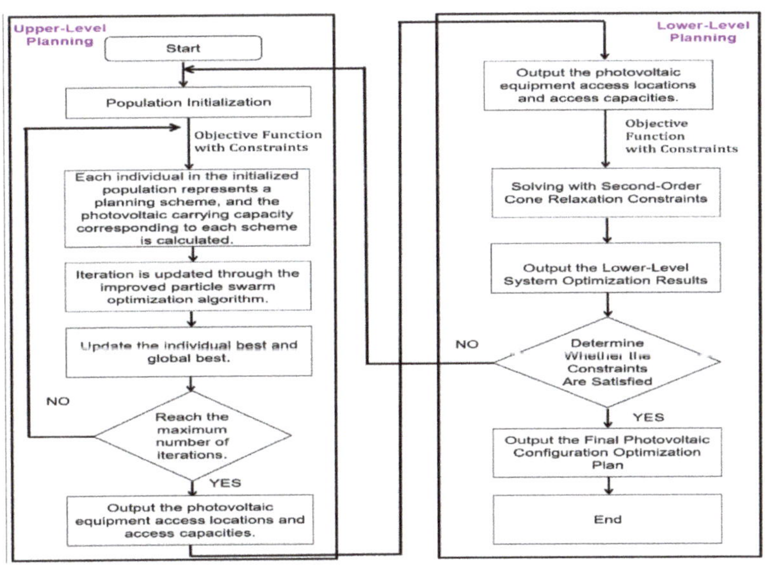

Figure 7. Model solution flowchart.

Bilevel Planning Model Solution Steps:

Step 1: Establish the upper-level planning model. Input the network component parameters, active, and reactive power outputs of conventional photovoltaic (PV) systems, SVC (static VAR compensator) devices, and conventional load parameters.

Step 2: Solve the upper-level model using the SCMPSO algorithm. Determine the PV equipment access locations and capacities at each node, and pass the results to the lower level.

Step 3: Establish the lower-level optimization model with the objectives of reducing active power losses, voltage stability margins, and voltage deviations.

Step 4: Solve the lower-level model using the SCMPSO algorithm to obtain the reactive power optimization values. Pass the optimized values back to the upper-level model for the next iteration.

Step 5: Evaluate the convergence conditions. If the preset number of iterations is reached, terminate the process and output the current optimal PV equipment access locations and capacities. If not, continue from Step 1 to Step 4 until the optimal conditions are achieved.

In this paper, the PV access capacity in the first layer is a continuous variable, and the PV access device location is a discrete variable. The upper-level objective function is challenging to solve linearly, so the improved particle swarm optimization (SCMPSO) is used for multi-objective solving.

The second layer of planning is essentially an optimal power flow problem. The non-convex power flow model is relaxed using the second-order cone relaxation method, with the following constraints:

$$W_{ij} = U_i * U_j \tag{39}$$

$$P_j = \sum_{k:j\to k} \mathrm{Re}(W_{ij}Y_{ij}^*) \tag{40}$$

$$Q_j = \sum_{k:j\to k} \mathrm{Im}(W_{ij}Y_{ij}^*) \tag{41}$$

$$\mathrm{min}P_{\mathrm{loss}}(w) = \sum_{(i,j)\in N} G_{ij}W_{ii} - 2\sum_{(i,j)\in N} G_{ij}\mathrm{Re}(Y_{ij}^*) \tag{42}$$

$$U_j^2 = U_i^2 - 2\left(r_{ij}P_{ij} + x_{ij}Q_{ij}\right) + \left(r_{ij}^2 + x_{ij}^2\right)I_{ij}^2 \tag{43}$$

$$P_{min} \le P_i \le P_{max} \tag{44}$$

$$Q_{min} \le Q_i \le Q_{max} \tag{45}$$

$$V_{min} \le V_i \le V_{max} \tag{46}$$

$$l_{min} \le l_{min} \le l_{max} \tag{47}$$

$$d_{pv,k}\% \le d_{max}\% \tag{48}$$

$$||W_{ii}||_2 \le W_{ii} \tag{49}$$

$$\left\|\begin{matrix} 2P_{ij}^t \\ 2Q_{ij}^t \\ l_{ij}^t - u_j^t \end{matrix}\right\| \le l_{ij}^t + u_j^t \tag{50}$$

Equation (40) to (41), Pj and Qj represent the active and reactive power at the node, respectively; Re denotes the real part of the expression, and Im denotes the imaginary part; Yij* represents the conjugate of the admittance matrix between branch i−j; Gij represents the real part of the admittance matrix; rij and xij are the resistance and reactance of the branch, respectively. Other variables used in the equations are annotated in Sections 2 and 3 of the paper and are not explained here again.

7. Case Analysis

This study uses the improved IEEE-33 node distribution network system, as shown in Figure 8.

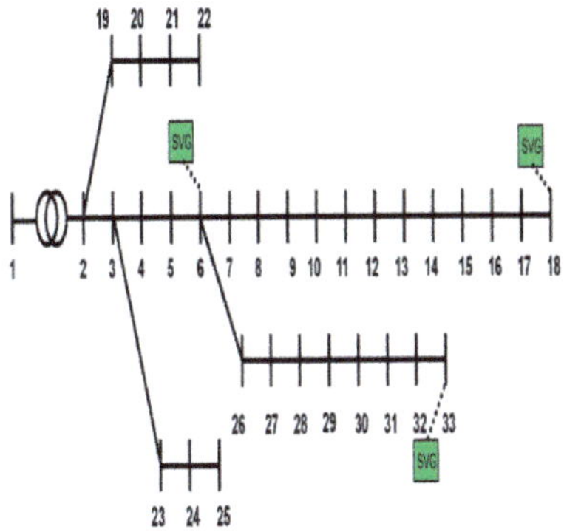

Figure 8. IEEE-33 node system.

The IEEE-33 node system has a distribution network load of 3.715 MW + 2.3 MVAR and a base voltage of 12.66 kV. The system node voltage safety range is set between 0.95 and 1.05 p.u. The distribution network uses LGJ-150 conductors(which was manufactured by Hebei Guangjie Cable Co., Ltd., located in Cangzhou, Hebei, China.), with a maximum allowable continuous current of 375 A. The unit investment cost for distributed PV is 3500 CNY/kW, and the average operation and maintenance cost per unit is 50 CNY/kWh. The annual depreciation rate is 0.05, and the service life is 25 years. At the beginning of the distribution network, there is an on-load tap-changing transformer (OLTC) with 11 tap positions and a voltage regulation range from 0.95 to 1.05 p.u.

Assuming that each node in the distribution network can be equipped with a distributed PV and energy storage, the number of distributed PV installations is set to 6. Additionally, three static VAR compensators (SVG) with an installed capacity of 200 kVar each are added and connected to nodes 6, 8, and 33. The number of energy storage installations is 4, with an initial charge/discharge efficiency of 75%. In the distribution network, nodes with PV system connections must ensure stability in multiple aspects, including voltage levels, reasonable load distribution, and stable power quality.

This study employs the K-means clustering method to analyze the annual solar intensity and basic load data for a location in Jiangsu, China. The year is divided into five scenarios, labeled from 1 to 5.

Scenario 1: Weak solar intensity, rainy days, low electricity consumption, and light load, mostly occurring during rainy days in spring and winter.

Scenario 2: Slightly stronger solar intensity compared to Scenario 1, low electricity consumption, and light load, common during clear days in spring and winter.

Scenario 3: A stronger solar intensity than Scenario 2, higher electricity consumption and moderate load, typical in late spring and early autumn.

Scenario 4: Strong solar intensity within a day, but reduced due to thunderstorms, high electricity consumption, and heavy load, often seen on rainy summer days.

Scenario 5: Very strong solar intensity throughout the day, high electricity consumption, and heavy load, usually occurring on clear summer days.

Figure 9 shows the division of the annual solar intensity and load status into five scenarios. Figure 10 illustrates the normalized load trend over 24 h on a specific day of the year. The load trend is influenced by factors such as seasons and holidays, but overall trends exhibit certain similarities, with differences mainly in the load magnitude.

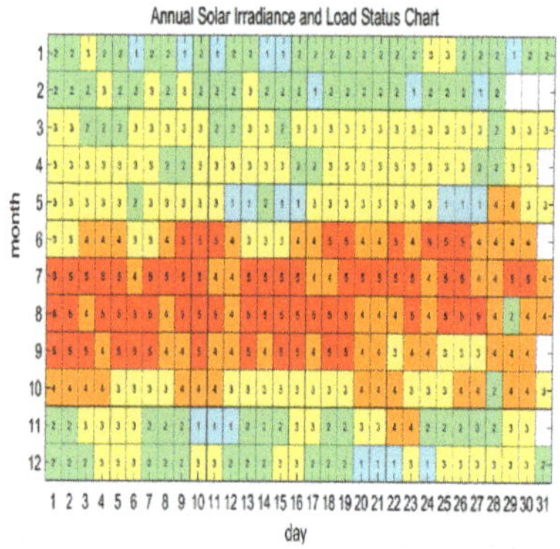

Figure 9. Annual solar intensity and load status.

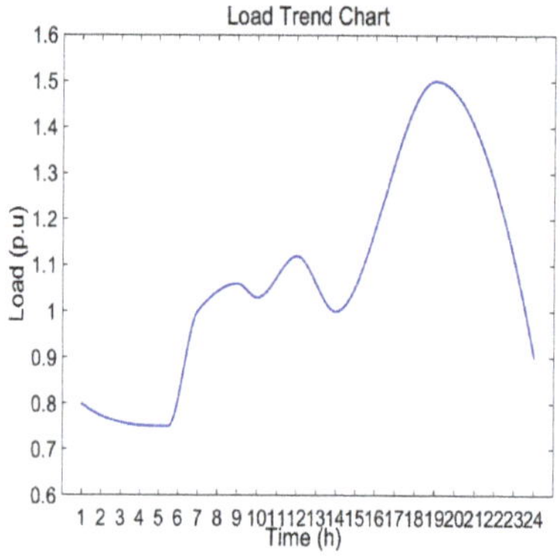

Figure 10. 24-hour load trend in Jiangsu.

7.1. Actual Simulation and Comparative Analysis

Based on Figure 9, which shows the annual solar intensity and load scenarios, and Figure 10, the 24 h load trend in Jiangsu, this study uses the SCMPSO algorithm for simulation and solution. By employing the proposed two-level optimization approach, this research aims to provide reasonable optimization strategies for PV integration into the distribution network system, thereby enhancing economic efficiency.

Table 2 provides the access locations and capacities of distributed PV for three different scenarios, along with the corresponding access locations, capacities, and power of the energy storage systems. The specific scenarios are as follows:

Scenario 1: Distributed PV access parameters under the condition of maximum voltage stability.

Scenario 2: Distributed PV access parameters under the condition of maximum economic efficiency.

Scenario 3: Distributed PV access parameters considering the system carrying capacity, economic efficiency, public interest, and grid stability margin.

Table 3 presents the cost analysis under different planning scenarios. After determining the configuration of distributed PV and energy storage equipment in Table 1, the upper-level optimization provides the investment cost and annual operation and maintenance cost of PV equipment, as well as the proportion of distributed PV generation in the overall system. In Scenario 1, the PV equipment access capacity is 1280 kwh with a total cost of CNY 87.1 million; in Scenario 2, the PV equipment access capacity is 1140 kWh with a total cost of CNY 74.645 million; in Scenario 3, the PV equipment access capacity is 1260 kWh with a total cost of CNY 83.116 million.

Table 2. Node location and capacity configuration of distributed photovoltaic under two-level planning scenarios.

Optimization Solution	Distributed PV Installation Node (Capacity/kwh)	Energy Storage System Installation Node (Capacity/kwh, Power/kw)
Voltage optimization	6 (130), 9 (170), 13 (150), 18 (210), 25 (200), 32 (420)	10 (300, 210), 18 (250, 160) 26 (180, 120), 33 (280,200)
Economic optimization	7 (150), 12 (220), 15 (170), 25 (220), 26 (170), 30 (210)	7 (200, 130), 18 (120, 70) 28 (210, 120), 33 (200, 140)
Comprehensive optimization	6 (110), 10 (190), 15 (240), 18 (230), 25 (300), 31 (190)	6 (100, 70), 15 (230, 170) 27 (280, 180), 31 (350, 250)

Table 3. Cost analysis under different planning scenarios.

Optimization Solution	Total Cost/ Million CNY	Investment Cost/ Million CNY	Operation and Maintenance Costs/ Million CNY	Depreciation Loss/ Million CNY	Distributed PV Penetration/%
Voltage optimization	87.1	63.2	20.74	3.16	83.69
Economic optimization	74.645	53.5	18.47	2.675	60.35
Comprehensive optimization	83.116	59.72	20.41	2.986	73.23

7.2. Operating Results under the Two-Level Planning Approach

7.2.1. Operation State of Energy Storage Devices

Figure 11 displays the optimal timing for charging and discharging under five scenarios within a comprehensive planning scheme.

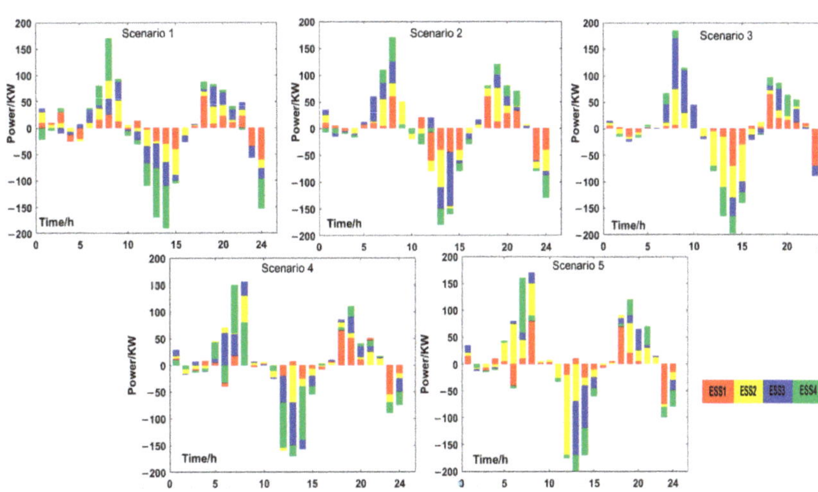

Figure 11. Charging and discharging schematic of energy storage devices across five scenarios.

Scenarios 1 and 2: These scenarios are characterized by lower photovoltaic (PV) outputs and lighter loads, which result in a more gradual distribution of energy storage charging during the afternoon periods. The reduced solar generation leads to less demand on the storage systems, allowing for smoother energy intake and less variability in charging requirements.

Scenarios 3, 4, and 5: In these scenarios, the PV output is considerably higher, and the load gradually increases, necessitating the more concentrated charging of distributed PV systems in the afternoon. This is indicative of peak solar hours aligning with higher energy consumption patterns, maximizing the use of solar power directly and reducing the reliance on stored energy during these periods.

Scenarios 4 and 5: The loads in these scenarios are nearly identical; however, due to meteorological variations, the PV output in Scenario 5 is greater than in Scenario 4. Despite similar timings for charging and discharging the energy storage devices, the power levels vary. This indicates that Scenario 5 may require the more dynamic management of energy storage due to higher PV outputs, potentially leading to more aggressive charging strategies or faster discharge rates to stabilize the grid.

These insights highlight the need for adaptive energy storage management strategies that can respond to fluctuations in both solar energy generation and consumption demands, ensuring stability and efficiency in the power supply.

7.2.2. Distribution Network Node Voltage Status

Table 4 presents comparative data on the voltage status at distribution network nodes across five scenarios under three operating modes: without any PV or energy storage devices, with only PV devices but no energy storage, and with both the PV and energy storage devices. Insights from Table 3 include the following.

Table 4. Voltage data for each scenario with and without PV and energy storage device integration.

Typical Scenarios		Scenario 1	Scenario 2	Scenario 3	Scenario 4	Scenario 5
Number of voltage overruns states	Without PV and ESS	4	6	7	8	11
	With PV but without ESS	8	11	13	17	22
	With PV and ESS	0	0	1	3	6
Maximum voltage/p.u	Without PV and ESS	1.0276	1.0269	1.027	1.0395	1.0455
	With PV but without ESS	1.0437	1.0428	1.0532	1.0595	1.0636
	With PV and ESS	1.0105	1.0112	1.0129	1.0217	1.0232
Minimum voltage/p.u	Without PV and ESS	0.9632	0.9611	0.9513	0.9479	0.9466
	With PV but without ESS	0.9513	0.9502	0.9499	0.9486	0.9473
	With PV and ESS	0.9836	0.9844	0.9867	0.9881	0.9885
Voltage Deviation	Without PV and ESS	0.0644	0.0658	0.0757	0.0916	0.0989
	With PV but without ESS	0.0924	0.0926	0.1033	0.1109	0.1163
	With PV and ESS	0.0269	0.0268	0.0262	0.0336	0.0347
Voltage Stability Margin	Without PV and ESS	0.1623	0.1611	0.1577	0.1564	0.1558
	With PV but without ESS	0.1425	0.1438	0.1247	0.1211	0.1195
	With PV and ESS	0.1873	0.1837	0.1868	0.1865	0.1875

With only PV devices and no energy storage, the system is the least stable, exhibiting the most instances of voltage violations, the greatest voltage deviations, and the poorest stability margins. In this mode, the electrical energy produced by PV cannot be effectively stored, leading to increased voltage fluctuations. With both PV and energy storage devices, the system operates in the most stable condition, with the fewest voltage violations, minimal voltage deviations, and the best stability margins. This indicates that the use of energy storage devices can effectively mitigate the voltage fluctuations and instability introduced by the integration of PV devices into the system. These data clearly demonstrate that the charging and discharging of energy storage devices not only support the stability of the grid operations but also effectively alleviate the adverse impacts caused by PV integration, thereby enhancing the overall operational efficiency and safety of the system.

7.2.3. Active Power Loss in the Distribution Network

Figure 12 shows the time-varying curves of active power loss across five scenarios under three operating modes. From these scenarios, it is evident that integrating the distributed photovoltaic (PV) and energy storage devices can effectively reduce the active power losses in the network. Night and early morning: During these periods, there is no sunlight, and the PV systems are inactive with no power output. As shown in Figure 11, energy storage devices are in charging mode during the night and early morning. However, due to the low load and minimal losses during these times, there is not much additional power loss. Thus, in the initial stages of all five scenario graphs, the curves for the three operating states almost coincide. Daytime with increasing load: As the system load increases, the actions of PV and energy storage devices cause changes in active power loss over time, leading to deviations among the three curves. Scenarios 1, 2, and 3 exhibit lower PV output and, consequently, lower system power loss compared to Scenarios 4 and 5. Scenarios 4 and 5 experience the highest system loads and the greatest PV output. As a result, there is a slight increase in the system power loss due to the higher energy throughput. In summary, the integration of PV and energy storage devices in the distribution network significantly mitigates active power losses, particularly during periods of high solar output and peak loads.

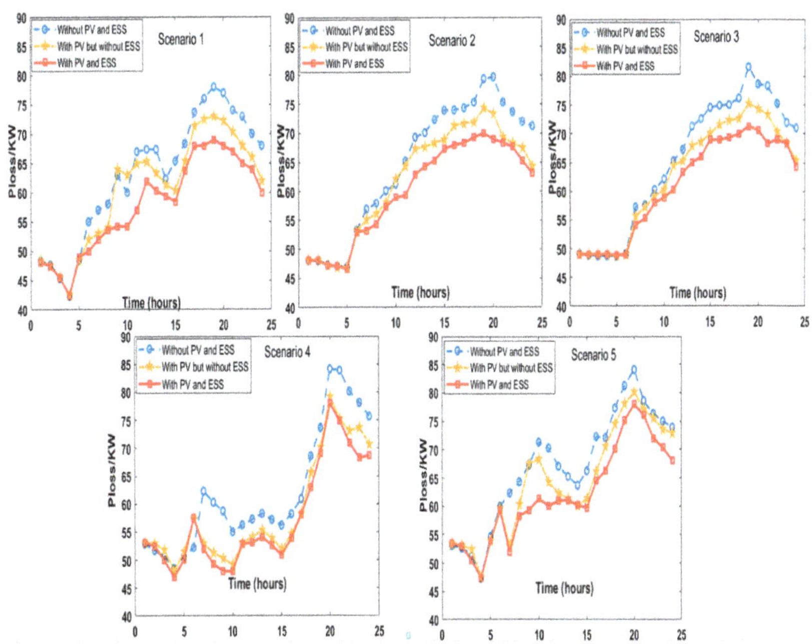

Figure 12. Active power loss over time across five scenarios.

From the above analysis, it can be seen that the integration of PV devices increases the instability of system node voltages, which is detrimental to the stable operation of the system. However, by reasonably planning the access nodes and capacities of distributed PV and energy storage devices, the aforementioned issues can be effectively avoided. As shown in Figures 11 and 12, and Table 3, the distributed PV two-level planning based on the proposed SCMPSO algorithm not only enhances system capacity but also reduces the active power losses, improves economic efficiency, and decreases the voltage deviations.

8. Conclusions

This paper utilizes the SCMPSO algorithm to solve the upper-level optimization model for distributed PV site selection and capacity determination. The simulation results lead to the following conclusions:

1. By fully considering the temporal characteristics of load and PV system output, a two-level optimization model for distributed PV site selection and capacity determination is proposed. Through upper and lower-level optimization, the optimal locations and capacities for distributed PV and energy storage devices were determined. The results indicate that the optimized distribution network significantly improves the power quality and reduces active power losses.

2. The SCMPSO algorithm was employed to solve the proposed model. The simulation results demonstrate that this algorithm effectively addresses the site selection and capacity determination problem for distributed PV. Comparative simulation results verify the efficiency and accuracy of the SCMPSO algorithm in handling high-dimensional nonlinear problems.

Author Contributions: Conceptualization, A.D.; methodology, A.D.; software, MATLAB; validation, A.D.; formal analysis, A.D.; investigation, S.-K.L.; resources, data curation, A.D.; writing—original draft preparation, A.D.; writing—review and editing, A.D.; visualization, A.D. All authors have read and agreed to the published version of the manuscript.

Funding: This research received no external funding.

Institutional Review Board Statement: Not applicable.

Data Availability Statement: The data presented in this study are available on request from the corresponding author. The data are not publicly available.

Conflicts of Interest: The authors declare no conflicts of interest.

References

1. Wen, Y.; Yang, W.; Wang, R.; Xu, W.; Ye, X.; Li, T. Review and prospects of constructing 100% renewable energy power system. *Proc. CSEE* **2020**, *40*, 1843–1856. (In Chinese)
2. Wang, Y.; He, J.; Chen, W. Distributed solar photovoltaic development potential and a roadmap at the city level in China. *Renew. Sustain. Energy Rev.* **2021**, *141*, 110772. [CrossRef]
3. da Silva, P.P.; Dantas, G.; Pereira, G.I.; Câmara, L.; De Castro, N.J. Photovoltaic distributed generation—An international review on diffusion, support policies, and electricity sector regulatory adaptation. *Renew. Sustain. Energy Rev.* **2019**, *103*, 30–39. [CrossRef]
4. Rahdan, P.; Zeyen, E.; Gallego-Castillo, C.; Victoria, M. Distributed photovoltaics provides key benefits for a highly renewable European energy system. *Appl. Energy* **2024**, *360*, 122721. [CrossRef]
5. Yang, F.-F.; Zhao, X.-G. Policies and economic efficiency of China's distributed photovoltaic and energy storage industry. *Energy* **2018**, *154*, 221–330. [CrossRef]
6. Jain, A.K.; Horowitz, K.; Ding, F.; Sedzro, K.S.; Palmintier, B.; Mather, B.; Jain, H. Dynamic hosting capacity analysis for distributed photovoltaic resources—Framework and case study. *Appl. Energy* **2020**, *280*, 115633. [CrossRef]
7. Zhou, X.; Chen, S.; Lu, Z.; Huang, Y.; Ma, S.; Zhao, Q. Technical Characteristics of China's New Generation Power System in Energy Transition. *Proc. Chin. Soc. Electr. Eng.* **2018**, *38*, 1893–1904+2205.
8. Hao, W.; Meng, Z.; Zhang, Y.; Xie, B.; Peng, P.; Wei, J. Research on the Evaluation Method of Distribution Network Carrying Capacity with Multiple Distributed Power Sources under the New Power System. *Power Syst. Prot. Control.* **2023**, *51*, 23–33.
9. Li, L.; Zhang, K.; Zeng, S.; Ji Feng, Z. Evaluation of Distributed Power Capacity in Distribution Networks Based on APDE Algorithm. *Acta Energiae Solaris Sin.* **2023**, *44*, 55–63.
10. Wang, L.; Zhang, F.; Kou, L.F. Large-scale distributed pv cluster division based on fast unfolding clustering algorithm. *Actal Energiae Solaris Sin.* **2021**, *42*, 29–34. (In Chinese)
11. Wang, H.; Zhang, Y.; Yin, S. The Risk of Voltage Violation and Harmonic Effects in Distributed Photovoltaic Grid-Connected Systems. *Power Syst. Clean Energy* **2024**, *40*, 128–146.
12. Pereira, L.D.L.; Yahyaoui, I.; Fiorotti, R. Optimal allocation of distributed generation and capacitor banks using probabilistic generation models with correlations. *Appl. Energy* **2022**, *307*, 118097. [CrossRef]
13. Cao, W.; Zhang, Y.; Zhang, A. Resonance Mechanism and Impact Characteristics of Distributed Photovoltaic Grid-Connected Systems under Weak Grid Conditions. *Electr. Power Constr.* **2024**, *45*, 149–159.
14. Zou, H.L.; Han, X.Y.; Liao, Q.F.; Di, C.L.; Zhen, Z. Analysis of the penetration capacity of distributed generation considering voltage quality and short- circuit capacity constraints. *Power Syst. Technol.* **2016**, *40*, 2273–2280.

15. Wang, Y.; Guo, L.; Lv, W.; Xiong, L.; Yan, T.Z. Research on Grid-Connected Control Strategies for Distributed Photovoltaic Power Sources. *Power Supply Technol.* **2019**, *43*, 637–640.
16. Hong, Z.; Lin, Q. Comprehensive Evaluation Method for the Carrying Capacity of Distribution Networks with High Proportion of Distributed Power Sources. *Large Electr. Mach. Technol.* **2023**, *289*, 77–82.
17. Liao, J.; Lin, J.; Wu, G. Two-layer optimization configuration method for distributed photovoltaic and energy storage systems based on IDEC-K clustering. *Energy Rep.* **2024**, *11*, 5172–5188. [CrossRef]
18. Yang, Z.; Zhu, B.; Cao, Y.; Dan, W. A Two-Level Planning Method for Distributed Photovoltaics to Improve Distribution Network Carrying Capacity. *J. Huazhong Univ. Sci. Technol. (Nat. Sci. Ed.)* **2024**, *03*, 110–117.
19. Lai, C.; Li, J.; Chen, B.; Huang, Y.J.; Wei, S.Y. A review of research on photovoltaic power generation output prediction technology. *Trans. Chin. Electr. Soc.* **2019**, *34*, 1201–1217. (In Chinese)
20. *GB/T12325—2008*; Power Quality–Voltage Deviation. China Zhijian Publishing House: Beijing, China, 2008.
21. *GB/T14549—1993*; Power Quality-Harmonics in Public Supply Network. China Zhijian Publishing House: Beijing, China, 1993.
22. Lu, Y.; Zhao, T.; Du, H.; Zhu, X.Y. Reactive power optimization of distribution network with DG based on improved particle swarm optimization. *Electr. Power Eng Technol.* **2018**, *37*, 69–74.
23. Li, L.; Wang, W.; Xu, X.; Li, W.K. Multi-objective particle swarm optimization algorithm based on grid sorting. *Comput. Res. Dev.* **2017**, *54*, 1012–1023.
24. Zhang, L.; Zhou, C.; Ma, M.; Chen, C. Solving Multi-Objective Optimization Problems Using Particle Swarm Optimization Algorithm. *J. Comput. Res. Dev.* **2024**, *7*, 1286–1291.
25. Bai, X.; Hou, Y.; Han, H. Adaptive knowledge transfer-based particle swarm optimization for constrained multitask optimization. *Swarm Evol. Comput.* **2024**, *87*, 101569. [CrossRef]
26. Pedrycz, A.; Dong, F.; Hirota, K. Nonlinear mappings in problem solving and their PSO-based development. *Inf. Sci.* **2011**, *181*, 4112–4123. [CrossRef]
27. Wan, Z.P.; Wang, G.M.; Sun, B. A hybrid intelligent algorithm by combining particle swarm optimization with chaos searching technique for solving nonlinear bilevel programming problems. *Swarm Evol. Comput.* **2013**, *8*, 26–32. [CrossRef]
28. Tian, D.; Xu, Q.; Yao, X.; Zhang, G.; Li, Y.; Xu, C. Diversity-guided particle swarm optimization with multi-level learning strategy. *Swarm Evol. Comput.* **2024**, *86*, 101533. [CrossRef]
29. Houssein, E.H.; Gad, A.G.; Hussain, K.; Suganthan, P.N. Major Advances in Particle Swarm Optimization: Theory, Analysis, and Application. *Swarm Evol. Comput.* **2021**, *63*, 100868. [CrossRef]
30. He, L.; Cun, F.R.; Jun, Y.D. Microgrid Group Optimization and Scheduling Based on Improved Second-Order Particle Swarm Optimization Algorithm. *J. Wuhan Univ. (Eng. Ed.)* **2022**, *55*.

energies

MDPI

Article

Grid-Connected Inverter Grid Voltage Feedforward Control Strategy Based on Multi-Objective Constraint in Weak Grid

Su'e Wang *, Kaiyuan Cui and Pengfei Hao

College of Electrical and Control Engineering, Shaanxi University of Science and Technology, Xi'an 710021, China
* Correspondence: wangsue@sust.edu.cn

Abstract: In weak grid, feedforward of grid voltage control is widely used to effectively suppress grid-side current distortion of inverters caused by harmonics in point of common coupling (PCC) voltage. However, due to its introduction of a positive feedback loop related to the grid impedance, it results in a significant reduction in the system phase margin. In view of this, in this paper, the output impedance of a three-phase LCL grid-connected inverter under a quasi-proportional resonant (QPR) controller is first modeled. Instead of the traditional grid voltage feedforward control strategy, a band-pass filter is added to the grid voltage feedforward channel. Secondly, a multi-objective constraint method is proposed to make improvements to the feedforward function. Then, a multi-objective constraint function is established with the constraints of base-wave current tracking performance, system stability margin, and low-frequency amplitude, and the feasibility of its function optimization design method is verified. Theoretical analysis shows that the optimized grid voltage feedforward control strategy can effectively reshape the phase characteristics of the system output impedance, which greatly broadens the adaptation range of the system to the grid impedance. Finally, the effectiveness of the proposed control strategy is verified by building a semi-physical simulation experimental platform based on RT-LAB OP4510.

Keywords: weak grid; grid-connected inverter; grid voltage feedforward; band-pass filter; multi-objective constraint

Citation: Wang, S.; Cui, K.; Hao, P. Grid-Connected Inverter Grid Voltage Feedforward Control Strategy Based on Multi-Objective Constraint in Weak Grid. *Energies* **2024**, *17*, 3288. https://doi.org/10.3390/en17133288

Academic Editors: Lin Jiang, Bo Yang and Zhijian Liu

Received: 12 June 2024
Revised: 30 June 2024
Accepted: 2 July 2024
Published: 4 July 2024

1. Introduction

In recent years, as the world attaches great importance to renewable energy sources such as solar energy and wind energy, the proportion of distributed generation systems in the future power grid will continue to increase [1,2]. Grid-connected inverters as an important interface for distributed generation and necessary equipment for power quality management, such as new energy grid-connected inverters [3], active power filter (APF) [4], and Static Var Generator (SVG) [5], etc., play a crucial role in the construction of the smart grid.

However, with the wide distribution of distributed power sources as well as users, low-voltage transmission and distribution lines tend to be long, and the grid impedance becomes non-negligible under the combined influence of line impedance and transformer leakage inductance, in addition to the fact that the actual grid increasingly exhibits weak grid characteristics due to the presence of severe background harmonics in the grid-connected voltage [6,7]. In order to reduce the grid-connected current harmonics and the size, weight and cost of grid-connected filters, LCL filters are usually used [8]. However, the LCL inverter has resonant characteristics, and the phase is prone to -180° jump at the resonant spike crossing 0 dB in the grid-side current feedback control, resulting in system instability, which is generally suppressed by passive damping [9–11] or active damping [12–15] methods. Although the passive damping method can effectively suppress the resonance peaks of the LCL filter, it introduces additional system losses and affects the service life of the inverter. The active damping method simply introduces an algorithm into the controller and does not require an actual resistor to be connected to the circuit, which does not generate additional system losses and also effectively suppresses resonance peaks. Therefore, in this paper, a capacitive current active damping strategy under QPR control is used [16], which not only enhances the fundamental gain of the system and reduces the steady state error, but also effectively suppresses the resonance peaks of the LCL filter [17].

Grid voltage feedforward control [18–20] can offset the influence of background harmonics of grid voltage on the grid-connected current, and it is widely used because it has the advantages of suppressing the inrush current during the startup of the grid-connected inverter and reducing the steady state error of the grid-connected current. However, in the weak grid case, the grid voltage feedforward control introduces an additional feedback loop related to the grid impedance, which drastically reduces the phase angle margin of the grid-connected inverter and poses a serious threat to the quality and stability of the grid-connected current of the grid-connected inverter. For this reason, there are some studies from this point of view that proposed to set different transfer functions on this positive feedback channel to weaken the influence of positive feedback and improve the stability of the grid-connected inverter. For example, in [21], a second-order generalized integrator (SOGI) centered on the fundamental frequency is proposed to be added to the grid voltage positive feedback channel. The system stability is improved due to the use of SOGI to compensate the phase margin of the grid-connected inverter. In [22], a new grid voltage feedforward adaptive algorithm is proposed to make it work well in the presence of grid impedance variations. A grid voltage weighted feedforward scheme based on multiple resonant components is proposed in [23], which introduces a series of quasi-resonant components in the full feedforward path of the grid voltage such that only the background harmonics in the grid voltage are fed forward. Additionally, in Reference [24], a weighted proportional-derivative (PD) PCC voltage feedforward scheme is proposed. By appropriately adjusting the weighting coefficients of the two feedforward parameters, the real part of the inverter output impedance becomes non-negative over a sampling frequency range of up to 0.47 times, harvesting satisfying passivity. In addition to this, [25] proposed to use grid voltage instead of PCC voltage for feedforward and considering that it is more difficult to detect the real grid voltage directly, a method to extract the grid voltage from the detected PCC voltage is proposed to realize the suggested feedforward scheme, which improves the situation that PCC voltage feedforward makes the grid-connected inverter unstable in weak grid due to the effect of delay.

In summary, although there are various methods to improve the grid voltage feedforward control, these control strategies are an impedance reshaping of the system output impedance. These improvement strategies mainly improve the grid voltage feedforward channel, which needs to be realized by adding various control links, but the design of the control parameters is seldom from a macro point of view to control and link the various indicators of the system as a whole; thus, the accuracy and universality of the control parameters are still to be solved urgently.

Based on the above problems, this paper takes the three-phase LCL grid-connected inverter as the research object. Firstly, the control system is modeled and secondly, a band-pass filter is added into the grid voltage feedforward channel [26] to perform impedance reshaping of the system to improve the robustness of the inverter against grid impedance variations. Furthermore, a multi-objective constraint method is proposed to optimize the band-pass filter parameters in terms of base-wave current tracking performance, system stability margin and low-frequency amplitude constraints for grid voltage feedforward in weak grid. The main contributions of this study are summarized as follows:

(1) Based on the impedance stability criterion analysis method, this study analyzes the causes of system instability caused by grid voltage feedforward under a weak grid, and explains through the bode plot analysis that the system instability is due to the fact that the equivalent shunt virtual impedance correction introduced by the grid voltage feedforward will change the originally designed system, which leads to the phase lag of the original output impedance of the inverter, and the system is destabilized.

(2) Improvements are made to the grid voltage feedforward channel by adding a band-pass filtering link to its positive feedback channel, analyzing the cumbersomeness of its traditional parameter design method, and proposing a multi-objective constraint method to optimize the design of its parameters, so that the optimized system can directly satisfy the various grid-connected requirements, and enhance the stability of the system under the weak grid without affecting the performance of current tracking.

The remainder of this paper is organized as follows: Section 2 models and briefly describes the three-phase LCL grid-connected inverter system and derives an expression for the output impedance of the inverter by means of a structure-diagram equivalent transformation. Section 3 elucidates the reasons why grid voltage feedforward destabilizes the system in weak grid. Section 4 describes and analyzes the inclusion of a band-pass filter in the feedforward channel. Section 5 proposes a multi-objective constrained optimization of the band-pass filter parameters and analyzes the stability of the optimized system. Section 6 verifies the feasibility and effectiveness of the proposed multi-objective constrained optimization method by building a semi-physical simulation experimental platform based on RT-LAB. Finally, Section 7 summarizes the whole passage.

2. Establishment of Mathematical Model for LCL Grid-Connected Inverter

Figure 1 shows a schematic diagram of a three-phase LCL grid-connected inverter and its control system. The inverter-side inductance L_1, the grid-side inductance L_2, and the filter capacitor C form the LCL filter. Since the grid-side grid resistance is beneficial to the stability of the system, only the role of the grid inductance L_g is considered when analyzing the effect of the grid impedance on the stability of the LCL grid-connected inverter. The grid-connected reference current I^* of the inverter is mainly given by the power control of the inverter or the DC-side voltage control loop. Here, the given value of the reference current is kept constant, and in the control system, in order to make the grid-connected current in phase with the grid voltage, it is necessary to utilize the phase-locked loop (PLL) to sample and analyze the PCC voltage, so as to obtain real-time phase information θ of the grid voltage. H_{i1} is the active damping coefficient of the capacitor current, H_{i2} is the sampling coefficient of grid current, and H_v is the grid voltage sampling coefficient, $G_{ff}(s)$ is the feedforward control transfer function, and $G_i(s)$ is the current controller transfer function.

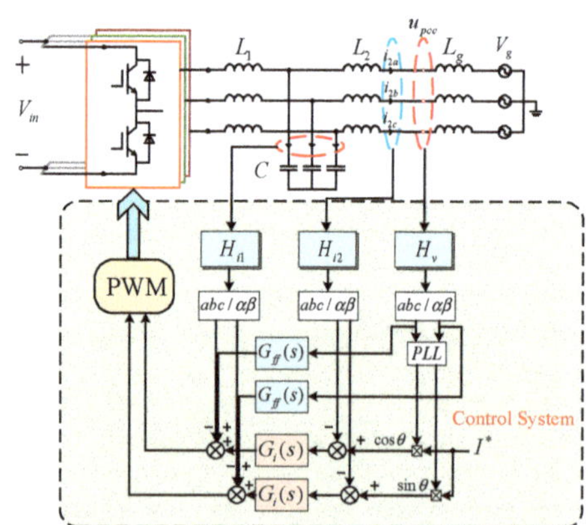

Figure 1. *LCL* grid-connected inverter and control system schematic diagram.

The three-phase LCL grid-connected inverter mathematical model is transformed into a two-phase stationary coordinate system after coordinate transformation. At this time, the two-phase currents are independent of each other, and there is no coupling relationship. Take the double closed-loop control structure with grid-connected current i_2 feedback as an example and, at this time, the control block diagram of Figure 1 on the α axis can be simplified as shown in Figure 2.

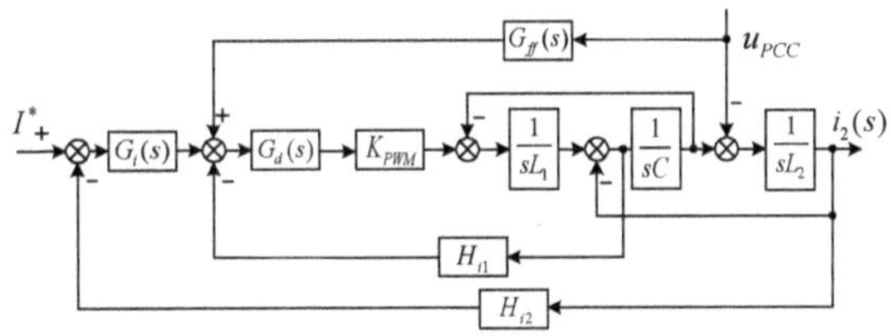

Figure 2. System control block diagram.

In Figure 2, K_{pwm} is the equivalent proportional gain of the inverter bridge, and there is $K_{pwm} \approx V_{in}$ (when the peak value of the triangular carrier is 1). In addition, most of the feedforward links use proportional feedforward, when the grid voltage proportional feedforward link $G_{ff}(s) \approx 1/K_{pwm}$.

The mathematical model of the inverter in the two-phase stationary coordinate system is independent of each other. There is no coupling relationship, but the control variables are sinusoidal and the traditional proportional-integral (PI) controller cannot be achieved without static control, so the QPR controller is used. The QPR controller can be in the two-phase stationary coordinate system to achieve the same control effect as the rotating coordinate system of the PI controller. The transfer function of the QPR controller is given by

$$G_i(s) = k_p + \frac{2k_r\omega_c s}{s^2 + 2\omega_c s + \omega_0^2} \tag{1}$$

In Equation (1), k_p, k_r are the proportional and resonance parameters of the QPR controller. ω_c is the control bandwidth of the controller. ω_0 is the fundamental angular frequency of the grid voltage.

When the grid-connected inverter is digitally controlled, there are delays in the system introduced by digital computation, sampling switches, and zero-order holder, and the equivalent transfer function is shown as

$$G_d(s) = e^{-(0.5+\lambda)sT_s}, \lambda = 1 \tag{2}$$

In Equation (2), T_s is the system sampling period.

A series of equivalent transformations are made to Figure 2 to obtain Figure 3:

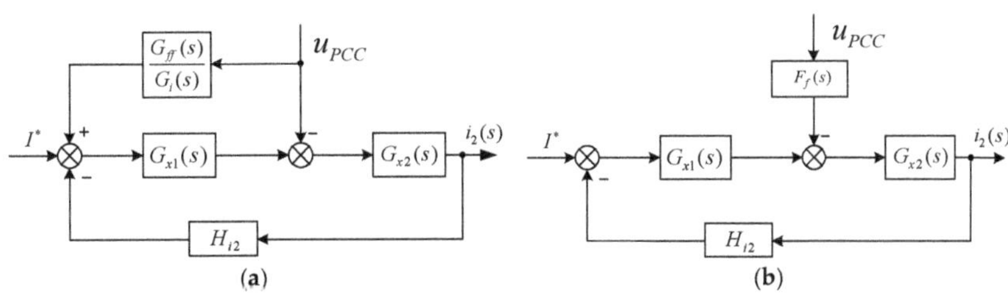

Figure 3. System control block diagram equivalent transformation. (**a**) with feedforward channel; (**b**) merge feedforward channel.

Where the transfer function $G_{x1}(s)$ and $G_{x2}(s)$ in Figure 3 can be deduced as

$$G_{x1}(s) = \frac{G_i(s)G_d(s)K_{pwm}}{s^2L_1C + sH_{i1}CK_{pwm}G_d(s) + 1} \tag{3}$$

$$G_{x2}(s) = \frac{s^2L_1C + sH_{i1}CK_{pwm}G_d(s) + 1}{s^3L_1L_2C + s^2L_2H_{i1}CK_{pwm}G_d(s) + s(L_1 + L_2)} \tag{4}$$

The loop gain T_0 and the grid-connected current i_2 of the system can be expressed as

$$T_0(s) = \frac{G_i(s)G_d(s)K_{pwm}H_{i2}}{s^3L_1L_2C + s^2L_2H_{i1}CK_{pwm}G_d(s) + s(L_1+L_2)} \tag{5}$$

$$\begin{aligned} i_2(s) &= \frac{1}{H_{i2}}\frac{T_0(s)}{1+T_0(s)}I^* - \frac{G_{x2}(s)}{1+T_0(s)}F_f(s)u_{pcc} \\ &= \frac{1}{H_{i2}}\frac{T_0(s)}{1+T_0(s)}I^* - \frac{1}{Z_0(s)}u_{pcc} \end{aligned} \tag{6}$$

$F_f(s)$ in Figure 3, is derived as shown in Equation (7).

$$F_f(s) = 1 - \frac{G_{ff}(s)}{G_i(s)}G_{x1}(s) \tag{7}$$

Then, the output impedance $Z_0(s)$ of the inverter containing the grid voltage feedforward control link is derived as

$$\begin{aligned} Z_0(s) &= \frac{1+T_0(s)}{G_{x2}(s)F_f(s)} \\ &= \frac{s^3L_1L_2C+s^2L_2H_{i1}CK_{pwm}G_d(s)+s(L_1+L_2)+G_i(s)G_d(s)K_{pwm}H_{i2}}{s^2L_1C+sCH_{i1}K_{pwm}G_d(s)+1-G_{ff}(s)K_{pwm}G_d(s)} \end{aligned} \tag{8}$$

when there is no grid voltage feedforward in the control system, the $F_f(s)$ link is taken to be 1. At this time, the inverter output impedance $Z_{01}(s)$ is shown as Equation (9).

$$Z_{01}(s) = \frac{s^3L_1L_2C + s^2L_2H_{i1}CK_{pwm}G_d(s) + s(L_1+L_2) + G_i(s)G_d(s)K_{pwm}H_{i2}}{s^2L_1C + sCH_{i1}K_{pwm}G_d(s) + 1} \tag{9}$$

In order to facilitate the analysis of the subsequent problems, this paper firstly gives a three-phase LCL grid-connected inverter with a rated power of 20 kW based on the relevant design guidelines, and the relevant technical parameters are shown in Table 1.

Table 1. Parameters of model.

Parameters	Value
DC voltage, V_{in}	800 V
Grid phase voltage, V_g	220 V
Rated power, P_0	20 Kw
Grid frequency, f_0	50 Hz
Rated current	40 A
Switching frequency, f_{sw}	40 kHz
Inverter-side inductance, L_1	5 mH
Grid-side inductance, L_2	2 mH
Filter capacitance, C	10 μF
Capacitor current feedback parameter, H_{i1}	0.18
Grid-connected current feedback parameter, H_{i2}	0.05
Grid voltage sampling parameter, H_v	1
QPR regulator proportional parameter, k_p	0.2
QPR regulator resonance parameter, k_r	1

3. System Stability Analysis in Weak Grid

When the grid is in weak grid conditions, the fluctuation range of its equivalent impedance is large, which can easily lead to inverter instability. Secondly, according to the above analysis, when the traditional grid voltage proportional feedforward is used to offset the influence of grid voltage on the incoming current, the feedforward channel introduces an additional positive feedback loop for the incoming current, and under the weak grid conditions, the introduction of this control strategy will lead to a drastic reduction in the phase angle margin of the system, which seriously threatens the stability of the grid-connected system. The causes of system destabilization will be discussed in the following passage.

Since the current controller is used in the paper, the actual output current is equal to the given value, thus when analyzing the stability of the grid-connected inverter under grid impedance, the inverter is equated to a constant current source i_0 and an equivalent output impedance Z_0 connected in parallel with it, and the specific equivalent circuit diagram is shown in Figure 4.

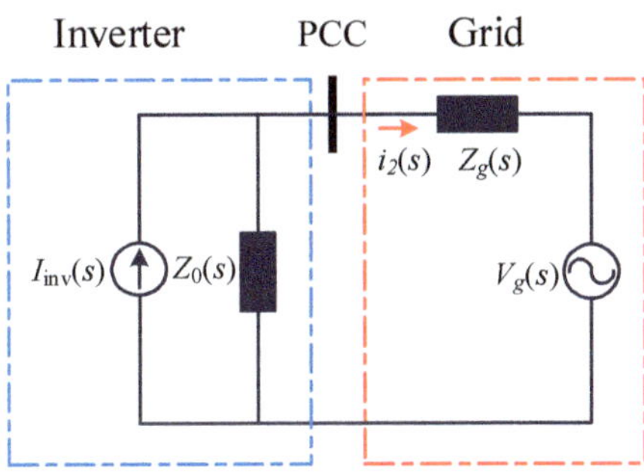

Figure 4. Impedance model of the grid-connected inverter.

Based on Figure 4, the grid-connected current i_2 can be rewritten as shown in Equation (10).

$$\begin{aligned} i_2(s) &= \frac{Z_0(s)}{Z_0(s)+Z_g(s)} I_{inv}(s) - \frac{V_g(s)}{Z_0(s)+Z_g(s)} \\ &= [I_{inv}(s) - \frac{V_g(s)}{Z_0(s)}] \frac{1}{1+\frac{Z_g(s)}{Z_0(s)}} \end{aligned} \quad (10)$$

In Equation (10), $Z_g(s)$ is the grid impedance. In order to better study the effect of grid impedance on the stability of the grid-connected inverter system, $Z_g(s)$ is only taken as the inductive part. From [27], the following two conditions need to be fulfilled in order to ensure that the inverter remains stable under conditions of varying grid impedance:

(1) When $Z_g(s) = 0$, it is necessary to ensure that the inverter is stabilized;

(2) When $Z_g(s) \neq 0$, it is necessary to ensure that $Z_g(s)/Z_0(s)$ satisfies the impedance stabilization criterion.

The system is stabilized if $Z_g(s)$ and $Z_0(s)$ do not intersect or if the phase angle margin between $Z_g(s)$ and $Z_0(s)$ at the intersection frequency f_s satisfies Equation (11).

$$PM = 180^\circ - (\angle Z_g(j2\pi f_s) - \angle Z_0(j2\pi f_s)) > 0^\circ \quad (11)$$

Since $Z_g(s)$ is inductive, the phase is 90°, and referring to (11), we can obtain the phase relation equation of $Z_0(s)$ when the system is stabilized, as shown in Equation (12).

$$\angle Z_0(j2\pi f_s)) > -90^\circ \quad (12)$$

Based on the results obtained in the previous section, the bode plot of the inverter output impedance with and without grid voltage feedforward is plotted as shown in Figure 5.

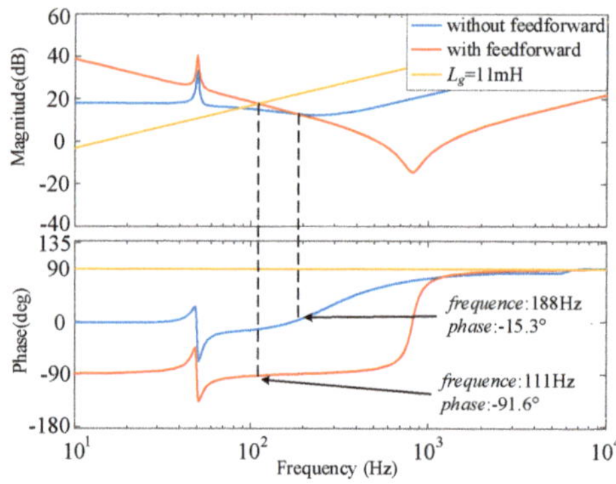

Figure 5. Bode plot of inverter output impedance with and without grid voltage feedforward.

As can be seen in Figure 5, the amplitude of the output impedance in the low-frequency band increases dramatically after the addition of the grid voltage feedforward, and the phase characteristics lag dramatically. However, the stability margin of the system changes in the weak grid case. For example, in Figure 5, when the grid impedance is 11 mH, corresponding to the short circuit ratio SCR = 2, the phase margin of the grid-connected inverter system before adding grid voltage feedforward is about 74.7°, while after adding grid voltage feedforward, the phase is −91.6°, which makes the system unstable at this time according to the analysis above. In conclusion, the increase in amplitude, while improving the inverter's ability to suppress background harmonics in the grid voltage, also poses a problem, which is that the lack of phase margin makes the grid-connected system much less stable. When the grid impedance in the weak grid case is non-negligible, the system is at risk of instability. In summary, regarding the reason for the instability of the grid-connected system caused by the traditional grid voltage feedforward method, the explanation from the perspective of output impedance is that the equivalent shunt virtual impedance correction introduced by the grid voltage feedforward will change the originally designed system, resulting in the phase lag of the original output impedance of the inverter.

4. Improved Feedforward Control Strategy Based on Band-pass Filter

According to the above analysis, the original intention of introducing proportional feedforward is to be able to effectively suppress the inverter grid-connected current distortion problem caused by the voltage harmonics at the point of common coupling (PCC), but after considering the weak grid, this control strategy will in turn reduce the overall phase margin of the system, resulting in system instability.

Therefore, to address the above problems, a band-pass filter is used in the proportional feedforward channel, so that this control can counteract the effect of grid voltage on the incoming current, while at the same time improve the overall phase margin of the system and broaden the system's adaptability to the grid impedance.

In this passage, instead of using a conventional voltage-controlled voltage-source second-order band-pass filter composed of a single integrated operational amplifier, subsequent experiments use a band-pass filter (BPF) composed of a low-pass filter (LPF) and a high-pass filter (HPF) connected in series. The band-pass filter transfer function is given below as shown in Equation (13).

$$\begin{aligned} G_{BPF}(s) &= \frac{K_1\omega_1^2}{s^2+\frac{\omega_1}{\zeta}s+\omega_1^2}\frac{K_2s^2}{s^2+\frac{\omega_2}{\zeta}s+\omega_2^2} \\ &= \frac{K\omega_1^2s^2}{(s^2+\frac{\omega_1}{\zeta}s+\omega_1^2)(s^2+\frac{\omega_2}{\zeta}s+\omega_2^2)} \end{aligned} \tag{13}$$

In Equation (13), K_1, K_2 are the gains of LPF and BPF, respectively, ω_2 and ω_1 are the upper and lower cutoff angular frequencies of BPF, K is its total gain, and ξ is the damping parameter (generally taking the value of 0.707).

The system control block diagram after adding BPF is shown in Figure 6.

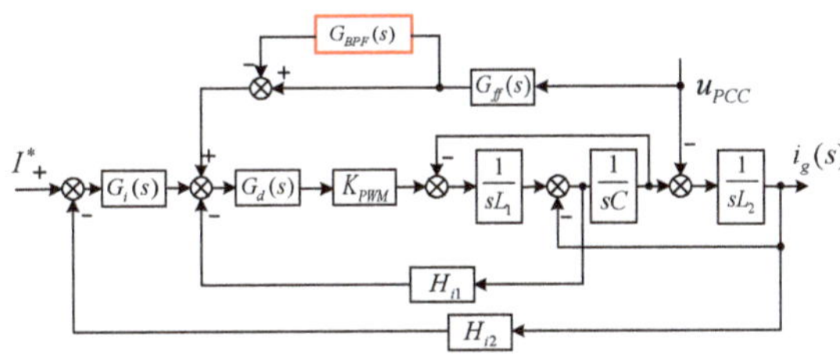

Figure 6. System control block diagram with BPF.

The improved inverter output impedance $Z_{03}(s)$ can be derived from Figure 6 as shown in Equation (14)

$$Z_{03}(s) = \frac{s^3L_1L_2C + s^2L_2H_{i1}CK_{pwm}G_d(s) + s(L_1 + L_2) + G_i(s)G_d(s)K_{pwm}H_{i2}}{s^2L_1C + sCH_{i1}K_{pwm}G_d(s) + 1 - (1 - G_{BPF}(s))G_{ff}(s)K_{pwm}G_d(s)} \quad (14)$$

Comparing Equations (8), (9), and (14), it can be seen that changing the feedforward link is actually changing the output impedance of the inverter for the purpose of impedance reshaping, and if the reshaped output impedance satisfies Equation (12), the original intention of the system design can be achieved.

In Table 1, the underlying parameters of the system have been given, but not the parameters of the BPF, which is now based on the basics of analog circuits to make a preliminary design of the BPF.

First, determine the center frequency f_0 = 50 Hz and the bandwidth B is also set to 50 Hz at this time. With the quality factor $Q = f_0/B$, determine the voltage amplification at all levels through the center frequency, so that the mode of the denominator of the voltage amplification multiplier is equal to 1.414 and obtain the lower cutoff frequency ω_1 and the upper cutoff frequency ω_2.

Here, only a simple design of the BPF is made. Usually, the gain of the BPF, if set too small cannot improve the phase margin of the inverter impedance in the mid-frequency band, while if set too large will lead to the amplitude of the amplitude–frequency curve of the inverter impedance amplitude reduction; the base wave gain is reduced, which is not conducive to the stability of the system. Here, the value of K is taken as 300. The upper and lower cutoff angular frequency is 4500 π and 100 π, respectively.

The given parameters are now brought into Equation (14) and the improved inverter output impedance bode plot is plotted as shown in Figure 7.

From Figure 7, it can be seen that the improved inverter output impedance $Z_{03}(s)$ possesses the same high impedance in the low-frequency band as that of $Z_{01}(s)$ without the addition of proportional feedforward, which retains the control system's immunity to grid harmonic voltages, whereas under weak grid conditions, $Z_{03}(s)$ has a phase of greater than $-90°$ at the frequency of intersection with the grid impedance and the system is in a stable state, which improves the system's instability in the case of proportional feedforward.

In the above introduction, the system impedance is changed by incorporating a BPF to adapt to a wide range of variations in the grid impedance and to improve the robustness of the system. However, from the beginning, the BPF was designed by applying only the knowledge of analog circuits to its most basic functions, such as setting the voltage signals within the desired frequency range to pass through, setting the signals outside the frequency to attenuate quickly, and setting the quality factor to improve its frequency selective characteristics. However, when the BPF parameters are changed, its inverter output impedance $Z_{03}(s)$ is also changed, and substituting the designed BPF into $Z_{03}(s)$ is not so ideal for every set of data.

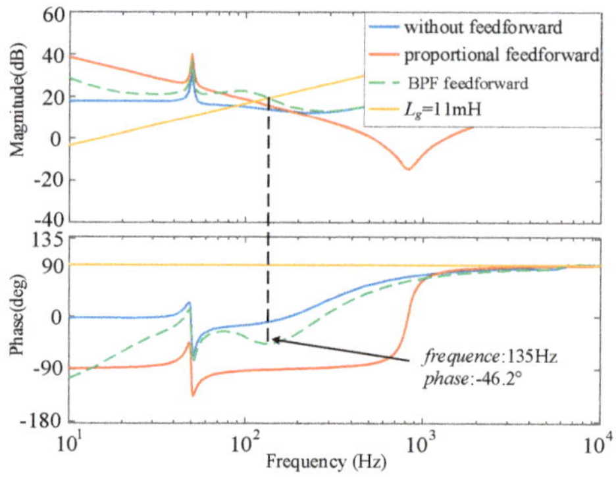

Figure 7. Bode plot of output impedance of the improved inverter.

Figure 8 shows the bode plot of $Z_{03}(s)$ with a K value of 300 and upper and lower cutoff angular frequency of 2500 π and 100 π, respectively, which reduces the passband range of the BPF, and it can be seen from Figure 8 that from the design point of view of the filter, the parameters obtained from the design satisfy the requirements of filtering but do not satisfy the requirements of system stability. At this time, the inverter output impedance will appear to intersect the grid impedance several times. Moreover, at f_s = 844 Hz, the phase is -88.4°, at which time the system is already in an unstable boundary state.

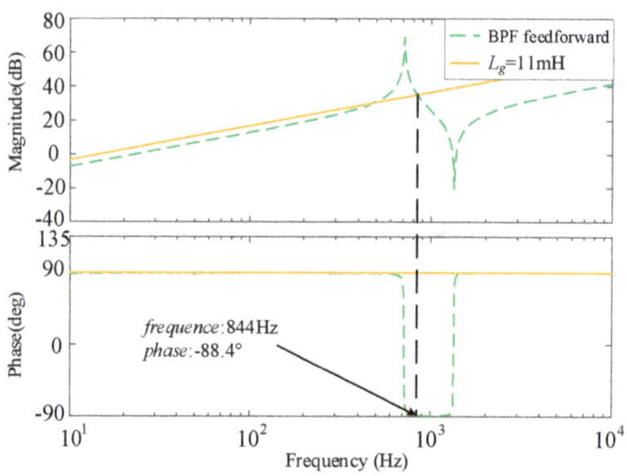

Figure 8. Bode plot of inverter output impedance after changing BPF parameter.

In addition to this, if a physical BPF is used, its resistance and capacitance need to be designed, and substituting each calculated parameter into $Z_{03}(s)$ if it is found to be inappropriate and then carrying out the calculation again will greatly increase the workload.

5. Multi-Objective Constraint

In the previous section of the analysis, the method of improving the system margin was actually an impedance remodeling of the system. Methods for impedance remodeling are well established, and there are many remodeling methods for this aspect of the feedforward channel. For example, low-pass, band-pass, high-pass, band-stop, and complex filters are strung in the feedforward channel, and a phase correction link is added to the current-controlled forward channel, and so on. But whenever a new link is added, the parameters of the new link need to be designed. For example, the key parameter cutoff frequency of a complex filter is designed based on its dominant eigenvalue and remains without a necessary connection to the system.

For grid connection, according to national standards, there are numerous criteria in addition to keeping the system stable such as unit power factor grid connection, grid current total harmonic distortion (THD), base-wave current tracking performance, and so on. If it is possible to use the grid-connected conditions as a baseline to establish the relationship between each of the required performance indicators and the currently incorporated improvements, it is possible to design the ideal parameters in one step. When the system is subjected to large disturbances and wide-range variations, the system can still meet the requirements and be connected to the grid in that case because the design leaves a certain margin. In order to solve the above problems and claims, the three important parameters K, ω_1, and ω_2 in BPF are jointly optimized in this passage. Firstly, let K, ω_1, and ω_2 be X, Y and Z, respectively, then Equation (13) can be rewritten as

$$G_{BPF}(s) = \frac{XY^2s^2}{(s^2 + \frac{Y}{\zeta}s + Y^2)(s^2 + \frac{Z}{\zeta}s + Z^2)} \quad (15)$$

Then, the multi-objective constraint functions between the BPF parameters and the base-wave current tracking performance, system stability margin, and low-frequency amplitude constraints are established, and these constraint functions are used to find the BPF parameters that satisfy the system requirements, so that the output impedance of the inverter directly establishes a direct relationship with the grid-connection index of the system, which makes the method of impedance reshaping more intuitive.

5.1. *Base-Wave Current Tracking Performance*

Since the following analyses are performed at both low frequency (50 Hz) and intermediate frequency (1 kHz), the filter capacitance C is neglected, and the QPR controller in term $G_i(s)G_d(s)K_{pwm}H_{i2}$ has a nearly infinite gain at the base-wave, and the $s(L_1 + L_2)$ term can be further neglected, and in this passage, the proportional feedforward is used for $G_{ff}(s)$.

$$G_{ff}(s) = \frac{1}{K_{pwm}} \quad (16)$$

Then, the output impedance of the improved inverter can be simplified as $Z_{04}(s)$.

$$Z_{04}(s) = \frac{G_i(s)G_d(s)K_{pwm}H_{i2}}{1 - (1 - G_{BPF}(s))G_d(s)} \quad (17)$$

In order to minimize the impact on the tracking performance of the base-wave current it is necessary to keep the characteristics at the base-wave unchanged before and after reshaping the output impedance of the inverter. For without grid voltage feedforward control, without grid voltage feedforward control in the strong grid case already meets the requirements of system stability and unit power factor grid connection. Therefore, for the introduction of the BPF feedforward, the base-wave characteristics of the feedforward control without grid voltage are used as a benchmark, and the base-wave is constrained in terms of both amplitude and phase.

5.1.1. Base-Wave Amplitude Constraint

Similarly, the inverter output impedance $Z_{01}(s)$ without grid voltage feedforward control can be simplified to $Z_{05}(s)$.

$$Z_{05}(s) = G_i(s)G_d(s)K_{pwm}H_{i2} \quad (18)$$

To keep the base-wave amplitude constant before and after impedance reshaping, (19) can be derived

$$|Z_{04}(j\omega_0)| = |Z_{05}(j\omega_0)| \quad (19)$$

Since the amplitude gain due to the delay link is one, it is negligible when considering the amplitude characteristic. Then, Equation (19) can be written as

$$|1 - G_{BPF}(j\omega_0)| \approx 0 \tag{20}$$

supposing that

$$|1 - G_{BPF}(j\omega_0)| \leq AM_{erorr-\max} \tag{21}$$

At this point, the base-wave amplitude can be approximated as unchanged. In Equation (21), $AM_{error\text{-}max}$ is the ratio of the maximum amplitude of the base-wave. Accordingly, the constraint based on the amplitude error of the base-wave can be obtained.

$$1 - G_{BPF}(j\omega_0) = \frac{\begin{array}{c}\omega_0^4 + \omega_0^2(XY^2 - Z^2 - Y^2 - 1.414^2 YZ) + Y^2Z^2 \\ +j(\omega_0 1.414(Z^2Y + Y^2Z) - \omega_0^3 1.414(Y+Z))\end{array}}{\begin{array}{c}\omega_0^4 - \omega_0^2(Z^2 + Y^2 + 1.414^2 YZ) + Y^2Z^2 \\ +j(\omega_0 1.414(Z^2Y + Y^2Z) - \omega_0^3 1.414(Y+Z))\end{array}} \tag{22}$$

For ease of reference below, Equations (23)–(25) are given

$$a = \omega_0^4 - \omega_0^2(Z^2 + Y^2 + 1.414^2 YZ) + Y^2Z^2 \tag{23}$$

$$b = \omega_0 1.414(Z^2Y + Y^2Z) - \omega_0^3 1.414(Y+Z) \tag{24}$$

$$c = \omega_0^4 + \omega_0^2(XY^2 - Z^2 - Y^2 - 1.414^2 YZ) + Y^2Z^2 \tag{25}$$

The constraint on the amplitude of the base-wave can then be written as

$$|1 - G_{BPF}(j\omega_0)| = \frac{\sqrt{c^2 + b^2}}{\sqrt{a^2 + b^2}} \leq AM_{error-\max} \tag{26}$$

5.1.2. Base-Wave Phase Constraint

Assuming that there is an error PF_{error} in the grid-connected power factor of the system before and after the introduction of the BPF, $\Delta\varphi$ is the corresponding power factor error angle. The inverter is considered to achieve unit power factor grid connection if PF_{error} satisfies $PF_{error} \leq PF_{error\text{-}max}$ ($PF_{error\text{-}max}$ is the maximum permissible power factor error). Then, the condition yields $|\Delta\varphi| = \Delta\varphi_{max}$. $\Delta\varphi_{max}$ as the power factor error angle corresponding to $PF_{error\text{-}max}$.

When the base-wave phase constraint is considered, the delay link can no longer be ignored. Then, Equation (17) is rewritten as

$$Z_{04}(j\omega_0) = \frac{G_i(j\omega_0)K_{pwm}H_{i2}e^{-j1.5T_s\omega_0}}{1 - (1 - G_{BPF}(j\omega_0))G_d(s)e^{-j1.5T_s\omega_0}} \tag{27}$$

The phase angle can be written as

$$\Phi_{Z_{04}(j\omega_0)} = \arctan\frac{b}{a} - \arctan\frac{b - b\cos(1.5T_s\omega_0) + c\sin(1.5T_s\omega_0)}{a - c\cos(1.5T_s\omega_0) - b\sin(1.5T_s\omega_0)} - 57.3(1.5T_s\omega_0) \tag{28}$$

For the case without BPF, the phase angle at the base-wave is counted as $\Phi_{Z01(j\omega 0)}$. According to Figure 5, its phase angle at the base-wave is $-15°$. To ensure that the phase angle offset of the base-wave after the introduction of the BPF is small at this point, it is necessary to satisfy Equation (29).

$$\begin{aligned}|\Delta\varphi| &= \left|\left|\Phi_{Z_{04}(j\omega_0)}\right| - \left|\Phi_{Z_{01}(j\omega_0)}\right|\right| \\ &= \left|\left|\Phi_{Z_{04}(j\omega_0)}\right| - 15°\right| \\ &\approx 0\end{aligned} \tag{29}$$

$|\Delta\varphi|$ should be as small as possible so that its maximum value does not exceed $\Delta\varphi_{max}$. When $|\tan(\Delta\varphi_{max})|$ is extremely small, the power factor error $PF_{error\text{-}max}$ can be considered to be approximately 0. At this point, the inverter unit power factor grid connection.

Based on the above constraint requirement, the base-wave phase error constraint on the output impedance of the inverter shown in Equation (30) can be obtained. There is no need to add a fundamental frequency phase correction link after considering this constraint.

$$|tan(\Delta\varphi_{\max})| \geq \left|\tan\left(\left|\Phi_{Z_{04}(j\omega_0)}\right| - \left|\Phi_{Z_{01}(j\omega_0)}\right|\right)\right| \tag{30}$$

$AM_{error\text{-}max} = 3\%$, $PF_{error\text{-}max} = 0.1\%$ (correspond to $\Delta\varphi_{max} = 2.56°$). Then, the constraint on the tracking performance of the base-wave current can be obtained.

5.2. System Stability Margin

In order to ensure that the system based on grid voltage feedforward control still has a high stability margin under weak and very weak grid when using the multi-objective constraint method, the stability margin constraint is performed under the condition of SCR = 2 for very weak grid at this time. In practical engineering, the system can be considered to be stabilized with PM = 30° considering that there may be some phase error before and after the simplification of Equation (17). To further ensure that the system has sufficient stabilization margin, the system stability margin constraint is set at PM = 40° in this paper.

When SCR = 2, the intersection frequency f_s of impedance $Z_{04}(s)$ with impedance $Z_g(s)$ is unknown. Therefore, the intersection frequency f_s = 110 Hz of impedance $Z_0(s)$ and impedance $Z_g(s)$ is used instead. To ensure the effectiveness of this alternative, the intersection frequency of impedance $Z_{04}(s)$ and impedance $Z_g(s)$ is first constrained to be near or greater than f_s.

The system output impedance intersection frequency constraint is given by

$$\left|\frac{Z_{04}(j\omega_g)}{Z_g(j\omega_g)}\right| \geq 1 \tag{31}$$

In Equation (31), ω_g is the intersection angle frequency between the inverter output impedance and the grid impedance. When Equation (31) is satisfied, the intersection frequency of impedance $Z_{04}(s)$ and impedance $Z_g(s)$ can be considered to be f_s or greater than f_s.

The constraint on the stability margin of the system is given by

$$PM = 90° + \angle Z_{04}(j\omega_g) \geq 40° \tag{32}$$

In Equation (32), $Z_{04}(j\omega_g)$ is shown as

$$\begin{aligned}\angle Z_{04}(j\omega_g) = \arctan & \frac{k_p(\omega_0^2 - \omega_g^2)^2 + 2k_r\omega_c\omega_g a_{\omega_g}(\omega_0^2 - \omega_g^2) + 4\omega_c^2\omega_g^2(k_p + k_r)b_{\omega_g}}{k_p a_{\omega_g}(\omega_0^2 - \omega_g^2)^2 - 2b_{\omega_g}k_r\omega_c\omega_g(\omega_0^2 - \omega_g^2) + 4\omega_c^2\omega_g^2(k_p + k_r)a_{\omega_g}} \\ & -\arctan\frac{b_{\omega_g} - b_{\omega_g}\cos(1.5T_s\omega_g) + c_{\omega_g}\sin(1.5T_s\omega_g)}{a_{\omega_g} - c_{\omega_g}\cos(1.5T_s\omega_g) - b_{\omega_g}\sin(1.5T_s\omega_g)} \\ & -57.3(1.5T_s\omega_g)\end{aligned} \tag{33}$$

In Equation (33), $a_{\omega g}$, $b_{\omega g}$, $c_{\omega g}$ are obtained by replacing ω_0 with ω_g in Equations (23)–(25).

Substituting Equation (33) into Equation (32) yields the constraint on the stability margin of the system.

5.3. Low-Frequency Amplitude Constraint

After impedance reshaping, the impedance amplitude in the frequency band below the fundamental frequency of the inverter output impedance is unknown and is likely to be below 0 dB, resulting in multiple intersections with $Z_g(s)$, which is not conducive to system stability. Therefore, further optimized design of the reshaped inverter output impedance with low-frequency amplitude as a constraint is required.

The amplitude of $G_{BPF}(s)$ is continuously increasing in the range [0–50 Hz]. And considering that the gain of the QPR controller is mainly determined by the proportionality parameter k_p, $G_i(s)$ is equated here to the proportionality link k_p. When considering amplitude below the fundamental frequency, $Z_{04}(s)$ can be simplified as

$$Z_{04}(s) = \frac{k_p K_{pwm} H_{i2}}{G_{BPF}(s)} \tag{34}$$

To ensure that the inverter output impedance below the fundamental frequency has no intersection with the amplitude of $Z_g(s)$, it is only necessary to ensure that the minimum value of the amplitude of $Z_{04}(s)$ below the fundamental frequency after reshaping is greater than the maximum value of the amplitude of $Z_g(s)$. From the above analysis, the minimum of the $Z_{04}(s)$ magnitude is obtained at ω_0.

$$\left| Z_{04}(j\omega_0) = \frac{k_p K_{pwm} H_{i2}}{G_{BPF}(j\omega_0)} \right| > A \tag{35}$$

In order to improve the applicability of the constraint, take SCR = 2 for very weak grid condition here. At this point the grid impedance has an amplitude of 10 dB at 50 Hz, so take A as 10.

By combining the above considerations, then the constraint on the low-frequency amplitude of the system can be obtained.

Based on the constraints of Equations (26), (30), (31), (32), and (35), then the constraint space of each constraint with respect to X, Y, and Z under the 3D coordinate system can be obtained. The points that satisfy these constraint spaces are the values of X, Y, and Z that satisfy the requirements.

5.4. System Stability Analysis Based on Grid Voltage Feedforward under Multi-Objective Constraint

Based on the constraint space that has been obtained, further stability analysis of the proposed grid voltage feedforward based on the multi-objective constraint.

It can be seen that the five constraint equations will form five constraint ranges in the space, and taking the intersection of these five constraint ranges is the value of X, Y, and Z that satisfies the requirements as shown in Figure 9.

In this paper, we take the value of X = 800, Y = 50, Z = 1750 as an example to verify and analyze. The three parameters are substituted into $Z_{03}(s)$ to obtain the BPF feedforward inverter output impedance $Z'_{03}(s)$ under multi-objective constraint. Also analyzed under very weak grid is the plotting of the bode plot of $Z_{01}(s)$ and $Z'_{03}(s)$, as shown in Figure 10.

At this point, for the constraint, the base-wave amplitude ratio is 2.56%, less than $AM_{error\text{-}max}$, and the base-wave amplitude constraint is satisfied; the phase difference in the base wave is 2.2°, less than $|\Delta\varphi| = \Delta\varphi_{max}$ = 2.56°. The grid-connected inverter can be considered to be grid-connected at a unit power factor and satisfies the base-wave phase constraint.

The bode plot of $Z'_{03}(s)$ and the inverter output impedance $Z_0(s)$ without the addition of BPF is plotted as shown in Figure 11 to verify the system stability margin constraint. From Figure 11, it can be seen that the intersection frequency of $Z'_{03}(s)$ with the grid impedance is larger than that of $Z_0(s)$ with the grid impedance, which satisfies the intersection frequency constraint. Secondly, the system stability margin after adding the multi-objective constraint reaches 71.6° greater than 40°, which satisfies the system stability margin constraint.

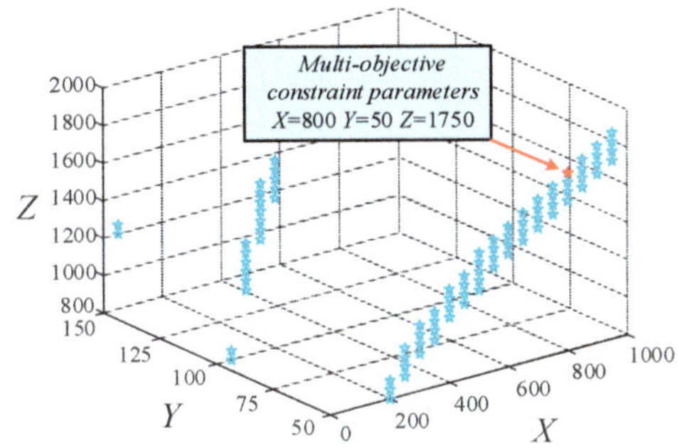

Figure 9. Distribution of parameter values.

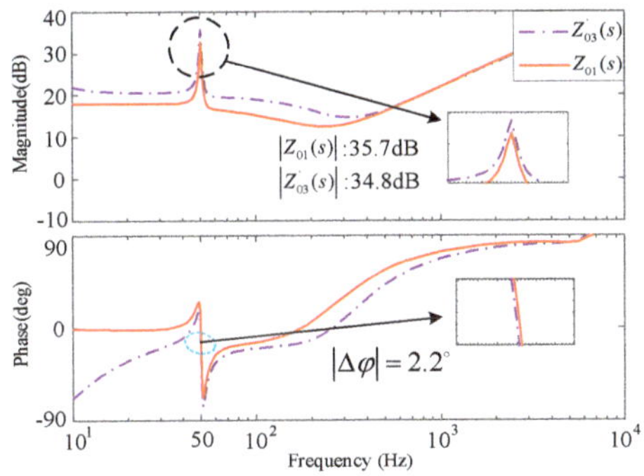

Figure 10. Bode plot of $Z_{01}(s)$ and $Z'_{03}(s)$.

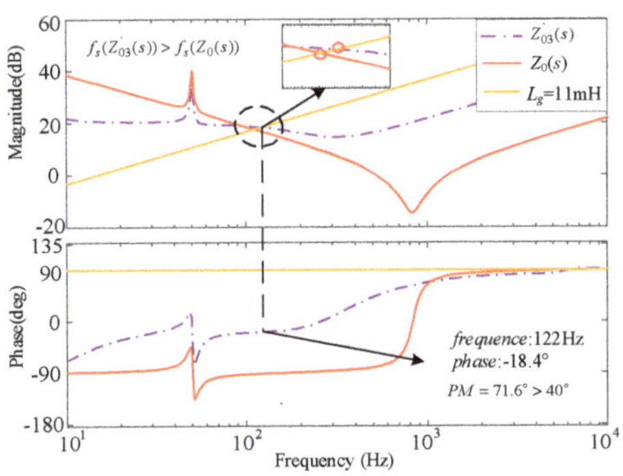

Figure 11. Bode plot of $Z_0(s)$ and $Z'_{03}(s)$.

Besides, below the fundamental frequency of the system, there is no intersection between the grid impedance and the inverter output impedance, which ensures the stability of the system and satisfies the low-frequency amplitude constraint. The correctness of the selected values is verified.

Figure 12 shows the bode plot of $Z'_{03}(s)$ and the inverter output impedance bode plot including BPF feedforward obtained according to the conventional design method in Section 3.

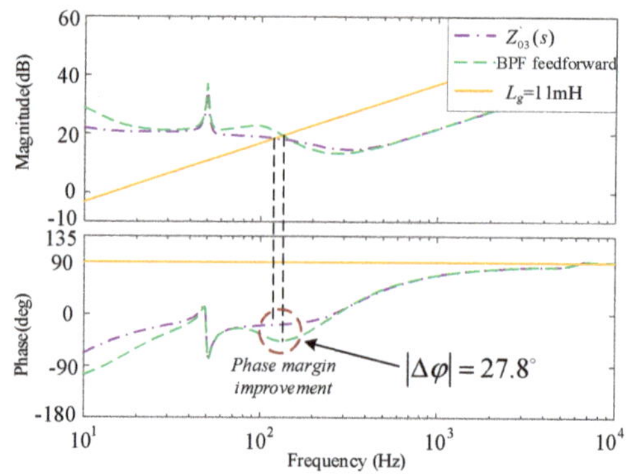

Figure 12. Bode plot of $Z_{03}(s)$ and $Z'_{03}(s)$.

As can be seen in Figure 12, the system margin is improved by 27.8° compared to the case when the multi-objective constraint is not used, which greatly improves the stability of the system.

Plotting the bode plot of $Z'_{03}(s)$ and the bode plot of the grid impedance as it varies is shown in Figure 13.

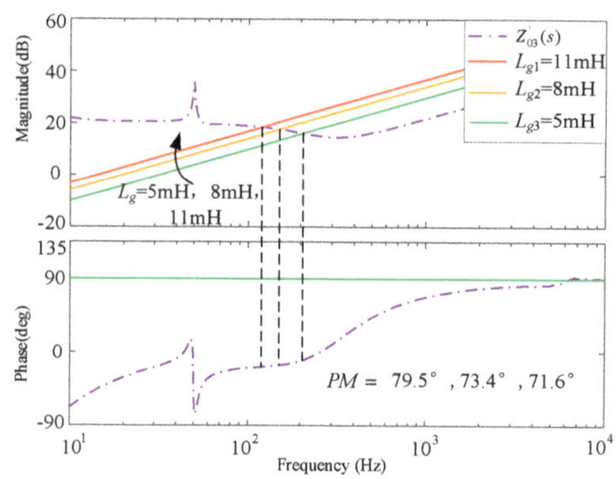

Figure 13. Bode plot of $Z'_{03}(s)$ and L_g.

From Figure 13, it can be seen that when the grid impedance is varied, L_g is 5 mH, 8 mH, and 11 mH, respectively, and the system stability margin is 79.5°, 73.4°, and 71.6°, respectively. It is shown that the system has excellent robustness after multi-objective constraint.

According to this section, it can be seen that the system is improved considerably after adding the multi-objective constraint feedforward control without affecting the system's base-wave current tracking performance. Once the various grid-connection requirements have been determined, combining the variable part of the feedforward with the grid-connection requirements makes impedance remodeling intuitive. This method, contrary to the idea of traditional methods, takes into account the holistic nature of the system before considering the innovative nature of the proposed strategy, making the method more flexible and versatile. Apart from this, the proposed method is also more universal than the feedforward scheme designed according to the traditional parameter design method, at the same time, it also greatly reduces the workload during the experiment.

For the above analysis, there could be an increase in the constraints. The constraints considered may also not satisfy the situation in a weaker grid. In addition to this, if the system has multiple inverters connected in parallel, then the applicability of this method is yet to be considered.

6. Hardware-in-the-Loop Experimental Results

To further validate the effectiveness of the grid voltage feedforward strategy based on multi-objective constraint proposed in this passage, the DSPTMS320F28335 (Texas Instruments, Xi'an, China) is used as the core control, and a semi-physical simulation platform based on RT-LAB is built to verify the effectiveness of the proposed control strategy. This experiment is conducted in a hardware-in-the-loop approach. The experimental platform is shown in Figure 14, and the system parameters are consistent with Table 1.

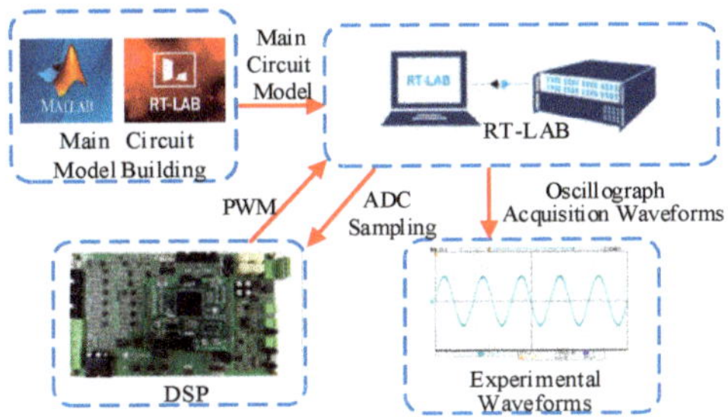

Figure 14. Hardware-in-the-loop experimental platform.

Figure 15 shows the experimental waveform of the grid-connected current of the system without grid voltage feedforward control under strong grid condition. At this time, the grid-connected current waveform remains stable, and the THD of the current is 0.76%, which meets the grid-connected condition.

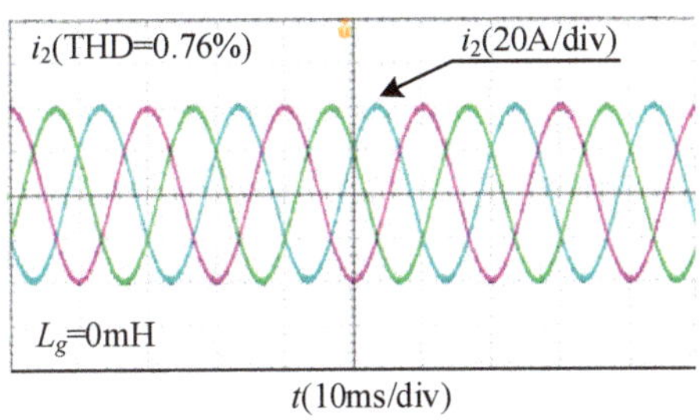

Figure 15. Hardware-in-the-loop experiment waveform of grid-connected current of the system under strong grid condition without grid voltage feedforward control.

Figure 16 shows the experimental waveforms of the grid-connected current after injecting each 3% of the 5th and 7th harmonics into the grid voltage in the case of a strong grid and without feedforward control of the grid voltage, at which time the grid-connected current undergoes a distortion, and the THD of the current is 5.97%.

Figure 17 shows the experimental waveform of the grid-connected current with the proportional feedforward control added after injecting grid voltage harmonic. From Figure 18, it can be seen that the added proportional feedforward control can suppress the effect of grid voltage fluctuation on the grid-connected current. The grid-connected current is improved and stabilized. The THD of the current is 3.59%.

Weak grid experiments are performed next, and the following experiment is conducted under a very weak grid condition, SCR = 2, L_g = 11 mH and all are injected with 3% of the 5th and 7th grid voltage harmonic.

Figure 18 shows the experimental waveform of the grid-connected current in weak grid, which only added proportional feedforward control, and it can be seen that the system grid-connected three-phase current occurs serious distortion at this time. The original system will no longer be stable.

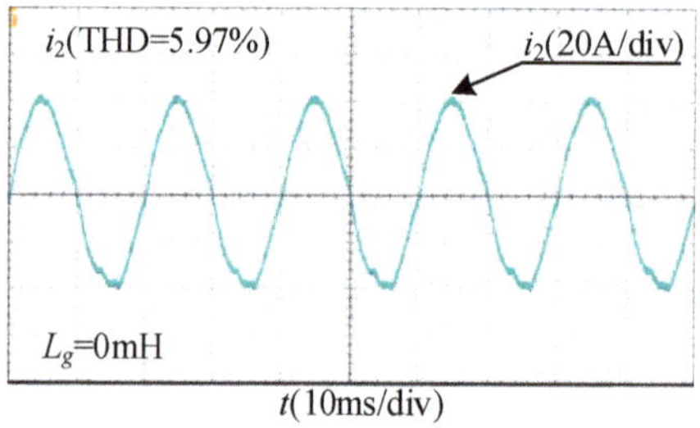

Figure 16. Hardware-in-the-loop experiment waveform of grid-connected current after harmonic injection.

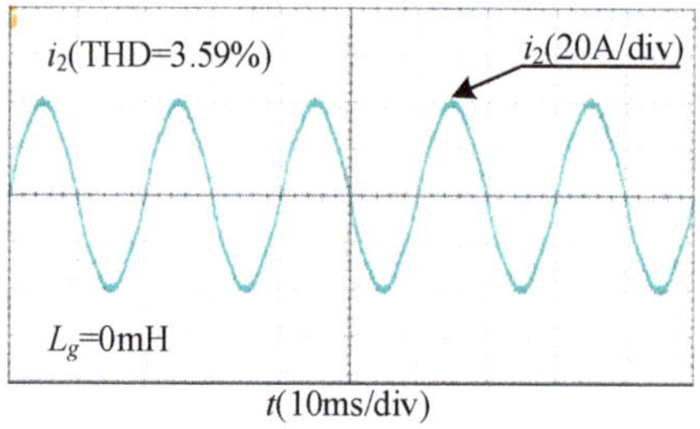

Figure 17. Hardware-in-the-loop experiment waveform of grid-connected current under proportional feedforward control.

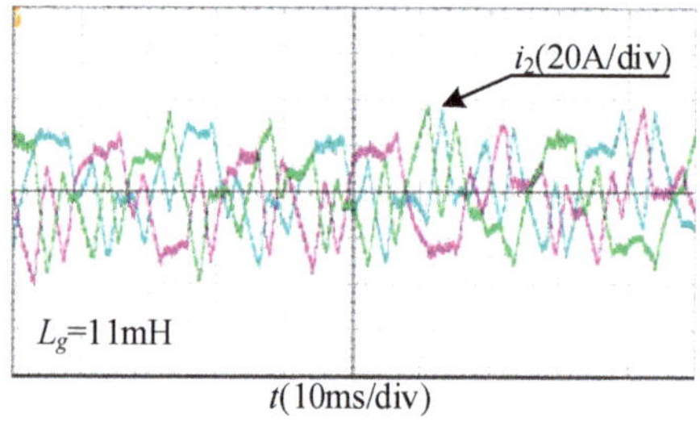

Figure 18. Hardware-in-the-loop experiment waveform of grid-connected current with proportional feedforward control in weak grid.

Figure 19 shows the experimental waveform of the grid-connected current for the weak grid case where the proportional feedforward control is replaced by the BPF feedforward control. The introduction of the BPF re-stabilizes the system and the THD of the grid-connected current is 1.07%.

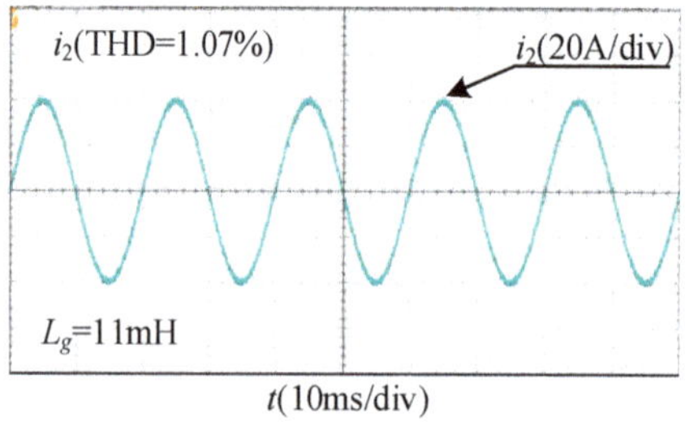

Figure 19. Hardware-in-the-loop experiment waveform of grid-connected current with BPF feedforward control in weak grid.

Figure 20 shows the experimental waveform of the grid-connected current under the proposed multi-objective constraint control. From Figure 21, it can be seen that the proposed multi-objective constraint method is able to link directly with the grid-connected requirements to realize the grid connection in the case of very weak grid. Its current THD is 0.8%.

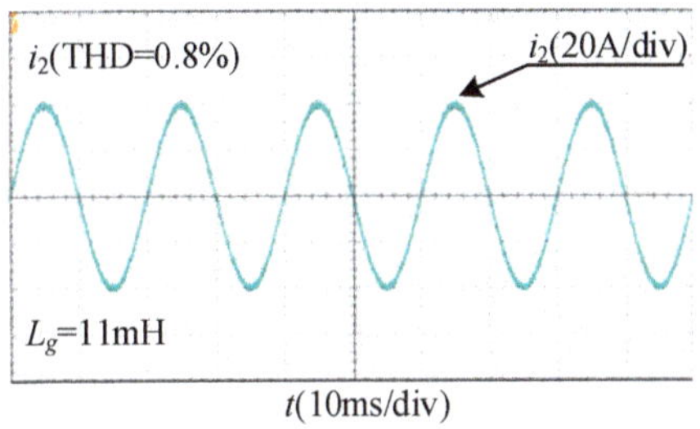

Figure 20. Hardware-in-the-loop experiment waveform of grid-connected current under multi-objective constraint control in weak grid.

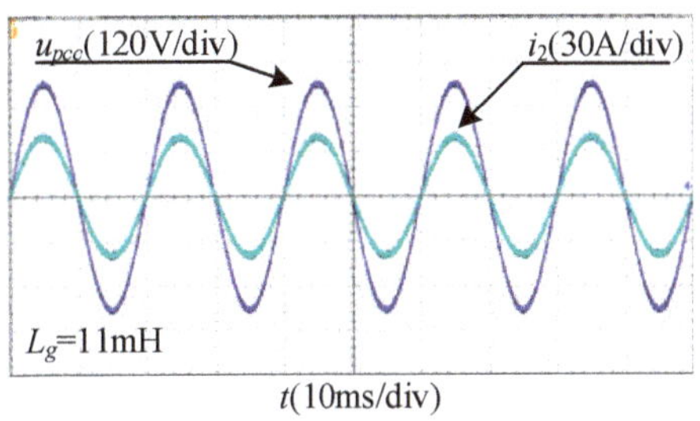

Figure 21. Hardware-in-the-loop experiment waveform of grid-connected voltage and grid-connected current under multi-objective constraint control in weak grid.

Figure 21 shows the experimental waveform of the system grid-connected current and grid-connected voltage when the multi-objective constraint feedforward control is used. Analyzing Figure 21, it can be seen that based on this control, the unit power factor grid connection can be guaranteed. There is no phase difference in grid-connected voltage and grid-connected current.

Figure 22 gives the dynamic experimental waveform when the grid-connected current jumps using the multi-objective constraint. The reference current is set to jump from full load to half load. Under this control, the grid-connected current undergoes a short regulation process at the jumping instant and enters into stabilization within half a cycle, which indicates that the system has excellent dynamic performance.

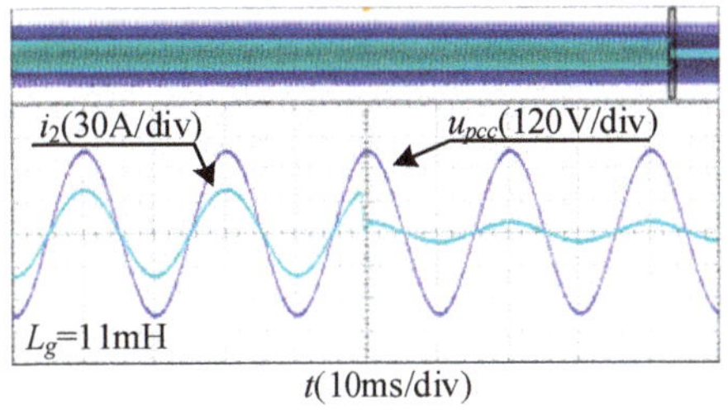

Figure 22. Hardware-in-the-loop experiment waveform of grid-connected current jump.

Based on the above hardware-in-the-loop experimental results, it is shown that the multi-objective constraint feedforward control strategy proposed in this passage can effectively solve the problems of system robustness degradation and instability in weak grid, and it also solves the problem of designing the BPF parameters redundantly, making the impedance reshaping method more intuitive, and improving the stability of the system in very weak grid.

7. Conclusions

In this paper, a grid voltage feedforward control strategy based on a multi-objective constraint method is proposed as an example for a three-phase *LCL* grid-connected inverter, and the proposed strategy effectively solves the situation of insufficient system stability margin in the case of a weak grid, and the work of this paper is summarized as follows:

(1) This paper analyzed the mechanisms and feedforward channels that can lead to system destabilization in weak grid. By adding BPF to feedforward control, the role of feedforward control in reducing the effect of grid voltage fluctuation on the grid-connected current was retained.

(2) For the BPF feedforward control used, the parameters design is too redundant and cannot be integrated with the final grid connection requirements. A multi-objective constraint method is proposed to solve this problem. The suggestions are as follows: relying on inverter output impedance stability criterion, considering the improvement of the feedforward channel as a reshaping of the inverter output impedance, detaching from the perspective of filter design, linking the added BPF to the system as a whole, optimizing the variables in the BPF, reshaping of the inverter output impedance, and making the system adapt to more severe conditions.

(3) The multi-objective constraint approach proposed in this paper has a wide range of applications. The multi-objective constraint method can still be used when the improvement for feedforward control is no longer adding filter but other methods. The method can combine the proposed improvements with the grid connection requirements, satisfying actual engineering requirements.

(4) For other feedforward control block diagrams, this study builds on the underlying feedforward control block diagram. Currently, there may be new control block diagrams proposed by scholars that are widely applied. So, for a new control block diagram, the modeling calculations have to be re-conducted. In addition to this, a multi-inverter parallel system will be investigated to verify whether this method is usable or not.

Author Contributions: Conceptualization, S.W. and P.H.; Methodology, K.C.; Software, K.C.; Validation, K.C.; Writing—original draft, K.C. All authors have read and agreed to the published version of the manuscript.

Funding: This research was funded by the Key Research and Development Plan of Shaanxi Province, grant number 2024GX-YBXM-468.

Data Availability Statement: The original contributions presented in the study are included in the article, further inquiries can be directed to the corresponding author.

Conflicts of Interest: The authors declare no conflicts of interest.

References

1. Blaabjerg, F.; Teodorescu, R.; Liserre, M.; Timbus, A.V. Overview of Control and Grid Synchronization for Distributed Power Generation Systems. *IEEE Trans. Ind. Electron.* **2006**, *53*, 1398–1409. [CrossRef]
2. Han, Y.; Zhang, K.; Li, H.; Coelho, E.A.A.; Guerrero, J.M. MAS-Based Distributed Coordinated Control and Optimization in Microgrid and Microgrid Clusters: A Comprehensive Overview. *IEEE Trans. Power Electron.* **2018**, *33*, 6488–6508. [CrossRef]
3. Karasala, C.; Lekkala, R.R.; Myneni, H.; Ganjikunta, S.K. PV Grid-Connected Inverter with DC Voltage Regulation in CCM and VCM Operation to Reduce Switching Losses. *IEEE Trans. Ind. Electron.* **2023**, *70*, 11262–11275. [CrossRef]
4. Ferreira, S.C.; Gonzatti, R.B.; Pereira, R.R.; da Silva, C.H.; da Silva, L.E.B.; Lambert-Torres, G. Finite Control Set Model Predictive Control for Dynamic Reactive Power Compensation with Hybrid Active Power Filters. *IEEE Trans. Ind. Electron.* **2018**, *65*, 2608–2617. [CrossRef]

5. Qi, J.; Zhao, W.; Bian, X. Comparative Study of SVC and STATCOM Reactive Power Compensation for Prosumer Microgrids with DFIG-Based Wind Farm Integration. *IEEE Access* **2020**, *8*, 209878–209885. [CrossRef]
6. Wang, X.; Blaabjerg, F.; Loh, P.C. Grid-Current-Feedback Active Damping for LCL Resonance in Grid-Connected Voltage-Source Converters. *IEEE Trans. Power Electron.* **2016**, *31*, 213–223. [CrossRef]
7. Mohamed, Y.A.-R.I. Suppression of Low- and High-Frequency Instabilities and Grid-Induced Disturbances in Distributed Generation Inverters. *IEEE Trans. Power Electron.* **2011**, *26*, 3790–3803. [CrossRef]
8. Zhang, H.; Ruan, X.; Lin, Z.; Wu, L.; Ding, Y.; Guo, Y. Capacitor Voltage Full Feedback Scheme for LCL-Type Grid-Connected Inverter to Suppress Current Distortion Due to Grid Voltage Harmonics. *IEEE Trans. Power Electron.* **2021**, *36*, 2996–3006. [CrossRef]
9. Ma, G.; Xie, C.; Li, C.; Zou, J.; Guerrero, J.M. Passivity-Based Design of Passive Damping for LCL-Type Grid-Connected Inverters to Achieve Full-Frequency Passive Output Admittance. *IEEE Trans. Power Electron.* **2023**, *38*, 16048–16060. [CrossRef]
10. Wu, W.; He, Y.; Tang, T.; Blaabjerg, F. A New Design Method for the Passive Damped LCL and LLCL Filter-Based Single-Phase Grid-Tied Inverter. *IEEE Trans. Ind. Electron.* **2013**, *60*, 4339–4350. [CrossRef]
11. Beres, R.N.; Wang, X.; Blaabjerg, F.; Liserre, M.; Bak, C.L. Optimal Design of High-Order Passive-Damped Filters for Grid-Connected Applications. *IEEE Trans. Power Electron.* **2016**, *31*, 2083–2098. [CrossRef]
12. Rodriguez-Diaz, E.; Freijedo, F.D.; Vasquez, J.C.; Guerrero, J.M. Analysis and Comparison of Notch Filter and Capacitor Voltage Feedforward Active Damping Techniques for LCL Grid-Connected Converters. *IEEE Trans. Power Electron.* **2019**, *34*, 3958–3972. [CrossRef]
13. Zhang, H.; Wang, X.; He, Y.; Pan, D.; Ruan, X. A Compensation Method to Eliminate the Impact of Time Delay on Capacitor-Current Active Damping. *IEEE Trans. Ind. Electron.* **2022**, *69*, 7512–7516. [CrossRef]
14. Zeng, C.; Wang, H.; Li, S.; Miao, H. Grid-Voltage-Feedback Active Damping with Lead Compensation for LCL-Type Inverter Connected to Weak Grid. *IEEE Access* **2021**, *9*, 106813–106823. [CrossRef]
15. Chen, W.; Zhang, Y.; Tu, Y.; Guan, Y.; Shen, K.; Liu, J. Unified Active Damping Strategy Based on Generalized Virtual Impedance in LCL-Type Grid-Connected Inverter. *IEEE Trans. Ind. Electron.* **2023**, *70*, 8129–8139. [CrossRef]
16. Bao, C.; Ruan, X.; Wang, X.; Li, W.; Pan, D.; Weng, K. Step-by-Step Controller Design for LCL-Type Grid-Connected Inverter with Capacitor–Current-Feedback Active-Damping. *IEEE Trans. Power Electron.* **2014**, *29*, 1239–1253.
17. Zhou, X.; Zhou, L.; Chen, Y.; Shuai, Z.; Guerrero, J.M.; Luo, A.; Wu, W.; Yang, L. Robust Grid-Current-Feedback Resonance Suppression Method for LCL-Type Grid-Connected Inverter Connected to Weak Grid. *IEEE J. Emerg. Sel. Top. Power Electron.* **2018**, *6*, 2126–2137. [CrossRef]
18. Wang, X.; Ruan, X.; Liu, S.; Tse, C.K. Full Feedforward of Grid Voltage for Grid-Connected Inverter with LCL Filter to Suppress Current Distortion Due to Grid Voltage Harmonics. *IEEE Trans. Power Electron.* **2010**, *25*, 3119–3127. [CrossRef]
19. Xu, F.; Zhu, M.; Ye, Y. The Stability of LCL-Type Grid-Tied Inverter Based on Repetitive Control and Grid Voltage Feed-Forward. *IEEE J. Emerg. Sel. Top. Power Electron.* **2023**, *11*, 1496–1506. [CrossRef]
20. Wang, X.; Qin, K.; Ruan, X.; Pan, D.; He, Y.; Liu, F. A Robust Grid-Voltage Feedforward Scheme to Improve Adaptability of Grid-Connected Inverter to Weak Grid Condition. *IEEE Trans. Power Electron.* **2021**, *36*, 2384–2395. [CrossRef]
21. Li, M.; Zhang, X.; Zhang, H.; Zhang, H.; Guan, W. An Improved Grid Voltage Feedforward Compensation Scheme of Grid-Connected Inverter in Weak Grid. In Proceedings of the IEEE International Power Electronics and Application Conference and Exposition, Shenzhen, China, 4–7 November 2018; pp. 1–6.
22. Xu, J.; Xie, S.; Qian, Q.; Zhang, B. Adaptive Feedforward Algorithm Without Grid Impedance Estimation for Inverters to Suppress Grid Current Instabilities and Harmonics Due to Grid Impedance and Grid Voltage Distortion. *IEEE Trans. Ind. Electron.* **2017**, *64*, 7574–7586. [CrossRef]
23. Lin, Z.; Ruan, X.; Wu, L.; Zhang, H.; Li, W. Multi resonant Component-Based Grid-Voltage-Weighted Feedforward Scheme for Grid-Connected Inverter to Suppress the Injected Grid Current Harmonics Under Weak Grid. *IEEE Trans. Power Electron.* **2020**, *35*, 9784–9793. [CrossRef]
24. Wang, C.; Wang, X.; He, Y.; Ruan, X. A Passivity-Based Weighted Proportional-Derivative Feedforward Scheme for Grid-Connected Inverters with Enhanced Harmonic Rejection Ability. *IEEE J. Emerg. Sel. Top. Power Electron.* **2023**, *11*, 3656–3668. [CrossRef]
25. Fang, T.; Liu, H.; Zhang, H.; Liu, C. Robust Technique Using the Optimized Grid Voltage Feedforward to Conquer the Impact Induced by Dual Factors for Grid-Connected Inverter in Weak Grid. *IEEE Trans. Ind. Electron.* **2024**, *71*, 7199–7209. [CrossRef]
26. Wu, X.; Li, X.; Yuan, X.; Geng, Y. Grid Harmonics Suppression Scheme for LCL-Type Grid-Connected Inverters Based on Output Admittance Revision. *IEEE Trans. Sustain. Energy* **2015**, *6*, 411–421. [CrossRef]
27. Sun, J. Impedance-Based Stability Criterion for Grid-Connected Inverters. *IEEE Trans. Power Electron.* **2011**, *26*, 3075–3078. [CrossRef]

electricity

MDPI

Article

Hardware-in-the-Loop Emulation of a SEPIC Multiplier Converter in a Photovoltaic System

Johnny Posada Contreras [1,*] and Julio C. Rosas-Caro [2,*]

[1] Programa de Ingeniería Eléctrica, Grupo de Investigación en Energía, GIEN, Facultad de Ingeniería, Universidad Autónoma de Occidente, Cll 25 # 115-85, Cali 760035, Colombia
[2] Facultad de Ingeniería, Universidad Panamericana, Álvaro del Portillo 49, Zapopan 45010, Mexico
* Correspondence: jposada@uao.edu.co (J.P.C.); crosas@up.edu.mx (J.C.R.-C.)

Abstract: This article presents the development and execution of a Single-Ended Primary-Inductor Converter (SEPIC) multiplier within a Hardware-in-the-Loop (HIL) emulation environment tailored for photovoltaic (PV) applications. Utilizing the advanced capabilities of the dSPACE 1104 platform, this work establishes a dynamic data exchange mechanism between a variable voltage power supply and the SEPIC multiplier converter, enhancing the efficiency of solar energy harnessing. The proposed emulation model was crafted to simulate real-world solar energy capture, facilitating the evaluation of control strategies under laboratory conditions. By emulating realistic operational scenarios, this approach significantly accelerates the innovation cycle for PV system technologies, enabling faster validation and refinement of emerging solutions. The SEPIC multiplier converter is a new topology based on the traditional SEPIC with the capability of producing a larger output voltage in a scalable manner. This initiative sets a new benchmark for conducting PV system research, offering a blend of precision and flexibility in testing supervisory strategies, thereby streamlining the path toward technological advancements in solar energy utilization.

Keywords: pulse-width-modulated (PWM) converter; SEPIC multiplier converter; power electronics; hardware-in-the-loop emulation; photovoltaic system

Citation: Posada Contreras, J.; Rosas-Caro, J.C. Hardware-in-the-Loop Emulation of a SEPIC Multiplier Converter in a Photovoltaic System. *Electricity* **2024**, *5*, 426–448. https://doi.org/10.3390/electricity5030022

Academic Editors: Bo Yang, Zhijian Liu and Lin Jiang

Received: 16 April 2024
Revised: 29 June 2024
Accepted: 2 July 2024
Published: 5 July 2024

1. Introduction

The growing concern over the exacerbation of the greenhouse effect, primarily attributed to the significant emissions of pollutants from human activities, has been a driving force behind the innovation of new models for energy generation [1–4]. This imperative has led to the development and implementation of renewable energy systems, including solar photovoltaic (PV), wind, hydraulic, and geothermal power systems, aimed at curtailing our dependency on fossil fuels and other non-renewable resources [4–6].

At the forefront of these advancements in power electronics is the strategic elimination of traditional fieldwork requirements. This objective is realized through the deployment of sophisticated emulation systems designed to replicate the dynamics of real-world power systems accurately. Such systems incorporate the precise characteristics of the energy systems under study, including the simulation of environmental atmospheric conditions. The adoption of these emulation techniques not only enhances the accuracy of energy system evaluations but also significantly contributes to the refinement of power generation technologies by providing a controlled yet realistic environment for testing and optimization. By leveraging these advanced emulation platforms, researchers and engineers are equipped with the tools necessary to conduct comprehensive analyses, thereby accelerating the development and adoption of efficient and sustainable energy solutions [7–11].

The employment of Hardware-in-the-Loop (HIL) simulation stands as a cornerstone in the realm of control systems testing, offering an innovative avenue for rigorous evaluation within a controlled virtual setting before establishing a direct interface with the actual converter. This methodology markedly diminishes the temporal and logistical demands traditionally associated with the setup and execution of empirical trials and validations using tangible hardware components. Presently, the integration of advanced simulation platforms equipped with high-performance multicore processors empowers researchers and engineers to meticulously design and test complex systems. These platforms enable the construction and manipulation of detailed virtual models, thereby accelerating the development process without compromising the fidelity or accuracy of the outcomes. The strategic utilization of HIL simulation not only streamlines the research and development cycle but also enhances the precision and reliability of the system under test, ensuring that results are both swift and precise [10–14].

Although solar panel emulation systems are available on the market [15,16], they are developed around closed (non-customizable by the user) platforms that do not allow for simple experimentation with different power electronics-based conversion systems. Programmable sources, such as those offered by CHROMA, are expensive compared to laboratory prototypes that can maximize the flexibility of emulating solar panel modules and the conversion systems used to implement MPPT algorithms.

This paper posits the utilization of Power Hardware-in-the-Loop (PHIL) simulation as a transformative approach for the integration of a DC/DC Single-Ended Primary-Inductor Converter (SEPIC) multiplier converter within a solar panel emulation framework managed by an advanced Maximum Power Point Tracking (MPPT) algorithm. An improved version of the work preliminarily introduced in [17], the proposed system aims to optimize power extraction, thereby enhancing the overall efficiency and reliability of photovoltaic systems. The SEPIC multiplier converter is a recently introduced and studied topology which can provide a larger voltage gain according to the number of diodes and capacitors added to the topology.

PHIL simulation facilitates a detailed and realistic emulation of solar energy systems, allowing for the thorough investigation and refinement of control strategies to ensure maximum energy yield and system performance. Through this approach, the research endeavors to bridge the gap between theoretical models and practical applications, fostering the development of more efficient and sustainable renewable energy solutions.

The PHIL system implemented in this work has the DS1104 data acquisition system from dSPACE as its central element. The DS1104 is a comprehensive data acquisition and processing card. With the aid of Matlab-Simulink – R2023b, detailed models can be created, which are then programmed on the DS1104. The general scheme of the data acquisition system is shown in Figure 1, where the software–hardware interaction that allows the emulation of the system is schematized. While the mathematical models are digitally processed in the DAQ system (DS1104), the output variables, voltages, currents, and powers are amplified by an electronic power source. At the same time, the voltages entering the model—the external command signals—are integrated into the process through the analog inputs of the DS1104 system.

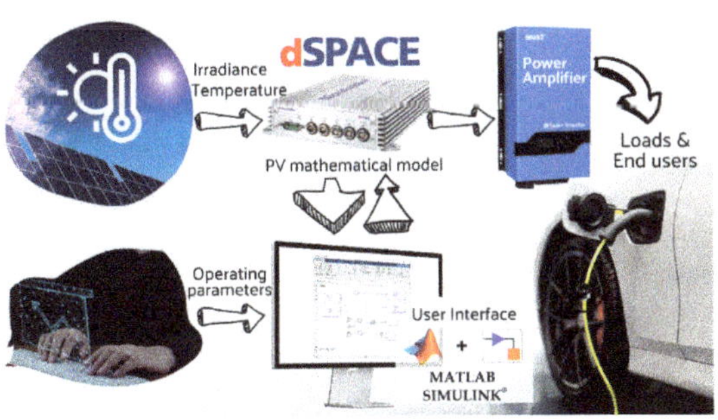

Figure 1. PV system emulation scheme.

2. Emulating a Photovoltaic Panel-Based Source

In this section, we start by covering the emulation of a solar energy source derived from a photovoltaic panel. A photovoltaic panel comprises multiple photovoltaic cells connected in series, which set the panel's output voltage, V_{pv}. At the same time, several panels are connected in series to form a string of panels in a photovoltaic system. In the photovoltaic system, multiple strings of panels can be connected in parallel, increasing the output power of the photovoltaic system.

A fundamental step in this process involves evaluating the device's output performance and solar efficiency. To achieve this, it is imperative to delve into the characteristics of the solar PV panel in question. These essential attributes can be gleaned from the solar cell I-V characteristic curve depicted in Figure 2. This initial examination sets the groundwork for understanding how to accurately replicate the behavior of solar panels, which is crucial for the development of an effective photovoltaic system emulation.

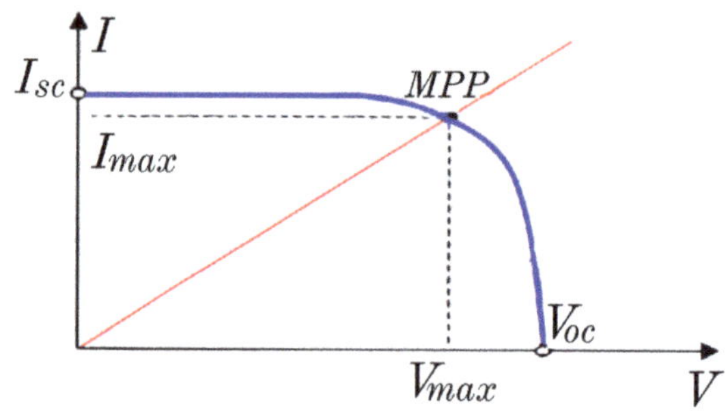

Figure 2. The characteristic solar cell I-V.

As depicted in Figure 2, to trace the curve comprehensively, a variable load is needed to identify all potential operating points along it. It is important to recognize that multiple I-V curves can be derived from a single panel by varying conditions such as temperature and irradiance. The point of maximum power occurs where the curve shows the highest combination of current and voltage the panel can provide. These are the critical values targeted by the MPPT (Maximum Power Point Tracking) algorithm [18,19].

2.1. Modeling and Characteristics

For the purposes of emulation, the model of one diode from the photovoltaic solar cell will be taken as a reference, as shown in Figure 3. In this circuit, it can be observed that a photovoltaic solar cell works as a controlled source, where the control variable is the irradiance or solar radiation.

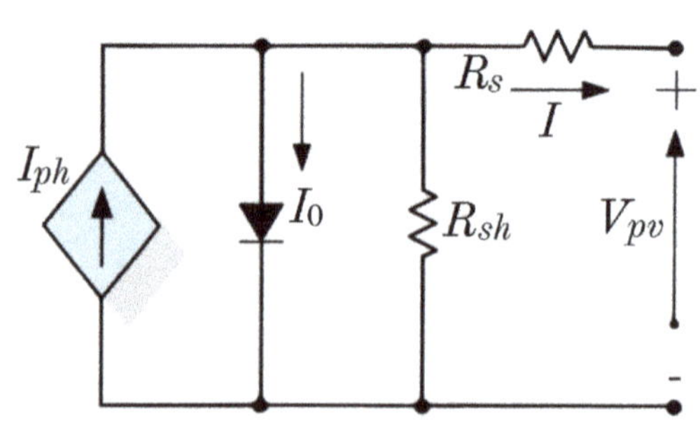

Figure 3. Solar cell diode model [20].

The basic diode equation for a photovoltaic solar cell is (1).

$$I = I_{ph}(G) - I_0\left(e^{\frac{qV}{k*T}} - 1\right) - \frac{V}{R_{sh}} \quad (1)$$

where

I represents the total circuit current;
I_{ph} is the photocurrent generated by illumination;
I_0 denotes the dark saturation current of the diode;
V stands for the voltage across the diode;
q is the charge of an electron (1.602×10^{-19} C);
k is the Stefan–Boltzmann constant (1.381×10^{-23} J/K);
T is the temperature in Kelvin.

The equation incorporating solar irradiance and internal resistances within a photovoltaic (PV) cell model, aiming to represent the dynamics of the cell comprehensively, involves parameters like $I_{ph}(G)$ for the photo-generated current, I_0 for the dark saturation current, V for the voltage across the cell, I for the output current, R_s for the series resistance, R_{sh} for the parallel (shunt) resistance, and factors for temperature and irradiance effects. This model accounts for the non-ideal behavior of real PV cells by including resistances that affect the panel's efficiency and performance. The $I_{ph}(G)$ increases with solar irradiance (G, in W/m^2).

The mathematical function representing I_{ph} as a function of solar irradiance (G) typically assumes direct proportionality between the irradiance and the current generated by the cell. A simplified form could be $I_{ph} = k \cdot G$, where k is a proportionality constant depending on the specific characteristics of the photovoltaic cell, including its efficiency and area. This model reflects how the photo-generated current increases linearly with irradiance under the assumption that all other factors remain constant.

A more detailed form of I_{ph} considers various factors, including the panel's efficiency dependency on temperature and specific material parameters. However, due to complexity and variability among different solar panel technologies, there is not a universally accepted 'non-simplified form' that does not consider specific panel model parameters. For precise details, consulting a solar panel's technical datasheet is necessary, where equations based on its characteristics and response to irradiance and temperature are provided.

A detailed and widely accepted expression for I_{ph}, considering both irradiance (G) and temperature (T), could be the following:

$$I_{ph} = \left(\frac{G}{G_{STC}}\right) I_{ph,STC}[1 + \alpha(T - T_{STC})]. \quad (2)$$

where $I_{ph,STC}$ is the photocurrent under standard test conditions (normally, 1000 W/m^2 irradiance and 25 °C temperature), G_{STC} is the standard test condition irradiance (1000 W/m^2), G is the current irradiance, T_{STC} is the standard test condition temperature (25 °C), T is the current module temperature, and α is the temperature coefficient of the current, indicating how the current varies with temperature. This formula reflects how the current generated by the cell increases with irradiance and adjusts with changes in module temperature. Reference [20] can be studied for a deeper analysis of the equations describing the operation of photovoltaic solar panels.

Let us discuss the emulated panel's characteristics. The panel is a YINGLI 60 CELL-series, 40 mm, model YL250p-29b. The YL250p-29b photovoltaic panel has 60 cells connected in series, with an open-circuit voltage of 38.40 Vdc. According to its datasheet, under standard test conditions (STCs), it possesses the following electrical characteristics: The maximum rated power is 250 W, and the efficiency panel is 15.3%. At full power, its voltage is 30.4 V, and its current is 8.24 A. The open circuit voltage is 38.4 V, while the short circuit current is 8.79 A.

Beyond the I-V solar panel characteristics, the emulator also needs to consider various technical parameters: the temperature coefficient of the current $\alpha = 0.00385$; the reference inverse saturation current (I_o), which in this case is 1.2723×10^{-10} A; the electron charge $q = 1.602 \times 10^{-19}$ C; the diode's ideality factor $n = 0.9985$; the Stefan–Boltzmann constant ($K = 1.3806 \times 10^{-23}$ J/K); the energy bandgap at the reference temperature ($EgRef = 1.212$ J); and the values of the resistances $R_{sh} = 364.52\ \Omega$ and $R_s = 0.42475\ \Omega$.

Figure 4 showcases the SIMULINK model representing the solar panel. As can be observed in the figure, two subsystems are implemented to determine the essential variables for solar panel voltage control according to weather and load conditions. The part marked in purple is a MATLAB script for the implementation of Equations (1) and (2). This script receives two variables, the panel temperature and the irradiance; the inputs to the model are highlighted in green. Since it is an emulation, the values for the mentioned signals are acquired through a data acquisition system that allows for the emulation of the PV array. The system marked in red is an electrical circuit similar to the single-diode model of the solar panel, the difference being that in the circuit there is no current source that depends on irradiance. Instead, the voltage across the panel terminals is used for its implementation, with 'V_D' being equal to the variable 'V' in Equation (1). The controlled current source, the gain, and the '*RLim*' resistance aim to recreate the current demand in the PV. When a current demand is simulated through the dependent current source, the voltage V_D drops, as shown in Figure 1. The value of the voltage V_D is determined using Equation (1), where the current I (I_PV in Figure 4) is the current delivered by the PV panel to the load emulated by the circuit marked in blue. The simulation of the current demand is also performed by one of the analog inputs of the data acquisition system, which is described in a subsequent section. The model's inputs and outputs are highlighted in green and orange, respectively. With these inputs and outputs, changes in the variables of voltages, currents, and powers in a PV system are obtained.

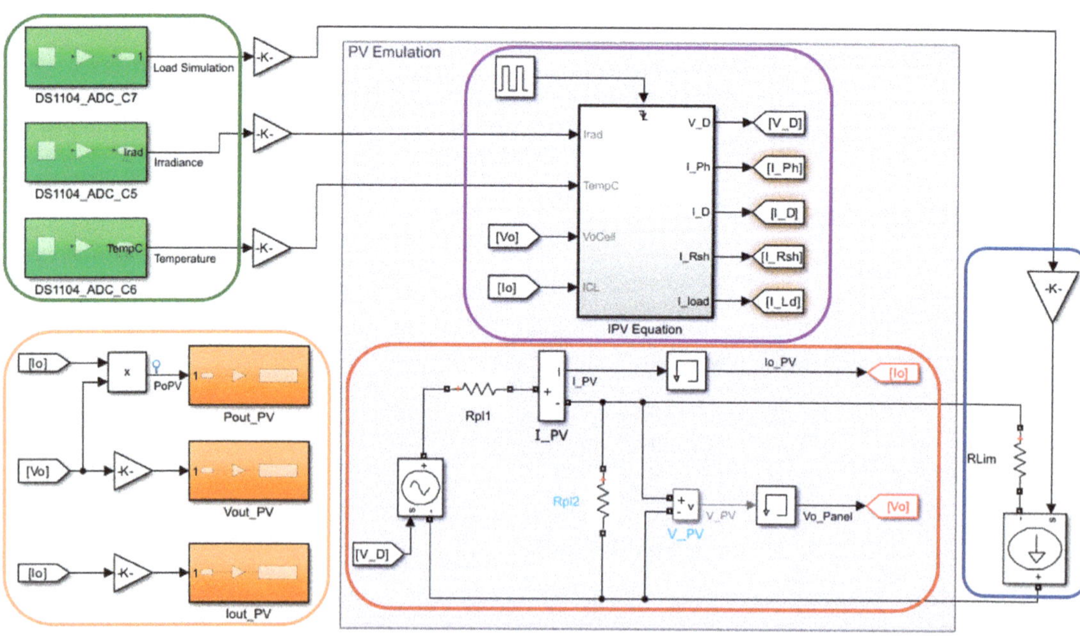

Figure 4. Solar panel model for emulation.

2.2. PV Simulation

To obtain the array I-V graph of the emulated panel array, it was necessary to create a script that simulated the current demand (I) from 0 A up to its short-circuit value. This behavior is achieved by manipulating the controlled current source shown in Figure 4. The manipulation of this source is performed using a DC voltage, which is injected through one of the analog inputs of the data acquisition card. Figure 5 demonstrates how this emulator automatically transitions from one curve to another, replicating the panel's array behavior, which is subjected to temperature and solar radiation. This ensures that the emulator maintains characteristics consistent with the referenced model.

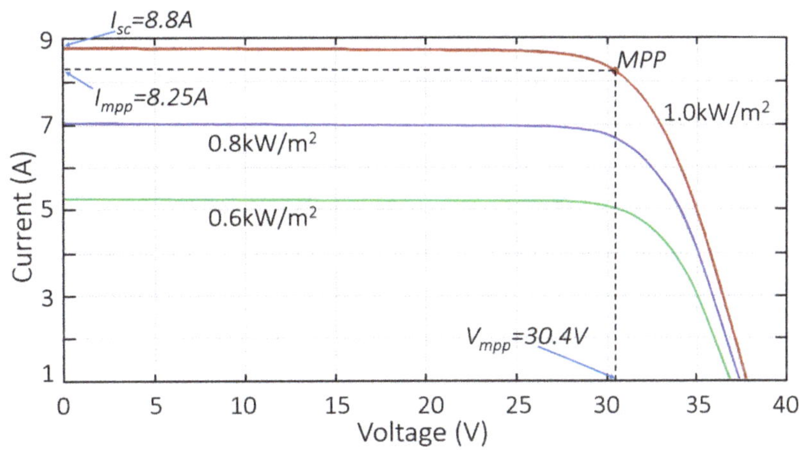

Figure 5. I-V characteristic curve of the solar panel.

Taking as a reference the panel's response curve at a power of 1 kW/m^2, we find that the values for current and voltage at the maximum power point differ slightly from those reported by the manufacturer, as do the open-circuit operating conditions. The comparison of these values can be seen in Table 1, where the error percentage is also recorded. The error percentage is attributed to calculating the diode current in the panel model.

Table 1. Panel parameters and simulation results.

Electrical Characteristics	Symbol	Manufacturer's Value	Simulation Result	%Error
Voltage at P_{max}	V_{mpp}	29.8 V	30.4 V	2.01%
Current at P_{max}	I_{mpp}	8.39 A	8.25 A	1.67%
Short-circuit current	I_{sc}	8.92 A	8.80 A	1.36%
Open-circuit voltage	V_{oc}	37.6 V	38.4 V	2.13%

To represent the solar panel's module, a programable voltage source was devised to accept a reference signal from the dSPACE control board, which receives the signal from the simulation. This setup will regulate the voltage specified by the model using a linear regulator. Additionally, sensors will provide feedback on voltage and the current of the voltage source to ensure correct operation.

A complete scheme of the emulator power supply is shown in Figure 6. The core objective was to enable voltage regulation through a control signal, leading to the selection of an LM317, which allows for voltage adjustment by varying the input current at the ADJUST pin. It is important to remember that the reference signal for the voltage comes from the dSPACE board; an isolating barrier featuring isolation amplifiers was incorporated for protection. To augment the output current capacity of the regulator, the inclusion of a pass transistor was necessary, ensuring stable system regulation. Furthermore, to supply adequate power to the system, three additional regulators were implemented, as indicated by the blue boxes. The green boxes represent a control stage designed to distribute the current across each regulator evenly.

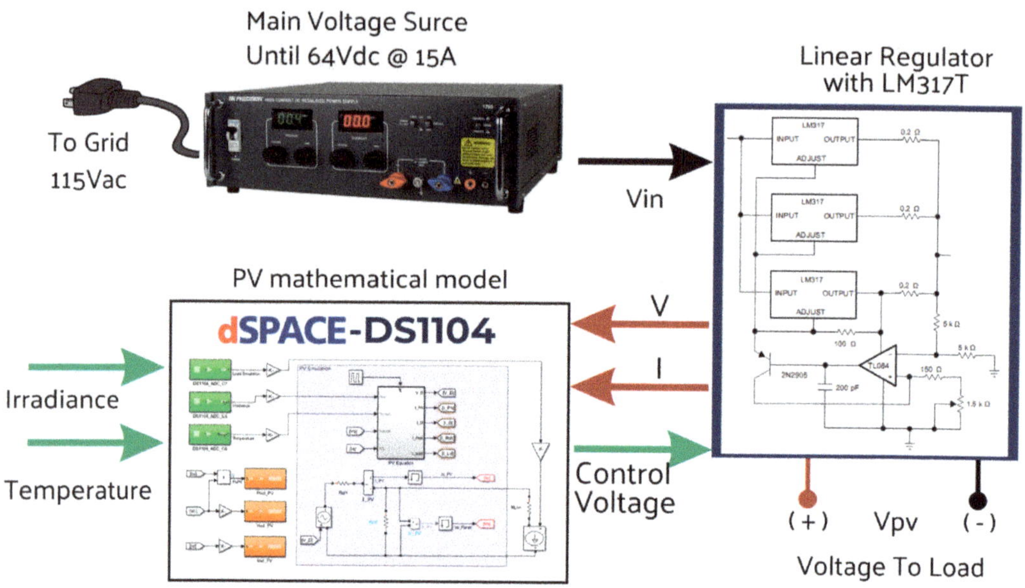

Figure 6. Schematic diagram of the panel's array emulator.

3. Multiplier SEPIC Converter

Power electronics is an essential discipline in the efficient conversion and control of electrical energy [21–23]. Its applications range from renewable energy generation systems to everyday consumer electronic devices. In the context of solar energy, power electronics play a crucial role in enabling the conversion of the energy captured by solar panels into usable and regulated forms of electrical power. The field of the DC-DC converter has many applications in PV panels [23–25]. The SEPIC (Single-Ended Primary Inductor Converter) is a power converter topology that offers advantages in terms of flexibility and efficiency, making it particularly suitable for solar panel emulators, where irradiance and environmental conditions can vary widely. Implementing a solar panel emulator based on a SEPIC converter allows for the recreation of the electrical characteristics of a real solar panel, facilitating the design and testing of photovoltaic systems under controlled and repeatable conditions.

With the aim of obtaining power from the PV-emulated circuit, a Multiplier SEPIC Converter was implemented. This converter will allow it to carry out the maximum power from the PV emulator using a traditional MPPT algorithm like Perturb and Observe (P&O). The strategy is to increase the voltage provided by the PV array with the SEPIC converter. At the same time, the SEPIC converter output voltage is supplied to a resistor, and the SEPIC voltage is changed to find the emulated PV array's maximum power point (MPP).

The Single-Ended Primary-Inductor Converter (SEPIC) is composed of a switch (s) (which can be synthesized with a transistor), a diode (d_1), inductors (L_1 and L_2), and capacitors (C_1 and C_2); the load can be represented as a resistor (R), as shown in Figure 7a. The SEPIC is capable of providing output voltages that are higher and lower than the input voltage.

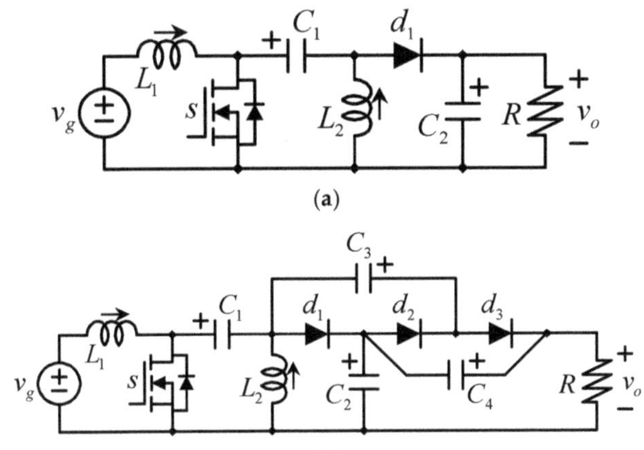

Figure 7. (**a**) SEPIC converter. (**b**) Multiplier SEPIC converter.

An improved version of the conventional SEPIC [26,27] includes a hybrid topology with diode-capacitor multipliers, known as the Dickson charge pump. Figure 7b illustrates a SEPIC that incorporates a $2x$ multiplier.

The transistor closes and opens periodically with a constant switching frequency (f_{SW}); the inverse of the switching frequency is called the switching period (T_{SW}). During a switching period, the transistor remains closed for a duration called t_{on}, and subsequently, it remains closed for a time called t_{off}. Evidently, the relation between the switching period and the times in which the transistor is closed and open can be expressed as in (3):

$$T_{SW} = \frac{1}{f_{SW}} = t_{on} + t_{off} \tag{3}$$

A duty cycle or duty ratio (D) and its complement can be defined as in (4) and (5).

$$D = \frac{t_{on}}{T_{SW}} \tag{4}$$

$$(1 - D) = \frac{t_{off}}{T_{SW}} \tag{5}$$

Figure 8 shows the equivalent circuits according to the switch state (open or closed). The fundamental operation of the SEPIC, as depicted in Figure 7a, can be explained with the following steps and by observing the equivalent circuits in Figure 8.

(i) Upon transistor s closing (see Figure 8a), the inductor L_1 gets in parallel connection to the source v_g, leading to an increment in the L_1 current characterized by a uniform incline and a positive direction (as per the orientation defined in Figure 7).
(ii) When the switch opens (see Figure 8b), it allows L_1's current to infuse C_1 with a positive voltage (in alignment with the designated signs) since they are connected in series, and the current gets through the positive voltage sign of C_1.
(iii) The switching action of the transistor is periodic; it inevitably closes again, but this time the capacitor C_1 already has some charge; it gets in parallel connection with the inductor L_2, facilitating C_1 in endowing L_2 with a positive flow of current.
(iv) Upon the transistor's subsequent opening, L_2's current moves through d_1, enriching C_2 with a positive charge.

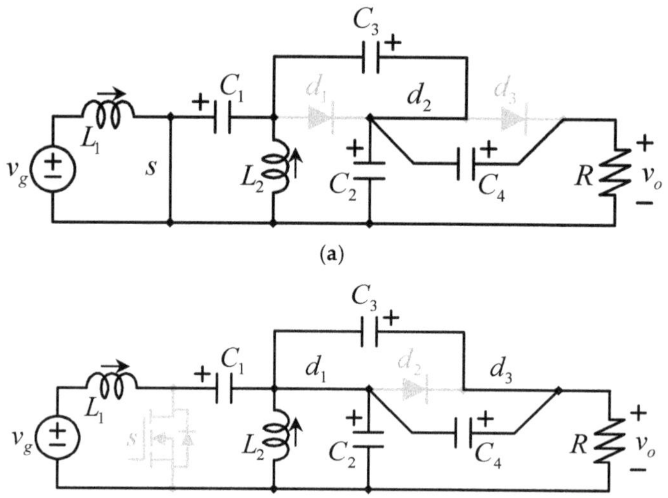

(**a**)

(**b**)

Figure 8. Switching states when (**a**) the transistor is on and (**b**) when the transistor is off.

This sequence of actions detailing the operation of the traditional SEPIC illustrated in Figure 7a is equally applicable to the configuration shown in Figure 7b, with the following additional steps:

(v) With the transistor open and L_2's current passing through d_1, as seen in Figure 8b, the negative terminals of C_3 and C_4 are conjoined at the same potential, subsequently allowing C_3 to engage d_3 to positively charge C_4.

(vi) With the transistor's closure, C_1 aligns in a series setup with C_2, as shown in Figure 8a, which then permits the charging of C_3 through the engagement of d_2.

The circuit in Figure 7b operates a diode-capacitor voltage multiplier, specifically a 2x multiplier in this scenario, which can be expanded through the addition of more diodes and capacitors. Calculating the average voltage across L_1 over a single switching cycle while employing the conventional small ripple approximation [23] results in (6).

$$L_1\frac{d\langle i_{L1}\rangle}{dt} = \frac{1}{T}\left(t_{on} * v_{L1ton} + t_{off} * v_{L1toff}\right) \tag{6}$$

where v_{L1ton} represents the voltage across L_1 with the switch closed and v_{L1toff} denotes the voltage across L_1 when the transistor is open. This formula can be expressed in relation to the duty ratio outlined in (2) as follows.

$$L_1\frac{d\langle i_{L1}\rangle}{dt} = D * v_{L1ton} + (1-D)v_{L1toff} \tag{7}$$

The voltage across L_1 during each equivalent circuit or switching state can be derived from the circuits depicted in Figure 8.

$$v_{L1ton} = V_g \tag{8}$$

$$v_{L1toff} = V_g - V_{C1} - V_{C2} \tag{9}$$

At equilibrium, the average voltage across the inductors equals zero, resulting in a steady current. Following Equation (7) and incorporating Equations (8) and (9), this can be articulated as (10).

$$D * V_g + (1-D)\left(V_g - V_{C1} - V_{C2}\right) = 0 \tag{10}$$

Similarly, the (average) voltage across L_2 can be expressed as (11).

$$L_2 \frac{d\langle i_{L2} \rangle}{dt} = D * v_{L2ton} + (1-D) v_{L2toff} \tag{11}$$

And the voltage in each state can also be determined by applying the KVL in Figure 8 as in (12) and (13).

$$v_{L2ton} = V_{C1} \tag{12}$$

$$v_{L2toff} = -V_{C12} \tag{13}$$

In the equilibrium condition or steady state, the voltage in L_2 must also be in that state, which can be written using (11) to (13) as (14).

$$D(V_{C1}) - (1-D)V_{C2} = 0 \tag{14}$$

From (10), the voltage in C_2 can be expressed as (15).

$$V_{C2} = V_{C1} \frac{D}{1-D} \tag{15}$$

Substituting (15) in (11) leads to the following:

$$V_{C1} = V_g \tag{16}$$

From Figure 8, C_3 gets charged by both C_1 and C_2 in a series connection, which leads to (17).

$$V_{C3} = V_{C1} + V_{C2} = V_g + V_g \frac{D}{1-D} = V_g \frac{1}{1-D} \tag{17}$$

From Figure 7, the voltage across C_4 is equal to the one across C_3, and then

$$V_{C4} = V_{C3} = V_g \frac{1}{1-D} \tag{18}$$

Finally, the load voltage is given by the series connection of C_2 and C_4, which can be expressed as (19).

$$V_o = V_{C2} + V_{C4} = V_g \left(\frac{D}{1-D} + \frac{1}{1-D} \right) = V_g \frac{1+D}{1-D} \tag{19}$$

The voltage gain is significant. Figure 9 illustrates the voltage gain relative to the duty cycle or duty ratio expressed in (19).

As can be seen in Figure 8b, the voltage that the transistor needs to block (when it is open) is relatively low in comparison to the output voltage. This attribute is advantageous, as it allows for the use of transistors rated for lower voltages in the construction of high-voltage converters, embodying the essence of multilevel converters. Contrary to this, in other configurations like the cascaded boost converter, the final transistor must withstand the entire output voltage. This requirement caps the maximum achievable output voltage to the voltage rating of the transistor. Moreover, transistors designed to handle higher voltages typically exhibit greater on-resistance than those intended for lower voltages.

A further benefit of the Multilevel SEPIC lies in the ability to enhance the voltage gain without augmenting the count of inductors. Inductors are bulky, costly, and challenging to encapsulate. Moreover, a singular transistor suffices irrespective of the level count, as adding more transistors necessitates additional circuitry.

Figure 10 displays the simulated model of the proposed Multiplier SEPIC in Matlab-Simulink R2023b software; the values of the Multiplier SEPIC components and the selected semiconductors are detailed in Table 2.

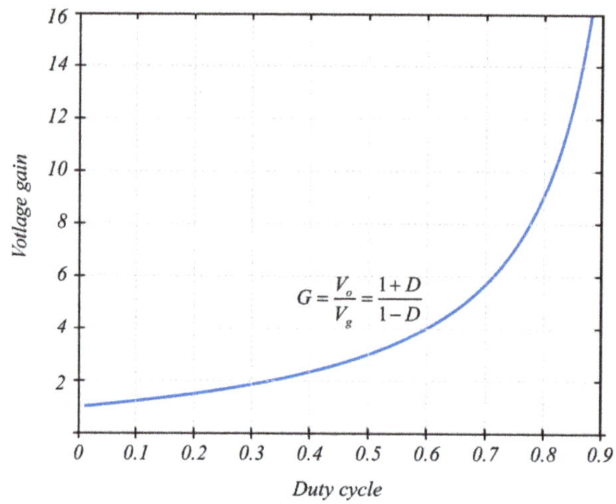

Figure 9. Voltage boost factor vs. duty cycle or duty ratio.

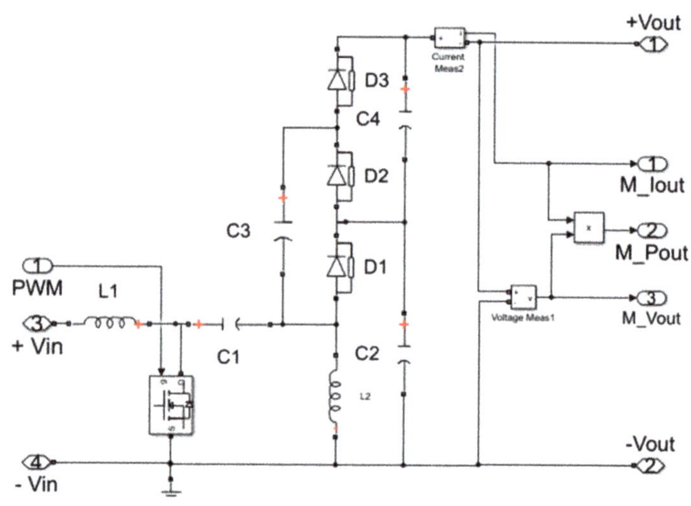

Figure 10. Schematic of the Multiplier SEPIC.

Table 2. Selected components of the Multiplier SEPIC.

Theoretical Component	Selected Component
$L_{1min} = L_{2min} \cong 800$ uH	Four 420 uH inductors for a current of 9 A
C_1, C_2, C_3, $C_4 \geq 60$ uF	One 60 uF film capacitor (C4ATDBW5600A30J), 250 V, ESR = 3.9 mOhms
Semiconductors	MOSFET mode IRFP250N with a max current equal to 30 A and a maximum power dissipation of 214 W at 175 °C, Ron = 1×10^{-5} Ω
	Diode 30CPF10: 30 A, 1000 V max

Multiplier SEPIC Simulation and Hardware Test

For the simulation and testing of the Multiplier SEPIC, an input voltage of 25 V, a useful cycle of 60%, and two different load resistances of 115 and 69 Ω were established in open loop. For the first case, with a load equal to 115 Ω, the output voltage should be

100 V, a theoretical current of approximately 0.88 A should be delivered, and the input current to the Multiplier SEPIC will be 3.5 A on average. For the second case, with a load equal to 69 ohms, the output voltage should be 100 V, a theoretical current of approximately 1.45 A should be delivered, and the input current to the Multiplier SEPIC will be 5.8 A on average.

Figure 11a shows the output voltages for the two cases, while Figure 11b shows the output and input currents. The hardware results for the Multiplier SEPIC are shown in Figure 12a,b. The over peaks and transients observed in the simulations are due to numerical approximations and initial conditions, which are not present in the real hardware. In boot cases, simulations, and hardware implementation, the output voltage was reduced by approximately 4 V. The current was measured with a sensing probe, with a 1 A/0.1 V reduction rate.

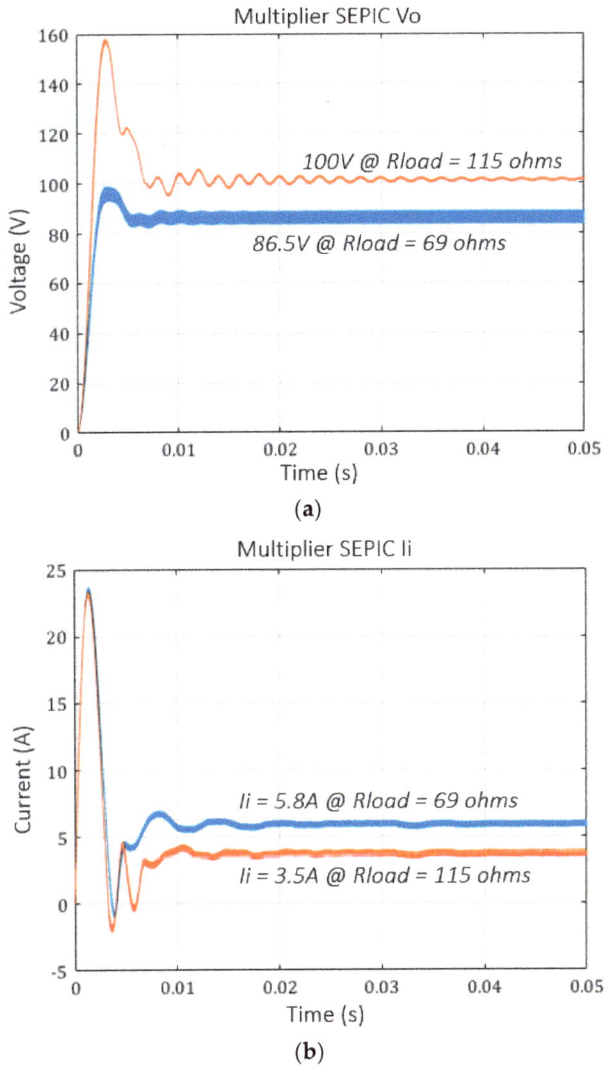

Figure 11. (**a**) Simulation results for Vo. (**b**) Simulation results for Ii.

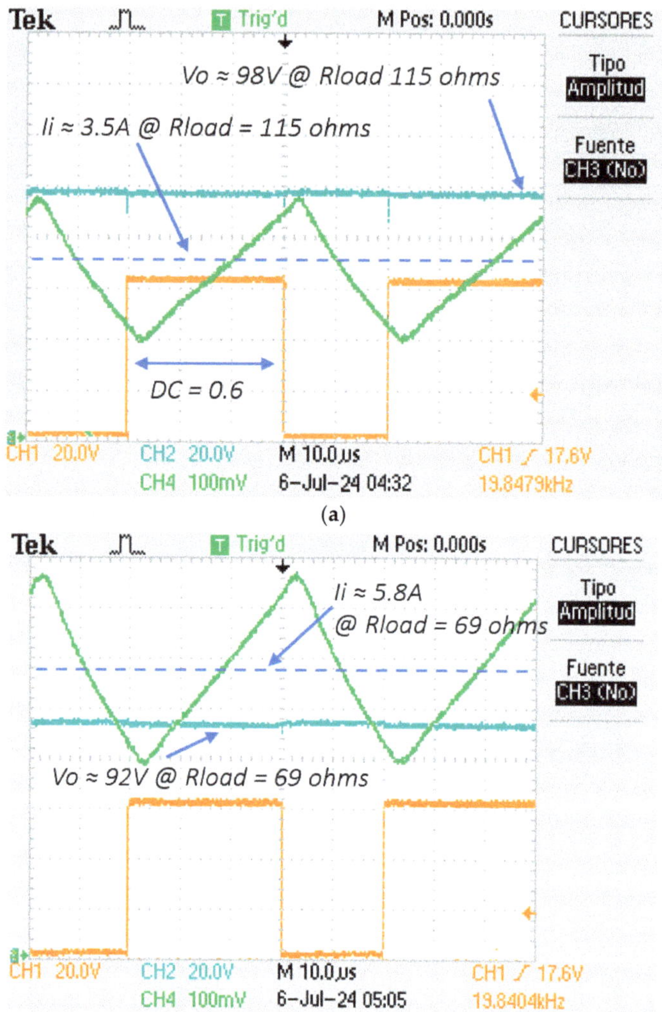

Figure 12. (**a**) Experimental results for the case with Rload = 115 ohms load. (**b**) Experimental results for the case with Rload = 69 ohms load.

4. Simulation Results for the PV + Multiplier SEPIC

Table 3 outlines the specifications of the panel array used in the simulation; they correspond to a temperature equal to 25 °C and an irradiance of 1000 W/m^2.

Table 3. Input and output converter parameters.

Voltage at the input	30.4 V	Power at the output	250 W
		Voltage at the output	80.3 V
Current at the input	8.25 A	Current at the output	3.11 A
Maximum duty ration	45%	Load resistor	25.8 Ω

Utilizing this information, one can refer to the I-V curve presented in Figure 5 to establish the necessary current and voltage values for designing the converter. The converter is designed to have an output power of 250 W and will operate in the linear region with a maximum duty ratio of 45% to ensure operation within a safe range. From this, the load resistance can be determined, which in turn allows for the calculation of the converter's output voltage and current.

After defining the input and output variables, the next step is to choose the components to be used. It is crucial to select an adequate transistor; in this case, a metal–oxide–semiconductor field-effect transistor (MOSFET) was chosen to minimize energy losses through heat dissipation.

The results of the simulation described above are shown, with Figure 13 revealing that the voltage achieved was very near the ideal 80.25 V value, with minimal losses of 6.75 V. Additionally, the voltage gain achieved was 2.42, whereas, theoretically, with a 45% duty ratio, the ideal value is three, indicating that the system's efficiency in terms of voltage gain is 91.6%. Furthermore, Figure 14 shows the current converter and illustrates that the emulator is not ideal and incurs power losses. When the output power is 209.5 W, the input power is about 237.1 W, resulting in an efficiency of 88.4%. The overpeaks and transients in Figures 13 and 14 are due to numerical approximations and initial conditions, which are not present in the real hardware.

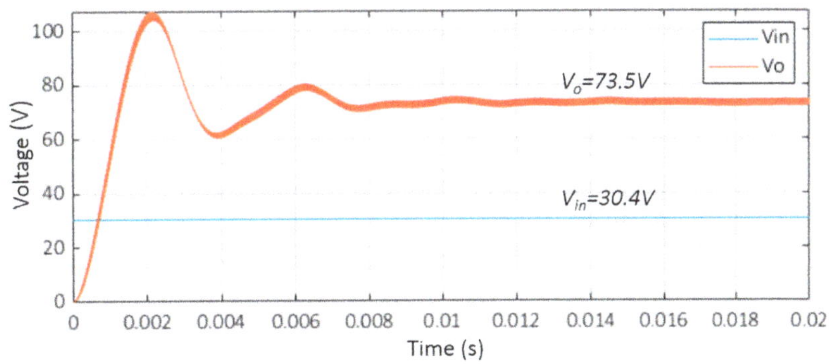

Figure 13. Input (red) and output (blue) converter voltage signals.

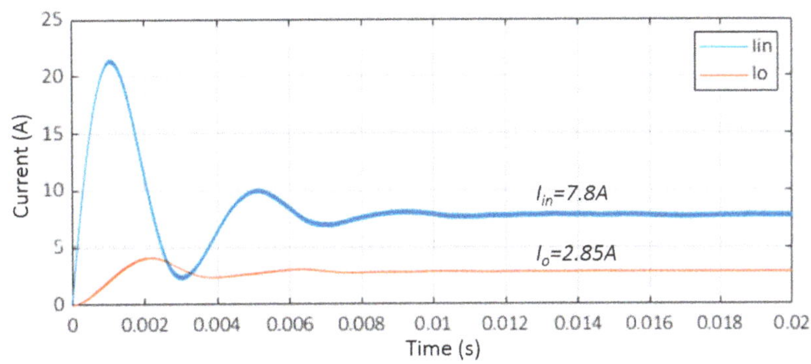

Figure 14. Input (red) and output (blue) converter currents.

The block diagram shown in Figure 15 corresponds to the complete system. The system was run with 0.286 uS of the sample period in the simulation, adhering to the Nyquist criterion. This adjustment was necessary because the SEPIC operated with a sample period of 28.5714 µS (35 KHz) and it is the fastest subsystem. For example, the panel's array emulator had a sample period of 1 mS and the Maximum Power Point Tracker (MPPT) algorithm ran with a sample period of 28 mS. Other parameters were the load resistance of 25.8 Ω over the total period simulation, which had a duration of 9 s.

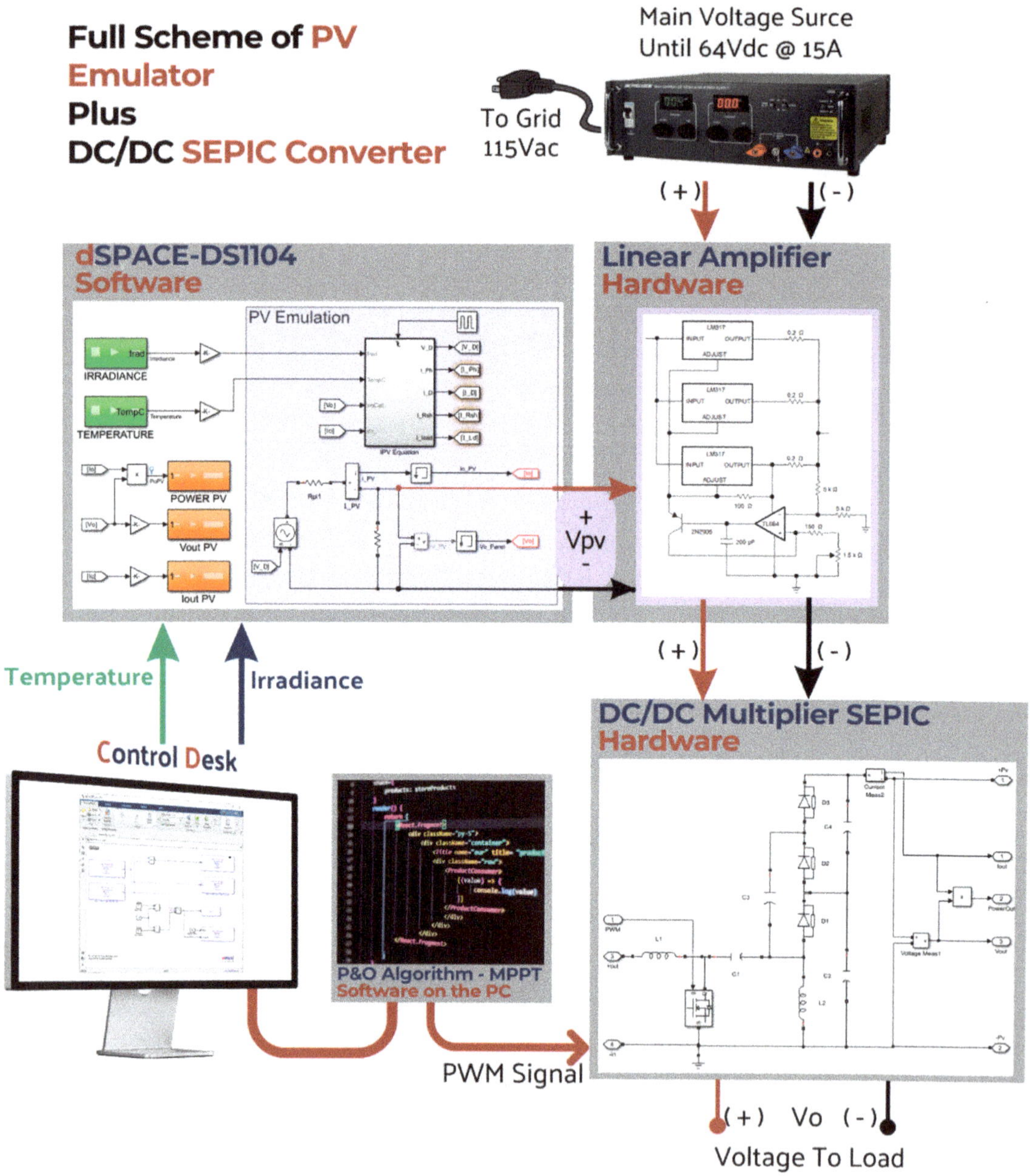

Figure 15. System simulation.

Among the various Maximum Power Point Tracking (MPPT) techniques, the Perturb and Observe (P&O) technique is the most widely used, being extensively implemented in commercial photovoltaic inverters due to the use of low-cost microcontrollers. However, despite its simplicity and reliability, the P&O algorithm has some major disadvantages. First, when the tracking reaches the vicinity of the MPP, the operating point tends to oscillate around it, resulting in a persistent fluctuation in the output power and, consequently, reducing the system quality. Secondly, P&O is prone to losing the search direction, meaning that it fails to adapt to changes in irradiance, causing the operating point to deviate from the MPP curve. This deviation also results in energy loss. In the work presented, the traditional Perturb and Observe technique was implemented to test the PHIL system rather than as an algorithm that excels in efficiency and effectiveness for finding and following the MPP. The P&O technique operates with a DC/DC or DC/AC converter, which helps extract power from the photovoltaic panel. When the panel's output power changes, the voltage across its output terminals, V_{pv}, also changes. The algorithm periodically increases or decreases the output voltage of the photovoltaic module to estimate a power value. Subsequently, the power obtained in the current cycle, $P(k)$, is compared with the power from the previous cycle, $P(k-1)$. If a perturbation is introduced in the system voltage, this will cause an increase or decrease in the output power. If the perturbation in the voltage causes an increase in power, the next perturbation is applied in the same direction as initially applied.

Conversely, if the power decreases, the next perturbation is applied in the opposite direction. After several perturbations, the algorithm converges to the maximum power point (MPP). Once the maximum power is reached, if the power in the next instant decreases, the perturbation should be applied in the opposite direction. Figure 16 shows the P&O algorithm. The emulation platform allows other algorithms to be used in the emulation process. Testing the platform with algorithms other than P&O was not within the scope of this article; rather, the aim was to demonstrate a platform development that will allow for their easy subsequent implementation.

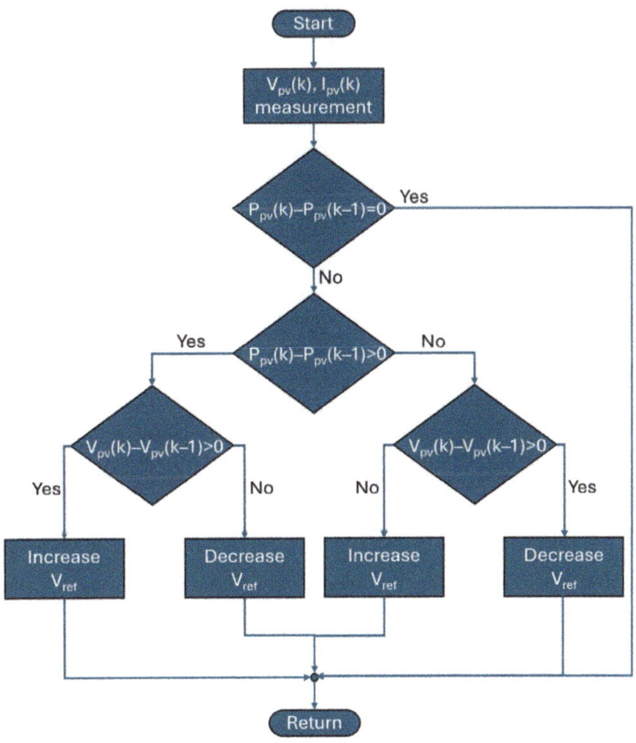

Figure 16. P&O algorithm.

To test the MPPT algorithm and the Multilevel SEPIC regulator working together, a simulation was first conducted with two scenarios of constant solar radiation: the first with continuous radiation of 1000 W/m^2, and the second with continual radiation of 600 W/m^2—both scenarios with a panel temperature of 25 °C.

The results of this simulation are presented in Figures 17–19. In Figure 17, the input power represents the maximum provided by the solar module. However, the output power demonstrates that the ripple caused by the MOSFET switching and the duty cycle adjustments enacted by the MPPT algorithm resulted in the system operating at an efficiency level below 100%. Figure 17 shows the power at the MPP. The peaks and transients in power are due to the typical oscillation around the MPP of the P&O algorithm; these variations can be reduced by adjusting the reference voltage increment of the P&O algorithm. Notably, under real environmental conditions, an increase in solar radiation or changes due to shadows are gradual, so the power transitions will be smoother.

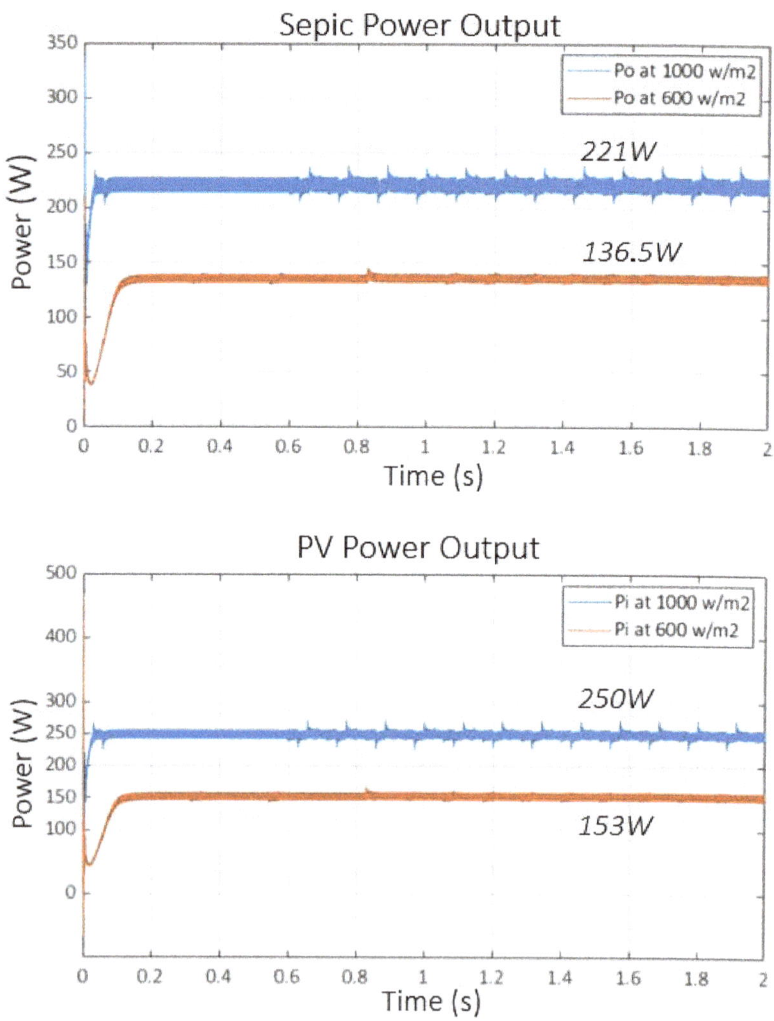

Figure 17. System power with the SEPIC running in the dynamic mode.

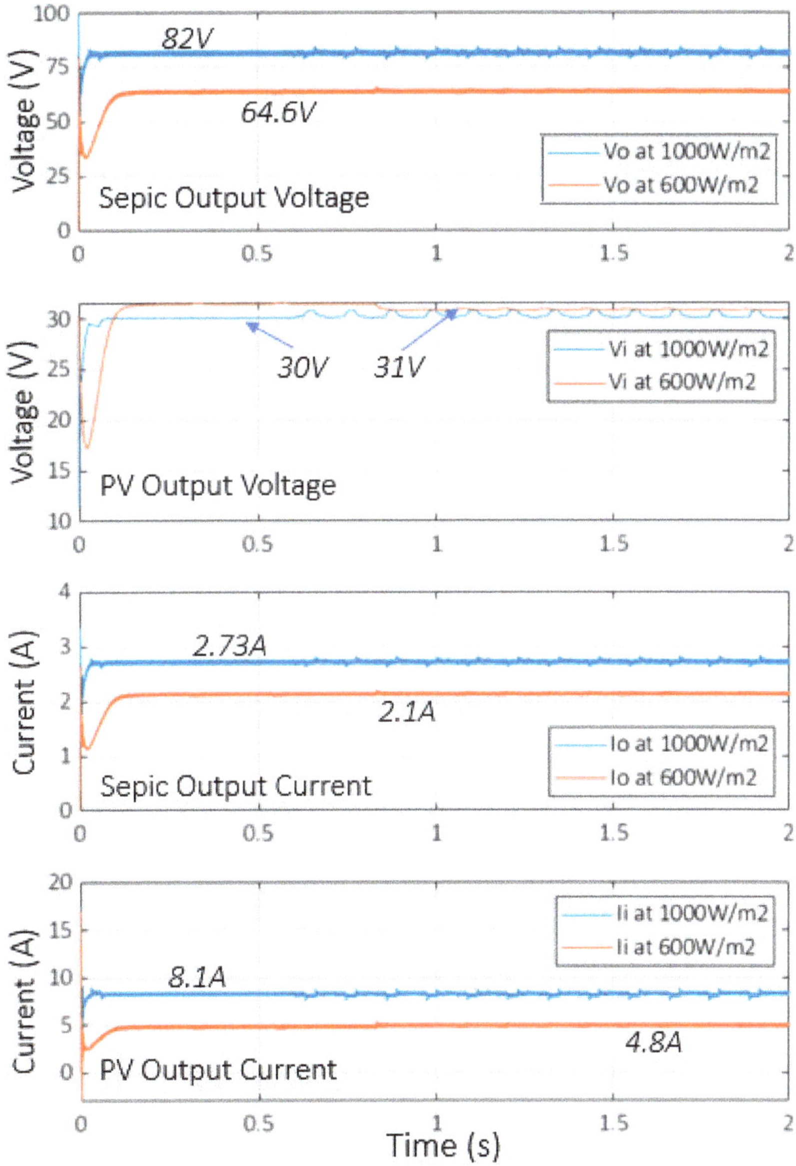

Figure 18. Voltage and current signals with the dynamic mode for the SEPIC.

Figure 18 shows the voltage evolution at the output of the Panel, V_{pv}, and at the output of the Multilevel SEPIC converter, *Vo*. The voltage and current variations can be observed as the result of changes in the reference voltage of the P&O algorithm.

Finally, Figure 19 shows the evolution of the system's efficiency under the two operating conditions. Since the initial conditions of the Multilevel SEPIC converter components were set to zero at the start of the simulation, oscillations occurred in the behavior of all the measured variables, which decreased once normal operating conditions were reached.

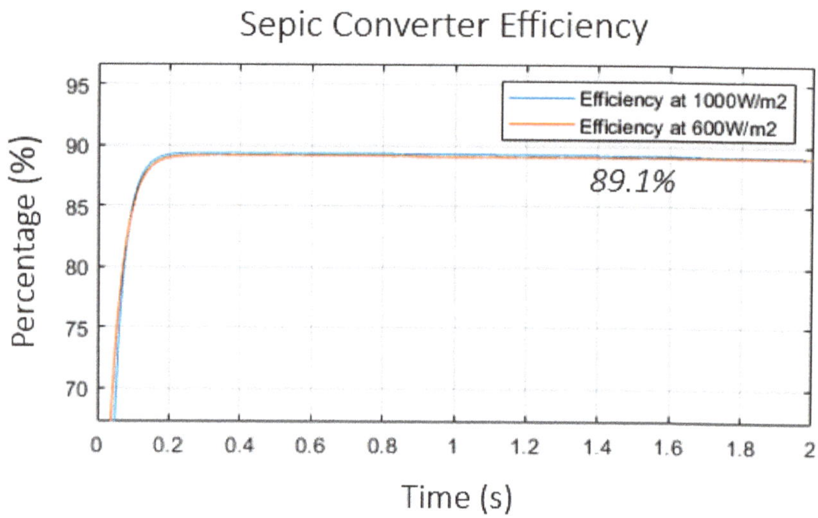

Figure 19. System efficiency during the dynamic mode of the SEPIC.

5. Hardware Results

Figure 20 shows the hardware implementation of the PHIL system. The figure highlights the interface in the control desk, the connection panel of the DS1104 data acquisition card, the linear source that emulates the behavior of the PV, the SEPIC Multiplier, and the electronic load in charge of emulating the system's power demand. The electronic load fulfills the load resistance functions in the implementation.

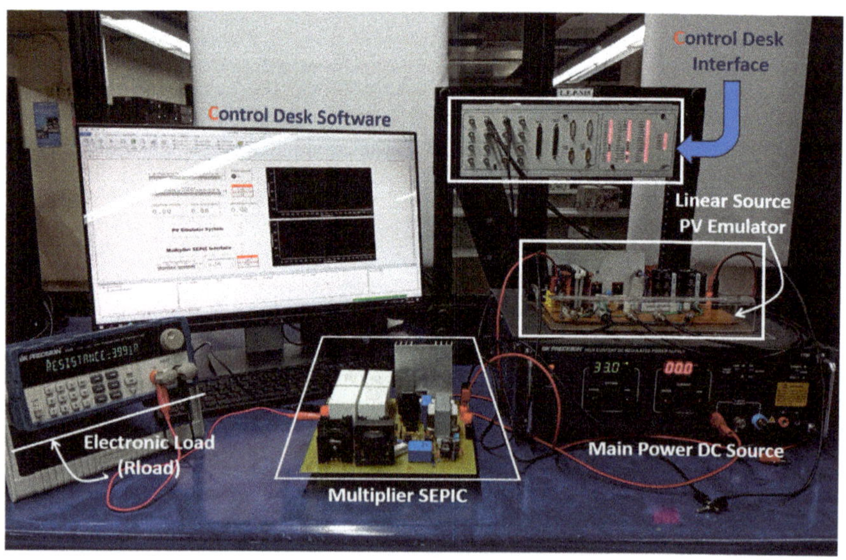

Figure 20. Hardware implementation of PV Emulator + Multiplier SEPIC.

In a photovoltaic solar panel, temperature directly impacts the power the panel can deliver, as evidenced by analyzing Equation (2), where the photocurrent shows a direct dependence on temperature. In a PV curve like the one in Figure 21, the influence of temperature on the MPP can be observed, which decreases in value as the temperature increases, directly affecting the panel's voltage. In Figure 21, the red line shows the PV curve for a temperature of 25 °C @ 1000 W/m^2 radiation, while the blue line shows the same curve for a temperature of 45 °C. The effect of temperature was considered for the Testbed used for the hardware validation of the designed prototype.

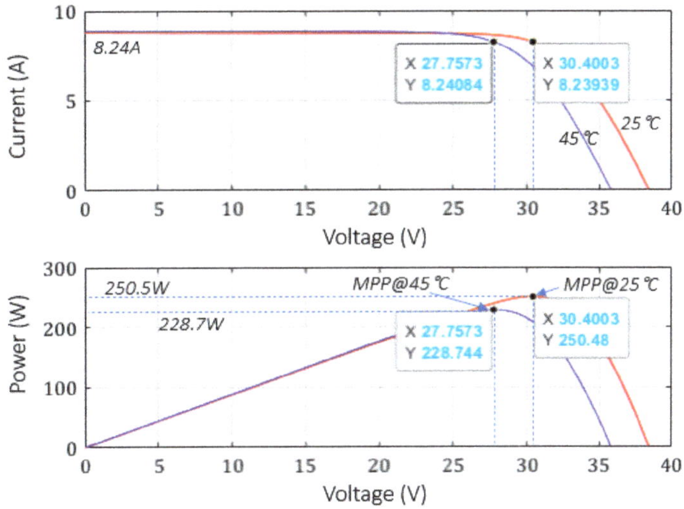

Figure 21. Modification of the MPP due to temperature change. The red curves are at 25 °C, and the blue curves are at 45 °C.

The DS1104 software 6.5 develops the Testbed for the laboratory prototype. At $t = 0$ s, a step in solar radiation of 600 W/m^2 and a temperature of 25 °C are generated, at which point the MPPT algorithm begins the search for the MPP. It takes approximately 6 s for the MPPT algorithm to reach the MPP. Ten seconds later, $t = 10$ s, a change in the panel temperature is generated, shifting from 25 °C to 45 °C. With this change, a decrease in the power provided by the panel should be observed, as shown in Figure 21.

Afterward, at $t = 14.5$ s, the solar radiation reference change is generated at 1000 W/m^2 and a temperature of 45 °C. Once again, after this change in solar radiation, an increase in the power provided by the solar panel is expected. Finally, at $t = 25$ s, the temperature reference changes from 45 °C to 25 °C, which would be expected to increase the panel's power due to the temperature decrease. On a normal day, the behavior of solar radiation does not experience sudden changes, so transients are not large. Furthermore, the MPPT algorithm can be executed with a much longer sampling time. In this scenario, the efficiency of the Multiplier SEPIC was around 88%, similar to the simulation results. It should be noted that the MPPT algorithm is executed every 200 mS, which is sufficient time for the Multiplier SEPIC converter to stabilize after a change in the duty cycle reference.

To calculate the power of the emulated panel, measurements of panel voltage and panel current are acquired via DS1104, which are measured at the output of the linear source using Hall Effect current sensors AMPLOC 25, LEM50, and LV25P voltage sensors. The PWM signal that activates the MOSFET of the Multiplier SEPIC also comes from the DS1104 system, where the MPPT algorithm is located. Once the measurement is performed, the data are transferred to Matlab for further processing and visualization.

Figure 22a shows the evolution of the power output from the panel, *Ppv*, and the power output from the Multilevel SEPIC, *Po*. As expected, the power provided by the photovoltaic panel decreases when the temperature reference changes at $t = 10$ s, consistent with the expected behavior observed in Figure 21. Additionally, when the solar radiation reference changes from 600 W/m^2 to 1000 W/m^2, there is also an increase in the panel's power, as anticipated. Finally, the panel's power increases when the panel temperature drops to 25 °C. Comparing the panel's output voltages, current, and power in the emulated system for radiation of 1000 W/m^2, they are very similar to those expected from the panel according to the manufacturer's parameters shown in Figure 21.

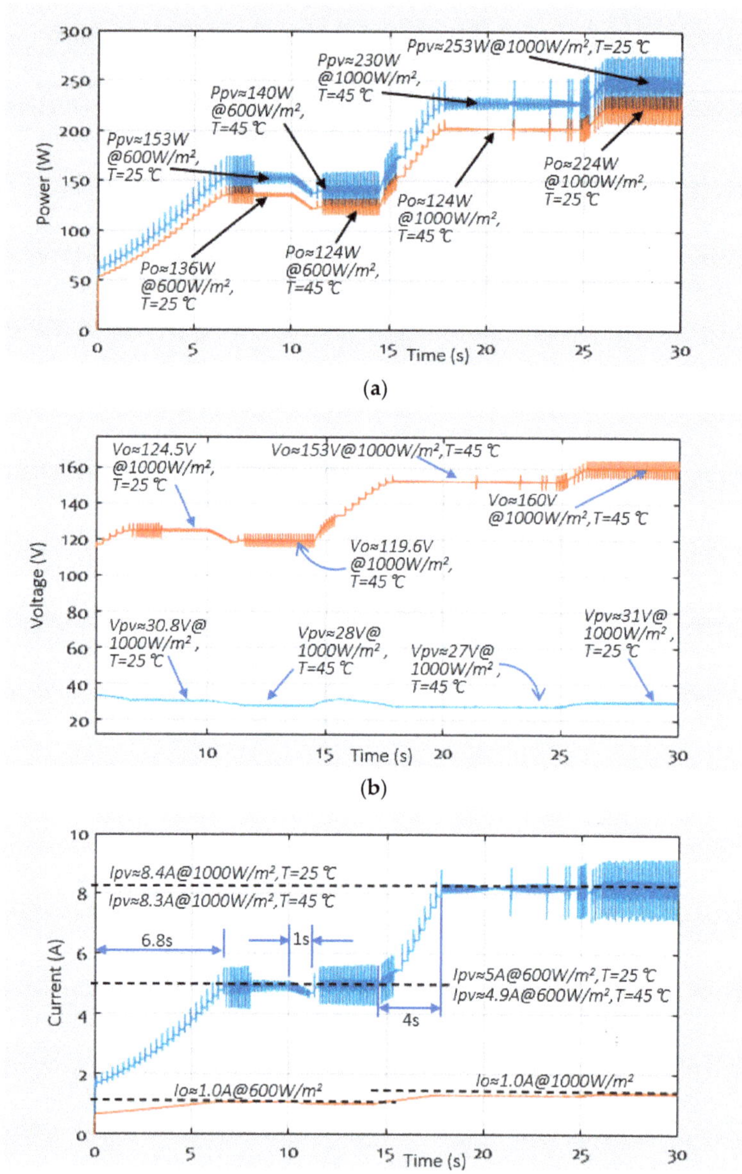

Figure 22. (**a**) Power evolution of P_{pv} panel (blue) and Po in the output of Multiplier SEPIC (red). (**b**) Voltage evolution of V_{pv} panel (blue) and Vo in the output of Multiplier SEPIC (red). (**c**) Current evolution of I_{pv} panel (blue) and Io in the output of Multiplier SEPIC (red).

Figure 22b,c show the evolution of the voltage and current in the laboratory hardware using an R_{load} of 115 Ω. As expected for a panel, the output voltage decreases when the temperature increases, causing the panel's power to change. When the temperature and irradiance change, the panel voltage also changes by around 30 V, as can be seen in Figure 22b, as expected, and, similarly, the output voltage of the Multilevel SEPIC changes when the algorithm searches for the MPP. Figure 22c shows the current evolution in the panel, whose short-circuit limit is constrained by the irradiance. Equations (1) and (2) illustrate this dependence, which can be graphically observed in Figure 5.

Figure 22c also shows the times at which the MPP is reached: between 1 s and 2 s, when temperature changes occur, and between 3 s and 6 s, when abrupt changes in irradiance occur.

6. Conclusions

This study uses a novel approach to model, design, and emulate a solar array feeding a DC-DC SEPIC multiplier converter with a Maximum Power Point Tracker (MPPT). The approach uses a Power Hardware-In-the-Loop (PHIL) DSpace platform for experimental validation along with a real DC-DC SEPIC multiplier converter. The article's contributions can be summarized as follows: (i) The application of an innovative emulation technique in a didactic PV + Multiplier SEPIC system: Our approach uniquely combines PHIL emulation with a real DC-DC converter, providing a versatile tool for emulating solar panel arrays under various conditions. This allows the simulation of renewable energy sources, offering a high degree of precision and flexibility in the emulation of different scenarios, the implementation of different MPPT algorithms, and experimentation with other types of DC/DC and DC/AC converters. (ii) The experiments conducted using the DSpace platform demonstrated the effectiveness and the reliability of our PHIL emulator in replicating the behavior of solar arrays connected to a DC-DC converter, including variations in sunlight intensity and other environmental factors, such that it can serve as a teaching and laboratory tool for experimentation at low cost. (iii) The successful development and validation of our PHIL emulator provides a powerful tool for researchers and engineers in the field of renewable energy. By allowing for detailed and accurate simulation of solar power systems, our work facilitates the exploration of new control strategies, the optimization of energy conversion processes, and the enhancement of system efficiency and stability.

Finally, our research contributes to the broader goal of advancing sustainable energy technologies. By improving our ability to simulate and study solar power systems, we are better equipped to develop innovative solutions that will increase the viability and efficiency of renewable energy, paving the way for a more sustainable and environmentally friendly future. New control algorithms can be tested with the proposed equipment in future work.

Author Contributions: J.P.C. contributed to the conceptualization of the article, J.C.R.-C. contributed to the methodology, J.P.C. contributed to the software, J.C.R.-C. contributed to the formal analysis, J.P.C. and J.C.R.-C. wrote the draft and contributed to the manuscript preparation. All authors have read and agreed to the published version of the manuscript.

Funding: The authors would like to thank the Universidad Panamericana and the Universidad Autonoma de Occidente.

Data Availability Statement: Data are contained within the article.

Acknowledgments: The authors would like to thank the Universidad Autónoma de Occidente in Colombia and the Universidad Panamericana in Mexico.

Conflicts of Interest: The authors declare no conflicts of interest.

References

1. Subbulakshmy, R.; Palanisamy, R.; Alshahrani, S.; Saleel, C.A. Implementation of Non-Isolated High Gain Interleaved DC-DC Converter for Fuel Cell Electric Vehicle Using ANN-Based MPPT Controller. *Sustainability* **2024**, *16*, 1335. [CrossRef]
2. Khan, M.R.; Haider, Z.M.; Malik, F.H.; Almasoudi, F.M.; Alatawi, K.S.S.; Bhutta, M.S. A Comprehensive Review of Microgrid Energy Management Strategies Considering Electric Vehicles, Energy Storage Systems, and AI Techniques. *Processes* **2024**, *12*, 270. [CrossRef]

3. Daccò, E.; Falabretti, D.; Ilea, V.; Merlo, M.; Nebuloni, R.; Spiller, M. Decentralised Voltage Regulation through Optimal Reactive Power Flow in Distribution Networks with Dispersed Generation. *Electricity* **2024**, *5*, 134–153. [CrossRef]
4. Bordbari, M.J.; Nasiri, F. Networked Microgrids: A Review on Configuration, Operation, and Control Strategies. *Energies* **2024**, *17*, 715. [CrossRef]
5. Min, C.; Kim, H. A Practical Framework for Developing Net-Zero Electricity Mix Scenarios: A Case Study of South Korea. *Energies* **2024**, *17*, 926. [CrossRef]
6. Gayathri, R.; Chang, J.-Y.; Tsai, C.-C.; Hsu, T.-W. Wave Energy Conversion through Oscillating Water Columns: A Review. *J. Mar. Sci. Eng.* **2024**, *12*, 342. [CrossRef]
7. Mihalič, F.; Truntič, M.; Hren, A. Hardware-in-the-Loop Simulations: A Historical Overview of Engineering Challenges. *Electronics* **2022**, *11*, 2462. [CrossRef]
8. Yousefzadeh, M.; Hedayati Kia, S.; Hoseintabar Marzebali, M.; Arab Khaburi, D.; Razik, H. Power-Hardware-in-the-Loop for Stator Windings Asymmetry Fault Analysis in Direct-Drive PMSG-Based Wind Turbines. *Energies* **2022**, *15*, 6896. [CrossRef]
9. Hermassi, M.; Krim, S.; Kraiem, Y.; Hajjaji, M.A.; Alshammari, B.M.; Alsaif, H.; Alshammari, A.S.; Guesmi, T. Design of Vector Control Strategies Based on Fuzzy Gain Scheduling PID Controllers for a Grid-Connected Wind Energy Conversion System: Hardware FPGA-in-the-Loop Verification. *Electronics* **2023**, *12*, 1419. [CrossRef]
10. Dini, P.; Saponara, S. Modeling and Control Simulation of Power Converters in Automotive Applications. *Appl. Sci.* **2024**, *14*, 1227. [CrossRef]
11. Lamo, P.; de Castro, A.; Sanchez, A.; Ruiz, G.A.; Azcondo, F.J.; Pigazo, A. Hardware-in-the-Loop and Digital Control Techniques Applied to Single-Phase PFC Converters. *Electronics* **2021**, *10*, 1563. [CrossRef]
12. Estrada, L.; Vázquez, N.; Vaquero, J.; de Castro, Á.; Arau, J. Real-Time Hardware in the Loop Simulation Methodology for Power Converters Using LabVIEW FPGA. *Energies* **2020**, *13*, 373. [CrossRef]
13. Sanchez, A.; Todorovich, E.; De Castro, A. Exploring the Limits of Floating-Point Resolution for Hardware-In-the-Loop Implemented with FPGAs. *Electronics* **2018**, *7*, 219. [CrossRef]
14. De Souza, I.D.T.; Silva, S.N.; Teles, R.M.; Fernandes, M.A.C. Platform for Real-Time Simulation of Dynamic Systems and Hardware-in-the-Loop for Control Algorithms. *Sensors* **2014**, *14*, 19176–19199. [CrossRef] [PubMed]
15. Regatron Programmable Power Supplies. Available online: https://www.regatron.com/programmable-power-supplies/en/#hardware-in-the-loop (accessed on 16 June 2024).
16. Chroma Solar Array Simulator Model 62000H-S Series. Available online: https://www.chromaate.com/en/product/solar_array_simulator_62000h_s_series_205 (accessed on 16 June 2024).
17. Martínez, J.R.; Rengifo, H.R.; Córdoba, J.S.; Palacios, J.; Posada, J. Design and Implementation of a Multiplier SEPIC Converter to Emulate a Photovoltaic System Using Power HIL. In Proceedings of the 2019 FISE-IEEE/CIGRE Conference—Living the energy Transition (FISE/CIGRE), Medellin, Colombia, 4–6 December 2019; pp. 1–7.
18. Olayiwola, T.N.; Hyun, S.-H.; Choi, S.-J. Photovoltaic Modeling: A Comprehensive Analysis of the I–V Characteristic Curve. *Sustainability* **2024**, *16*, 432. [CrossRef]
19. Pan, W.; Zhang, Y.; Jin, W.; Liang, Z.; Wang, M.; Li, Q. Photovoltaic-Based Residential Direct-Current Microgrid and Its Comprehensive Performance Evaluation. *Appl. Sci.* **2023**, *13*, 12890. [CrossRef]
20. Duffie, J.A.; Beckman, W.A. *Solar Engineering of Thermal Processes, Photovoltaics and Wind*; Wiley: New York, NY, USA, 1991; pp. 762–772.
21. Rashid, M.H. *Power Electronics: Circuits, Devices, and Applications*, 3rd ed.; Pearson Education: Hoboken, NJ, USA, 2009; pp. 289–292.
22. Mohan, N.; Undeland, T.M.; Robbins, W.P. *Power Electronics in Converters, Applications, and Design*, 3rd ed.; Whiley: Hoboken, NJ, USA, 2002.
23. Erickson, R.W.; Maksimovic, D. *Fundamentals of Power Electronics*, 3rd ed.; Springer: New York, NY, USA, 2020.
24. Valdez-Resendiz, J.E.; Mayo-Maldonado, J.C.; Alejo-Reyes, A.; Rosas-Caro, J.C. Double-Dual DC-DC Conversion: A Survey of Contributions, Generalization, and Systematic Generation of New Topologies. *IEEE Access* **2023**, *11*, 38913–38928. [CrossRef]
25. Rosas-Caro, J.C.; Mayo-Maldonado, J.C.; Valdez-Resendiz, J.E.; Alejo-Reyes, A.; Beltran-Carbajal, F.; López-Santos, O. An Overview of Non-Isolated Hybrid Switched-Capacitor Step-Up DC–DC Converters. *Appl. Sci.* **2022**, *12*, 8554. [CrossRef]
26. Rosas-Caro, J.C.; Mayo-Maldonado, J.C.; Valdez-Resendiz, J.E.; Salas-Cabrera, R.; Gonzalez-Rodriguez, A.; Salas-Cabrera, E.N.; Cisneros-Villegas, H.; Gonzalez-Hernandez, J.G. Multiplier SEPIC converter. In Proceedings of the CONIELECOMP 2011, 21st International Conference on Electrical Communications and Computers, Cholula, Puebla, Mexico, 28 February–2 March 2011; pp. 232–238.
27. Rosas-Caro, J.C.; Sanchez, V.M.; Vazquez-Bautista, R.F.; Morales-Mendoza, L.J.; Mayo-Maldonado, J.C.; Garcia-Vite, P.M.; Barbosa, R. A novel DC-DC multilevel SEPIC converter for PEMFC systems. *Int. J. Hydrogren Energy* **2016**, *41*, 23401–23408. [CrossRef]

energies

Article

Multi-Site Wind Speed Prediction Based on Graph Embedding and Cyclic Graph Isomorphism Network (GIN-GRU)

Hongshun Wu and Hui Chen *

School of Electrical and Electronic Engineering, Hubei University of Technology, Wuhan 430068, China; w2379439736@gmail.com
* Correspondence: chenhui@hbut.edu.cn

Abstract: Accurate and reliable wind speed prediction is conducive to improving the power generation efficiency of electrical systems. Due to the lack of adequate consideration of spatial feature extraction, the existing wind speed prediction models have certain limitations in capturing the rich neighborhood information of multiple sites. To address the previously mentioned constraints, our study introduces a graph isomorphism-based gated recurrent unit (GIN-GRU). Initially, the model utilizes a hybrid mechanism of random forest and principal component analysis (PCA-RF) to discuss the feature data from different sites. This process not only preserves the primary features but also extracts critical information by performing dimensionality reduction on the residual features. Subsequently, the model constructs graph networks by integrating graph embedding techniques with the Mahalanobis distance metric to synthesize the correlation information among features from multiple sites. This approach effectively consolidates the interrelated feature data and captures the complex interactions across multiple sites. Ultimately, the graph isomorphism network (GIN) delves into the intrinsic relationships within the graph networks and the gated recurrent unit (GRU) integrates these relationships with temporal correlations to address the challenges of wind speed prediction effectively. The experiments conducted on wind farm datasets for offshore California in 2019 have demonstrated that the proposed model has higher prediction accuracy compared to the comparative model such as CNN-LSTM and GAT-LSTM. Specifically, by modifying the network layers, we achieved higher precision, with the mean square error (MSE) and root mean square error (RMSE) of wind speed at a height of 10 m being 0.8457 m/s and 0.9196 m/s, respectively.

Keywords: multisite wind prediction; PCA-RF; graph embedding; graph neural network; convolutional neural network

Citation: Wu, H.; Chen, H. Multi-Site Wind Speed Prediction Based on Graph Embedding and Cyclic Graph Isomorphism Network (GIN-GRU). *Energies* **2024**, *17*, 3516. https://doi.org/10.3390/en17143516

Academic Editor: Frede Blaabjerg

Received: 31 May 2024
Revised: 27 June 2024
Accepted: 10 July 2024
Published: 17 July 2024

1. Introduction

The growing global energy crisis and the critical issue of environmental pollution have highlighted the need for clean and renewable energy sources. Wind energy has garnered significant attention due to its relatively short construction cycle, minimal environmental prerequisites, and vast reserves [1]. This has led to its widespread adoption and rapid development globally. Many nations have recognized the potential of wind power and are actively promoting its generation to capture wind energy. Consequently, the wind power sector has undergone rapid growth and expansion, firmly establishing wind power generation as a field with promising current prospects. Presently, wind speed prediction plays a pivotal role in formulating control strategies for wind farms, which stands as a cornerstone technology that enhances the operational efficiency of wind turbines [2].

Accurate wind speed prediction is crucial to optimizing wind power generation. It enhances wind energy utilization, mitigates wind power's grid impact, and ensures efficient operation of wind farms. To refine wind speed forecasting, researchers worldwide have pioneered various innovative approaches. For instance, data decomposition techniques like wavelet transform [3], empirical modal decomposition [4–6], and variational modal decomposition [7] are employed to analyze wind speed's multiscale attributes. Additionally, data fusion strategies such as Kalman filtering, data assimilation, and multi-source data fusion are utilized [8–10]. These methods provide a comprehensive analysis of various meteorological elements and data sources, which improves the accuracy of weather forecasting and monitoring.

Traditional statistical methods often focus on specific learning facets. With the accumulation of a large amount of historical operational data in the wind power industry, deep learning methods based on large-scale and multi-dimensional data have been increasingly applied to the field of wind speed prediction due to their powerful nonlinear mapping ability. H Xu et al. developed a hybrid deep learning model for wind speed prediction that integrates grid search optimization with a recurrent neural network [11]. This approach effectively corrected the errors of numerical weather prediction to enhance both the precision and dependability of wind speed forecasts. Ai X et al. proposed a wind speed prediction model based on data augmentation and deep learning. The results showed that empirical mode decomposition and data augmentation techniques generated more training data which improved the generalization ability as well as robustness of the wind speed prediction model [12]. B Shao et al. utilized a configuration of long short-term memory (LSTM) networks with diverse architectural complexities to uncover the underlying patterns in wind speed time series data [13]. The researchers integrated the predictive outputs using a nonlinear regression layer, which consisted of support vector regression machines. This integration resulted in a combined prediction model that exhibited high accuracy. However, the abovementioned wind speed prediction methods mostly use the historical and real-time data of wind speed at a single point. Thus, the prediction accuracy of the models still requires further improvement and augmentation.

The integration of comprehensive spatio-temporal data from multiple locations can provide a novel approach to predicting wind speeds. Q Zhu et al. leveraged the spatio-temporal correlation characteristics across various spatial scales to improve wind speed prediction [14]. By employing convolutional neural networks (CNN) for spatial modeling, the method captures the intricate spatial correlations between the macroscopic and microscopic levels, which offers a more comprehensive approach than traditional one-to-one modeling methods. Trebing K et al. employ a multiscale convolutional neural network filter to extract the multi-features influencing long-term wind speed distribution [15]. It introduces a deep convolutional recurrent neural network model for wind speed prediction at various points within a wind farm. This model excels at feature extraction and time-series analysis, which enables simultaneous ultra-short-term wind speed forecasts for each turbine. By integrating spatial and temporal flow field data, the model enhances overall prediction accuracy. W Tuerxun et al. proposed a novel spatio-temporal neural network that employs a deep convolutional neural network coupled with a bidirectional gated recurrent unit [16]. This model comprehensively captures the spatio-temporal dynamics between wind speed and direction across various altitudes within a wind farm, as well as pertinent NWP data. MM Yuan et al. combine CNN with LSTM networks to propose a multifactor spatio-temporal correlation model [17]. It introduces a data representation technique that uses three-dimensional matrices. This innovative model demonstrates enhanced predictive performance in wind speed forecasting. Therefore, by considering not only the direct spatio-temporal distribution of wind speed within the wind farm but also the potential spatio-temporal relationships of wind speed at multiple sites using deep learning algorithms, it will be probable to further improve the prediction accuracy.

This paper introduces a deep learning-based integrated multi-site wind speed prediction model for wind farms, designated as RPCA-GIN-GRU. This model enhances the predictive accuracy for each site by leveraging the feature correlations across multiple sites. Firstly, we apply PCA-RF to the feature set, which captures the essential features and extracts key insights by reducing the dimensions of the remaining features. Secondly, we use a method that combines graph embedding techniques with the Mahalanobis distance metric to construct the network graph's edges, which facilitates a thorough analysis of the relationships between each site's features. Ultimately, the GIN model is leveraged for its proficient learning of graph structure similarities. By conducting deep learning on the spatial attributes of these graph structures, we can then feed the deeply learned feature data into a GRU network for subsequent prediction. This approach generates wind speed predictions that considers the characteristics of each site and their interconnectedness to generate wind speed predictions. Additionally, it summarizes the interactive effects between sites on the predicted wind speeds.

2. Materials and Methods

2.1. Construction of Graph Networks

2.1.1. PCA-RF Fusion Model

Raw meteorological data are complex and multifaceted. However, not all variables are pertinent to changes in wind speed. An overabundance of predictive variables can introduce redundancy, thereby diminishing the model's generalization capabilities. Some residual features have some key information that needs to be extracted. Therefore, we focus on primary feature extraction and dimensionality reduction of residual features from the original meteorological elements. This approach streamlines the dataset and enhances model interpretability and efficiency. Consequently, this research leverages the intrinsic feature extraction capabilities of the random forest algorithm in conjunction with PCA to further diminish the dimensionality of residual features. This approach enables the separate processing of primary and residual features to achieve maximal efficiency. Subsequently, the primary features and the dimensionality-reduced residual features are concatenated to form the dataset used for constructing graph networks.

2.1.2. Random Forests

Random forest is a kind of machine learning algorithm that makes predictions from multiple decision trees and integrates their results. Random forests construct each decision tree using random samples and features, which imparts the model with robustness against overfitting [18]. MDI is a measure of feature importance based on the reduction in Gini impurity of each feature at the split point in the decision tree [19]. Given a dataset containing nodes from C categories, where the probability that node j belongs to category c is denoted as p_j, the Gini impurity of node j is as follows:

$$Gini(j) = 1 - \sum_{c=1}^{C} p_j^2 \tag{1}$$

Gini impurity is a statistical metric that measures the probability of a random classification error for an element chosen from a dataset, given that the classification is random and reflects the distribution of classes in the dataset. Owing to its simplicity and direct computation, MDI is selected for evaluating the significance of feature variables within this study. The MDI value signifies a feature's greater relevance in strengthening the predictive accuracy of the model. The steps for calculating MDI are given as follows:

Step 1. Each decision tree generated through bootstrap sampling on the training set constitutes a random forest.

Step 2. For every tree, the Gini impurity of each node is computed.

Step 3. MDI for each feature is ascertained by averaging the reduction in Gini impurity across all trees.

2.1.3. Principal Component Analysis

PCA is a prevalent technique for reducing data dimensionality. It employs orthogonal transformations to convert correlated variables into a new set of uncorrelated variables, known as principal components [20]. PCA is designed to preserve critical information from the original dataset and reduce dimensionality by decreasing the number of variables. PCA streamlines complex datasets by isolating pivotal features to reduce noise and safeguard essential information.

The fundamental method involves computing the eigenvalues and eigenvectors from the covariance matrix. The eigenvalues represent the core information of the original variables, which are crucial to the remaining features. Subsequently, the dimensionality reduction of the feature matrix facilitates the isolation of distinctive meteorological factors that exert a significant influence on wind speed variations.

2.1.4. PCA-RF Feature Fusion

In this study, the primary and residual feature matrices are based on the ordering of features according to their MDI scores. The residual feature matrix is then subjected to PCA for dimensional reduction. For a wind field with N sites, the feature matrix of the n-th site is expressed as $\mathbf{X}_n = \left[\mathbf{V}_1^T, \mathbf{V}_2^T \cdots \mathbf{V}_M^T\right]$, where $n = 1, \ldots, N$, M is the number of features per site, the total number of features is $M \times D$ and $V_i = [v_i(1), v_i(2) \cdots v_i(t)]$ $(i = 1, \ldots, M\ t = 1, \ldots, D)$ is the feature vector of the site. This dataset primarily includes measurements taken at 15-m intervals for wind speed, wind direction, temperature, pressure, and other relevant characteristics. The steps of wind speed feature processing for the n-th site are given as follows:

(1) *Build a random forest.* Bootstrap sampling and random feature selection techniques are utilized to generate a training dataset and a corresponding subset of features. These elements form the basis for constructing an individual decision tree. The process of building a decision tree involves randomly selecting a subset of the dataset and recursively splitting the nodes based on optimal segmentation criteria until the stopping criteria of the random forest are met. Through the repetition of this process, a random forest model is assembled with I different decision trees, where the i-th decision tree has T nodes and C categories.

(2) *Calculate the Gini impurity and the amount of its variation.* In tree i, at node j, it is essential to calculate the proportion of category c, represented as $p(c|(i,j),k)$, against the total categories. The variation in Gini impurity is then determined by evaluating the Gini impurity before and after the branching of node j. The calculation formula is as follows:

$$Gini(i,j,k) = 1 - \sum_{c=1}^{C} p(c|(i,j),k)^2 \tag{2}$$

$$\Delta Gini(i,j,k) = Gini(i,j,k) - Gini(i,j_b,k) - Gini\left(i,j_f,k\right) \tag{3}$$

where $j = 1, 2 \ldots T$, $i = 1, 2 \ldots I$, $k = 1, 2 \ldots M$, $c = 1, 2 \ldots C$. j_b and j_f represent the two new nodes after node j is branched. $\Delta Gini(i,j,k)$ are feature vector $\mathbf{V_k}$ reduction of Gini impurity after node splitting in decision tree i.

(3) *Calculate the MDI.* MDI for feature $\mathbf{V_k}$ is as follows:

$$M(\mathbf{V_k}) = \frac{1}{I} \sum\nolimits_{t=1}^{T} \Delta Gini(i,j,k) \tag{4}$$

(4) *Feature selection.* In the feature selection phase, features are ranked by their MDI values. Subsequently, a threshold denoted by γ is determined. Features with an MDI value of $M(V_k)$ are selected to comprise the primary feature matrix X_f, which includes M_f features. Conversely, features with an MDI value of $M(V_k) < \gamma$ are used to construct the auxiliary matrix X_b, which incorporates the remaining $M_b = M - M_f$ features.

(5) *Feature decentralization.* The auxiliary matrix X_b is decentralized to yield the matrix X_b'.

$$X_b' = X_b - \overline{X_b} \tag{5}$$

where $\overline{X_b}$ is the auxiliary matrix X_b.

(6) *Compute eigenvalues and eigenvectors.* The eigenvalue decomposition is performed on the covariance matrix. The feature values are ranked in descending order and

sequentially aggregated until the cumulative contribution rate, denoted by $\eta \geq 0.85$. Consequently, in adherence to this criterion, the primary r features are chosen to constitute the matrix R of $r \times D$.

$$\lambda l = \frac{1}{D-1} X_b{}'^T X_b{}' l \tag{6}$$

$$\eta = \frac{\sum_{i=1}^{r} \lambda_i}{\sum_{k=1}^{M} \lambda_k} \tag{7}$$

where $\lambda = \{\lambda_1, \lambda_2 \ldots \lambda_{M_b}\}$ is a set of eigenvalues, $l = [l_1, l_2 \ldots l_{M_b}]$ is the matrix of eigenvectors.

(7) *Data dimensionality reduction.* We calculate the dimensionality reduction matrix Y utilizing the subsequent formula:

$$Y = RX_b \tag{8}$$

(8) *Matrix splicing.* By horizontally concatenating the matrices X_f and Y, we obtain a matrix W_n of dimension $\left(M_f + r\right) \times D$.

$$W_n = \left[X_f Y\right] \tag{9}$$

PCA-RF processes all sites to yield a sequence of feature matrices $W_1, W_2 \ldots W_N$, which captures the pivotal features of the dataset. Finally, all the feature matrices are horizontally spliced to obtain a feature matrix W of $g \times D$.

$$W = [W_1 W_2 \ldots W_N] \tag{10}$$

where $g = N * \left(M_f + r\right)$. Figure 1 illustrates the steps of PCA-RF. Initially, the raw datasets from different sites are processed through RF algorithm to extract the primary feature matrix and the auxiliary feature matrix. Subsequently, the auxiliary feature matrix is subjected to PCA for dimensionality reduction to yield the dimensionality reduction matrix. Finally, the final feature matrix is constructed by concatenating the primary feature matrix with the dimensionality reduction matrix.

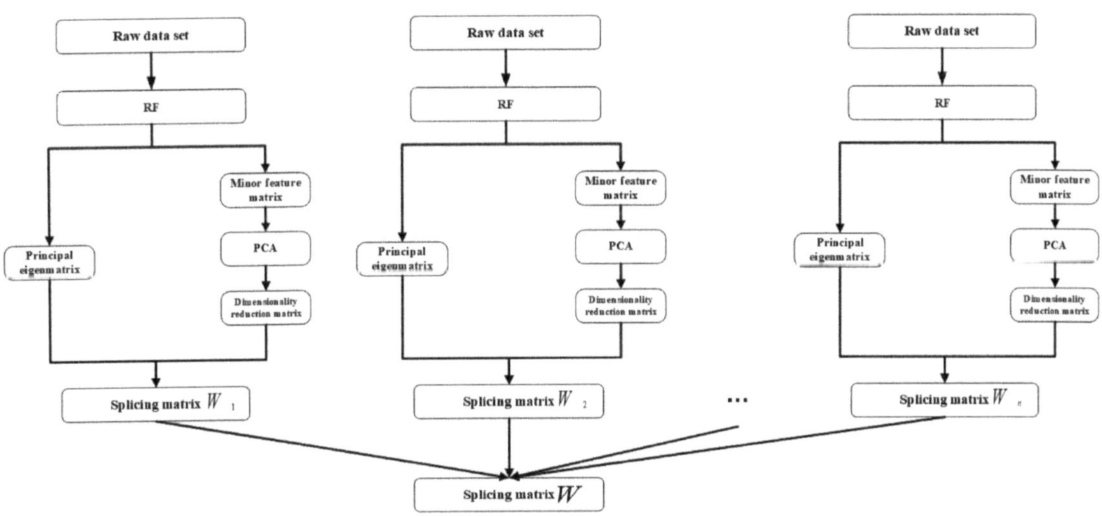

Figure 1. The flow chart of PCA-RF.

2.1.5. Integrative Modeling of Graph Embeddings and Mahalanobis Distances

Graph neural networks combined with graph networks on some feature correlations have achieved excellent results in wind speed prediction [21,22]. However, the performance of graph neural networks is highly dependent on predefined graphs to characterize the relationships between site features, such as those based on the Pearson correlation coefficients. Predefined graphs often do not reflect the complex dynamics of wind speed features, with their quality significantly depending on expert judgment and accurate wind speed measurements. Considering these factors, this paper introduces a novel compositional algorithm to analyze the spatio-temporal attributes of wind speed. The algorithm acquires node embedding vectors by employing graph embedding techniques on a predefined graph, which elucidate the graph's topology and the inter-node relationships [23]. Subsequently, it quantifies the distance between node embedding vectors by employing the Mahalanobis distance [24,25]. The final graph network results from the adjustment of edge relationships in the predefined graph through a filter based on a threshold Mahalanobis distance.

2.1.6. Construction of Predefined Graphs

In this paper, we use the matrix W obtained above to construct the predefined graph $G = (\boldsymbol{V}, E)$, where $\boldsymbol{V} = \left\{ \boldsymbol{V}_1^T, \boldsymbol{V}_2^T, \boldsymbol{V}_3^T, \ldots \boldsymbol{V}_g^T \right\}$ represents the set of nodes. We consider each column of data of matrix W as the feature vector of a node and the dimension is $g \times D$ where g is the number of nodes and D is the dimension of each node feature. The set of edges $E = \left(\boldsymbol{e_{ij}}\right)_{i,j=1}^{v}$ embodies the connectivity relationships among nodes within the network, such that there exists an edge connecting any two nodes $\boldsymbol{V}_i$ and $\boldsymbol{V}_j$. In the predefined graphs, the existence of an edge is denoted by $\boldsymbol{e_{ij}} = 1$, which is presented as follows:

$$e_{ij} = \begin{cases} 1, & \frac{\sum_{t=1}^{D} (v_i(t)-\overline{v}_i)\left(v_j(t)-\overline{v}_j\right)}{\sqrt{\sum_{t=1}^{D} (v_i(t)-\overline{v}_i)^2}\sqrt{\sum_{t=1}^{D} \left(v_j(t)-\overline{v}_j\right)^2}} < \delta \\ 0, & \frac{\sum_{t=1}^{D} (v_i(t)-\overline{v}_i)\left(v_j(t)-\overline{v}_j\right)}{\sqrt{\sum_{t=1}^{D} (v_i(t)-\overline{v}_i)^2}\sqrt{\sum_{t=1}^{D} \left(v_j(t)-\overline{v}_j\right)^2}} > \delta \end{cases} \tag{11}$$

where δ is the correlation threshold. $\overline{v}_i$ represents the mean value of the i-th node.

Graph embedding is a process that maps graph data into low-dimensional, dense vectors, preserving the graph's structure and properties. This process facilitates node classification, clustering, link prediction, graph reconstruction, and visualization [26]. In this study, graph embedding techniques are employed to transform a predefined graph into a continuous vector space. This conversion enables the effective learning of features and facilitates subsequent data analysis tasks. Certain shallow graph embedding techniques are initiated by randomly selecting neighboring nodes within the network to create a fixed-length random walk sequence. This sequence is then utilized by the skip-gram model to project the sequence of nodes into a low-dimensional space, which results in an embedding vector that captures the essential structural features of the graph [27]. This method effectively reduces the dimensionality and maintains node contextuality. The specific process is shown in Figure 2. However, the shallow model is unable to capture the highly nonlinear structure, which in turn leads to the generation of non-optimal solutions. Recent breakthroughs in deep learning have profoundly impacted graph analysis, as deep neural network techniques are increasingly employed to advance graph embedding methodologies [28]. Then we use deep neural networks to perform a nonlinear transformation of node features and neighborhood information, which leads to the generation of node embedding vectors that capture the graph's higher-order dependencies. This approach allows for a more profound understanding of node characteristics, which optimizes the graph's structural representation.

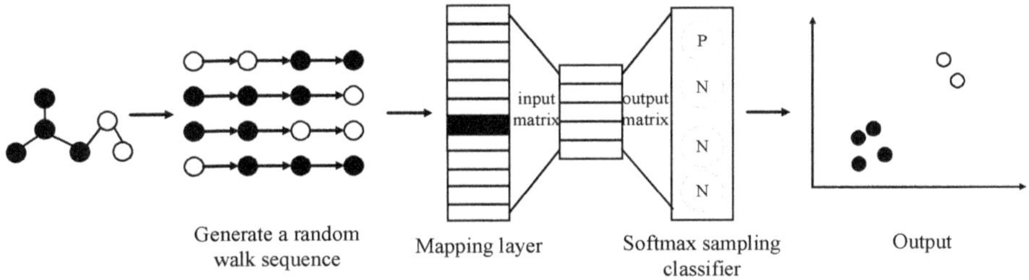

Figure 2. The flow chart of graph embedding.

2.1.7. GraphSAGE-Based Graph Embedding

We employ a deep learning approach grounded in GraphSAGE for graph embedding. This method leverages node feature information to generate embedding vectors for new nodes or subgraphs via an enhanced graph convolutional network. It facilitates incremental updates to node embeddings and preserves the graph's features and structural information. Based on this groundwork, the algorithm establishes 'edges' that precisely delineate the relationships between nodes. The specific process is given as follows:

A. *Sampling*: for node V_i, we randomly sample a subset of its neighboring nodes $N(V_i)$ from the set of its neighbor nodes S_k to create a subgraph. This method decreases computational demands yet preserves the heterogeneity among adjacent nodes.
B. *Aggregation*: for node V_i, we employ an aggregation function $AGGREGATE^k$ that compresses and transforms the feature vectors of its neighboring nodes $\left\{h_u^{k-1}, \forall u \in N(V_i)\right\}$ to generate a new feature vector $h_{V_i}^k$ through aggregation. The formula for aggregation is expressed as

$$h_{V_i}^k = \sigma\left(Q^k \cdot CONCAT\left(h_{V_i}^{k-1}, AGGREGATE^k\left(\left\{h_u^{k-1}, \forall u \in N(V_i)\right\}\right)\right)\right) \tag{12}$$

where $h_{V_i}^k$ denotes the node V_i in the first k layer of the embedding vector, and $h_{V_i}^{k-1}$ is the feature vector of node V_i. σ denotes the sigmoid activation function. The matrix Q^k represents the learnable weights. $CONCAT$ signifies the splicing operation. $AGGREGATE^k$ is the aggregation function at the first k layer and $N(V_i)$ refers to the node V_i neighboring set.
C. *Update*: for node V_i, the feature vector $h_{V_i}^{k-1}$ is concatenated with the aggregated feature vector $h_{V_i}^k$. This combined vector then passes through a fully connected layer followed by an activation function, which results in the embedding vector $h_{V_i}^{k+1}$ for node V_i. This process facilitates the integration and nonlinear transformation of the features represented by node V_i with those of its neighboring nodes. The updated formula is expressed as follows:

$$h_{V_i}^{k+1} = \sigma\left(Q^{k+1} \cdot COMBINE\left(h_{V_i}^{k-1}, h_{V_i}^k\right)\right) \tag{13}$$

where $COMBINE$ denotes the splicing or summing operation and Q^{k+1} denotes the learnable weight matrix.

2.1.8. Modification of Graph Network Edges

The node embedding vectors obtained through GraphSAGE graph embedding capture the features of the nodes and the relationships between the nodes. However, they are not directly used as components of the graph network. In this study, the edge connections are determined by setting a similarity threshold based on the Mahalanobis distance, which dictates the connectivity between nodes.

The Mahalanobis distance quantifies the similarity or divergence between two data samples by incorporating their covariance matrix. This metric excels at handling features with varying scales and interdependencies, which effectively neutralizes the impact of scale disparities and feature correlations. Unlike traditional distance measures (such as Euclidean distance), which often assume that individual features are independent, real-world data often exhibit correlations between features. The Mahalanobis distance accounts for these correlations by utilizing the covariance matrix. Consequently, we employ the Mahalanobis distance method for optimizing node embedding vectors. This method is particularly suitable for high-dimensional data influenced by numerous meteorological factors and can accommodate the conditions of non-independent as well as identically distributed dimensions. Its calculation formula is expressed as follows:

$$d_{mahal} = \sqrt{\left(h_{V_i}^{k+1} - h_{V_j}^{k+1}\right)^T \Sigma^{-1} \left(h_{V_i}^{k+1} - h_{V_j}^{k+1}\right)} \tag{14}$$

where $h_{V_i}^{k+1}$ denotes the node embedding vector. The covariance matrix is represented by Σ, and its inverse is denoted as Σ^{-1}. Finally, we can compute the dimension d_{mahal} and the correlation threshold α to determine the connectivity of the optimized edges $E' = (e_{ij})_{i,j=1}^{v}$ within the predefined graph network $G = (V, E)$.

$$e_{ij} = \begin{cases} 1, & d_{mahal} < \alpha \\ 0, & d_{mahal} > \alpha \end{cases} \tag{15}$$

A new graph network $G' = (V, E')$ is thus created. The process of embedding the graph network is illustrated in Figure 3.

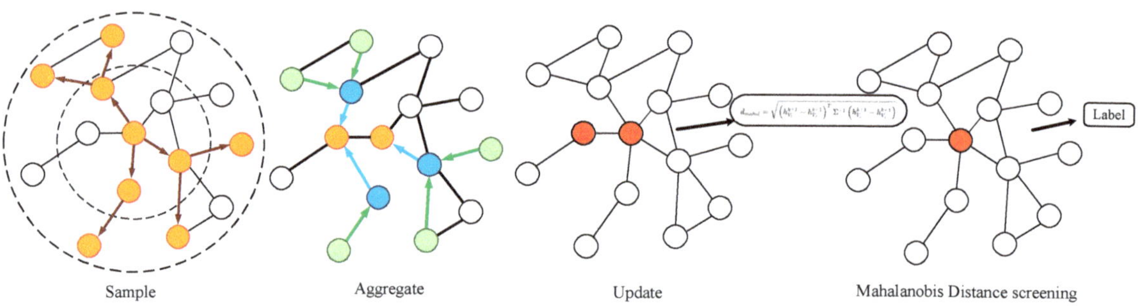

Figure 3. The flow chart of graph creation.

2.2. Spatio-Temporally Integrated Forecasting Model Based on GIN and GRU

2.2.1. Graph Isomorphism Network

GIN is a robust method for learning graph representations, which closely approximates the graph isomorphism test (WL-test) performance [29,30]. It effectively distinguishes between non-isomorphic graphs. GIN excels at learning node embedding vectors, discerning patterns within graphs to enhance efficiency and capturing structural dependencies to improve performance. The framework of GIN proceeds as follows:

(1) *Aggregation*. The GIN model employs a summation aggregation function to compile the feature vectors h_u^{k-1} from the neighboring nodes $u \in N(V_i)$ of node V_i, which aims to integrate the information from all adjacent nodes.

$$a_{V_i}^k = \sum\nolimits_{u \in N(V_i)} h_u^{k-1} \tag{16}$$

where $a_{V_i}^k$ is the temporary aggregation result at layer k, which contains only the information of neighboring nodes.

(2) *Combination*. The aggregated features of neighboring nodes are combined with the target node V_i and the features from the previous layer $h_{V_i}^{k-1}$ to form a new node feature $h_{V_i}^{k}$. This process, facilitated by learnable parameters and the nonlinear transformations of a multilayer perceptron (MLP), enhances the model's ability to learn and represent complex patterns.

$$h_{V_i}^{k} = MLP^{k}\left(\left(1+\epsilon^{k}\right)\cdot h_{V_i}^{k-1} + a_{V_i}^{k}\right) \quad (17)$$

where ϵ^{k} is a trainable parameter that allows the model to adjust the self-loop when updating the node features with the contribution. $h_{V_i}^{k}$ is the final node feature representation after combination.

2.2.2. Gated Recurrent Unit

GRU is a variant of the recurrent neural network (RNN) designed for processing sequential data, including natural language, speech, and video. Characterized by its dual gating mechanisms—the reset and update gates—GRU regulates the information flow and memory within the network [31,32]. This architecture effectively addresses the vanishing gradient problem, preserves long-term dependencies, and enhances model performance. GRU is particularly adept at extracting temporal correlations, which makes it a prime choice for the second layer in deep learning neural networks. The procedural steps are as follows:

(1) *Determine the values of the update and reset gates*. The update gate assesses the degree to which the hidden state from the preceding timestep is retained in the current timestep. Conversely, the reset gate regulates the proportion of the previous timestep's hidden state that is incorporated into the computation of the current state. The formula for this step is given by

$$\begin{aligned} z_t &= \sigma(U_z[h_{t-1}, x_t] + b_z) \\ r_t &= \sigma(U_r[h_{t-1}, x_t] + b_r) \end{aligned} \quad (18)$$

where z_t and r_t denote the values of the update gate and the reset gate, respectively. The sigmoid activation function, represented by σ, is utilized to facilitate the computation of gradients and enable effective backpropagation. The weight matrices U_z and U_r, along with the bias vectors b_z and b_r, play a pivotal role in this mechanism. Additionally, b_r represents the hidden state from the previous timestep, while x_t corresponds to the input at the current timestep.

(2) *Compute the candidate hidden state*. This represents the candidate hidden state, obtained by applying the hyperbolic tangent ($tanh$) activation function to the current input and the previously reset hidden state. The formulas are expressed as follows:

$$\tilde{h}_t = tanh(U_h[r_t \odot h_{t-1}, x_t] + b_h) \quad (19)$$

where h_{t-1} is the candidate hidden state, U_h is the weight matrix, b_h is the bias vector, and $\odot$ represents the element-wise multiplication operation.

(3) *Compute the current hidden state*. The current hidden state is computed as the weighted average of the previous hidden state and the candidate hidden state, where the weights are governed by the update gate. The formulas are expressed as follows:

$$h_t = (1 - z_t) \odot h_{t-1} + z_t \odot \tilde{h}'_t \quad (20)$$

where h_t represents the hidden state at the current moment, z_t denotes the value of the update gate, and $\tilde{h}'_t$ is the candidate hidden state.

These equations collectively illuminate the dynamic update mechanism of the hidden state at each timestep, as orchestrated by gated recurrent units (GRUs). This update methodology is intricately designed to capture long-term dependencies and complex patterns inherent in sequential data. Illustrated in Figure 4, the GRU's input–output structure includes the current input x_t, as well as the transmitted hidden state h_{t-1} from the previous timestep, which retains essential historical information. By amalgamating x_t and h_{t-1}, the GRU calculates the hidden node output y^t for the current timestep and the hidden state h_t, which will be forwarded to the next timestep.

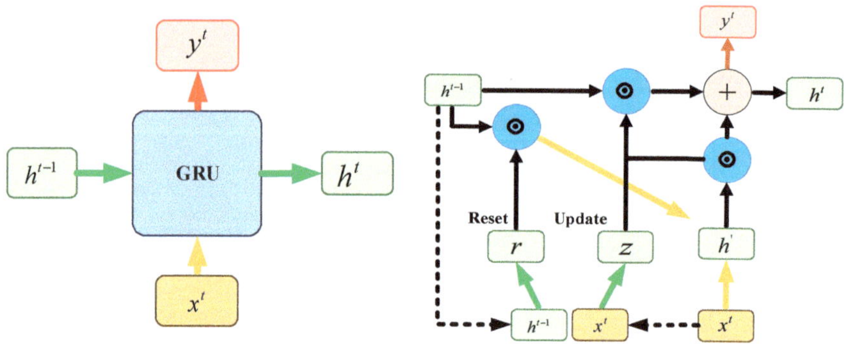

(**a**) Structure of GRU (**b**) Flow of GRU

Figure 4. GRU visual flow chart.

2.3. Hybrid Model of GE-GIN-GRU Network

In our study, we explore wind speed prediction models based on GIN and GRU. These models capture both multi-site wind speed spatio-temporal correlations and wind direction spatio-temporal correlations, which facilitate accurate local wind speed predictions across various time domains. The data such as wind direction data, wind speed information, temperature information, and barometric pressure information are independent time series from each other. To incorporate feature information affecting loads and prepare it as input for the GIN, we need to process this information to construct a network graph. In our study, we use the PCA-RF method to optimize the nodes. Furthermore, more accurate edge representations are obtained by employing graph embedding techniques in conjunction with Mahalanobis distance. Graph embedding techniques are intended to augment the accuracy of graph network within the GIN framework.

Firstly, at the node level, primary features are extracted using PCA-RF and residual features are dimensionally reduced to retain essential information. This process yields the $g \times D$ fusion matrix W, which serves as the nodes in the predefined graph $G = (V,E)$.

Secondly, at the edge level, the predefined graph $G = (V,E)$ is subject to aggregation and subsequent updating to obtain the node embedding vector. Subsequently, the Mahalanobis distance between node embedding vectors is employed as a discriminative criterion to establish the edges within the newly formulated graph network $G' = (\mathbf{V}, E')$.

Finally, we use the new graph network $G' = (\mathbf{V}, E')$ as the input for the GIN-GRU network. The structure of the proposed GIN-GRU network model is illustrated in Figure 5. Within this model, the GIN component primarily handles feature extraction; furthermore, the GRU network focuses on load prediction. The GIN is structured with two convolutional layers, which sequentially perform the aggregation and combination operations. In the GRU network component, we observed that a greater number of GRU network units, which adds to the model's depth, enhances its prediction capability. Consequently, the proposed model includes two layers of GRU networks, with 128 neurons in each layer. In each layer of the GRU network, random deactivation is employed to prevent overfitting. Ultimately, the wind speed prediction vector is generated through the fully connected layer (Dense).

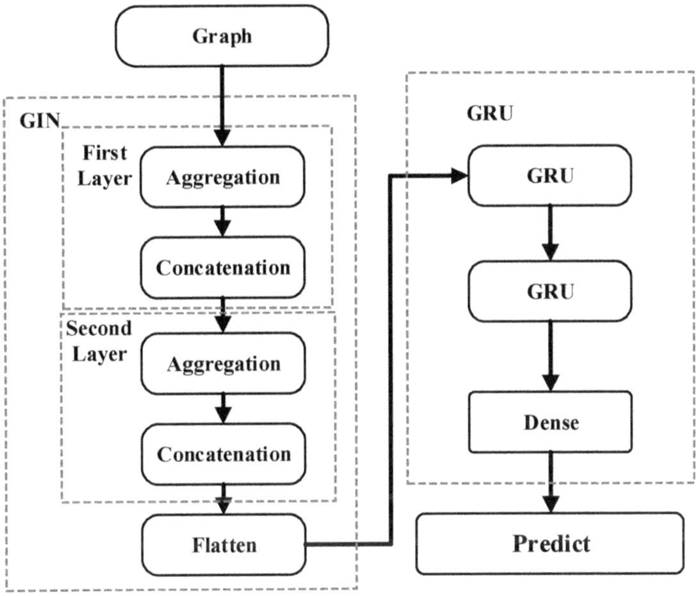

Figure 5. The flow chart of GIN-GRU.

3. Results

This subsection is structured into four different parts: the first part introduces the dataset, the second part analyzes the dimension reduction results of PCA-RF on the dataset, the third part discusses the results of graph networks for graph embedding, and the fourth part examines the prediction results of different models on the same test data.

3.1. *Datasets and Settings*

3.1.1. Dataset

The dataset is historical data for 20 sites along the California coast. Site IDs are 0, 22, 23, 24, 25, 26, 27, 28, 29, 30, 31, 32, 69, 70, 71, 72, 73, 74, 75, and 76 for 200 days at 15 min intervals. The size of the preprocessed dataset is set to 5000 and the ratio of the training, validation, and test sets is set to 6:2:2. Table 1 describes the characteristics of the dataset for each site.

3.1.2. Experimental Equipment

This experiment implements the TensorFlow 2.10.0 framework in Python 3.10.9 and accelerates computation through Compute Unified Device Architecture (CUDA). The simulation hardware platform features an Intel Core i7-10875H CPU (manufactured by Intel Corporation, Santa Clara, CA, USA) running at 2.30 GHz with 32GB of RAM, complemented by an Nvidia GeForce RTX 2070 GPU (manufactured by NVIDIA Corporation, Santa Clara, CA, USA).

3.1.3. Error Assessment Criteria

For the multi-site local wind speed prediction with graph network input in the wind farm, we evaluate its effectiveness using several key metrics. These metrics serve as indicators of prediction accuracy and performance. The following evaluation metrics are employed in this simulation: mean squared error (MSE), mean absolute error (MAE), normalized (RMSE), and normalized mean absolute error (MAPE) [33]. The formulas are expressed as follows:

$$E_{MSE} = \frac{1}{n}\sum_{i=1}^{n}(y_i - \hat{y}_i)^2, \tag{21}$$

$$E_{RMSE} = \sqrt{\frac{1}{n}\sum_{i=1}^{n}(y_i - \hat{y}_i)^2}, \tag{22}$$

$$E_{MAE} = \frac{1}{n}\sum_{i=1}^{n}|y_i - \hat{y}_i|, \tag{23}$$

$$E_{MAPE} = \frac{1}{n}\sum_{i=1}^{n}\left|\frac{y_i - \hat{y}_i}{y_i}\right| \times 100\%, \tag{24}$$

where n represents the number of samples, y_i denotes the i-th observed value, and $\hat{y}_i$ represents the i-th predicted value. MSE and RMSE quantify the goodness of fit of the model. Smaller values indicate better performance and reduced prediction error. However, it's important to note that MSE and RMSE are sensitive to outliers and can reflect the distribution of prediction errors. The MAE is the mean of the absolute differences between the predicted values and the true values. Unlike MSE, MAE is not influenced by outliers and provides a robust measure of prediction accuracy. MAPE remains unaffected by outliers and provides insight into the relative error across different data points. MSE and RMSE are commonly utilized for evaluating model performance, while MAE and MAPE offer alternative perspectives that take outliers into account, providing a more comprehensive assessment.

Table 1. Site dataset classification.

Category	Variables
Velocity and friction	friction_velocity_2 m
Monin–Obukhov length	inversemoninobukhovlength_2 m
Roughness	roughness_length
Sea temperature	surface_sea_temperature
Pressure	pressure_0 m, pressure_100 m, pressure_200 m
Humidity	relativehumidity_2 m
Precipitation rate	precipitationrate_0 m
Wind speed (at various heights)	windspeed_10 m, windspeed_40 m, windspeed_60 m, windspeed_80 m, windspeed_100 m, windspeed_120 m, windspeed_140 m, windspeed_160 m, windspeed_180 m, windspeed_200 m
Wind direction (at various heights)	winddirection_10 m, winddirection_20 m, winddirection_40 m, winddirection_60 m, winddirection_80 m, winddirection_100 m, winddirection_120 m, winddirection_140 m, winddirection_160 m, winddirection_180 m, winddirection_200 m
Temperature (at various heights)	temperature_2 m, temperature_10 m, temperature_20 m, temperature_40 m, temperature_60 m, temperature_80 m, temperature_100 m, temperature_120 m, temperature_140 m, temperature_160 m, temperature_180 m, temperature_200 m
General weather parameters	wind_speed, wind_direction, pressure, temperature

3.2. *Experimental Results and Analysis*

3.2.1. Analysis of PCA-RF

In our study, we address the challenge of extracting meaningful meteorological features from a dataset containing 46 variables. Previous attempts using individual methods yielded inaccurate results and exhibited bias. Therefore, we employ random forest to identify weather elements that capture essential information from the original variables.

By doing so, we filter out noise and focus on the most relevant features. Subsequently, we employ PCA to downsize the remaining minor weather elements. Finally, we fuse the main features with the dimensionality-reduced data to extract the characteristic weather elements that significantly impact wind changes. Figure 6 shows the MDI of the first 13 features of windspeed_10 m, which is the feature with the largest contribution to the other features of the site numbered 0. This paper observed a significant decline in MDI for all sites after the 10th feature. Consequently, the first 10 features are selected as the main features. Subsequently, the remaining 36 residual features undergo downsizing via PCA, which results in a new downscaled matrix with five principal components. Finally, the main feature matrix and the downscaling matrix are spliced as our feature set to build the graph network.

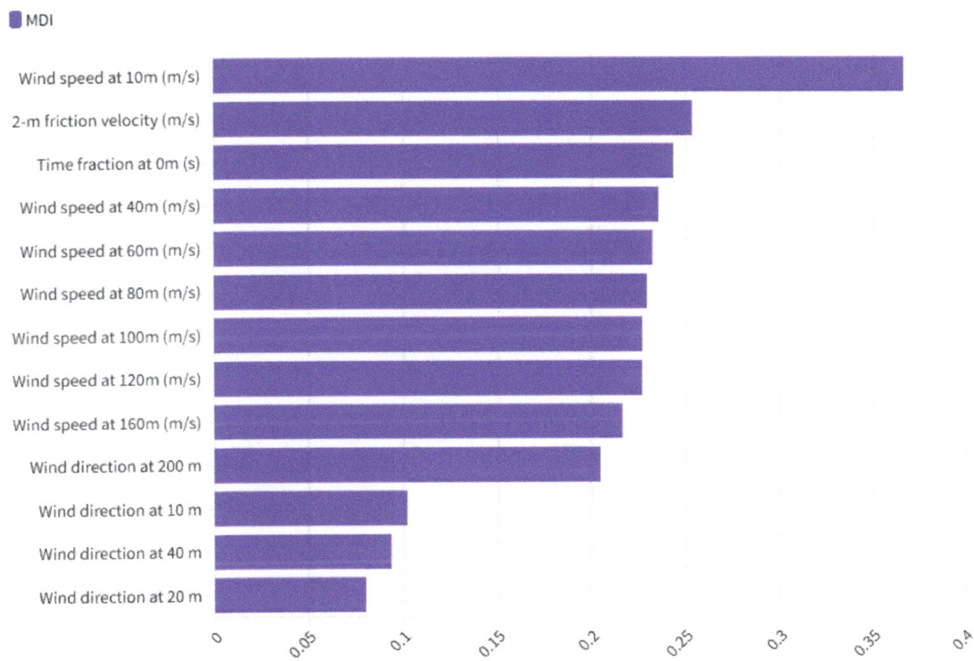

Figure 6. Features of the top thirteen MDI scores.

3.2.2. Evaluation of Graph Embedding for Graph Networks

To verify the effectiveness of graph embedding to build graph networks, we will compare the method of building graph networks solely using the Pearson correlation coefficient, i.e., the PS-GIN-GRU model. Due to the variability in the number of edges resulting from distinct threshold settings, we employ the optimally constructed network graph by both models for training purposes. The constructed graph networks were put into the same GIN-GRU model training and wind speed prediction at 10 m height for 40 days was performed. The model's performance was assessed using four key metrics: MSE, RMSE, MAE, and MAPE. We compared the predicted wind speeds from several stations with the actual measurements to validate the model's effectiveness. The error metrics for wind speed prediction, as obtained from graph networks constructed by different methods, are presented in Table 2.

Our focus was on evaluating the stability of predictions across multiple sites. The GE-GIN-GRU model integrates various neural network algorithms. Notably, this model demonstrates robustness and strong generalization ability. Models constructed solely using Pearson correlation coefficients rely exclusively on the correlation threshold method for building graph networks. While this approach is straightforward, it lacks the adaptability and complexity inherent in the GE-GIN-GRU model. We evaluated both models to assess the stability of wind speed forecasts across various sites. Especially, the GE-GIN-GRU model consistently outperformed the single correlation threshold model in terms of stability. Additionally, the GE-GIN-GRU model demonstrated robust prediction quality across diverse wind speed datasets. Our study underscores the critical role of constructing a graph that captures intricate relationships among wind speed features. By embedding vectors within this graph, we enhance accuracy in wind speed prediction. Particularly, the fusion of graph embedding techniques with neural networks significantly contributes to the model's performance and reliability.

Table 2. Comparison results of different methods of constructing graph networks.

Graph Foundation Models	MSE	RMSE	MAE	MAPE
①PS-GIN-GRU	2.5422	1.5944	1.4224	0.1151
①GE-GIN-GRU	0.8457	0.9196	0.7612	0.0636
②PS-GIN-GRU	2.7184	1.6488	1.4988	0.1231
②GE-GIN-GRU	1.0712	1.0350	0.8645	0.0726
③PS-GIN-GRU	2.7143	1.6475	1.4856	0.1209
③GE-GIN-GRU	0.9258	0.9622	0.8027	0.0685
④PS-GIN-GRU	2.7724	1.6651	1.4912	0.1208
④GE-GIN-GRU	0.9222	0.9603	0.7897	0.0666
⑤PS-GIN-GRU	2.5713	1.6035	1.4319	0.1160
⑤GE-GIN-GRU	0.8507	0.9223	0.7455	0.0624
⑥PS-GIN-GRU	2.1951	1.4816	1.3085	0.1062
⑥GE-GIN-GRU	1.0044	1.0022	0.8317	0.0705
⑦PS-GIN-GRU	2.2201	1.4900	1.3143	0.1066
⑦GE-GIN-GRU	0.9340	0.9665	0.8060	0.0684
⑧PS-GIN-GRU	2.2993	1.5163	1.3373	0.1079
⑧GE-GIN-GRU	1.1580	1.0761	0.9188	0.0761
⑨PS-GIN-GRU	2.2724	1.5074	1.3321	0.1074
⑨GE-GIN-GRU	0.7793	0.8828	0.7233	0.0613
⑩PS-GIN-GRU	2.1376	1.4620	1.2833	0.1032
⑩GE-GIN-GRU	1.0573	1.0283	0.8545	0.0719

Figure 7 presents the graph networks constructed by various models. With 20 sites in total, the diagram is divided into 20 corresponding groups. Within this diagram, each feature is depicted as a node, and features belonging to the same site are denoted by a uniform color. Figure 7 reveals that the graph networks generated through the Pearson correlation coefficient possess fewer edges compared to the one constructed via graph embeddings. Additionally, the node density within the correlation-based graph networks is comparatively lower, which leads to a more uniform distribution of edge connections. In contrast, the graph networks constructed by graph embeddings demonstrate a concentrated focus on specific nodes to capture their profound connections.

3.2.3. Evaluation and Analysis of GIN-GRU Neural Networks

To further validate the effectiveness of the GIN-GRU neural network, we compared it with recent popular wind speed prediction models: CNN-LSTM proposed by W. Tuerxun [16], GAT-GRU proposed by D. Aykas [34], and GAT-LSTM proposed by A. Flores [35]. For these models, we also constructed a network using the TensorFlow 2.10.0 framework in Python for training and fine-tuning. We evaluated these models using a test set of 20 stations over 40 days for each height wind speed prediction. The input graph network, which is identical across various models, is constructed utilizing consistent graph embedding techniques. We employed four key metrics (MSE, RMSE, MAE, and MAPE) to summarize the results. The results were summarized using four metrics: MSE, RMSE, MAE, and MAPE. The error metrics for the wind speed predictions made by different neural network models are displayed in Tables 3 and 4.

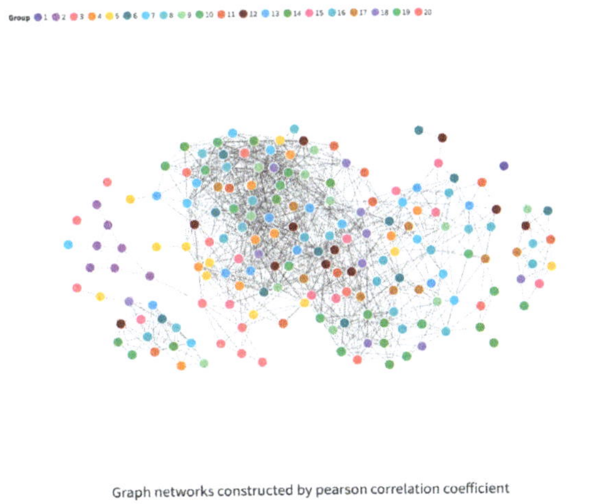

(**a**) Graph networks constructed by pearson correlation coefficient

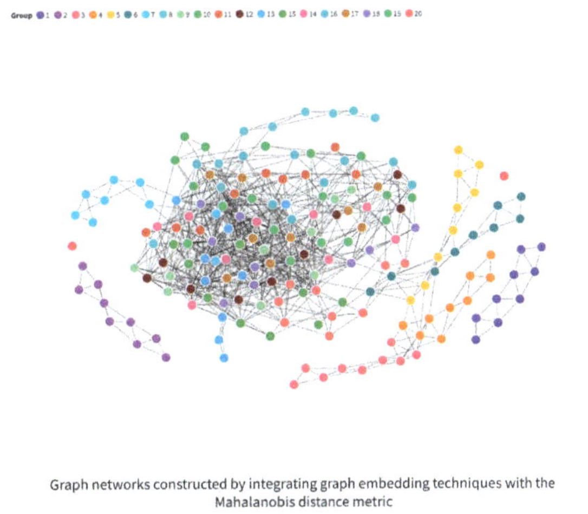

(**b**) Graph networks constructed by integrating graph embedding techniques with the Mahalanobis distance metric

Figure 7. Graph networks constructed by different models.

Table 3. Wind speed error metrics predicted by different neural networks at 10 m height.

Error Metrics	CNN-LSTM	GAT-GRU	GAT-LSTM	GE-GIN-GRU
MSE	7.7851	2.1808	2.3968	0.8457
RMSE	2.7902	1.4768	1.5482	0.9196
MAE	2.6851	1.3567	1.4009	0.7612
MAPE	0.2283	0.1152	0.1213	0.0636

Table 4. Wind speed error metrics predicted by different neural networks at 30 m height.

Error Metrics	CNN-LSTM	GAT-GRU	GAT-LSTM	GE-GIN-GRU
MSE	4.9849	1.3163	1.4214	0.6400
RMSE	2.2327	1.1473	1.1922	0.8000
MAE	2.0489	0.9414	0.9726	0.6076
MAPE	0.1629	0.0759	0.0787	0.0503

As shown in Figure 8, we compare the wind speed predictions for heights of 10 m and 30 m using the GIN-GRU model.

Despite variations in altitude, the GIN-GRU model consistently predicts wind speed with higher accuracy compared to other models. This result validates the effectiveness of the proposed method.

An examination of wind speed forecasts using various models at the designated site for elevations of 10 m and 30 m uncovers notable trends, as illustrated in Figure 9. While other networks generally align with measured wind speed signals, they often fail to grasp the nuanced graphical structural relationships. In contrast, GIN excels in this regard by leveraging additional spatio-temporal nodes as a predictive foundation.

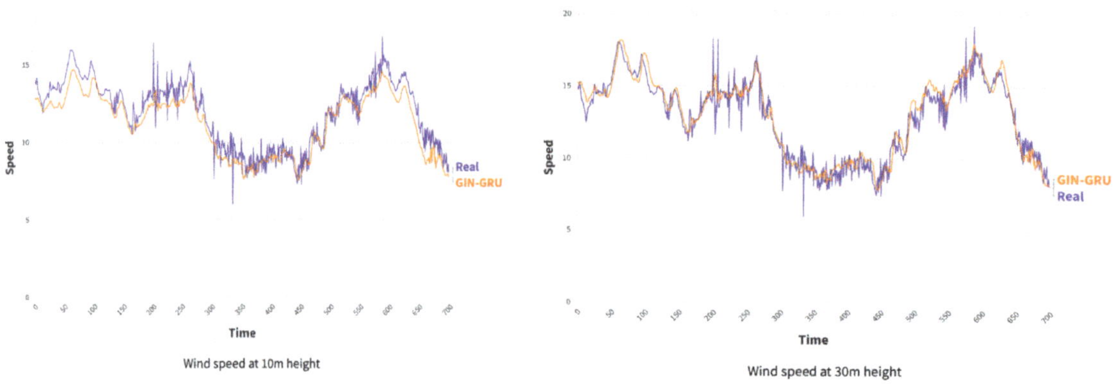

(**a**) wind speed prediction at 10 m height (**b**) wind speed prediction at 30 m height

Figure 8. Result of comparison between GE-GIN-GRU wind speed prediction and real value.

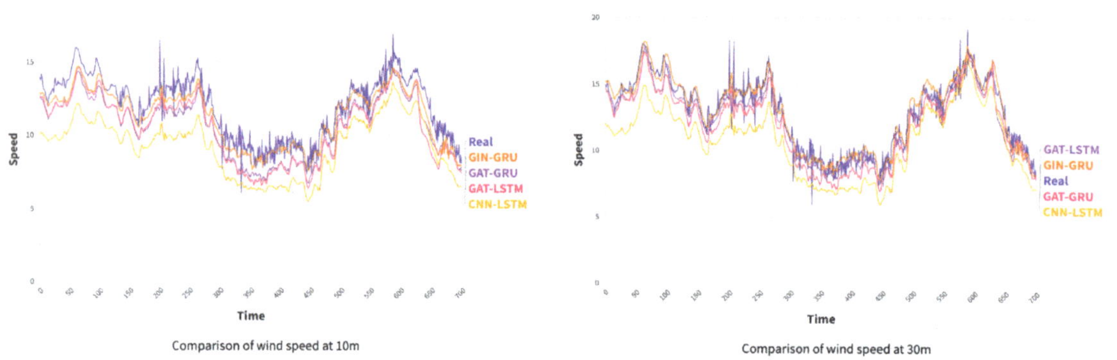

(**a**) wind speed prediction at 10 m height (**b**) wind speed prediction at 30 m height

Figure 9. GE-GIN-GRU wind speed prediction compared with other models.

Overall, both GE-GIN-GRU and other models effectively track wind speed trends. However, due to the limitations of other models in data processing and their sensitivity to critical information, the prediction error tends to be larger during wind speed fluctuations. GE-GIN-GRU stands out by leveraging information from neighboring stations. This approach better simulates unique wind speed patterns across wind of different heights. Notably, the predicted trend of GE-GIN-GRU closely aligns with the measured wind speed data, particularly during smooth wind speed changes and gradual increases. Additionally, during wind speed fluctuation periods, GE-GIN-GRU leverages its nonlinear fitting ability to approximate the wind speed more accurately than the continuous method.

4. Discussion

To delve deeper into the impact of network graph alterations on wind speed prediction outcomes within graph neural networks, this study will explore two primary dimensions: the variations in nodes within the input graph networks and the modifications in edges within the input graph networks.

4.1. *Effect of the Different Nodes on Graph Networks*

In this study, we construct the graph nodes using the primary site features obtained through PCA-RF processing. We focus exclusively on the principal features identified after

RF processing as the graph's nodes and deliberately omit the residual feature matrix that is obtained following PCA dimensionality reduction for our comparative analysis. Tables 5 and 6 present the error metrics for wind speed predictions, following the application of different feature selection methods to the nodes.

Table 5. Different feature selection methods are utilized to predict the wind speed error metrics at 10 m height.

Error Metrics	RF	PCA-RF
MSE	2.5422	0.8457
RMSE	1.5944	0.9196
MAE	1.4224	0.7612
MAPE	0.1151	0.0636

Table 6. Different feature selection methods are utilized to predict the wind speed error metrics at 30 m height.

Error Metrics	RF	PCA-RF
MSE	0.6945	0.6400
RMSE	0.8334	0.8000
MAE	0.6354	0.6076
MAPE	0.0510	0.0503

The distinction between the RF and PCA-RF models is evident in the diminished quantity of nodes and edges within the graph networks, as presented in Figure 10. To ascertain the influence of varying nodes, we modulate the threshold to align the number of edges as closely as possible. Our empirical observations reveal that both models exhibit parallel accuracy levels in predicting wind speeds at an altitude of 30 m. However, at a 10-m elevation, the performance of the RF model is significantly inferior to that of the PCA-RF model. This discrepancy may stem from the presence of nodes in the residual feature matrix, obtained through the PCA dimensionality reduction, which significantly enhances the precision of wind speed predictions at the 10-m mark. Conversely, the RF model which relies solely on the primary feature matrix yields suboptimal predictions due to an inadequate feature set.

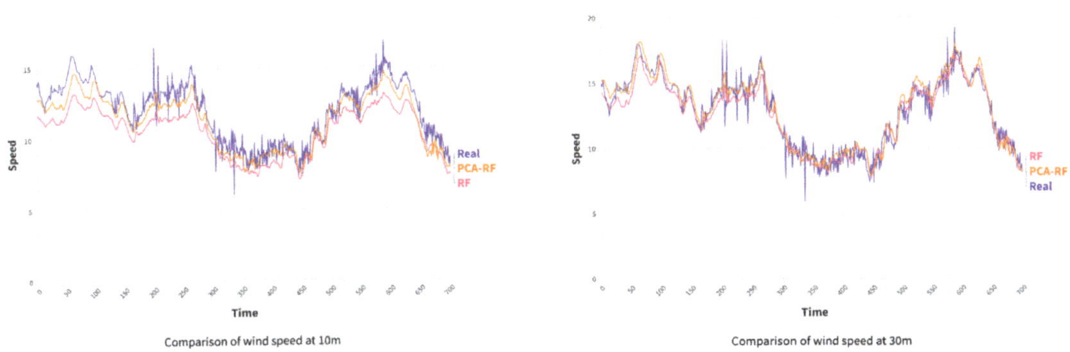

(**a**) wind speed prediction at 10 m height (**b**) wind speed prediction at 30 m height

Figure 10. Comparison between PCA-RF feature screening and RF feature screening.

4.2. *Effect of the Different Edges on Graph Networks*

The construction of additional edges within the input graph networks do not necessarily correlate with the improvement of prediction accuracy. By varying the threshold value, we manipulate the number of edges via graph embedding and the predictive outcomes are expressed as Table 7.

Table 7. The impact of varying the number of network edges on the error metrics of wind speed prediction.

Error Metrics	GE-GIN-GRU (362)	GE-GIN-GRU (1724)	GE-GIN-GRU (3124)
MSE	1.6184	0.8457	1.4014
RMSE	1.2722	0.9196	1.1838
MAE	1.1254	0.7612	1.0025
MAPE	0.0924	0.0636	0.0857

The experimental data suggest that the network achieves peak predictive accuracy when it maintains a balanced number of edges, as illustrated in Figure 11. A paucity of edges compromises accuracy due to the neural network's deficiency in requisite information for effective learning. On the other hand, an overabundance of edges leads to computational inefficiency by introducing unnecessary data, which does not translate to improved accuracy.

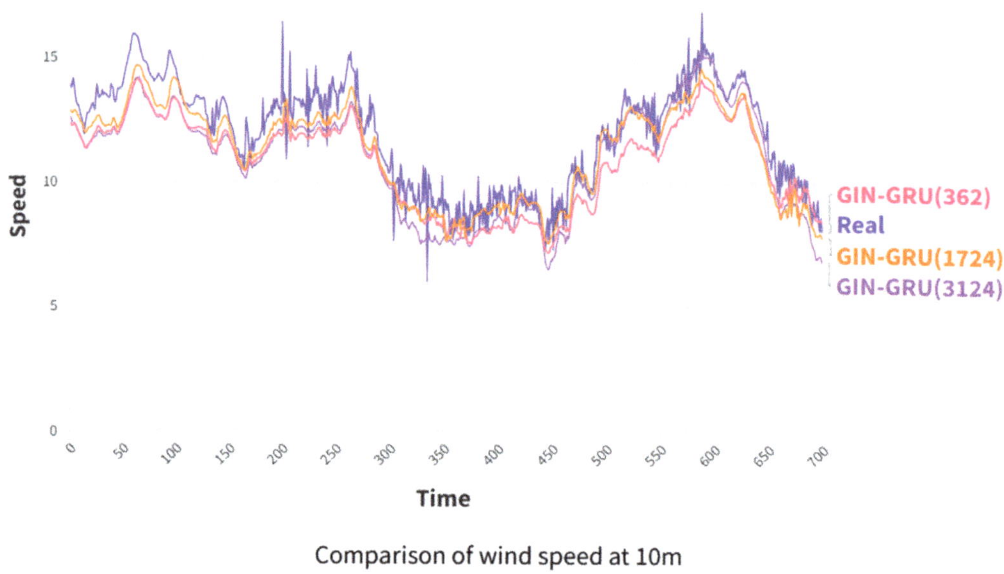

Figure 11. Comparison of varying the number of network edges on the error metrics of wind speed prediction.

Furthermore, the capacity for information processing and filtration varies across graph networks constructed via disparate methodologies. Notably, when a graph network constructed via correlation amasses 1000 edges, there is a significant decline in the accuracy of neural network processing. Comparatively, when the correlation and graph embedding models are used to construct graph networks with an equivalent number of edges, the latter outperforms the former in terms of quality. This superiority is attributed to the graph embedding model's advanced proficiency in feature extraction, understanding of the dataset's topology, and advanced analysis of node relationships, which facilitates the construction of a network graph optimally suited for graph neural network learning. Moreover, the methods by which various neural networks process graphical networks differ significantly. GIN exhibits superior performance over other graph neural networks in managing intricate network edges, which translates to enhanced prediction accuracy.

5. Conclusions

This paper presents the GE-GIN-GRU model, which synthesizes graph embeddings and neural network techniques to streamline wind speed prediction for stations within the study area. The proposed PCA-RF algorithm effectively reduces the number of features involved in the computation and improves the computational efficiency of the model. Subsequently, we employed deep learning-based graph embedding techniques to construct graph networks that capture the interrelationships among the sites. Our graph embedding methods capitalize on the strengths of both GraphSAGE and the Mahalanobis distance. The former excels at extracting intricate connections within wind speed features, forming feature vectors. Meanwhile, the latter demonstrates advantages in processing high-dimensional feature vectors. By fully leveraging the strengths of these two methods, we construct optimized graph networks. The GIN-GRU model seamlessly integrates diverse neural network algorithms to enhance generalization capabilities and improve prediction accuracy. Consequently, it consistently maintains excellent prediction quality and stability across wind speed datasets at varying heights. By fully leveraging the strengths of both models, we achieve deep extraction of spatio-temporal relationship features. In subsequent research, our primary goal is to construct new graph networks by combining the location coordinates of the sites with their corresponding nodes.

Author Contributions: Conceptualization, H.W. and H.C.; methodology, H.W.; resources, H.W.; writing—original draft preparation, H.W.; writing—review and editing, H.W. and H.C.; funding acquisition, H.W. and H.C. All authors have read and agreed to the published version of the manuscript.

Funding: This research received no external funding.

Data Availability Statement: The wind field data were generated using the Wind Toolkit API (version 2) provided by the National Renewable Energy Laboratory (NREL). Available at https://developer.nrel.gov/docs/wind/wind-toolkit/ (accessed on 15 January 2024).

Conflicts of Interest: The authors declare no conflicts of interest.

References

1. Valdivia-Bautista, S.M.; Domínguez-Navarro, J.A.; Pérez-Cisneros, M.; Vega-Gómez, C.J.; Castillo-Téllez, B. Artificial Intelligence in Wind Speed Forecasting: A Review. *Energies* **2023**, *16*, 2457. [CrossRef]
2. Chandra, D.R.; Kumari, M.S.; Sydulu, M. A detailed literature review on wind forecasting. In Proceedings of the 2013 International Conference on Power, Energy and Control (ICPEC), Dindigul, India, 6–8 February 2013; pp. 630–634. [CrossRef]
3. Zhu, C.; Zhu, L. Wind Speed Short-Term Prediction Based on Empirical Wavelet Transform, Recurrent Neural Network and Error Correction. In *Journal of Shanghai Jiaotong University (Science)*; Springer: Berlin/Heidelberg, Germany, 2022; pp. 1–12.
4. Bokde, N.; Feijóo, A.; Villanueva, D.; Kulat, K. A review on hybrid empirical mode decomposition models for wind speed and wind power prediction. *Energies* **2019**, *12*, 254. [CrossRef]
5. Zhang, Y.; Zhang, C.; SUN, J.B.; Guo, J.J. Improved Wind Speed Prediction Using Empirical Mode Decomposition. *Adv. Electr. Comput. Eng.* **2018**, *18*, 3–10. [CrossRef]
6. Ren, Y.; Suganthan, P.N. Empirical mode decomposition-k nearest neighbor models for wind speed forecasting. *J. Power Energy Eng.* **2014**, *2*, 176–185. [CrossRef]
7. Wang, L.; Liao, Y. A short-term hybrid wind speed prediction model based on decomposition and improved optimization algorithm. *Front. Energy Res.* **2023**, *11*, 1298088. [CrossRef]
8. Qu, Z.; Hou, X.; Hu, W.; Yang, R.; Ju, C. Wind power forecasting based on improved variational mode decomposition and permutation entropy. *Clean Energy* **2023**, *7*, 1032–1045. [CrossRef]
9. Mohapatra, M.R.; Radhakrishnan, R.; Shukla, R.M. A Hybrid Approach using ARIMA, Kalman Filter and LSTM for Accurate Wind Speed Forecasting. *arXiv* **2023**, arXiv:2311.08550.
10. Che, Y.; Xiao, F. An integrated wind-forecast system based on the weather research and forecasting model, Kalman filter, and data assimilation with nacelle-wind observation. *J. Renew. Sustain. Energy* **2016**, *8*, 053308. [CrossRef]
11. Xu, H.; Chang, Y.; Zhao, Y.; Wang, F. A novel hybrid wind speed interval prediction model based on mode decomposition and gated recursive neural network. *Environ. Sci. Pollut. Res.* **2022**, *29*, 87097–87113. [CrossRef]

12. Ai, X.; Li, S.; Xu, H. Wind speed prediction model using ensemble empirical mode decomposition, least squares support vector machine and long short-term memory. *Front. Energy Res.* **2023**, *10*, 1043867. [CrossRef]
13. Shao, B.; Song, D.; Bian, G.; Zhao, Y. Wind speed forecast based on the LSTM neural network optimized by the firework algorithm. *Adv. Mater. Sci. Eng.* **2021**, *2021*, 4874757. [CrossRef]
14. Zhu, Q.; Chen, J.; Zhu, L.; Duan, X.; Liu, Y. Wind speed prediction with spatio–temporal correlation: A deep learning approach. *Energies* **2018**, *11*, 705. [CrossRef]
15. Trebing, K.; Mehrkanoon, S. Wind speed prediction using multidimensional convolutional neural networks. In Proceedings of the 2020 IEEE symposium series on computational intelligence (SSCI), Canberra, ACT, Australia, 1–4 December 2020; IEEE: Piscataway, NJ, USA, 2020; pp. 713–720.
16. Tuerxun, W.; Xu, C.; Guo, H.; Guo, L.; Zeng, N.; Cheng, Z. An ultra-short-term wind speed prediction model using LSTM based on modified tuna swarm optimization and successive variational mode decomposition. *Energy Sci. Eng.* **2022**, *10*, 3001–3022. [CrossRef]
17. Yuan, M.M.; Gong, F.M.; Li, X. Multifactor Spatio-Temporal Wind Speed Prediction Based on CNN-LSTM. *Comput. Syst. Appl.* **2021**, *30*, 133–141. (In Chinese)
18. Louppe, G. Understanding random forests: From theory to practice. *arXiv* **2014**, arXiv:1407.7502.
19. Li, X.; Wang, Y.; Basu, S.; Kumbier, K.; Yu, B. A debiased MDI feature importance measure for random forests. *Adv. Neural Inf. Process. Syst.* **2019**, *32*, 713–720.
20. Abdi, H.; Williams, L.J. Principal component analysis. *Wiley Interdiscip. Rev. Comput. Stat.* **2010**, *2*, 433–459. [CrossRef]
21. Zhou, J.; Cui, G.; Hu, S.; Zhang, Z.; Yang, C.; Liu, Z.; Sun, M. Graph neural networks: A review of methods and applications. *AI open* **2020**, *1*, 57–81. [CrossRef]
22. Waikhom, L.; Patgiri, R. Graph neural networks: Methods, applications, and opportunities. *arXiv* **2021**, arXiv:2108.10733.
23. Cai, H.; Zheng, V.W.; Chang, K.C.C. A comprehensive survey of graph embedding: Problems, techniques, and applications. *IEEE Trans. Knowl. Data Eng.* **2018**, *30*, 1616–1637. [CrossRef]
24. De Maesschalck, R.; Jouan-Rimbaud, D.; Massart, D.L. The mahalanobis distance. *Chemom. Intell. Lab. Syst.* **2000**, *50*, 1–18. [CrossRef]
25. Leys, C.; Klein, O.; Dominicy, Y.; Ley, C. Detecting multivariate outliers: Use a robust variant of the Mahalanobis distance. *J. Exp. Soc. Psychol.* **2018**, *74*, 150–156. [CrossRef]
26. Goyal, P.; Ferrara, E. Graph embedding techniques, applications, and performance: A survey. *Knowl. Based Syst.* **2018**, *151*, 78–94. [CrossRef]
27. Ju, W.; Fang, Z.; Gu, Y.; Liu, Z.; Long, Q.; Qiao, Z.; Zhang, M. A comprehensive survey on deep graph representation learning. *Neural Netw.* **2024**, *173*, 106207. [CrossRef] [PubMed]
28. Jiang, W. Graph-based deep learning for communication networks: A survey. *Comput. Commun.* **2022**, *185*, 40–54. [CrossRef]
29. Kim, B.H.; Ye, J.C. Understanding graph isomorphism network for rs-fMRI functional connectivity analysis. *Front. Neurosci.* **2020**, *14*, 545464. [CrossRef]
30. Chen, Z.; Villar, S.; Chen, L.; Bruna, J. On the equivalence between graph isomorphism testing and function approximation with gnns. *Adv. Neural Inf. Process. Syst.* **2019**, *32*.
31. Chung, J.; Gulcehre, C.; Cho, K.; Bengio, Y. Empirical evaluation of gated recurrent neural networks on sequence modeling. *arXiv* **2014**, arXiv:1412.3555.
32. Dey, R.; Salem, F.M. Gate-variants of gated recurrent unit (GRU) neural networks. In Proceedings of the 2017 IEEE 60th international midwest symposium on circuits and systems (MWSCAS), Boston, MA, USA, 6–9 August 2017; IEEE: Piscataway, NJ, USA, 2017; pp. 1597–1600.
33. Chicco, D.; Warrens, M.J.; Jurman, G. The coefficient of determination R-squared is more informative than SMAPE, MAE, MAPE, MSE and RMSE in regression analysis evaluation. *Peerj Comput. Sci.* **2021**, *7*, e623. [CrossRef]
34. Aykas, D.; Mehrkanoon, S. Multistream graph attention networks for wind speed forecasting. In Proceedings of the 2021 IEEE Symposium Series on Computational Intelligence (SSCI), Orlando, FL, USA, 5–7 December 2021; IEEE: Piscataway, NJ, USA, 2021; pp. 1–8.
35. Flores, A.; Tito-Chura, H.; Yana-Mamani, V. An ensemble GRU approach for wind speed forecasting with data augmentation. *Int. J. Adv. Comput. Sci. Appl.* **2021**, *12*, 569–574. [CrossRef]

energies

MDPI

Article

Adaptability Evaluation of Power Grid Planning Scheme for Novel Power System Considering Multiple Decision Psychology

Yuqing Wang [1,2,*], Chaochen Yan [2], Zhaozhen Wang [2] and Jiaxing Wang [2]

[1] State Key Laboratory of Alternate Electrical Power System with Renewable Energy Sources, North China Electric Power University, Beijing 102206, China

[2] Department of Economic Management, North China Electric Power University, Baoding 071000, China; ycc@ncepu.edu.cn (C.Y.); 220232218003@ncepu.edu.cn (Z.W.); 220222218095@ncepu.edu.com (J.W.)

* Correspondence: yuqingwang@ncepu.edu.cn

Abstract: With a substantial fraction of renewable energy integrated into the electrical grid, the new power system urgently requires grid planning scheme displaying adaptability to different energy types and their volatility. Considering the indeterminacy of renewable energy generation output and the different attitudes of decision-makers towards its risk, this paper proposes an adaptability assessment methodology for power grid planning schemes considering multiple decision psychology. First, an evaluation indicator framework is established based on the adaptive requirements of the grid planning for novel power system, and the weights of indicators are calculated based on an improved AHP-CRITIC combination weighting method. Second, improved cumulative prospect theory (ICPT) is adopted to improve to the calculation method of the distance between the evaluation program and the positive and negative ideal programs in the GRA and TOPSIS, which effectively characterize the different decision-making psychologies, and a combination evaluation model is constructed based on a cooperative game (CG), namely, an adaptability evaluation model of grid planning schemes for novel power systems based on GRA-TOPSIS integrating CG and ICPT. Finally, the proposed model serves to evaluate grid planning schemes of three regions in China's 14th Five-Year Plan. The evaluation results show that the adaptability of the schemes varies under different decision-making psychologies, and under the risk-aggressive and loss-sensitive decision-making psychologies, grid planning scheme of Region 1 with the greatest accommodation capacity of renewable energy is preferable.

Keywords: novel power system; adaptability evaluation; grid planning; decision psychology; combination evaluation

Citation: Wang, Y.; Yan, C.; Wang, Z.; Wang, J. Adaptability Evaluation of Power Grid Planning Scheme for Novel Power System Considering Multiple Decision Psychology. *Energies* **2024**, *17*, 3672. https://doi.org/10.3390/en17153672

Academic Editor: Javier Contreras

Received: 31 May 2024
Revised: 8 July 2024
Accepted: 22 July 2024
Published: 25 July 2024

1. Introduction

With the increasing energy crisis and environmental pollution, China is committed to constructing a novel power system mainly composed of new energy to promote the clean and low-carbon transformation of energy [1]. In the future, the substantial integration of renewable energy generation will emerge as a pivotal characteristic of the novel power system, but its stochastic and intermittent nature will pose a serious challenge to the existing power grid [2,3]. In view of the central position of the power grid in power transmission and distribution, the adaptability of its planning scheme is decisive for ensuring the stable operation and economic benefits of the novel power system [4,5]. Therefore, there is an immediate need to assess the adaptability of the grid planning approaches for novel power system to efficiently identify the weaknesses, rationally construct the grid, and offer guidance for ensuring the system's future economic and stable functioning [6].

Recently, academics have conducted extensive research on evaluating the adaptability of grid planning schemes for emerging power systems. In [7], the authors pointed out that the grid planning for a novel power system should adapt to the new challenges brought by the development of new energy sources, and more attention should be paid to the adaptability of different energy types and the volatility of different power sources. In [8], the authors believed that upon the interconnection of a substantial proportion of renewable energy sources to the grid, a comprehensive assessment of the grid planning scheme's adaptability must encompass both the grid's intrinsic properties and its interactions with external factors. This constituted a vital aspect in evaluating and appraising the grid's construction quality, serving as a feedback mechanism for enhancing subsequent construction quality and operational performance of the grid planning.

In assessing the adaptability of distribution network planning, the technical indicators for instance capacity expansion margin, power supply capacity margin, and expandability were proposed to construct an evaluation index system, and then AHP methodology was employed to derive the comprehensive score of the planning scheme [9]. In [8], some technical indicators, for instance, load ratio, current, power quality, operating life, and new energy utilization rate, were considered to construct an adaptability evaluation index system of distribution network planning schemes; entropy weight and the AHP method were adopted to calculate weight; then, TOPSIS was used to construct the evaluation model. In [10], the complexity of the grid structure after the significant integration of renewable energy sources was considered, and for the planning of distribution network under the big data environment, the adaptability evaluation index system was established by selecting technical indicators from five perspectives: grid structure, power supply capacity, equipment level, load characteristics, and grid integration of new elements; then, the planning scheme was evaluated employing the back propagation neural network (BNPP) method. On this basis, in [11], the authors considered the economic and environmental benefits brought about by the substantial integration of renewable energy sources into the grid, proposed an evaluation index system of distribution grid planning adaptability, which contains equipment operation status, power supply reliability, economy, and environmental friendliness, and adopted a variety of empowerment methods for combined empowerment; evaluation results were obtained through the AHP method.

In assessing the adaptability of transmission grid planning, in [12], the efficiency benefits of transmission grid planning scheme were considered to propose the evaluation index system of system scale, development, and environment. Then G1 method and GRA method were adopted for combination assignment, and the fuzzy comprehensive evaluation approach was employed to derive the assessment outcomes. In [13], the evaluation index system was formulated encompassing the dimensions of economy, technology, reliability, and new energy acceptance capacity, and the "Over-Average Penalty" entropy weight method was employed to allocate weights to the indicators; then, a comprehensive evaluation method based on high penetration of new energy (HPNE) was proposed. Moreover, in [14], the impact of flexibility and vulnerability on the grid planning scheme was considered, and the IFAHP method was adopted for the assignment of indicators and comprehensive evaluation.

In summary, the current study exhibits the following weaknesses: (1) mainly, studies are focused separately on the adaptability evaluation of the transmission or distribution network, while few studies consider the coordination factors among all levels of a power grid, and the new needs generated by the grid planning for a novel power system, to construct the evaluation index system; (2) in the process of grid planning for novel power system, due to the significant uncertainty in the renewable energy power, it is necessary to fully consider the different attitudes of decision-makers towards risk when constructing the evaluation model. However, current studies generally ignore the influence of decision-making psychological factors on the evaluation results, and tend to use a single evaluation model, which may lead to a large limitation in the results. Thus, it is difficult to comprehensively reflect the complexity of the real decision-making environment, which may lead to bias or conflict in the practical application.

Therefore, this paper introduces a methodology for assessing the adaptability of power grid planning schemes for a novel power system considering multiple decision psychology, and the key contributions of this paper are outlined as follows:

(1) A comprehensive evaluation index system of power grid planning adaptability is established, which comprehensively addresses the emerging requirements of grid planning for the novel power system, including economy adaptability, energy structure adaptability, power grid structure adaptability, reliability adaptability, and environment adaptability.
(2) An improved cumulative prospect theory (ICPT) is introduced into the evaluation model to effectively characterize the different decision-making psychologies, which enhances the adaptability to the uncertainty of renewable energy and makes the evaluation results more realistic.
(3) A combination evaluation method based on a cooperative game (CG) is constructed, fully contributing to the advantages of different evaluation models to make the evaluation results fairer.

The subsequent sections of this paper are structured as follows: In Section 2, the adaptive requirement of grid adaptive planning for novel power system are analyzed. Section 3 describes the adaptability evaluation index system of power grid planning scheme for novel power system. Section 4 presents the weighting method and adaptability evaluation model of grid planning scheme for novel power system based on GRA-TOPSIS integrating CG and ICPT. Section 5 presents case studies and comparative analysis of the results. Lastly, Section 6 concludes this study.

2. Adaptive Requirements of Grid Planning for Novel Power System

In the context of developing a novel power system characterized by a high integration of renewable energy sources, the characteristics of the power grid have undergone tremendous changes: from passive network to active network, from one-way to two-way power flow, from pure consumption to both production and consumption, from rigid demand to adjustable and controllable, and from source–network coordination to source–network–load–storage coordination [15]. Therefore, due to the transformation of the distinctive features of the power grid, power grid planning is facing many new demands, as shown in Figure 1:

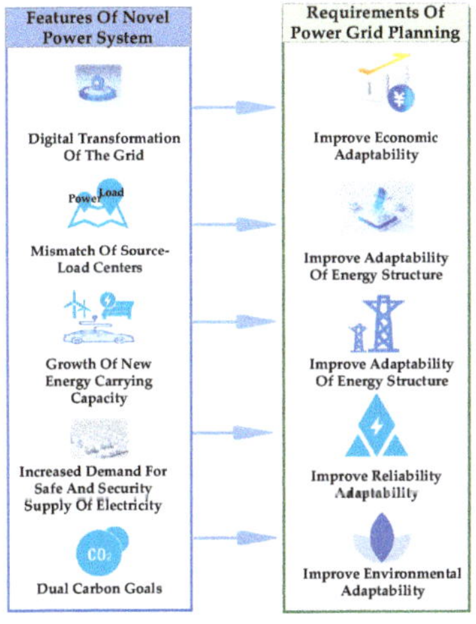

Figure 1. Adaptive demand of grid planning for novel power system.

2.1. *Adaptive Requirements for Economic*

To propel the transformation and upgrading of the power system, the power grid needs to accelerate its digital transformation urgently. This will enable it to optimize the allocation of multiple factors, effectively fulfill its platform role, and facilitate the in-depth development of the energy revolution [16]. In this process, the first step in power grid planning is to focus on the economic aspect of each link, including sourcing, networking, loading, and storage. It is crucial to fully consider the impact of various power sources, electric vehicles, and energy storage on load forecasting. This comprehensive approach enhances the overall planning concept, shifting the focus from the main power grid to the broader power system extension. The second consideration is to anticipate the increase in electricity demand and load, implementing proactive planning and construction of the power grid. This foresighted approach ensures efficient scalability and adaptability of the grid network. Furthermore, enhancing the deployment of intelligent terminals and improving the distribution communication network are essential. These actions boost the observability, measurability, adjustability, and controllability of the power grid, thereby enhancing its internal rate of return. Ultimately, this translates into a reduction in the overall lifecycle expenditure of power grid development.

2.2. Adaptive Requirements for Energy Structure

In the novel power system, non-fossil energy sources, notably hydroelectric, wind, and solar power, will progressively emerge as the primary sources of installed capacity and electricity generation [17]. Nevertheless, China faces a persistent challenge of mismatched distribution between clean energy resources and demand, with water resources concentrated in the southwest, wind and solar resources predominantly in the "Three-North" regions, and electricity demand heavily skewed towards the eastern, central, and southern regions [18]. Ultrahigh-voltage transmission emerges as a pivotal solution for facilitating long-distance, large-scale power transmission, thereby enhancing new energy integration capabilities and mitigating wind and solar curtailment issues [19]. As new energy sources integrate on a massive scale, cross-regional power transmission will inevitably escalate. Consequently, power grid planning necessitates the development of ultra-high-voltage and various levels of power grids to enhance the power grid's capacity to accept new energy sources.

2.3. Adaptive Requirements for Grid Structure

Under the "Peak Carbon and Carbon Neutral" goal, the penetration rate of distributed power generation and the proportion of electricity to end energy consumption will continue to increase [20]. Emerging new loads, represented by electric vehicles, will scale up significantly, leading to the normalization of the integrated production and sales model. As new energy sources are increasingly integrated into the power grid on a large scale, issues such as equipment overload and declining power quality are gradually becoming prominent, placing higher demands on the substation capacity. When planning the grid, the flexible coordination between the transmission and distribution networks should be considered, and the potential for transformation and intelligent upgrade of grid construction should be enhanced at the planning stage to facilitate the harmonious integration of large-scale new energy sources with the power grid [21]. At the same time, substation full stop and turn rate should be improved. When the power outage is caused by special circumstances, the power grid structure should respond positively enough to transfer the power load to other normally operating substation buses and quickly restore power supply.

2.4. Adaptive Requirements for Reliability

Considering economic globalization, the establishment of a contemporary industrial system is speeding up, the eco-friendly transition of conventional industries is picking up pace, and the high-tech manufacturing sector is gradually emerging as the primary catalyst for progress. As people's expectations for an improved quality of life rise, there arises a critical requirement for power grid design to holistically enhance the electricity supply quality [22]. Moreover, the substantial share of renewable energy power fluctuations transforms the initial unidirectional random demand alteration system into a bidirectional random modification system, resulting in challenges such as diminished inertia and insufficient voltage support capabilities. To safeguard grid security and elevate power quality standards, it is imperative to augment investments in reliability during the planning phase.

2.5. Adaptive Requirements for Environment

Traditional power grid planning aims to enhance the economic efficiency of the system while meeting specific stability and reliability criteria. Under the "Peak Carbon and Carbon Neutral" goal, power grid planning should prioritize safety and environmental friendliness. Building upon the existing standards for reliability and cost-effectiveness, power grid planning must now incorporate heightened environmental protection measures. The focus of planning and design needs to shift from solely ensuring safety to achieving a balance between safety and sustainability.

3. Construction of Evaluation Index System

This paper adheres to the principles of scientific rigor, comprehensiveness, independence, applicability, and operability in the construction of the index system. An adaptability evaluation index system of power grid planning scheme for novel power system is established by integrating the adaptive requirement of grid adaptive planning for novel power system. This index system encompasses five key dimensions, including economic adaptability, energy structure adaptability, grid structure adaptability, reliability adaptability, and environment adaptability. The adaptability evaluation index system are shown in Table 1.

Table 1. Adaptability evaluation index system of power grid planning scheme for novel power system.

First-Level Indicators	Second-Level Indicators
Economic Adaptability (C_1)	Elasticity coefficient of power production (C_{11})
	Investment revenue expansion ratio (C_{12})
	Additional load capacity per unit investment (C_{13})
	Additional electricity supply per unit investment (C_{14})
Energy Structure Adaptability (C_2)	Proportion of clean energy (C_{21})
	Capacity to accommodate renewable energy (C_{22})
Grid Structure Adaptability (C_3)	Substation full stop and turn rate (C_{31})
	Capacity ratio of transformer (C_{32})
	Capacity expansion margin of substation (C_{33})
	Remaining interval ratio (C_{34})
	Line capacity-to-load ratio (C_{35})
	Line loss rate (C_{36})
Reliability Adaptability (C_4)	N-1 pass rate of power lines (C_{41})
	N-1 pass rate of transformers (C_{42})
	Voltage compliance rate (C_{43})
	Mean power supply reliability (C_{44})
Environment Adaptability (C_5)	CO_2 emission reduction (C_{51})
	NO_X emission reduction (C_{52})
	SO_2 emission reduction (C_{53})

3.1. Economic Adaptability

Economic adaptability refers to the adaptability and support capacity of a new power grid project to future economic development. The load growth rate of the region is directly affected by the level of local economic development. As the load increases, the maximum transmission capacity of the grid will be insufficient to supply the load demand. So, the capacity construction and economic cost of spare capacity for future development should be balanced in the grid planning, which enhance the grid's ability to resist uncertainties in the future. Therefore, this paper proposes that secondary indicators of economic adaptability include elasticity coefficient of power production, investment revenue expansion ratio, additional load capacity per unit investment, and additional electricity supply per unit investment [23].

3.2. Energy Structure Adaptability

In the assessment of power grid planning, energy structure adaptability constitutes a pivotal evaluation dimension, assessing the grid's capacity to accommodate shifts in energy composition. Amid the ongoing transformation of energy mix and the proliferation of renewable energy sources, grid planning necessitates a comprehensive consideration of clean energy's share within the energy structure, along with the grid's resilience to integrate new energy forms, ultimately ensuring grid stability and optimizing energy utilization efficiency. Therefore, this paper proposes that secondary indicators of energy structure adaptability include the proportion of clean energy and capacity to accommodate renewable energy.

3.3. Grid Structure Adaptability

The evaluation framework for energy grid structural adaptability primarily aims to secure that the grid can flexibly and effectively respond to various load changes, resource allocations, and emergencies. These measures guarantee a stable supply of electricity and improve a power supply reliability, thereby fostering optimal resource allocation and enabling sustainable grid development. Therefore, this paper proposes that secondary indicators of grid structure adaptability include substation full stop and turn rate, capacity ratio of transformer, capacity expansion margin of substation, remaining interval ratio, line capacity-to-load ratio, and line loss rate [24].

3.4. Reliability Adaptability

In the evaluation of grid planning, reliability adaptability is essential metric for measuring whether the grid system can provide power supply to customers in a continuous and stable manner when facing various uncertainties. The integration of large-scale renewable energy into the grid presents new challenges to the reliability of the grid, especially considering the randomness in both power supply and load during summer and winter load peaks. Therefore, this paper proposes that secondary indicators of reliability adaptability include the N-1 pass rate of power lines, N-1 pass rate of transformers, voltage compliance rate, and mean power supply reliability [8,25].

3.5. Environment Adaptability

As global attention to climate change and environmental protection continues to grow, the power sector, as one of the major areas of energy consumption and greenhouse gas emissions, bears an important responsibility for reducing environmental impacts. Considering environmental constraints, the impact of high-ratio renewable energy integration on mitigating grid-connected emissions must be assessed. This assessment is crucial for evaluating the planning scheme's compatibility with impending environmental demands. Therefore, this paper proposes that secondary indicators of environment adaptability include CO_2 emission reduction, NO_X emission reduction, and SO_2 emission reduction.

4. Construction of Evaluation Method Considering Multiple Decision Psychology

4.1. Framework of the Evaluation Method

The volatility, intermittency, and randomness in renewable energy output in a novel power system have a profound impact on grid planning. The risk attitude of decision-makers directly leads to the extent to which the grid planning scheme can adapt to the grid-connected capacity of renewable energy in advance. Therefore, this paper considers the impact of multiple psychological factors of decision-makers on evaluation results and expands the cumulative prospect theory into an improved method that includes multiple risk attitudes and multiple profit and loss attitudes. On this basis, in order to weaken the one-sidedness of the results caused by a single evaluation method, this paper combines the subjective and objective factors in the evaluation process and the impact of different measures on the evaluation results. The evaluation results of the GRA method are more subjective and based on geometric similarity measures, while the evaluation results of the

TOPSIS method are more objective and based on distance similarity. We combined these two methods with ICPT to build two different evaluation models, namely, ICPT-GRA and ICPT-TOPSIS. Then, in order to ensure the fairness and rationality of the evaluation results, the results of the two evaluation models were scientifically coupled based on the ideas of CG and overall difference maximization. In addition, in order to improve the accuracy of the weights, this paper considers the impact of subjective and objective factors on the weights and adopts an improved AHP and CRITIC indicator combination weighting method. Based on the above, this paper proposes an adaptability evaluation method of grid planning scheme for a novel power system considering multiple decision psychology.

Figure 2 shows the flow chart of the adaptability evaluation method of the grid planning scheme for a novel power system considering multiple decision psychology. Firstly, the subjective weight of the improved AHP method is obtained based on expert scores, the objective weight of the CRITIC method is obtained based on the original evaluation data, and the optimal combination weight is obtained based on the minimum deviation combination weighting method. Secondly, the comprehensive prospect values are calculated based on the ICPT-GRA method and the ICPT-TOPSIS method. Finally, based on CG and overall difference maximization, the combined weight coefficient of the results of the two evaluation methods is obtained, and the combined comprehensive prospect value is calculated, which is the evaluation result.

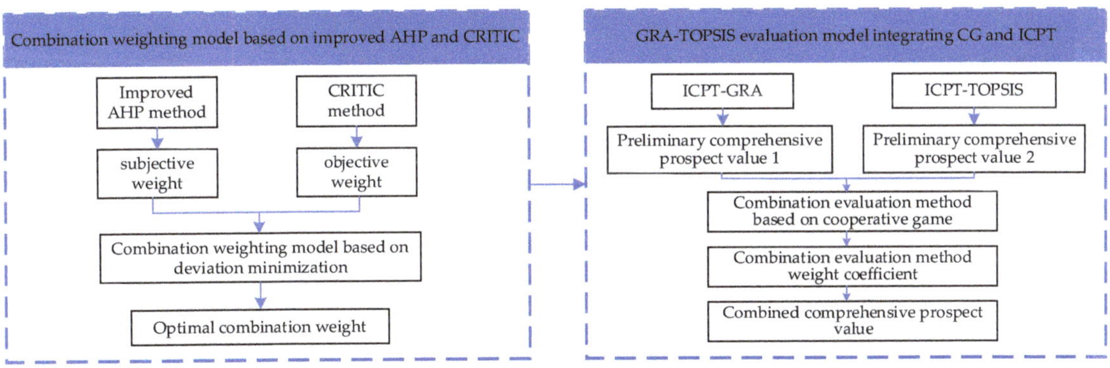

Figure 2. Flow chart of adaptability evaluation method of grid planning scheme for novel power system considering multiple decision psychology.

4.2. *Weighting Method Based on Improved AHP-CRITIC*

4.2.1. Subjective Weight Calculation Based on Improved AHP

At present, AHP is a method widely used in power grid planning evaluation, and mostly uses a 1–9 scale to weigh the significance between indicators to determine the weight value. However, in actual engineering, it is difficult for decision-makers to make such detailed distinctions between the differences in indicators, resulting in errors; and when calculating weights, decisions may not be made when the judgment matrix dissatisfies the consistency check. In response to the above problems, this paper proposes a method by assigning a three-scale value based on the significance of every two indicators to improve the judgment matrix. And by constructing a consistency matrix to omit the consistency check step, the calculation process is simplified, and the decision-making efficiency and accuracy are improved. The key steps for improvement are as follows:

(1) Improved judgment matrix

A three-scale evaluation method on the basis of the importance of elements is introduced in this paper, categorized as significant, equally significant, and insignificant. This approach requires only the comparison of whether elements are important, without the need to compare their relative importance. This makes for a more intuitive matrix construction, omitting the step of the consistency test of judgment matrix, and eventually simplifying subsequent calculations. Additionally, it becomes easier to determine the degree of importance between indicators. The specific steps of matrix construction are as follows:

(a) Experts select the importance of each indicator;
(b) Based on the opinions of experts, a judgment matrix $A = \left(a_{ij}\right)_{n\times n}$ is formed, with the following parameters:

$$a_{ij} = \begin{cases} 1, Element\ j\ is\ less\ significant\ than\ i \\ 0, The\ significance\ of\ element\ i\ and\ j\ is\ same \\ -1, Element\ j\ is\ more\ significant\ than\ i \end{cases} \tag{1}$$

where a_{ij} represents the value obtained from comparing the element i with the element j. When $i = j$, it is stipulated that holds the same level of importance when compared to itself, that is, $a_{ii} = 0$. a_{ii} is the comparison of the element itself. When $i \neq j$, the element j is less significant than the element i and the value is assigned as 1, otherwise as -1.

(2) Improve matrix consistency

One approach to determine the significance of each element in AHP is through the empirical method, but because people's understanding of things is subjective, it may fail to capture objective facts accurately. Traditional AHP requires consistency testing of the judgment matrix because of discontinuity in expert judgment on multi-indicator. If it is inconsistent, mathematical methods need to be used to adjust it, thereby increasing the computational complexity of the problem. However, if a consistent matrix can be constructed from the start, the consistency test can be omitted, allowing for the matrix to inherently satisfy the consistency requirements and thus simplifying the process of matrix calculation.

(3) Construct an antisymmetric matrix

Let there be n-order real matrices $A = \left[a_{ij}\right]_{n\times n}$ and $B = \left[b_{ij}\right]_{n\times n}$, where $i = 1, 2, \cdots, n$; $j = 1, 2, \cdots, n$.

Definition 1. *For a real matrix A, if $j = 1, 2, \cdots, n$, and there is always $a_{ij} = -a_{ji}$, then it is called an antisymmetric matrix A.*

Based on Equation (1) and incorporating expert opinions, the judgment matrices A is constructed. From definition 1, it can be seen that the matrix A must be an antisymmetric matrix, and the size of the matrix varies according to the number of indicators.

$$A = \begin{bmatrix} a_{11} & a_{12} & \cdots & a_{1j} & \cdots & a_{1n} \\ a_{21} & a_{22} & \cdots & a_{2j} & \cdots & a_{2n} \\ \vdots & \vdots & & \vdots & & \vdots \\ a_{i1} & a_{i} & \cdots & a_{ij} & \cdots & a_{in} \\ \vdots & \vdots & & \vdots & & \vdots \\ a_{n1} & a_{n2} & \cdots & a_{ij} & \cdots & a_{nn} \end{bmatrix} \tag{2}$$

(4) Solve the optimal transfer matrix

Definition 2. *If the antisymmetric matrix A satisfies $a_{ij} = a_{ik} + a_{kj}$, then the matrix A must be a transfer matrix. If matrix A is a transfer matrix, it is necessary to fulfill the above condition for all k less than or equal to the dimension of matrix. Among them, a_{ik} a is the element in ith row and kth column of matrix A, and a_{kj} is the element in kth row and j kth column of matrix A.*

Definition 3. *For transfer matrix A, if B is the optimal transfer matrix of A, then $\sum_{i=1}^{n}\sum_{j=1}^{n}\left(b_{ij} - a_{ij}\right)$ must obtain the minimum value, where b_{ij} is the element in row ith row and column jth column of the transfer matrix B.*

Theorem 1. *If A is an antisymmetric matrix, then the optimal transfer matrix B satisfies the following:*

$$b_{ij} = \frac{1}{n}\sum_{k=1}^{n}\left(a_{ik} - a_{jk}\right) \tag{3}$$

Reasoning 1. *Because of the property of antisymmetric matrix A, the optimal transfer matrix B must satisfy the following:*

$$b_{ij} = \frac{1}{n}\sum_{k=1}^{n}\left(a_{ik} + a_{kj}\right) \tag{4}$$

From Theorem 1 or Reasoning 1, the optimal transfer matrix of B can be obtained:

$$B = \begin{bmatrix} b_{11} & b_{12} & \cdots & b_{1j} & \cdots & b_{1n} \\ b_{21} & b_{22} & \cdots & b_{2j} & \cdots & b_{2n} \\ \vdots & \vdots & & \vdots & & \vdots \\ b_{i1} & b_{i2} & \cdots & b_{ij} & \cdots & b_{in} \\ \vdots & \vdots & & \vdots & & \vdots \\ b_{n1} & b_{n2} & \cdots & b_{ij} & \cdots & b_{nn} \end{bmatrix} \tag{5}$$

where $b_{ij} = \frac{1}{n}\sum_{k=1}^{n}\left(a_{ik} - a_{jk}\right) = \frac{1}{n}\sum_{k=1}^{n}\left(a_{ik} + a_{kj}\right)$.

(5) Solve the consistency matrix

Definition 4. *For matrix A, if* $\forall i, j, k \in N$, *there is* $a_{ik}a_{kj} = a_{ij}$, *which is called a completely consistent matrix A.*

Reasoning 2. *For the antisymmetric matrix A, if matrix B is an optimal transfer matrix of A, when* $A^* = e^B$, A^* *is a completely consistent matrix of A.*

It can be derived from Theorem 2 that the matrix B can be converted into a completely consistent matrix A^*.

$$A^* = \begin{bmatrix} a^*_{11} & a^*_{12} & \cdots & a^*_{1j} & \cdots & a^*_{1n} \\ a^*_{21} & a^*_{22} & \cdots & a^*_{2j} & \cdots & a^*_{2n} \\ \vdots & \vdots & & \vdots & & \vdots \\ a^*_{i1} & a^*_{i2} & \cdots & a^*_{ij} & \cdots & a_{in} \\ \vdots & \vdots & & \vdots & & \vdots \\ a^*_{n1} & a^*_{n2} & \cdots & a^*_{ij} & \cdots & a^*_{nn} \end{bmatrix} \tag{6}$$

where $a^*{}_{ij}$ is the element in ith row and jth column of A^*, $a^*_{ij} = \exp(b_{ij})$. A^* is the completely consistency matrix of A, which satisfies consistency requirements and guarantees the information of A to the maximum extent.

(6) Calculate the weight value

The weight value of indicators signifies the significance of elements in this layer relative to the previous layer. Determining these values can be simplified to computing the principal eigenvalues and eigenvectors of the matrix. The eigenvectors of the consistency matrix A^*, which corresponds to the eigen roots, must satisfy $A^*W = \lambda W$. In this equation, A^* is the eigenvector; λ is the eigen root. This paper identifies the eigenvectors associated with the largest eigenvalues through the method of the square root, and the specific steps are as follow:

(a) The nth root of the product for the elements in each row of A^* are calculated.

$$\overline{W}_i = \sqrt[n]{\prod_{j=1}^{n} a_{ij}^*},\ i = 1, 2, \cdots, n; j = 1, 2, \cdots, n \tag{7}$$

where $\overline{W_i}$ is the nth root of the product of the elements of the i row.

(b) The values of the elements of each row are processed to the nth power and recorded as vectors

$$\overline{W} = \left[\overline{W}_1, \overline{W}_2, \cdots, \overline{W}_n\right]^{\mathrm{T}} \tag{8}$$

(c) Through the step of normalizing $\overline{W}$, the weight obtained:

$$W_i = \frac{\overline{W}_i}{\sum\limits_{j=1}^{n} \overline{W}_j} \tag{9}$$

where $\overline{W_j}$ is for the elements of the jth column of the nth root of the product

$$W = \left[W_1, W_2, \cdots, W_n\right]^{\mathrm{T}} \tag{10}$$

W in Equation (10) is the eigenvector which is corresponding to the maximum eigenvalue λ, that is, the eventual weight value.

4.2.2. Objective Weight Calculation Based on CRITIC

This paper uses the CRITIC method to calculate the objective weight of indicators based on the amount of information in the indicator data [26]. When weights are determined by this method, not only is the amount of information contained in the indicator considered, but the contrast and the conflict between different solutions and indicators are also regarded. Therefore, the results are more objective and reasonable.

4.2.3. Combination Weight Calculation Based on Deviation Minimization

Using a single method of subjective empowerment or objective empowerment will lead to differences and defects in evaluation results. The objective function aims to minimize the sum of the squares of the differences between "the deviation between the improved AHP weight and the combination weight" and "the CRITIC weight and the deviation between the combination weight". The combination weight and the subjective and objective weights are solved, respectively, when the sum of squares of deviations is minimized and the optimal combined weight result is solved, in which the minimization problem is solved with respect to the variable β.

$$\min z = \sum_{i=1}^{m} \left[(u_i - W_i)^2 + (u_i - V_i)^2\right] \tag{11}$$

$$u_{\mathrm{i}} = \beta W_{\mathrm{i}} + (1 - \beta) V_i \tag{12}$$

Among them, u_i is the comprehensive weight of the *ith* indicator after combining the two weighting methods that are represented as a linear combination of W_i and V_i; β is the proportion of the subjective preference coefficient weight in the combination weight; W_i is the improved analytic hierarchy process weight of the *ith* indicator; $1 - \beta$ is the proportion of the objective preference coefficient in the combination weight; and V_i is the CRITIC weight of the *ith* indicator.

4.3. Adaptability Evaluation Model Based on GRA-TOPSIS Integrating CG and ICPT

4.3.1. ICPT Method

Cumulative prospect theory focuses on the irrational behavior of decision-makers and reflects bounded rational behavior by establishing a value function. However, the existing value function does not distinguish the risk preference type and profit and loss attitude of decision-makers [27]. This paper expands the value range of the decision-making risk preference coefficient; proposes an improved prospect value function for three risk attitudes, radical, balanced, and cautious; and adds parameters δ to adjust the decision-maker's outlook profit and loss attitude [28]. The details are as follows:

$$v(\Delta x) = \begin{cases} \delta(\Delta x)^{\alpha}, \Delta x \geqslant 0 \\ -\theta(-\Delta x)^{\beta}, \Delta x < 0 \end{cases} \tag{13}$$

where $v(\Delta x)$ is the prospect value; Δx is the difference between the evaluation plan value and the reference plan value -1~1, which is the value under standard circumstances. If $\Delta x \geqslant 0$, then the prospect value is the income value $\Delta x \geqslant 0$; otherwise, it is the loss value v^{-}; $\alpha\beta$ are the parameters of risk attitude from different decision-maker; δ is the decision-maker's sensitivity coefficient to returns; θ is the sensitivity coefficient of the decision-maker to the loss.

The improved traditional prospect theory is shown in Figure 3. The value range of $\alpha\beta$ is expanded and decision-makers are divided into three types. If $0 < \alpha, \beta < 1$, then the decision-maker is a radical type; if $\alpha, \beta = 1$, then the decision-maker is a balanced type; if $\alpha, \beta > 1$, then the decision-maker is a cautious type. Traditional prospect theory is only a cautious decision-making model.

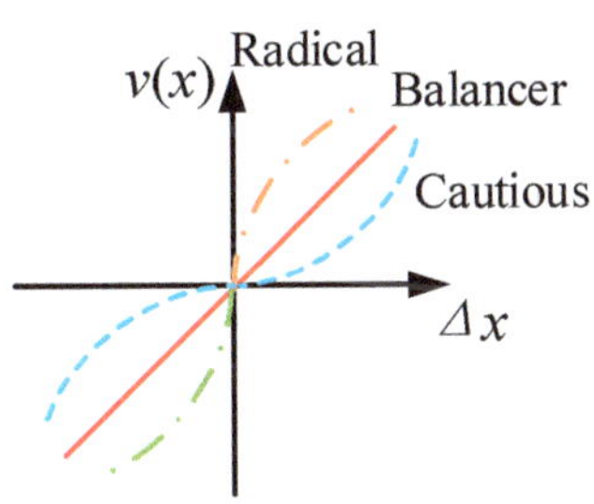

Figure 3. Improved prospect value function.

A new prospect value function parameter δ is added to adjust the decision-maker's attitude towards profit and loss. If the decision-maker is more sensitive to prospect losses than to prospect losses, then let $\delta > \theta = 1$; if the decision-maker is more sensitive to prospective losses than to prospective gains, then $\theta > \delta = 1$; if the decision-maker is equally sensitive to prospective gains and losses, then make $\theta = \delta = 1$.

Based on the improved prospect value function and using the cumulative functional to optimize the decision weight, the comprehensive prospect value of ICPT can be obtained:

$$V_j = \sum_{i=1}^{n} v_{ij}^{+} \pi^{+}(\omega_i) + \sum_{i=1}^{n} v_{ij}^{-} \pi^{-}(\omega_i) \tag{14}$$

where V_j is the comprehensive prospect value of the jth plan; $v_{ij}^{+} v_{ij}^{-}$ are, respectively, the positive and negative prospect values of the ith plan under the jth indicator; $\pi^{+}(\omega_i)$, $\pi^{-}(\omega_i)$ are, respectively, the decision weight functions of the positive and negative prospect values corresponding to the ith indicator weight ω_i; $i = 1,2,\ldots,n$, $j = 1,2,\ldots,m$. The decision weight function is as below:

$$\begin{cases} \pi^+(\omega_i) = \dfrac{\omega_i^{\gamma^+}}{\left[\omega_i^{\gamma^+} + (1-\omega_i)^{\gamma^+}\right]^{\frac{1}{\gamma^+}}} \\ \pi^-(\omega_i) = \dfrac{\omega_i^{\gamma^-}}{\left[\omega_i^{\gamma^-} + (1-\omega_i)^{\gamma^-}\right]^{\frac{1}{\gamma^-}}} \end{cases} \tag{15}$$

where $\gamma^+\gamma^-$ are the fitting parameters; usually, the values are $\gamma^+ = 0.61, \gamma^- = 0.69$ [29].

4.3.2. ICPT-GRA Method

The GRA method determines the closeness based on the geometric similarity between the comparison sequence curve and the reference sequence curve. The greater the gray correlation degree, the closer the comparison sequence is to the reference sequence [30]. According to the standardized evaluation matrix, the positive and negative ideal schemes are established as reference schemes, and the Dun's gray correlation coefficient between each evaluation scheme and the positive and negative ideal schemes under each evaluation index is calculated [31].

The gray correlation coefficient ξ_{ij} between the reference plan sequence x_0' and the plan sequence x_j' to be evaluated with respect to the ith indicator is as follows:

$$\xi_{ij} = \frac{\min_j \min_i \left|x_{i0}' - x_{ij}'\right| + \rho \max\max_i \left|x_{i0}' - x_{ij}'\right|}{\left|x_{i0}' - x_{ij}'\right| + \rho \max_i \min_{i0} x \left|x_{i0}' - x_{ij}'\right|} \tag{16}$$

where $x_{ij}'x_{i0}'$ are the evaluation values of the index $x_j'x_0'$, respectively; ρ is the resolution coefficient, which generally takes the value 0.5 [32].

Any evaluation plan should be a gain compared with the negative ideal plan, so its prospect value should be a positive number; similarly, any evaluation plan should be a loss compared with the positive ideal plan, so its prospect value should be a negative number.

The prospect value function of the ICPT-GRA method takes the positive ideal solution as a reference:

$$\begin{cases} v_{1ij}^+ = \delta \cdot \left[1 - \xi_{ij}^-\right]^\alpha, \textit{Take the negative ideal solution as a reference;} \\ v_{1ij}^- = -\theta \cdot \left\{-\left[\xi_{ij}^+ - 1\right]\right\}^\beta, \textit{Take the positive ideal solution as a reference.} \end{cases} \tag{17}$$

where $\xi_{ij}^+\xi_{ij}^-$ are the coefficients of gray correlation for the ith plan and the positive and negative ideal plans under the jth index, respectively; v_{1ij}^+ is the positive prospect value of the jth plan and the negative ideal plan with respect to the ith index; v_{1ij}^- is the positive prospect value of the jth plan and the positive ideal plan with respect to the ith index negative prospect value. Combining Equation (14), we can obtain the comprehensive prospect value V_{1j} of the first solution using the ICPT-GRA method (the first evaluation method in this paper).

4.3.3. ICPT-TOPSIS Method

The basic idea of the TOPSIS method is to use Euclidean distance to measure the distance between the evaluation object and the positive and negative ideal solutions [33]. As shown in Table 1, there is a correlation between the secondary indicators in environment adaptability. If calculated using the Euclidean TOPSIS method, it will lead to biased ranking results. Mahalanobis distance can eliminate the impact of indicator correlation, but it requires that the number of evaluation objects must be greater than the number of indicators, so it is not suitable for power grid planning adaptability evaluation, while cosine similarity (CS) is not interfered with by the correlation of indicators, and there is no requirement for the relationship between the number of indicators and the number of evaluation objects [34]. Therefore, this paper uses cosine similarity as the ranging algorithm of the TOPSIS method.

$$\begin{cases} v_{2ij}^{+} = \delta \cdot \left(x'_{ij} - x'_{i\min}\right)^{\alpha}, \textit{Take the negative ideal solution as a reference;} \\ v_{2ij}^{-} = -\theta \cdot \left[-\left(x'_{ij} - x'_{i\max}\right)\right]^{\beta}, \textit{Take the postive ideal solution as a reference;} \end{cases} \tag{18}$$

$$v_{2ij} = v_{2ij}^{+} \cdot \pi^{+}(\omega_i) + v_{2ij}^{-} \cdot \pi^{-}(\omega_i) \tag{19}$$

where v_{2ij}^{+} is the positive prospect value of the jth plan and the negative ideal plan with respect to the ith index; v_{2ij}^{-} is the negative prospect value of the jth plan and the positive ideal plan with respect to the ith index; $x'_{i\min}$, $x'_{i\max}$ are, respectively, the negative and positive ideal plans in the standard evaluation matrix index value; v_{2ij} is the comprehensive prospect value of the jth plan with regard to the ith index. The comprehensive prospect matrix is $V = \left(v_{2ij}\right)_{n\times m} = \left(v_{2j}\right)_{m}$.

$$\operatorname{sim}(v_{2a}, v_{2b}) = \cos\theta_{ab} = \frac{\sum_{i=1}^{n} v_{2ia} \cdot v_{2ib}}{\sqrt{\sum_{i=1}^{n} (v_{2ia})^2} \cdot \sqrt{\sum_{i=1}^{n} (v_{2ib})^2}} \tag{20}$$

$$v_{2avg} = 0.5(v_{2\max} + v_{2\min}) \tag{21}$$

$$\begin{cases} d(v_{2j}, v_{2\max}) = \log_{\frac{1}{2}}\left(\frac{\sin(v_{2j} - v_{2avg}, v_{2\max} - v_{2avg}) + 1}{2}\right) \\ d(v_{2j}, v_{2\min}) = \log_{\frac{1}{2}}\left(\frac{\sin(v_{2j} - v_{2avg}, v_{2\min} - v_{2avg}) + 1}{2}\right) \end{cases} \tag{22}$$

$$V_{2j} = \frac{d(v_{2j}, v_{2\min})}{d(v_{2j}, v_{2\min}) + d(v_{2j}, v_{2\max})} \tag{23}$$

where v_{2j} is the comprehensive prospect column vector of the jth plan; $v_{2\max}$, $v_{2\min}$ are the comprehensive prospect column vectors of the positive and negative ideal plans, respectively; V_{2j} is the comprehensive prospect value of the jth ICPT-TOPSIS method (the second evaluation method in this paper).

4.3.4. Integrating CG and ICPT GRA-TOPSIS Method

Using the TOPSIS method to evaluate the adaptability of power grid planning schemes, if the evaluation index data are limited, it may lead to large errors in the evaluation results [35]. The GRA method is suitable for comprehensive evaluation in gray environments with incomplete information [36]. The evaluation results of the TOPSIS method are more objective, while the GRA method has the subjective color of dividing the optimal value of the gray index; the TOPSIS method is based on distance measurement, while the GRA method is based on geometric similarity [37]. In view of the advantages and disadvantages of the TOPSIS method and the GRA method, this article adopts the combined evaluation method of ICPT-GRA and ICPT-TOPSIS (ICPT-GRA-TOPSIS method). In order to determine the combined evaluation coefficient, this paper uses the CG method, which has a relatively small total system error. However, the feasibility of using the average value as the benchmark value of this method needs further study, and it is not suitable for the combination of two evaluation methods. The combination coefficient has no solution [38]. Therefore, this paper improves the CG combination evaluation method, uses variance maximization as a measurement standard, and constructs an integrated CG and ICPT GRA-TOPSIS combination evaluation model and overall difference maximization to solve the problem of artificial setting of benchmark values, the limited number of combination evaluation methods, and other issues.

The comprehensive prospect values obtained by the ICPT-GRA method and the ICPT-TOPSIS method are standardized as follows:

$$V_{kj}^{*} = \frac{V_{kj} - V_{k\min}}{V_{k\max} - V_{k\min}} \quad (24)$$

$$e_{kj} = \frac{V_{kj}^{*} - \overline{V_k^{*}}}{s_k} \quad (25)$$

$$\begin{cases} \overline{V_k^{*}} = \dfrac{1}{m}\sum\limits_{j=1}^{m} V_{kj}^{*} \\ s_k = \sqrt{\dfrac{1}{m-1}\sum\limits_{j=1}^{m}\left(V_{kj}^{*} - \overline{V_k^{*}}\right)^2} \end{cases} \quad (26)$$

where V_{kj} is the comprehensive prospect value of the kth evaluation method for the jth plan; $V_{k\min}$, $V_{k\max}$ are, respectively, the minimum and maximum comprehensive prospect values of the kth evaluation method; V_{kj}^{*}, V_{kj} are the standardized processing results; $\overline{V_k^{*}}$, s_k are the average values of all plans of the kth evaluation method's standard deviations, respectively.

$$maxJ = LHL^{\mathrm{T}} = L(ee^{\mathrm{T}})L^{\mathrm{T}} \quad (27)$$

where $L = [l_1 l_2], l_1 + l_2 = 1, l_k > 0(k = 1,2)$; H is the variance information matrix of the combined evaluation model; $e = \left(e_{kj}\right)_{2\times m}$.

Let $G = \{1,2,\ldots,g\}, f \subset G, u(f)$ be a real-valued function defined on the set 2^G, and let $u(f) = J(f)$.

$$\begin{cases} u(\phi) = 0 \\ u(G) \leqslant \sum\limits_{k=1}^{g} u(\{k\}) \end{cases} \quad (28)$$

If $u(f)$ satisfies the above conditions, it is called the characteristic function of the cooperative game $[G,u]$, where $J(f)$ is the variance information matrix of the alliance f for combined evaluation.

$$\varphi_k(u) = \sum_{f} \frac{(g-|f|)!(|f|-1)!}{g!}[u(f) - u(f - \{k\})] \quad (29)$$

φ_k is the Shapely value, which represents the average contribution of the kth evaluation method in the cooperative game. After normalizing the obtained Shapely values, the combined evaluation weight coefficient determined by the cooperative exchange is as follows:

$$l_k = \frac{\varphi_k(u)}{u(G)} \Big/ \sum_{k=1}^{g} \frac{\varphi_k(u)}{u(G)} \quad (30)$$

5. Example Analysis

5.1. Basic Data and Standardized Processing

In this paper, grid planning schemes from three regions encompassed by China's 14th Five-Year Plan are designated as the evaluation subjects. The adaptability evaluation model of grid planning scheme for novel power system based on GRA-TOPSIS integrating CG and ICPT is applied to assess the degree of adaptability and identify the limitations inherent in each regional grid planning scheme.

The region-specific data were transformed into isotropic indicators, which were then converted to positive indicators and rendered dimensionless prior to presentation in Table 2.

Table 2. Standardized data.

Indicators	Region 1	Region 2	Region 3
C_{11}	0.6200	0.6000	0.9800
C_{12}	0.8627	0.8013	0.8480
C_{13}	0.8460	0.5331	0.7840
C_{14}	0.7578	0.6525	0.8867
C_{21}	0.6421	0.5262	0.5575
C_{22}	0.5846	0.6048	0.5408
C_{31}	0.7541	0.8230	0.6997
C_{32}	0.4710	0.9058	0.7609
C_{33}	1.0000	0.6750	0.9300
C_{34}	1.0000	0.9555	1.0000
C_{35}	0.8036	0.8794	0.7851
C_{36}	0.6584	0.6569	0.6996
C_{41}	0.9000	1.0000	0.8300
C_{42}	0.7456	0.6616	0.7144
C_{43}	0.7546	0.8750	0.8449
C_{44}	0.9709	0.7750	0.9340
C_{51}	0.6121	0.5538	0.5645
C_{52}	0.6411	0.6554	0.3992
C_{53}	0.5839	0.5209	0.6227

5.2. Weighting of Indicators

Using the improved AHP method, expert opinions are collected and scores for each indicators are obtained according to Equations (1)–(10); the average value is taken as subjective weights W = (0.0672, 0.037, 0.0579, 0.0774, 0.0761, 0.0525, 0.0395, 0.0903, 0.0168, 0.0254, 0.054, 0.0373, 0.084, 0.0703, 0.0782, 0.0821, 0.0301, 0.0117, 0.0122)^T. Meanwhile, we used the RANCOM method [39], which takes into account the inaccuracy of expert judgment, to calculate the value of indicator weights as W′ = (0.0654, 0.0354, 0.0498, 0.0787, 0.0759, 0.0498, 0.0327, 0.1057, 0.0215, 0.0265, 0.0504, 0.0363, 0.0873, 0.0669, 0.0683, 0.0928, 0.0239, 0.0193, 0.0134)^T, where the difference in weight under the two methods is minimal and the results are consistent.

The CRITIC method was employed to compute the index information quantity and corresponding objective weights, with the outcomes presented in Table 3.

Table 3. Objective weights.

Evaluation Indicators	G_i	V_i	Evaluation Indicators	G_i	V_i
C_{11}	0.8402	0.0498	C_{35}	1.3243	0.0785
C_{12}	0.6686	0.0397	C_{36}	0.8448	0.0501
C_{13}	0.6666	0.0395	C_{41}	1.2951	0.0768
C_{14}	0.7340	0.0435	C_{42}	0.6796	0.0403
C_{21}	0.7453	0.0442	C_{43}	1.2499	0.0741
C_{22}	1.2320	0.0731	C_{44}	0.6662	0.0395
C_{31}	0.6853	0.0406	C_{51}	0.7662	0.0454
C_{32}	1.2692	0.0753	C_{52}	1.1607	0.0688
C_{33}	0.6674	0.0396	C_{53}	0.6995	0.0415
C_{34}	0.6653	0.0395			

Table 4 displays the comprehensive weights of each indicator, derived from the integration of subjective and objective weights. Notably, reliability and grid structure exhibit a heightened influence in assessing grid planning adaptability, particularly through indicators like N-1 pass rate of power lines and mean power supply reliability, which are intimately tied to this metric. Conversely, environmental adaptability demonstrates a lesser impact.

Table 4. Combined weights.

First-Level Indicators	Combined Weights	Second-Level Indicators	Combined Weights
C_1	0.1969	C_{11}	0.0601
		C_{12}	0.0383
		C_{13}	0.0476
		C_{14}	0.0509
C_2	0.1210	C_{21}	0.0492
		C_{22}	0.0718
C_3	0.2936	C_{31}	0.0388
		C_{32}	0.0873
		C_{33}	0.0364
		C_{34}	0.0341
		C_{35}	0.0548
		C_{36}	0.0422
C_4	0.03065	C_{41}	0.0825
		C_{42}	0.0672
		C_{43}	0.0764
		C_{44}	0.0804
C_5	0.07940	C_{51}	0.0315
		C_{52}	0.0235
		C_{53}	0.0244

5.3. Adaptability Evaluation of Grid Planning Scheme for Novel Power System Based on GRA-TOPSIS Integrating CG and ICPT

5.3.1. Evaluating Based on ICPT-GRA

Using the geometric similarity of the GRA method, the grey correlation coefficient between the planning scheme and the positive and negative ideal schemes is separately calculated. To investigate the influence of the limited psychological behaviors of the decision-makers in grid scheduling on the evaluation results, the ICPT-GRA method, which takes into account the psychology of decision-making, introduces the cumulative prospect theory on the basis of the grey correlation coefficient, and the comprehensive prospect value of each evaluation scheme is calculated. The final evaluation scores and ranking results of each scheme, as shown in Table 5. This section takes the more radical loss-sensitive psychological evaluation of decision-makers as an example for analysis. It is consistent with the most conservative mindset held by grid planning decision-makers to ensure reliable and safe operation of the system, for which $\alpha = \beta = 1, \delta = 1, \theta = 2.25$.

Table 5. Overall scores and ranking results in ICPT-GRA method.

Sample	Positive Ideal Solution Distance	Negative Ideal Solution Distance	Comprehensive Prospect Value of ICPT-GRA	Ranking Results in ICPT-GRA
Region 1	0.6555	0.5990	1.7874	1
Region 2	0.5998	0.6037	0.5395	3
Region 3	0.6312	0.5659	1.6623	2

5.3.2. Evaluating Based on ICPT-TOPSIS

According to the TOPSIS method, the Euclidean distance between the decision matrix and the positive and negative ideal schemes is separately acquired. Combined with the cumulative prospect theory, the ICPT-TOPSIS method considering the decision psychology applies cosine similarity to obtain the comprehensive prospect value of each index of the schemes, as shown in Table 6.

Table 6. Overall scores and ranking results in ICPT-TOPSIS method.

Sample	Positive Ideal Solution Distance	Negative Ideal Solution Distance	Comprehensive Prospect Value of ICPT-TOPSIS	Ranking Results in ICPT-TOPSIS
Region 1	0.6138	0.6873	0.33364	2
Region 2	0.46540	0.5548	0.1974	3
Region 3	0.6973	0.4213	0.4064	1

5.3.3. Combination Evaluation Based on CG

Considering the advantages and disadvantages of the ICPT-GRA and ICPT-TOPSIS evaluation methods, their combined prospective values are normalized to obtain the variance information matrix. Based on the variance maximization principle, the optimal weights for the combined evaluation methods are derived by calculating the shapely values of individual evaluation techniques. Combined with the results of the improved combination assignment, the comprehensive evaluation scores and ranking results of GRA-TOPSIS integrating CG and ICPT methods are shown in the Table 7.

Table 7. Overall scores and ranking results in GRA-TOPSIS integrating CG and ICPT method.

Sample	Comprehensive Prospect Value of GRA-TOPSIS Integrating CG and ICPT Method	Ranking Results in GRA-TOPSIS Integrating CG and ICPT Method
Region 1	0.9879	1
Region 2	0.2983	3
Region 3	0.9100	2

Comparison of the results in Tables 5–7 reveals that the combination evaluation not only evaluates the optimal scheme explicitly from the overall dimension, but also widens the gap between different schemes, which makes up for the shortcomings of a single evaluation method. Figure 4 shows the comprehensive evaluation results of ICPT-GRA, ICPT-TOPSIS, and GRA-TOPSIS integrating CG and ICPT. Figures 5 and 6 show the first-level indicator scores for each scheme under the ICPT-GRA and ICPT-TOPSIS methods. Although both evaluation methods are based on the relative distances of positive and negative ideal scenarios, there are large differences in their overall scenario and first-level indicator scores. As can be seen from Figures 2 and 3, the scores for each first-level indicator under the ICPT-GRA method of evaluation are relatively balanced, while the ICPT-TOPSIS method makes a more significant distinction between the strengths and weaknesses of the first-level indicators. For Region 1, economic adaptability and adaptability of grid structure are evaluated poorly in the ICPT-GRA method, but they are significantly worse in the ICPT-TOPSIS method. That is, when evaluating the overall scenario, the ICPT-TOPSIS method highlights the worse structures in the scenario, while the ICPT-GRA method can distinguish the better overall scenario more intuitively. The evaluation approach that incorporates the combination of CG assigns weights based on the marginal contribution of each evaluation method, thereby addressing the limitations of single-method evaluations and quantitatively fusing the outcomes from two individual evaluation techniques.

Combined with the results of the scenarios in Table 3, it can be seen that Region 1 has the highest overall rating, Region 3 occupies second place, and Region 2 is the worst. The safe and reliable operation of the grid is the prerequisite for evaluating the adaptability of the grid planning scheme, which has the largest weight, as high as 0.3065. Among them, the two indicators of mean power supply reliability and N-1 pass rate of power lines determine the safety and reliability of the grid, the combined weight of two accounting for more than 53.15%. The strengths of Region 1 in terms of reliability adaptability, in particular the mean power supply reliability and N-1 pass rate of lines, make its overall score stand out among the three scenarios. Region 1 has significant advantages in terms of adaptability of energy structure, reliability adaptability, and environmental adaptability. This region is equipped with a certain scale of installed renewable energy capacity and realizes the transfer of power resources in time through large-capacity energy storage to ensure the controllability of grid fluctuations, so that the proportion of clean energy and the mean reliability of power supply are increased by 18.78% and 30.17%, respectively. Compared with other regions, the significant increase in the proportion of its clean energy makes the environmental adaptability of the region better than that of other regions. At the same time, limited by the scale of renewable energy installed capacity and line capacity ratio, the ability to increase load and power supply by subsequent unit investment is insufficient in Region 1. The planning scheme of Region 2 has the largest transformer capacity–load ratio and large available capacity, which is more conducive to the access of a high proportion of new energy sources. Despite the capacity to integrate new energy sources, there is a mismatch between the scale of renewable energy sources and the grid structure planning, resulting in grid line redundancy. The inefficiency can lead to increased operational costs and underutilized infrastructure. The primary cause is the lack of synchronization between renewable energy development and grid planning, leading to oversupply in certain grid lines and insufficient supply in others. Region 3 focuses on wind power, due to the large differences in the seasonal distribution of wind resources, resulting in the overall power supply of the project being overly dependent on the generation of traditional thermal power units. This reliance on thermal power is primarily due to the intermittent nature of wind energy and the lack of sufficient storage or backup renewable sources to compensate for periods of low wind availability. The efficient use of existing thermal power plants and strategic deployment of renewable resources make it outstanding in terms of economic adaptability and grid structure adaptability, and relatively balanced power supply reliability capacity.

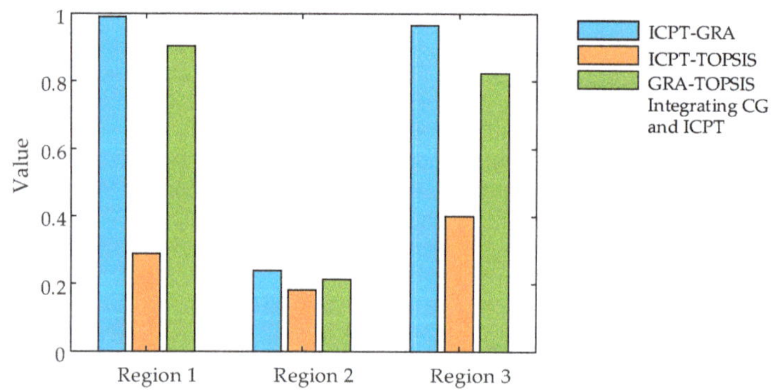

Figure 4. Comparison of the assessment results of 3 evaluation methods.

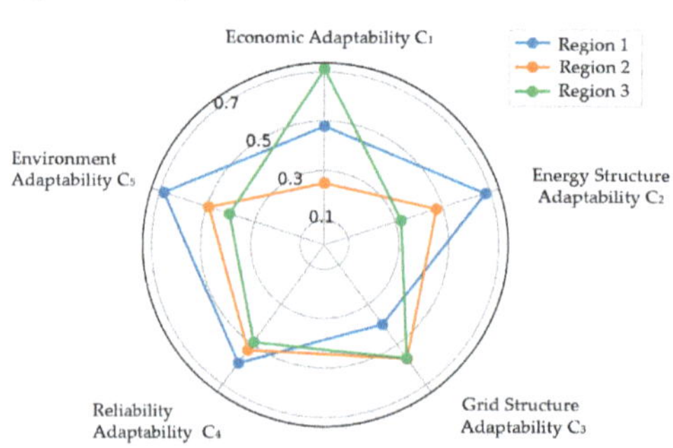

Figure 5. The scores of the first-level indicators in ICPT-GRA.

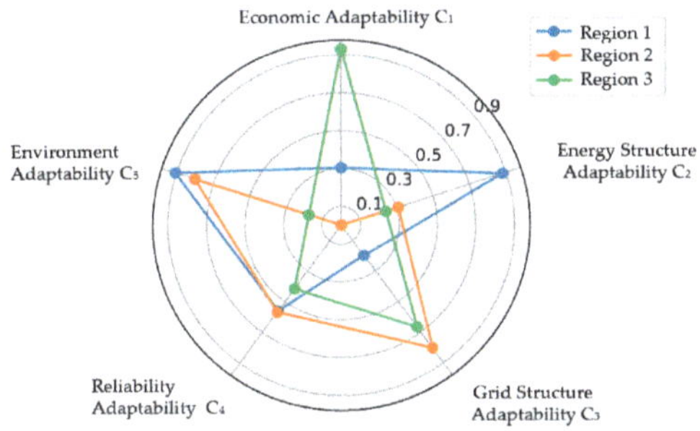

Figure 6. The scores of the first-level indicators in ICPT-TOPSIS.

Considering the limited psychological behavior of the planning decision-makers, the results of the evaluation of the fist and secondary indicators under the planning scenarios for the three districts based on the ICPT-GRA-TOPSIS methodology are shown in Figures 7 and 8.

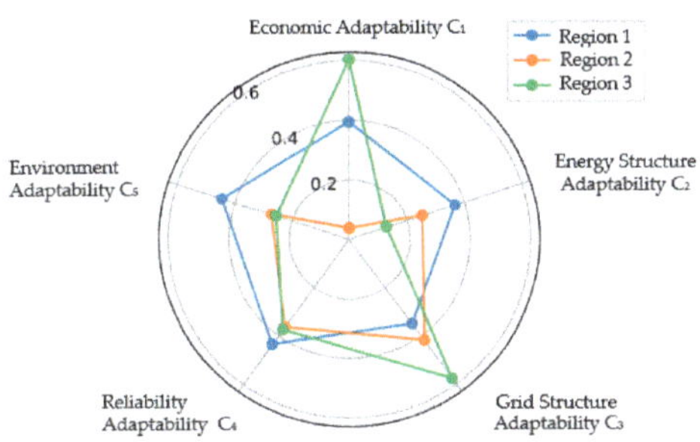

Figure 7. The scores of the first-level indicators in GRA-TOPSIS integrating CG and ICPT.

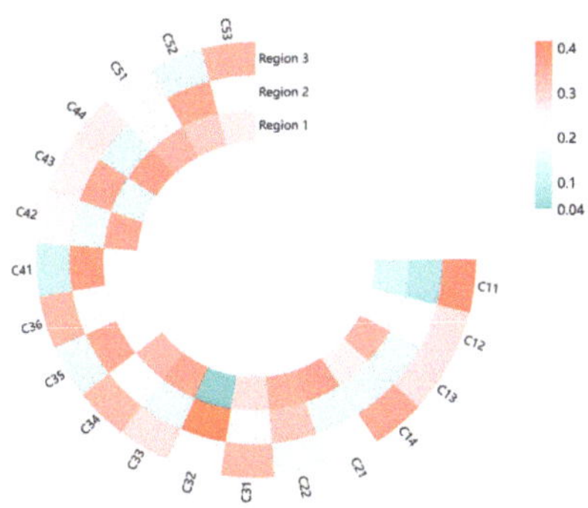

Figure 8. The scores of the second-level indicators in GRA-TOPSIS integrating CG and ICPT.

5.4. Sensitivity Analysis Based on Multiple Psychology of Decision-Maker on the Evaluation of Planning Schemes

In this paper, decision-making psychology is taken as a sensitive factor that is introduced into the GRA-TOPSIS integrating CG and ICPT method combined evaluation model for the adaptive evaluation of grid planning schemes. The personalities of different decision-makers are classified as radical, balanced, or cautious based on the improved cumulative prospect theory. Combining these three psychological behaviors as well as profit–loss attitudes to explore their influence on the comprehensive evaluation and ranking results of regional planning schemes, the constructed six combinations of psychological parameters for grid planning decision-making are shown in Table 8, Figure 9 shows the results of the comprehensive rating evaluation of the six combinations, and Figure 10 shows the results of the evaluation of the first-level indicators under multiple psychology in the case of Region 1.

Table 8. Combination of mental parameters for decision-makers.

Sequence	α	β	δ	θ	Risk Attitudes of Decision-Maker	Profit–Loss Attitude of Decision-Maker
1	0.4	0.4	1	2.25	Radical	Loss-sensitive
2	1	1	1	2.25	Balanced	
3	1.9	1.9	1	2.25	Cautious	
4	0.4	0.4	2.25	1	Radical	Profit-sensitive
5	1	1	2.25	1	Balanced	
6	1.9	1.9	2.25	1	Cautious	

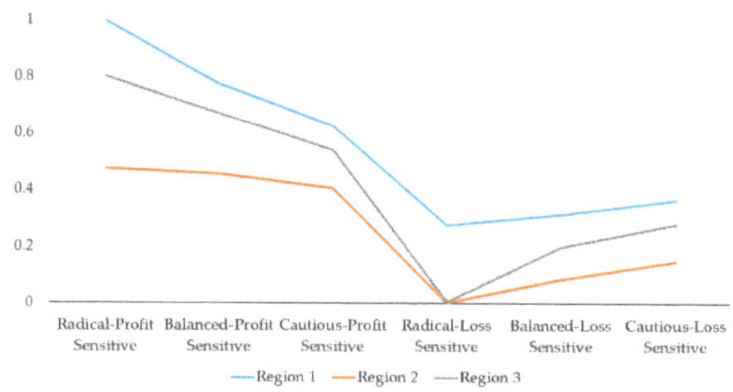

Figure 9. Results of multiple psychological comprehensive evaluation of decision-makers.

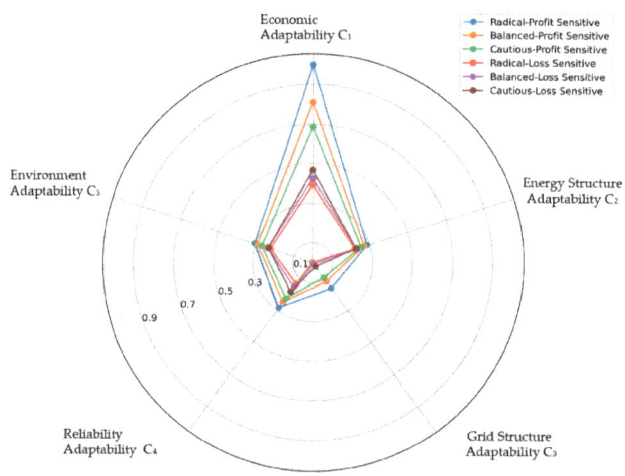

Figure 10. Results of the evaluation of first-level indicators under multiple psychology in Region 1.

From the Figure 9, it can be seen that in the case of the decision-maker having the same profit–loss attitude, the three different decision-making mindsets of radical, balanced, and cautious evaluate the value of the integrated prospect value of the same scheme differently, but the rank of three regions remains consistent, which proves the validity in accurately evaluating the value of different schemes and informing decision-makers. Furthermore, for the more cautious, with a loss-sensitive attitude, the higher the prospect value of the scheme is assessed at, while for the less aggressive, with a lower prospect value of the scheme under the profit-sensitive attitude, it suggests that a difference in the personality/mindset of grid planning decision-makers will produce different selection outcomes for scenario evaluation. Under the same risk attitude, the profit-sensitive type evaluates the prospect value of the scheme 42.5% higher than the loss-sensitive type on average. When the prospect value of Regions 2 and 3 tends to zero, the evaluation value of Region 1 remains around 0.3, which means that under the most conservative and negative planning mindset proposed in this paper, the planning scheme of Region 1 still has a great prospect of revenue and planning application value. That is, the decision-makers believe that the planning scheme in Region 1 has a larger prospect of gain and can bear a controllable risk of loss. The more radical the loss-sensitive decision-maker is, the lower the prospective value of the scenario is considered for the same profit–loss attitude, while the opposite is true for the gain-sensitive decision-maker. In region 1, the difference in the evaluation of the indictor in adaptability of energy structure across risk attitudes for loss-sensitive attitudes is less than 0.02, which shows that the adaptability of the energy structure is less considered to be significantly affected by the psychology of decision-makers. Nevertheless, the prospect value of economic adaptability varies by as much as 0.57 across different decision-making mindsets, which means this level of indicator is the most sensitive to the mentalities of decision-makers.

6. Conclusions

To foster the efficient utilization and consumption of renewable energy sources, this paper proposes an adaptability evaluation of power grid planning scheme for a novel power system considering multiple decision psychology. Compared with the existing studies, this paper analyzes the demand for the adaptability of grid planning for a novel power system in depth, constructs an index system based on it to characterize the adaptability of the planning scheme more comprehensively, and builds an adaptability evaluation model of grid planning scheme for a novel power system based on GRA-TOPSIS integrating CG and ICPT, so as to more accurately reflect the evaluation results of the experts based on different decision-making mentalities in the real environment. After the simulation and analysis of the algorithms, subsequent conclusions can be drawn, as follows:

(1) In the evaluation of the adaptability of grid planning for novel power system, grid structure adaptability and reliability adaptability have a greater impact.
(2) It is considered that the different risk and loss attitudes of decision-makers can effectively improve the accuracy of the evaluation results, and radical profit-sensitive decision-making psychology pays more attention to economic adaptability.
(3) The ICPT-TOPSIS method can better identify the weakness of the evaluation scheme, while the ICPT-GRA method can distinguish the better overall scenario more intuitively, and the combination evaluation method based on CG effectively combines the advantages of these two different methods.

In the future, we will consider the different identities and backgrounds of decision-makers to build a more credible evaluation model, and carry out extensive empirical evaluations to guide power grid planning and promote the high-quality development of new energy.

Author Contributions: Y.W. and C.Y. contributed to conception and design of this study; methodology and validation, C.Y. and Z.W.; investigation, Z.W. and J.W.; writing—original draft preparation, Y.W. and C.Y.; writing—review and editing, Z.W. and J.W.; visualization, Z.W. and J.W. All authors have read and agreed to the published version of the manuscript.

Funding: This research was funded by Natural Science Foundation of Hebei Province, grant number G2022502004, Specia Fund for Basic Scientific Research Business Expenses of Central Universities, grant number 2023MS156.

Data Availability Statement: Data are contained within the article.

Conflicts of Interest: The authors declare no conflicts of interest.

References

1. Wu, J.; Chen, Y.; Yu, L.; Li, G.; Li, J. Has the evolution of renewable energy policies facilitated the construction of a new power system for China? A system dynamics analysis. *Energy Policy* **2023**, *183*, 113798. [CrossRef]
2. Li, R.; Hu, Y.; Wang, X.; Zhang, B.; Chen, H. Estimating the impacts of a new power system on electricity prices under dual carbon targets. *J. Clean. Prod.* **2024**, *438*, 140583. [CrossRef]
3. Bo, H.; Xie, K.G.; Shao, C.Z.; Pan, C.; Lin, C.; Zhao, Y. Commentary on Risk of New Power System Under Goals of Carbon Emission Peak and Carbon Neutrality: Characteristics, indices and Assessment Methods. *Autom. Electr. Power Syst.* **2023**, *47*, 1–15.
4. Liao, Z.; Kally, J.; Ru, S. Probabilistic modeling of renewable energy sources in smart grids: A stochastic optimization perspective. *Sustain. Cities Soc.* **2024**, *109*, 105522. [CrossRef]
5. Wu, Y.; Fang, J.; Ai, X.; Xue, X.; Cui, S.; Chen, X.; Wen, J. Robust co-planning of AC/DC transmission network and energy storage considering uncertainty of renewable energy. *Appl. Energy* **2023**, *339*, 120933. [CrossRef]
6. Li, F.; Yang, J.; Shen, S.; Zhu, K.; Qian, J.; Yan, C. Adaptability Evaluation of Power Grid Planning Scheme Based on Improved AHP-CRITIC-TOPSIS with High Proportion of Renewable Energy. In Proceedings of the 2023 IEEE International Conference on Power Science and Technology (ICPST), Kunming, China, 5–7 May 2023; pp. 623–630. [CrossRef]
7. Yin, L.; Wei, X. Multigroup differential evolutionary and multilayer Taylor dynamic network planning for zero-carbon grid extension model with user satisfaction. *Energy Convers. Manag.* **2023**, *297*, 117753. [CrossRef]
8. Huang, Y.S.; Jiang, Y.Q.; Wang, J. Research on Adaptability Evaluation of Distribution Network Based on Improved TOPSIS-PSO-SVM. *Electron. Sci. Technol.* **2022**, *35*, 54–63.
9. Lu, X.Q.; Ye, Y.; Cao, C.; Meng, J.H.; Tang, H.; He, J. Comprehensive evaluation method of distribution network planning for distributed photovoltaic access. *J. North China Electr. Power Univ. (Nat. Sci. Ed.)* **2022**, *51*, 1–10.
10. Zhang, Z.; Yang, H.Y.; Gao, X.T.; Wang, J.; Wang, Q. Research on Evaluation Method of Distribution Network Planning Scheme Adaptability Based on BPNN Model. *Distrib. Util.* **2021**, *38*, 56–63+88.
11. He, X.; Gao, C.; Cao, H.Z.; Li, Y.; Yu, T. Index evaluation of distribution network based on improved analytic hierarchy process. *Electr. Meas. Instrum.* **2022**, *59*, 93–99.
12. An, Z.; Wei, N.; Liu, S.; Chen, Q.F.; Xing, D. An Efficiency and Benefit Evaluation Method for New Transmission Network Planning Based on Production Simulation. *Power Syst. Clean Energy* **2024**, *40*, 73–83.
13. Zhang, J.; Gao, C.; Wang, T.; Duan, Y.; Xu, M.; Guo, Z. A Dynamic Evaluation Method for High-Permeability New Energy Distribution Network Planning Considering Multistage Development Trends. *Front. Energy Res.* **2022**, *10*, 958892. [CrossRef]
14. Lu, L.; Zhou, H.; Cai, S.; Liao, Y.; Jiang, L.; Wang, Y. Comprehensive Evaluation of Transmission Network Planning Schemes Based on IFAHP. In Proceedings of the 2020 4th International Conference on HVDC (HVDC), Xi'an, China, 6–9 November 2020; pp. 375–381.
15. Si, J.D.; Wu, X.; Guo, Q.S.; Cai, H.; Cheng, L. Review of Flexible Interconnection of Regional Grids Interconnection Planning and Operation Techniques for High Percentage of Renewable Energy Consumption. *Power Syst. Technol.* **2024**, *48*, 2272–2286.
16. Gao, Y.; Gao, Q.C. Serve the high-quality development of the power grid with standard digital innovation. *State Grid News China* **2024**, 1. [CrossRef]
17. Fan, W.; Fan, Y.; Tan, Z.; Ju, L.; Yao, X. Distributionally robust optimization model for virtual power plant participation in electricity carbon market based on multi-layer benefit sharing. *Syst. Eng.-Theory Pract.* **2024**, *44*, 661–683.
18. Wang, Z.; Lu, X.; Zhuang, M.; Zhang, C.; Chen, S. Spatial Optimization of Wind-PV Hybrid Energy Systems for the Three-North Region in China. *J. Glob. Energy Interconnect.* **2020**, *3*, 97–104.
19. Gu, J.M.; Song, Y.T.; Liu, X.Y.; Zhu, S.X.; Lei, Y. Design of Security Defense System for UHV Power Grids under the Dual Carbon Background. *Electr. Drive* **2024**, 1–6. [CrossRef]
20. Tian, C.; Liu, Y.; Zhang, G.; Yang, Y.; Yan, Y.; Li, C. Transfer learning based hybrid model for power demand prediction of large-scale electric vehicles. *Energy* **2024**, *300*, 131461. [CrossRef]
21. Wang, J.H. Exploration of Power Grid Planning Method Based on New Energy Consumption Under the "Dual Carbon" Goal. *Electr. Eng.* **2023**, *S1*, 237–239.
22. Zhang, Y.; Xiang, R.M.; Zheng, Z.H. Industrial User Decision Model Considering Differentiated Power Quality Services in the Context of Carbon Market. *Proc. CSEE* **2024**, 1–14.
23. Lu, Y.F.; Liu, D.; Zhang, Y. Construction Sequence Planning of Power Transmission and Transformation Projects Based on Game Theory Combined Weighting and Improved Technique for Order Preference by Similarity to Ideal Solution. *Sci. Technol. Eng.* **2024**, *24*, 4124–4131.
24. Fang, H.; Shang, L.; Dong, X.; Tian, Y. High Proportion of Distributed PV Reliability Planning Method Based on Big Data. *Energies* **2023**, *16*, 7692. [CrossRef]
25. Zhang, K.; Zhao, Q.; Wang, S.; Liang, F.; Ni, J.F. Construction and calculation method of comprehensive evaluation index reflecting distributed photovoltaic hosting capacity of distribution network. *Distrib. Util.* **2024**, *41*, 3–11.
26. Zhao, S.Q.; Tang, S.F. Comprehensive evaluation of transmission network planning scheme based on improved analytic hierarchy process, CRITIC method and TOPSIS. *Electr. Power Autom. Equip.* **2019**, *39*, 143–148+162.
27. Ilbahar, E.; Kahraman, C.; Cebi, S. Risk assessment of renewable energy investments: A modified failure mode and effect analysis based on prospect theory and intuitionistic fuzzy AHP. *Energy* **2022**, *239*, 121907. [CrossRef]

28. Cao, Y.; Li, G.J. Comprehensive evaluation of island isolated microgrid dispatch considering multiple decision psychology. *Control Decis.* **2022**, *37*, 1591–1600.
29. Liu, Y.; Gaun, X.; Li, Y.C. Multi-attribute decision making based on comprehensive hesitation fuzzy entropy. *Control Decis.* **2024**, *39*, 2022–2030.
30. Qu, K.Q.; Qiao, J.M.; Mao, L.; Zhu, S.J.; Zhao, J.B. Optimal Configuration and Site-selection Evaluation Method for Shared Energy Storage Stations. *Mod. Electr. Power* **2024**, 1–9. [CrossRef]
31. He, L.J.; Li, W.F.; Zhang, Y. Multi-objective optimization method based on grey synthetic incidence analysis. *Control Decis.* **2020**, *35*, 1134–1142.
32. Zhao, H.; Hao, X. Location decision of electric vehicle charging station based on a novel grey correlation comprehensive evaluation multi-criteria decision method. *Energy* **2024**, *299*, 131356. [CrossRef]
33. Wang, L.; Yang, J.; Qu, B.; Pang, C. Multi-Objective Optimization of an Organic Rankine Cycle (ORC) for a Hybrid Solar–Waste Energy Plant. *Energies* **2024**, *17*, 1810. [CrossRef]
34. Akram, F.; Ahmad, T.; Sadiq, M. An integrated fuzzy adjusted cosine similarity and TOPSIS based recommendation system for information system requirements selection. *Decis. Anal. J.* **2024**, *11*, 100443. [CrossRef]
35. Aiello, G.; Quaranta, S.; Inguanta, R.; Certa, A.; Venticinque, M. A Multi-Criteria Decision-Making Framework for Zero Emission Vehicle Fleet Renewal Considering Lifecycle and Scenario Uncertainty. *Energies* **2024**, *17*, 1371. [CrossRef]
36. Çakiroğlu, R.; Çinici, O.K.; Asal, Ş.; Acır, A. Multi-objective optimization of the hydrogen fuel production of a solar-based cobalt-chlorine (Co–Cl) thermochemical cycle with grey relational analysis. *Int. J. Hydrogen Energy* **2024**, *68*, 360–373. [CrossRef]
37. Zhang, X.; Duan, X. Evaluating water resource carrying capacity in Pearl River-West River economic Belt based on portfolio weights and GRA-TOPSIS-CCDM. *Ecol. Indic.* **2024**, *161*, 111942. [CrossRef]
38. Chen, Y.T.; Chen, G.H.; Li, M.J. Research on Determining Weights of Combination Evaluation Method Using Cooperative Games. *Chin. J. Manag. Sci.* **2005**, *13*, 89–94.
39. Więckowski, J.; Kizielewicz, B.; Shekhovtsov, A.; Sałabun, W. RANCOM: A novel approach to identifying criteria relevance based on inaccuracy expert judgments. *Eng. Appl. Artif. Intell.* **2023**, *122*, 106114. [CrossRef]

sustainability

MDPI

Article

Maximization of Total Profit for Hybrid Hydro-Thermal-Wind-Solar Power Systems Considering Pumped Storage, Cascaded Systems, and Renewable Energy Uncertainty in a Real Zone, Vietnam

Phu Trieu Ha [1,2], Dao Trong Tran [3], Tan Minh Phan [4] and Thang Trung Nguyen [4,*]

1 Faculty of Electrical and Electronics Engineering, Ton Duc Thang University, Ho Chi Minh City 700000, Vietnam; hatrieuphu.st@tdtu.edu.vn or phu.ht@sgu.edu.vn
2 Faculty of Electronics and Telecommunications, Saigon University, Ho Chi Minh City 700000, Vietnam
3 Division of MERLIN, Faculty of Electrical and Electronics Engineering, Ton Duc Thang University, Ho Chi Minh City 700000, Vietnam; trantrongdao@tdtu.edu.vn
4 Power System Optimization Research Group, Faculty of Electrical and Electronics Engineering, Ton Duc Thang University, Ho Chi Minh City 700000, Vietnam; phanminhtan@tdtu.edu.vn
* Correspondence: nguyentrungthang@tdtu.edu.vn; Tel.: +84-986515857

Abstract: The study maximizes the total profit of a hybrid power system with cascaded hydropower plants, thermal power plants, pumped storage hydropower plants, and wind and solar power plants over one operation day, considering the uncertainty of wind speed and solar radiation. Wind speed and solar radiation in a specific zone in Vietnam are collected using the wind and solar global atlases, and the maximum data are then supposed to be 120% of the collection for uncertainty consideration. The metaheuristic algorithms, including the original Slime mould algorithm (SMA), Equilibrium optimizer, and improved Slime mould algorithm (ISMA), are implemented for the system. ISMA is a developed version of SMA that cancels old methods and proposes new methods of updating new solutions. In the first stage, the cascaded system with four hydropower plants is optimally operated by simulating two cases: simultaneous optimization and individual optimization. ISMA is better than EO and SMA for the two cases, and the results of ISMA from the simultaneous optimization reach greater energy than individual optimization by 154.8 MW, equivalent to 4.11% of the individual optimization. For the whole system, ISMA can reach a greater total profit than EO and SMA over one operating day by USD 6007.5 and USD 650.5, equivalent to 0.12% and 0.013%. The results indicate that the optimization operation of cascaded hydropower plants and hybrid power systems can reach a huge benefit in electricity sales

Keywords: pumped storage hydropower plant; cascaded hydropower plant; wind speed; solar radiation; uncertainty; total profit

Citation: Ha, P.T.; Tran, D.T.; Phan, T.M.; Nguyen, T.T. Maximization of Total Profit for Hybrid Hydro-Thermal-Wind-Solar Power Systems Considering Pumped Storage, Cascaded Systems, and Renewable Energy Uncertainty in a Real Zone, Vietnam. *Sustainability* **2024**, *16*, 6581. https://doi.org/10.3390/su16156581

Academic Editors: Bo Yang, Zhijian Liu and Lin Jiang

Received: 19 June 2024
Revised: 26 July 2024
Accepted: 27 July 2024
Published: 1 August 2024

1. Introduction

In recent years, there has been a growing need for cheaper, more sustainable, and cleaner power sources due to the high cost and environmental impact of fossil fuels [1]. Wind power plants (WPs) and solar power plants (SPs) are promising solutions, but they face challenges due to the uncertain nature of wind and solar radiation [2]. This has led to a focus on energy storage, with pumped storage hydropower plants (PSHPs) emerging as a popular option. PSHPs consist of upper and lower reservoirs connected by a tunnel, with turbines that can function as both generators and pumps [3]. They play a significant role in reducing the total electric generation cost of Thermal Power Plants (ThPs) in a hybrid power system. The study also explores the optimal scheduling of a hybrid power system comprising various power plants, including PSHPs, cascaded hydropower plants (CasHPs), WPs, SPs, and ThPs, to maximize the total electricity sale profit from these power plants.

Hourly electric price, load demand, wind speed, and solar radiation over one day are given. Then, the maximum data of wind and solar radiations are collected for an uncertainty case with 120% of the maximum given data, which allows for maximizing the total electric sale profit. PSHPs use the power from other plants to pump water back to the upper reservoirs and the stored water volume can be used for many hours with a high price. The study applies metaheuristic algorithms to determine power of WPs, SPs, and ThPs, discharge and volume of CasHPs, and pumping hours, generation hours, and discharge of generation hours for PSHPs. So, the importance of the study is to reach optimal operation solutions for the system that can supply enough electricity to loads and reach a high total electric sale profit. In addition, the use of suitable energy storage can produce more energy for power systems, and ThPs can continuously operate.

The earliest studies have applied deterministic algorithms, such as the linear programming algorithm [4] and Gradient algorithm (GRA) [5], for simple systems and have reached the same solutions for different runs. On the contrary, metaheuristic algorithms, such as improved particle swarm optimization (MPSO) [6] and Evolutionary Programming (EP) [7], have found different solutions for different runs. In general, the goal of these studies was to compare costs and demonstrate the contribution of PSHPs in meeting load demand, even without inflows to the reservoir. In recent years, studies have focused on complex problems related to PSHPs in various countries, including Turkey [8], China [9], Spain [10], Portugal [11], Iran [12], and Saudi Arabia [13]. These studies have integrated PSHPs into other power sources to compensate for power shortages and reduce peak load demand. They have also examined the effectiveness of PSHPs in hybrid systems, and their potential for increasing total revenue. Additionally, this research has been conducted on optimizing the operation of PSHP to alleviate peak shaving in power systems. However, the studies did not address the cost of constructing PSHPs in these power systems. Several studies have investigated the profitability and benefits of pumped storage hydropower (PSHP) in various electric markets in Turkey [14], China [15], and benchmark systems [16]. While some studies [17,18] found that PSHP could not bring enough profit for the power systems, others suggested that it could compensate for insufficient power and consume surplus power from renewable plants. Additionally, studies have proposed transforming existing Conventional Hydropower Plants (CoHPs) into PSHPs and showed that PSHP could reduce generation costs and cut emissions in hybrid systems [19].

On the other hand, cascaded hydropower plants and pumped storage hydropower plants could be combined in the same system by adding a pumping function into the cascaded hydropower plants or letting them work separately [20–22]. The cooperation among cascaded hydropower plants is driven by the goal of maximizing energy production or profit [23]. Coordinated operations of these plants can enhance the benefits of the entire river system [24]. However, allocating these benefits fairly and efficiently presents a significant challenge for river system operators or managers [25]. The operation combination of PSHPs and CasHPs is very complicated, but the effectiveness of the successful cooperation can lead to a huge benefit. In [20], researchers investigated a larger electric power grid comprising two PSHPs, twenty-nine TPs, and four cascaded power plants. The study aimed to achieve a balance between demand and supply, taking into consideration energy purchases from external power plants. The results revealed that all system constraints were met, but the efficacy of the PSHPs was inconclusive. Another study [21] focused on optimizing the generation over five days for three cascaded hydroelectric units, with the lowest unit capable of running pumps for water storage. The study demonstrated the system's potential to achieve the highest energy output, but it lacked comparisons with other studies or different methods. In the study [22], researchers determined the optimal location for constructing a PSHP in a six-cascaded hydroelectric power plant system. In general, each previous study had significant contribution in power systems; however, most of the studies have not proved their solution method and metaheuristic algorithms to be highly effective. In addition, data of test systems were almost neglected for replication. All types of power plants were not combined, and the uncertainty of renewable energies was not considered for the large integrated system.

In two early studies regarding renewable energies, real power output can be obtained by substituting certain data into mathematical equations for WPs [26] and SPs [27]. Considering the uncertainty of wind, the cumulative distribution function (cdf) and probability density function (pdf) were proposed to calculate the power output and the costs [28]. Similarly, the electricity generation costs of SPs were also calculated by using the probability function [29]. The costs of wind and solar power considering uncertainty were studied for small systems with wind, solar, and small hydro plants [30], a large system considering Electric Vehicles [31], and transmission power systems [32]. In the study, we investigate the positive impact of pumped storage hydropower plants (PSHPs) on maximizing the total electric sale profit of a hybrid power system with PSHPs, CasHPs, WPs, SPs, and ThPs. Real data of wind speed and solar radiations in a specific zone are collected by using the wind global atlas [33] and solar global atlas [34]. On the other hand, their uncertain characteristics are also considered by using probability functions in calculating generation costs as shown in previous studies [30–32]. Three metaheuristic algorithms, including the original Slime mould algorithm (SMA) [35], Equilibrium optimizer [36], and a Slime mould algorithm (ISMA), are applied to find the optimal scheduling of the system in two cases: certain and uncertain wind and solar. The novelty of the paper can be summarized as follows:

- Develop an improved version of SMA to improve the search performance of SMA.
- Select a real zone in Vietnam and then access the wind and solar global atlases to collect real wind speed and solar radiation. In addition, the maximum possible data are supposed to be 120% of the collected data to run uncertainty case of wind and solar.
- Apply a variety of major power plants, such as CasHP, PSHP, WPs, SPs, and ThPs, which are considered as power sources. Their operation principles and characteristics are obtained from previous studies.

The main contributions of the whole paper can be clarified as follows:

- The PSMA can find better solutions and provide more stable searchability than SMA and EO for two cases: considering certain and uncertain characteristics for wind and solar power plants.
- Optimal operation solutions for the PSHP can lead to higher profits for hybrid systems with renewable energies. The total profit can be greater for the system with the PSHP when considering both certain and uncertain wind and solar power.
- For the PSHP, optimal solutions can lead to significant profits after accounting for electric purchases. The PSHP needs to buy electricity to run pumps for storing water, and then it can sell the energy generated by discharging the stored water. The difference between the sale and purchase is the profit.
- CasHPs can reach the maximum energy and it can lead to a great profit for the whole system with many power plant types.

Other parts of the study are as follows: Section 2 presents the problem formulation with objective functions and constraints. Section 3 shows an improved version of SMA. Section 4 presents numerical results to show the contributions. Finally, Section 5 summarizes the whole study with obtained results, contributions, shortcomings, and future work.

2. The Problem Description

In this paper, the contribution of pumped storage hydropower plants to the cost reduction for thermal power plants is investigated. A typical power system is considered with the presence of power plant types, including N_{ThP} thermal power plants, N_{PSHP} pumped storage hydropower plants, N_{WP} wind power plants, N_{SP} solar power plants, and N_{CasHP} cascaded hydropower plants. The system is plotted in Figure 1. In the figure, CasHP1, CasHP2, CasHP3, and CasHP4 are the four cascaded hydropower plants, in which CasHP1 and CasHP2 are, respectively, the upper plants of CasHP4 and CasHP3.

In addition, CasHP3 is the upper plant of CasHP4. Discharges 1–4 and flows 1–4 are, respectively, the discharge and inflow of the four CasHPs. ThP1 and ThP2 are two thermal power plants.

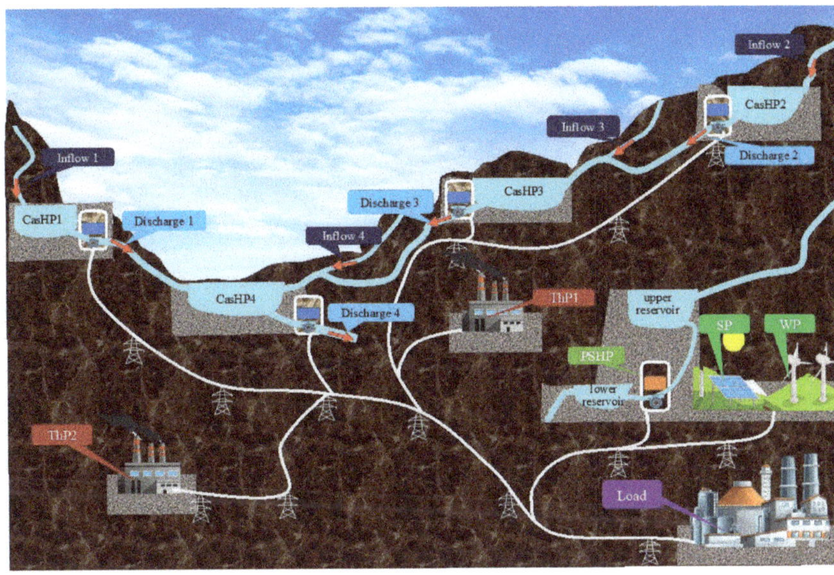

Figure 1. The structure of the considered hybrid system.

2.1. The Total Profit Maximization

The objective function of the study is to maximize the total electric sales profit for hybrid power systems throughout one day. Basically, ThPs have a cost function modeled as a second-order function based on active power output and specific coefficients; meanwhile, the cost function of hydropower plants is supposed to be zero due to the free water. Regarding the costs of WPs and SPs, generation costs are considered in two aspects: neglecting generation costs and taking them into account. In this paper, we explore both cases of renewable power plants. This objective function is calculated as the total revenue minus the total costs. The objective is mathematically presented as follows:

$$Maximize\ Profit_{sys} = Sale_{Load} - Cost_{all} \tag{1}$$

$$Sale_{Load} = \sum_{h=1}^{N_h} (Price_h.Load_h) \tag{2}$$

$$Reduce\ Cost_{all} = C_{ThP} + C_{WP} + C_{SP} \tag{3}$$

where GC_{TP}, GC_{WP}, and GC_{PVP} are the total generation costs of all THPPs, WPPs, and PVPPs in the considered power system. The costs of thermal power plants [37], wind, and solar power plants [29] are respectively obtained by

$$C_{ThP} = \sum_{h=1}^{N_h} \sum_{th=1}^{N_{ThP}} \left(\delta_{1th} + \delta_{2th} P_{th,h} + \delta_{3th} P_{th,h}^2 \right) \tag{4}$$

$$C_{WP} = \sum_{h=1}^{N_h} \sum_{w=1}^{N_{wp}} (DC_{w,h} + PC_{w,h} + RC_{w,h}) \tag{5}$$

$$C_{SP} = \sum_{h=1}^{N_h} \sum_{pv=1}^{N_{SP}} \left(DC_{pv,h} + PC_{pv,h} + RC_{pv,h} \right) \quad (6)$$

where $DC_{w,h}$, $RC_{w,h}$, and $PC_{w,h}$ are the direction, reserve, and penalty costs of the wth WP at the hth hour; and $DC_{pv,h}$, $RC_{pv,h}$, and $PC_{pv,h}$ are the direction, reserve, and penalty costs of the $pvth$ SP at the hth hour.

Here, the total electric sale profit belongs to the whole power plants in the system. It is supposed that all power plants working in the power system belong to one generation company and the sharing profit is in charge of the CEO of the company. The study acts as an optimization operation expert and proposes the optimal generation schedule for the total profit maximization.

2.2. Constraints

2.2.1. Active Power Balance

The power systems need to balance total generated power and total consumed power, as expressed in the following model [38]:

$$\sum_{th=1}^{N_{Thp}} P_{th,h} + \sum_{sp=1}^{N_{SP}} P_{sp,h} + \sum_{w=1}^{N_{WP}} P_{w,h} + \sum_{cas=1}^{N_{CasHP}} P_{cas,h} + \sum_{ps=1}^{N_{PSHP}} \left(1 - K_{ps,h}\right) P_{ps,h} - \sum_{ps=1}^{N_{PSHP}} K_{ps,h} \,.Pump_{ps,h} - Load_h = 0; h = 1, \ldots, N_h \quad (7)$$

where $K_{ps,h}$ is the operating status of the $psth$ PSHP at the hth hour, it has two values: 1 for the pumping action, and 0 for the generating action. Its values and actions are summarized as follows:

$$K_{ps,h} = \begin{cases} 0 \; for \; generating \; action \\ 1 \; for \; pumping \; action \end{cases} \quad (8)$$

2.2.2. Generation Limits of Power Plants

All power plants have limits of generating capacity so that their generators can work stably and effectively. The power output must be within the minimum and maximum limits as follows:

$$P_{th}^{Min} \leq P_{th,h} \leq P_{th}^{Max} \quad (9)$$

$$P_{sp}^{Min} \leq P_{sp,h} \leq P_{sp}^{Max} \quad (10)$$

$$P_{w}^{Min} \leq P_{w,h} \leq P_{w}^{Max} \quad (11)$$

$$P_{cas}^{Min} \leq P_{cas,h} \leq P_{cas}^{Max} \quad (12)$$

$$P_{ps}^{Min} \leq P_{ps,h} \leq P_{ps}^{Max} \quad (13)$$

$$Pump_{ps,h} = \begin{cases} 0 \; for \; generating \; action \\ P_{ps}^{Max} \; for \; pumping \; action \end{cases} \quad (14)$$

2.2.3. Pumped Storage Hydropower Plant's Hydraulic Constraints

Inequality constraints: the upper reservoir and hydro turbines of the power plants have an operating range, called the minimum and maximum limits as the following model [39]:

$$V_{ps}^{Min} \leq V_{ps,h} \leq V_{ps}^{Max} \quad (15)$$

$$Q_{ps}^{Min} \leq Q_{ps,h} \leq Q_{ps}^{Max} \quad (16)$$

Here, $Q_{ps,h}$ is the $psth$ PSHP's discharge at the hth hour, and it is function of power output and discharge coefficients as follows [16]:

$$Q_{ps,h} = m_{1ps} \left(P_{ps,h}\right)^2 + m_{2ps} P_{ps,h} + m_{3ps} \quad (17)$$

Equality constraints: Equality constraints are about the continuity of water in reservoirs, the relationship between discharge and pump flow, and the reservoir volumes at different hours. These constraints are as follows [40]:

$$V_{ps,h-1} + I_{ps,h} - V_{ps,h} - \left(K_{ps,h} - 1\right).Q_{ps,h} + K_{ps,h} .Q_{pumps} = 0; k = 1, \ldots, N_5; h = 1, \ldots, N_1 \quad (18)$$

$$Q_{pumps} = \eta_{ps}.Q_{ps}^{Max} \quad (19)$$

$$V_{ps,0} = V_{ps}^{Initial} \quad (20)$$

$$V_{ps,N_h} = V_{ps}^{Finall} \quad (21)$$

where $V_{ps,0}$ and $V_{ps}^{Initial}$ are the *psth* PSHP's reservoir at the 0th hour and the beginning of the scheduled day. V_{ps,N_h} and V_{ps}^{Finall} are the *psth* PSHP's reservoir at the last hour and the end of the scheduled day. $V_{ps,0}$ is the special value of $V_{ps,h-1}$ when considering $h = 1$.

2.2.4. Cascaded Hydropower Plants' Constraints

The cascaded hydropower plant system has a diagram as shown in Figure 2. The cascaded system has the constraints as follows:

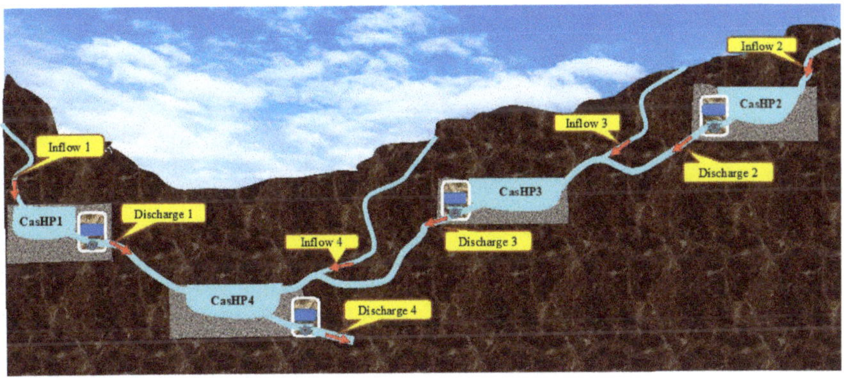

Figure 2. The configuration of the four cascaded hydroelectric plants.

Equality constraint: each CasHP is subject to the water continuity constraint as follows [41]:

$$V_{cas,h-1} + I_{cas,h} - V_{cas,h} - Q_{cas,h} + \sum_{cas'=1}^{NU_{cas}} \left(Q_{cas',h-T_{cas',cas}}\right) = 0 \quad (22)$$

where $T_{cas',cas}$ is the travelling water time from the *cas'th* to the *casth* CasHP; $Q_{cas',h-T_{cas',cas}}$ is the discharge of the *cas'th* upper CasHP of the *casth* CasHP at the $(h - T_{cas',cas})th$ hour; and NU_{cas} is the *casth* CasHP's upstream CasHP number.

In addition, the CasHPs are also constrained by the initial volume and end volume as indicated in Equations (20) and (21) of the PSHPs.

For each CasHP, power output is a function of discharge, volume, and other coefficients as follows:

$$P_{cas,h} = \gamma_{1cas}(V_{cas,h})^2 + \gamma_{2cas}(Q_{cas,h})^2 + \gamma_{3cas}V_{cas,h}Q_{cas,h} + \gamma_{4cas}V_{cas,h} + \gamma_{5cas}Q_{cas,h} + \gamma_{6cas} \quad (23)$$

Inequality constraints: For each CasHP, volume and discharge also have their operating range [42] as shown in Equations (15) and (16).

3. Improved Slime Mould Algorithm

3.1. Slime Mould Algorithm

In the Slime mould algorithm (SMA), each control variable has two ways of updating new values depending on the comparison result between a random number and the fitness computation. The two ways are summarized in the following equation [35]:

$$cv_{x,y}^{new} = \begin{cases} r_1.cv_{x,y} & if\ rd \geq tanh(FN_y - FN_{b1}) \\ cv_{x,b1} + r_2(r_3.cv_{x,r1} - cv_{x,r2})\ else \end{cases} \quad (24)$$

$$Range_{r1} = \left[\frac{G_c}{G_M} - 1;\ 1 - \frac{G_c}{G_M}\right] \quad (25)$$

$$Range_{r2} = \left[-arctanh\left(1 - \frac{G_c}{G_M}\right);\ arctanh\left(1 - \frac{G_c}{G_M}\right)\right] \quad (26)$$

$$r_3 = \begin{cases} 1 - rd.\log\left(1 + \frac{FN_{b1} - FN_y}{FN_{b1} - FN_{w1}}\right) & if\ y > 0.5N_s \\ 1 + rd.\log\left(1 + \frac{FN_{b1} - FN_y}{FN_{b1} - FN_{w1}}\right) & else \end{cases} \quad (27)$$

3.2. Improved Slime Mould Algorithm

The upper model in Equation (24) is a shortcoming of SMA in searching for new values for the *xth* control variable. In fact, the upper model uses a random number between 0 and 1 to multiply the old value, this result can lead to a zero solution if the random value is zero. This model can cause an ineffective search process if the condition of $[rd \geq tanh(FN_y - FN_{b1})]$ happens. So, in the ISMA, the upper model is replaced with another model, which searches around the current value of the *xth* control variable. The change can avoid finding the zero solutions with all zero control variables. This is the first modification, and it is expressed as follows:

$$cv_{x,y}^{new} = cv_{x,y} + r_2(r_3.cv_{x,b1-4} - cv_{x,y})\ if\ rd \geq tanh(FN_y - FN_{b1}) \quad (28)$$

The lower model in Equation (24) finds new values of the *xth* control variable by finding a close value of the *xth* control variable in the best solution. On the other hand, the lower model uses an increased interval by using two random solutions. The search cannot lead to the best control variable because one interval cannot enlarge the search space, and using two random solutions cannot find the most effective increased interval. So, the lower model is modified as follows:

$$cv_{x,y}^{new} = \begin{cases} cv_{x,b1} + r_2(r_3.cv_{x,b1-4} - cv_{x,b1})\ if\ 0.5 \leq rd < tanh(FN_y - FN_{b1}) \\ cv_{x,b1} + r_2(r_3.cv_{x,b1-4} - cv_{x,b1}) + r_2(r_3.cv_{x,r3} - cv_{x,b1})\ if\ rd < 0.5\ \&\ tanh(FN_y - FN_{b1}) \end{cases} \quad (29)$$

In Equation (29), two models exploit the search space around the best solution with different scales: one increased interval in the upper model and two increased intervals in the lower model. The condition is the comparison of the random number and 0.5 to balance the increased steps in one or two intervals.

4. Numerical Results

4.1. Data, Parameter Settings, and Simulation Scebnario

In this section, the total profit of one hybrid power system with four thermal power plants, four cascaded hydropower plants, one solar power plant, one wind power plant, and one pumped storage hydropower plant is maximized under the consideration of renewable power source uncertainty. In the first simulation scenario, three algorithms, SMA [35], EO [36], and ISMA are run to reach the maximum energy of the cascaded hydropower plant system. Then, the most effective solution of the system is used to maximize the total profit of the whole system in the second simulation scenario. The whole study is summarized as follows:

(1) The first simulation scenario: energy maximization of the cascaded hydropower plants:
Case 1: Optimization operation of each plant;
Case 2: Simultaneous optimization operation of the whole plants.
(2) The second simulation scenario: maximization of the total profit of the whole system.

The whole study is performed on a personal computer with the central processing unit (CPU) Core i7, 2.6 GHz, and 8 GB of random accessing memory. (RAM). MATLAB software with the version of R2018a is used to code the program of algorithms. For running the three algorithms, the iteration number and population size are set to 50 and 200 for each plant of Case 1, 50, and 1000 for the whole cascaded system of Case 2 in the first scenario, and 200 and 2000 in the second scenario.

The whole system is plotted in Figure 1. The system consists of four THPPs, one PSHP, one WPP, and one PVPP. The ThPs' cost function and generation limits are given in Table A1 in the Appendix A. The PSHP's whole data, including discharge function, generation and pump limits, and pump efficiency, are collected from [5]. Three costs of the WP and SP, including direct cost, penalty cost, and reserve cost, in the second simulation scenario are reported in Figures A1 and A2 in the Appendix A. In the system, a location with geographical coordinates of 11°12′39″ and 108°40′53″ in Binh Thuan province is selected to collect wind and solar data by referring to the websites for the wind global atlas [29] and solar global atlas [37]. This location is shown in Figure 3. The wind power plant has fifty turbines, and each turbine has a rated power of 2.5 MW. The hourly wind speed and power are plotted in Figure 4, and it is supposed that the wind speed can be greater than the maximum index by 20% for uncertainty consideration. Finally, the wind power plant's maximum power is given in Figure 5.

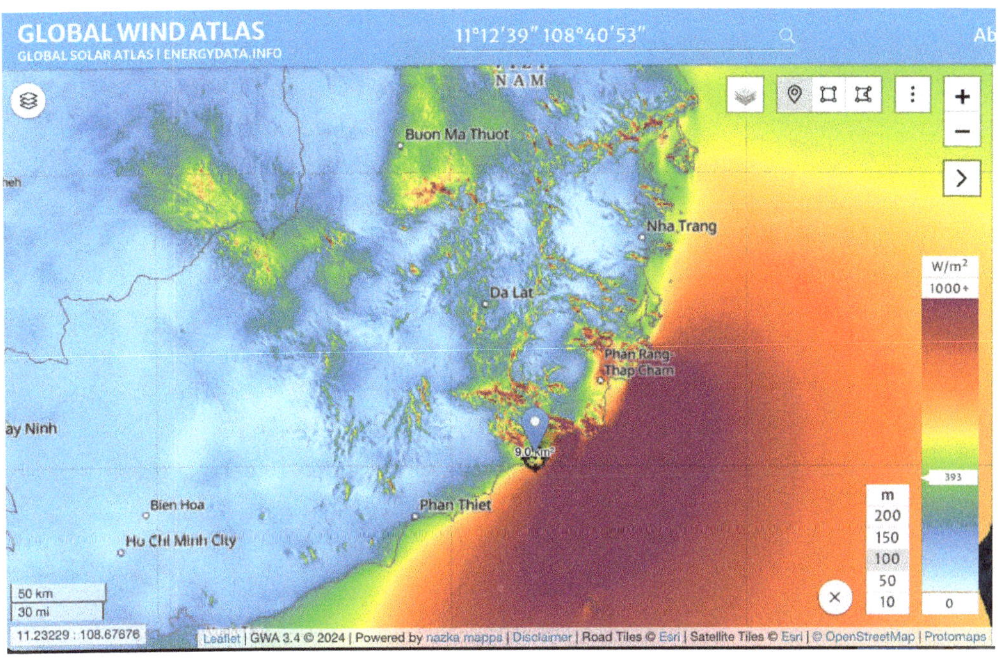

Figure 3. Location of Binh Thuan province in Vietnam.

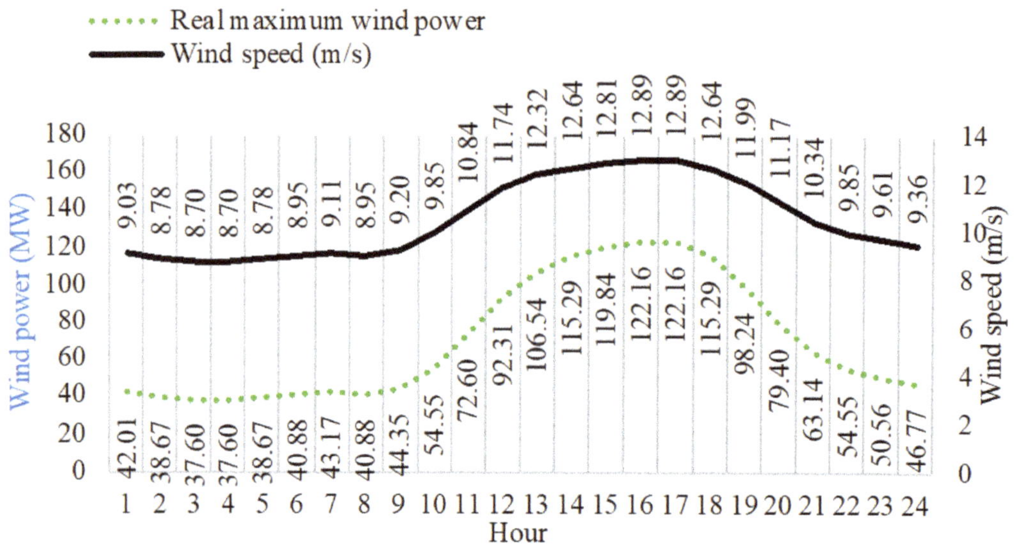

Figure 4. Real wind speed and power in a location of Binh Thuan province in Vietnam.

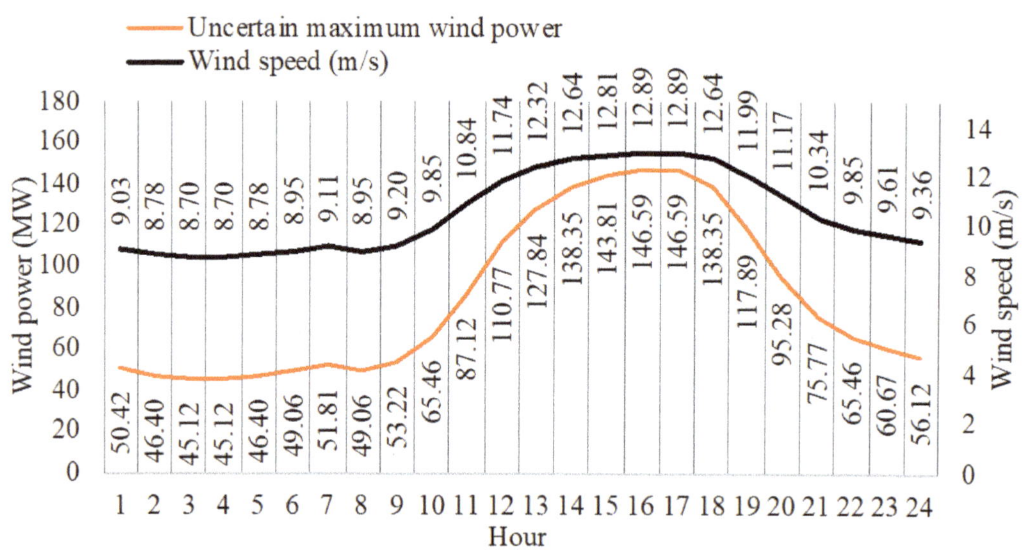

Figure 5. Wind power applied for uncertainty.

The solar power plant's rated power is selected to be 150,000-kWp using Tilt and Azimuth of solar panels with 12° and 180°. The maximum solar power is given in Figure 6. The maximum power of wind and solar plants for uncertainty and the electric price are reported in Figure 7.

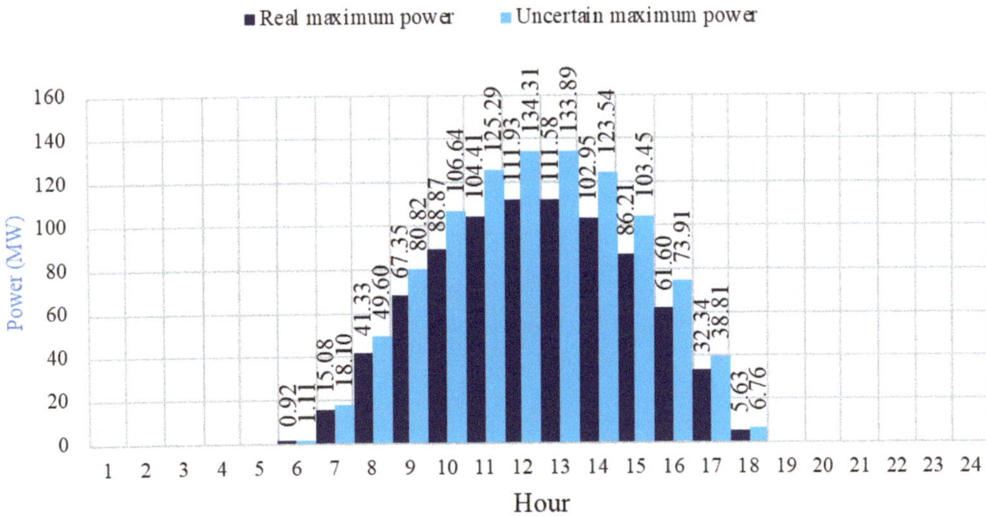

Figure 6. Maximum hourly solar power.

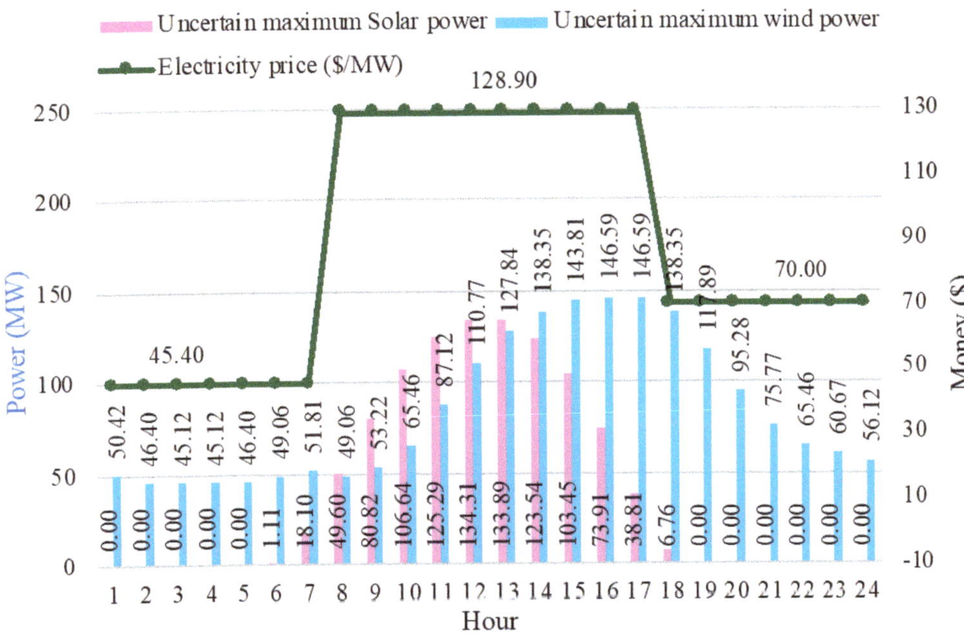

Figure 7. Renewable powers for uncertainty and electric price.

4.2. Results of the First Simulation Scenario

The one-day energy of the four cascaded power plants is given in Figure 8. Three algorithms can reach greater one-day energy in Case 2 than in Case 1. The greatest energy for each case is found by ISMA, which is 3767.86 MWh for Case 1 and 3922.66 MWh for Case 2. So, Case 2 finds a greater energy than Case 1 by (3922.66–3767.86) = 154.81 MWh, which is about 4.1% of Case 1's energy. The comparison indicates that the simultaneous optimization operation of the whole system can reach greater energy than planning the optimal operation of each plant. In addition, ISMA is the most suitable algorithm for the first scenario.

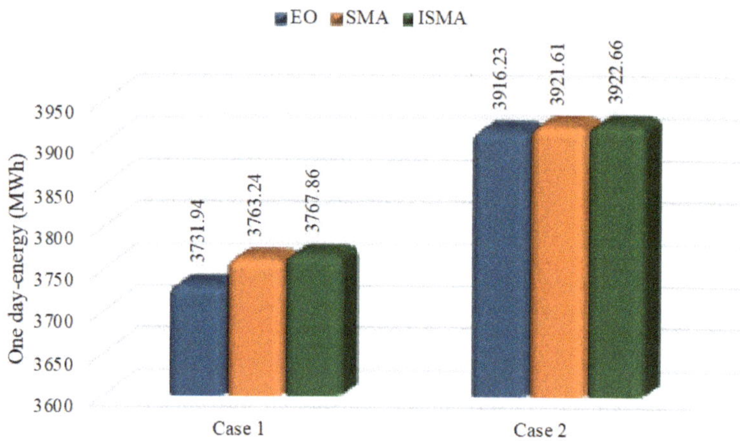

Figure 8. One-day energy of the whole cascaded system.

The hourly power of each cascaded power plant is given in Figure 9. The sum of hourly energy from Case 2 is greater than Case 1 by 6.89 MW for the first CasHP (CasHP1), 42.09 MW for the second CasHP (CasHP2), −81.76 MW for the third CasHP (CasHP3), and 187.59 MW for the fourth CasHP (CasHP4). Clearly, CasHP3 can reach greater energy in Case 1 than in Case 2, which is different from other plants.

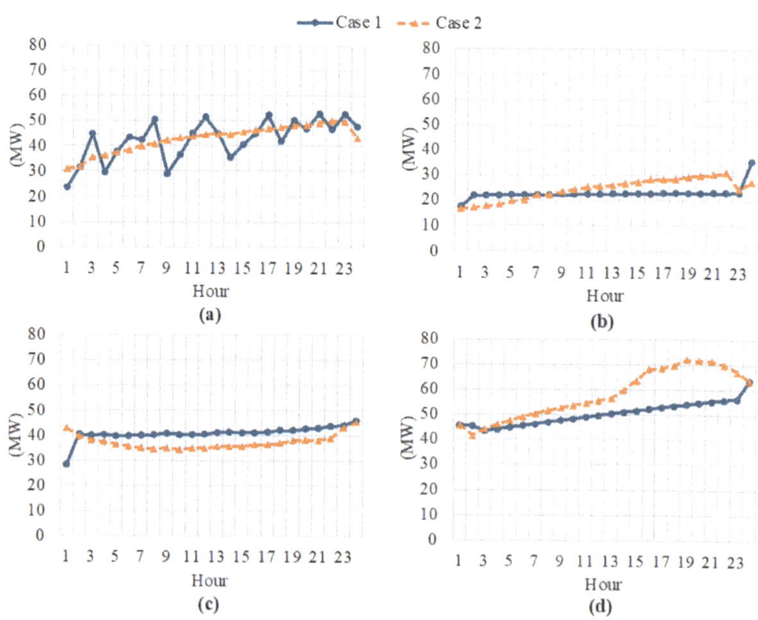

Figure 9. Comparison of generation: (**a**) CasHP1, (**b**) CasHP 2, (**c**) CasHP3, (**d**) CaseHP4.

4.3. Results of the Second Simulation Scenario

The results obtained from three algorithms are reported in Figure 10. The maximum profit of ISMA is USD 5,196,221.4, while that of EO and SMA is USD 5,190,213.9 and USD

5,195,570.9, respectively. The exact computation indicates that ISMA can reach a greater total profit than EO and SMA over one operating day by USD 6007.5 and USD 650.5, equivalent to 0.12% and 0.013%, respectively. The lowest total profit of EO, SMA, and ISMA is, respectively, USD 5,176,953.8, USD 5,174,086.4, and USD 5,176,953.8. EO and ISMA have the same lowest profit. The comparisons indicate that ISMA is more suitable than EO and SMA for the study.

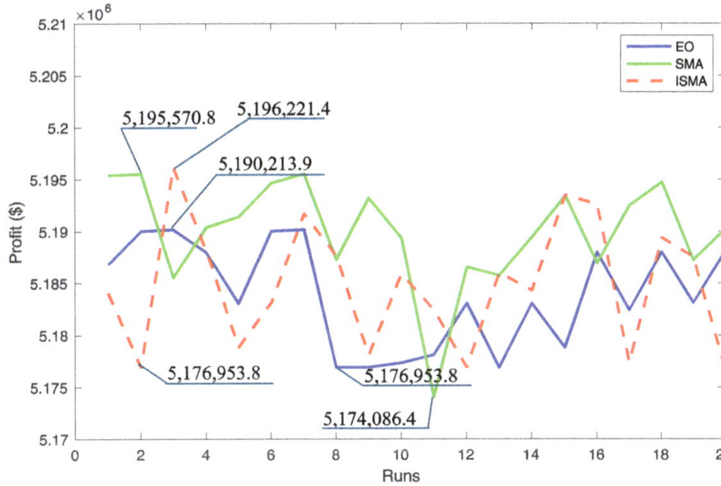

Figure 10. Total profit values collected from twenty trials.

The optimal generation and pump power of power plants are reported in Table A2 in Appendix A.

The cost, revenue, and profit of each thermal power plant are shown in Figure 11. The four thermal power plants have the same manner in that revenue and profit are approximately the same from hours 1 to 8 and from hours 22 to 24. For the other remaining hours, the total revenue and profit have a bigger deviation. Checking the optimal solution in Table A2 in the Appendix A, the generation of these power plants is under 350 MW for hours 1–6 and 280 MW for hours 22–24, whereas it is much higher than these values for hours 7–21, which are about from 700 to under 1000 MW. Clearly, these hours cannot reach high profit for the power system. Paying attention to the pump power of PSHP in the last column in the table, we can see the values of—300 MW. It indicates that the PSHP runs pumps over the first six hours and the last three hours. So, the system achieves some profit.

The cost, revenue, and profit of the WP and SP are plotted in Figure 12. The optimal solutions have reached great profit and revenue at hours with high solar radiation and wind speed. The optimal generations in Table A2 in the Appendix A indicate that the plants have low generations at hours 1–6 and 22–24, where SP has 0 MW at the hours. At other hours with high load demand, the generations of the plants are highly used. The profit is directly proportional to the generation. The system's total cost, revenue, and profit are reported in Figure 13. Thanks to the zero cost of PSHP, the total profit and revenue are high at hours 8–17. The generation of PSHP is about 40 to 140 MW in the period and approximately zero MW at others.

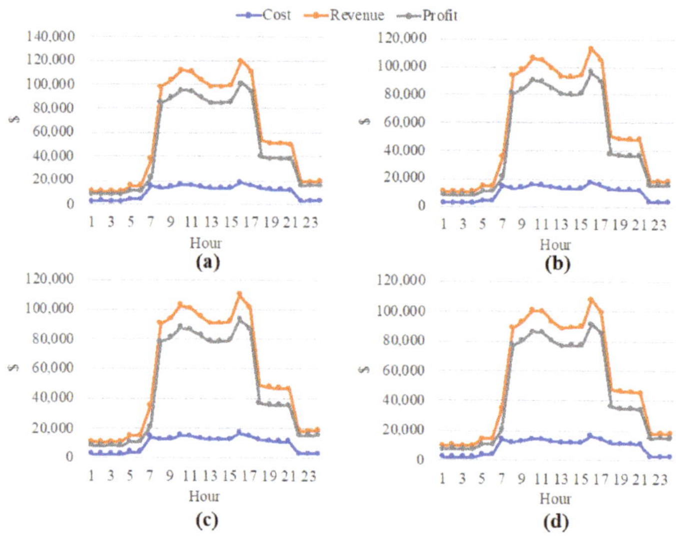

Figure 11. Cost, revenue, and profit of thermal power plants: (**a**) the first ThP, (**b**) the second ThP, (**c**) the third ThP, (**d**) the fourth ThP.

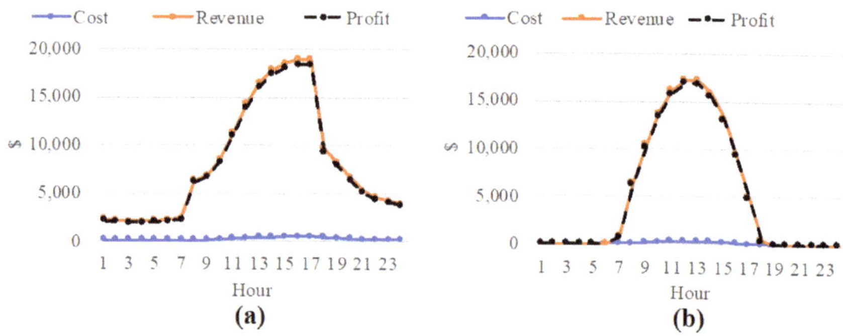

Figure 12. Cost, revenue, and profit: (**a**) wind power plant, (**b**) solar power plant.

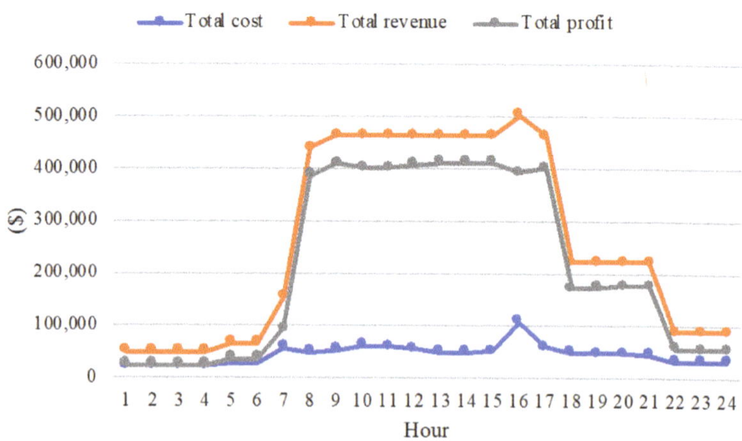

Figure 13. Total cost, revenue and profit of whole system.

4.4. Discussion on the Advantages and Disadvantages

In this section, the advantages and disadvantages of the study are discussed. The study applied three algorithms, and valid and optimal solutions satisfying all constraints could be found. The proposed ISMA could converge to better solutions than EO and SMA. However, the algorithms need a high setting for control parameters, such as 100 for the population size and 2000 for the iteration number. The solution method to handle all constraints is powerful, reaching a 100% success rate. So, the study reached the advantages of powerful applied algorithms and a high-success-rate solution method.

On the contrary, the study also copes with disadvantages that need to be improved in future work. The study used data obtained from the wind and solar global atlases, which reported the mean wind speed and mean solar radiation of each hour in every month. So, we collected the mean data for each hour for 12 days, where one was represented for one month. There are twelve months, and each hour has twelve wind speed values. The best value among the twelve values of each hour was selected and reported in Figure 6. Similarly, we have values for 24 h in Figure 6. Clearly, the data details for each hour on each day of each month are not provided in the wind and solar global atlases. For solving the uncertainty case, we have supposed that the maximum value for uncertainty case is 120% of the highest real value of each hour, as shown in Figures 4 and 5. The best way to reach more practical data of wind and solar radiation is to use the best day of the year when the wind speed is practically stable during the day. On the other hand, the worst day should be used to collect the minimum data for the uncertainty case. Thus, the practical data are not 100% accurate.

In addition, the study ignored practical factors regarding control technology for pumps and generators in pumped storage hydropower plants and operation change time from pumping mode to generation mode or from generation mode to pumping mode. The status change time has an impact on the generation of the pumped storage hydropower plants to electric power systems, but the problem is not considered in the paper. In this study, the time was supposed to be 0 s. The problem can influence the fluctuation of power, and the power balance constraint cannot be satisfied exactly. The power systems last a few minutes, fluctuating power throughout the process. About the thermal power plants, we have also supposed that the start-up time and costs or power changing time were zero. So, the exact cost was not obtained by using the assumption. Basically, turbine technology in thermal power plants is a factor related to fuel cost, power changing cost, power changing time, and life cycle. The operation schedule was one day for calculating the total profit. Thus, the payback period of the whole system could not be found, and the decision to build the hybrid system was not concluded in this study.

5. Conclusions

The study optimized the maximum total electric sale profit of hybrid power systems with the different power plant types and real data on renewable energies in a specific zone of Vietnam. The generation costs of these thermal power plants and renewable power plants were taken into account, while those of the hydropower plant were neglected. The wind and solar global atlases were employed to collect wind and solar data, and the maximum data for the uncertainty case was supposed to be 120% of the reported data from the wind and solar global atlases. The study ran a hybrid power system with four cascaded hydropower plants, four thermal power plants, one pumped storage hydropower plant, one wind power plant, and one solar power plant over one day. The inflows of the pumped storage hydropower plant were supposed to be zero, and it must consume power generated by other plants to run the pumps. The operation of the cascaded hydropower plant was complicated, and they needed the highest power when operating in the hybrid power system. The plants were optimally operated under two considerations: simultaneous and separate. Then, the most effective solution was employed to run other power plants. The simulation results were obtained by running three algorithms, including the original Slime mould algorithm (SMA), Equilibrium optimizer, and improved Slime mould algorithm (ISMA). The results and contributions are summarized as follows:

- For the one-day energy of the cascaded hydropower plants, ISMA reached greater energy than EO and SMA by 35.92 MWh (0.96%) and 4.62 MWh (0.122%) for the simultaneous operation, and 6.43 MWh (0.164%) and 1.05 MWh (0.027%) for the separate operation.
- For the whole system's profit, ISMA could reach a greater total profit than EO and SMA by USD 6007.5 (0.12%) and USD 650.5 (0.013%).

From the results above, ISMA performed more effectively than EO and SMA, and the contribution of the study was to provide a high-performance metaheuristic for the optimization operation of the hybrid power system. In addition, the proposed ISMA has achieved a great profit for the whole system. However, the study has faced limits that need to be improved. The study has ignored the power loss on lines connecting power plants together, and power flows on the transmission power grid where power plants are located. In addition, the electric prices in an electric market one day ahead were not considered, but electric prices in distribution power grids were used to optimize the hybrid power system. The investment cost and other operating costs of power plants were not considered to evaluate the chance to recover the investment cost. In practice, wind speed and solar radiation normally changes every five minutes, and the use of hourly data is not as practical. So, the limitations will be considered in future work. All power plant types in the study or in Vietnam will be integrated in a real transmission power grid in Vietnam or standard IEEE transmission power grid. The electric price of one day ahead will be applied to change the price data. The real transmission power grids, real power plants, and real electric market will form a more practical study that can tackle the shortcomings of the current study.

Author Contributions: Conceptualization, P.T.H.; methodology, P.T.H.; software, P.T.H. and T.M.P.; validation, D.T.T.; formal analysis, T.T.N.; investigation, D.T.T.; data curation, D.T.T.; writing—original draft preparation, P.T.H.; writing—review and editing, T.T.N.; visualization, P.T.H. and T.M.P.; supervision, D.T.T. All authors have read and agreed to the published version of the manuscript.

Funding: This study received no external funding.

Informed Consent Statement: Not applicable.

Data Availability Statement: Data were available in the study.

Conflicts of Interest: The authors declare no conflict of interest.

Nomenclature

$Profit_{sys}$	Total profit of the whole system
$Sale_{Load}$	Total electricity sale revenue
$Cost_{all}$	Total electricity generation cost
$Load_h$, $Price_h$	Load demand and electric price at the *hth* hour
$P_{sp,h}$, $P_{w,h}$	Power of the *spth* solar power plant and *wth* wind power plant at the *hth* hour
$P_{cas,h}$	Power of the *casth* cascaded power plant at the *hth* hour
$P_{ps,h}$, $Pump_{ps,h}$	Generation and pump power of the *psth* pumped storage hydropower plant
P_{th}^{Min}, P_{th}^{Max}	Minimum and maximum generation limits of the *thth* thermal power plant
P_{sp}^{Min}, P_{sp}^{Max}	Minimum and maximum generation limits of the *spth* solar power plant
P_{w}^{Min}, P_{w}^{Max}	Minimum and maximum generation limits of the *wth* wind power plant
P_{cas}^{Min}, P_{cas}^{Max}	Minimum and maximum generation limits of the *casth* cascaded hydropower plant
P_{ps}^{Min}, P_{ps}^{Max}	Minimum and maximum generation limits of the *psth* pumped storage hydropower plant
$V_{ps,h}$, $Q_{ps,h}$	The *psth* PSHP's reservoir volume and discharge at the *hth* hour
V_{ps}^{Min}, V_{ps}^{Max}	The *psth* PSHP's minimum and maximum reservoir volumes
Q_{ps}^{Min}, Q_{ps}^{Max}	The *psth* PSHP's minimum and maximum discharges
m_{1ps}, m_{2ps}, m_{3ps}	The *psth* PSHP's coefficients in discharge function
$I_{ps,h}$	The *psth* PSHP's inflow at the *hth* hour
η_{ps}	The *psth* PSHP's water storage efficiency

γ_{1cas}, γ_{2cas}, γ_{3cas}, γ_{4cas}, γ_{5cas}, γ_{6cas}	Coefficients in power output function of the *casth* CasHP
$V_{cas,h}$, $Q_{cas,h}$	The *casth* CasHP's reservoir volume and discharge at the *hth* hour
$I_{cas,h}$	The *casth* CasHP's inflow at the *hth* hour
$cv_{x,y}^{new}$, $cv_{x,y}$	The new and old *xth* control variable in the *yth* solution
rd	The random number within 0 and 1
FN_y, FN_{b1}	Fitness value of the *yth* and the best solutions
$cv_{x,r1}$, $cv_{x,r2}$, $cv_{x,r3}$	The *xth* control variable in three randomly selected solutions
$cv_{x,b1}$	The *xth* control variable in the best solution
G_c, G_M	the current and maximum iteration numbers
FN_{w1}	The worst solution's fitness
N_s	Number of solutions or population size
$cv_{x,b1-4}$	The *xth* control variable randomly selected in the four best solutions

Appendix A

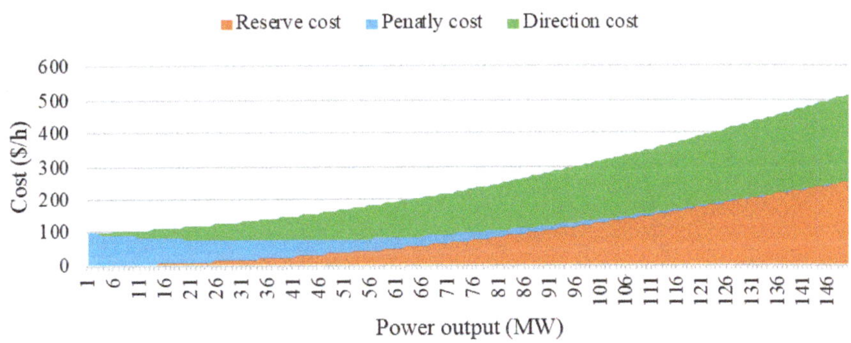

Figure A1. Costs of the wind power plant.

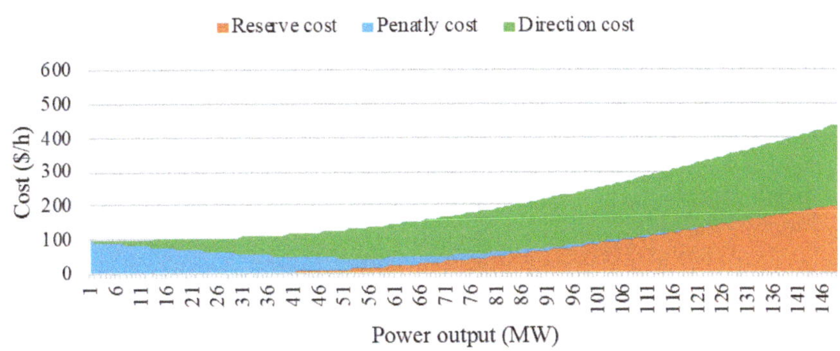

Figure A2. Costs of the solar power plant.

Table A1. The four ThPs' cost function's coefficients.

th	δ_{1th}	δ_{2th}	δ_{3th}	P_{th}^{Min}	P_{th}^{Max}
1	38.5	7.959	0.0127	50	1000
2	39	7.8	0.0135	50	1000
3	35	7.4	0.0142	50	1000
4	36	7.6	0.0143	50	1000

Table A2. Optimal generation and pump power of power plants obtained by ISMA.

h	$P_{th,h}$ ($th = 1$)	$P_{th,h}$ ($th = 2$)	$P_{th,h}$ ($th = 3$)	$P_{th,h}$ ($th = 4$)	$P_{w,h}$ ($w = 1$)	$P_{sp,h}$ ($sp = 1$)	$P_{ps,h}$ ($ps = 1$)	$Pump_{ps,h}$ ($ps = 1$)
1	236.51	225.24	234.38	217.14	50.42	0.00	0.00	−300.00
2	240.93	227.27	227.87	226.88	46.40	0.00	0.00	−300.00
3	233.84	225.82	238.36	221.12	45.12	0.00	0.00	−300.00
4	235.68	226.29	236.94	217.69	45.12	0.00	0.00	−300.00
5	341.05	327.89	322.39	320.77	46.40	0.00	0.00	−300.00
6	342.51	318.14	324.27	321.35	49.06	1.01	0.00	−300.00
7	843.62	796.28	776.71	765.58	51.81	18.10	0.00	0.00
8	762.84	729.79	699.98	691.31	49.06	49.60	267.68	0.00
9	802.53	757.44	727.04	725.24	53.22	80.82	300.00	0.00
10	866.87	825.57	797.89	781.57	65.46	106.64	0.00	0.00
11	855.24	812.89	783.02	777.58	87.12	125.29	0.00	0.00
12	804.08	767.77	741.10	725.04	110.77	134.31	155.68	0.00
13	762.50	722.87	701.76	688.43	127.84	133.89	299.38	0.00
14	760.98	720.29	703.61	694.09	138.35	123.54	291.76	0.00
15	771.23	729.18	711.64	695.04	143.81	103.45	272.74	0.00
16	927.48	879.32	853.25	839.84	146.59	73.91	0.00	−300.00
17	859.88	815.26	785.28	773.73	146.59	38.81	0.00	0.00
18	763.69	716.82	703.63	688.06	138.35	6.76	0.00	0.00
19	730.23	694.42	676.96	661.29	117.89	0.00	130.98	0.00
20	734.71	688.88	668.91	659.80	95.28	0.00	163.82	0.00
21	721.41	686.83	660.51	647.79	75.77	0.00	217.96	0.00
22	269.25	259.14	258.56	256.86	65.38	0.00	0.00	−300.00
23	274.27	262.64	257.51	259.54	60.67	0.00	0.00	−300.00
24	277.17	264.37	264.72	257.66	56.12	0	0	−300.00

References

1. Gao, J.; Zheng, Y.; Li, J.; Zhu, X.; Kan, K. Optimal Model for Complementary Operation of a Photovoltaic-Wind-Pumped Storage System. *Math. Probl. Eng.* **2018**, *2018*, 5346253. [CrossRef]
2. Alam, M.S.; Al-Ismail, F.S.; Salem, A.; Abido, M.A. High-Level Penetration of Renewable Energy Sources into Grid Utility: Challenges and Solutions. *IEEE Access* **2020**, *8*, 190277–190299. [CrossRef]
3. Ha, P.T.; Dinh, B.H.; Phan, T.M.; Nguyen, T.T. Jellyfish Search Algorithm for Optimization Operation of Hybrid Pumped Storage-Wind-Thermal-Solar Photovoltaic Systems. *Heliyon* **2024**, *10*, e29339. [CrossRef] [PubMed]
4. Nanda, J.; Bijwe, P.R.; Kothari, D.P. Application of Progressive Optimality Algorithm to Optimal Hydrothermal Scheduling Considering Deterministic and Stochastic Data. *Int. J. Electr. Power Energy Syst.* **1986**, *8*, 61–64. [CrossRef]
5. Wood, A.J.; Wollenberg, B.F.; Sheblé, G.B. *Power Generation, Operation, and Control*; John Wiley & Sons: Hoboken, NJ, USA, 2013.
6. Fakhar, M.S.; Rehman Kashif, S.A.; Saqib, M.A.; Mehmood, F.; Hussain, H.Z. Non-Cascaded Short-Term Pumped-Storage Hydro-Thermal Scheduling Using Accelerated Particle Swarm Optimization. In Proceedings of the 2018 International Conference on Electrical Engineering (ICEE), Lahore, Pakistan, 15–16 February 2018; IEEE: Piscataway, NJ, USA, 2018; pp. 1–5.
7. Khandualo, S.K.; Barisal, A.K.; Hota, P.K. Scheduling of Pumped Storage Hydrothermal System with Evolutionary Programming. *J. Clean. Energy Technol.* **2013**, *1*, 308–312. [CrossRef]
8. Petrakopoulou, F.; Robinson, A.; Loizidou, M. Simulation and Analysis of a Stand-Alone Solar-Wind and Pumped-Storage Hydropower Plant. *Energy* **2016**, *96*, 676–683. [CrossRef]
9. Cheng, C.; Su, C.; Wang, P.; Shen, J.; Lu, J.; Wu, X. An MILP-Based Model for Short-Term Peak Shaving Operation of Pumped-Storage Hydropower Plants Serving Multiple Power Grids. *Energy* **2018**, *163*, 722–733. [CrossRef]
10. Pérez-Díaz, J.I.; Jiménez, J. Contribution of a Pumped-Storage Hydropower Plant to Reduce the Scheduling Costs of an Isolated Power System with High Wind Power Penetration. *Energy* **2016**, *109*, 92–104. [CrossRef]
11. Correia, P.F.; Ferreira de Jesus, J.M.; Lemos, J.M. Sizing of a Pumped Storage Power Plant in S. Miguel, Azores, Using Stochastic Optimization. *Electr. Power Syst. Res.* **2014**, *112*, 20–26. [CrossRef]
12. Karimi, A.; Heydari, S.L.; Kouchakmohseni, F.; Naghiloo, M. Scheduling and Value of Pumped Storage Hydropower Plant in Iran Power Grid Based on Fuel-Saving in Thermal Units. *J. Energy Storage* **2019**, *24*, 100753. [CrossRef]
13. Awan, A.B.; Zubair, M.; Sidhu, G.A.S.; Bhatti, A.R.; Abo-Khalil, A.G. Performance Analysis of Various Hybrid Renewable Energy Systems Using Battery, Hydrogen, and Pumped Hydro-based Storage Units. *Int. J. Energy Res.* **2019**, *43*, 6296–6321. [CrossRef]
14. Ercan, E.; Kentel, E. Optimum Daily Operation of a Wind-Hydro Hybrid System. *J. Energy Storage* **2022**, *50*, 104540. [CrossRef]

15. He, Y.; Liu, P.; Zhou, L.; Zhang, Y.; Liu, Y. Competitive Model of Pumped Storage Power Plants Participating in Electricity Spot Market——In Case of China. *Renew. Energy* **2021**, *173*, 164–176. [CrossRef]
16. Kwon, J.; Levin, T.; Koritarov, V. Optimal Market Participation of Pumped Storage Hydropower Plants Considering Hydraulic Short-Circuit Operation. In Proceedings of the 2020 52nd North American Power Symposium (NAPS), Tempe, AZ, USA, 11–13 April 2021; IEEE: Piscataway, NJ, USA, 2021; pp. 1–6.
17. Yuan, W.; Xin, W.; Su, C.; Cheng, C.; Yan, D.; Wu, Z. Cross-Regional Integrated Transmission of Wind Power and Pumped-Storage Hydropower Considering the Peak Shaving Demands of Multiple Power Grids. *Renew. Energy* **2022**, *190*, 1112–1126. [CrossRef]
18. Xu, B.; Chen, D.; Venkateshkumar, M.; Xiao, Y.; Yue, Y.; Xing, Y.; Li, P. Modeling a Pumped Storage Hydropower Integrated to a Hybrid Power System with Solar-Wind Power and Its Stability Analysis. *Appl. Energy* **2019**, *248*, 446–462. [CrossRef]
19. Bayazıt, Y.; Bakış, R.; Koç, C. A Study on Transformation of Multi-Purpose Dams into Pumped Storage Hydroelectric Power Plants by Using GIS Model. *Int. J. Green. Energy* **2021**, *18*, 308–318. [CrossRef]
20. Sawa, T.; Sato, Y.; Tsurugai, M.; Onishi, T. Daily Integrated Generation Scheduling for Thermal, Pumped-storage, and Cascaded Hydro Units and Purchasing Power Considering Network Constraints. *Electr. Eng. Jpn.* **2011**, *175*, 25–34. [CrossRef]
21. Jia, J. Mixed-Integer Linear Programming Formulation for Short-Term Scheduling of Cascaded Hydroelectric Plants with Pumped-Storage Units. *Electr. Power Compon. Syst.* **2013**, *41*, 1456–1468. [CrossRef]
22. Rajšl, I.; Ilak, P.; Delimar, M.; Krajcar, S. Dispatch Method for Independently Owned Hydropower Plants in the Same River Flow. *Energies* **2012**, *5*, 3674–3690. [CrossRef]
23. Shen, J.; Cheng, C.; Zhang, X.; Zhou, B. Coordinated Operations of Multiple-Reservoir Cascaded Hydropower Plants with Cooperation Benefit Allocation. *Energy* **2018**, *153*, 509–518. [CrossRef]
24. Feng, Z.; Niu, W.; Cheng, C.; Liao, S. Hydropower System Operation Optimization by Discrete Differential Dynamic Programming Based on Orthogonal Experiment Design. *Energy* **2017**, *126*, 720–732. [CrossRef]
25. Cheng, C.-T.; Shen, J.-J.; Wu, X.-Y.; Chau, K. Operation Challenges for Fast-Growing China's Hydropower Systems and Respondence to Energy Saving and Emission Reduction. *Renew. Sustain. Energy Rev.* **2012**, *16*, 2386–2393. [CrossRef]
26. Yao, F.; Dong, Z.Y.; Meng, K.; Xu, Z.; Iu, H.H.-C.; Wong, K.P. Quantum-Inspired Particle Swarm Optimization for Power System Operations Considering Wind Power Uncertainty and Carbon Tax in Australia. *IEEE Trans. Ind. Inform.* **2012**, *8*, 880–888. [CrossRef]
27. Augusteen, W.A.; Geetha, S.; Rengaraj, R. Economic Dispatch Incorporation Solar Energy Using Particle Swarm Optimization. In Proceedings of the 2016 3rd International Conference on Electrical Energy Systems (ICEES), Chennai, India, 17–19 March 2016; IEEE: Piscataway, NJ, USA, 2016; pp. 67–73.
28. Liu, X. Economic Load Dispatch Constrained by Wind Power Availability: A Wait-and-See Approach. *IEEE Trans. Smart Grid* **2010**, *1*, 347–355. [CrossRef]
29. Biswas, P.P.; Suganthan, P.N.; Amaratunga, G.A.J. Optimal Power Flow Solutions Incorporating Stochastic Wind and Solar Power. *Energy Convers. Manag.* **2017**, *148*, 1194–1207. [CrossRef]
30. Biswas, P.P.; Suganthan, P.N.; Qu, B.Y.; Amaratunga, G.A.J. Multiobjective Economic-Environmental Power Dispatch with Stochastic Wind-Solar-Small Hydro Power. *Energy* **2018**, *150*, 1039–1057. [CrossRef]
31. Reyes, E.D.; Bretas, A.S.; Rivera, S. Marginal Uncertainty Cost Functions for Solar Photovoltaic, Wind Energy, Hydro Generators, and Plug-In Electric Vehicles. *Energies* **2020**, *13*, 6375. [CrossRef]
32. Pham, L.H.; Dinh, B.H.; Nguyen, T.T. Optimal Power Flow for an Integrated Wind-Solar-Hydro-Thermal Power System Considering Uncertainty of Wind Speed and Solar Radiation. *Neural Comput. Appl.* **2022**, *34*, 10655–10689. [CrossRef]
33. Available online: https://Globalwindatlas.Info/En (accessed on 1 May 2024).
34. Available online: https://Globalsolaratlas.Info/Map (accessed on 1 May 2024).
35. Li, S.; Chen, H.; Wang, M.; Heidari, A.A.; Mirjalili, S. Slime Mould Algorithm: A New Method for Stochastic Optimization. *Future Gener. Comput. Syst.* **2020**, *111*, 300–323. [CrossRef]
36. Faramarzi, A.; Heidarinejad, M.; Stephens, B.; Mirjalili, S. Equilibrium Optimizer: A Novel Optimization Algorithm. *Knowl. Based Syst.* **2020**, *191*, 105190. [CrossRef]
37. Nguyen, T.T.; Nguyen, T.T.; Duong, M.Q.; Doan, A.T. Optimal Operation of Transmission Power Networks by Using Improved Stochastic Fractal Search Algorithm. *Neural Comput. Appl.* **2020**, *32*, 9129–9164. [CrossRef]
38. Ha, P.T.; Tran, D.T.; Nguyen, T.T. Electricity Generation Cost Reduction for Hydrothermal Systems with the Presence of Pumped Storage Hydroelectric Plants. *Neural Comput. Appl.* **2022**, *34*, 9931–9953. [CrossRef]
39. Nguyen, T.T.; Vu Quynh, N.; Duong, M.Q.; Van Dai, L. Modified Differential Evolution Algorithm: A Novel Approach to Optimize the Operation of Hydrothermal Power Systems While Considering the Different Constraints and Valve Point Loading Effects. *Energies* **2018**, *11*, 540. [CrossRef]
40. Özyön, S. Optimal Short-Term Operation of Pumped-Storage Power Plants with Differential Evolution Algorithm. *Energy* **2020**, *194*, 116866. [CrossRef]

41. Wang, C.; Chen, S. Planning of Cascade Hydropower Stations with the Consideration of Long-Term Operations under Uncertainties. *Complexity* **2019**, *2019*, 1534598. [CrossRef]
42. Nguyen, T.T.; Vo, D.N. The Application of an Effective Cuckoo Search Algorithm for Optimal Scheduling of Hydrothermal System Considering Transmission Constraints. *Neural Comput. Appl.* **2019**, *31*, 4231–4252. [CrossRef]

MDPI AG
Grosspeteranlage 5
4052 Basel
Switzerland
Tel.: +41 61 683 77 34
www.mdpi.com

MDPI Books Editorial Office
E-mail: books@mdpi.com
www.mdpi.com/books

www.ingramcontent.com/pod-product-compliance
Lightning Source LLC
Chambersburg PA
CBHW061457220926
51850CB00004B/114
9783725819232